BRIEF VERSION

Human Anatomy and Physiology Laboratory Manual

Elaine N. Marieb, R.N., Ph.D.
Holyoke Community College

THIRD EDITION

The Benjamin/Cummings Publishing Company, Inc.

Redwood City, California · Menlo Park, California
Reading, Massachusetts · New York · Don Mills, Ontario
Wokingham, U.K. · Amsterdam · Bonn · Sydney
Singapore · Tokyo · Madrid · San Juan

Sponsoring Editor: Melinda Adams
Production Coordinator: Anne Friedman
Manufacturing Coordinator: Casi Kostecki
Assistant Editor: Diane Honigberg
Editorial Assistant: Sami Iwata
Photo Editor: Cecilia Mills
Cover and Book Design: Gary Head
Copy Editor: Mary Prescott
Proofreaders: Laura Kenney and Anita Wagner
Art to accompany photomicrographs: Linda McVay
Cover art: Top right (temporal bone) from *A Stereoscopic Atlas of Human Anatomy,* David L. Bassett, M.D.; bottom left (xs of scalp w/ follicles) © Manfred Kage/Peter Arnold, Inc.; bottom right (resin cast) © Martin Dohrn, Royal College of Surgeons/Science Photo Library/Photo Researchers, Inc.
Typesetter: GTS Graphics, Commerce, CA
Printer: Von Hoffman Press, Jefferson City, MO

The Author and Publisher believe that the lab experiments described in this publication, when conducted in conformity with the safety precautions described herein and according to the school's laboratory safety procedures, are reasonably safe for the students to whom this manual is directed. Nonetheless, many of the described experiments are accompanied by some degree of risk, including human error, the failure or misuse of laboratory or electrical equipment, mismeasurement, spills of chemicals, and exposure to sharp objects, heat, bodily fluids, blood, or other biologics. The Author and Publisher disclaim any liability arising from such risks in connection with any of the experiments contained in this manual. If students have any questions or problems with materials, procedures, or instructions on any experiment, they should *always* ask their instructor for help before proceeding.

ISBN 0-8053-4807-7

6 7 8 9 10-VH-96 95 94

The Benjamin/Cummings Publishing Company, Inc.
390 Bridge Parkway
Redwood City, California 94065

The Benjamin/Cummings Series in Human Anatomy and Physiology:

E.N. Marieb and J. Mallatt
Human Anatomy (1992)

R. A. Chase
The Bassett Atlas of Human Anatomy (1989)

S. W. Langjahr and R. D. Brister
Coloring Atlas of Human Anatomy, Second Edition (1992)

E. N. Marieb
Human Anatomy and Physiology, Second Edition (1992)

E. N. Marieb
Human Anatomy and Physiology, Study Guide, Second Edition (1992)

E. N. Marieb
Human Anatomy and Physiology Laboratory Manual: Cat Version, Fourth Edition (1993)

E. N. Marieb
Human Anatomy and Physiology Laboratory Manual: Fetal Pig Version, Fourth Edition (1993)

E. N. Marieb
Essentials of Human Anatomy and Physiology, Third Edition (1991)

E. N. Marieb
The A & P Coloring Workbook: A Complete Study Guide, Third Edition (1991)

A. P. Spence
Basic Human Anatomy, Third Edition (1991)

R. L. Vines and A. Hinderstein
California State University, Sacramento
Human Musculature Videotape (1989)

R. L. Vines and University Media Services,
California State University, Sacramento
Human Nervous System Videotape (1992)

Histology insert photos by Victor Eroschenko, University of Idaho, except as noted. Plate 10: Courtesy of Churchill Livingstone. Plate 20: ©Carolina Biological Supply Company. Plate 30: Courtesy of Marian Rice; Table 10.11, 13.3c, 13.4c, 13.5c, 13.6a, 13.8d, 13.10b, 13.11b, 13.12c: From *A Stereoscopic Atlas of Human Anatomy*, by David L. Bassett, M.D.;

1.1: BioMed Arts Associates, Inc.; 2.1–2.03: Courtesy of Jack Scanlon; 3.1: Courtesy of Leica Inc.; 4.3a–f: Courtesy of Victor Eroschenko; 5.1: Richard Megna/Fundamental Photographs; 5.2a: ©K.R. Porter/Photo Researchers, Inc.; 5.2b: ©David M. Phillips/Photo Researchers, Inc.; 5.2c: Courtesy of Dr. Mohandas Narla, Berkeley Livermore Laboratory; 6.2a: Marian Rice; 6.2b: ©Ed Reschke/Peter Arnold, Inc.; 6.2c: Marian Rice; 6.2d–f: ©Ed Reschke; 6.2g: ©John D. Cunningham/Visuals Unlimited; 6.2h: ©Bruce Iverson; 6.3a: ©Ed Reschke; 6.3b: Marian Rice; 6.3c–i: © Ed Reschke; 6.3j: ©Biology Media/Photo Researchers; 6.3k,l: ©Ed Reschke; 6.4a: ©Eric Grave/Photo Researchers, Inc.; 6.4b: ©Ed Reschke; 6.4c: Department of Anatomy & Histology, University of California, San Francisco; 6.5: ©Ed Reschke; 7.2: Marian Rice; 7.3: Courtesy of Victor Eroschenko, University of Idaho; 8.3b: Marian Rice; 10.7b: Courtesy of Richard Hastings, Redwood City Medical Center; 12.2: Marian Rice; 16.8b, 16.9, 16.10: Courtesy of Ann Allworth; 17.1a: ©Alexander Tsiaras/Photo Researchers, Inc.; 20.3: ©William Thompson; 21.4: Courtesy of Jack Scanlon; 21.11: Courtesy of Christopher S. Sherman, Swedish Hospital Medical Center; 26.4, 26.5: Courtesy of Jack Scanlon; 26.6: Marian Rice; 27.3a,b: Photo courtesy of Ann Allworth; 27.6: Marian Rice; 32.4: ©Science Photo Library/Photo Researchers, Inc.; 33.5: ©Alfred Pasieka/Custom Medical Stock Photo; 33.7, 37.3: Photo courtesy of Ann Allworth; 41.1: Courtesy of T.T. Puck. 1972, *The Mammalian Cell as Microorganism.* San Francisco: Holden-Day; 42.1–42.22: BioMed Arts Associates, Inc.

Table of Contents

Exercises denoted by a red (*) contain physiology experiments.

The Respiratory System

The Digestive System

The Urinary System

The Reproductive System, Development, and Heredity

Surface Anatomy

Preface to the Instructor

With the writing of each new edition of a book comes a matured sensibility for the way teachers teach and students learn. This sensibility is achieved not just through years of teaching the subject, but also by listening to the suggestions of other instructors as well as those of students enrolled in the multifaceted health-care programs. The third edition of *Human Anatomy and Physiology Laboratory Manual: Brief Version* has been developed in a continuing effort to facilitate the job of both teachers and students.

As with previous editions of this manual, the current edition maintains as its audience the introductory courses in human anatomy and physiology. It presents a wide range of laboratory experiences for students concentrating in nursing, physical therapy, occupational therapy, respiratory therapy, dental hygiene, pharmacology, health and physical education, as well as biology and premedical programs. It differs from the Cat and Fetal Pig Versions (Fourth Editions, 1993) in that it does not include detailed dissection guidelines for any particular laboratory animal; the descriptive material for each experiment is streamlined for easier assimilation; and the laboratory review sheets require less time to complete and correct. Thus, the manual's coverage is broad enough to serve two-semester courses and yet, because of the shorter exercises and streamlined review sheets, it should serve one semester courses equally well.

ORGANIZATION

The variety of both anatomical studies and physiological experiments provides flexibility that enables instructors to gear their courses to specific academic programs, or to their own teaching preferences. The manual is still independent of any textbook, and contains all the background discussion and terminology necessary to perform all experiments. Such a self-contained learning aid eliminates the need for students to bring a textbook into the laboratory.

Each of the 42 exercises leads students toward a coherent understanding of the structure and function of the human body. The manual begins with anatomical terminology and an orientation to the body, which provides the necessary tools for studying the various body systems. The exercises that follow reflect the dual focus of the manual—both anatomical and physiological aspects receive considerable attention. As the various organ systems of the body are introduced, the initial exercises focus on organization from the cellular to the organ level. As indicated by the Table of Contents, the anatomical exercises are followed by physiological experiments that familiarize students with various aspects of body functioning and promote an appreciation for the critical fact that function follows structure. Homeostasis is continually emphasized, and these discussions can be recognized by the homeostasis imbalance logo within the descriptive material of each exercise. This holistic approach ultimately gives students an integrated understanding of the human body.

NEW TO THIS EDITION

In this revision, I really tried to respond to reviewers' and users' feedback concerning trends that are having an impact on the anatomy and physiology laboratory experience, most importantly:

- the pedagogical advantage of using four-color illustrations
- the growing reluctance of students to perform experiments using living laboratory animals and the declining popularity of animal dissection exercises
- the increased use of computers in the laboratory and, hence, the subsequent desire for more computer simulation exercises
- the demand for clearer safety guidelines for students, especially when working with blood and other body fluids.

Among the specific changes implemented to address these trends are the following:

1. The use of full four-color illustrations aids student learning in the Classification of Tissues exercise (Exercise 6), for the Axial Skeleton exercise (Exercise 9), for the Identification of Human Muscles (Exercise 13), and in the new Surface Anatomy exercise (Exercise 42). Additionally, the skull art in Exercise 9 is color-coded for easy identification.
2. The addition of a new eight-page Histology Atlas (following p. 176) and new colored plates on the front and back endpapers brings the total number of colored histology plates in this edition to sixty. Great care has been taken to select histology photomicrographs that will be the most helpful to students, corresponding closely with what they will view in the laboratory. While most tissues are stained with hematoxylin and eosin, a few of those depicted in the Histology Atlas have special stains, such as the photomicrograph of a pancreatic islet of Langerhans that has been differentially stained to distinguish between its glucagon-producing alpha cells and its insulin-secreting beta cells. For

some organs studied, a low-power view is presented first to orient the student to the tissue slide before presenting the anatomical detail of the high-power view.

3. In response to the trend towards less animal dissection, a computer simulation exercise, Exercise 14B, Muscle Physiology (Computerized Simulations), has been added to this edition. This exercise provides an *alternative* method to study skeletal muscle physiology for those who choose not to use frogs to conduct the traditional laboratory experiments. The standard frog gastrocnemius exercise (Exercise 14A) is still provided for the traditionalists among us.

   In addition, Exercise 12, the Microscopic Anatomy, Organization, and Classification of Skeletal Muscle, now suggests using chicken meat purchased from the butcher shop for the study of skeletal muscle tissue histology rather than sacrificing a frog to obtain the needed fresh tissue.

4. Five logos, which are visual mnemonics to alert students to a special feature or instructions, appear in this edition.

   The homeostatic imbalance logo described on p. v indicates clinical implications sections as applicable.

   A blood and body fluid logo, which appears when blood or other body fluids (saliva, urine) must be handled, signifies where special self-protective measures are to be taken.

   A safety logo notifies students that specific safety precautions are to be observed when using certain equipment or conducting particular lab procedures (for example, a hood is to be used when working with ether, and so on).

   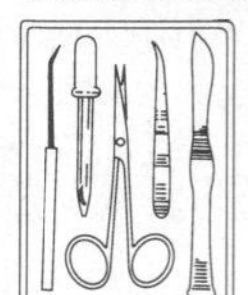
   The dissection tray logo appears at the beginning of lab activities to be conducted by the students.

   The cooperative learning logo is seen before experiment heads or procedures where learning would be enhanced and/or time saved by having students work together (in pairs or teams).

5. Two new exercises are included—the computer simulation exercise for muscle physiology (Exercise 14B) mentioned above, and a new surface anatomy lab that invites students to apply the anatomy concepts they have learned, involving them in active observation and manipulation of surface features of their own bodies and those of other students.

6. Several exercises have been expanded, modified, or rewritten:

   - Exercise 6, the Classification of Tissues, has been expanded to include all the common tissue types, in four-color.
   - Exercise 11 on joints, has been rewritten with a *structural classification* focus in response to users' requests.
   - Exercise 15, Neuron Anatomy and the Nerve Impulse, has been expanded to include an exercise using the oscilloscope to study properties of the nerve impulse.
   - Exercise 26 on blood reflects current CDC guidelines for handling human blood. (Since it is important that nursing students, in particular, learn how to safely handle blood-stained articles, the human focus has been retained. In other cases, the use of animal blood for testing purposes is suggested.)
   - Exercise 31, Frog Cardiovascular Physiology, now contains additional set-up instructions for use of the physiograph.
   - Exercise 36 on enzyme activity in the digestive system has been entirely revised. The simplified experiments are designed to provide greater student success in this difficult-to-conduct set of experiments.
   - Exercise 38, Urinalysis, has been rewritten to provide for greater safety in handling human urine. In addition, the *Instructor's Guide* provides the source of a published "recipe" for making safe-to-handle artificial urine.

7. In response to user feedback, three infrequently used exercises—Food Propulsion Mechanisms, Kidney Regulation of Fluid and Electrolyte Balance, and Pregnancy Tests—have been deleted.

8. The anatomical terminology in this manual has been updated to match the updated terminology in *Human Anatomy and Physiology, Second Edition* (by this author) to avoid student confusion over conflicting terminology.

9. The questions in the laboratory review sheets are expanded and now include more labeling exercises.

10. The newly added Appendix A: The Metric System is a useful reference for the student, giving conversion factors to and from the English system to the metric system.

11. All laboratory instructions and procedures have been revised to incorporate the latest precautions as recommended by the Centers for Disease Control (CDC), including recommendations for safely handling body fluids. These guidelines are reinforced by the laboratory safety procedures described in the front section of this text and the *Instructor's Guide.*

12. Finally, because art plays such a critical role in helping students visualize anatomical and physiological concepts, every effort has been made in this manual to offer a figure, in either line art or photographs, for virtually everything students will examine in the laboratory.

SPECIAL FEATURES

Virtually all the special features appreciated by the adopters of the second edition are retained.

- Each exercise begins with learning objectives.
- Key terms appear in boldface print, and each term is defined when introduced.
- Illustrations are large and of exceptional quality. The use of full color is expanded in this edition to highlight, to differentiate, and to focus student attention on important structures.
- Body structures are studied from simple-to-complex levels, and physiological experiments allow ample opportunity for student observation, manipulation, and experimentation.
- There are numerous physiological experiments for each organ system, ranging from simple experiments that can be performed without specialized tools to more complex experiments using laboratory equipment and instrumentation techniques.
- Tear-out laboratory review sheets, located towards the end of the manual, are designed to accompany the laboratory exercises. They require students to label diagrams and answer multiple-choice, short-answer, and review questions. As noted, the current edition has more review questions and more labeling exercises.
- Space is provided for recording and interpreting experimental results.
- Isolated animal organs such as the sheep heart and pig kidney are employed because of their exceptional similarity to human organs, but figures and instructions for using major dissection animals are not included. If the fetal pig or cat is used, one of the other two versions of this manual is recommended.
- The prologue, "Getting Started—What to Expect, The Scientific Method, Scientific Notation, and Metrics," explains the scientific method, the logical, practical, and reliable way of approaching and solving problems in the laboratory. It also reviews the use of exponents, metric units, and interconversions.

SUPPLEMENTS

The *Instructor's Guide* that accompanies the *Human Anatomy & Physiology Laboratory Manual: Brief Version* was formerly just a solutions manual for the review sheets. Now it features a wealth of information for anyone involved in teaching this course. Instructors can find help in planning the experiments, ordering equipment and supplies, anticipating pitfalls and problem areas, and locating audiovisual material. The probable in-class time required for each lab is indicated by a clock logo. Other useful resources are the Trends in Instrumentation section that describes the latest laboratory equipment and technological teaching tools available, and the Anatomy and Physiology Laboratory Safety Procedures section that gives current Centers for Disease Control (CDC) guidelines for handling body fluids and other precautions recommended by the CDC. These procedures, located in the front section of the *Instructor's Guide,* can be photocopied and posted in the lab. By adhering to CDC-recommended measures, the students of this course will be well trained in laboratory safety.

To obtain Mechanical Properties of Active Muscle software that accompanies Exercise 14B, Muscle Physiology (Computerized Simulations), please contact your sales representative.

ACKNOWLEDGMENTS

I wish to thank the following people for their contributions to this edition: Jennifer Breckler (San Francisco State University), Mark B. Frasier (Colorado State University), William J. Higgins (University of Maryland at College Park), Paul Holmgren (Northern Arizona University), John A. Knesel (Northeast Louisiana University), Linda S. Kollett (Massasoit Community College), Donn D. Martin (Texas Wesleyan University), Randall McKee (University of Wisconsin-Parkside), Connie Vader-Lindholm (Colorado State University), and Peter Zao (North Idaho College). These instructors reviewed the manuscript and provided invaluable assistance with this revision.

My continued thanks to my colleagues at Benjamin/Cummings who helped in the production of this edition, especially Melinda Adams, my Sponsoring Editor, and Anne Friedman, Senior Production Editor. Also not to be forgotten is the diligent work of Diane Honigberg, Assistant Editor, and the Editorial Assistant, Sami Iwata, and Cecilia Mills who was responsible for photo research.

Finally, since preparations for the next edition begin well in advance of its publication, I invite users of this edition to send me their comments and suggestions for subsequent editions.

Elaine N. Marieb
Department of Biology
Holyoke Community College
303 Homestead Avenue
Holyoke, MA 01040

Preface to the Student

Laboratory Activities

The A&P laboratory exercises within this manual are designed to help you gain a broad understanding of both anatomy and physiology. So, you can anticipate that you will be examining models, dissecting isolated animal organs, and using a microscope to look at tissue slides (anatomical approaches). You will also investigate chemical conditions or observe changes in both living and nonliving systems, and conduct experiments that examine responses of living organisms, such as frogs or yourself, to various stimuli (physiological approaches).

Four suggestions to help make your lab experience more rewarding are:

1. Scan the scheduled laboratory exercise and the questions in the review section in the back of the manual that pertain to it *before* going to lab.
2. Be on time. Most instructors explain what the lab is about, pitfalls to avoid, and the sequence or format to be followed at the beginning of the lab session. If you are late, not only will you miss this information, you will not endear yourself to the instructor.
3. Keep your work area clean and neat. This reduces confusion and accidents.
4. Assume that all laboratory chemicals and equipment are sources of potential danger to you, and follow directions for equipment use and rules of lab safety provided in this manual and by your instructor.

Anatomy and Physiology laboratory safety guidelines are provided on the next page.

Because some students question the use of animals in the laboratory setting, their concerns need to be addressed. Be assured that the commercially available preserved organ specimens used in the A&P lab are *not* harvested from animals raised specifically for dissection purposes. Instead, organs that are of no use to the meat packing industry (such as the brain, heart, or lungs) are sent from slaughter houses to qualified biological supply houses, where they are prepared for laboratory use according to USDA guidelines.

Relative to using live animals for experimentation, every effort is being made to find alternative methods that do not use living animals to study physiological concepts. For example, new to this edition is Exercise 14B, a computer simulation that studies the properties of contracting muscles. This is provided as an alternative in addition to the traditional exercise (14A) that uses frogs. Without a doubt, computer simulations offer certain advantages: (1) they allow you to experiment at length without time constraints of traditional animal experiments, in which fragile living tissues need to be kept alive and viable for the duration of the experiment; and (2) they make it possible to investigate certain concepts (such as isometric versus isotonic contraction in this case) that would be difficult (or impossible) to explore in traditional exercises. Yet, the main disadvantage of computer simulations is that the real-life aspects of experimentation are sacrificed; an animated frog muscle or heart on a computer screen is not really a substitute for observing the responses of actual muscle tissue.

Although it might appear that the choice is simply to decide which approach offers fewer disadvantages, unfortunately adequate software applications are still unavailable. Consequently, living animal experiments remain an important part of the approach of this manual to the study of human A&P. However, wherever possible, the minimum number of animals needed to demonstrate a particular point are used. Futhermore, more instructor-delivered demonstrations of live animal experiments are suggested.

If you use living animals for experiments, you will be expected to handle them humanely. Inconsiderate treatment of laboratory animals will not be tolerated in your A&P laboratory.

Logos/Visual Mnemonics

I have tried to make this manual very easy for you to use, and to this end, five different types of logos (visual mnemonics) are described below:

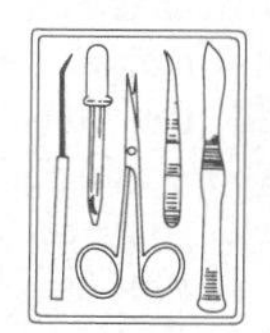

The **dissection tray logo** appears at the beginning of lab activities. Since most exercises have some explanatory background provided before the experiment(s), this visual cue alerts you that your involvement is imminent.

The **blood/body fluid logo** appears where blood or other body fluids (saliva, urine) must be handled, and it signifies that you should take special measures to protect yourself.

The **safety logo** alerts you to special precautions in handling lab equipment or conducting certain procedures, e.g., use a ventilating hood when using volatile chemicals.

The **cooperative learning logo** is seen before procedures where your learning will be enhanced (or time saved) if you work with a partner (or team).

The **homeostatic imbalance logo** appears where a clinical disorder is described to indicate what happens when there is a structural abnormality or physiological manfunction, i.e., a loss of homeostasis.

Anatomy and Physiology Laboratory Safety Guidelines*

1. Upon entering the laboratory, locate exits, fire extinguisher, fire blanket, chemical shower, eyewash station, first aid kit, broken glass containers, and cleanup materials for spills.
2. Do not eat, drink, smoke, store food, or apply cosmetics in the laboratory.
3. Students who are pregnant, taking immunosuppressive drugs, or who have any other medical conditions (e.g., diabetes, immunological defect) that might necessitate special precautions in the laboratory must inform the instructor immediately.
4. Restrain long hair, loose clothing, and dangling jewelry.
5. Use safety glasses in all experiments in which solutions or chemicals are heated over a Bunsen burner. Never leave heat sources unattended.
6. Decontaminate work surfaces at the beginning and end of every lab period, using a commercially prepared disinfectant or 10% sodium hypochlorite solution (household bleach). After labs involving dissection of preserved material, use hot soapy water or disinfectant.
7. Keep all liquids away from the edge of the lab bench to avoid spills. Clean up spills of viable materials using disinfectant or 10% sodium hypochlorite solution.
8. Properly label glassware and slides.
9. Use mechanical pipetting devices; mouth pipetting is prohibited.
10. Place glassware and plasticware contaminated by blood and other body fluids in a disposable autoclave bag for decontamination by autoclaving, or place them directly into a 10% sodium hypochlorite solution before reuse or disposal. Place disposable materials such as gloves, mouthpieces, swabs, and toothpicks that have come into contact with body fluids into a disposable autoclave bag.
11. Wear disposable gloves when handling blood and other body fluids, mucous membranes, and nonintact skin, and/or when touching items or surfaces soiled with blood or other body fluids. Change gloves between procedures. Wash hands immediately after removing gloves. (Note: cover open cuts or scrapes with a sterile bandage before donning gloves.)
12. To prevent contamination by needle stick injuries, use only disposable needles and lancets. Do not bend the needles and lancets, replace them in sheaths, or remove them from syringes following use. Needles and lancets should be placed promptly in a freshly prepared 10% sodium hypochlorite solution or placed in a puncture-resistant container and decontaminated, preferably by autoclaving.
13. Report all spills or accidents, no matter how minor, to the instructor.
14. Never work alone in the laboratory.
15. Wash hands and remove protective clothing before leaving the laboratory.

* Adapted from:

Biosafety in Microbiological and Biomedical Laboratories. 1988. U.S. Government Printing Office, Washington D.C. 20402.

Centers for Disease Control. 1989. "Guidelines for Prevention of Transmission of Human Immunodeficiency Virus and the Hepatitis B Virus to Health-Care and Public-Safety Workers." *MMWR:* 38 (S6).

———.1987. "Recommendations for Prevention of HIV Transmission in Health-Care Settings." *MMWR:*36 (2s).

Johnson, Ted, and Christine Case. 1992. *Laboratory Experiments in Microbiology, Brief Version, Third Edition.* Redwood City, CA: Benjamin Cummings Publishing Co.

School Science Laboratories: A Guide to Some Hazardous Substances. 1984. U.S. Consumer Product Safety Commission. Washington D.C. 20207.

Getting Started—What to Expect, The Scientific Method, Scientific Notation, and Metrics

Two hundred years ago science was largely a plaything of wealthy patrons, but today's world is dominated by science and its technology. Whether or not we believe that such domination is desirable, we all have a responsibility to try to understand the goals and methods of science that have seeded this knowledge and technological explosion.

The biosciences are very special and exciting because they open the doors to an understanding of all the wondrous workings of living things. A course in human anatomy and physiology (a minute subdivision of bioscience) provides such insights in relation to your own body. Although some experience in scientific studies is helpful when beginning a study of anatomy and physiology, perhaps the single most important prerequisite is curiosity.

Gaining an understanding of science is a little like becoming acquainted with another person. Even though a written description can provide a good deal of information about the person, you can never really know another unless there is personal contact. And so it is with science—if you are to know it well, you must deal with it intimately.

The laboratory is the setting for "intimate contact" with science. It is where scientists test their ideas (do research), the essential purpose of which is to provide a basis from which predictions about scientific phenomena can be made. Likewise, it will be the site of your "intimate contact" with the subject of human anatomy and physiology as you are introduced to the methods and instruments used in biological research.

For many students, human anatomy and physiology is taken as an introductory-level course; and their scientific background exists, at best, as a dim memory. If this is your predicament, this prologue may be just what you need to fill in a few gaps and to get you started on the right track before your actual laboratory experiences begin. So—let's get to it!

THE SCIENTIFIC METHOD

Science would quickly stagnate if new knowledge were not continually derived from and added to it. The approach commonly used by scientists when they investigate various aspects of their respective disciplines is called the **scientific method.** This method is *not* a single rigorous technique that must be followed in a lockstep manner. It is nothing more or less than a logical, practical, and reliable way of approaching and solving problems of every kind—scientific or otherwise—to gain knowledge. It comprises five major steps.

Step 1: Observation of Phenomena

The crucial first step involves observation of some phenomenon of interest. In other words, before a scientist can investigate anything, he or she must decide on a *problem* or focus for the investigation. In most college laboratory experiments, the problem or focus has been decided for you. However, to illustrate this important step, we will assume that you want to investigate the true nature of apples, particularly green apples. In such a case you would begin your studies by making a number of different observations concerning apples.

Step 2: Statement of the Hypothesis

Once you have decided on a focus of concern, the next step is to design a significant question to be answered. Such a question is usually posed in the form of a **hypothesis,** an unproven conclusion that attempts to explain some phenomenon. (At its crudest level, a hypothesis can be considered to be a "guess" or an intuitive hunch that tentatively explains some observation.) Generally, scientists do not restrict themselves to a single hypothesis; instead, they usually pose several and then test each one systematically.

We will assume that, to accomplish step one, you go to the supermarket and randomly select apples from several bins. When you later eat the apples, you find that the green apples are sour, but the red and yellow apples are sweet. From this observation, you might conclude (*hypothesize*) that "green apples are sour." This statement would represent your current understanding of green apples. You might also reasonably predict that, if you were to buy more apples, any green ones you buy will be sour. Thus, you would have gone beyond your initial observation that "these" green apples are sour to the prediction that "all" green apples are sour.

Any good hypothesis must meet several criteria. First, *it must be testable.* This characteristic is far more important than its being correct. The tests may prove the hypothesis incorrect; or new information may require that the hypothesis be modified. Clearly the accuracy of a prediction in the green apple example or in any scientific study depends on the accuracy of the initial information on which it is based.

In our example, no great harm will come from an inaccurate prediction—that is, were we to find that some green apples are sweet. However, in some cases human life may depend on the accuracy of the prediction. Take the case of testing drugs for their effectiveness in treating disease. If one set of observations erroneously indicates that the drugs are risky but very effective, such a conclusion could lead to the death of the subsequent drug recipient(s). This illustrates two points: (1) Repeated testing of scientific ideas is important, particularly because scientists working on the same problem do not always agree in their conclusions. The studies on the use of saccharin and amino acid sweeteners are only two examples. (2) Conclusions drawn from scientific tests are only as accurate as the information on which they are based; therefore, careful observation is essential, even at the very outset of a study.

A second criterion is that, even though hypotheses are guesses of a sort, *they must be based on measurable, describable facts. No mysticism can be theorized.* We cannot conjure up, to support our hypothesis, forces that have not been shown to exist. For example, as scientists, we cannot say that the tooth fairy took Johnny's tooth unless we can prove that the tooth fairy exists!

Third, a hypothesis *must not be anthropomorphic.* Human beings tend to anthropomorphize—that is, to relate all experiences to human experience. Because man is a social animal influenced by culture, these two characteristics tend to promote biased thinking. Whereas we could state that bears instinctively protect their young, it would be anthropomorphic to say that bears love their young, because love is a human emotional response. Thus, the initial hypothesis must be stated without interpretation.

Step 3: Data Collection

Once the initial hypothesis has been stated, scientists plan experiments that will provide data (or evidence) to verify or disprove their hypotheses—that is, they *test* their hypotheses. Data are accumulated by making qualitative or quantitative observations of some sort. The observations are often aided by the use of various types of equipment such as cameras, microscopes, stimulators, or various electronic devices that allow chemical and physiologic measurements to be made.

Observations referred to as **qualitative** are those we can make with our senses—that is, by using our vision, hearing, or sense of taste, smell, or touch. The color of an object, its texture, the relationship of one part to another, and its relative size (large versus small) may all be part of a qualitative description. For some quick practice in qualitative observation, compare and contrast* an orange and an apple.

Whereas the differences between an apple and an orange are obvious, this is not always the case in biological observations. Quite often a scientist tries to detect very subtle differences that cannot be determined by qualitative observations; and data must be derived from measurements made using a variety of scientific equipment. Such observations based on precise measurements of one type or another are **quantitative observations.** Examples of quantitative observations include careful measurements of body or organ dimensions such as mass, size, and volume; measurement of volumes of oxygen consumed during metabolic studies; determination of the concentration of glucose and other chemicals in urine; and determination of the differences in blood pressure and pulse under conditions of rest and exercise. An apple and an orange could be compared quantitatively by performing chemical measurements of the relative amounts of sugar and water in a given volume of fruit flesh, by analyzing the pigments and vitamins present in the apple skin and orange peel, and so on.

A valuable part of data gathering is the use of experiments to verify or disprove a hypothesis. An **experiment** is a procedure designed to describe the factors in a given situation that affect one another (that is, to discover cause and effect) under certain conditions.

Two general rules govern experimentation. The first of these rules is that the experiment(s) should be conducted in such a manner that every **variable** (any factor that might affect the outcome of the experiment) is under the control of the experimenter. The experimenter manipulates the **independent variables** and observes the effects of this manipulation on the **dependent** (or **response**) **variable.** For example, if the goal is to determine the effect of body temperature on breathing rate, the value measured (breathing rate) is called the dependent variable because it "depends on" the value chosen for the independent variable (body temperature). The ideal way to perform such an experiment is to set up and run a series of tests that are all identical, except for one specific factor that is varied.

One specimen (or group of specimens) is used as the **control** against which all other experimental samples are compared. The importance of the control sample cannot be overemphasized. It is essential to know how the system you are investigating works under normal circumstances before you can be sure that the results obtained from experimentation are due solely to the manipulation of the independent variable(s). Taking our example one step further, if we wanted to investigate the effects of body temperature (the independent variable) on breathing rate (the dependent variable), we could collect data on the breathing rate of individuals with "normal" body temperature (the implicit control group), and compare these data to breathing-rate measurements obtained from groups of individuals with higher and lower body temperatures. The control group would provide the "normal standard" against which all other samples would be compared relative to the dependent variable.

The second rule governing experimentation is that valid results require that testing be done on large numbers of subjects. It is essential to understand that it is nearly impossible to control all possible variables in biological tests. Indeed, there is a bit of scientific wisdom that mirrors this truth—that is, that laboratory animals, even in the most rigidly controlled and carefully designed experiments, "will do as they damn well please." Thus, stating that the testing of a drug for its pain-killing effects was successful after having tested it on only one postoperative patient would be scientific suicide. Large numbers of patients would have to receive the drug and be monitored for a decrease in postopera-

* *Compare* means to emphasize the similarities between two things, whereas *contrast* means that the differences are to be emphasized.

tive pain before such a statement could have any scientific validity. Then, other researchers would have to be able to uphold those conclusions by running similar experiments. *Repeatability* is an important part of the scientific method, and is the primary basis for acceptance or rejection of many hypotheses.

During experimentation and observation, data must be carefully recorded. Usually, such initial, or raw, data are recorded in tabular (table) form. The table should be labeled to show the variables investigated and the results for each sample. At this point, *accurate recording* of observations is the primary concern. Later, these raw data will be reorganized and manipulated to show more explicitly the outcome of the experimentation.

Some of the observations that you will be asked to make in the anatomy and physiology laboratory will require that a drawing be made. Don't panic! The purpose of making drawings (in addition to providing a record) is to force you to observe things very closely. You need not be an artist (most biological drawings are simple outline drawings), but you do need to be neat and as accurate as possible. It is advisable to use a 4H pencil to do your drawings, because it is easily erased and doesn't smudge. Before beginning to draw, you should examine your specimen closely, studying it as though you were going to have to draw it from memory. For example, when looking at cells you should ask yourself questions such as "What is their shape—the relationship of length and width? How are they joined together?" Then decide precisely what you are going to show and how large the drawing must be to show the necessary detail. After making the drawing, add labels in the margins; and connect them, by straight lines (leader lines), to the structures being named.

Step 4: Manipulation and Analysis of Data

The form of the final data varies, depending on the nature of the data collected. Usually, the final data represent information converted from the original measured values (raw data) to some other form. This may mean that averaging or some other statistical treatment must be applied, or it may require conversions from one kind of units to another. In other cases, graphs may be needed to display the data.

ELEMENTARY TREATMENT OF DATA Only very elementary statistical treatment of data is required in this manual. For example, you will be expected to understand and/or compute an average (mean), percentages, and a range.

Two of these statistics, the average and the range, are useful in describing the *typical* case among a large number of samples evaluated. Let us use a simple example. We will assume that the following heart rates (in beats/min) were recorded during an experiment: 64, 70, 82, 94, 85, 75, 72, 78. If you put these numbers in numerical order, the **range** is easily computed, because the range is the difference between the highest and lowest numbers obtained (highest number minus lowest number). What is the range of the set of numbers just provided?

1. ______________________________ *

The **average,** or **mean,** is obtained by summing the items and dividing the sum by the number of items. Compute the average for the set of numbers just provided:

2. ______________________________

The word *percent* comes from the Latin meaning "for 100"; thus *percent,* indicated by the percent sign, %, means parts per 100 parts. Thus, if we say that 45% of Americans have type O blood, what we are really saying is that among each group of 100 Americans, 45 (45/100) can be expected to have type O blood.

It is very easy to convert any number (including decimals) to a percent. The rule is to move the decimal point two places to the right and add the percent sign. If no decimal point appears, it is *assumed* to be at the end of the number; and zeros are added to fill any empty spaces. Two examples follow:

0.25 = 0.2 5 = 25%

5 = 5 = 500%

Change the following numbers to percents:

3. 38.2 = __________ 5. 1.6 = __________

4. 402 = __________

Note that although you are being asked here to convert numbers to percents, percents by themselves are meaningless. We always speak in terms of a percentage *of* something.

To change a percent to a whole number (or decimal), remove the percent sign, and move the decimal point two places to the left. Change the following percents to whole numbers or decimals:

6. 36% = __________ 8. 25777% = __________

7. 800% = __________ 9. 0.05% = __________

MAKING AND READING LINE GRAPHS For some laboratory experiments you will be required to show your data (or part of them) graphically. Simple line graphs allow relationships within the data to be shown interestingly and allow trends (or patterns) in the data to be demonstrated. An advantage of properly drawn graphs is that they save the reader's time because the essential meaning of large numbers of statistical data can be visualized at a glance.

To aid in making accurate graphs, graph paper (or a printed grid in the manual) is used. Line graphs have both horizontal and vertical scales. Each scale should have uniform intervals—that is, each unit measured on

* Answers are given on page xvii.

the scale should require the same distance along the scale as any other. Variations from this rule may be misleading and result in false interpretations of the data. By convention, the condition that is manipulated (the independent variable) in the experimental series is plotted on the X-axis (the horizontal axis); and the value that we then measure (the dependent variable) is plotted on the Y-axis (the vertical axis). To plot the data, a dot or a small *x* is placed at the precise point where the two variables (measured for each sample) meet; and then a line (this is called the **curve**) is drawn to connect the plotted points.

Sometimes, you will see the curve on a line graph extended beyond the last plotted point. This is (supposedly) done to predict "what comes next." When you see this done, be skeptical. The information provided by such a technique is only slightly more accurate than that provided by a crystal ball!

To read a line graph, pick any point on the line, and match it with the information directly below on the horizontal scale and with that directly to the left of it on the vertical scale. Figure P.1 is a graph that illustrates the relationship between breaths per minute (respiratory rate) and body temperature. Answer the following questions about this graph:

10. What was the respiratory rate at a body temperature of 96°F?

____

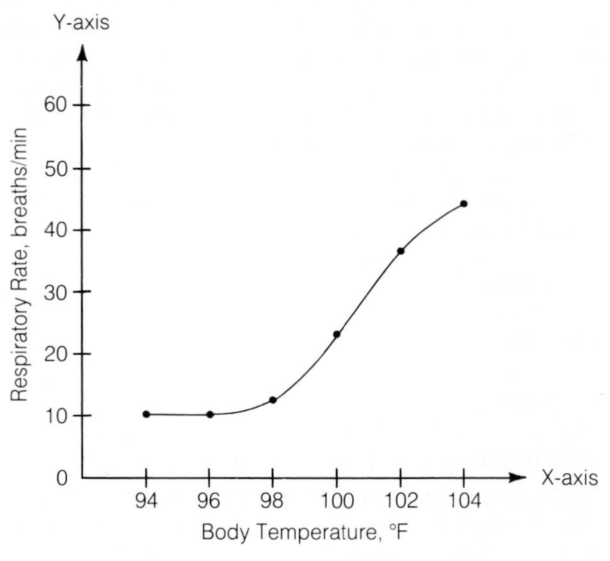

Example of graphically presented data.

11. Between 98° and 102°F, the respiratory rate increased from

_______ to _______ breaths per minute.

12. Between which two body temperature readings was the increase in breaths per minute greatest?

____

13. Are the intervals on each scale uniform?

____

Step 5: Reporting Conclusions of the Study

Drawings, tables, and graphs alone do not suffice as the final presentation of scientific results. The final step requires that you provide a straightforward description of the conclusions drawn from your results. If possible, your findings should be compared to those of other investigators working on the same problem. (For laboratory investigations conducted by students, these comparative figures are provided by classmates.)

It is important to realize that scientific investigations do not always yield the anticipated results. If there are discrepancies between your results and those of others, or what you expected to find based on your class notes or textbook readings, this is the place to try to explain those discrepancies.

Results are often only as good as the observation techniques used. Depending on the type of experiment conducted, several questions may need to be answered. Did you weigh the specimen carefully enough? Did you balance the scale first? Was the subject's blood pressure actually as high as you recorded it, or did you record it hastily (and inaccurately)? If you did record it accurately, is it possible that the subject was emotionally upset about something, which (even though the matter of concern had nothing to do with the experiment) might have given falsely high data for the variable being investigated? Attempting to explain an unexpected result will often teach you more than you would have learned from anticipated results.

When the experiment produces results that are consistent with the hypothesis, then the hypothesis can be said to have reached a higher level of certainty. There is now a greater probability that the hypothesis is correct.

A hypothesis that has been validated by many different investigators is called a **theory.** Theories are useful in two important ways. First, they link sets of data; and second, they make predictions that may lead to additional avenues of investigation. (Okay, we know this with a high degree of certainty; what's next?)

When a theory has been repeatedly verified and appears to have wide applicability in biology, it may assume the status of a **biological principle.** A principle is a statement that applies with a high degree of probability to a range of events. For example, "Living matter is made of cells or cell products" is a principle stated in many biology texts. It is a sound and useful principle, and will continue to be used as such—unless new findings prove it wrong.

We have been through quite a bit of background concerning the scientific method and what its use entails. Because it is important that you remember the phases of the scientific method, they are summarized here:

1. Observation of some phenomenon
2. Statement of a hypothesis (based on the observations)
3. Collection of data (testing the hypothesis with controlled experiments)
4. Manipulation and analysis of the data
5. Reporting of the conclusions of the study

SCIENTIFIC NOTATION AND METRICS

No matter how highly developed our ability to observe, observations have scientific value only if they can be communicated to others. This necessitates the use of scientific notation and the widely accepted system of metric measurements.

Scientific Notation

Because quantitative measurements often yield very large or very small numbers, you are quite likely to encounter numbers such as 3.5×10^{12} or 10^{-3}. It is important that you understand what this **scientific notation** means.

Scientific notation is dependent on the properties of exponents and on the movement of the decimal point when multiplying or dividing by 10. When you multiply 10 by itself, you get a product that is one followed by zeros. The number of zeros (two, in this case) in the product is equal to the number of times you have used 10 as a factor and is shown as an **exponent.** Thus, the following notation

$$\text{base} \rightarrow 10^{2 \leftarrow} \text{exponent}$$

translates to "the base 10 multiplied by itself (10 × 10)."

The powers of 10 are represented as follows:

$10^0 = 1$	(Any number followed by a zero exponent is one.)
$10^1 = 10$	$(10 \times 1 = 10)$
$10^2 = 100$	$(10 \times 10 = 100)$
$10^3 = 1000$	$(10 \times 10 \times 10 = 1000)$
$10^4 = 10{,}000$	$(10 \times 10 \times 10 \times 10 = 10{,}000)$

As you can see, each time the exponent is increased by one, another zero (× 10) is added to the answer.

When you multiply any number by a power of 10 written with exponents, the decimal point is moved to the right the number of times shown in the exponent. Thus:

$3.25 \times 10^1 = 3.2\,5 = 32.5$

$3.25 \times 10^3 = 3.2\,5\_\_ = 3250$

$3.25 \times 10^5 = 3.2\,5\_\_\_\_ = 325{,}000$

By using such exponential notation, very large numbers may be written in far simpler form.

Write the following numbers using the proper scientific notation:

14. 140,000 = 1.4 × _______

15. 9,650,000 = 9.65 × _______

16. 852 = 8.52 × _______

17. 10 = 1.0 × _______

Notice that proper scientific notation entails only one number to the left of the decimal point. Thus 1.03×10^3 is correct, but 10.3×10^2 is not.

In the above examples, all of the numbers used were greater than one. Scientific notation can also be used to report numbers less than one. To do this, negative exponents are used. For example, in

$$3.25 \times 10^2$$

the positive exponent means that the decimal point is to be moved two places to the right, and the number designated is 325 (3.25 × 10 × 10). However, in

$$3.25 \times 10^{-2}$$

the negative exponent means that the number is to be divided by the power of 10 indicated by the exponent and the decimal point is to be moved two places to the left. The number so designated is 0.0325 [3.25 ÷ (10 × 10)].

Thus, the rule for converting scientific notation (using powers of 10) to decimal notation is to move the decimal point the number of places indicated by the exponent. When the exponent is positive (with or without a plus sign), the decimal point is moved to the right. When the exponent is negative (always provided with a minus sign), the decimal point is moved to the left.

For a little practice, write the following numbers in scientific notation: (18–23)

140,000 = 1.4 × ______	45,000 = 4.5 × ________
0.0000063 = 6.3 × ______	0.265 = 2.65 × ________
0.00054 = 5.4 × ______	0.10 = 1.0 × ________

Metrics

Without measurement, we would be limited to qualitative description. However, with a system of measurement, quantitative description becomes possible.

Anyone can establish a system of measurement. All that is required is a reference point; and, historically, much of our common (the British) system of measurement evolved from units based on objects everyone knew. For example, horses were measured in "hands," and a "fathom" was the distance between outstretched arms. However, the variability in such measurements is immediately apparent—for example, an infant's hand is substantially smaller than that of an adult. Therefore, for precise and repeatable communication of information,

TABLE G.1 Commonly Used Units of the Metric System, and Their Fractions and Multiples

Measurement	Unit	Fraction or multiple		Prefix	Symbol
Length	Meter (m)	10^6	one million	mega	M
Volume	Liter (l)	10^3	one thousand	kilo	k
Mass	Gram (g)	10^{-1}	one tenth	deci	d
Time*	Second (s)	10^{-2}	one hundredth	centi	c
Temperature	Degree Celsius (°C)	10^{-3}	one thousandth	milli	m
		10^{-6}	one millionth	micro	μ
		10^{-9}	one billionth	nano	n

*The accepted standard for time is the second; and thus hours and minutes are used in scientific, as well as everyday, measurement of time. The prefixes used in the designation of units of length, mass, and volume are also used in specifying units of time. However, because minutes and hours are terms that indicate *multiples* of seconds, the only prefixes generally used are those indicating *fractional portions* of seconds—for example, millisecond and microsecond.

the agreed-upon system of measurement used by scientists is the **metric system,** a nonvarying standard of reference.

A major advantage of the metric system is that it is based on units of 10. This allows rapid conversion to workable numbers so that neither very large nor very small figures need be used in calculations. Fractions or multiples of the standard units of length, volume, mass, time, and temperature have been assigned specific names. Table G.1 shows the commonly used units of the metric system, along with the prefixes used to designate fractions and multiples thereof.

To change from smaller units to larger units, you must *divide* by the appropriate factor of 10 (because there are fewer of the larger units). For example, a milliunit (milli = one thousandth), such as a milliliter or millimeter, is one step smaller than a centiunit (centi = one hundredth), such as a centiliter or centimeter. Thus to change milliunits to centiunits, you must divide by 10. On the other hand, when converting from larger units to smaller ones, you must *multiply* by the appropriate factor of 10 (because there will be more of the smaller units). A partial scheme for conversions between the metric units is shown below. (See Table G.1.)

Students studying a science or preparing for a profession in the health-related fields find that certain of the metric units are encountered and dealt with more frequently than others. Thus, the objectives of the sections that follow are to provide a brief overview of these most-used measurements and to help you gain some measure of confidence in dealing with them. (A listing of the most frequently used conversion factors, for conversions between British and metric system units, is provided in Appendix A.)

LENGTH MEASUREMENTS The metric unit of length is the **meter (m)**. In addition to measuring things in meters, you will be expected to measure smaller objects in centimeters or millimeters. Subcellular structures are measured in micrometers.

To help you picture these units of length, some equivalents follow:

One meter (m) is slightly longer than one yard (1 m = 39.37 in.).

One centimeter (cm) is approximately the width of a piece of chalk. (Note: there are 2.54 cm in 1 in.)

One millimeter (mm) is approximately the thickness of the wire of a paper clip or of a mark made by a No. 2 pencil lead.

One micrometer (μm) is extremely tiny and can be measured only microscopically.

Make the following conversions between metric units of length: (24–28)

352 cm = ________ mm   12 cm = ________ mm

150 km = ________ m   1 mm = ________ m

2000 μm = ________ mm

Now, circle the answer that would make the most sense in each of the following statements:

29. A match (in a matchbook) is (0.3, 3, 30) cm long.
30. A standard-size American car is about 4 (mm, cm, m, km) long.
31. John pole-vaults a height of 5 meters, whereas

microunit ⇄ (÷ 1000 / × 1000) milliunit ⇄ (÷ 10 / × 10) centiunit ⇄ (÷ 100 / × 100) unit ⇄ (÷ 1000 / × 1000) kilounit

smallest ⇄ largest

Gerry vaults a height of 5 yards. Does John or Gerry make the more difficult vault?

____

VOLUME MEASUREMENTS The metric unit of volume is the liter (l). A **liter (l)** is slightly more than a quart (1 l = 1.057 quarts). Liquid products, measured in liters, are becoming more common, and laboratory solutions are often prepared in 1-liter quantities. Liquid volumes measured out for laboratory experiments are usually measured in milliliter (ml) volumes. (The terms *ml* and *cc,* cubic centimeter, are used interchangeably in laboratory and medical settings.)

To help you visualize metric volumes, the equivalents of some common substances follow:

A 12-oz can of soda is just slightly more than 360 ml.
A cup of coffee is approximately 180 ml.
A fluid ounce is 30 ml (cc).
A teaspoon of vanilla is about 5 ml (cc), and many drug injections are given in 5-ml volumes.

Compute the following:

32. How many 5-ml injections can be prepared from 1 liter of a medicine?

____

33. A 450-ml volume of alcohol is ________ l.

34. The volume of one grape is approximately 0.004 l. What is the volume of the grape in milliliters?

____

MASS MEASUREMENTS Although many people use the terms *mass* and *weight* interchangeably, this usage is inaccurate. **Mass** is the amount of matter in an object; and an object has a constant mass, regardless of where it is—that is, at sea level, on a mountaintop, or in outer space. However, weight varies with gravitational pull; the greater the gravitational pull, the greater the weight. Thus, our astronauts are said to be weightless* when in outer space, but they still have the same mass as they do on earth.

The metric unit of mass is the **gram (g)**, and most objects weighed in the laboratory will be measured in terms of this unit or fractions thereof. Medical dosages are usually prescribed in milligrams (mg) or micrograms (μg); and, in the clinical agency, body weight (particularly of infants) is typically specified in kilograms (kg) (1 kg = 2.2 lb).

The following examples are provided to help you become familiar with the masses of some common objects:

* Astronauts are not *really* weightless. It is just that they and their surroundings are being pulled toward the earth at the same speed; and so, in reference to their environment, they appear to float.

Two aspirin tablets have a mass of approximately 1 g.
A nickel has a mass of 5 g.
The mass of an average woman (132 lb) is 60 kg.

Make the following conversions:

35. 300 g = ________ mg = ________ μg

36. 4000 μg = ________ mg = ________ g

37. A nurse must administer to her patient, Mrs. Smith, 5 mg of a drug per kg of body mass. Mrs. Smith weighs 140 lb. How many grams of the drug should the nurse administer to her patient?

________ g

TEMPERATURE MEASUREMENTS In the laboratory and in the clinical agency, temperature is measured both in metric units (degrees Celsius, °C) and in British units (degrees Fahrenheit, °F). Thus it helps to be familiar with both temperature scales.

The temperatures of boiling and freezing water can be used to compare the two scales:

The boiling point of water is 100°C and 212°F.
The freezing point of water is 0°C and 32°F.

As you can see, the range from the freezing point to the boiling point of water on the Celsius scale is 100 degrees, whereas the comparable range on the Fahrenheit scale is 180 degrees. Hence, one degree on the Celsius scale represents a greater change in temperature. Normal body temperature is approximately 98.6°F and 37°C.

To convert from the Fahrenheit scale to the Celsius scale (or vice versa), the following equation is used:

$$°C = 5(°F - 32)/9$$

For example, to convert 180°C to °F:

$$°C = 5(°F - 32)/9$$
$$180 = 5(°F - 32)/9$$
$$1620 = 5(°F - 32)$$
$$1620 = 5(°F) - 160$$
$$1780 = 5(°F)$$
$$356 = °F$$

and to convert 72°F to °C:

$$°C = 5(°F - 32)/9$$
$$°C = 5(72 - 32)/9$$
$$°C = 5(40)/9$$
$$°C = 200/9$$
$$°C = 22.2$$

Perform the following temperature conversions:

38. Convert 38°C to °F: ________

39. Convert 158°F to °C: ________

Answers

1. range of 94–64; 30 beats/min
2. average 77.5
3. 3820%
4. 40200%
5. 160%
6. 0.36
7. 8
8. 257.77
9. 0.0005
10. 10 breaths/min
11. 12 to 36
12. interval between 100–102° (went from 22 to 36 breaths/min)
13. yes
14. 10^5
15. 10^6
16. 10^2
17. 10^1
18. 1.4×10^5
19. 6.3×10^{-6}
20. 5.4×10^{-4}
21. 4.5×10^4
22. 2.65×10^{-1}
23. 1.0×10^{-1}
24. cm = 3520 mm
25. km = 150,000 m
26. μm = 2 mm
27. cm = 120 mm
28. mm = 0.001 m
29. 3 cm
30. m long
31. John
32. 200
33. 0.45 l
34. 4 ml
35. 300 g = $\underline{3 \times 10^5}$ mg = $\underline{3 \times 10^8}$ μg
36. 4000 μg = $\underline{4}$ mg = $\underline{4 \times 10^{-3}}$ g ($\underline{0.004}$)
37. 0.32 g
38. 100.4°F
39. 70°C

1 EXERCISE

The Language of Anatomy

OBJECTIVES	MATERIALS
1. To describe the anatomical position verbally or by demonstration. 2. To use proper anatomical terminology to describe body directions, planes, and surfaces. 3. To name the body cavities and note the important organs in each.	Human torso model (dissectible) Human skeleton Demonstration: sectioned and labeled kidneys (three separate kidneys uncut or cut so that a. entire, b. transverse section, and c. longitudinal section views are visible)

Scientific terminology is a precise tool that allows a great deal of information to be communicated clearly with a minimum of words. Thus it is not surprising that physicians' orders and progress notes, therapists' records, and nurses' notes use anatomical terminology to describe body parts, regions, positions, and activities. The ability to understand and use correct anatomical terminology is a skill that distinguishes health care personnel who are successful and comfortable in their chosen profession from those perpetually unsure of just what is expected of them.

When anatomists or doctors discuss the human body, they refer to specific areas in accordance with a universally accepted standard position called the **anatomical position.** In the anatomical position the human body is erect, with feet together, head and toes pointed forward, and arms hanging at the sides with palms facing forward (Figure 1.1).

Assume the anatomical position, and note that it is not particularly comfortable. The hands are held unnaturally forward rather than hanging partially cupped and facing toward the thighs.

F1.1

Anatomical position.

BODY PLANES AND SECTIONS

The body is three-dimensional and, in order to observe internal structures, it is often helpful and necessary to make use of a **section,** or cut. When the section is made through the body wall or through an organ, it is made along an imaginary surface or line called a **plane.** Anatomists commonly refer to three planes (Figure 1.2), or sections, which lie at right angles to one another:

Sagittal plane: A plane that runs longitudinally, dividing the body into right and left parts, is referred to as a sagittal plane. If it divides the body into equal parts, right down the median plane of the body, it is called a **midsagittal,** or **median, plane.** All other

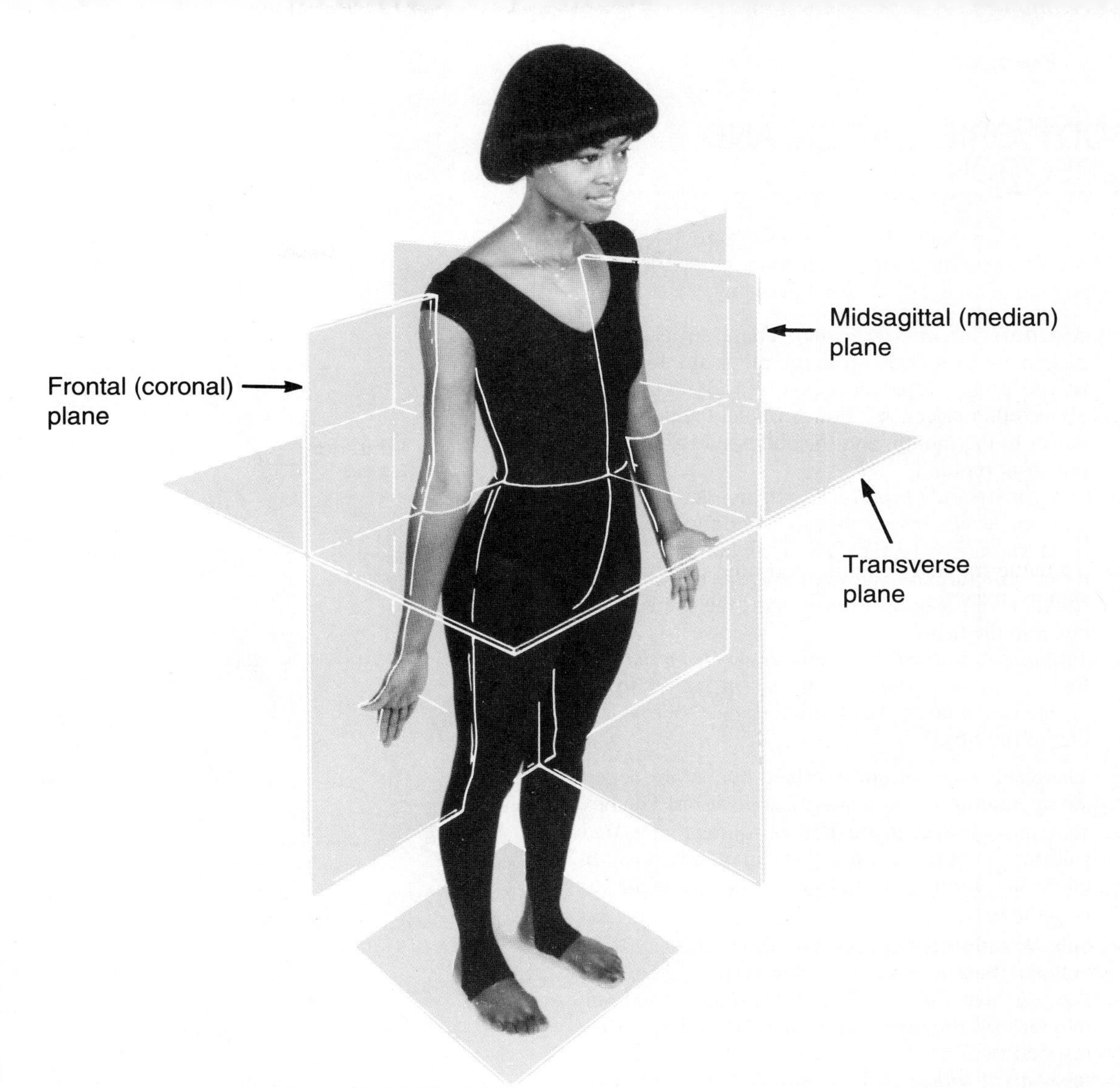

F1.2

Planes of the body.

sagittal planes are referred to as **parasagittal planes.**

Frontal plane: Sometimes called a **coronal plane,** a frontal plane is a longitudinal plane that divides the body (or an organ) into anterior and posterior parts.

Transverse plane: A transverse plane runs horizontally, dividing the body into superior and inferior parts. When organs are sectioned along the transverse plane, the sections are commonly called **cross sections.**

As shown in Figure 1.3, an organ cut along the sagittal or frontal plane provides quite a different view from one cut along the transverse plane.

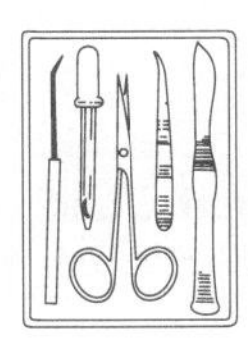

Go to the demonstration area, and observe the transversely and longitudinally cut organ specimens. Pay close attention to the different details of structure seen in the samples provided.

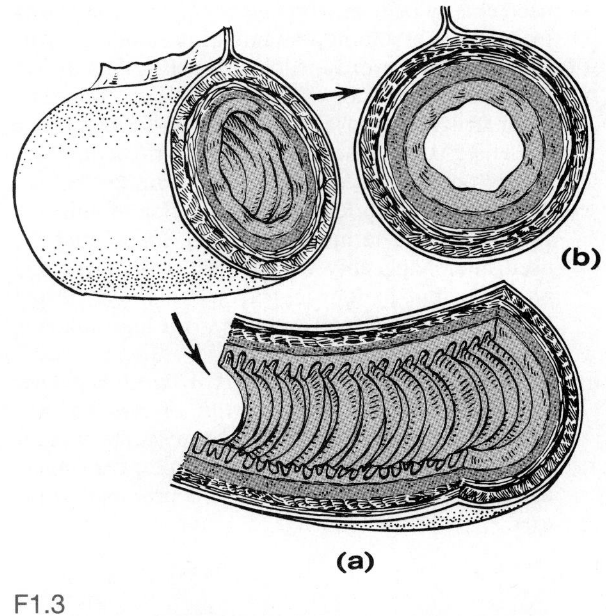

F1.3

Segment of the small intestine. (a) Cut longitudinally; (b) cut transversely.

BODY ORIENTATION AND DIRECTION

Study the terms below, referring to Figure 1.4. Notice that certain terms have a different connotation for a four-legged animal than they do for a human.

Superior/inferior (*above/below*): These terms refer to placement of a body structure along the long axis of the body. Superior structures always appear above other structures. For example, the nose is superior to the mouth, and the abdomen is inferior to the chest region.

Anterior/posterior (*front/back*): In humans the most anterior structures or surfaces are those that are most forward—the face, chest, and abdomen. Posterior structures or surfaces are those toward the backside of the body. For instance, the spine is posterior to the heart.

Medial/lateral (*toward the midline/away from the midline or median plane*): The ear is lateral to the bridge of the nose; the sternum (breastbone) is medial to the ribs.

The terms of position described above are dependent on an assumption of anatomical position. The next four term pairs, however, are more absolute; that is, their applicability is not relative to a particular body position, and they consistently have the same meaning in all vertebrate animals.

Cephalad/caudal (*toward the head/toward the tail*): In humans these terms are used interchangeably with *superior* and *inferior*. But in four-legged animals they are synonymous with *anterior* and *posterior,* respectively.

Dorsal/ventral (*backside/belly side*): These terms are used chiefly in discussing the comparative anatomy of animals, assuming the animal is standing. *Dorsum* is a Latin word meaning "back"; thus *dorsal* refers to the backside of the animal's body or of any other structures. For instance, the posterior surface of the leg is its dorsal surface. The term *ventral* derives from the Latin term *venter,* meaning "belly," and thus always refers to the belly side of animals. In humans the terms *ventral* and *dorsal* may be used interchangeably with the terms *anterior* and *posterior,* but in four-legged animals *ventral* and *dorsal* are synonymous with *inferior* and *superior,* respectively.

Proximal/distal (*nearer the trunk or attached end/farther from the trunk or point of attachment*): These terms are used primarily to locate various areas of the body limbs. For example, the fingers are distal to the elbow; the knee is proximal to the toes.

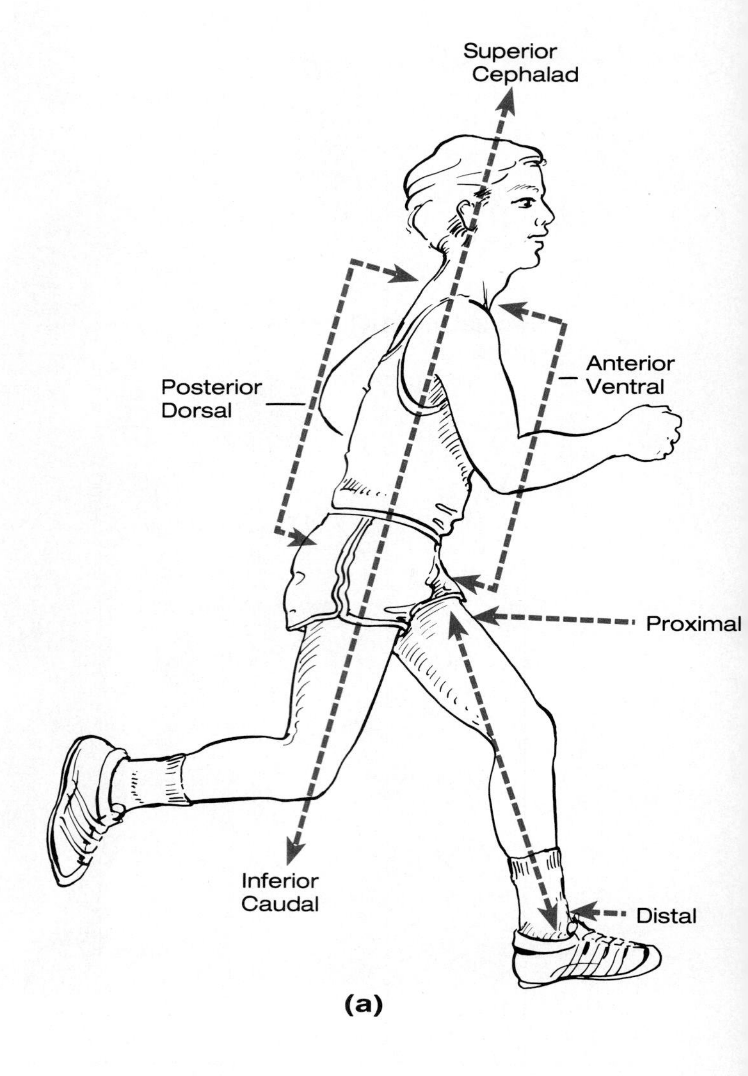

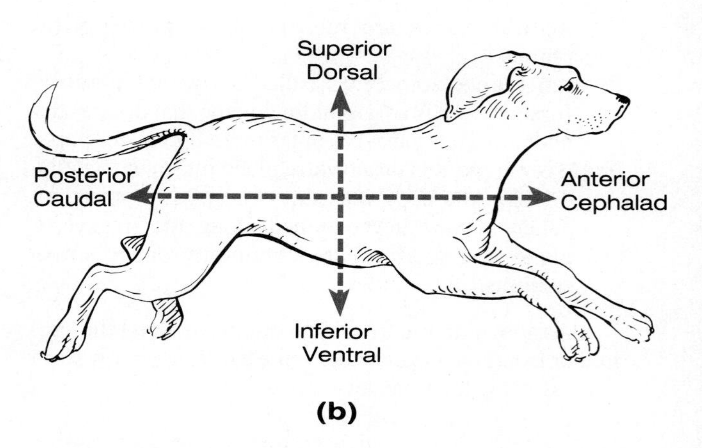

F1.4

Anatomical terminology describing body orientation and direction. (a) With reference to a human; (b) with reference to a four-legged animal.

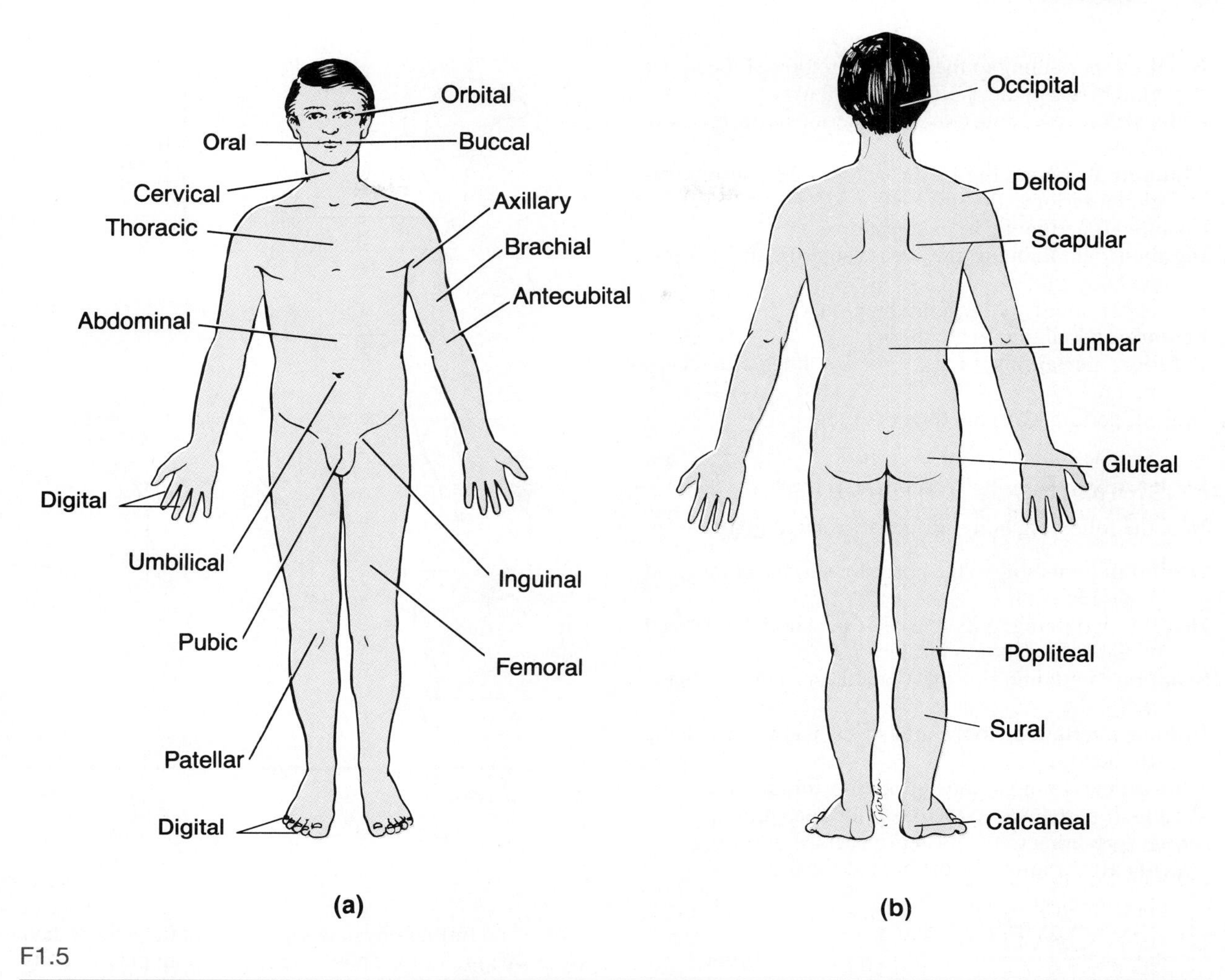

F1.5

Surface anatomy. (a) Anterior body landmarks; (b) posterior body landmarks.

Superficial/deep (*toward or at the body surface/away from the body surface or more internal*): These terms locate body organs in terms of their *relative* closeness to the body surface. For example, the lungs are deep to the rib cage, and the skin is superficial to the skeletal muscles.

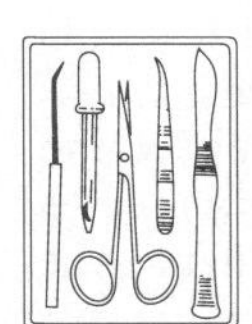

Before continuing, use a human torso model, a skeleton, or your own body to specify the relationship between the following structures. Use the correct anatomical terminology:

The wrist is ______________ to the hand.

The trachea (windpipe) is ______________ to the spine.

The brain is ______________ to the spinal cord.

The kidneys are ______________ to the liver.

The tip of the nose is ______________ to the cheekbones.

SURFACE ANATOMY

Body surfaces provide visible landmarks for study of the body.

Anterior Body Landmarks

Note the following regions in Figure 1.5a:

Oral: pertaining to the mouth
Orbital: pertaining to the bony eye socket (orbit)
Buccal: pertaining to the cheek
Cervical: pertaining to the neck region
Thoracic: pertaining to the chest
Axillary: pertaining to the armpit

Brachial: pertaining to the arm (the region of the upper limb between the shoulder and elbow)
Antecubital: pertaining to the anterior surface of the elbow
Abdominal: pertaining to the anterior body trunk region inferior to the ribs
Umbilical: pertaining to the navel
Inguinal: pertaining to the area where the thigh meets the body trunk
Pubic: pertaining to the genital region
Femoral: pertaining to the thigh
Patellar: pertaining to the anterior knee (kneecap) region
Digital: pertaining to the fingers or toes

Posterior Body Landmarks

Note the following body surface regions in Figure 1.5b.

Occipital: pertaining to the posterior surface of the head or base of skull
Deltoid: pertaining to the curve of the shoulder formed by the large deltoid muscle
Scapular: pertaining to the scapula or shoulder blade area
Lumbar: pertaining to the area of the back between the ribs and hips
Gluteal: pertaining to the buttocks or rump
Popliteal: pertaining to the posterior knee region
Sural: pertaining to the posterior surface of the leg
Calcaneal: pertaining to the heel of the foot

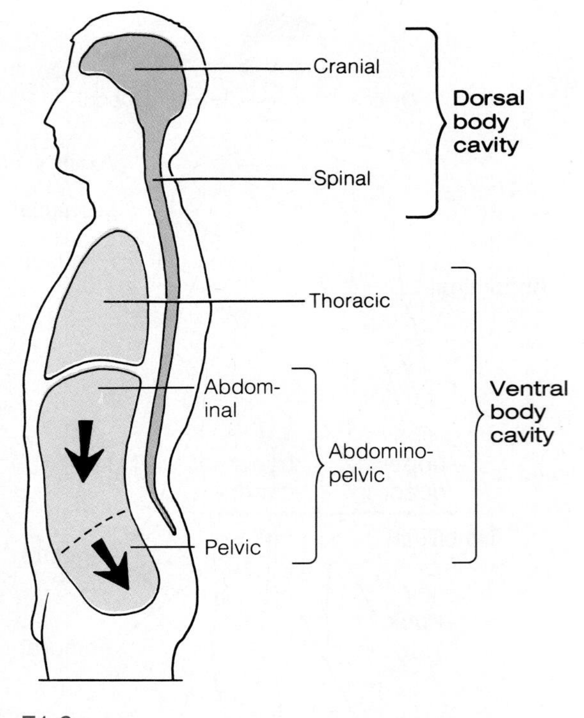

F1.6

Body cavities.

BODY CAVITIES

The body has two sets of cavities, which provide quite different degrees of protection to the organs within them (Figure 1.6).

Dorsal Body Cavity

The dorsal body cavity can be subdivided into the **cranial cavity,** in which the brain is enclosed within the rigid skull, and the **spinal cavity,** in which the delicate spinal cord is protected within a bony vertebral column. Because the spinal cord is a continuation of the brain, these cavities are continuous with each other.

Ventral Body Cavity

Like the dorsal cavity, the ventral body cavity is subdivided. The superior **thoracic cavity** is separated from the rest of the ventral cavity by the dome-shaped diaphragm. The heart and lungs, located in the thoracic cavity, are afforded some measure of protection by the bony rib cage. The cavity inferior to the diaphragm is often referred to as the **abdominopelvic cavity,** since there is no further physical separation of the ventral cavity. Some prefer to subdivide the abdominopelvic cavity into a superior **abdominal cavity,** which houses the stomach, intestines, liver, and other organs, and an inferior **pelvic cavity,** partially enclosed by the bony pelvis and containing the reproductive organs, bladder, and rectum. Note in Figure 1.6 that the abdominal and pelvic cavities are not continuous with each other in a straight plane but that the pelvic cavity is tipped away from the perpendicular.

ABDOMINOPELVIC QUADRANTS & REGIONS

Because the abdominopelvic cavity is quite large and contains many organs, it is helpful to divide it up into smaller areas for discussion or study. The scheme used most by physicians and nurses divides the abdominopelvic cavity into four more or less equal regions called **quadrants;** the quadrants are then simply named according to their relative positions—that is, *right upper quadrant, right lower quadrant, left upper quadrant,* and *left lower quadrant* (see Figure 1.7a).

Another system commonly used by anatomists divides the abdominopelvic cavity into nine separate regions by four planes, as shown in Figure 1.7b. Al-

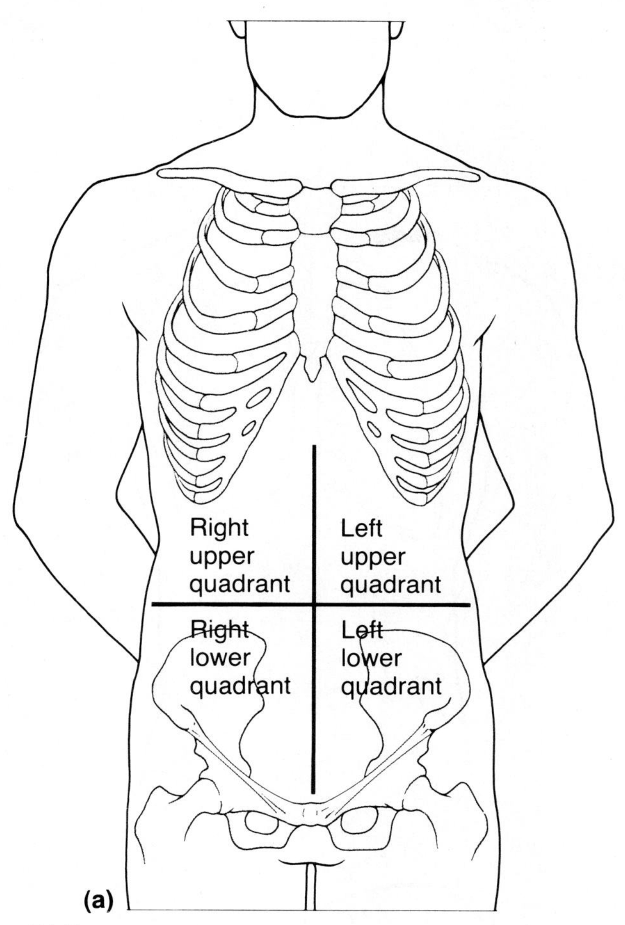

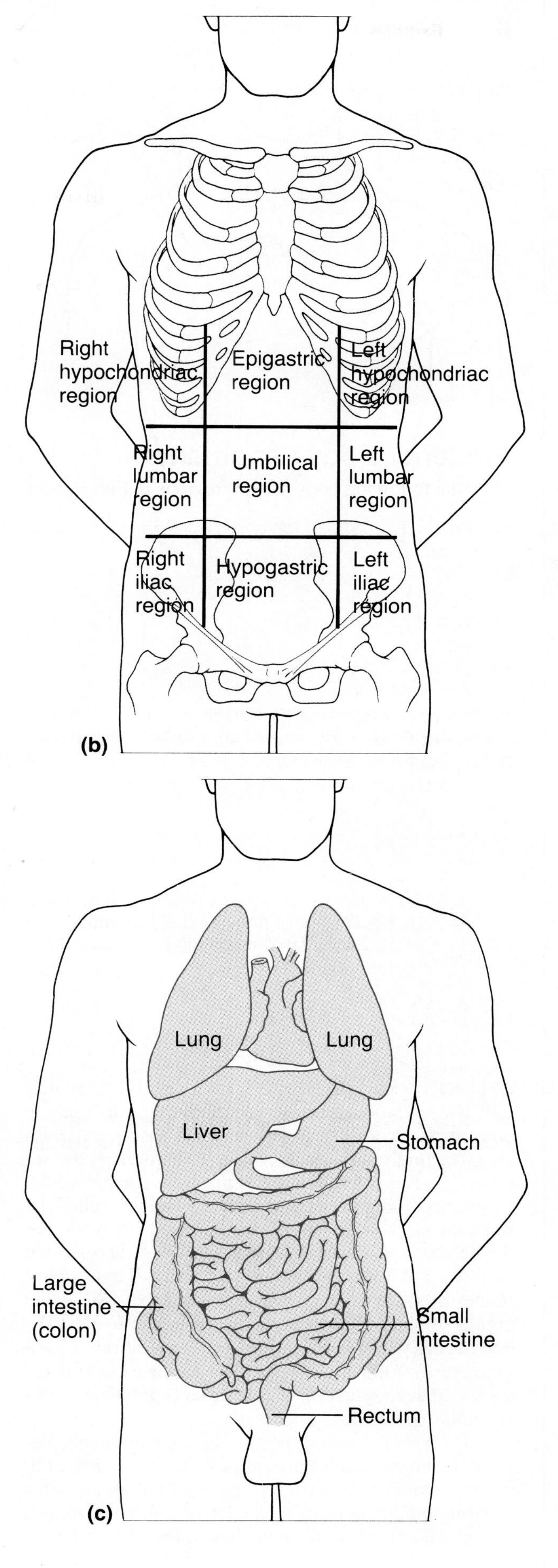

F1.7

Abdominopelvic surface and cavity. (a) The four quadrants. (b) Nine regions delineated by four planes. The superior horizontal plane is just inferior to the ribs, the inferior horizontal plane is at the superior aspect of the hip bones, and the vertical planes are just medial to the nipples. (c) Anterior view of the abdominopelvic cavity showing superficial organs.

though the names of the nine regions are unfamiliar to you now, with a little patience and study they will become easier to remember. As you read through the descriptions of these nine regions below and locate them in the figure, notice the organs they contain by referring to Figure 1.7c.

Umbilical region: the centermost region, which includes the umbilicus

Epigastric region: immediately superior to the umbilical region; overlies most of the stomach

Hypogastric region: immediately inferior to the umbilical region; encompasses the pubic area

Iliac regions: lateral to the hypogastric regions and overlying the inferior portion of the hip bones

Lumbar regions: between the ribs and the flaring portions of the hip bones

Hypochondriac regions: flanking the epigastric regions and overlying the lower ribs

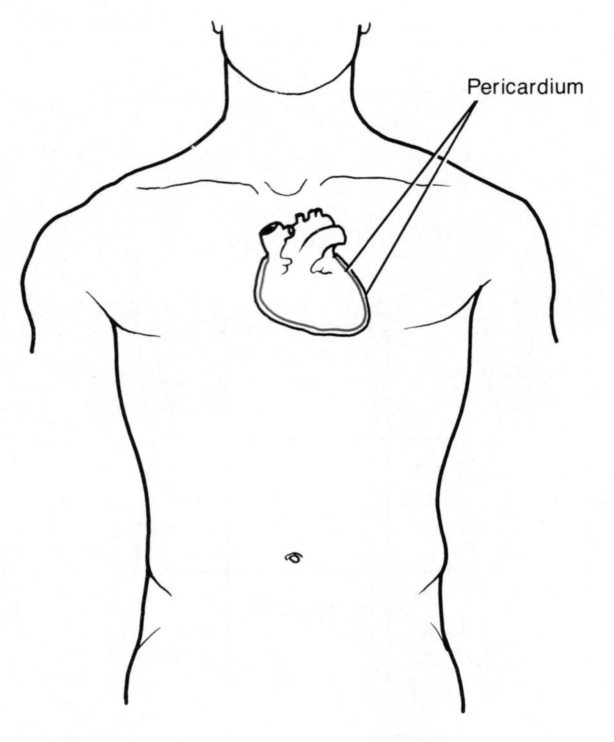

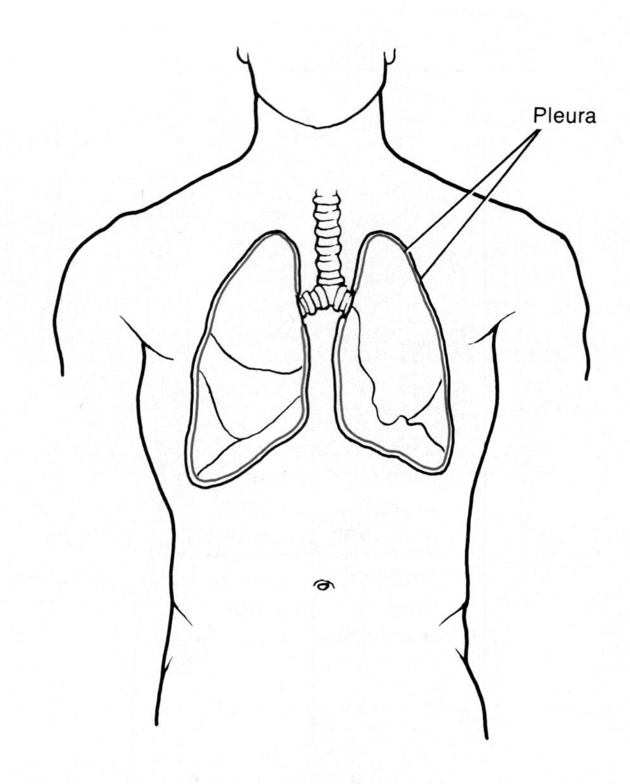

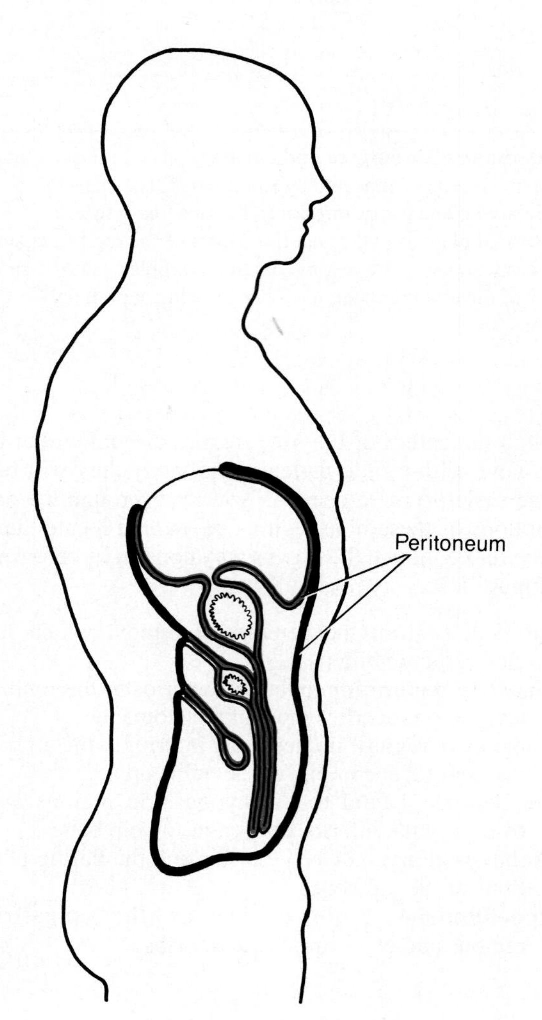

F1.8

Serous membranes. The visceral layer is shown in color; the parietal layer is shown in black.

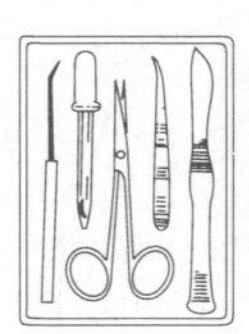

Locate the regions of the abdominal surface on a torso model and on yourself before continuing.

SEROUS MEMBRANES OF THE VENTRAL BODY CAVITY The walls of the ventral body cavity and the outer surfaces of the organs it contains are covered with an exceedingly thin, double-layered membrane, the **serosa,** or **serous membrane** (Figure 1.8). The part of the membrane lining the cavity walls is generally called the **parietal serosa,** and it is continuous with a similar membrane, the **visceral serosa,** covering the external surface of the organs within the cavity. These membranes produce a thin lubricating fluid that allows the organs to slide over one another or to rub against the body wall without friction. Serous membranes also function to compartmentalize the various visceral organs so that an infection of one organ is prevented from spreading to others.

The specific names of the serous membranes depend on the structures they envelop. Thus the serosa lining the abdominal cavity and covering its organs is the **peritoneum,** that enclosing the lungs is the **pleura,** and that around the heart is the **pericardium** (Figure 1.8).

Organ Systems Overview

OBJECTIVES

1. To name the human organ systems and state the major functions of each.
2. To list two or three organs of each system, and categorize the various organs by organ system.
3. To identify these organs in a dissected rat or on a dissectible human torso model or human cadaver.
4. To identify the correct organ system for each organ when presented with a list of organs (as studied in the laboratory).

MATERIALS

Freshly killed or preserved rat, predissected by instructor as a demonstration or for dissection by students (one for every two to four students), or dissected human cadaver
Dissecting pans and pins
Scissors
Forceps
Twine
Disposable plastic gloves
Human torso model (dissectible)

The basic unit, or building block, of all living things is the **cell.** Cells fall into four different categories according to their structures and functions. Each of these corresponds to one of the four **tissue** types: epithelial, muscular, nervous, and connective. An **organ** is a structure composed of two or more tissue types that performs a specific function for the body. For example, the small intestine, which digests and absorbs nutrients, is composed of all four tissue types. An **organ system** is a group of organs that act together to perform a particular body function. For example, the organs of the digestive system work together to assure that food moving through the digestive system is properly broken down and that the end products are absorbed into the bloodstream to provide nutrients and fuel for all the body's cells. In all, there are 10 organ systems, which are described in Table 2.1. In addition to these 10 organ systems, there is a *functional system,* the immune system, which is composed of an army of mobile cells (rather than organs) that act to protect the body from foreign substances. Read through this tabular summary before beginning the rat dissection.

RAT DISSECTION

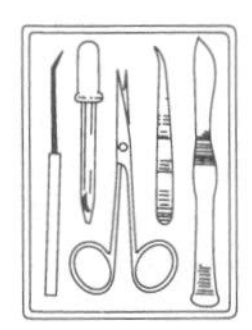

Now you will have a chance to observe the size, shape, location, and distribution of the organs and organ systems. Many of the external and internal structures of the rat are quite similar in structure and function to those of the human, so a study of the gross anatomy of the rat should help you understand your own physical structure.

The following instructions have been written to complement and direct the student's dissection and observation of a rat, but the descriptions for organ observations from procedure 4 under **Ventral Body Cavity** (p. 11) apply as well to superficial observations of a previously dissected human cadaver. In addition, the general instructions for observation of external or oral cavity structures can easily be extrapolated to serve human cadaver observations.

Note that four of the organ systems listed in Table 2.1 will not be studied at this time (integumentary, nervous, skeletal, and muscular), as they require microscopic study or more detailed dissection.

External Structures

1. Obtain a preserved or freshly killed rat (one for every two to four students), a dissecting pan, dissecting pins, scissors, forceps, and disposable gloves.

2. Don the gloves before beginning your observations. This precaution is particularly important when handling freshly killed animals which may harbor internal parasites.

3. Observe the major divisions of the animal's body—head, trunk, and extremities. Compare these divisions to those of humans.

Oral Cavity

1. Examine the structures of the oral cavity. Identify the teeth and tongue. Observe the extent of the hard palate (the portion underlain by bone) and the soft palate (immediately posterior to the hard palate, with no bony support). Note that the posterior end of the oral cavity leads into the throat, or pharynx. The pharynx is a passageway used by both the digestive and respiratory systems.

TABLE 2.1 Overview of Organ Systems of the Body

Organ system	Major component organs	Function
Integumentary (Skin)	Epidermal and dermal regions; cutaneous sense organs and glands	• Protects deeper organs from mechanical, chemical, and bacterial injury, and desiccation (drying out) • Excretion of salts and urea • Aids in regulation of body temperature • Produces vitamin D
Skeletal	Bones, cartilages, tendons, ligaments, and joints	• Body support and protection of internal organs • Provides levers for muscular action • Cavities provide a site for blood cell formation
Muscular	Muscles attached to the skeleton	• Primarily function to contract or shorten; in doing so, skeletal muscles allow locomotion (running, walking, etc.), grasping and manipulation of the environment, and facial expression • Generate heat
Nervous	Brain, spinal cord, nerves, and sensory receptors	• Allows body to detect changes in its internal and external environment and to respond to such information by activating appropriate muscles or glands • Helps maintain homeostasis of the body via rapid transmission of electrical signals
Endocrine	Pituitary, thyroid, parathyroid, adrenal, and pineal glands; ovaries, testes, and pancreas	• Helps maintain body homeostasis, promotes growth and development; produces chemical "messengers" (hormones) that travel in the blood to exert their effect(s) on various "target organs" of the body
Cardiovascular	Heart, blood vessels, and blood	• Primarily a transport system that carries blood containing oxygen, carbon dioxide, nutrients, wastes, ions, hormones, and other substances to and from the tissue cells where exchanges are made; blood is propelled through the blood vessels by the pumping action of the heart • Antibodies and other protein molecules in the blood act to protect the body
Lymphatic	Lymphatic vessels, lymph nodes, spleen, thymus, tonsils, and scattered collections of lymphoid tissue	• Picks up fluid leaked from the blood vessels and returns it to the blood • Cleanses blood of pathogens and other debris • Houses lymphocytes that act in body immunity
Respiratory	Nasal passages, pharynx, larynx, trachea, bronchi, and lungs	• Keeps the blood continuously supplied with oxygen while removing carbon dioxide
Digestive	Oral cavity, esophagus, stomach, small and large intestines, and accessory structures (teeth, salivary glands, liver, and pancreas)	• Acts to break down ingested foods to minute particles, which can be absorbed into the blood for delivery to the body cells • Undigested residue removed from the body as feces
Urinary	Kidneys, ureters, bladder, and urethra	• Rids the body of nitrogen-containing wastes (urea, uric acid, and ammonia), which result from the breakdown of proteins and nucleic acids by body cells • Maintains water, electrolyte, and acid-base balance of blood
Reproductive	Male: testes, scrotum, penis, and duct system, which carries sperm to the body exterior Female: ovaries, uterine tubes, uterus, and vagina	• Provides germ cells (sperm and eggs) for perpetuation of the species • Female uterus houses the developing fetus until birth
Immune (functional system)	Major components are cells (lymphocytes and macrophages) that inhabit the lymphoid tissues and circulate in blood and lymph	• Protects the body via the immune response from foreign substances (antigens)

Ventral Body Cavity

1. Pin the animal to the wax of the dissecting pan by placing its dorsal side down and securing its extremities to wax. If the dissecting pan is not waxed, you will need to secure the animal with twine as shown in Figure 2.1a. (Some may prefer this method in any case.) Obtain the roll of twine. Make a loop knot around one upper limb, pass the twine under the pan, and secure the opposing limb. Repeat for the lower extremities.

2. Lift the abdominal skin with a forceps, and cut through it with the scissors (Figure 2.1b). Close the scissor blades and insert them under the cut skin. Moving in a cephalad direction, open and close the blades to loosen the skin from the underlying connective tissue and muscle. Once this skin-freeing procedure has been completed, cut the skin along the midline, from the pubic region to the lower jaw (Figure 2.1c). Make a lateral cut about halfway down the ventral surface of each limb. Complete the job of freeing the skin with the scissor tips, and pin the flaps to the tray. The underlying tissue that is now exposed is the skeletal musculature of the body wall and limbs. It allows voluntary body movement. Note that the muscles are packaged in sheets of pearly white connective tissue (fascia), which protect the muscles and bind them together.

3. Carefully cut through the muscles of the abdominal wall in the pubic region, avoiding the underlying organs. Remember, to *dissect* means "to separate"—not mutilate! Now, hold and lift the muscle layer with a forceps and cut through the muscle layer from the pubic region to the bottom of the rib cage. Make two lateral cuts through the base of the rib cage (Figure 2.2). A thin membrane attached to the inferior boundary of the rib cage should be obvious; this is the **diaphragm,** which separates the thoracic and abdominal cavities. Cut the diaphragm away to loosen the rib cage. You can now lift the ribs to view the contents of the thoracic cavity.

4. Examine the structures of the thoracic cavity, starting with the most superficial structures and working deeper. As you work, refer to Figure 2.3, which shows the superficial organs.

Thymus: an irregular mass of glandular tissue overlying the heart

Push the thymus to the side to view the heart.

Heart: median oval structure enclosed within the pericardium (serous membrane sac)
Lungs: flanking the heart on either side

Now observe the throat region to identify the trachea.

Trachea: tubelike "windpipe" running medially down the throat; part of the respiratory system

Follow the trachea into the thoracic cavity; note where it divides into the bronchi.

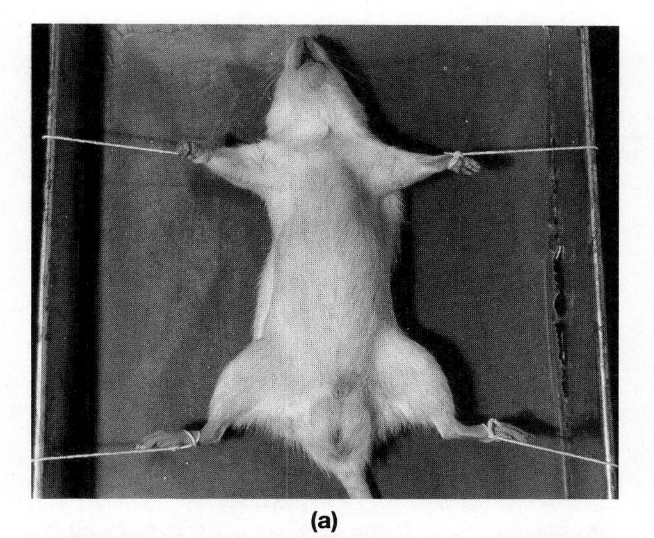

(a)

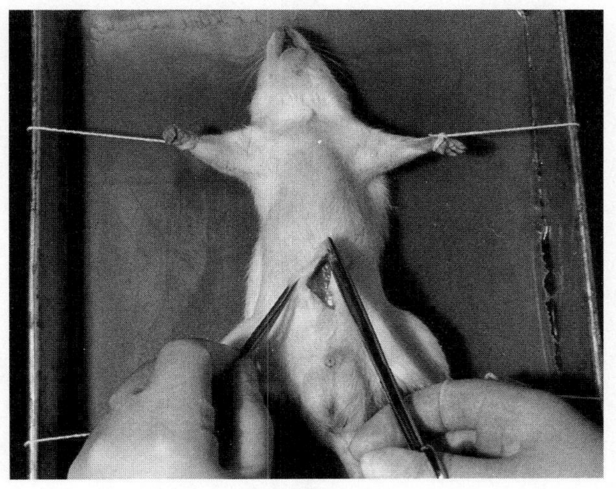

(b)

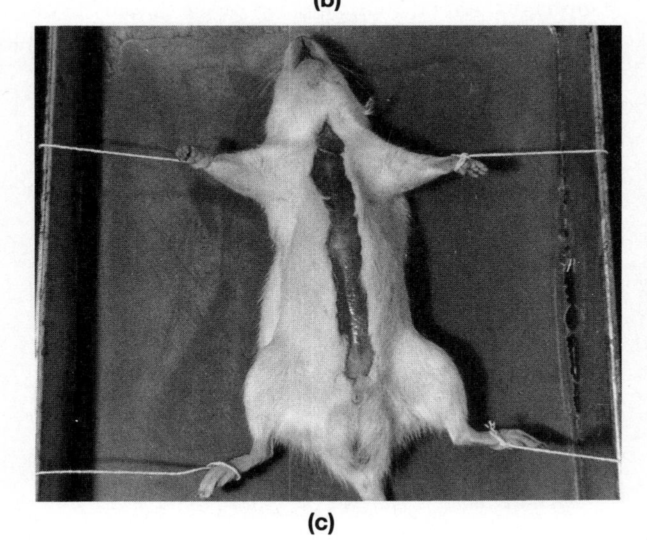

(c)

F2.1

Rat dissection. Securing for dissection and the initial incision. (a) Securing the rat to the dissection tray with twine; (b) using scissors to make the incision on the medial line of the abdominal region; (c) completed incision from the pelvic region to the lower jaw.

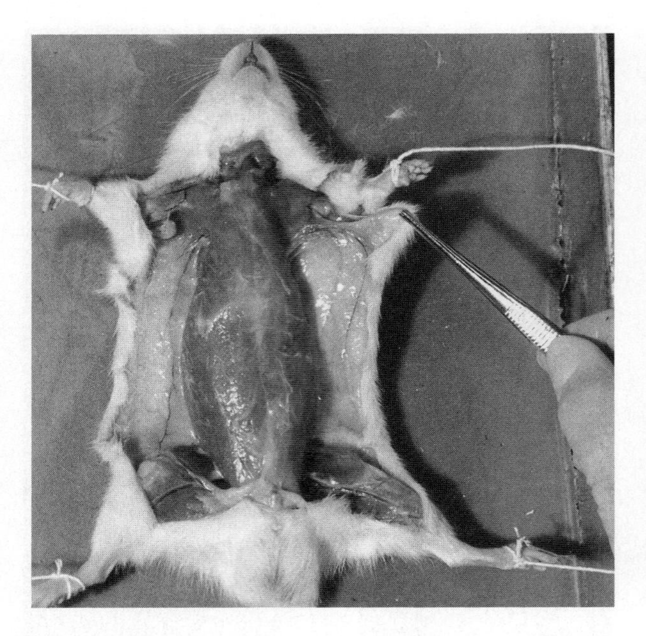

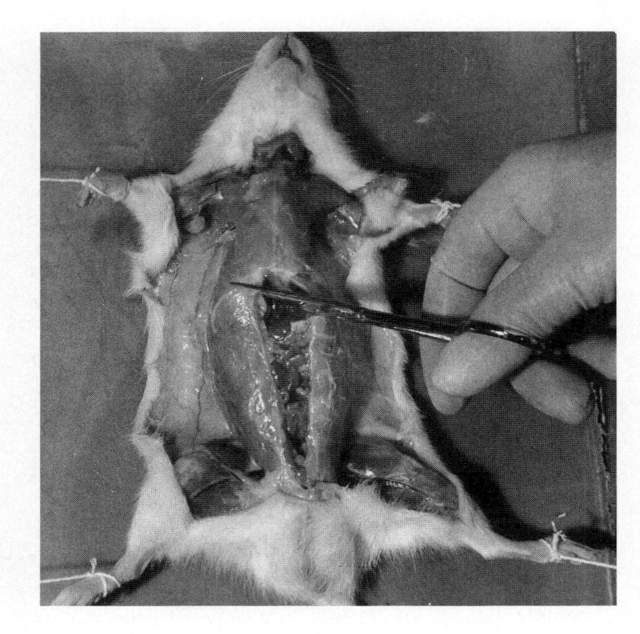

F2.2

Rat dissection. Exposing the skeletal musculature and making lateral cuts at the base of the rib cage.

Bronchi: two passageways that plunge laterally into the tissue of the two lungs

Now push the trachea to one side to expose the esophagus.

Esophagus: a food chute; the part of the digestive system that transports food from the pharynx (throat) to the stomach

Follow the esophagus through the diaphragm to its junction with the stomach.

Stomach: a C-shaped organ important in food digestion and temporary food storage

5. Examine the superficial structures of the abdominopelvic cavity. Beginning with the stomach, trace the rest of the digestive tract.

Small intestine: connected to the stomach and ending just before a large saclike cecum
Cecum: the initial portion of the large intestine
Large intestine: a large muscular tube coiled within the abdomen

Follow the course of the large intestine to the rectum, which is partially covered by the urinary bladder.

Rectum: terminal portion of the large intestine; continuous with the anal canal
Anus: the opening of the digestive tract (anal canal) to the exterior

Now lift the small intestine with the forceps to view the mesentery.

Mesentery: an apronlike serous membrane; suspends many of the digestive organs in the abdominal cavity. Notice that it is heavily invested with blood vessels and, more likely than not, riddled with large fat deposits.

Locate the remaining abdominal structures.

Pancreas: a diffuse gland; rests in and behind the mesentery between the first portion of the small intestine and the stomach
Spleen: a dark red organ curving around the left lateral side of the stomach; considered part of the lymphatic system and often called the red blood cell graveyard
Liver: large and brownish red; the most superior organ in the abdominal cavity, directly beneath the diaphragm

6. To locate the deeper structures of the abdominopelvic cavity, cut through the superior margin of the stomach and the distal end of the large intestine and lay them aside. (Refer to Figure 2.4 as you work.)

Examine the posterior wall of the abdominal cavity to locate the two kidneys.

Kidneys: bean-shaped organs; retroperitoneal (behind the peritoneum)

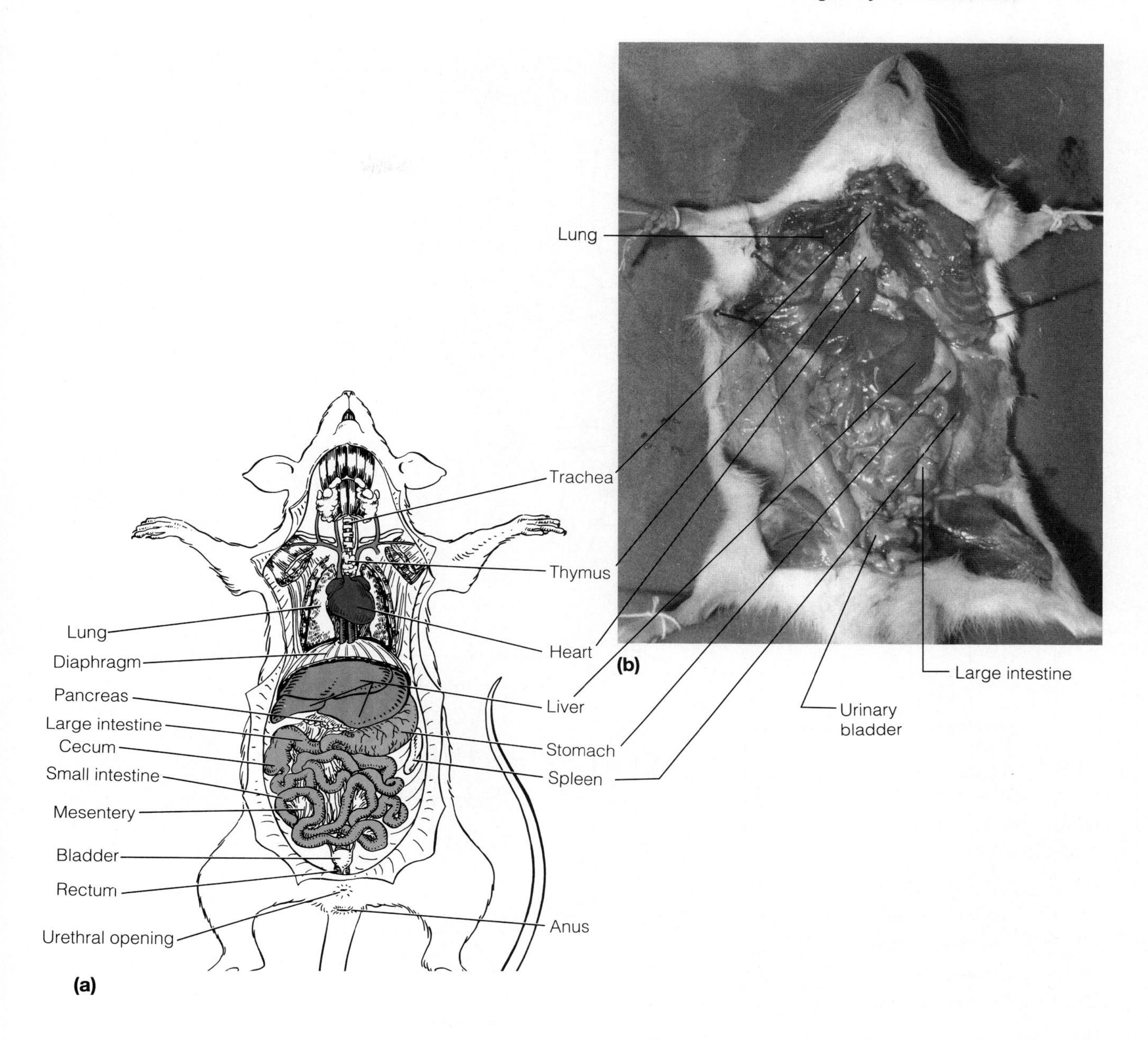

F2.3

Rat dissection. Superficial organs of the thoracic and abdominal cavities (a) Diagrammatic view; (b) photograph.

Adrenal glands: large glands that sit astride the superior margin of each kidney; part of the endocrine system

Carefully strip away part of the peritoneum and attempt to follow the course of one of the ureters to the bladder.

Ureter: tube running from the indented region of a kidney to the urinary bladder

Urinary bladder: the sac that serves as a reservoir for urine

7. In the midline of the body cavity, lying between the kidneys, are the two principal abdominal blood vessels. Identify each.

Inferior vena cava: the large vein that returns blood to the heart from the lower regions of the body

Descending aorta: deep to the inferior vena cava; the largest artery of the body; carries blood away from the heart, down the midline of the body

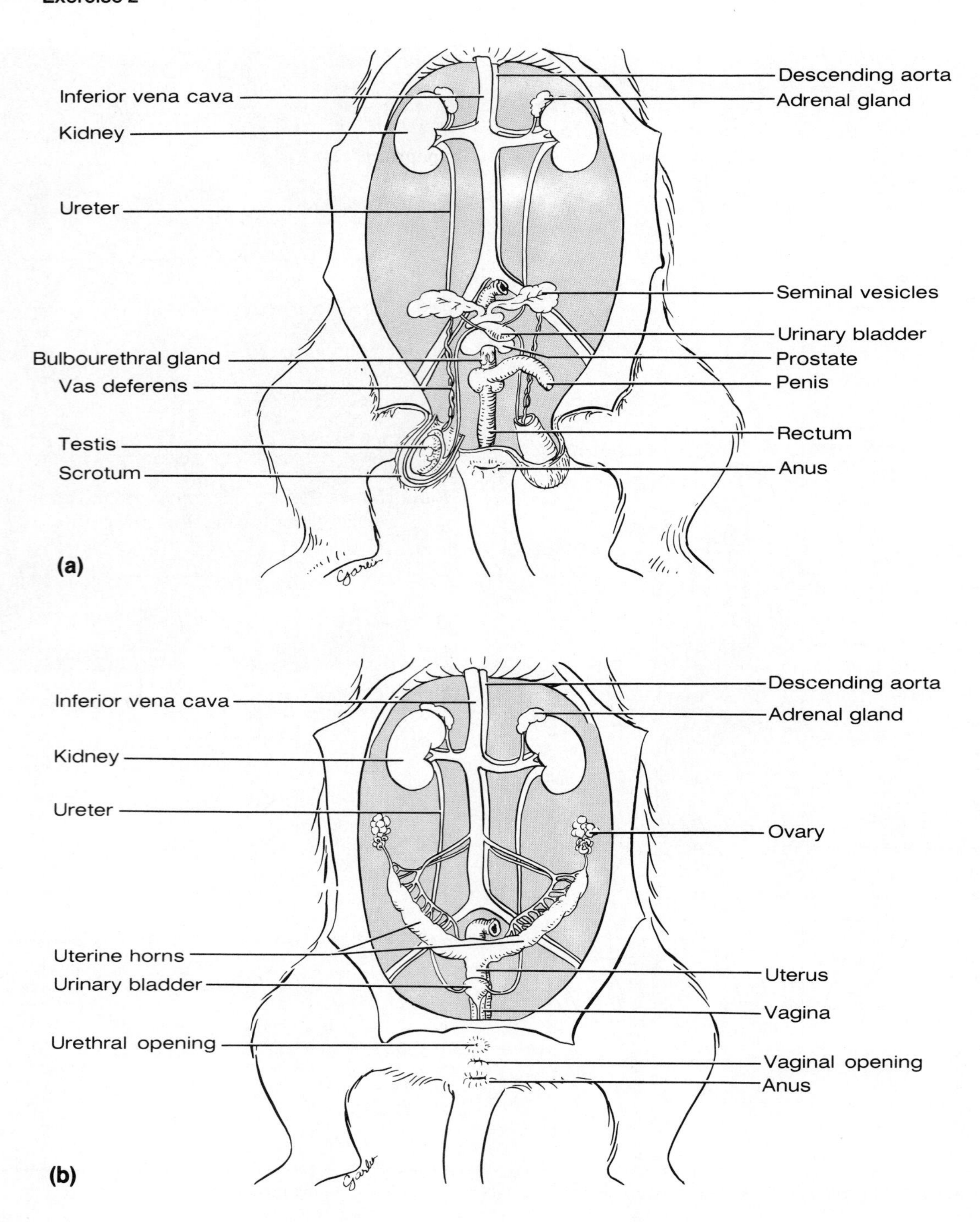

F2.4

Rat dissection. Deeper organs of the abdominal cavity and the reproductive structures. (a) Male; (b) female.

8. Only a cursory examination of reproductive organs will be done. First determine if the animal is a male or female. Observe the ventral body surface beneath the tail. If a saclike scrotum and a single body opening are visible, the animal is a male. If three body openings are present, it is a female. (See Figure 2.4.)

MALE ANIMAL Make a shallow incision into the scrotum. Loosen and lift out the oval **testis.** Exert a gentle pull on the testis to identify the slender **vas deferens,** or sperm duct, which carries sperm from the testis superiorly into the abdominal cavity and joins with the urethra. The urethra runs through the penis of the male, and carries both urine and sperm out of the body. Identify the **penis,** extending from the bladder to the ventral body wall. Figure 2.4a also shows some glands of the male reproductive system, but they need not be identified at this time.

FEMALE ANIMAL Inspect the pelvic cavity to identify the Y-shaped **uterus** lying against the dorsal body wall and beneath the bladder (Figure 2.4b). Follow one of the uterine horns superiorly to identify an **ovary,** a small oval structure at the end of the uterine horn. (The rat uterus is quite different from the uterus of human females, which is a single-chambered organ about the size and shape of a pear.) The inferior undivided part of the rat uterus is continuous with the vagina, which leads to the body exterior. Identify the **vaginal orifice** (external vaginal opening).

9. When you have finished your observations, store or dispose of the rat according to your instructor's directions. Wash the dissecting pan and tools with laboratory detergent. Dispose of the gloves. Then wash and dry your hands before continuing with your examination of the torso model.

EXAMINING THE HUMAN TORSO MODEL

Examine a human torso model to identify the following organs:

Dorsal cavity: brain, spinal cord
Thoracic cavity: heart, lungs, bronchi, trachea, esophagus, diaphragm
Abdominopelvic cavity: liver, stomach, pancreas, spleen, small intestine, large intestine, rectum, kidneys, ureters, bladder, adrenal glands, descending aorta, inferior vena cava

As you observe these structures, locate the nine abdominopelvic areas studied earlier, and determine which organs would be found in each area.

umbilical region ______________________

epigastric region ______________________

hypogastric region ______________________

right iliac region ______________________

left iliac region ______________________

right lumbar region ______________________

left lumbar region ______________________

right hypochondriac region ______________________

left hypochondriac region ______________________

Would you say that the shape and location of the human organs are similar or dissimilar to those of the rat?

____

Assign each of the organs just identified to one of the organ system categories below.

Digestive: ______________________

Urinary: ______________________

Cardiovascular: ______________________

Reproductive: ______________________

Respiratory: ______________________

Lymphatic: ______________________

Nervous: ______________________

3 EXERCISE

The Microscope

OBJECTIVES

1. To identify the parts of the microscope and list the function of each.
2. To describe and demonstrate the proper techniques for care of the microscope.
3. To define *total magnification* and *resolution.*
4. To demonstrate the proper focusing technique.
5. To define *parfocal, field,* and *depth of field.*
6. To estimate the size of objects in a field.

MATERIALS

Compound microscope
Millimeter ruler
Prepared slides of the letter *e* or newsprint
Immersion oil
Lens paper
Prepared slide of grid ruled in millimeters (grid slide)
Prepared slide of 3 crossed colored threads
Clean microscope slide and coverslip
Toothpicks (flat-tipped)
Physiologic saline in a dropper bottle
Methylene blue stain (dilute) in a dropper bottle
Filter paper
Forceps
Large beaker containing fresh 10% household bleach solution for wet-mount disposal
Disposable autoclave bag

With the invention of the microscope, biologists gained a valuable tool to observe and study structures (like cells) that are too small to be seen by the unaided eye. As a result, many of the theories basic to the understanding of the biological sciences have been established. This exercise will familiarize you with the workhorse of microscopes—the compound microscope—and provide you with the necessary instructions for its proper use.

CARE AND STRUCTURE OF THE COMPOUND MICROSCOPE

The compound microscope is a precision instrument and should always be handled with care. At all times you must observe the following rules for its transport, cleaning, use, and storage:

- When transporting the microscope, hold it in an upright position with one hand on its arm and the other supporting its base. Avoid jarring the instrument when setting it down.
- Use only special grit-free lens paper to clean the lenses. Clean all lenses before and after use.
- Always begin the focusing process with the lowest-power objective lens in position, changing to the higher-power lenses if necessary.
- *Never* use the coarse adjustment knob with the high-power or oil immersion lens.
- A coverslip must always be used with temporary (wet-mount) preparations.
- Before putting the microscope in the storage cabinet, remove the slide from the stage, rotate the lowest-power objective lens into position, and replace the dust cover.
- Never remove any parts from the microscope; inform your instructor of any mechanical problems that arise.

Note to the Instructor: The slides and coverslips used for viewing cheek cells are to be soaked for two hours (or longer) in 10% bleach solution and then drained. The slides, coverslips, and disposable autoclave bag (containing used toothpicks) are to be autoclaved for 15 min at 121°C and 15 pounds pressure to insure sterility. After autoclaving, the disposable autoclave bag may be discarded in any disposal facility and the glassware washed with laboratory detergent and reprepared for use. These instructions apply as well to any blood-stained glassware or disposable items used in other experimental procedures.

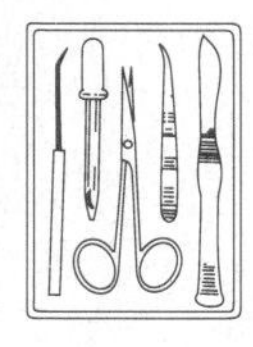

1. Obtain a microscope, and bring it to the laboratory bench. (Use the proper carrying technique!) Compare your microscope with the illustration in Figure 3.1, and identify the following microscope parts:

Base: supports the microscope. (Note: some microscopes are provided with an inclination joint, which allows the instrument to be tilted backward for viewing dry preparations.)

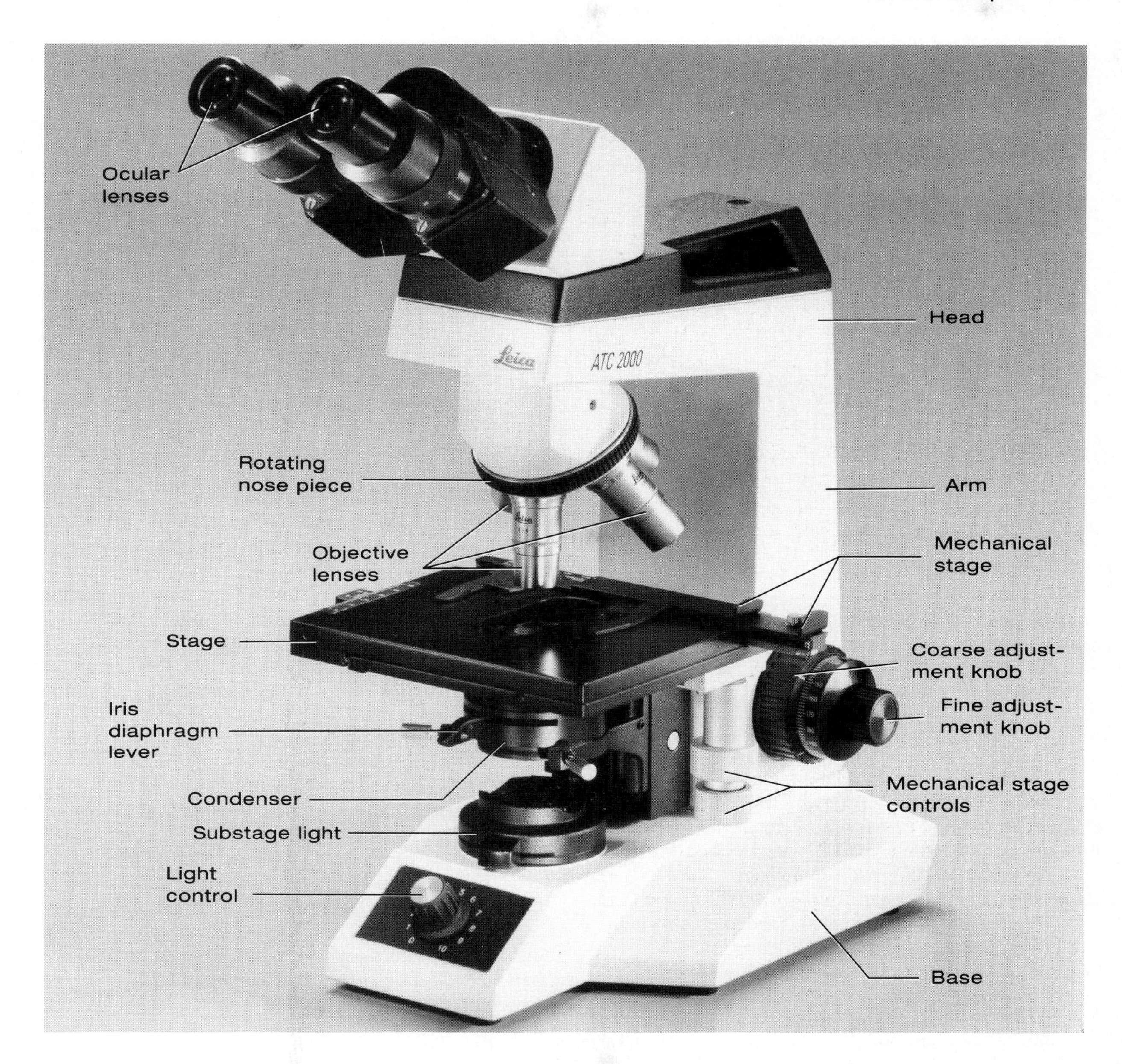

F3.1

Compound microscope and its parts.

Substage light (or *mirror*): located in the base. In microscopes with a substage light source, the light passes directly upward through the microscope. If a mirror is used, light must be reflected from a separate free-standing lamp.

Stage: the platform the slide rests on while being viewed. The stage always has a hole in it to permit light to pass through both it and the specimen. Some microscopes have a stage equipped with *spring clips*; others have a clamp-type *mechanical stage.* Both hold the slide in position for viewing; in addition, the mechanical stage permits precise movement of the specimen.

Condenser: concentrates the light on the specimen. The condenser may be equipped with a height-adjustment knob that raises and lowers the condenser to vary the delivery of the light. Generally, the best position for the condenser is close to the inferior surface of the stage.

Iris diaphragm lever: arm attached to the condenser

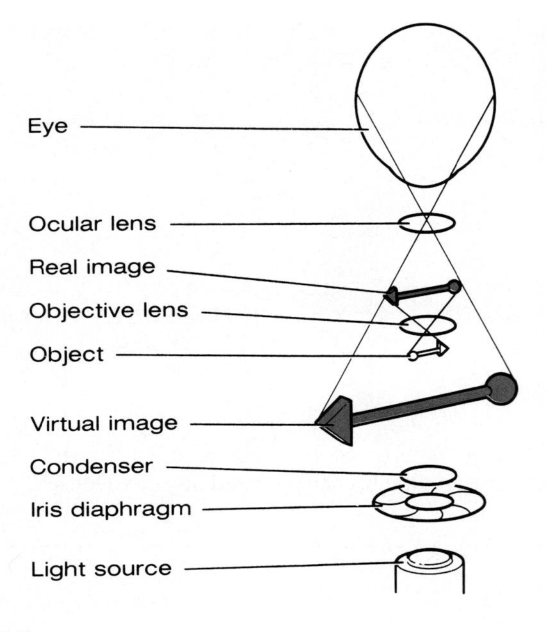

F3.2

Optical system of the compound microscope. Note the real and virtual images.

that regulates the amount of light passing through the condenser. The iris diaphragm permits the best possible contrast when viewing the specimen.

Coarse adjustment knob: used to focus the specimen.

Fine adjustment knob: used for precise focusing once coarse focusing has been completed.

Head (also called the *barrel* or *body tube*): supports the objective lens system (which is mounted on a movable nosepiece) and the ocular lens.

Ocular (or *eyepiece*): lens contained at the superior end of the head or body tube. Observations are made through the ocular. This lens has a magnification of 10× (it increases the apparent size of the object by 10 times or 10 diameters). If your microscope has a *pointer* (used to indicate a specific area of the viewed specimen), it is attached to the ocular and can be positioned by rotating the ocular lens.

Nosepiece: the movable (revolving) nosepiece generally carries three objective lenses and permits the sequential positioning of these lenses over the light passing through the hole in the stage.

Objective lenses: adjustable lens system that permits the use of a **low-power lens,** a **high-power lens,** or an **oil immersion lens.** The objective lenses have different magnifying and resolving powers.

2. Examine the objectives carefully, noting their relative lengths and the numbers inscribed on their sides. On most microscopes, the low-power (l.p.) objective is the shortest and generally has a magnification of 10×. The high-power (h.p.) objective is of intermediate length and has a magnification range from 40× to 50×, depending on the microscope. The oil immersion objective is usually the longest of the objectives and has a magnifying power of 95× to 100×. (Note: some microscopes lack the oil immersion lens but have a very low magnification lens called the **scanning lens,** which is a very short objective with a magnification of 4× to 5×.)

3. Rotate the low-power objective into position, and turn the coarse adjustment knob about 180 degrees. Notice how far the stage (or objective) travels during this adjustment. Move the fine adjustment knob 180 degrees, noting again the distance that the stage (or the objective) moves.

Magnification and Resolution

The microscope is an instrument of magnification. In the compound microscope, magnification is achieved through the interplay of two lenses—the ocular lens and the objective lens. The objective lens magnifies the specimen to produce a **real image** that is projected to the ocular. This real image is magnified by the ocular lens to produce the **virtual image** seen by your eye (Figure 3.2).

The **total magnification** of any specimen being viewed is equal to the power of the ocular lens multiplied by the power of the objective lens used. For example, if the ocular lens magnifies 10× and the objective lens being used magnifies 45×, the total magnification is 450×.

Determine the total magnification you may achieve with each of the objectives on your microscope, and record the figures on the chart on p. 21. (Note: At this time, also cross out the column relating to the lens that your microscope does not have, and record the number of your microscope at the top of the chart.)

The compound light microscope has certain limitations. Although the level of magnification is almost limitless, the **resolution** (or resolving power), the ability to discriminate two close objects as separate, is not. The human eye can resolve objects about 100 μm apart, but the compound microscope has a resolution of 0.2 μm under ideal conditions.

Resolving power (RP) is determined by the amount and physical properties of the visible light that enters the microscope. In general, however, the greater the amount of light delivered to the objective lens, the greater the resolution. The size of the objective lens aperture (opening) decreases with increasing magnification, allowing less light to enter the objective. Thus, you will probably find it necessary to increase the light intensity at the higher magnifications.

VIEWING OBJECTS THROUGH THE MICROSCOPE

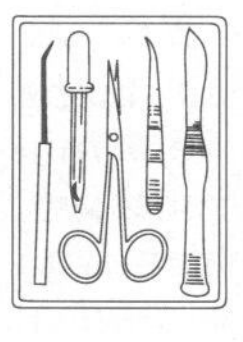

1. Obtain a millimeter ruler and a prepared slide of the letter *e* or newsprint. Secure the slide on the stage so that the letter *e* is centered over the hole, and switch on the light source. (If the light source is not built into the base, use the curved surface

of the mirror to reflect the light up into the microscope.) The condenser should be in its highest position.

2. With your lowest-power (scanning or low-power) objective in position over the stage, use the coarse adjustment knob to bring the objective and stage as close together as possible.

3. Looking through the ocular, adjust the light for comfort. Now use the coarse adjustment knob to focus slowly away from the *e* until it is as clearly focused as possible. Complete the focusing with the fine adjustment knob.

4. Sketch the letter in the circle just as it appears in the **field** (the area you see through the microscope).

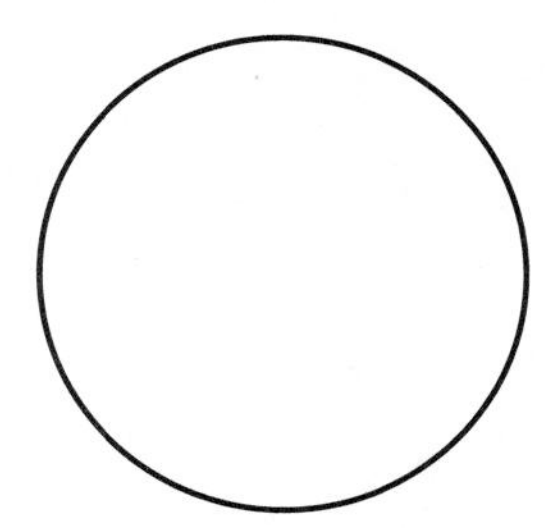

What is the total magnification? ________________ ×

How far is the bottom of the objective from the specimen? In other words, what is the **working distance**? (Use a millimeter ruler to make this measurement, and record it here and in the chart on page 21.)

______________ mm

How has the apparent orientation of the *e* changed top to bottom, right to left, and so on?

____

____

5. Move the slide slowly away from you on the stage as you look through the ocular. In what direction does the image move?

____

Move the slide to the left. In what direction does the image move?

____

At first this change in orientation will confuse you, but with practice you will learn to move the slide in the desired direction with no problem.

6. Without touching the focusing knobs, increase the magnification by rotating the next higher magnification lens (low-power or high-power) into position over the stage. Using the fine adjustment only, sharpen the focus.* What new details become clear?

____

____

What is the total magnification now? ____________ ×

Measure the distance between the objective and the slide (the working distance) and record it here and on the chart.

______________ mm

Why should the coarse focusing knob *not* be used when focusing with the higher-powered objective lenses?

____

____

Is the image larger or smaller? ____________________

Approximately how much of the letter is visible now?

____

Is the field larger or smaller? ____________________

Why is it necessary to center your object (or the portion of the slide you wish to view) before changing to a higher power?

____

____

Move the iris diaphragm lever while observing the field. What happens?

____

____

Is it more desirable to increase *or* decrease the light when changing to a higher magnification?

___________________ Why? ___________________

____

____

7. If you have just been using the low-power objective, repeat the steps given in direction 6, using the high-

* Today most good laboratory microscopes are **parfocal;** that is, the object should be in focus (or nearly so) at the higher magnifications once you have properly focused in l.p. If you are unable to swing the objective into position without raising the objective, your microscope is not parfocal. Consult your instructor.

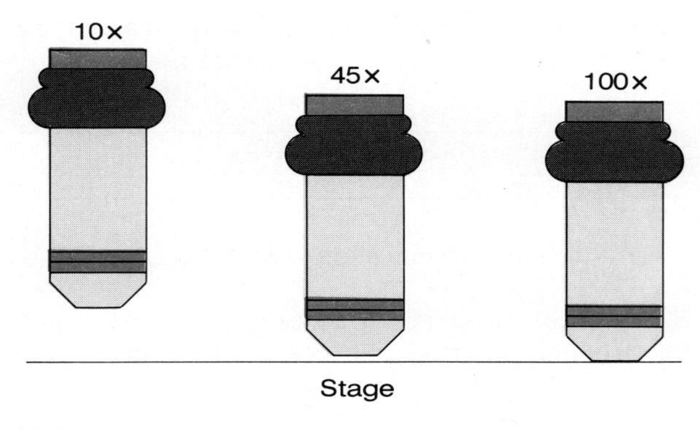

F3.3

Relative working distance of the 10×, 45×, and 100× objectives.

power objective lens. Record the total magnification, approximate working distance, and information on detail observed on the chart on p. 21.

8. Without touching the focusing knob, rotate the high-power lens out of position so that the area of the slide over the opening in the stage is unobstructed. Place a drop of immersion oil over the *e* on the slide and rotate the immersion lens into position. Adjust the fine focus and the light for the best possible resolution. Is the field again decreased in size?

____

What is the total magnification with the immersion lens?

__________ ×

Is the working distance less or greater than it was when the high-power lens was focused?

____

Compare your observations on the relative working distances of the objective lenses with the illustration in Figure 3.3. Explain why it is desirable to begin the focusing process in low power.

____

____

9. Rotate the immersion lens slightly to the side and remove the slide. Clean the oil immersion lens carefully with lens paper, and then clean the slide in the same manner.

DETERMINING THE SIZE OF THE MICROSCOPE FIELD

By this time you should know that the size of the microscope field decreases with increasing magnification. For future microscope work, it will be useful to determine the diameter of each of the microscope fields. This information will allow you to make a fairly accurate estimate of the size of the objects you view in any field. For example, if you have calculated the field diameter to be 4 mm and the object being observed extends across half this diameter, you can estimate the length of the object to be approximately 2 mm.

Microscopic specimens are usually measured in micrometers and millimeters, both units of the metric system. You can get an idea of the relationship and meaning of these units from Table 3.1.

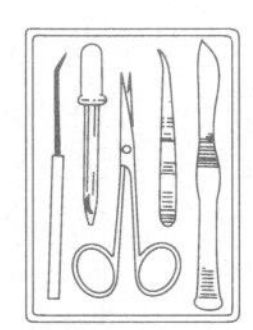

1. Return the letter *e* slide, and obtain a grid slide, a slide prepared with graph paper ruled in millimeters. Each of the squares in the grid is 1 mm on each side. Use your lowest-power objective to bring the grid lines into focus.

2. Move the slide so that one grid line touches the edge of the field on one side, and then count the number of squares you can see across the diameter of the field. If you can see only part of a square, as in the accompanying diagram, estimate the part of a millimeter that the partial square represents.

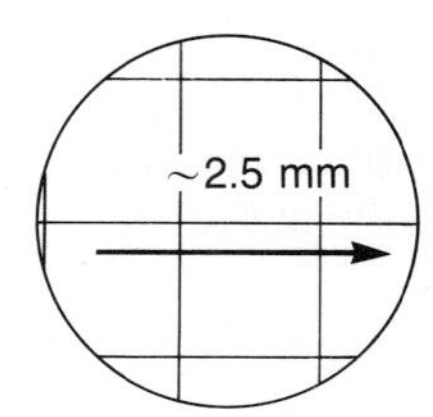

TABLE 3.1 Comparison of Metric Units of Length

Metric unit	Abbreviation	Equivalent
Meter	m	(about 39.3 in.)
Centimeter	cm	10^{-2} m
Millimeter	mm	10^{-3} m
Micrometer (or micron)	μm (μ)	10^{-6} m
Nanometer (or millimicrometer, or millimicron)	nm (mμ)	10^{-9} m
Angstrom	Å	10^{-10} m

For future reference, record this figure in the space marked "field size (mm)" on the summary chart below. (If you have been using the scanning lens, repeat the procedure with the low-power objective lens.) Complete the chart by computing the approximate diameter of the high-power and oil immersion fields. Say the diameter of the low-power field (total magnification of 50×) is 2 mm. You would compute the diameter of a high-power field with a total magnification of 100× as follows:

$$2 \text{ mm} \times 50 = Y \text{ (diameter of h.p. field)} \times 100$$
$$100 \text{ mm} = 100Y$$
$$1 \text{ mm} = Y \text{ (diameter of the h.p. field)}$$

The formula is

Diameter of the l.p. field (mm) × total magnification of the l.p. field = diameter of field Y × total magnification of field Y

3. Estimate the length of the following microscopic objects. *Base your calculations on the field sizes you have determined for your microscope.*

Object seen in low-power field:

Estimated length:

_____mm

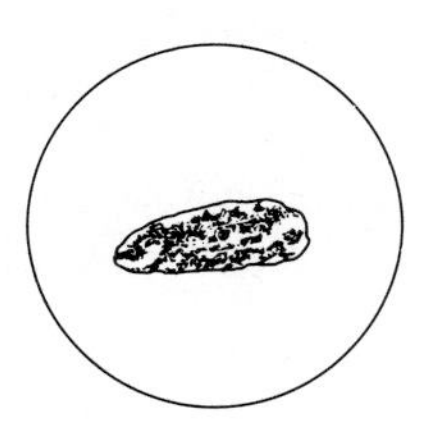

Object seen in high-power field:

Estimated length:

_____mm

or

_____μm

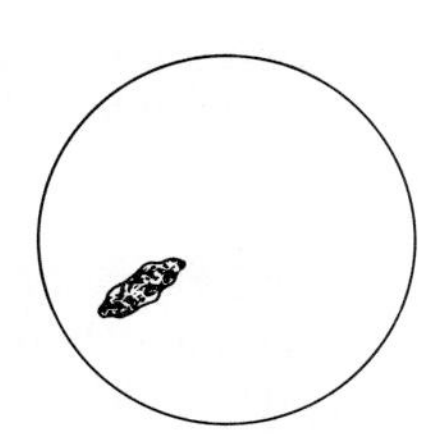

Object seen in oil immersion field:

Estimated length:

_____mm

or

_____μm

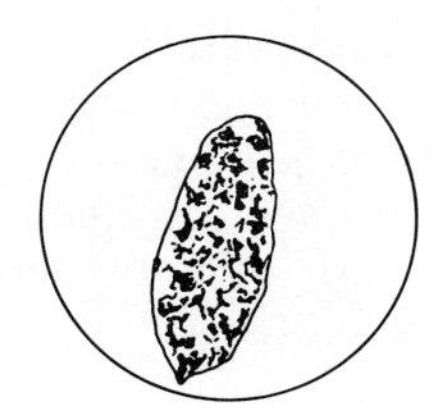

4. If an object viewed with the oil immersion lens looked like the field depicted below, could you determine its approximate size from this view?

____

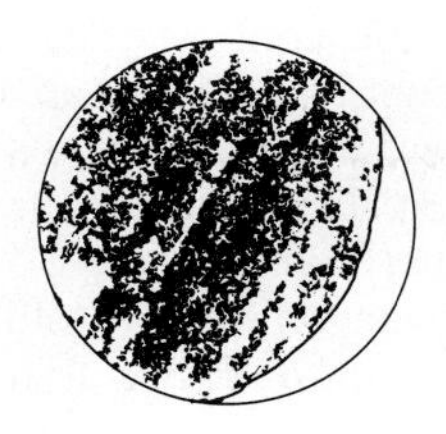

If not, then how could you determine it?

____

____

____

____

Summary Chart for Microscope #________

	Scanning	Low power	High power	Oil immersion
Magnification of objective lens	×	×	×	×
Total magnification	×	×	×	×
Detail observed				
Field size (diameter)	mm μm	mm μm	mm μm	mm μm
Working distance	mm	mm	mm	mm

PERCEIVING DEPTH

Any specimen mounted on a slide has depth as well as length and width; it is rare indeed to view a tissue slide with just one layer of cells. Normally you can see two or three cell thicknesses. Therefore, it is important to learn how to determine relative depth with your microscope.*

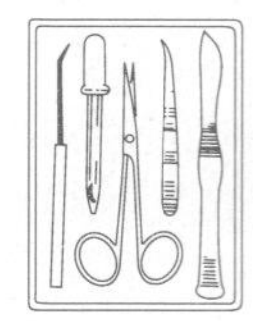

1. Return the grid slide and obtain a slide of colored crossed threads. Focusing at a low magnification, locate the point where the three threads cross each other.

2. Use the iris diaphragm lever to cut the light way down to increase the contrast. Focus down with the coarse adjustment until the threads are out of focus, then slowly focus upward again, noting which thread comes into clear focus first. This one is the lowest, or most inferior, thread. (Note: You will see two or even all three threads, so you must be very careful in determining which one comes into focus first.) Record your observation.

______________ thread over ____________________

Continue to focus upward until the uppermost thread is clearly focused. Again record your observation.

______________ thread over ____________________

Which of the threads is uppermost?

____

Lowest?

____

PREPARING AND OBSERVING A WET MOUNT

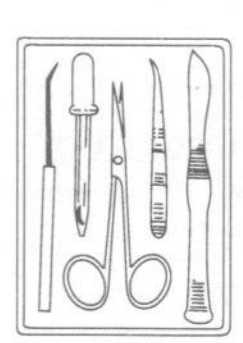

1. Obtain the following: a clean microscope slide and coverslip, a flat-tipped toothpick, a dropper bottle of physiologic saline, a dropper bottle of methylene blue stain, forceps, and filter paper.

2. Place a drop of physiologic saline in the center of the slide. Using the flat end of the toothpick, gently scrape the inner lining of your cheek. Agitate the end of the toothpick containing the cheek scrapings in the drop of saline (Figure 3.4a).

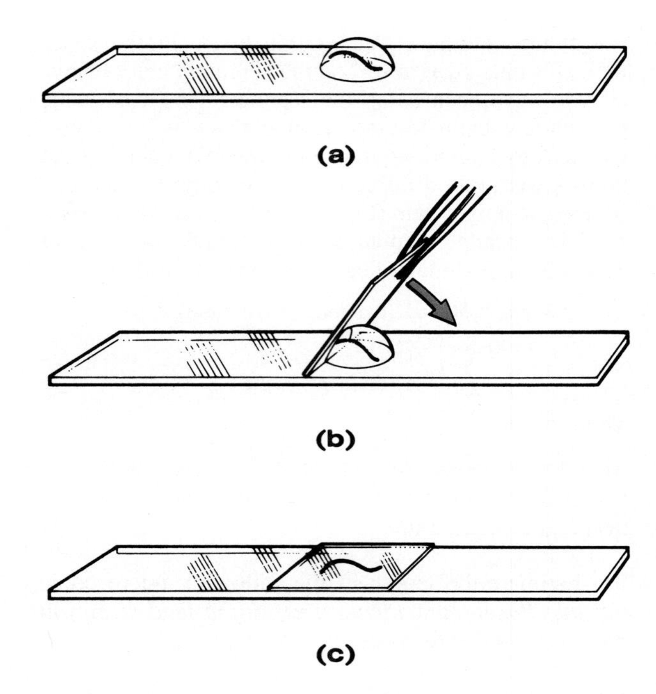

(a)

(b)

(c)

F3.4

Procedure for the preparation of a wet mount. (a) The object is placed in a drop of water (or saline) on a clean slide. (b) A coverslip is held at a 45-degree angle with forceps and (c) lowered carefully over the water and the object.

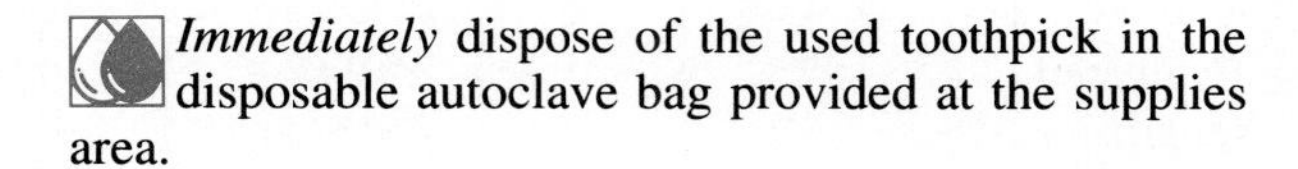

Immediately dispose of the used toothpick in the disposable autoclave bag provided at the supplies area.

3. Add a tiny drop of the methylene blue stain to the preparation. (These epithelial cells are nearly transparent and thus difficult to see without the stain, which colors the nuclei of the cells and makes them look much darker than the cytoplasm.) Stir again and then dispose of the toothpick as described above.

4. Hold the coverslip with the forceps so that its bottom edge touches one side of the fluid drop (Figure 3.4b), then *carefully* lower the coverslip onto the preparation (Figure 3.4c). *Do not just drop the coverslip,* or you will trap large air bubbles under it, which will obscure the cells. *A coverslip should always be used with a wet mount,* to prevent soiling the lens if you should misfocus.

5. Examine your preparation carefully. The coverslip should be closely opposed to the slide. If there is excess fluid around its edges, obtain a piece of filter paper. Fold the filter paper in half, and use the folded edge to absorb the excess fluid.

* In microscope work, the **depth of field** (the depth of the specimen clearly in focus) is greater at lower magnifications.

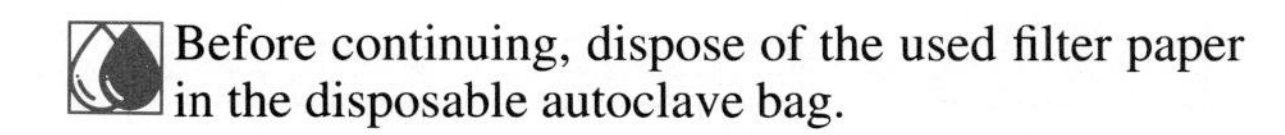

Before continuing, dispose of the used filter paper in the disposable autoclave bag.

6. Place the slide on the stage and locate the cells in low power. You will probably want to dim the light with the iris diaphragm to provide more contrast for viewing the lightly stained cells. (Furthermore, a wet mount will dry out quickly in bright light, because a bright light source is hot.)

7. Cheek epithelial cells are very thin, six-sided cells. In the cheek, they provide a smooth, tilelike lining, as shown in Figure 3.5.

8. Make a sketch of epithelial cells that you observed. Approximately how wide are the cheek epithelial cells?

_______________ mm

Why do *your* cheek cells look different from those illustrated in Figure 3.5? (Hint: what did you have to *do* to your cheek to obtain them?)

____

9. When you have completed your observations, dispose of your wet-mount preparation in the beaker of bleach solution.

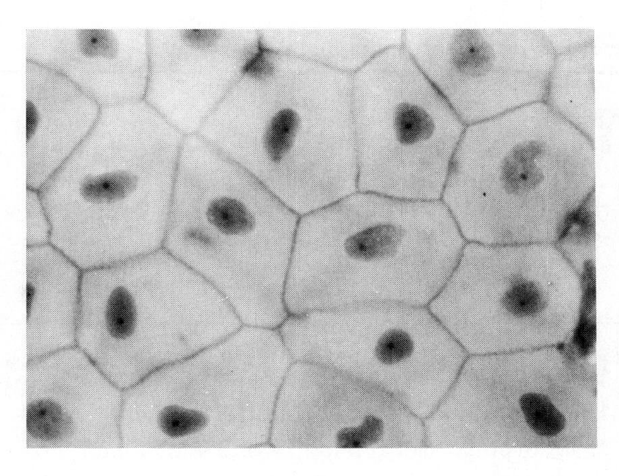

F3.5

Photomicrograph of cheek cells, surface view (150×).

10. Before leaving the laboratory, make sure all other materials are properly disposed of or returned to the proper laboratory station: clean the microscope lenses and slide, and put the dust cover on the microscope before you return it to the storage cabinet.

4 EXERCISE

The Cell—Anatomy and Division

OBJECTIVES

1. To define *cell, organelle,* and *inclusion.*
2. To identify on a cell model or diagram the following cellular regions, and to list the major function of each: nucleus, cytoplasm, and plasma membrane.
3. To identify and list the major functions of the various organelles studied.
4. To compare and contrast specialized cells with the concept of the "generalized cell."
5. To define *interphase, mitosis,* and *cytokinesis.*
6. To list the stages of mitosis and describe the events of each stage.
7. To identify the mitotic phases on projected slides or appropriate diagrams.
8. To explain the importance of mitotic cell division and its product.

MATERIALS

Three-dimensional model of the "composite" animal cell or laboratory chart of cell anatomy
Prepared slides of simple squamous epithelium ($AgNO_3$ stain), teased smooth muscle, human blood cell smear, and sperm
Compound microscope
Prepared slides of whitefish blastulae
Three-dimensional models of mitotic stages

The **cell**, defined as the structural and functional unit of all living things, is a very complex entity. The cells of the human body are highly diverse; and their differences in size, shape, and internal composition reflect their specific roles in the body. Yet cells do have many common anatomical features, and there are some functions that all must perform to maintain life. For example, all cells have the ability to maintain their boundaries, to metabolize, to digest foodstuffs and dispose of wastes, to grow and reproduce, to move, and to respond to a stimulus. Most of these functions are considered in detail in later exercises. This exercise focuses on structural similarities that typify the "composite," or "generalized," cell and considers only the function of cell reproduction (cell division). Cell membrane transport (the means by which substances cross the plasma membrane) is dealt with separately in Exercise 5.

ANATOMY OF THE COMPOSITE CELL

In general, all cells have three major regions, or parts, that can readily be identified with a light microscope: the **nucleus,** the **plasma membrane,** and the **cytoplasm.** The nucleus is usually seen as a round or oval structure near the center of the cell. It is surrounded by cytoplasm, which in turn is enclosed by the plasma membrane. Since the advent of the electron microscope, even smaller cell structures—organelles—have been identified. Figure 4.1 is a diagrammatic representation of the fine structure of the composite cell as revealed by the electron microscope.

Nucleus

The nucleus is often described as the control center of the cell and is necessary for cell reproduction. A cell that has lost or ejected its nucleus (for whatever reason) is literally programmed to die because the nucleus is the site of the "genes," or genetic material—DNA.

When the cell is not dividing, the genetic material is loosely dispersed throughout the nucleus in a thread-like form called **chromatin.** When the cell is in the process of dividing to form daughter cells, the chromatin coils and condenses to form dense, darkly staining rod-like bodies called **chromosomes**—much in the way a stretched spring becomes shorter and thicker when relaxed. (Cell division is discussed later in this exercise.)

The nucleus also contains one or more small round bodies, called **nucleoli,** composed primarily of proteins and ribonucleic acid (RNA). The nucleoli are believed to be storage sites for RNA and/or assembly sites for ribosomal particles (particularly abundant in the cytoplasm), which are the actual "factories" for synthesizing proteins.

The nucleus is bound by a double-layered porous membrane, the **nuclear membrane,** which is similar in composition to other cellular membranes.

- ***Label*** the nuclear membrane, chromatin threads, and a nucleolus in Figure 4.1.

Plasma Membrane

The **plasma membrane** separates cell contents from the surrounding environment. It is made up of protein and lipid (fat), and appears to have a bimolecular lipid core in which protein molecules float. (See enlargement in

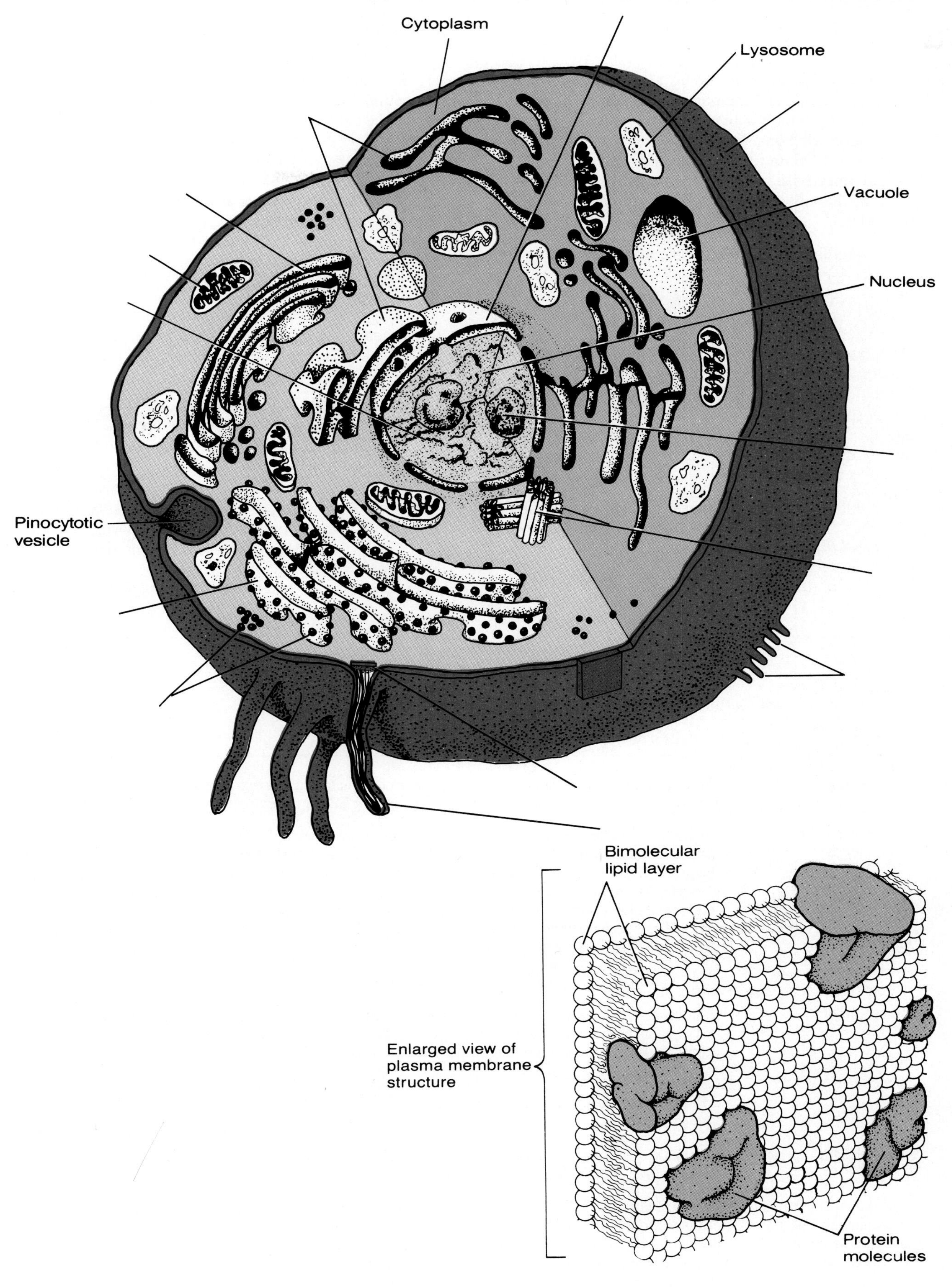

F4.1

Anatomy of the composite animal cell. The inset shows the structural details of the plasma membrane.

Figure 4.1.) Besides providing a protective barrier for the cell, the plasma membrane plays an active role in determining which substances may enter or leave the cell and in what quantity. In some cells the membrane is thrown into minute fingerlike projections, or folds, called **microvilli,** which greatly increase the surface area of the cell available for absorption or passage of materials.

- ***Label*** the plasma membrane and microvilli in Figure 4.1.

Cytoplasm and Organelles

The cytoplasm consists of the cell contents outside the nucleus. It is the major site of most activities carried out by the cell. Suspended in the cytoplasmic material are many small structures called **organelles.** The organelles are the metabolic machinery of the cell, and they are highly organized to carry out specific functions for the cell as a whole.

- Table 4.1 briefly describes the cytoplasmic organelles. Read through the table, and then ***correctly label the organelles in Figure 4.1.***

In addition to these cell structures, some cells have projections called **flagella** or **cilia** which, respectively, propel the cells or allow cells to sweep substances along a tract. ***Label*** the cilia in Figure 4.1.

The cell cytoplasm contains various other substances and structures, including stored foods (glycogen granules, lipid droplets, and other nutrients), pigment granules, crystals of various types, water vacuoles, and ingested foreign materials. But these are not part of the active metabolic machinery of the cell and are therefore called **inclusions.**

- Once you have located and labeled all of these structures in Figure 4.1, examine the cell model (or cell chart) to reinforce your identifications.

OBSERVING DIFFERENCES AND SIMILARITIES IN CELL STRUCTURE

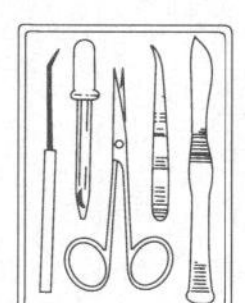

1. Obtain prepared slides of simple squamous epithelium, sperm, smooth muscle cells (teased), and human blood.

TABLE 4.1 Cytoplasmic Organelles

Organelle	Location and function
Ribosomes	Tiny spherical bodies composed of RNA and protein; actual sites of protein synthesis; seen floating free or attached to a membranous structure (the rough ER) in the cytoplasm
Endoplasmic reticulum (ER)	Membranous system of tubules that extends throughout the cytoplasm; two varieties: rough or granular ER—studded with ribosomes (tubules of the rough ER provide an area for storage and transport of the proteins made on the ribosomes to other cell areas; external face synthesizes phospholipids and cholesterol); smooth or agranular ER—no protein synthesis–related function (believed to be a site of lipid metabolism and for synthesis of lipids and steroid–based hormones)
Golgi apparatus	Stack of flattened sacs with bulbous ends and associated small vesicles; found close to the nucleus; role in packaging proteins or other substances for export from the cell or incorporation into the plasma membrane and in packaging lysosomal enzymes
Lysosomes	Various-sized membranous sacs containing powerful digestive enzymes; function to digest worn-out cell organelles and foreign substances that enter the cell; since they have the capacity of total cell destruction if ruptured, referred to as "suicide sacs of the cell"
Mitochondria	Generally rod-shaped bodies with a double membrane wall; inner membrane is thrown into folds, or cristae; contain enzymes that oxidize foodstuffs to produce cellular energy (ATP); often referred to as "powerhouses of the cell"
Centrioles	Paired, cylindrical bodies lying at right angles to each other, close to the nucleus; direct the formation of the mitotic spindle during cell division
Cytoskeletal elements: microtubules, intermediate filaments, and microfilaments*	Provide cellular support; function in intracellular transport. Microtubules form the internal structure of the centrioles and help to determine cell shape; intermediate filaments are stable elements, composed of a variety of proteins, that resist mechanical forces acting on cells. Microfilaments are formed largely of contractile proteins, and thus are important in cell mobility (particularly in muscle cells)

*Cytoskeletal elements are *not* depicted in Figure 4.1.

2. Observe each slide under the microscope, carefully noting similarities and differences in the cells. (The oil immersion lens will be needed to observe blood and sperm.) Distinguish the limits of the individual cells, and note the shape and position of the nucleus in each case. When you look at the human blood smear, direct your attention to the disc-shaped red blood cells only. Sketch your observations in the circles provided.

Simple squamous epithelium

Sperm cells

Teased smooth muscle cells

Human red blood cells

3. How do these four cell types differ in shape and size?

____

____

____

____

How might cell shape affect cell function?

____

____

____

____

Which cells have visible projections?

____

How do these projections relate to the function of this cell?

____

____

____

Do any of these cells lack a cell membrane? _______

A nucleus? _______________

In the cells with a nucleus, can you discern nucleoli?

____

Were you able to observe any of the organelles in these cells? _______________ Why or why not?

____

____

CELL DIVISION: MITOSIS AND CYTOKINESIS

Cell division in all cells other than bacteria consists of a series of events collectively called mitosis and cytokinesis. **Mitosis** is nuclear division; **cytokinesis** is the division of the cytoplasm, which begins after mitosis is nearly complete. Although mitosis is usually accompanied by cytokinesis, in some instances cytoplasmic division does not occur, leading to the formation of binucleate (or multinucleate) cells. This is relatively common in the human liver and during embryonic development of skeletal muscle cells.

The process of mitosis results in the formation of two daughter nuclei that are genetically identical to the mother nucleus. This distinguishes mitosis from **meiosis,** a specialized type of nuclear division that occurs only in the reproductive organs (testes or ovaries). Meiosis, which yields four daughter nuclei that differ genetically and in composition from the mother nucleus, is used only for the production of eggs and sperm (gametes) for sexual reproduction. The function of cell division, including mitosis and cytokinesis in the body, is to increase the number of cells for growth and repair while maintaining their genetic heritage.

In cells about to divide, an important event precedes cell division. The genetic material (the DNA molecules composing part of the chromatin strands) is replicated (duplicated exactly) during the portion of the cell's life cycle called **interphase.** Interphase is *not* part of mitosis; it represents the time when a cell is not actively involved in cell division. Although some people refer to interphase as the cell's resting period, this is an inaccurate description because the cell is quite active in its daily activities and is resting only from cell division. As shown in the diagrams of Figure 4.2 and in the photomicrographs of Figure 4.3, the stages of mitosis include the following events:

Prophase: At the onset of cell division, the chromatin threads coil and shorten to form densely staining, short, barlike **chromosomes.** By the middle of prophase the chromosomes appear as double-stranded structures (each strand is a **chromatid**) connected by a small median body called a **centromere.** The centrioles separate from one another and act as focal points for the assembly of a system of microtubules called the **mitotic spindle** which forms between them. The spindle acts as a scaffolding for the attachment and movement of the chromosomes during later mitotic stages. The nuclear membrane and the nucleolus break down and disappear, and the chromosomes randomly attach to the spindle fibers by their centromeres.

Metaphase: A brief stage, during which the chromosomes migrate to the central plane, or equator, of the spindle and align along that plane in a straight line from the superior to the inferior region of the spindle (lateral view). Viewed from the poles of the cell (end view), the chromosomes appear to be arranged in a "rosette," or circle, around the widest dimension of the spindle.

Anaphase: During anaphase, the centromeres break, and the chromatids (now called chromosomes again) separate from one another and then progress slowly toward opposite ends of the cell. The chromosomes appear to be pulled by their centromere attachment, with their "arms" dangling behind them. Anaphase is complete when poleward movement ceases.

Telophase: During telophase, the events of prophase are essentially reversed. The chromosomes at the poles begin to uncoil and resume the chromatin form, the spindle disappears, a nuclear membrane forms around each chromatin mass, and nucleoli appear in each of the daughter nuclei.

Mitosis is essentially the same in all animal cells; but depending on the type of tissue, it takes from 5 minutes to several hours to complete. In most cells, centriole replication is deferred until interphase of the next cell cycle.

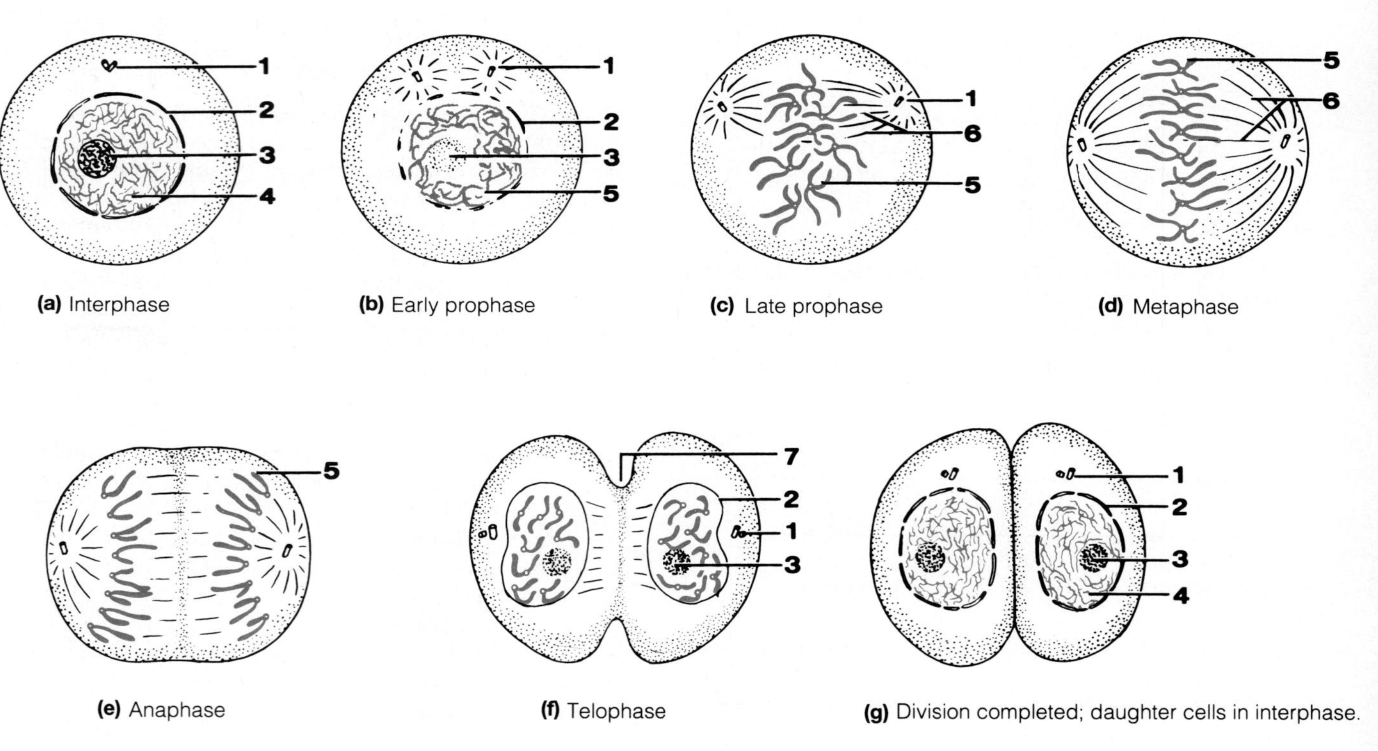

F4.2

Cell life cycle. (a) Mother cell in interphase. (b–f) The stages of mitosis, diagrammatic views (1 = centriole; 2 = nuclear membrane; 3 = nuceolus; 4 = chromatin; 5 = chromosomes; 6 = spindle; 7 = cleavage furrow). (g) Daughter cells in interphase, division completed.

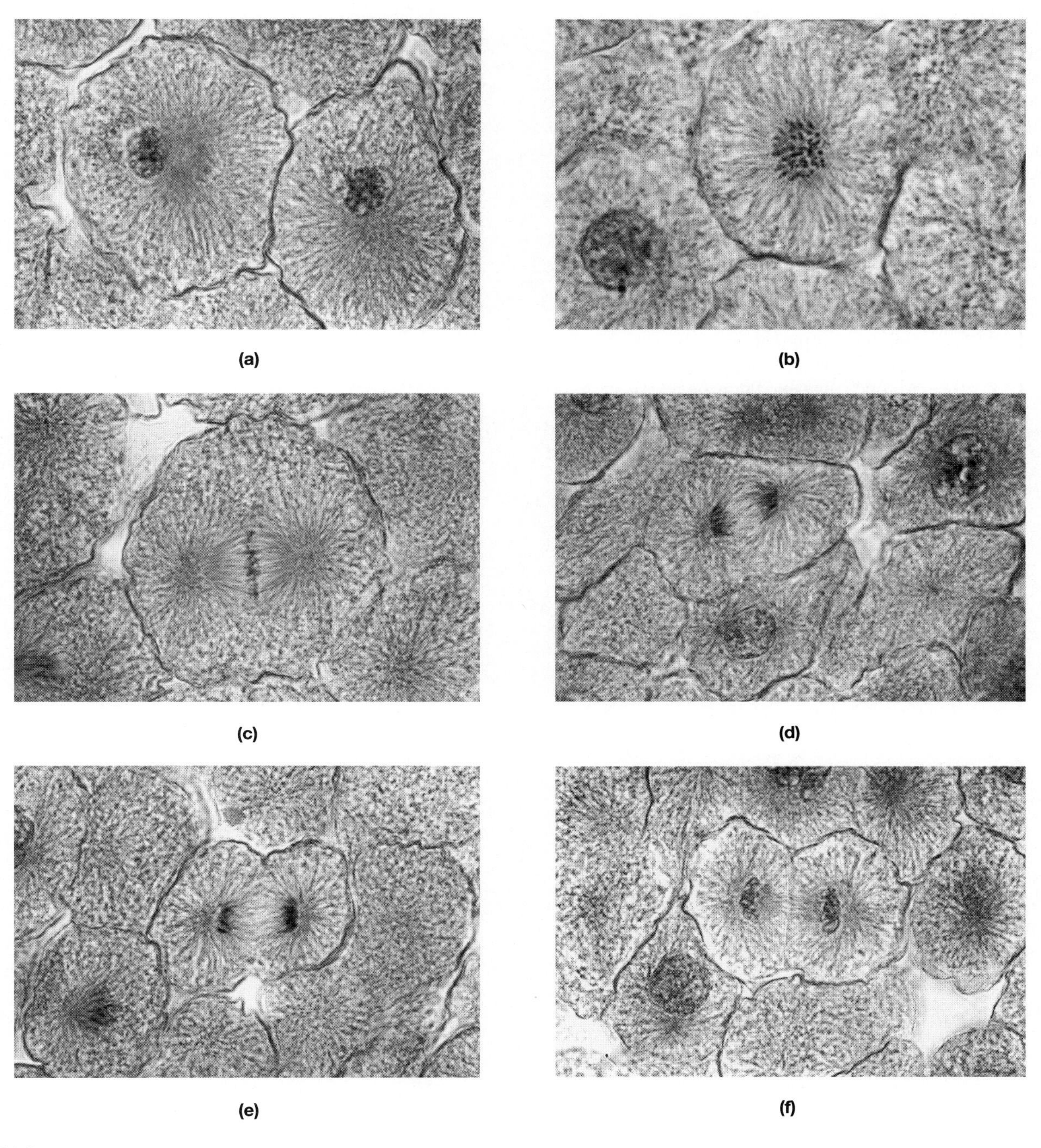

F4.3

Photomicrographs of mitotic phases in whitefish blastula (570×). (a) Early prophase; (b) late prophase; (c) metaphase; (d) anaphase; (e) early telophase; (f) late telophase.

Cytokinesis, or the division of the cytoplasmic mass, begins during telophase and provides a good guideline for where to look for the mitotic figures of telophase. In animal cells, a cleavage furrow begins to form approximately over the equator of the spindle and eventually splits or pinches the original cytoplasmic mass into two portions. Thus at the end of cell division two daughter cells exist, each smaller in cytoplasmic mass than the mother cell but genetically identical to it. The daughter cells grow and carry out the normal spectrum of metabolic processes until it is their turn to divide.

Cell division is extremely important during the body's growth period. Most cells (excluding nerve cells) undergo mitosis until puberty, when normal body size is achieved and overall body growth ceases. After this time in life, only certain cells routinely carry out cell division—for example, cells subjected to abrasion (epithelium of the skin and lining of the gut). Other cell populations—such as liver cells—stop dividing, but retain this ability should some of them be removed or damaged. Skeletal muscle and nervous tissue completely lose the ability to divide, and thus are severely handicapped by injury. Throughout life, the body retains its ability to repair cuts and wounds and to replace some of its aged cells. Mitosis that is uncontrolled, or "gone wild," is the basis of all tumors and cancers.

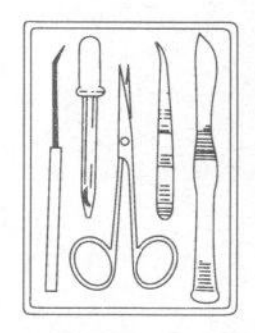

Obtain a prepared slide of whitefish blastulae to study the stages of mitosis. The cells of each blastula (a stage of embryonic development consisting of a hollow ball of cells) are at approximately the same mitotic stage, so it may be necessary to observe more than one blastula to view all the mitotic stages. The exceptionally high rate of mitosis observed in this tissue is typical of embryos; but if it occurs in specialized tissues, it can be an indication of cancerous cells, which also have an extraordinarily high mitotic rate. Examine the slide carefully, identifying the four mitotic stages and the process of cytokinesis. Compare your observations with Figures 4.2 and 4.3, and verify your identifications with your instructor. Then sketch your observations of each stage in the circles provided here.

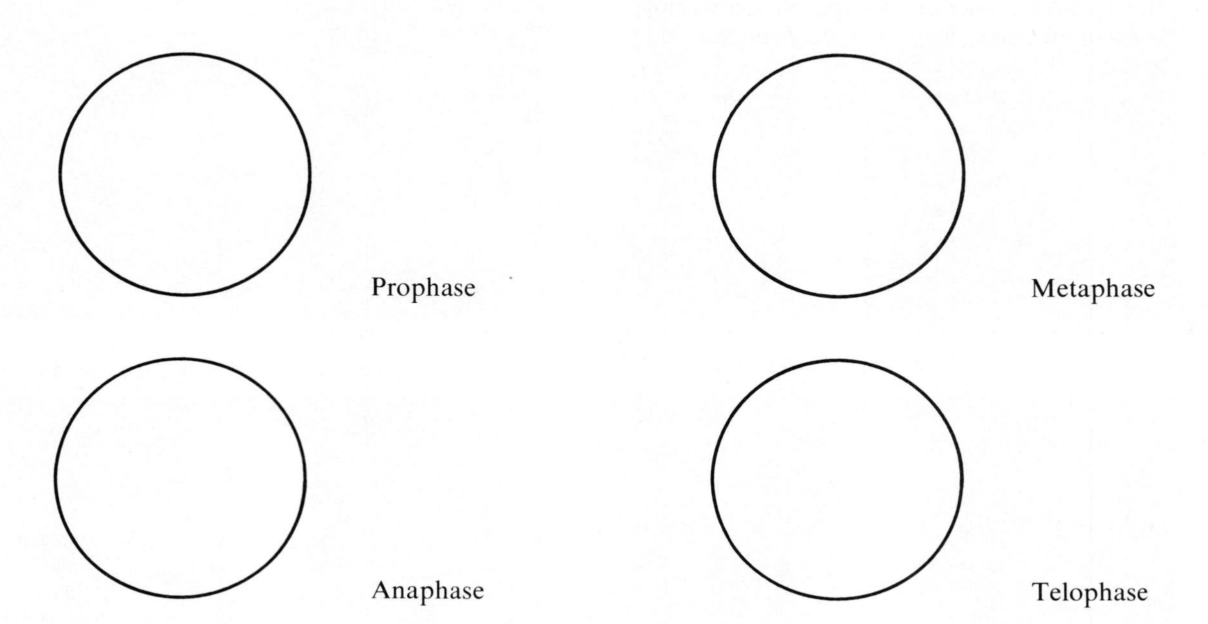

The Cell—Transport Mechanisms and Cell Permeability

OBJECTIVES

1. To define *differential,* or *selective, permeability; diffusion* (*dialysis* and *osmosis*); *Brownian motion, isotonic, hypotonic,* and *hypertonic; passive transport; active transport, pinocytosis, phagocytosis,* and *solute pumps.*
2. To explain the processes that account for the movement of substances across the cell membrane and to note the driving force for each.
3. To determine which way substances will move passively through a selectively permeable membrane (given appropriate information on concentration differences).

MATERIALS

For Passive Transport Experiments:
Clean slides and coverslips
Forceps
Glass stirring rods
15-ml graduated cylinders
Compound microscopes

Brownian motion:
Milk in dropper bottles
Hot plate

Diffusion:
Demonstration 1:
Potassium permanganate crystals are placed in a 1000-ml graduated cylinder, which is then carefully filled to the 1000-ml mark with distilled water. Demonstration is prepared the morning of the laboratory with setup time noted.
Demonstration 2:
Thistle tube containing molasses, broad end (bulb) closed with a selectively permeable membrane. Thistle tube bulb is then immersed in a beaker containing distilled water. Level of water–molasses mixture in the thistle tube is marked with a wax pencil at the beginning of the laboratory session.

Petri plate containing 12 ml of 1.5% agar-agar
Methylene blue dye crystals
Potassium permanganate dye crystals
Millimeter rulers

Four dialysis sacs (or small Hefty "alligator" sandwich bags)
Beakers (250 ml)
40% glucose solution
Fine twine or dialysis tubing clamps
10% NaCl solution, boiled starch solution
Laboratory balance
Benedict's solution in dropper bottle
Test tubes in racks, test tube holder
Wax marker
Small funnel
Silver nitrate ($AgNO_3$) in dropper bottle
Lugol's iodine solution in dropper bottle

Filter paper
Physiologic (mammalian) saline solution in dropper bottle
1.5% sodium chloride (NaCl) solution in dropper bottle
Distilled water in dropper bottle
Medicine dropper
Vials of animal blood obtained from a commercial source (veterinarian or biological supply house)
Disposable plastic gloves
Battery jar or basin containing freshly prepared 10% household bleach solution
Disposable autoclave bag

Filtration:
Ring stand, ring, clamp
Filter paper, funnel
Flask containing a mixture of uncooked starch, powdered charcoal, and copper sulfate ($CuSO_4$)

For Active Transport:
Culture of starved amoeba (*Amoeba proteus*)
Depression slide
Coverslip
Tetrahymena pyriformis culture
Compound microscope

Because of its molecular composition, the plasma membrane is selective about what passes through it. It allows nutrients to enter the cell but keeps out undesirable substances. By the same token, valuable cell proteins and other substances are kept within the cell, and excreta or wastes pass to the exterior. This property is known as **selective permeability.** Transport through the plasma membrane occurs in two basic ways. In **active transport,** the cell provides the energy (ATP) to power the transport process; in the other, **passive transport,** the transport process is driven by concentration or pressure differences.

PASSIVE TRANSPORT

All molecules vibrate randomly (because of their inherent kinetic energy) at all temperatures above absolute zero (about −460°F). In general, the smaller the particle, the greater the kinetic energy it possesses and the faster its molecular motion. This random movement may be detected indirectly by observing a suspension. The larger particles can be seen moving randomly as they are hit and deflected by the smaller, more rapidly moving particles. The zigzag movement of the larger particles is known as **Brownian motion.**

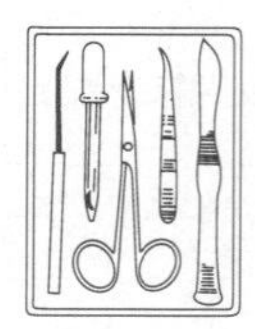

1. Make a wet mount of milk; that is, place a small drop of milk on a slide and cover carefully with a coverslip. Allow the slide to stand on the microscope stage for about 10 minutes before observing.

2. Keeping the light low, observe the slide with high power and then with the oil immersion lens. As the minute solvent (water) molecules collide with the fat globules of the milk, you can see the larger fat globules ricochet in an erratic manner (Brownian motion).

3. Place the preparation on a warm hot plate for a few seconds. Observe again. How has the *rate* of Brownian motion changed?

____

____

What can you conclude about the effect of increased temperature on the kinetic energy of molecules?

____

____

Diffusion

If a **concentration gradient** (difference in concentration) exists, molecules eventually become evenly distributed through random molecular motion. **Diffusion** is the movement of molecules from a region of their higher concentration to a region of their lower concentration. The driving force is the kinetic energy of the molecules themselves.

There are many examples of diffusion in nonliving systems; for example, if a bottle of ether were uncorked in the front of the laboratory, very shortly thereafter you would be nodding, as the ether molecules became distributed throughout the room. The ability to smell a friend's cologne shortly after he or she has entered the room is another such example.

The diffusion of particles into and out of cells is modified by the plasma membrane, which constitutes a physical barrier. In general, molecules diffuse passively through the plasma membrane if they are small enough to pass through its pores (and are aided by an electrical gradient) or if they can dissolve in the lipid portion of the membrane (as in the case of CO_2 and O_2). The diffusion of solutes through a semipermeable membrane is called **dialysis.** The diffusion of water through a semipermeable membrane is called **osmosis.** Both dialysis and osmosis, examples of diffusion phenomena, involve the movement of a substance from an area of its higher concentration to one of its lower concentration, i.e., down its concentration gradient.

DIFFUSION OF A DYE THROUGH AN AGAR GEL The relationship between molecular weight and the rate of diffusion can be examined simply by observing the diffusion of the molecules of two different types of dye through an agar gel. The dyes used in this experiment are methylene blue, which has a molecular weight of 320 and is deep blue in color, and potassium permanganate, a deep purple dye with a molecular weight of 158. Although the agar gel appears quite solid, it is primarily (98.5%) water and allows free movement of the diffusing dye molecules through it.

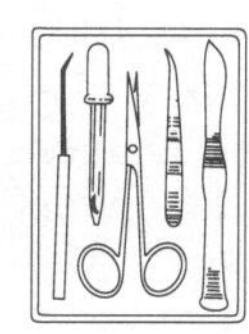

1. Obtain a petri dish containing agar gel, a forceps, a millimeter ruler, and containers of methylene blue crystals and potassium permanganate crystals, and bring them to your bench.

⚠ 2. *Avoid contact between your skin and the dye crystals. Using the forceps,* select approximately equal-size crystals of each dye, and place them gently on the agar gel surface, approximately 10 centimeters apart (Figure 5.1). Record the time.

____

3. At 15-minute intervals, use the millimeter ruler to measure the distance the dye has diffused from each crystal. These observations should be continued for 1½ hours, and the results recorded on p. 33.

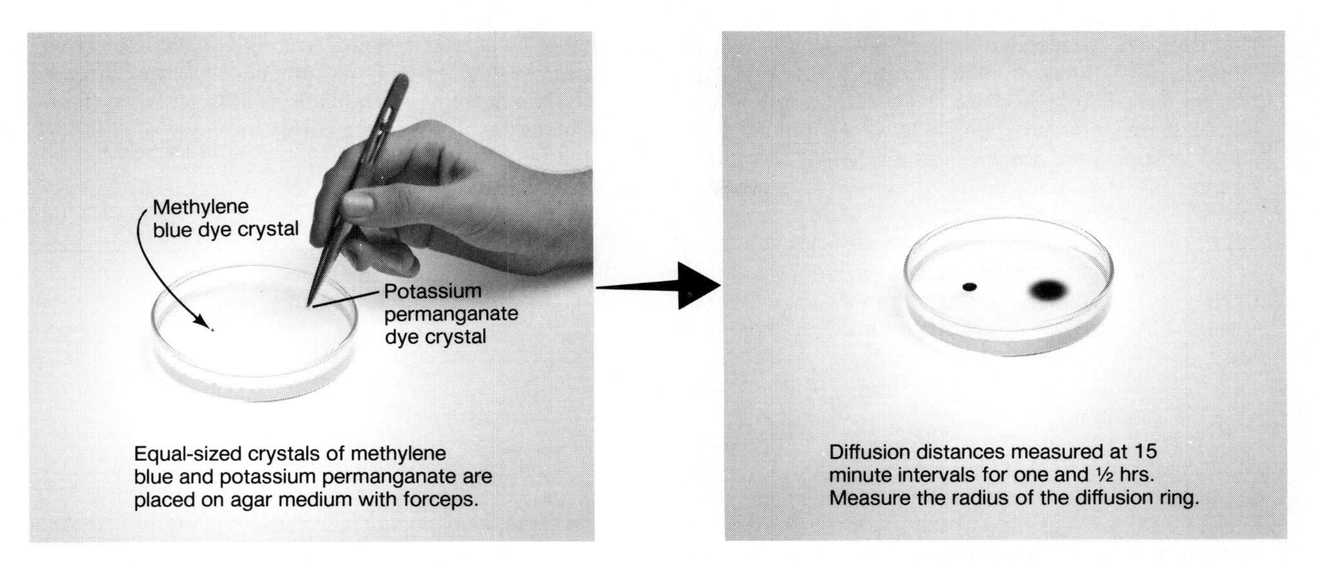

F5.1

Setup for comparing the diffusion rates of molecules of methylene blue and potassium permanganate.

Time (min)	Diffusion of methylene blue (mm)	Diffusion of potassium permanganate (mm)
15		
30		
45		
60		
75		
90		

Which dye diffused more rapidly?

____

What is the relationship between molecular weight and rate of molecular movement (diffusion)?

____

Why did the dye molecules move?

____

____

Compute the rate of diffusion of the potassium permanganate molecules in millimeters per minute and record.

_______________ mm/min

Compute the rate of diffusion of the methylene blue molecules in mm/min and record.

_______________ mm/min

DIFFUSION OF A DYE THROUGH WATER Make a mental note to yourself to go to demonstration area 1 at the end of the laboratory session to observe the extent of diffusion of potassium permanganate dye through water. At that time, follow the directions given next.

1. Measure the number of millimeters from the bottom of the graduated cylinder the dye has diffused and record.

_______________ mm

2. Note the time the demonstration was set up and the time of your observation. Compute the rate of the dye's diffusion through the water and record below.

_______________ Time of setup

_______________ Time of observation

_______________ mm/min (rate of diffusion)

3. Does the potassium permanganate dye move (diffuse) more rapidly through water or the agar gel? (Explain your answer.)

____

____

____

DIFFUSION THROUGH NONLIVING MEMBRANES A diffusion experiment providing information on the passage of water and solutes through semipermeable membranes, which may be applied to the study of transport mechanisms in living membrane-bound cells, is outlined below.

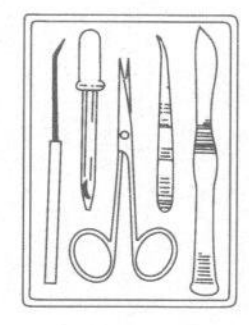

1. Obtain four dialysis sacs,* a small funnel, a graduated cylinder, and four beakers (250 ml). Number the beakers 1 through 4; and half fill all of them with distilled water, except for beaker 2, to which you should add 40% glucose solution.

2. Prepare the dialysis sacs one at a time. Using the small funnel, half fill each sac with 20 ml of the specified liquid. Press out the air, fold over the open end of the sac, and tie it tightly with fine twine. Before proceeding to the next sac, rinse each sac in tap water and then quickly and carefully blot the sac dry by rolling it on a paper towel. Weigh the sac with a laboratory balance, and record the weight below. Then drop each sac into the corresponding beaker. Be sure the sac is completely covered by the beaker solution, adding more solution if necessary.

- Sac 1: 40% glucose solution. Weight: __________ g
- Sac 2: 40% glucose solution. Weight: __________ g
- Sac 3: 10% NaCl solution. Weight: __________ g
- Sac 4: boiled starch solution. Weight: __________ g

Allow sacs to remain undisturbed in the beakers for 1 hour. (Use this time to continue with other experiments.)

3. After an hour, quickly and gently blot sac 1 dry and weigh it. (Note: Do not squeeze the sac during the blotting process.)

__________ g

Has there been any change in weight? __________

Conclusions? __________

____

____

Place 5 ml of Benedict's solution in each of two test tubes. Put 4 ml of the beaker fluid into one test tube, and 4 ml of the sac fluid into the other. Mark the tubes for identification, and then place them in a beaker containing boiling water. Boil 2 minutes. Cool slowly. If a green, yellow, or rusty red precipitate forms, the test is positive, meaning that glucose is present. If the solution remains the original blue color, the test is negative.

Was glucose still present in the sac? __________

In the beaker? __________

Conclusions? __________

____

____

4. Blot gently and weigh sac 2: __________ g

Was there an *increase* or *decrease* in weight? __________

With 40% glucose in the sac and 40% glucose in the beaker, would you expect to see any net movements of water (osmosis) or of glucose molecules (dialysis)?

__________ Why or why not? __________

____

____

____

5. Blot gently and weigh sac 3: __________ g

Was there any change in weight? __________

Conclusions? __________

____

____

Take a 5-ml sample of beaker 3 solution and put it into a test tube. Add a drop of silver nitrate. The appearance of a white precipitate or cloudiness indicates the presence of AgCl, which is formed by the reaction of $AgNO_3$ with NaCl (sodium chloride).

Results? __________

____

Conclusions? __________

____

6. Blot gently and weigh sac 4: __________ g

Was there any change in weight? __________

Conclusions? __________

____

* Dialysis sacs are selectively permeable membranes with pores of a particular size. The selectivity of living membranes depends on more than just pore size, but using the dialysis sacs will allow you to examine selectivity due to this factor.

Take a 5-ml sample of beaker 4 solution and add a couple of drops of Lugol's iodine solution. The appearance of a black color is a positive test for the presence of starch. Did any starch diffuse from the sac into the beaker? Explain.

___

___

7. In which of the test situations did net osmosis occur?

___

In which of the test situations did net dialysis occur?

___

What conclusions can you make about the relative size of glucose, starch, NaCl, and water molecules?

___

___

___

___

With what cell structure can the dialysis sac be compared?

___

8. Before leaving the laboratory observe the **osmometer** demonstration, which follows the movement of *water* through a membrane (i.e., osmosis). Measure the distance the water column has moved during the laboratory period and record below. (The position of the meniscus in the thistle tube at the beginning of the laboratory period is marked with wax pencil.)

Distance the meniscus has moved: ___ mm

DIFFUSION THROUGH LIVING MEMBRANES

To examine permeability properties of plasma membranes, conduct the following two experiments.

Experiment 1: 1. Obtain a clean slide and coverslip, a vial of animal blood, physiologic saline, 1.5% sodium chloride solution, 3 test tubes, test tube rack, glass stirring rod, 15-ml graduated cylinder, filter paper, and plastic gloves.

2. Label the three test tubes A, B, and C, and prepare them as follows:
   - A: add 2 ml physiologic saline
   - B: add 2 ml 1.5% sodium chloride solution
   - C: add 2 ml distilled water

3. Don the gloves and, using a medicine dropper, add 5 drops of blood to test tube A. Repeat this procedure for samples B and C. Stir each test tube with the glass rod, rinsing between each sample.

4. Hold each test tube in front of this printed page. ***Record*** the clarity of print seen through the fluid in each tube.

Test tube A ___

Test tube B ___

Test tube C ___

Experiment 2: Now you will examine red blood cells, suspended in the same three solutions, under the microscope to determine if these solutions have any effect on cell shape by promoting net osmosis.

1. Place a very small drop of physiologic saline on a slide and, using the medicine dropper, add a small drop of animal blood to the saline on the slide. Tilt the slide to mix, cover with a coverslip, and immediately examine the preparation under the high-power lens. Notice that the red blood cells retain their normal smooth disc-like shape. This is because the physiologic saline is **isotonic** to the cells; i.e., it contains a concentration of nonpenetrating solutes equal to that in the cells. Consequently, the cells neither gain nor lose water by osmosis.

2. Prepare another wet mount of the animal blood, but this time use 1.5% saline solution as the suspending medium. After 5 minutes, carefully observe the red blood cells under high power. What has begun to happen to the normally smooth disc shape of the red blood cells?

___

___

___

This crinkling-up process, called **crenation,** is due to the fact that the 1.5% saline solution is hypertonic to the cell sap of the red blood cells. A **hypertonic** solution contains more nonpenetrating solutes than are present in the cell. Under these circumstances, water tends to leave the cells by osmosis. Compare your observations to Figure 5.2b.

3. Add a drop of distilled water to the edge of the coverslip. Fold a piece of filter paper in half, and place its folded edge at the opposite edge of the coverslip; it will absorb the saline solution and draw the distilled water across the cells. Watch the red blood cells as they float across the field. After about 5 minutes have passed, describe the change in their appearance.

___

___

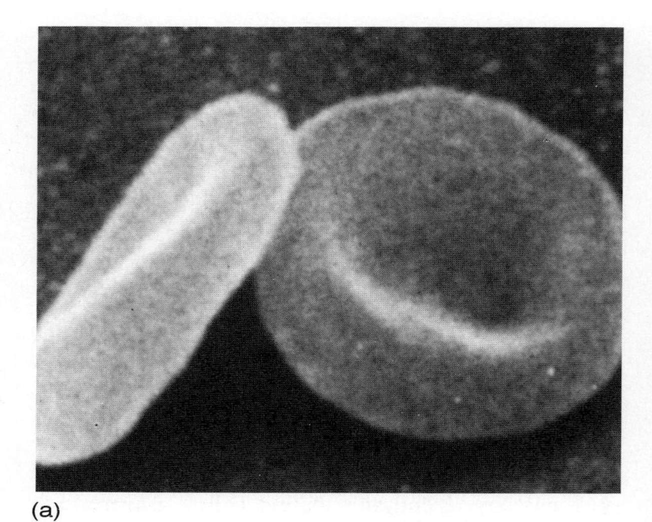
(a)

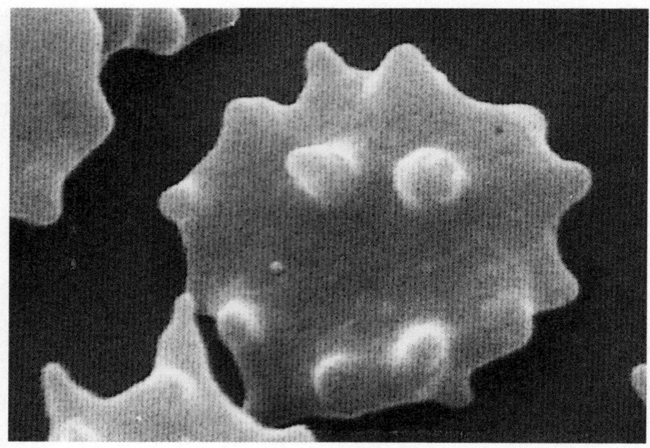
(b)

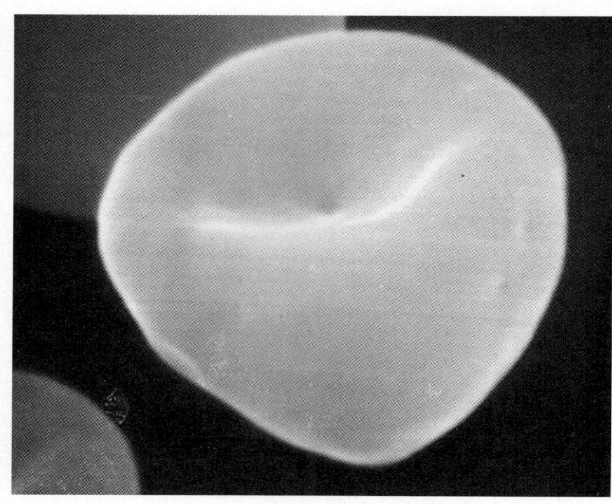
(c)

F5.2

Influence of hypertonic and hypotonic solutions on red blood cells. (a) Red blood cells suspended in an isotonic solution, where the cells retain their normal size and shape. (b) Red blood cells suspended in a hypertonic solution. As the cells lose water to the external environment, they shrink and become prickly; this phenomenon is called crenation. (c) Red blood cells suspended in a hypotonic solution. Notice their spherical bloated shape, a result of excessive water intake.

Distilled water contains *no* solutes (it is 100% water). Distilled water and *very* dilute solutions (that is, those containing less than 0.9% nonpenetrating solutes) are **hypotonic** to the cell. In a hypotonic solution, the red blood cells first "plump up," but then they suddenly start to disappear. The red blood cells burst as the water floods into them, leaving "ghosts" in their wake (see Figure 5.2c). This phenomenon is called **hemolysis.**

How do your observations of test tube C in experiment 1 correlate with what you have just observed under the microscope?

___

4. Place the blood-soiled slides in the bleach-containing beaker at the supply area and put the gloves you used in the disposable autoclave bag.

Filtration

Filtration is a physical process by which water and solutes pass through a membrane from an area of higher hydrostatic (fluid) pressure into an area of lower hydrostatic pressure. Like diffusion, it is a passive process. For example, fluids and solutes filter out of the capillaries in the kidneys into the kidney tubules because the blood pressure in the capillaries is greater than the fluid pressure in the tubules. Filtration is not a selective process. The amount of filtrate (fluids and solutes) formed depends almost entirely on the difference in pressure on the two sides of the membrane and on the size of the membrane pores.

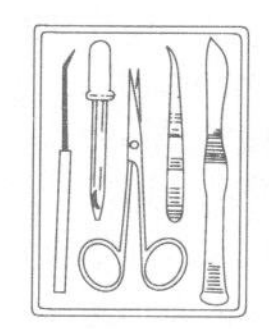

1. Obtain the following equipment: a ring stand, ring, and ring clamp; a piece of filter paper; a beaker; and the flask containing the mixture of uncooked starch, powdered charcoal, and copper sulfate. Attach the ring to the ring stand with the clamp.

2. Fold tne filter paper in half twice, open it into a cone, and place it in a funnel. Place the funnel in the ring of the ring stand, and place a beaker under the funnel. Shake the starch mixture, and fill the funnel with it to just below the top of the filter paper. When the steady stream of filtrate changes to countable filtrate drops, count the number of drops formed in 10 seconds.

___ drops

When the funnel is half empty, again count the number of drops formed in 10 seconds, and record the count.

___ drops

3. After all the fluid has passed through the filter, check the filtrate and paper to see which materials were

retained by the paper. (Note: If the filtrate is blue, the copper sulfate passed. Check both the paper and filtrate for black particles to see if the charcoal passed. Finally, add Lugol's iodine to a 2-ml filtrate sample in a test tube. If the sample turns blue/black when the iodine is added, starch is present in the filtrate.)

Passed: ______________________________

Retained: ______________________________

What does the filter paper represent? ______________

During which counting interval was the filtration rate greatest?

____

Explain: ______________________________

____

____

____

What characteristic of the three solutes determined whether or not they passed through the filter paper?

____

ACTIVE TRANSPORT

Whenever a cell expends metabolic energy (ATP) to move substances across its boundaries, the process is referred to as active transport. The substances moved by active means are generally unable to pass by diffusion. They may be too large to pass through the pores; they may not be lipid soluble; or they may have to move against, rather than with, a concentration gradient.

In one type of active transport, substances move across the plasma membrane by combining with carrier molecules (proteins) located in the membrane; the combining process resembles an enzyme-substrate interaction. ATP is required, and in many cases the substances move against concentration or electrochemical gradients or both. Substances that are moved into the cells by such carriers, commonly called **solute pumps,** are amino acids and some sugars. Neither is lipid soluble, and amino acids are too large to pass through the pores, but both are necessary for cell life. On the other hand, sodium ions (Na^+) are moved out of cells by active transport. There is more Na^+ outside the cell than there is inside; so the Na^+ tends to remain in the cell unless actively transported out.

Phagocytosis and pinocytosis also require ATP. In **pinocytosis** (cell drinking), the plasma membrane seems to sink beneath the material to form a small vesicle, which then pinches off into the cell interior (see Figure 5.3). Pinocytosis is most common for taking in liquids containing protein or fat.

In **phagocytosis** (cell eating), parts of the plasma membrane flow around a relatively large or solid material (for example, bacteria or cell debris) and engulf it,

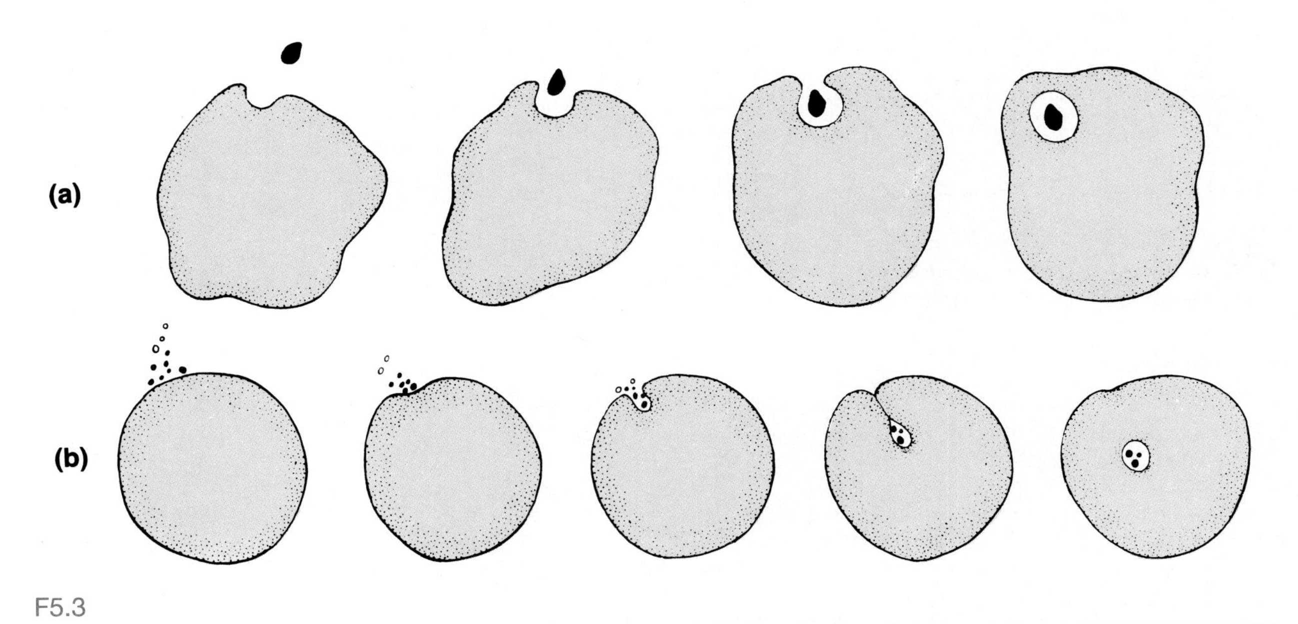

F5.3

Phagocytosis and pinocytosis. (a) In phagocytosis, extensions of the cytoplasm (pseudopodia) flow around the external particle and enclose it within a vacuole. (b) In pinocytosis, solutes (such as dissolved proteins) at the external surface of the plasma membrane stimulate the membrane to invaginate and incorporate a droplet of the fluid.

enclosing it within a sac. This is an important function of the body's phagocytic or scavenger cells, such as some white blood cells and macrophages. Such cells protect the body from pathogens such as viruses, bacteria, and fungi, and dispose of dead tissue cells by phagocytizing them and then digesting the contents of the phagocytic vesicle (usually by fusing it with a lysosome).

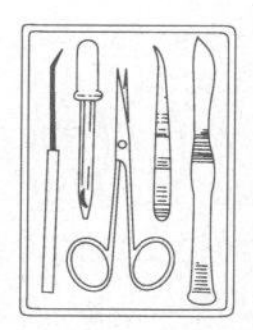

1. Obtain a drop of starved *Amoeba proteus* and place it on a coverslip. Add a drop of Tetrahymena pyriformis culture (an amoeba "meal") to the amoeba-containing drop, and then quickly but gently invert the coverslip over the well of a depression slide.

2. Locate an amoeba under low power. Keep the light as dim as possible; otherwise the amoeba will "ball up" and begin to disintegrate.

3. Observe as the amoeba phagocytizes the *Tetrahymena* by forming pseudopods that engulf it. In unicellular organisms like the amoeba, phagocytosis is an important food-getting mechanism; but in higher organisms, its more important function is in protection as described above.

4. Return all equipment to the appropriate supply areas and rinse glassware used before leaving the lab.

Classification of Tissues

OBJECTIVES

1. To name the four major types of tissues in the human body and the major subcategories of each.
2. To identify the tissue subcategories through microscopic inspection or inspection of an appropriate diagram or projected slide.
3. To indicate the location of the various tissue types in the body.
4. To state the general functions and structural characteristics of each of the four major tissue types.

MATERIALS

Compound microscope
Prepared slides of simple squamous, simple cuboidal, simple columnar, stratified squamous (nonkeratinized), stratified cuboidal, stratified columnar, pseudostratified ciliated columnar, and transitional epithelium
Prepared slides of mesenchyme; adipose, areolar, and dense connective tissue regular (tendon) and irregular (dermis) varieties; elastic connective tissue (in wall of large artery); hyaline and elastic cartilage; fibrocartilage, bone (cross section); and blood
Prepared slides of skeletal, cardiac, and smooth muscle (longitudinal sections)
Prepared slide of nervous tissue (spinal cord smear)

Exercise 4 describes cells as the building blocks of life and the all-inclusive functional units of unicellular organisms. But in higher organisms, cells do not usually operate as isolated, independent entities. In humans and other multicellular organisms, cells depend on one another and cooperate to maintain homeostasis in the body.

With a few exceptions (parthenogenic organisms), even the most complex animal starts out as a single cell, the fertilized egg, which divides almost endlessly. The thousands of cells that result become specialized for a particular function; some become supportive bone, others the transparent lens of the eye, still others skin cells, and so on. Thus a division of labor exists, with certain groups of cells highly specialized to perform functions that benefit the organism as a whole.

Groups of cells that are similar in structure and function are called **tissues.** The four primary tissue types—epithelium, connective tissue, nervous tissue, and muscle—have distinctive structures, patterns, and functions. Each of the four primary tissues are further divided into subcategories, as described shortly.

To perform specific body functions, the tissues are organized into **organs** such as the heart, kidneys, and lungs. Most organs contain several representatives of the primary tissues, and the arrangement of these tissues determines the organ's structure and function. Thus **histology,** the study of tissues, complements a study of gross anatomy and provides the structural basis for a study of organ physiology.

The main objective of this exercise is to familiarize you with the major similarities and dissimilarities of the primary tissues so that when the tissue composition of an organ is described, you will be able to more easily understand (and perhaps even predict) the organ's major function. Because epithelium and some types of connective tissue will not be considered again, they are emphasized more than muscle, nervous tissue, and bone (a connective tissue), which are covered in more depth in later exercises.

EPITHELIAL TISSUE

Epithelial tissue, or **epithelium,** covers surfaces, and since glands almost invariably develop from epithelial membranes, they too are logically classed as epithelium. Epithelium covers the external body surface (as the epidermis), lines its cavities and tubules, and composes the various edocrine (hormone-producing) and exocrine glands of the body.

Epithelial functions include *protection, absorption, filtration,* and *secretion.* For example, the epithelium covering the body protects against bacterial and chemical damage; that lining the respiratory tract is ciliated to sweep mucus away from the lungs. Epithelium specialized to absorb substances lines the stomach and small intestine. In the kidney tubules, the epithelium absorbs, secretes, and filters. Secretion is a specialty of the glands.

Epithelium generally exhibits the following characteristics:

- Cells fit closely together to form membranes, or sheets of cells, and are bound together by specialized junctions.

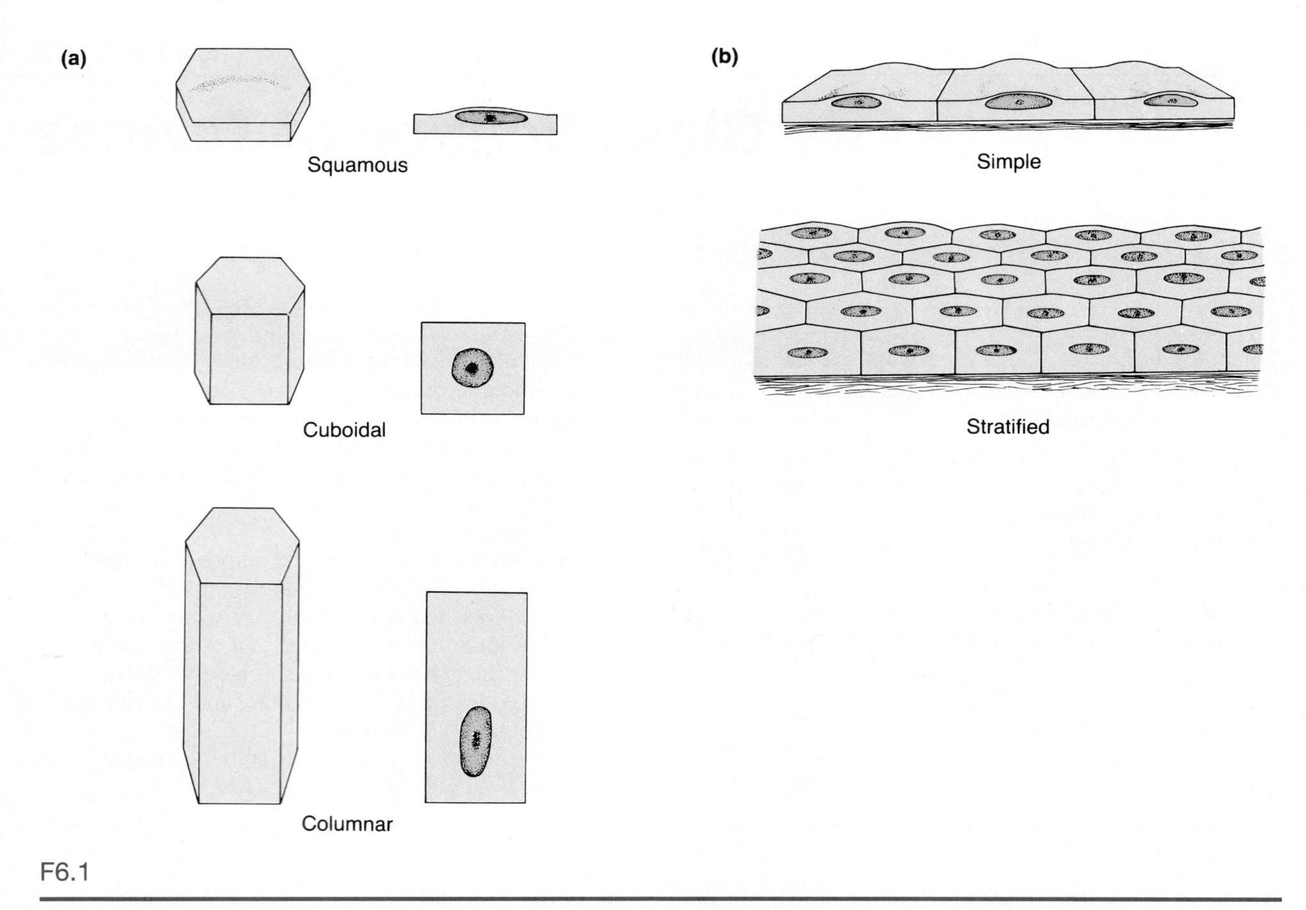

Classification of epithelia. (a) Classification on the basis of cell shape. For each category, a whole cell is shown on the left and a longitudinal section is shown on the right. (b) Classification on the basis of arrangement (layers).

- The membranes always have one free surface, called the *apical surface.*
- The cells are attached to an adhesive **basement membrane,** an amorphous material secreted by the epithelial and connective tissue cells that lie adjacent to each other.
- Epithelial tissues have no blood supply of their own (are avascular) but depend on diffusion of nutrients from capillaries in the underlying connective tissue.
- If well nourished, epithelial cells can easily regenerate themselves. This is an important characteristic because many epithelia are subjected to substantial friction.

The covering and lining epithelia are classified according to two criteria—cell shape and arrangement, or relative number of layers (Figure 6.1). **Squamous** (scalelike), **cuboidal** (cubelike), and **columnar** (column-shaped) epithelial cells are the general types based on shape. On the basis of arrangement, there are **simple** epithelia, consisting of one layer of cells attached to a basement membrane, and **stratified** epithelia, consisting of more than one layer of cells. The terms denoting shape and arrangement of the epithelial cells are combined to describe the epithelium fully. *Stratified epithelia are named according to the cells at the apical surface of the epithelial membrane,* not those resting on the basement membrane.

There are, in addition, two less easily categorized types of epithelia. **Pseudostratified columnar epithelium** is actually a simple epithelium (one layer of cells); but because its cells extend varied distances from the basement membrane, it gives the false appearance of being stratified. This epithelium is often ciliated. **Transitional epithelium** is a rather peculiar stratified squamous epithelium formed of rounded, or "plump," cells with the ability to slide over one another to allow the organ to be stretched. Transitional epithelium is found only in urinary system organs subjected to periodic distention, such as the bladder. The superficial cells are scalelike (like true squamous cells) when the organ is distended and rounded when the organ is empty.

The most common types of epithelia and their most common locations in the body are illustrated and described in Figure 6.2.

Epithelial cells forming glands are highly specialized to remove materials from the blood and to manufacture them into new materials, which they then secrete. There are two types of glands. The **endocrine glands** lose their surface connection (duct) as they develop; thus they are referred to as *ductless glands.* Their secretions (all hormones) are extruded directly into the

Text continues on p. 45

(a) Simple squamous epithelium

Description: Single layer of flattened cells with disk-shaped central nuclei and sparse cytoplasm; the simplest of the epithelia.

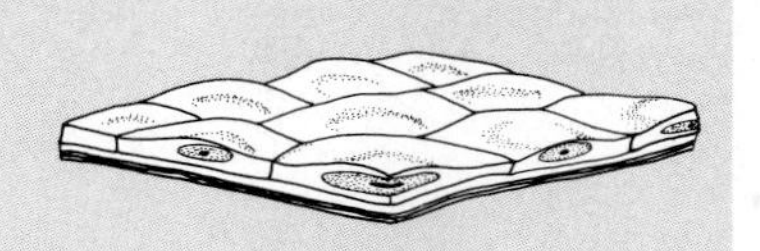

Location: Air sacs of lungs; kidney glomeruli; lining of heart, blood vessels, and lymphatic vessels; lining of ventral body cavity (serosae).

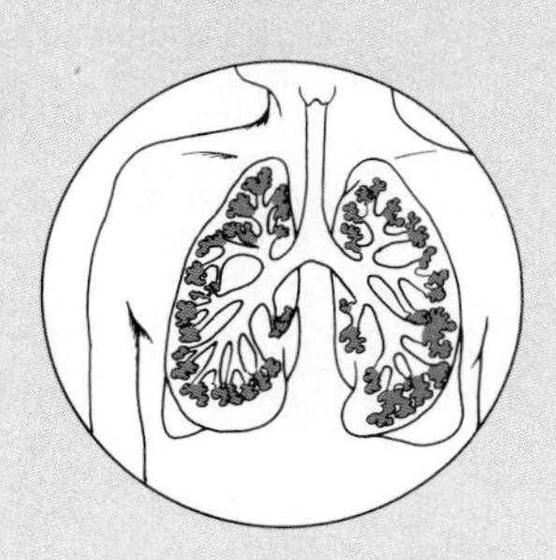

Function: Allows passage of materials by diffusion and filtration in sites where protection is not important; secretes lubricating substances in serosae.

Photomicrograph: Simple squamous epithelium forming walls of alveoli (air sacs) of the lung (280 ×).

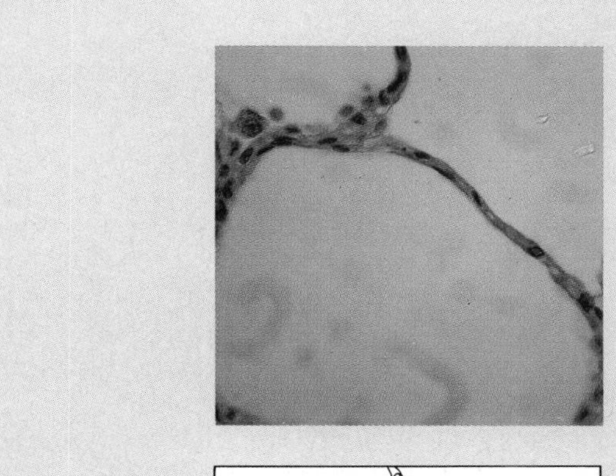

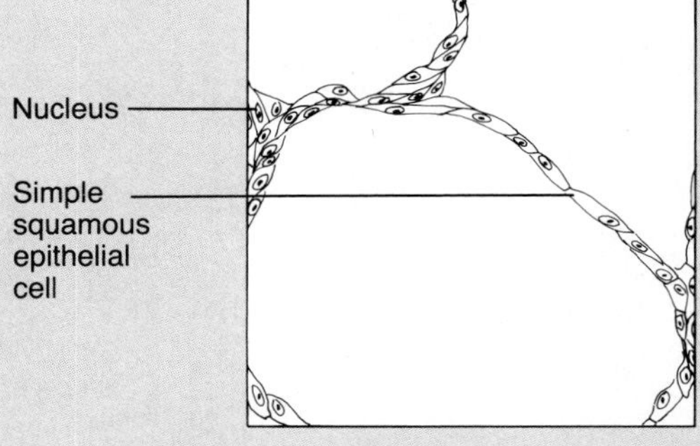

(b) Simple cuboidal epithelium

Description: Single layer of cubelike cells with large, spherical central nuclei.

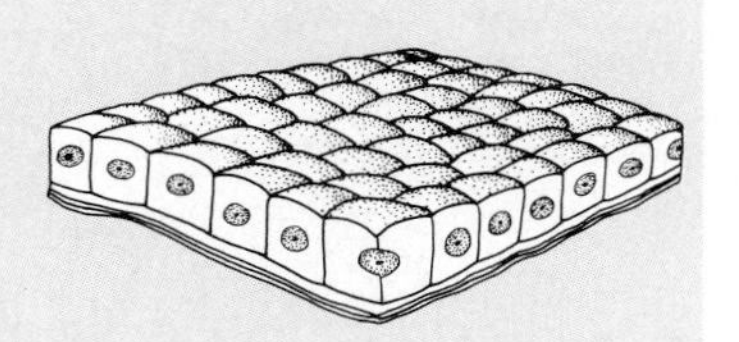

Location: Kidney tubules; ducts and secretory portions of small glands; ovary surface.

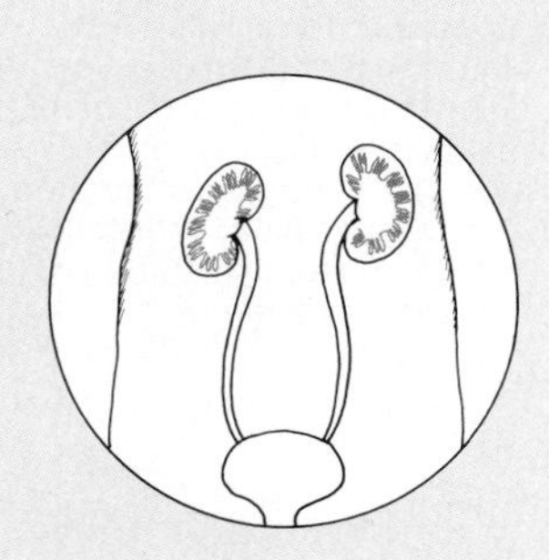

Function: Secretion and absorption.

Photomicrograph: Simple cuboidal epithelium in kidney tubules (260X)

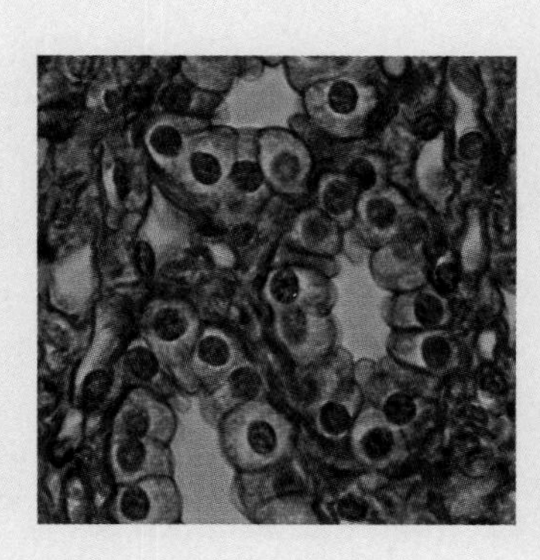

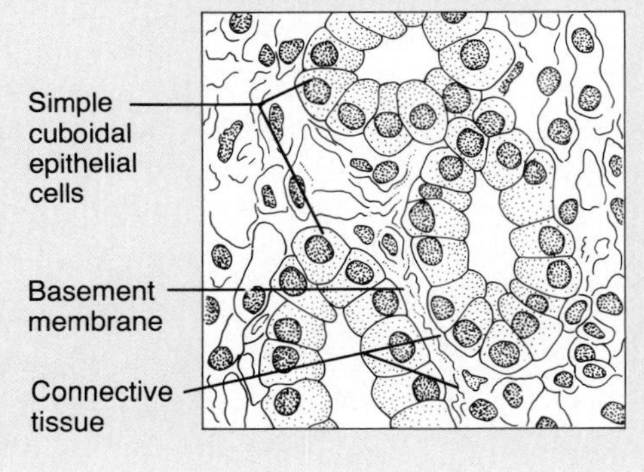

F6.2

Epithelial tissues. Simple epithelia (a and b).

(c) Simple columnar epithelium

Description: Single layer of tall cells with *oval* nuclei; some cells bear cilia; layer may contain mucus-secreting glands (goblet cells).

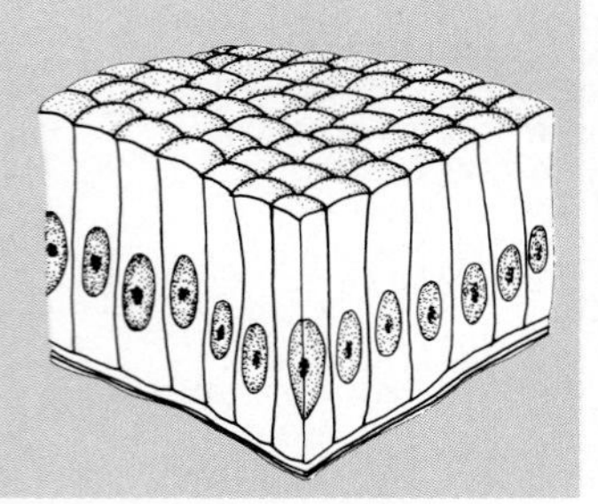

Location: Nonciliated type lines most of the digestive tract (stomach to anal canal), gallbladder and excretory ducts of some glands; ciliated variety lines small bronchi, uterine tubes, and some regions of the uterus.

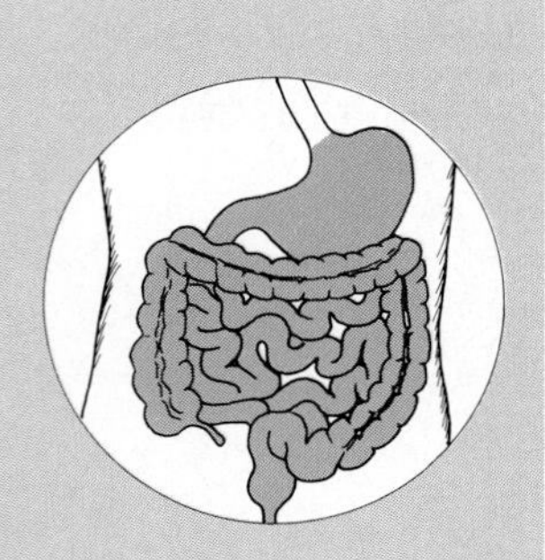

Function: Absorption; secretion of mucus, enzymes, and other substances; ciliated type propels mucus (or reproductive cells) by ciliary action.

Photomicrograph: Simple columnar epithelium of the stomach mucosa (280×).

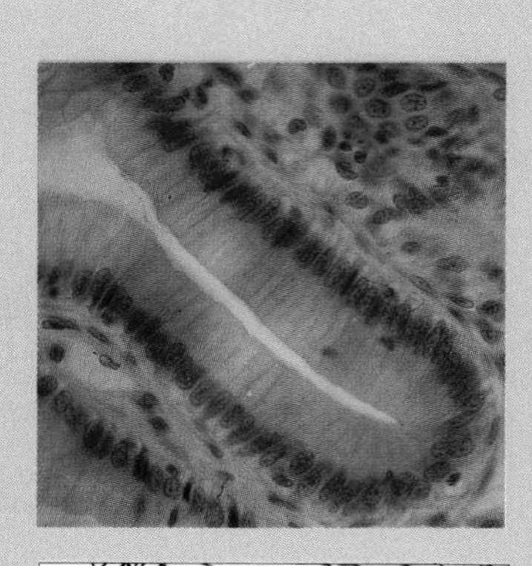

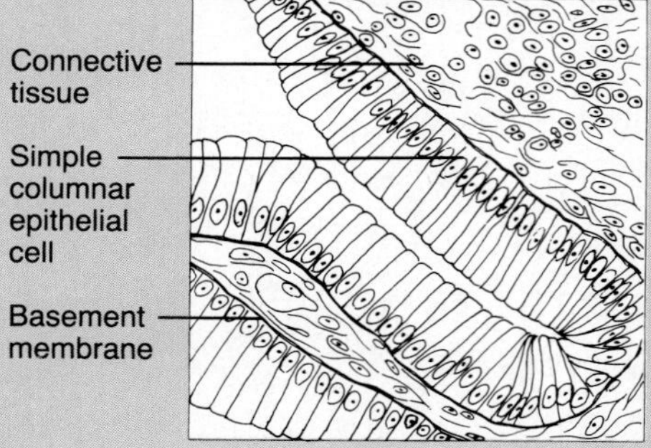

(d) Pseudostratified columnar epithelium

Description: Single layer of cells of differing heights, some not reaching the free surface; nuclei seen at different levels; may contain goblet cells and bear cilia.

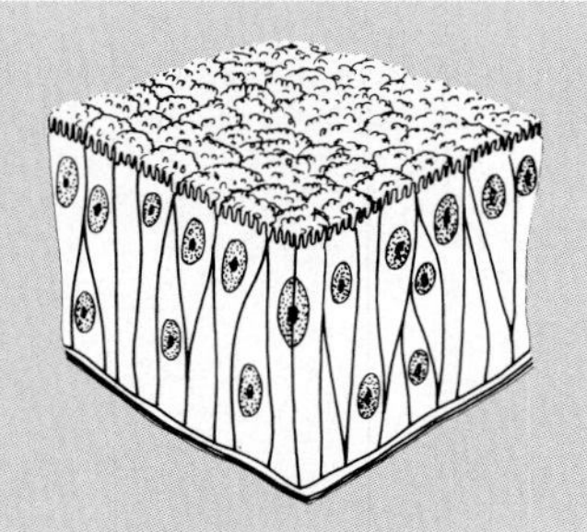

Location: Nonciliated type in ducts of large glands, parts of male urethra; ciliated variety lines the trachea, most of the upper respiratory tract.

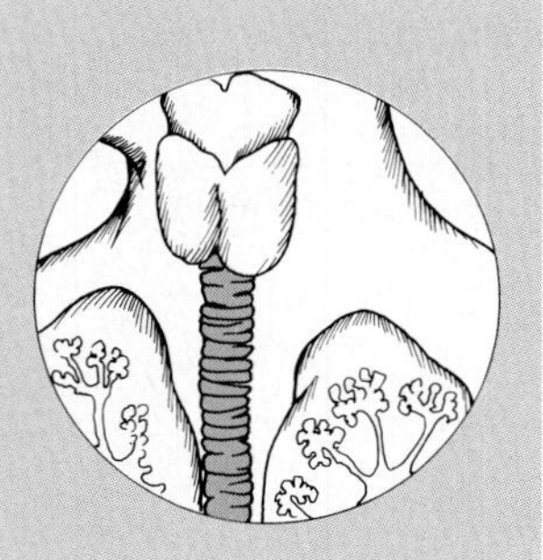

Function: Secretion, particularly of mucus; propulsion of mucus by ciliary action.

Photomicrograph: Pseudostratified ciliated columnar epithelium lining the human trachea (430×).

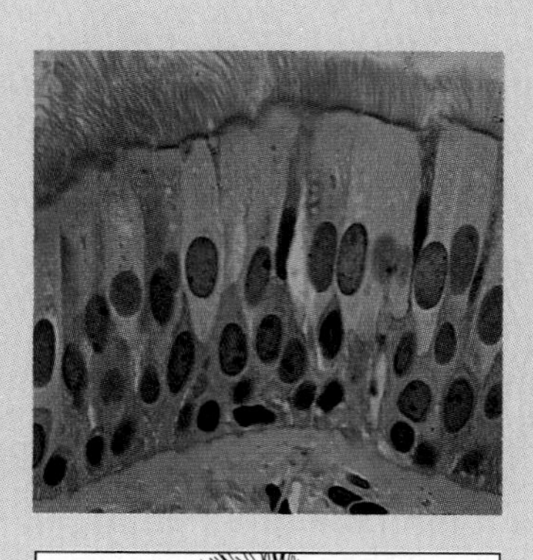

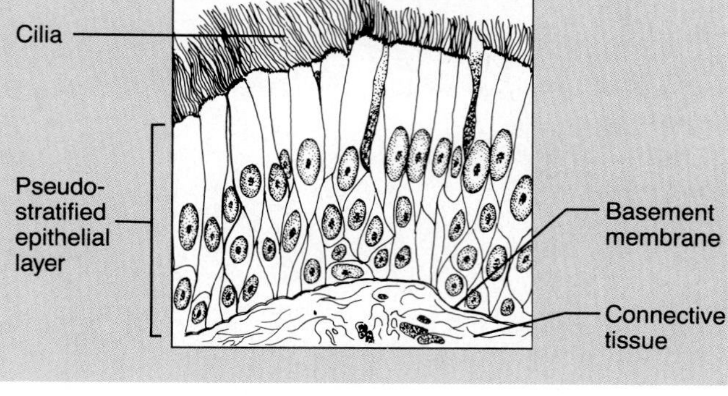

F6.2 (Continued)

Simple epithelia (c and d).

(e) Stratified squamous epithelium

Description: Thick membrane composed of several cell layers; basal cells are cuboidal or columnar and metabolically active; surface cells are flattened (squamous); in the keratinized type, the surface cells are full of keratin and dead; basal cells are active in mitosis and produce the cells of the more superficial layers.

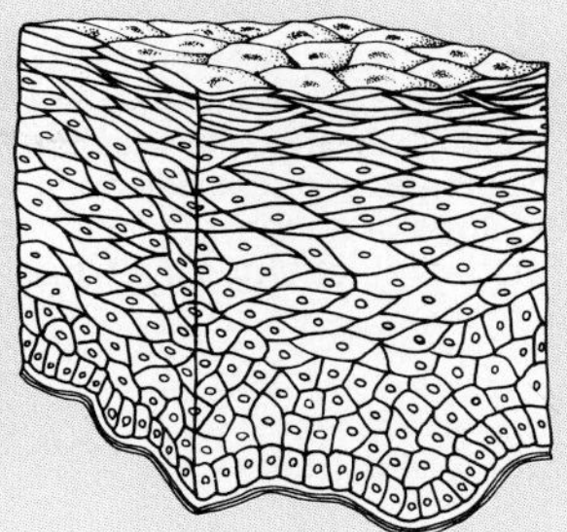

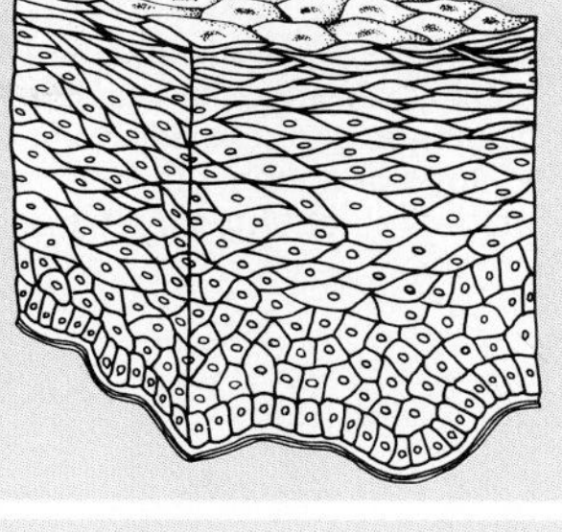

Location: Nonkeratinized type forms the moist linings of the esophagus, mouth, and vagina; keratinized variety forms the epidermis of the skin, a dry membrane.

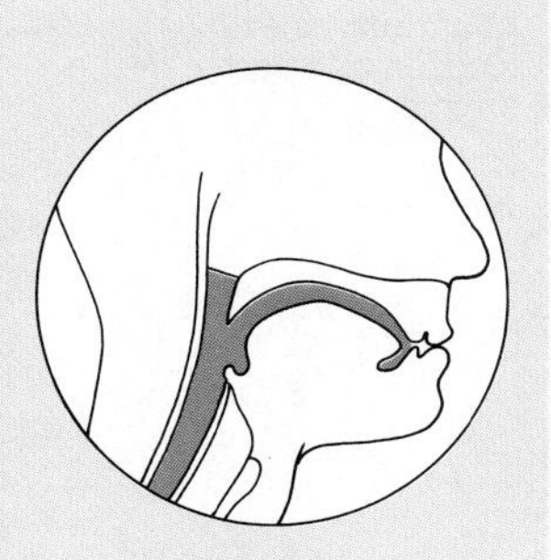

Function: Protects underlying tissues in areas subjected to abrasion.

Photomicrograph: Stratified squamous epithelium lining of the esophagus (173 ×).

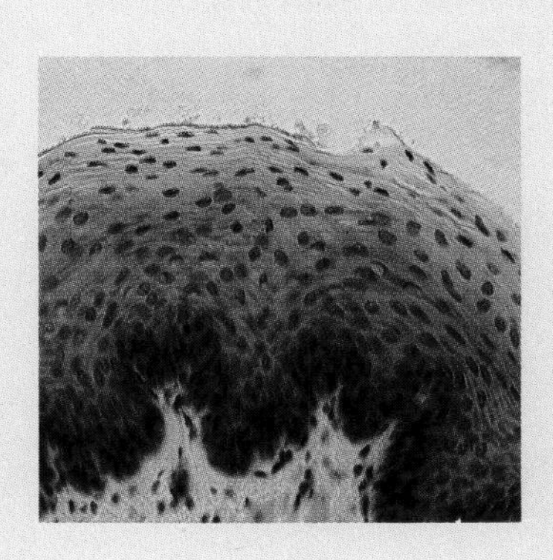

Stratified squamous epithelium

Nuclei

Basement membrane

Connective tissue

(f) Stratified cuboidal epithelium

Description: Generally two layers of cube-like cells.

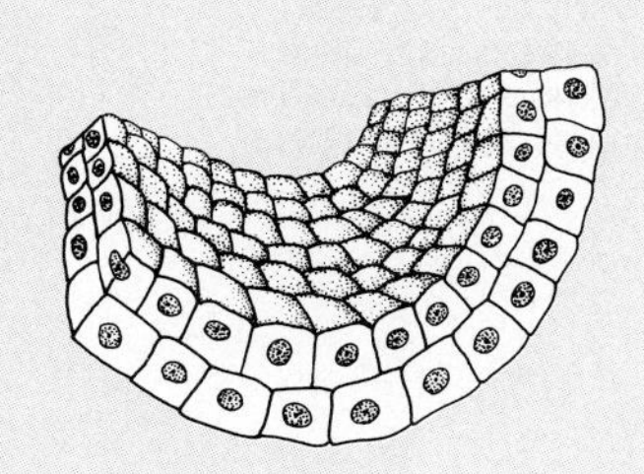

Location: Largest ducts of sweat glands, mammary glands, and salivary glands.

Function: Protection.

Photomicrograph: Stratified cuboidal epithelium forming a salivary gland duct (400 ×).

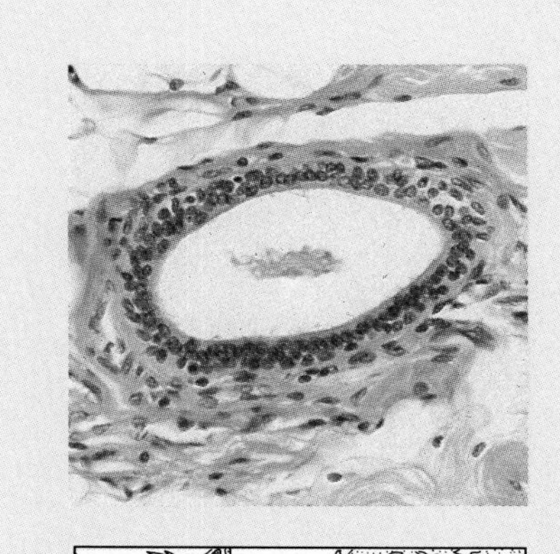

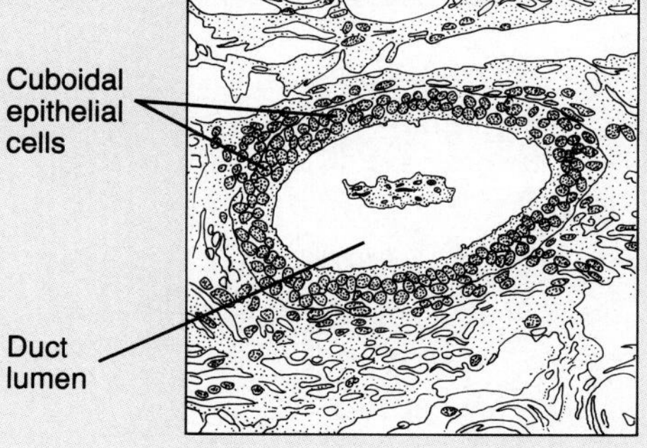

F6.2 (Continued)

Stratified epithelia (e and f).

(g) Stratified columnar epithelium

Description: Several cell layers; basal cells usually cuboidal; superficial cells elongated and columnar.

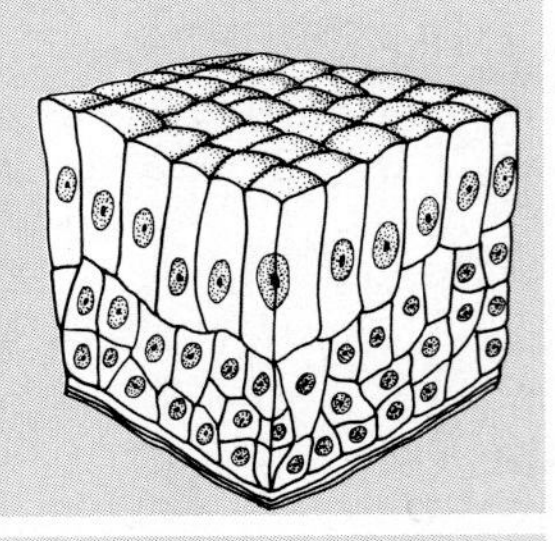

Location: Rare in the body; small amounts in male urethra and in large ducts of some glands.

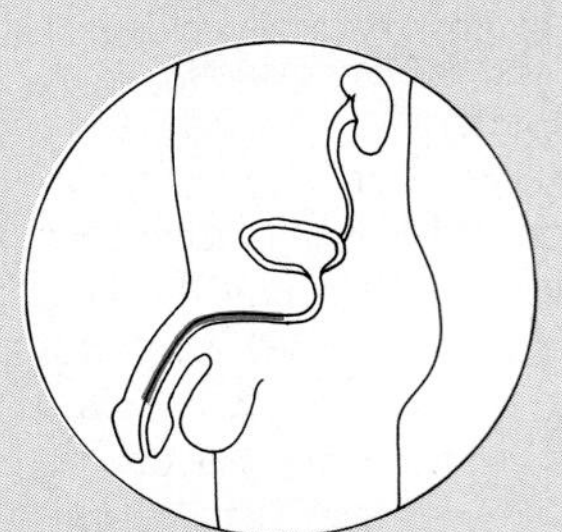

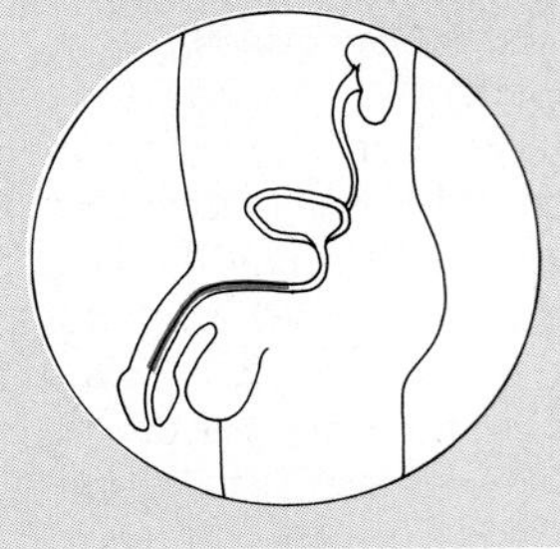

Function: Protection; secretion.

Photomicrograph: Stratified columnar epithelium lining of the male urethra (360 ×).

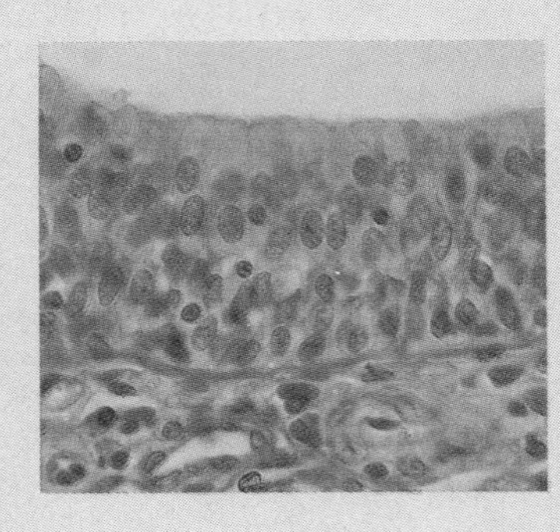

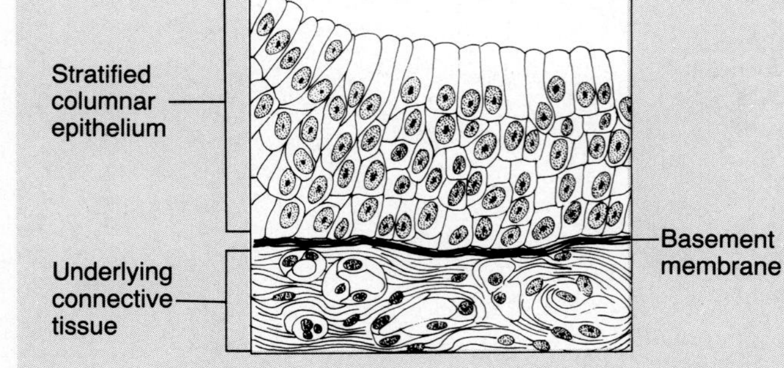

(h) Transitional epithelium

Description: Resembles both stratified squamous and stratified cuboidal; basal cells cuboidal or columnar; surface cells dome-shaped or squamous-like, depending on degree of organ stretch.

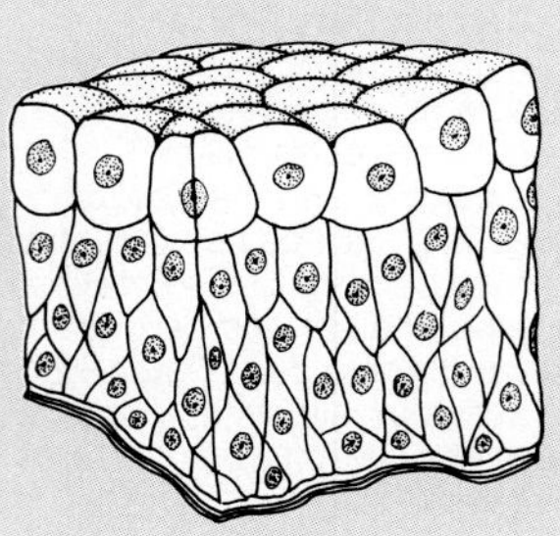

Location: Lines the ureters, bladder, and part of the urethra.

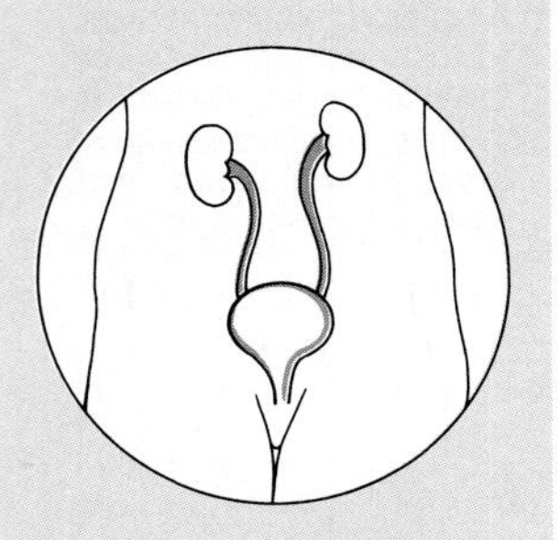

Function: Stretches readily and permits distension of urinary organ by contained urine.

Photomicrograph: Transitional epithelium lining of the bladder, relaxed state (170 ×); note the bulbous, or rounded, appearance of the cells at the surface; these cells flatten and become elongated when the bladder is filled with urine.

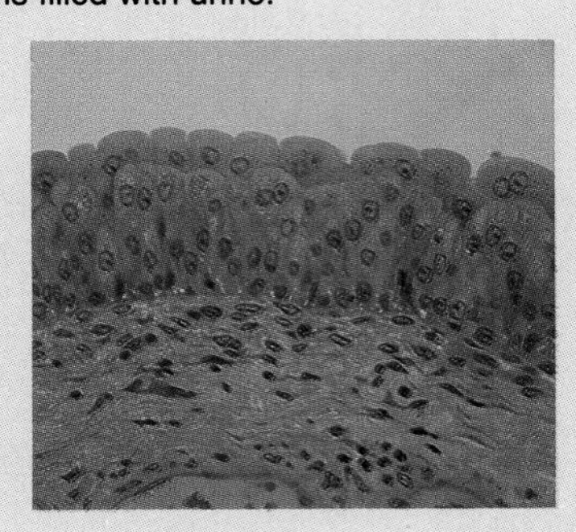

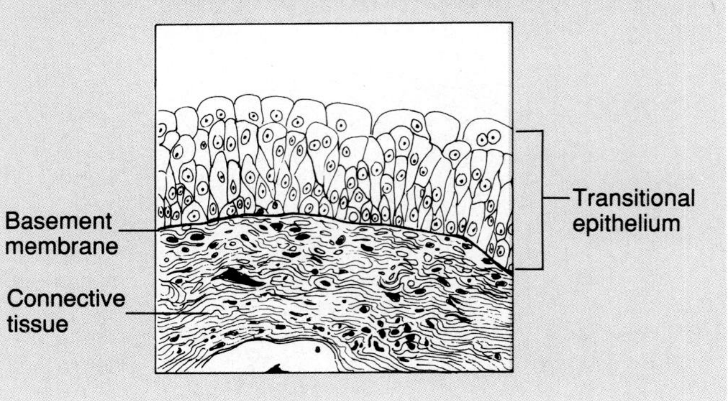

F6.2 (Continued)

Stratified epithelia (g and h).

extracellular fluid and ultimately enter the blood or the lymphatic vessels that weave through the glands. The **exocrine glands** retain their ducts, and their secretions empty through these ducts to an epithelial surface. The exocrine glands—including the sweat and oil glands, liver, and pancreas—are both external and internal; they will be discussed in conjunction with the organ systems to which their products are functionally related.

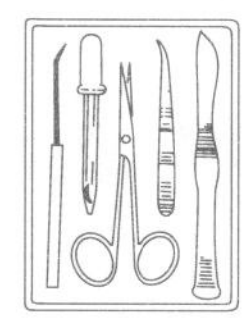

Obtain slides of simple squamous, simple cuboidal, simple columnar, stratified squamous (nonkeratinized), pseudostratified ciliated columnar, stratified cuboidal, stratified columnar, and transitional epithelia. Examine each carefully, and note how the epithelial cells fit closely together forming intact sheets of cells, a necessity for a tissue that forms linings or covering membranes. Compare your observations with the photomicrographs in Figure 6.2. Scan each epithelial type for any modifications for specific functions, such as cilia (motile cell projections which help to move substances along the cell surface) and microvilli, which increase the surface area for absorption. Also be on the alert for goblet cells which secrete lubricating mucus (see plate 1 of the histology atlas).

While working, check the questions in the laboratory review section for this exercise. Some of the questions there refer to some of the observations you are asked to make here.

CONNECTIVE TISSUE

Connective tissue is found in all parts of the body, as discrete structures or as part of various body organs. It is the most abundant and widely distributed of the tissue types.

The connective tissues perform a variety of functions; but they primarily *protect, support,* and *bind together* other tissues of the body. For example, bones are composed of connective tissue (**osseous tissue**), and they protect and support other body tissues and organs. The ligaments and tendons (**dense connective tissue**) bind the bones together or bind skeletal muscles to bones.

Areolar connective tissue is a soft packaging material that cushions and protects body organs. Fat (**adipose**) tissue provides insulation for the body tissues and a source of stored food. Blood-forming (**hematopoietic**) tissue replenishes the body's supply of red blood cells. In addition, connective tissue serves a vital function in the repair of all body tissues since many wounds are repaired by connective tissue in the form of scar tissue.

The characteristics of connective tissue include the following:

- With a few exceptions (cartilages, tendons, and ligaments), connective tissues are well vascularized.
- Connective tissues are composed of many types of cells.
- There is a great deal of noncellular, nonliving material (matrix) between the cells of connective tissue.

The nonliving material between the cells—the **extracellular matrix**—deserves a bit more explanation. It is produced by the cells and then extruded. The matrix is primarily responsible for the strength associated with connective tissue, but there is variation. At one extreme, hematopoietic and adipose tissues are composed mostly of cells. At the opposite extreme, bone and cartilage have very few cells and large amounts of matrix.

The matrix has two components—ground substance and fibers. The **ground substance** is composed largely of glycoproteins and large charged polysaccharide molecules. Depending on its specific composition, the ground substance may be liquid, semisolid, gel-like, or very hard. The fibers include **collagenic** (white), **elastic** (yellow), and **reticular** (fine collagenic) fibers. Of these, the collagenic fibers are most abundant.

Generally speaking, the ground substance functions as a molecular "sieve," or medium, through which nutrients and other dissolved substances can diffuse between the blood capillaries and the cells. The fibers in the matrix impede diffusion somewhat and make the ground substance less pliable. The relative durability and strength of the various connective tissues depend on the relative firmness of their ground substance and the number and kinds of fibers deposited in it.

There are four main types of adult connective tissue, all of which typically have large amounts of matrix—these are connective tissue proper, cartilage, bone, and blood. All of these derive from an embryonic tissue called **mesenchyme.** Figure 6.3 lists the general characteristics, location, and function of the connective tissues found in the body.

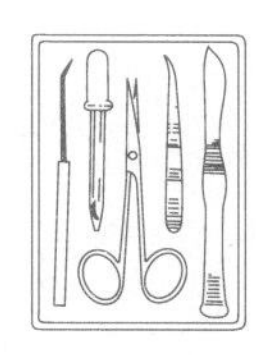

Obtain prepared slides of mesenchyme, adipose, areolar, reticular, and dense fibrous connective tissue (tendon and dermis of skin); elastic connective tissue; fibrocartilage; elastic and hyaline cartilage; osseous tissue (bone); and blood. Compare your observations with the views in Figure 6.3.

Distinguish between the living cells and the matrix and pay particular attention to the denseness and arrangement of the matrix in these tissues. For example, notice how the dense fibrous connective tissue seen in tendons and in the skin dermis is "chock full" of collagenic fibers and that in the regular variety the fibers are all running in the same direction.

While examining the areolar connective tissue, a soft "packing tissue," notice how much "empty space" there appears to be and distinguish between the collagen fibers and the coiled elastic fibers. Also, try to locate a **mast cell,** which has large, darkly staining granules in its cytoplasm. This cell type releases histamine that makes capillaries highly permeable during inflammatory reactions and allergies, and thus is partially responsible for the "runny nose" of some allergies.

In adipose tissue, locate a cell (signet ring cell) in which the nucleus can be seen pushed to one side by the large fat-filled vacuole, which appears to be a large empty space, and notice how little matrix there is in fat or adipose tissue.

Text continues on p. 52

Embryonic connective tissue

(a) Mesenchyme

Description: Embryonic connective tissue; gel-like ground substance containing fine fibers; star-shaped mesenchymal cells.

Location: Primarily in embryo.

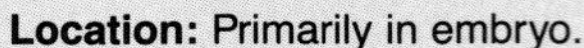

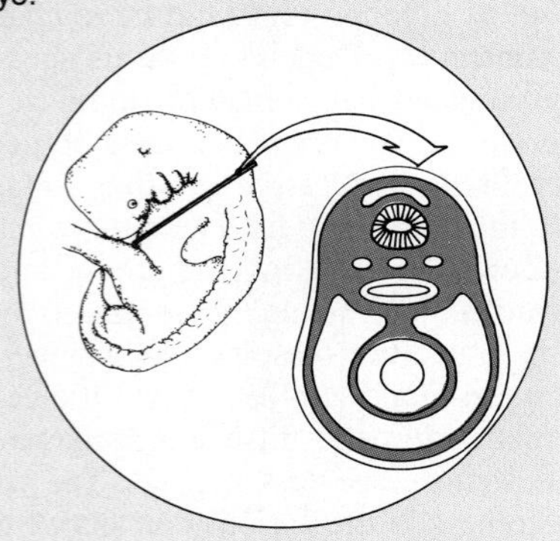

Function: Gives rise to all other connective tissue types.

Photomicrograph: Mesenchymal tissue, an embryonic connective tissue (475×); the clear-appearing background is the fluid ground substance of the matrix; notice the fine, sparse fibers.

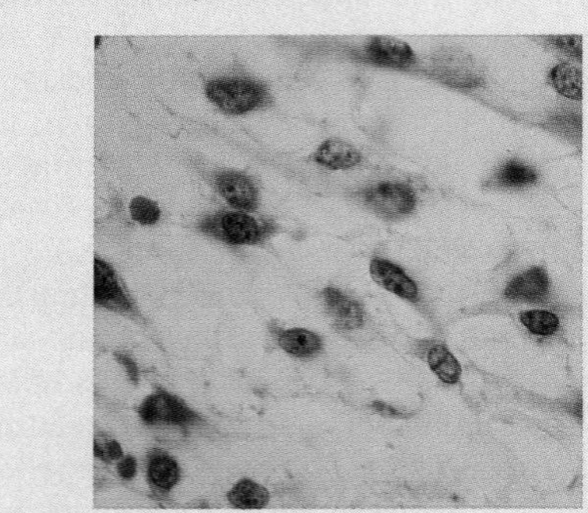

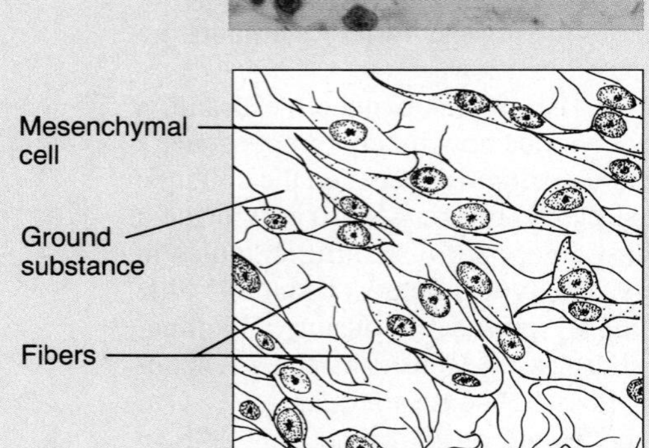

Connective tissue proper: Loose connective tissue (b to d)

(b) Areolar connective tissue

Description: Gel-like matrix with all three fiber types; cells: fibroblasts, macrophages, mast cells, and some white blood cells.

Location: Widely distributed under epithelia of body, e.g., forms lamina propria of mucous membranes; packages organs; surrounds capillaries.

Epithelium

Lamina propria

Function: Wraps and cushions organs; its macrophages phagocytize bacteria; plays important role in inflammation; holds and conveys tissue fluid.

Photomicrograph: Areolar connective tissue, a soft packaging tissue of the body (170×).

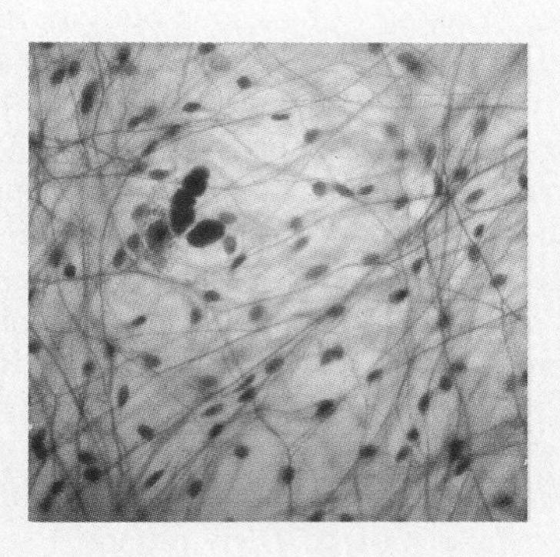

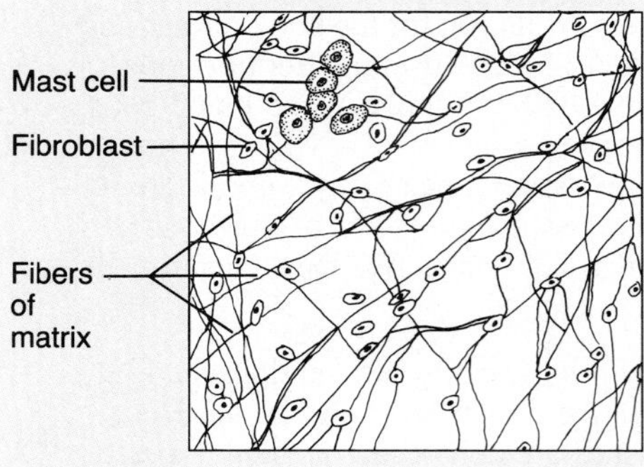

F6.3

Connective tissues. Embryonic connective tissue (a) and connective tissue proper: loose connective tissue (b–d).

(c) Adipose tissue

Description: Matrix as in areolar, but very sparse; closely packed adipocytes, or fat cells, have nucleus pushed to the side by large fat droplet.

Location: Under skin; around kidneys and eyeballs; in bones and within abdomen; in breasts.

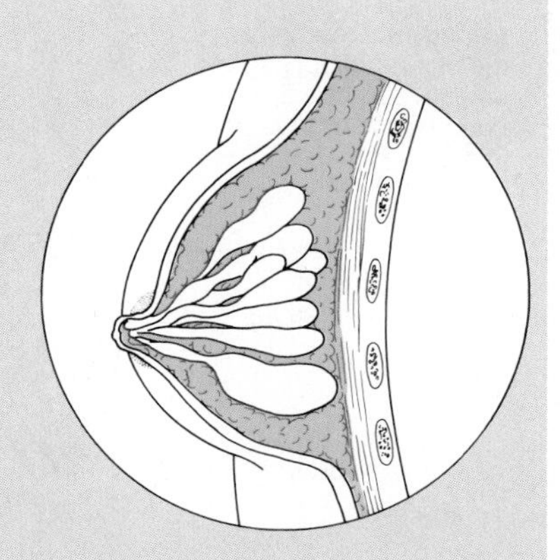

Function: Provides reserve food fuel; insulates against heat loss; supports and protects organs.

Photomicrograph: Adipose tissue from the subcutaneous layer under the skin (500×).

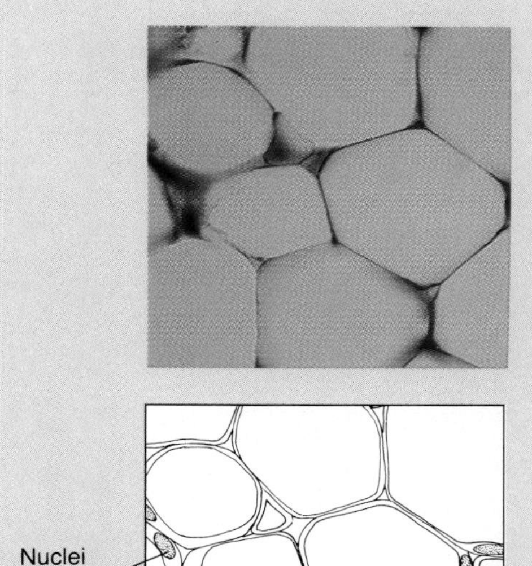

(d) Reticular connective tissue

Description: Network of reticular fibers in a typical loose ground substance; reticular cells predominate.

Location: Lymphoid organs (lymph nodes, bone marrow, and spleen).

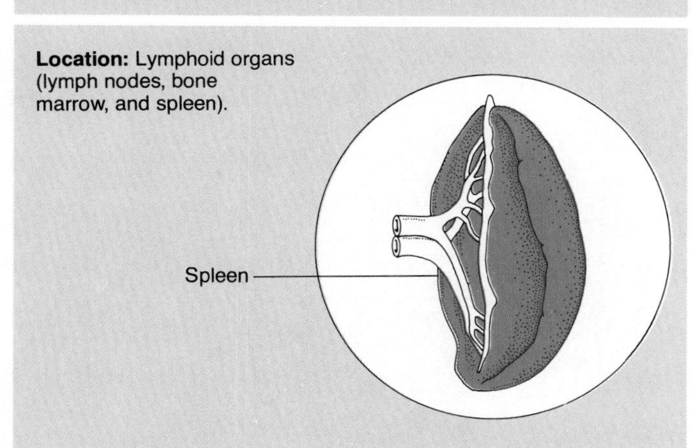

Function: Fibers form a soft internal skeleton that supports other cell types.

Photomicrograph: Dark-staining network of reticular connective tissue fibers forming the internal skeleton of the spleen (625×).

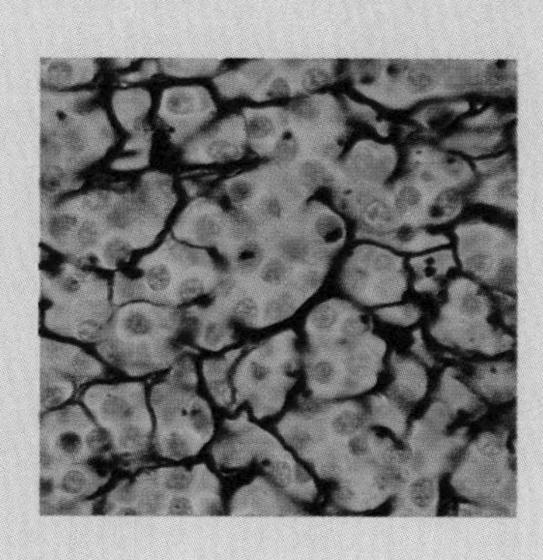

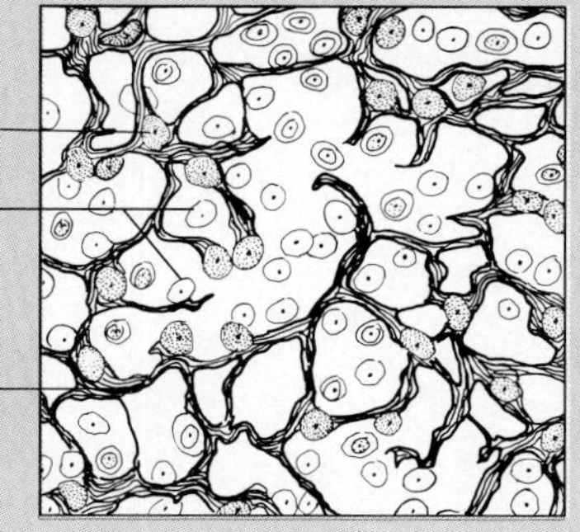

F6.3 (Continued)

Connective tissue proper: Dense connective tissue (e to g)

(e) Dense regular connective tissue

Description: Primarily parallel collagen fibers; a few elastin fibers; major cell type is the fibroblast.

Location: Tendons, most ligaments, aponeuroses.

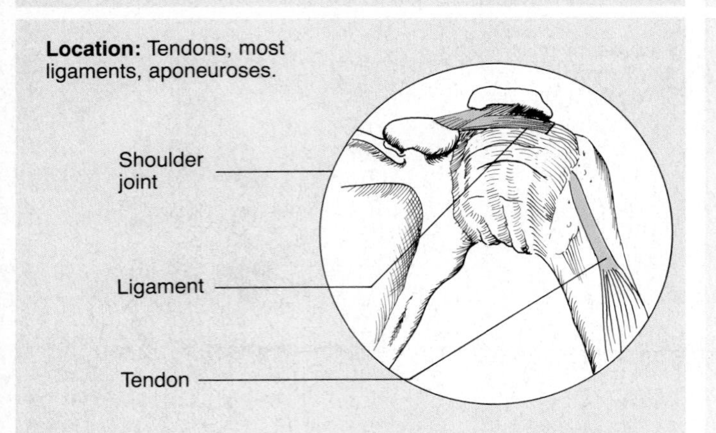

Function: Attaches muscles to bones or to muscles; attaches bones to bones; withstands great tensile stress when pulling force is applied in one direction.

Photomicrograph: Dense regular connective tissue from a tendon (200×).

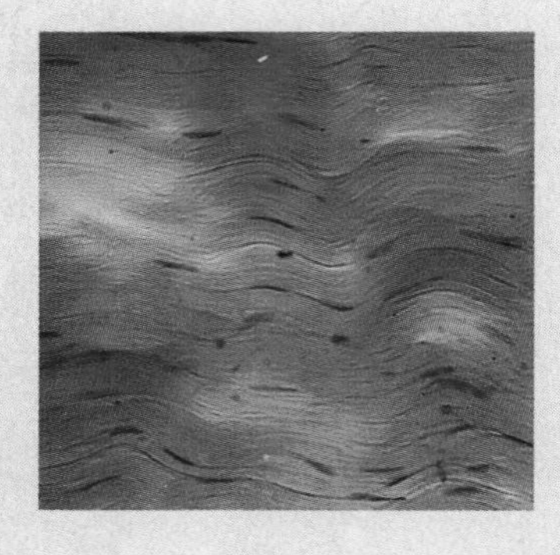

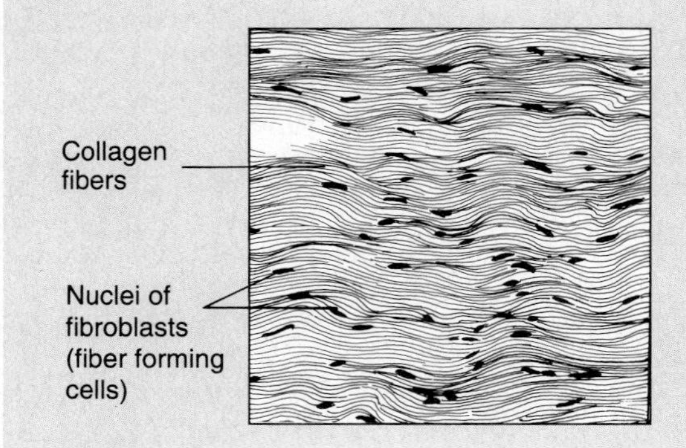

(f) Dense irregular connective tissue

Description: Primarily irregularly arranged collagen fibers; some elastic fibers; major cell type is the fibroblast.

Location: Dermis of the skin; submucosa of digestive tract; fibrous capsules of organs and of joints.

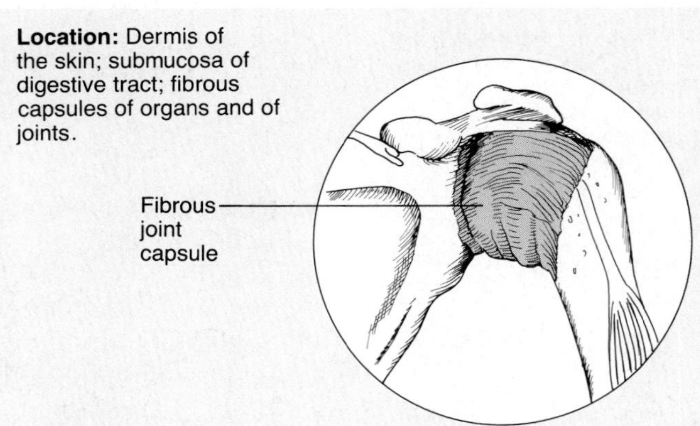

Function: Able to withstand tension exerted in many directions; provides structural strength.

Photomicrograph: Dense irregular connective tissue from the dermis of the skin (475×).

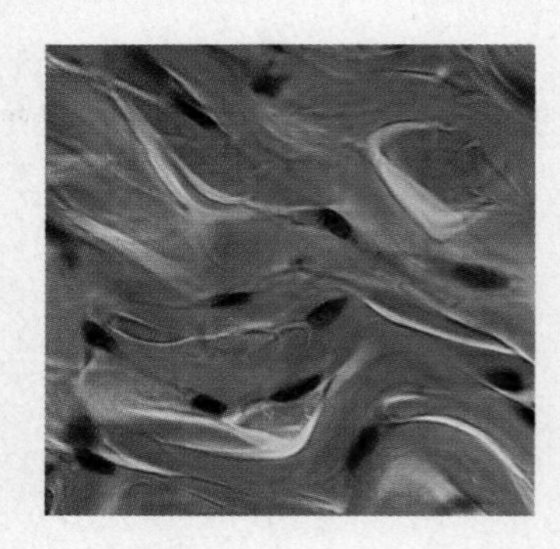

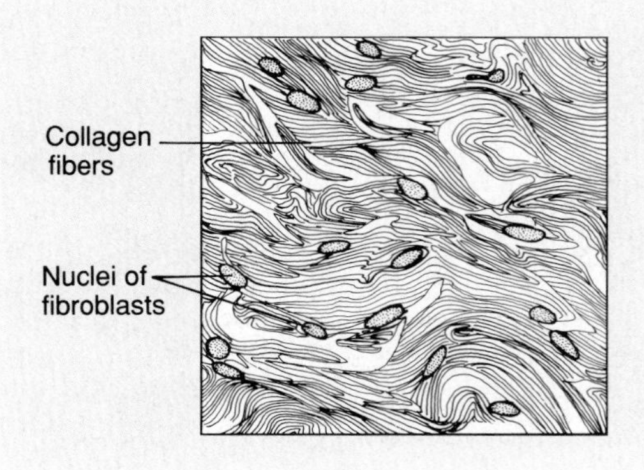

F6.3 (Continued)

Connective tissue proper: dense connective tissue (e–g). Cartilage (h).

Cartilage: (h to j)

(g) Elastic connective tissue

Description: Same as for the other dense connective tissues, but predominant fiber type is elastin.

Location: Walls of the aorta, some parts of trachea and bronchi; forms the vocal cords and the ligamenta flava connecting the vertebrae.

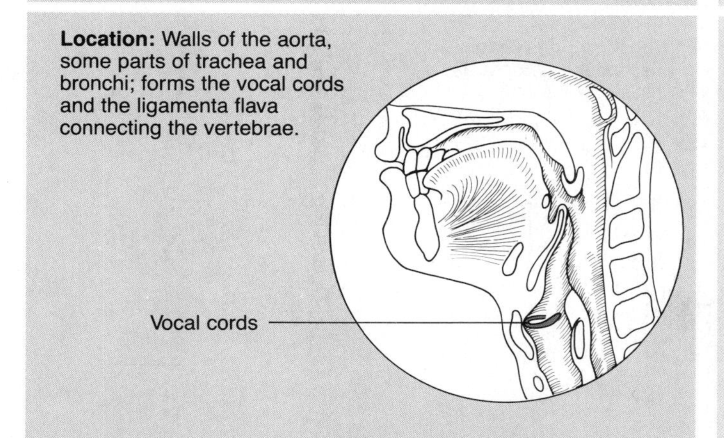

Function: Provides durability with stretch.

Photomicrograph: Elastic connective tissue in a wall of the aorta (190×); notice the wavy appearance of the elastin fibers.

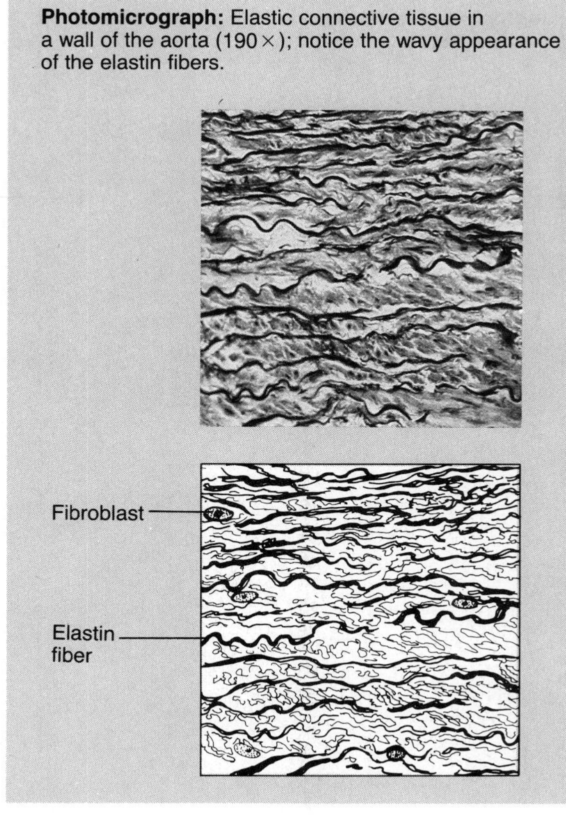

(h) Hyaline cartilage

Description: Amorphous but firm matrix; collagen fibers form an imperceptible network; chondroblasts produce the matrix and when mature (chondrocytes) lie in lacunae.

Location: Forms most of the embryonic skeleton; covers the ends of long bones in joint cavities; forms costal cartilages of the ribs; cartilages of the nose, trachea, and larynx.

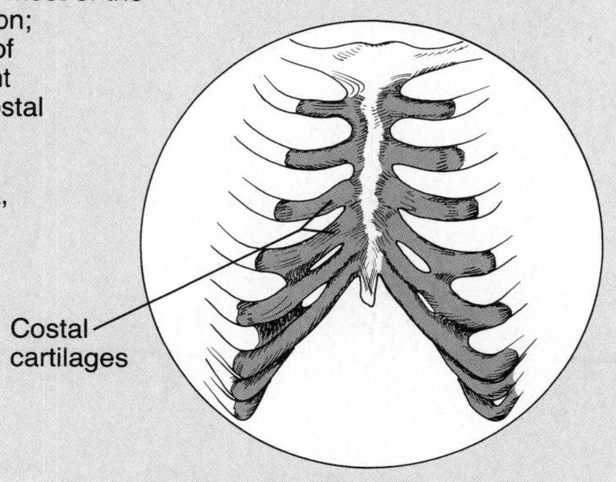

Function: Supports and reinforces; has resilient cushioning properties; resists compressive stress.

Photomicrograph: Hyaline cartilage from the trachea (475×).

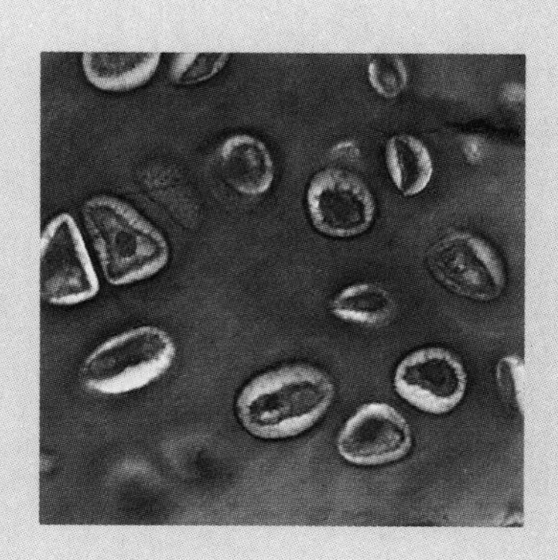

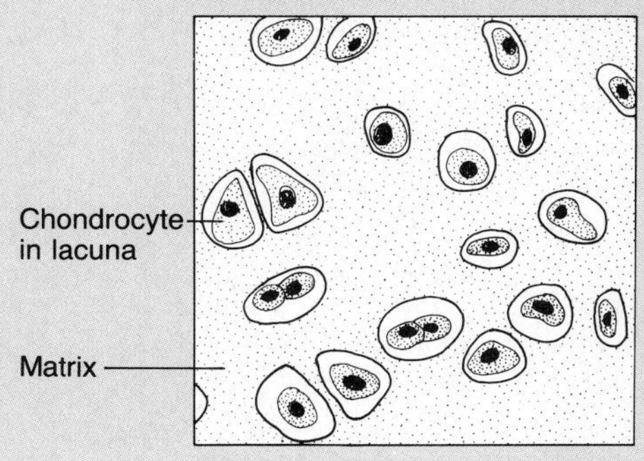

F6.3 (Continued)

(i) Elastic cartilage

Description: Similar to hyaline cartilage, but more elastic fibers in matrix.

Location: Supports the external ear (pinna); epiglottis.

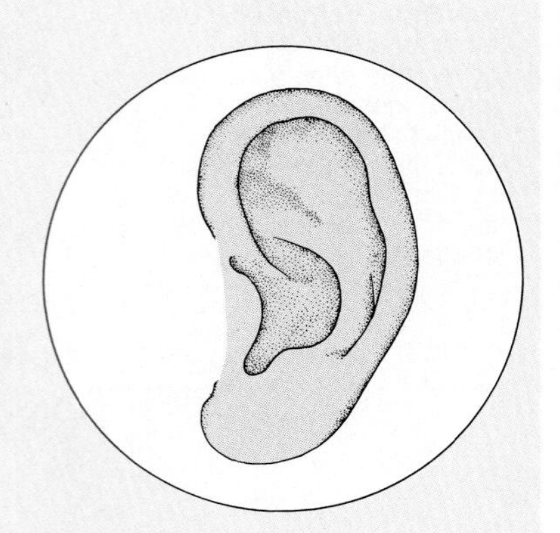

Function: Maintains the shape of a structure while allowing great flexibility.

Photomicrograph: Elastic cartilage from the human ear pinna; forms the flexible skeleton of the ear (250x).

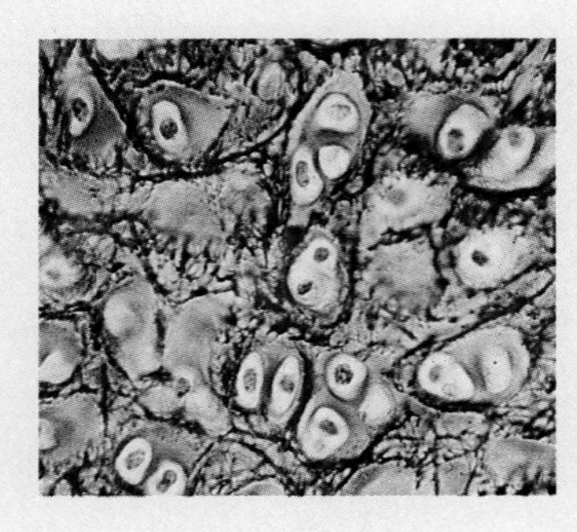

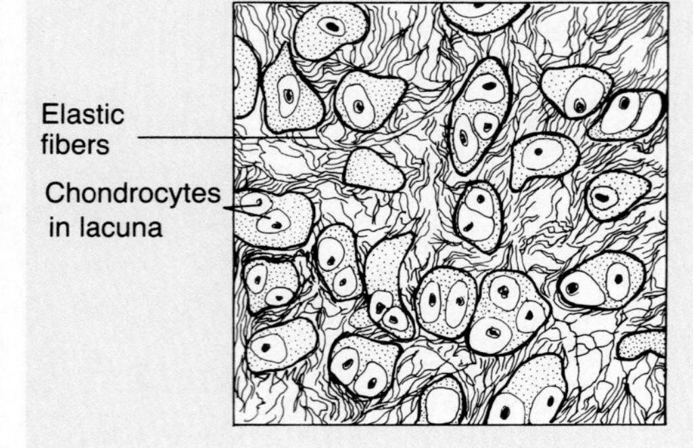

(j) Fibrocartilage

Description: Matrix similar but less firm than in hyaline cartilage; thick collagen fibers predominate.

Location: Intervertebral discs; pubic symphysis; discs of knee joint.

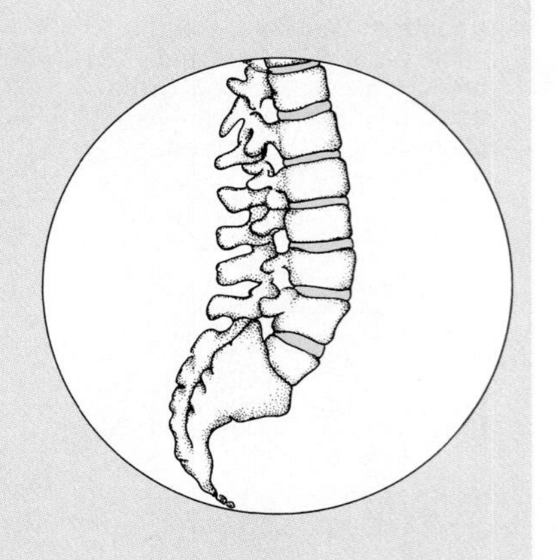

Function: Tensile strength with the ability to absorb compressive shock.

Photomicrograph: Fibrocartilage of an intervertebral disc (500x).

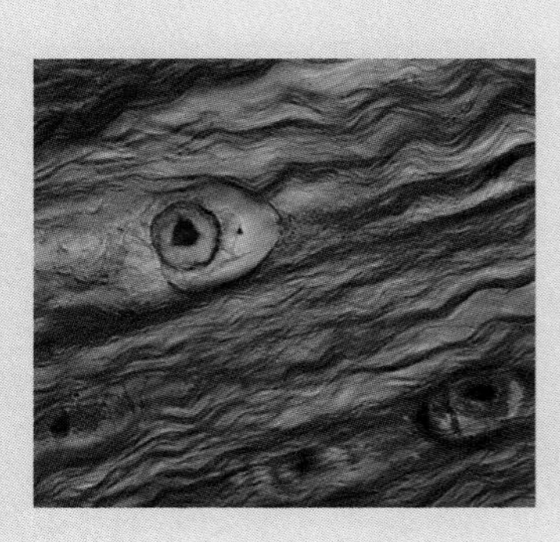

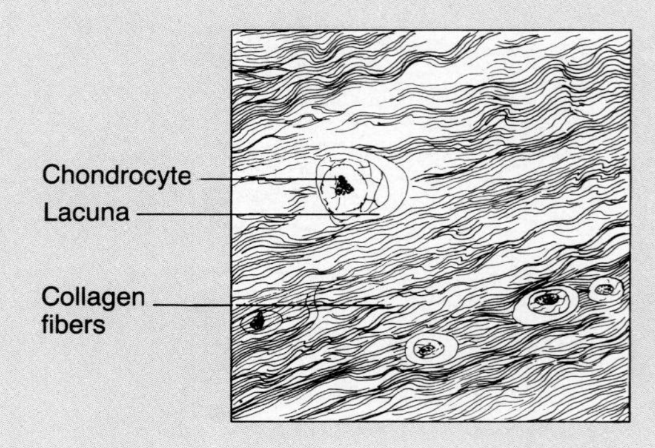

F6.3 (Continued)

Cartilage (i and j).

Others: (k and l)

(k) Bone (osseous tissue)

Description: Hard, calcified matrix containing many collagen fibers; osteocytes lie in lacunae. Very well vascularized.

Location: Bones

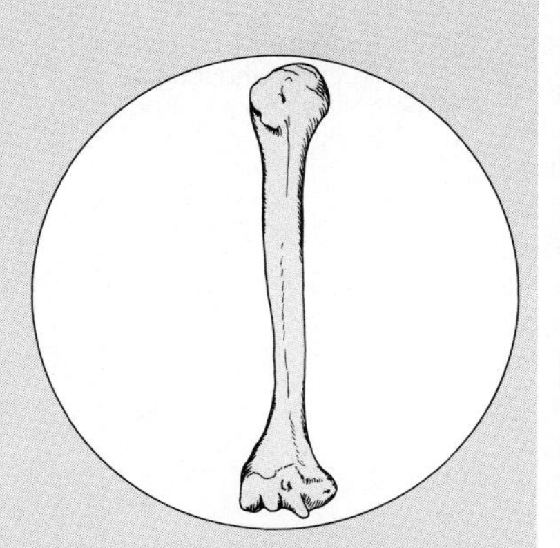

Function: Bone supports and protects (by enclosing); provides levers for the muscles to act on; stores calcium and other minerals and fat; marrow inside bones is the site for blood cell formation (hematopoiesis).

Photomicrograph: Cross-sectional view of bone (100 ×).

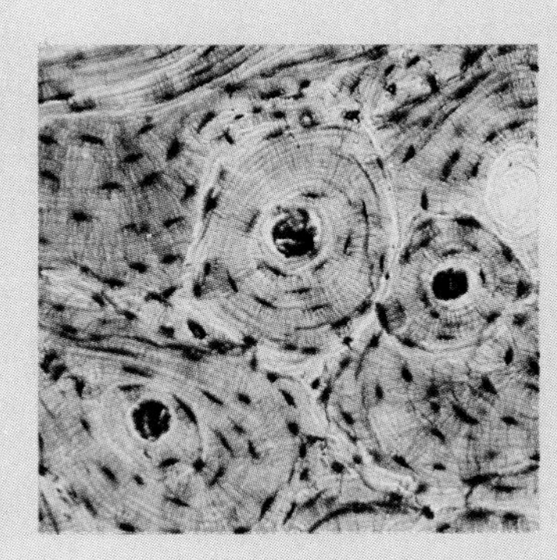

Osteocytes in lacunae

(l) Blood

Description: Red and white blood cells in a fluid matrix (plasma).

Location: Contained within blood vessels.

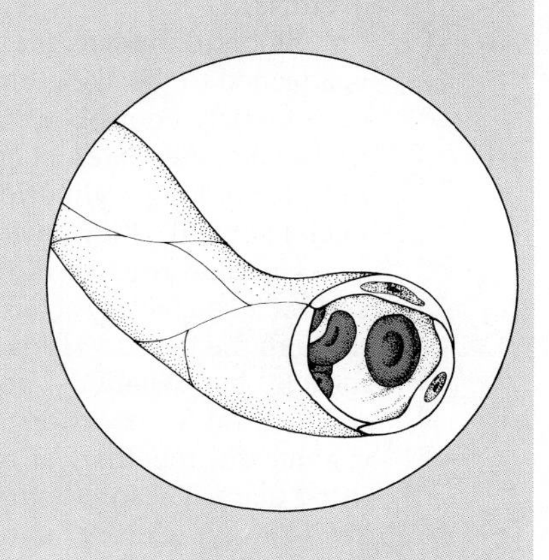

Function: Transport of respiratory gases, nutrients, wastes, and other substances.

Photomicrograph: Smear of human blood (1000 ×); two white blood cells (neutrophil in upper left and lymphocyte in lower right) are seen surrounded by red blood cells.

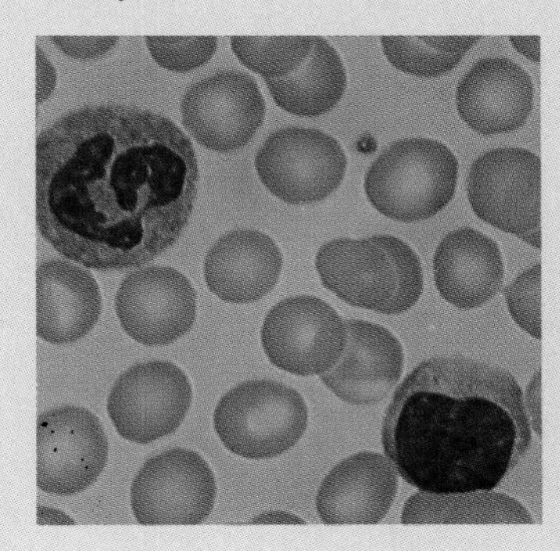

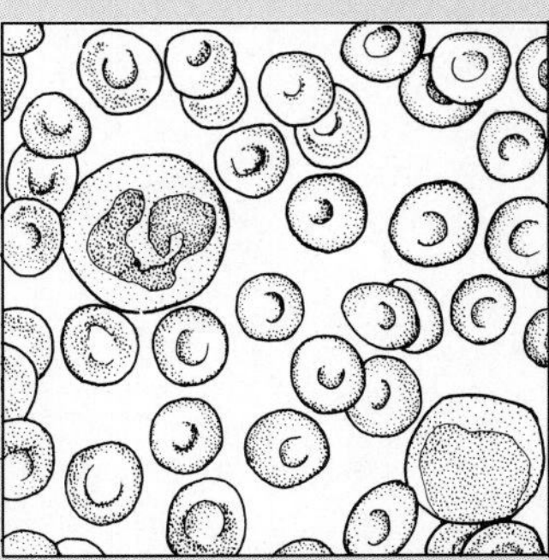

F6.3 (Continued)

Others (k and l).

MUSCLE TISSUE

Muscle tissue is highly specialized to contract (shorten) in order to produce movement of some body parts. As you might expect, muscle cells tend to be quite elongated, providing a long axis for contraction. The three basic types of muscle tissue are described briefly here; skeletal muscle is treated more completely in a later exercise.

Skeletal muscle, the "meat," or flesh, of the body, is attached to the skeleton. It is under voluntary control (consciously controlled), and its contraction moves the limbs and other external body parts. The cells of skeletal muscle are long, cylindrical, and multinucleate (several nuclei per cell); they have obvious striations (stripes).

Cardiac muscle is found only in the heart. As it contracts, the heart acts as a pump, propelling the blood through the blood vessels. Cardiac muscle, like skeletal muscle, has striations, but cardiac cells are branching uninucleate (or occasionally binucleate) cells that interdigitate (fit together) at tight junctions called **intercalated discs.** These structures allow the cardiac muscle to act as a unit. Cardiac muscle is under involuntary control, which means that we cannot voluntarily or consciously control the operation of the heart.

Smooth muscle, or visceral muscle, is found in the walls of hollow organs and blood vessels. Typically there are two layers that run at right angles to each other, hence its contraction can constrict or dilate the lumen (cavity) of the organ, thus propelling substances along predetermined pathways. Smooth muscle cells are quite different in appearance from those of skeletal or cardiac muscle. No striations are visible, and the uninucleate cells are spindle shaped.

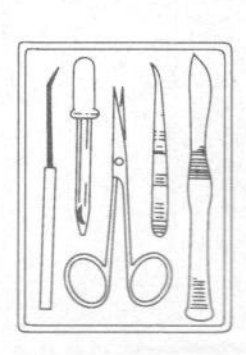

Obtain and examine prepared slides of skeletal, cardiac, and smooth muscle. Note their similarities and dissimilarities in both your observations and in the illustrations in Figure 6.4.

(a) Skeletal muscle

Description: Long, cylindrical, multinucleate cells; obvious striations.

Location: In skeletal muscles attached to bones or occasionally to skin.

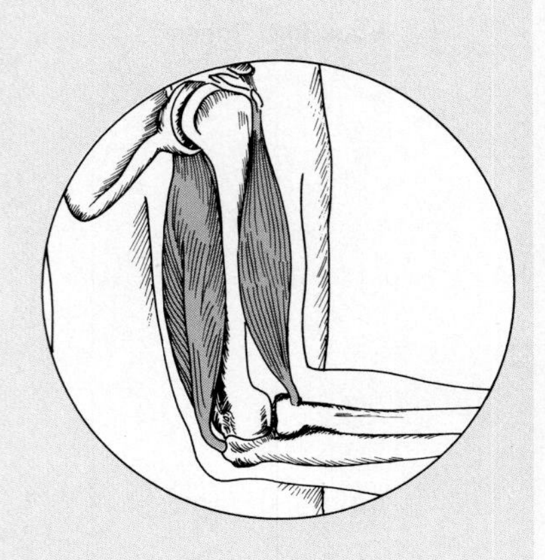

Function: Voluntary movement; locomotion; manipulation of the environment; facial expression. Voluntary control.

Photomicrograph: Skeletal muscle (approx. 30x). Notice the obvious banding pattern and the fact that these large cells are multinucleate.

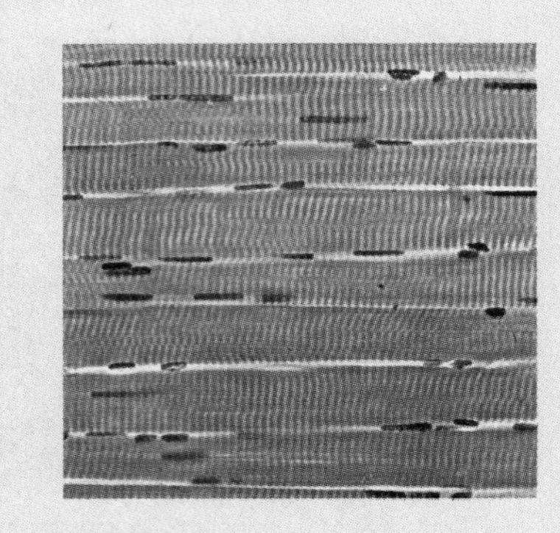

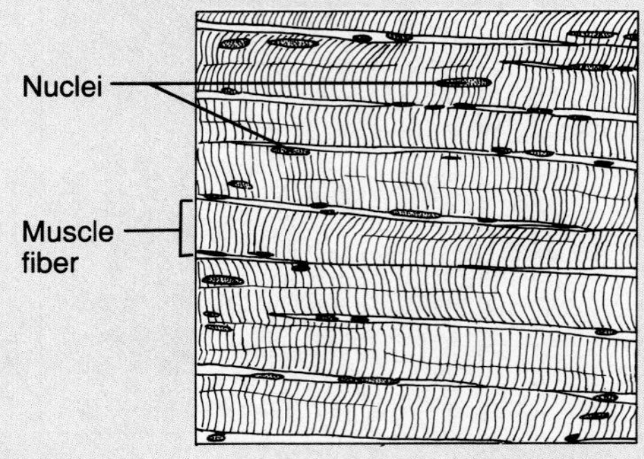

F6.4

Muscle tissues.

(b) Cardiac muscle

Description: Branching, striated, generally uninucleate cells that interdigitate at specialized junctions (intercalated discs).

Location: The walls of the heart.

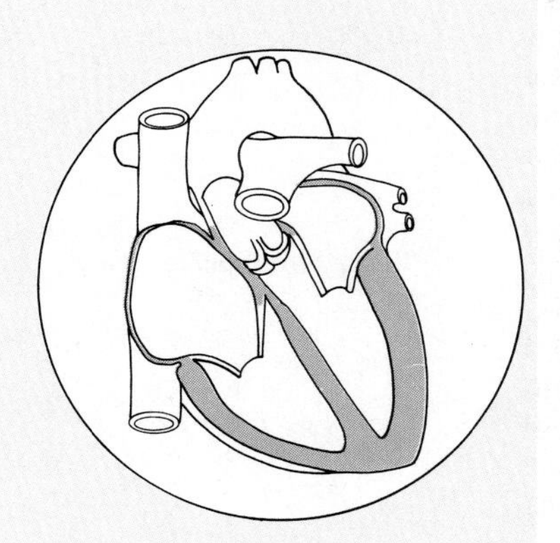

Function: As it contracts, it propels blood into the circulation; involuntary control.

Photomicrograph: Cardiac muscle (250x); notice the striations, branching of fibers, and the intercalated discs.

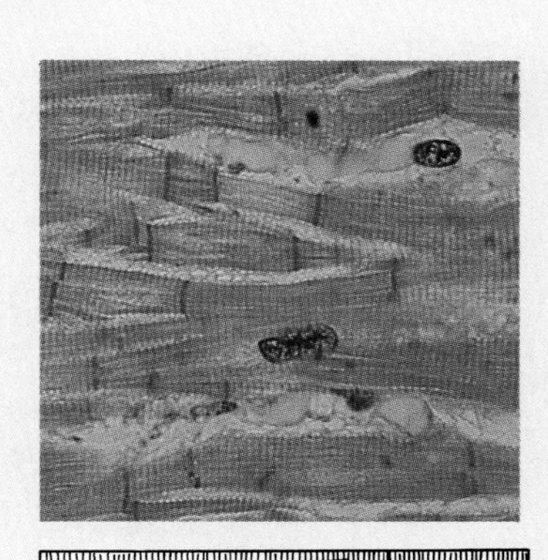

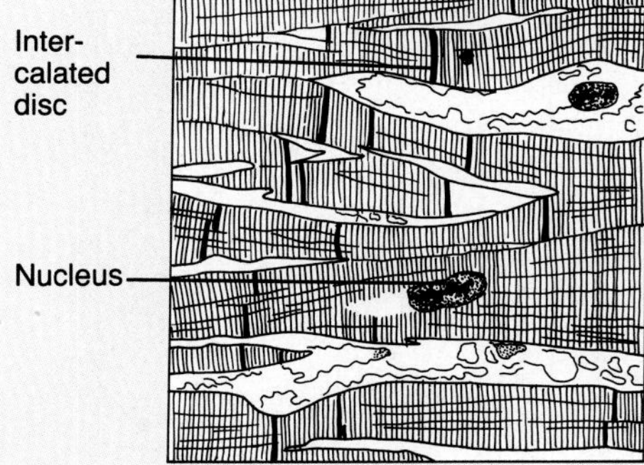

(c) Smooth muscle

Description: Spindle-shaped cells with central nuclei; cells arranged closely to form sheets; no striations.

Location: Mostly in the walls of hollow organs.

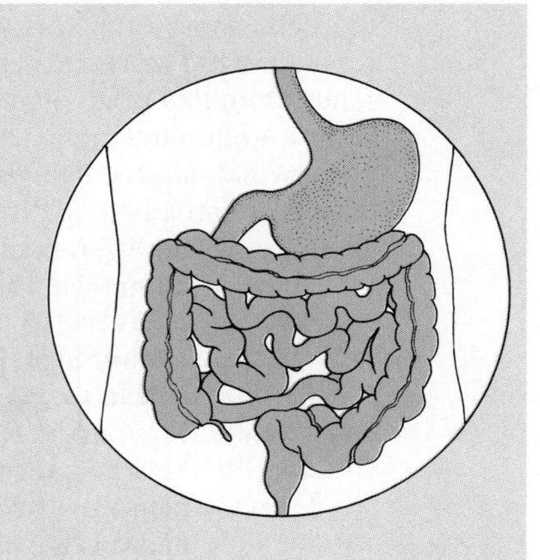

Function: Propels substances or objects (foodstuffs, urine, a baby) along internal passageways; involuntary control.

Photomicrograph: Sheet of smooth muscle (approx. 300x).

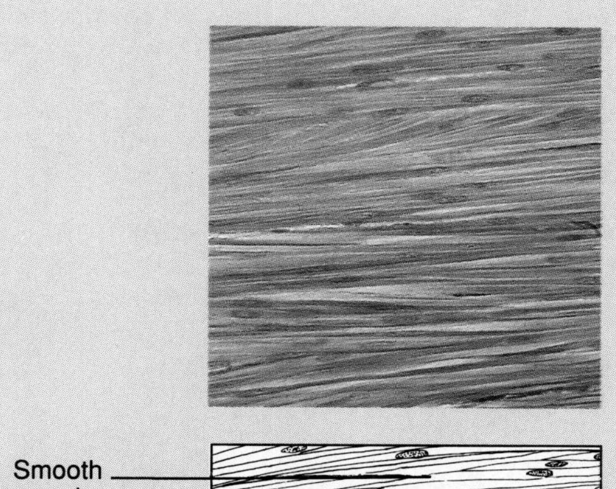

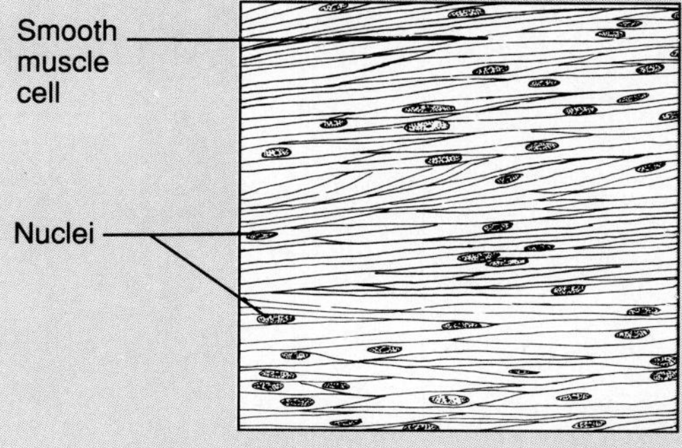

F6.4 (Continued)

NERVOUS TISSUE

Nervous tissue is composed of two major cell populations. The **neuroglia** are special supporting cells that *protect, support,* and *insulate* the more delicate neurons. The **neurons** are highly specialized to receive stimuli (*irritability*) and to conduct waves of excitation, or impulses, to all parts of the body (*conductivity*). They are the cells most often associated with nervous system functioning. The structure of neurons is markedly different from that of all other body cells. Their cytoplasm is drawn out into long extensions as much as 3 ft (about 1 m), which allows a single neuron to conduct a stimulus over relatively long distances. More detail about the anatomy of the different classes of neurons and neuroglia appears in Exercise 15.

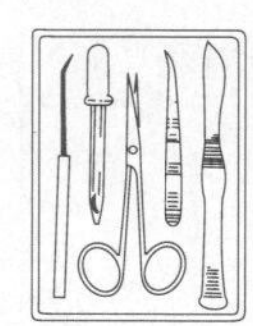

Obtain a prepared slide of a spinal cord smear. Locate a neuron, and compare it to Figure 6.5. Keep the light dim—this will help you see the cellular extensions of the neurons. The smaller cells surrounding the neurons are neuroglia cells.

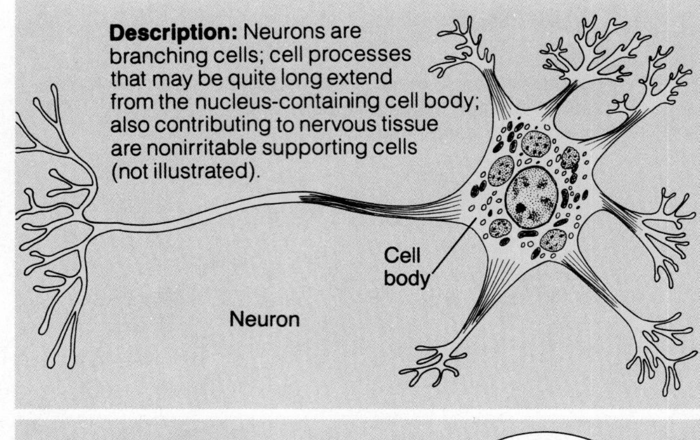

Description: Neurons are branching cells; cell processes that may be quite long extend from the nucleus-containing cell body; also contributing to nervous tissue are nonirritable supporting cells (not illustrated).

Location: Brain, spinal cord, and nerves.

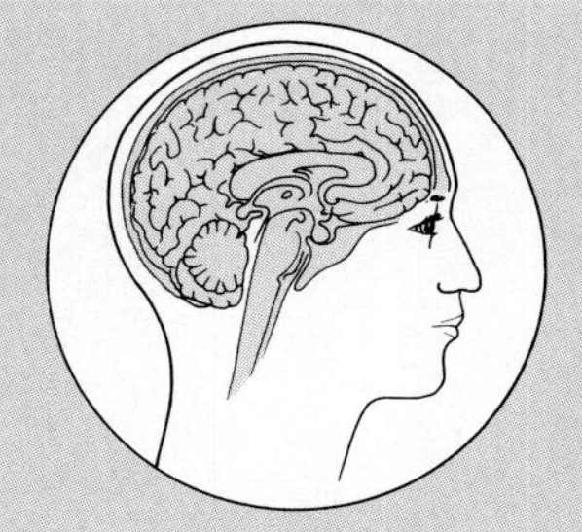

Function: Transmit electrical signals from sensory receptors and to effectors (muscles and glands) which control their activity.

Photomicrograph: Neuron (170×).

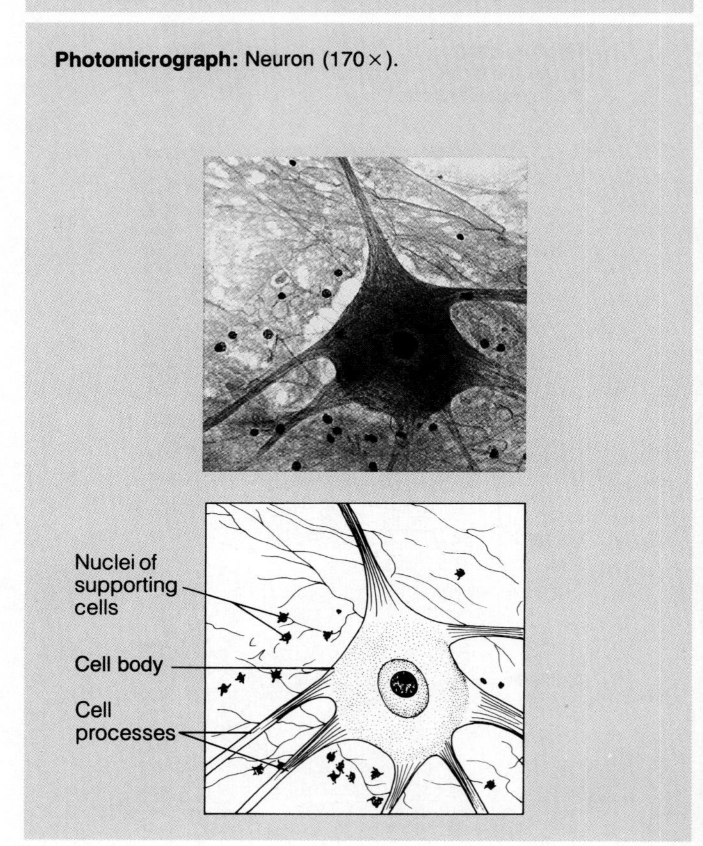

F6.5

Nervous tissue.

The Skin and Other Body Membranes

OBJECTIVES

1. To recount several important functions of the skin, or integumentary system.
2. To recognize and name from observation of an appropriate model, diagram, projected slide, or microscopic examination the following skin structures: epidermis (also note relative positioning of its various strata), dermis (papillary and reticular layers), hair follicles and hair, sebaceous glands, and sweat glands.
3. To compare the properties of the epidermis to those of the dermis.
4. To describe the distribution and function of the skin derivatives—sebaceous glands, sweat glands, and hair.
5. To differentiate between eccrine and apocrine sweat glands.
6. To enumerate the factors determining skin color.
7. To describe the function of melanin.
8. To compare the structure and function of the major membrane types (epithelial and synovial).
9. To list the general functions of each membrane type and note its location in the body.
10. To recognize by microscopic examination epidermal, mucous, and serous membranes.

MATERIALS

Skin model (three-dimensional, if available)
Compound microscope
Prepared slide of human skin with hair follicles
Sheet of #20 bond paper ruled to mark off cm^2 areas
Scissors
Betadine swabs, or Lugol's iodine and cotton swabs
Adhesive tape

Prepared slides of trachea (cross section) and small intestine (cross section)
Prepared slide of serous membrane (e.g., mesentery)
Longitudinally cut fresh beef joint (if available)

The **skin,** or **integument,** is often considered an organ system because of its extent and complexity, and one thing is certain, it is much more than an external body covering. Architecturally, the skin is a marvel; it is tough yet pliable. This characteristic enables it to withstand constant insult from outside agents.

The skin has many functions, most (but not all) concerned with protection. It *insulates* and *cushions* the underlying body tissues, and *protects* the entire body from mechanical damage (bumps and cuts), chemical damage (acids, alkalis, and the like), thermal damage (heat), and bacterial invasion (by virtue of its acid mantle and continuous surface). The uppermost layer of the skin (the cornified layer), which is hardened, prevents water loss from the body surface. The skin's abundant capillary network (under the control of the nervous system) plays an important role in regulating heat loss from the body surface.

The skin has other functions as well. For example, it acts as a miniexcretory system; urea, salts, and water are lost through the skin pores. The skin is also the site of vitamin D synthesis for the body. Finally, the skin is a large, diffuse sensory organ because the cutaneous sense organs are located in the dermis.

BASIC STRUCTURE OF THE SKIN

Structurally, the skin consists of two kinds of tissue. The more superficial **epidermis** is made up of stratified squamous epithelium; the underlying **dermis** is dense irregular connective tissue. These layers are firmly cemented together. Immediately deep to the dermis is the subcutaneous tissue (primarily areolar connective tissue with abundant fat cells), which is not considered part of the skin. The main areas and structures of the skin are described below.

As you read, locate the following structures on Figure 7.1 and on a skin model.

Epidermis

The avascular epidermis consists of several layers or *strata* of cells; from deep to superficial these are the germinativum, granulosum, lucidum, and corneum.

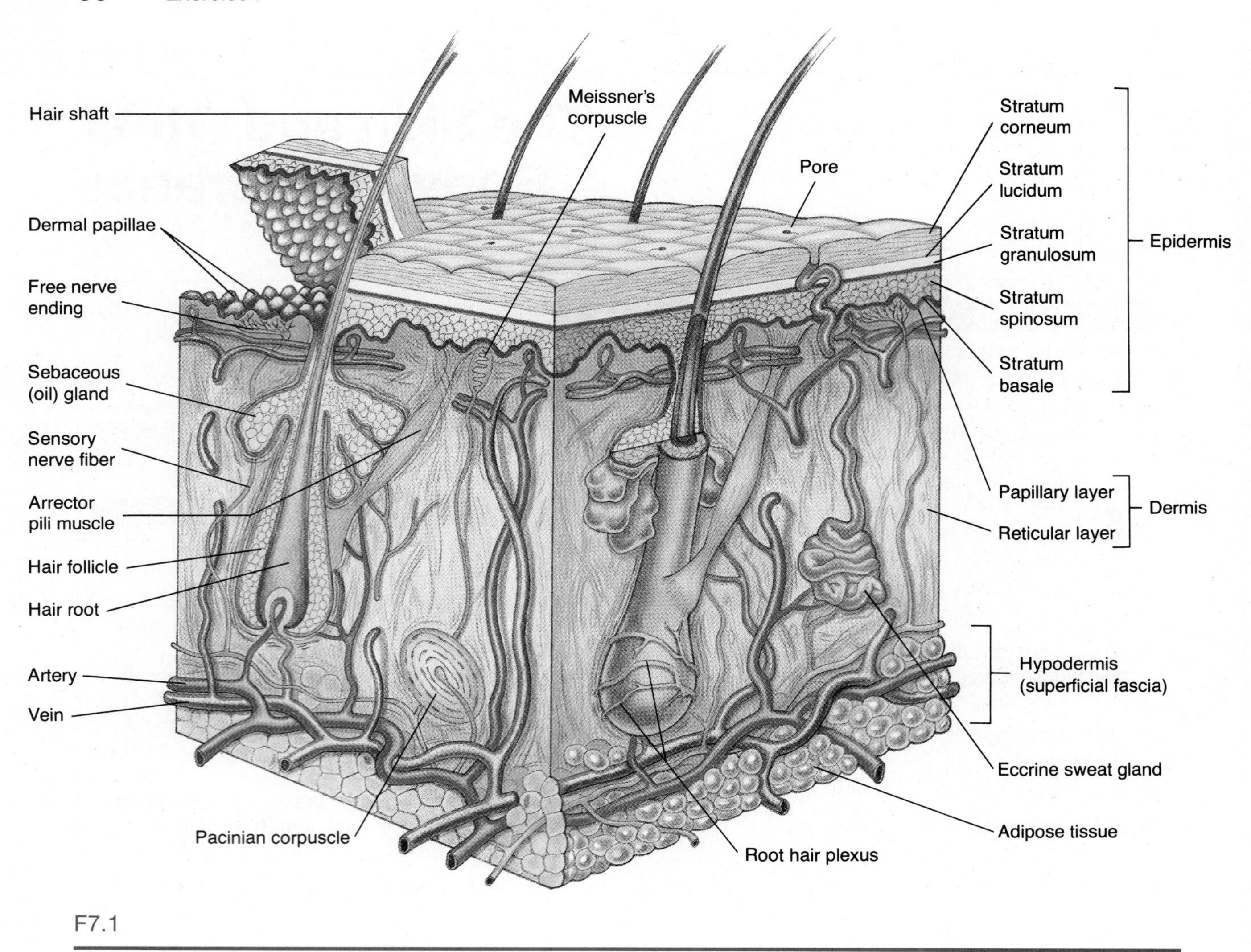

F7.1

Three-dimensional view of the skin.

The **stratum germinativum,** consisting of the **stratum basale** (basal layer) and **stratum spinosum** (spiny layer), is immediately adjacent to the dermis. This region contains the only epidermal cells that receive adequate nourishment (via diffusion of nutrients from the dermis). These cells, particularly those in the more inferior basal layer, are constantly undergoing cell division; millions of new cells are produced daily. As the daughter cells are pushed upward, away from the source of nutrition, they gradually die, and their soft protoplasm becomes increasingly keratinized.

Melanin, a brown pigment, is produced by special cells (*melanocytes*) found in the stratum germinativum. The skin tans because of an increase in melanin production when the skin is exposed to sunlight. Melanin provides a protective pigment "umbrella" over basal cell nuclei, thus shielding their genetic material (DNA) from the damaging effects of ultraviolet radiation. A concentration of melanin in one spot is commonly called a *freckle.*

The **stratum granulosum** (granular layer), immediately above the stratum spinosum, is the area in which the cells begin to die owing to their accumulation of keratohyalin granules (a keratin precursor) and their increasing distance from the dermal blood supply. Superficial to the stratum granulosum in thick skin (but not in thin skin) is the **stratum lucidum** (clear layer). This thin layer, which appears as a pale band, contains fully keratinized cells.

The uppermost epidermal layer is the **stratum corneum,** sometimes called the *horny layer* (*cornu* = horn) because it consists of tough flattened keratinized cells. Keratin is a protein with waterproofing properties; thus this layer provides a natural "raincoat" for the body and prevents water loss from the deeper tissues. Keratinized cells are dead cells. They constantly rub and flake off and are replaced by the division of deeper cells.

Dermis

The connective tissue making up the dermis consists of two principal regions—the papillary and reticular areas—and like the epidermis, it varies in thickness. For example, the dermis, and skin as a whole, is particularly thick on the palms of the hands and soles of the feet and is quite thin on the eyelids.

The **papillary layer** is the more superficial dermal region. It is very uneven and has fingerlike projections, the **dermal papillae,** from its superior surface which attach it to the epidermis above. These projections are reflected in fingerprints, which are unique patterns of ridges that remain unchanged throughout life. Abundant capillary networks in the papillary layer furnish nutrients to the epidermis and allow heat to radiate to the skin surface. The pain and touch receptors (Meissner's corpuscles) are also found here.

The **reticular layer** is the deepest skin layer. It contains many arteries and veins, sweat and sebaceous glands, and pressure receptors.

Both the papillary and reticular layers are heavily invested with collagenic and elastic fibers. There are substantial amounts of both fiber types in the subcutaneous tissue as well. The elastic fibers give skin its exceptional elasticity in youth, but in old age their number decreases and the subcutaneous layer loses fat, which leads to wrinkling and inelasticity of the skin. Fibroblasts, adipose cells, various types of macrophages (which are important in the body's defense), and other cell types are found throughout the dermis.

The dermis has an abundant blood supply, which allows it to play a role in the regulation of body temperature. When body temperature is high, the arterioles dilate and the capillary network of the dermis becomes engorged with the heated blood. Thus body heat is allowed to radiate from the skin surface. If the environment is cool and body heat must be conserved, the arterioles constrict so that blood bypasses the capillary networks of the skin.

Any restriction of the normal blood supply to the skin results in cell death and, if severe enough, skin ulcers (Figure 7.2). *Bedsores* (*decubitus ulcers*) occur in bedridden patients who are not turned regularly enough. The weight of the body exerts pressure on the skin, especially over bony projections, which leads to restriction of the blood supply and death of tissue. ■

The dermis is also richly provided with a nerve supply. Many of the nerve endings bear highly specialized receptor organs (touch, temperature, pressure, and so on) that, when stimulated by environmental changes, transmit messages to the central nervous system for interpretation. One of these, the deep pressure receptor called a *Pacinian* receptor, is illustrated in Figure 7.1. (These receptors are discussed in depth in Exercise 19.)

Skin Color

Skin color is the result of three factors—the relative amount of two pigments (melanin and carotene) and the degree of oxygenation of the blood. People who produce large amounts of melanin have brown-toned skin. In light-skinned people, who have less melanin, the dermal blood supply flushes through the rather transparent cell layers above, giving the skin a rosy glow. Carotene is a yellow-orange pigment present primarily in the stratum corneum and in adipose tissue of the hypodermis.

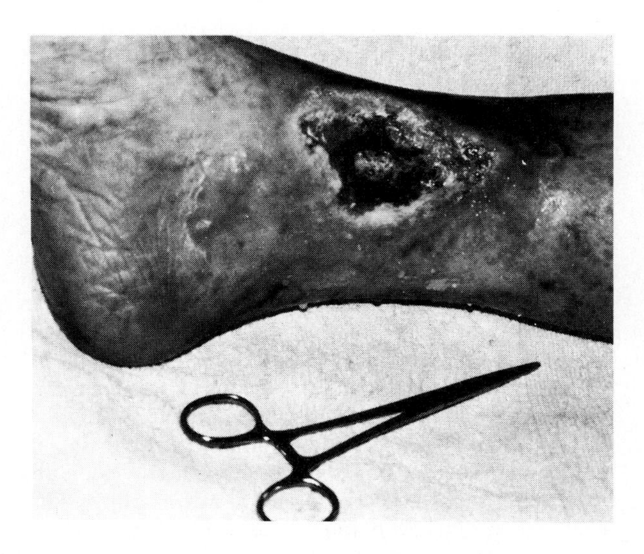

F7.2

A decubitus ulcer on the ankle of a patient.

Skin color may be an important diagnostic tool. For example, flushed skin may indicate hypertension, fever, or embarrassment, whereas pale skin is common in anemic individuals. When the blood is inadequately oxygenated, as during asphyxiation and serious lung disease, the skin takes on a bluish or *cyanotic* cast. *Jaundice,* in which the tissues become yellowed, is almost always diagnostic for liver disease, whereas a bronzing of the skin hints that a person's adrenal cortex is hypoactive (*Addison's disease*). ■

APPENDAGES OF THE SKIN

The appendages of the skin—hair, nails, and cutaneous glands—are all derivatives of the epidermis, but they reside almost entirely in the dermis. They originate from the stratum germinativum and grow downward into the deeper skin regions.

Cutaneous Glands

The cutaneous glands fall primarily into two categories: sebaceous glands and sweat glands. The **sebaceous glands** are found nearly all over the skin, except for the palms of the hands and soles of the feet. Their ducts usually empty into a hair follicle, but some open directly onto the skin surface.

The product of the sebaceous glands, called **sebum,** is a mixture of oily substances and fragmented cells. Sebum is a lubricant that keeps the skin soft and moist (a natural skin cream) and keeps the hair from becoming brittle. The sebaceous glands become particularly active during puberty, and the skin tends to become oilier during this period of life. Blackheads are accumulations of dried sebum and bacteria; acne is due to active infection of the sebaceous glands.

Epithelial openings, called pores, are the outlets for the **sweat glands** (*sudoriferous glands*). These exocrine glands are widely distributed in the skin. Sweat glands are subcategorized on the basis of the composition of their secretions. The **eccrine glands,** which are distributed all over the body, produce clear perspiration, consisting primarily of water, salts (NaCl), and urea. The **apocrine glands,** found predominantly in the axillary and genital areas, secrete a milky protein-based substance (also containing water, salts, and urea) that is an excellent nutrient medium for the microorganisms generally found on the skin.

The sweat glands, under the control of the nervous system, are an important part of the body's heat-regulating apparatus. They secrete perspiration when the external temperature or body temperature is high. When this water-based substance evaporates, it carries excess body heat with it. Thus evaporation of greater amounts of perspiration provides an efficient means of ridding the body of excess heat when the capillary cooling system is not sufficient or is unable to maintain body temperature homeostasis.

Hair

Hairs are found over the entire body surface, except for the palms of the hands, the soles of the feet, parts of the external genitalia, the nipples, and the lips. A hair, enclosed in a **hair follicle,** is also an epithelial structure. The portion of the hair enclosed within the follicle is called the **root;** the portion projecting from the scalp surface is called the **shaft.** The hair is formed by mitosis of the well-nourished germinal epithelial cells at the basal end of the follicle (the **hair bulb**). As the daughter cells are pushed farther away from the growing region, they become keratinized and die; thus the bulk of the hair shaft, like the bulk of the epidermis, is dead material.

Small bands of smooth muscle cells—**arrector pili**—connect each hair follicle to the papillary layer of the dermis. When these muscles contract (during cold or fright), the hair follicle (normally in a slanted position) is pulled upright, dimpling the skin surface with "goosebumps."

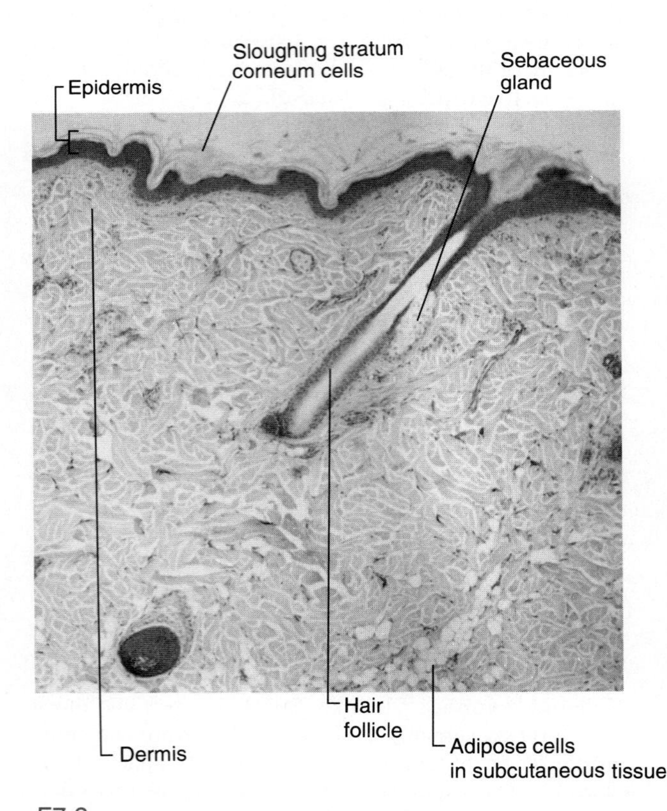

F7.3

Longitudinal section through skin of scalp showing hair follicles and cutaneous glands (21×).

Nails

Nails, the hornlike derivatives of the epidermis, consist of a root and a body. The germinal cells in the nail root divide to produce its growth. Like the hair shaft, the nail is composed of nonliving material. The nails are nearly colorless but appear pink because of the blood supply in the nail bed. When a person is cyanotic because of a lack of oxygen in the blood, the nails take on a blue cast.

EXAMINATION OF THE MICROSCOPIC STRUCTURE OF THE SKIN

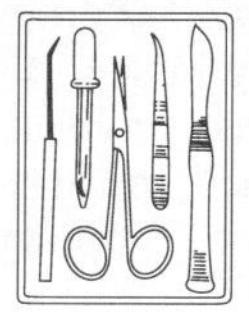

Obtain a prepared slide of human skin, and study it carefully under the microscope. Compare your tissue slide to the view shown in Figure 7.3, and identify as many of the structures diagramed in Figure 7.1 as possible. Diagram and label your observations below.

How is this stratified squamous epithelium different from that observed in Exercise 6?

____

____

How do these differences relate to the functions of these two similar epithelia?

____

____

____

PLOTTING THE DISTRIBUTION OF SWEAT GLANDS

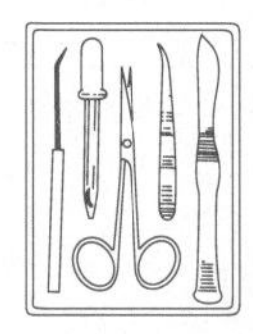

1. For this simple experiment you will need two squares of bond paper (each 1 cm × 1 cm), adhesive tape, and a betadine (iodine) swab *or* Lugol's iodine and a cotton-tipped swab. (The bond paper has been preruled in cm^2—just cut along the lines to obtain the required squares.)

2. Paint the medial aspect of your left palm (avoid the crease lines) and a region of your left forearm with the iodine solution, and allow it to dry thoroughly. The painted area in each case should be slightly larger than the paper squares to be used.

3. Have your lab partner *securely* tape a square of bond paper over each iodine-painted area, and leave them in place for 20 minutes. (Note: If it is very warm in the laboratory while this test is being conducted, good results may be obtained within 10 to 15 minutes.)

4. After 20 minutes, remove the paper squares and count the number of blue-black dots on each square. The presence of a blue-black dot on the paper indicates an active sweat gland. (The iodine in the pore is dissolved in the sweat and reacts chemically with the starch in the bond paper to produce the blue-black color.) Thus "sweat maps" have been produced for the two skin areas.

5. Which skin area tested has the greater density of sweat glands?

CLASSIFICATION OF BODY MEMBRANES

The body membranes, which cover surfaces, line body cavities, and form protective (and often lubricating) sheets around organs, fall into two major categories. These are the so-called *epithelial membranes* and the *synovial membranes.*

Epithelial Membranes

The term "epithelial membrane" is used in various ways. Here we will define an **epithelial membrane** as a simple organ consisting of an epithelial sheet bound to an underlying layer of connective tissue. Most of the covering and lining epithelia take part in forming one of three common types of epithelial membranes: cutaneous, mucous, or serous. The **cutaneous membrane** is the skin, a dry membrane with a keratinizing epithelium (the epidermis), which has just been described in some detail. We will focus on the mucous and serous membranes here.

MUCOUS MEMBRANES **Mucous membranes (mucosae)** are composed of epithelial cells resting on a layer of loose connective tissue called the **lamina propria.** They line all body cavities that open to the body exterior—the respiratory, digestive, and urinary tracts. All mucous membranes are "wet" membranes because they are continuously bathed by secretions (or in the case of urinary tract mucosae, urine). Although mucous membranes often secrete mucus, this is not a requirement. The mucous membranes of both the digestive and respiratory tracts secrete mucus, but that of the urinary tract does not.

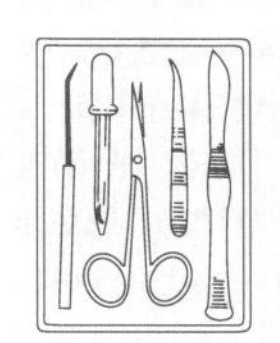

Examine a slide made from a cross section of the trachea and another of the small intestine (see plates 31 and 36 in the histology atlas). Draw the mucosa of each in the appropriate circle, and fully identify each epithelial type. Remember to look for the epithelial cells at the free surface. Search the epithelial sheets for goblet cells. **Goblet cells** are columnar epithelial cells with a large mucus-containing vacuole (goblet) in the cytoplasm. Which of these mucous membranes contains goblet cells?

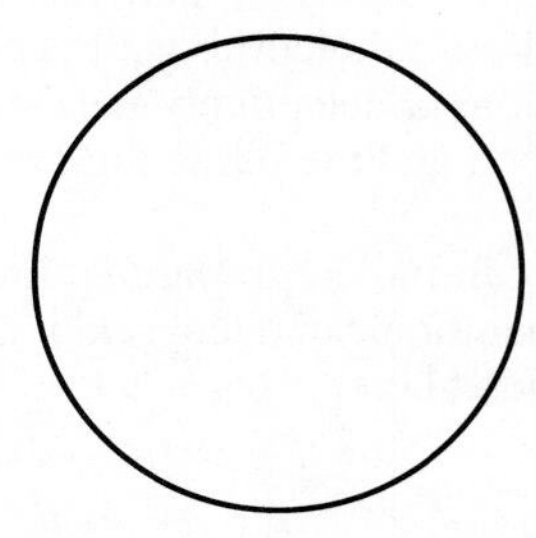

Mucosa of trachea

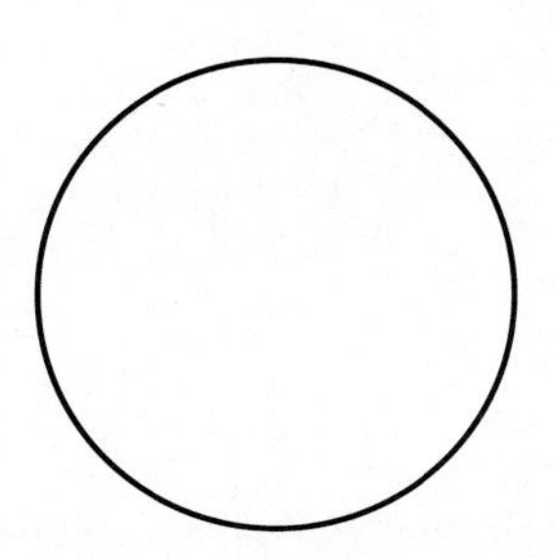

Mucosa of small intestine

How do the roles of these two mucous membranes differ?

How are they the same? ______________________

____

____

Would you say that mucous membranes have a high regenerative capacity? Why or why not?

____

____

____

SEROUS MEMBRANES The **serous membranes** (**serosae**) are also epithelial membranes. They are composed of a layer of simple squamous epithelium on a scant amount of loose connective tissue. The serous membranes generally occur in twos. The parietal layer lines a body cavity, and the visceral layer covers the outside of the organs in that cavity. In contrast to the mucous membranes, which line open body cavities, the serous membranes line body cavities that are closed to the exterior (with the exception of the female peritoneal cavity and the dorsal body cavity). The serosae secrete a thin fluid (serous fluid), which lubricates the organs and body walls and thus reduces friction as the organs slide across one another and against the body cavity walls. A serous membrane also lines the interior of blood vessels (endothelium) and the heart (endocardium). In capillaries, the entire wall is composed of serosa that serves as a selectively permeable membrane between the blood and the tissue fluid of the body.

Examine a prepared slide of a serous membrane and diagram it in the space provided here.

What kind of cells that you have seen many times before are these shaped like?

____

What are the specific names of the serous membranes covering the heart and lining the cavity in which it resides (respectively)?

____

The lungs and thoracic cavity (respectively)?

____

The abdominal viscera and visceral cavity (respectively)?

____

____

Synovial Membranes

Synovial membranes, unlike the mucous and serous membranes, are composed entirely of connective tissue; they contain no epithelial cells. These membranes line the cavities surrounding the joints, where they provide a smooth surface and secrete a lubricating fluid. They also line smaller sacs of connective tissue (bursae) and tendon sheaths, both of which cushion structures moving against each other, as during muscle activity. Figure 7.4 illustrates the positioning of a synovial membrane in the joint cavity.

- If a freshly sawed beef joint is available, visually examine the interior surface of the joint capsule to observe the smooth texture of the synovial membrane.

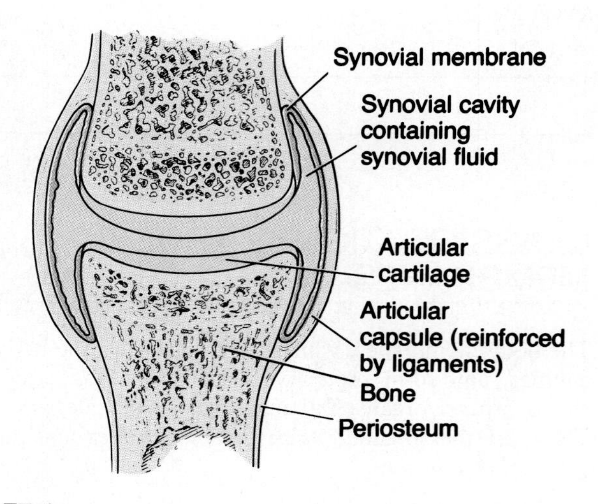

F7.4

A typical synovial joint.

Bone Classification, Structure, and Relationships: An Overview

OBJECTIVES

1. To list at least three functions of the skeletal system.
2. To identify the four main kinds of bones.
3. To identify surface bone markings and their function.
4. To identify the major anatomical areas on a longitudinally cut long bone (or diagram of one).
5. To identify the major regions and structures of an osteon in a histological specimen of compact bone (or diagram of one).
6. To explain the role of the inorganic salts and organic matrix in providing flexibility and hardness to bone.

MATERIALS

Disarticulated bones, identified by name or number, that demonstrate classic examples of the four bone classifications (long, short, flat, and irregular)
Long bone sawed longitudinally (beef bone from a slaughterhouse, if possible, or prepared laboratory specimen)
Compound microscope
Prepared slide of ground bone (cross section)
3-D model of microscopic structure of compact bone
Long bone soaked in 10% nitric acid (or vinegar) until flexible
Long bone baked at 250°F for more than 2 hours
Disposable plastic gloves

The skeleton is constructed of two of the most supportive tissues found in the human body—cartilage and bone. In embryos, the skeleton is predominantly composed of hyaline cartilage; but in the adult, most of the cartilage is replaced by more rigid bone. Cartilage persists only in such isolated areas as the bridge of the nose, the larynx, the trachea, the joints, and parts of the rib cage.

Besides supporting and protecting the body as an internal framework, the skeleton also provides a system of levers the skeletal muscles use to move the body. In addition, the bones store such substances as lipids and many minerals (most importantly calcium). Finally, the red marrow cavities of bones provide a site for hematopoiesis (blood cell formation).

The skeleton is made up of bones that are connected at joints, or articulations. The skeleton is subdivided into two divisions: the **axial skeleton** (those bones that lie around the body's center of gravity) and the **appendicular skeleton** (those of the limbs, or appendages) (Figure 8.1).

Before beginning your study of the skeleton, imagine for a moment that your bones have turned to putty. What if you were running when this metamorphosis took place? Now imagine your bones forming a continuous metal framework within your body, somewhat like a network of plumbing pipes. What problems could you envision with this arrangement? These images should help you understand how well the skeletal system provides support and protection, as well as facilitates movement.

BONE MARKINGS

Even a casual observation of the bones will reveal that bone surfaces are not featureless smooth areas but are scarred with an array of bumps, holes, and ridges. These bone markings reveal where bones form joints with other bones; where muscles, tendons, and ligaments were attached; and where blood vessels and nerves passed. Bone markings fall into two categories: projections, or processes, which grow out from the bone and serve as sites of muscle attachments or help to form joints, and depressions, or cavities, which are indentations or openings in the bone that often serve as conduits for nerves and blood vessels. The bone markings are summarized in Table 8.1.

CLASSIFICATION OF BONES

The 206 bones of the adult skeleton are composed of two basic kinds of osseous tissue that differ in their texture. **Compact bone** looks smooth and homogeneous; **spongy** (or *cancellous*) **bone** is composed of small spicules (bars) of bone and lots of open space. Bones may be further classified into four groups based on their relative gross anatomy: long, short, flat, or irregular.

The **long bones** are much longer than they are wide, generally consisting of a shaft with heads at either end. Long bones are composed predominantly of com-

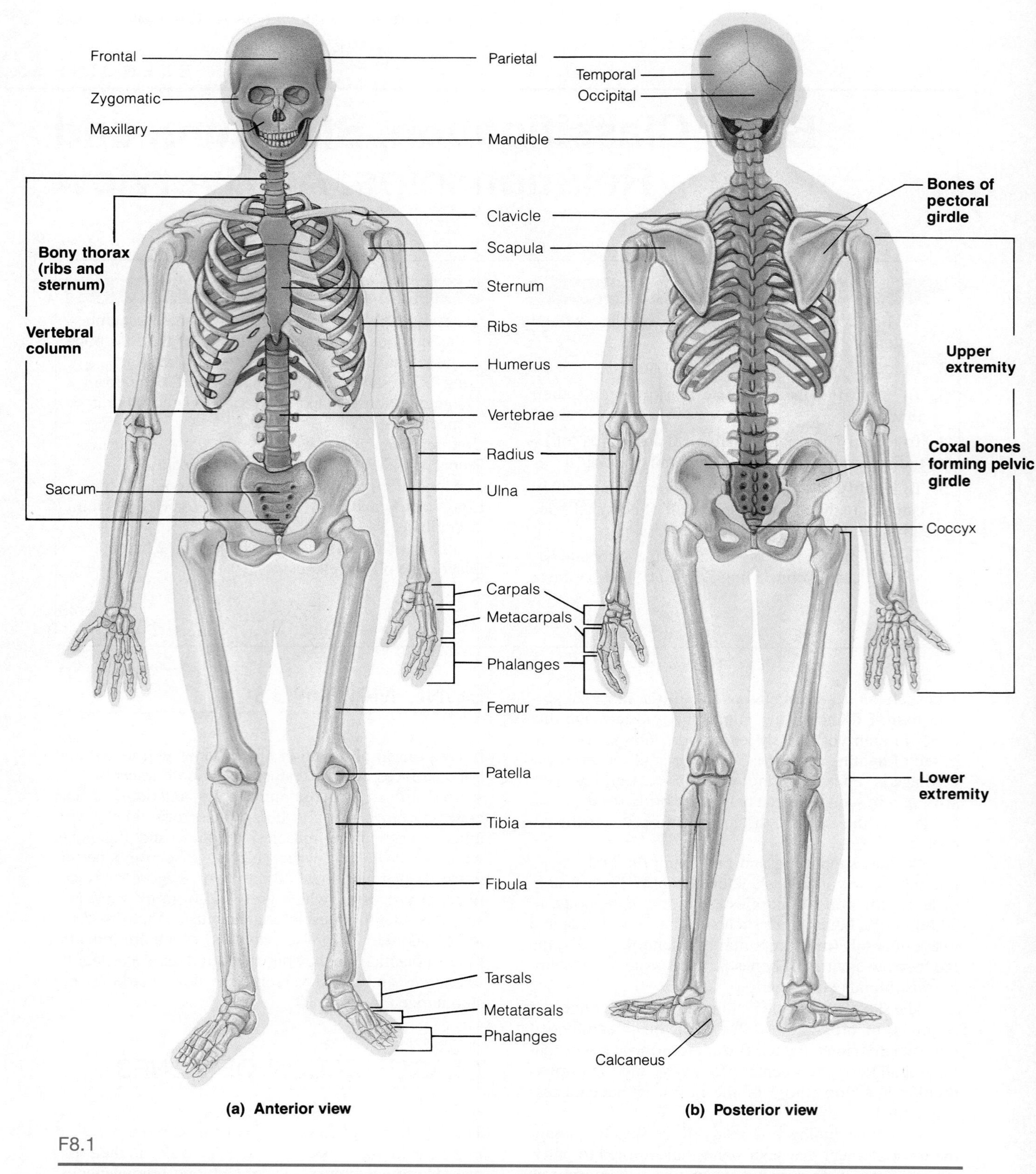

F8.1

The human skeleton. The bones of the axial skeleton are shaded green; the bones of the appendicular skeleton are shaded gold.

pact bone. The **short bones** are generally cube shaped, and they contain more spongy bone than compact bone.

The **flat bones** are typically thin, with two thin layers of compact bone sandwiching a layer of spongy bone between them. Although the name "flat bone" implies a structure that is level (or horizontal), many flat bones are curved. Bones that do not fall into one of the preceding categories are classified as **irregular bones.**

Some anatomists also recognize two other subcategories of bones. **Sesamoid bones** are small bones

TABLE 8.1 Bone Markings

Name of bone marking	Description	Illustration
Projections that are sites of muscle attachment		
Tuberosity	Large rounded projection; may be roughened	
Crest	Narrow ridge of bone; usually prominent	
Trochanter	Very large, blunt, irregularly shaped process. (The only examples are on the femur.)	
Line	Narrow ridge of bone; less prominent than a crest	
Tubercle	Small rounded projection or process	
Epicondyle	Raised area on or above a condyle	
Spine	Sharp, slender, often pointed projection	
Projections that help to form joints		
Head	Bony expansion carried on a narrow neck	
Facet	Smooth, nearly flat articular surface	
Condyle	Rounded articular projection	
Ramus	Armlike bar of bone	
Depressions and openings allowing blood vessels and nerves to pass		
Meatus	Canal-like passageway	
Sinus	Cavity within a bone, filled with air and lined with mucous membrane	
Fossa	Shallow, basinlike depression in a bone, often serving as an articular surface	
Groove	Furrow	
Fissure	Narrow, slitlike opening	
Foramen	Round or oval opening through a bone	

formed in tendons. The patellae (kneecaps) are sesamoid bones. **Wormian bones** are tiny bones between cranial bones. The sesamoid and wormian bones are not included in the bone count given above because they vary in number in different individuals.

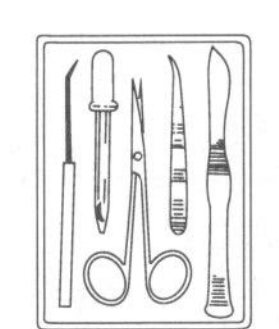

Examine the isolated (disarticulated) bones on display. See if you can find specific examples of the bone markings described in Table 8.1. Then classify each of the bones into one of the four anatomical groups by recording its name or number in the chart on the right.

Long	Short	Flat	Irregular

GROSS ANATOMY OF THE TYPICAL LONG BONE

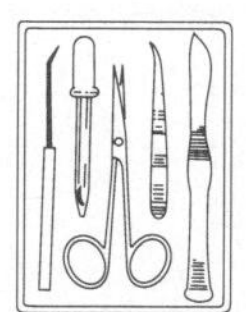

1. Obtain a long bone that has been sawed along its longitudinal axis. If a cleaned dry bone is provided, no special preparations need be made.

If the bone provided is a *fresh* beef bone, don gloves before beginning your observations.

2. With the help of Figure 8.2, identify the shaft, or **diaphysis.** Note its smooth surface, which is composed of compact bone. If you are using a fresh specimen, pull away the **periosteum,** or fibrous membrane covering, to view the bone surface. Notice that many fibers of the periosteum penetrate into the bone. These fibers are called **Sharpey's fibers.** The periosteum is the source of the blood vessels and nerves that invade the bone. **Osteoblasts** (bone-forming cells) on its inner face produce the bony matrix for increase in girth of the long bones.

3. Now inspect the **epiphysis,** the end of the long bone. Note that it is composed of a thin layer of compact bone filled with spongy bone. If present, identify the **articular cartilage** covering the epiphyseal surface. Because it is composed of glassy hyaline cartilage, it provides a smooth surface to prevent friction at joint surfaces.

4. If the source animal was still young and growing, you will be able to see a thin area of hyaline cartilage, the **epiphyseal plate,** which provides for longitudinal growth of the bone during youth. Once the long bone has stopped growing, these areas are replaced with bone and may appear as thin, barely discernible remnants—the **epiphyseal lines.**

5. In an adult animal, the interior **medullary cavity,** or **yellow marrow cavity,** of the shaft is essentially a storage region for adipose tissue. **Red marrow,** which is involved in blood cell formation, occupies the spaces between the spicules of spongy bone within the epiphyses.

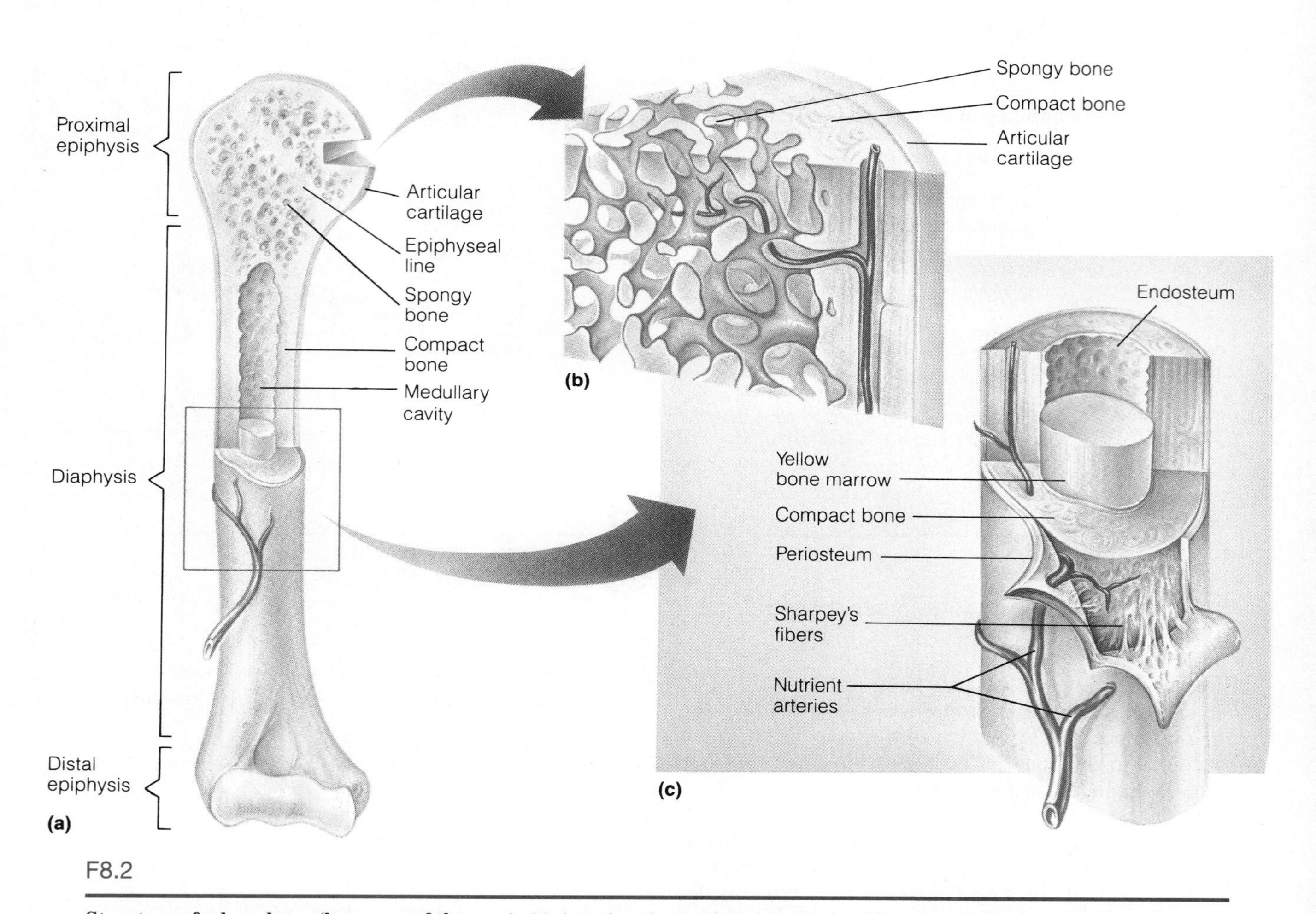

F8.2

Structure of a long bone (humerus of the arm). (a) Anterior view with longitudinal section cut away at the proximal end. (b) Pie-shaped, three-dimensional view of spongy bone and compact bone of the epiphysis. (c) Cross section of shaft (diaphysis). Note that the external surface of the diaphysis is covered by a periosteum, but the articular surface of the epiphysis is covered with hyaline cartilage.

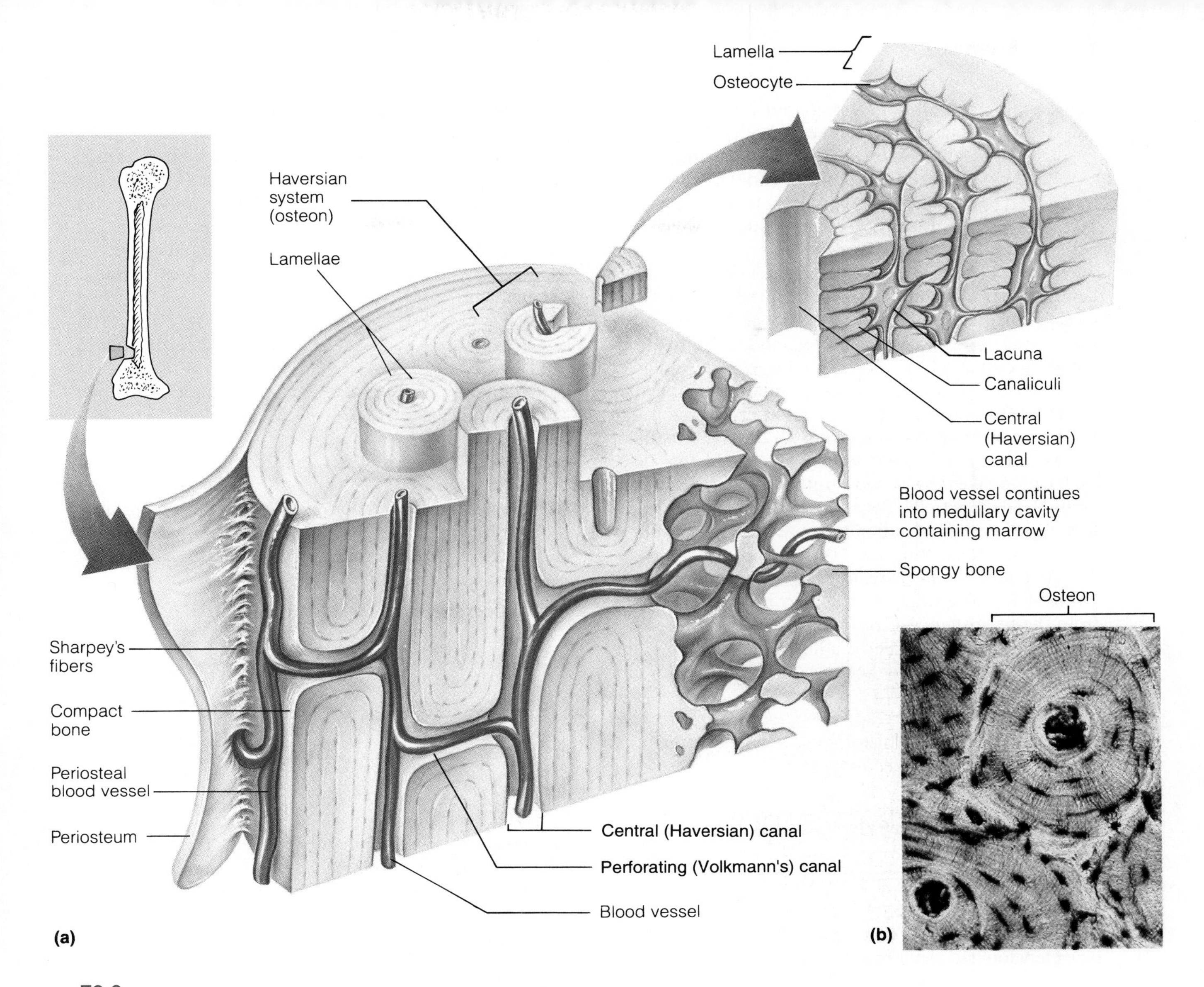

F8.3

Microscopic structure of compact bone. (a) Diagrammatic view of a pie-shaped segment of compact bone, illustrating its structural units (osteons). The inset shows a more highly magnified view of a portion of one osteon. Note the position of osteocytes in lacunae (cavities in the matrix). (b) Photomicrograph of a cross-sectional view of one osteon.

6. If you are examining a fresh bone, look carefully to see if you can distinguish the delicate **endosteum** lining the shaft. In a living bone, *osteoclasts* (bone-destroying cells) are found on the inner surface of the endosteum, against the compact bone of the diaphysis. As the bone grows in diameter on its external surface, it is constantly being broken down on its inner surface. Thus the thickness of the compact bone layer composing the shaft remains relatively constant.

Longitudinal bone growth at epiphyseal plates follows a predictable sequence and provides a reliable indicator of the age of children exhibiting normal growth. When problems of long-bone growth are suspected (for example, pituitary dwarfism), X-rays are taken to view the width of the growth plates. An abnormally thin epiphyseal plate indicates growth retardation. ■

MICROSCOPIC STRUCTURE OF COMPACT BONE

As you have seen, spongy bone has a spiky, openwork appearance resulting from the arrangement of the spicules of bony material, or *trabeculae,* that compose it, whereas compact bone appears to be dense and homogeneous. Microscopic examination of compact bone, however, reveals that it is riddled with passageways (Figure 8.3) for blood vessels, nerves, and lymphatic vessels that provide the living bone cells with needed substances and a way to eliminate wastes. Indeed, bone histology is much easier to understand when you recognize that bone tissue is organized around its blood supply.

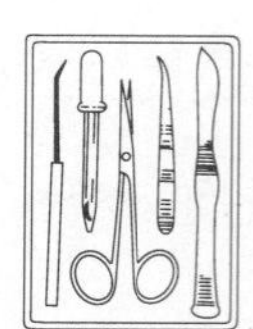

1. Obtain a prepared slide of ground bone and examine it under low power. Using Figure 8.3 as a guide, focus on a **central (Haversian) canal,** which contains the vascular supply in living bone tissue. Identify the **osteocytes** (mature bone cells) in **lacunae** (chambers), which are arranged in 3 to 5 concentric circles (concentric **lamellae**) around the central canal. A central canal and all the concentric lamellae surrounding it is referred to as an **osteon,** or **Haversian system.** Also identify **canaliculi,** tiny canals radiating outward from a central canal to the lacunae of the first lamella and then from lamella to lamella. The canaliculi form a dense transportation network through the hard bone matrix, connecting all the living cells of the osteon to the nutrient supply. The canaliculi allow each cell to take what it needs for nourishment and pass along the excess to the next osteocyte. You may need a higher-power magnification to see the fine canaliculi.

2. Also note the **perforating (Volkmann's) canals** in Figure 8.3a. These canals run into the compact bone from the periosteum, at right angles to the shaft. With the central canals, the perforating canals complete the communication pathway between the bone interior and its external surface.

3. If a model of bone histology is available, reidentify the same structures on the model.

CHEMICAL COMPOSITION OF BONE

Bone is one of the hardest materials in the body. Although relatively light, bone has a remarkable ability to resist tension and shear forces that continually act on it. The hardness of bone is due to the inorganic calcium salts deposited in its ground substance. Its flexibility comes from the organic elements of the matrix, particularly the collagenic fibers.

Obtain a bone sample that has been soaked in acid and one that has been baked. Heating removes the organic part of bone, while acid dissolves out the minerals. Do the treated bones retain the structure of untreated specimens?

____

Gently apply pressure to each bone sample. What happens to the heated bone?

____

The bone treated with acid? ______________________________

____

What does the acid appear to remove from the bone?

____

What does baking appear to do to the bone?

____

____

In rickets, the bones are not properly calcified. Which of the demonstration specimens would more closely resemble the bones of a child with rickets?

____

The Axial Skeleton

OBJECTIVES

1. To name the three bone groups composing the axial skeleton (skull, bony thorax, and vertebral column).
2. To identify the bones composing the axial skeleton, either by examining the isolated bones or by pointing them out on an articulated skeleton or skull, and to name the important bone markings on each.
3. To distinguish by examination the different types of vertebrae.
4. To discuss the importance of the intervertebral discs and spinal curvatures.
5. To differentiate among lordosis, kyphosis, and scoliosis.
6. To define *fontanel* and discuss its function and fate in the fetus.

MATERIALS

Intact skull and Beauchene skull
Fetal skull
Colored pencils
X-rays of individuals with scoliosis, lordosis, and kyphosis (if available)
Articulated skeleton
Isolated cervical, thoracic, and lumbar vertebrae; sacrum; and coccyx

The axial skeleton (the green-colored portion of Figure 8.1 on p. 62) can be divided into three parts: the skull, the vertebral column, and the bony thorax.

THE SKULL

The skull is composed of two sets of bones: The **cranium,** or *cranial vault,* encloses and protects the fragile brain tissue; the **facial bones** present the eyes in an anterior position and form the base for the facial muscles, which make it possible for us to present our feelings to the world. All but one of the bones of the skull are joined by interlocking joints called *sutures.* The mandible, or lower jawbone, is attached to the rest of the skull by a freely movable joint.

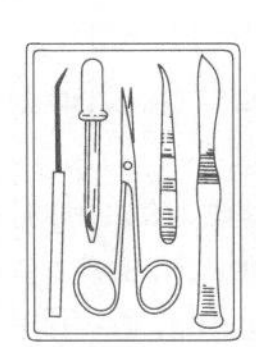

The bones of the skull, shown in Figures 9.1 through 9.4, are described below. As you read through this material, identify each bone on an intact (and/or Beauchene) skull. Note that important bone markings are listed beneath the bones on which they appear and that a color-coding dot before each bone name indicates its color in the figures.

The Cranium

The cranium is composed of eight large flat bones. *With the exception of two paired bones (the parietals and the temporals), all are single bones.* Sometimes the six ossicles of the middle ear are considered part of the cranium. Because the ossicles are functionally part of the hearing apparatus, their consideration is deferred to Exercise 22, Special Senses: Hearing and Equilibrium.

○ FRONTAL See Figures 9.1, 9.2, and 9.4. Anterior portion of cranium; forms the forehead, superior part of the orbit, and floor of anterior cranial fossa.

Supraorbital foramen: opening above each orbit allowing blood vessels and nerves to pass.
Glabella: smooth area between the eyes.

○ PARIETAL See Figures 9.1 and 9.2. Posterolateral to the frontal bone, forming sides of cranium.

Sagittal suture: midline articulation point of the two parietal bones.
Coronal suture: point of articulation of parietals with frontal bone.

○ TEMPORAL See Figures 9.1 through 9.4. Inferior to parietal bone on lateral skull.

Squamous suture: point of articulation of temporal (squamous region) with parietal bone.
External auditory meatus: canal leading to eardrum and middle ear.
Styloid (*stylo* = stake, pointed object) **process:** needle-like projection inferior to external auditory meatus; attachment point for muscles and ligaments. (This

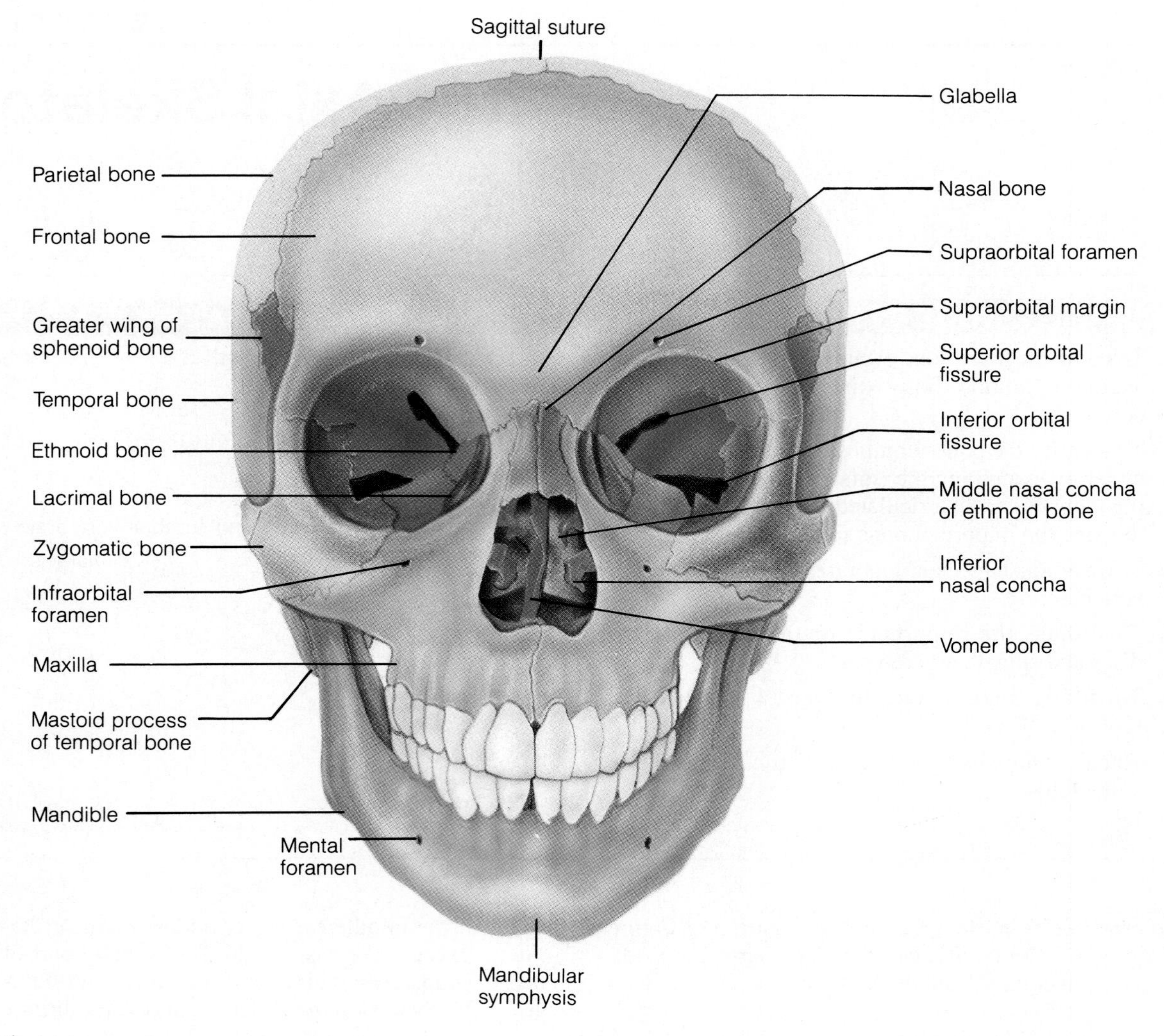

F9.1

Anatomy of the anterior aspect of the skull.

process is often missing from [broken off] demonstration skulls.)

Zygomatic process: ridgelike projection joining the zygomatic bone anteriorly. Together these two bones form the *zygomatic arch.*

Mastoid process: rough projection inferior and posterior to external auditory meatus; attachment site for muscles.

Mandibular fossa: rounded depression anterior to external auditory meatus; forms the socket for the mandibular condyle, the point where the mandible (lower jaw) joins the cranium.

Jugular foramen: opening medial to styloid process through which blood vessels and cranial nerves IX, X, and XI pass.

Carotid canal: opening through which the internal carotid artery passes into the petrous portion of the temporal bone.

Stylomastoid foramen: tiny opening between the mastoid and styloid processes through which the seventh cranial nerve leaves the cranium.

Internal acoustic meatus: opening on posterior aspect (petrous portion) of temporal bone allowing passage of two cranial nerves (VII and VIII).

OCCIPITAL See Figures 9.2, 9.3, and 9.4. Most posterior bone of cranium—forms floor and back wall. Joins sphenoid bone anteriorly via its narrow **basioccipital** region.

Lambdoidal suture: site of articulation of occipital bone and parietal bones.

Foramen magnum: large opening in base of occipital that allows the spinal cord to join with the brain.

Occipital condyles: rounded projections lateral to the foramen magnum that articulate with the first cervical vertebra (atlas).

External occipital protuberance: midline prominence posterior to the foramen magnum.

Superior and **inferior nuchal lines:** inconspicuous ridges that serve as sites of muscle attachment.

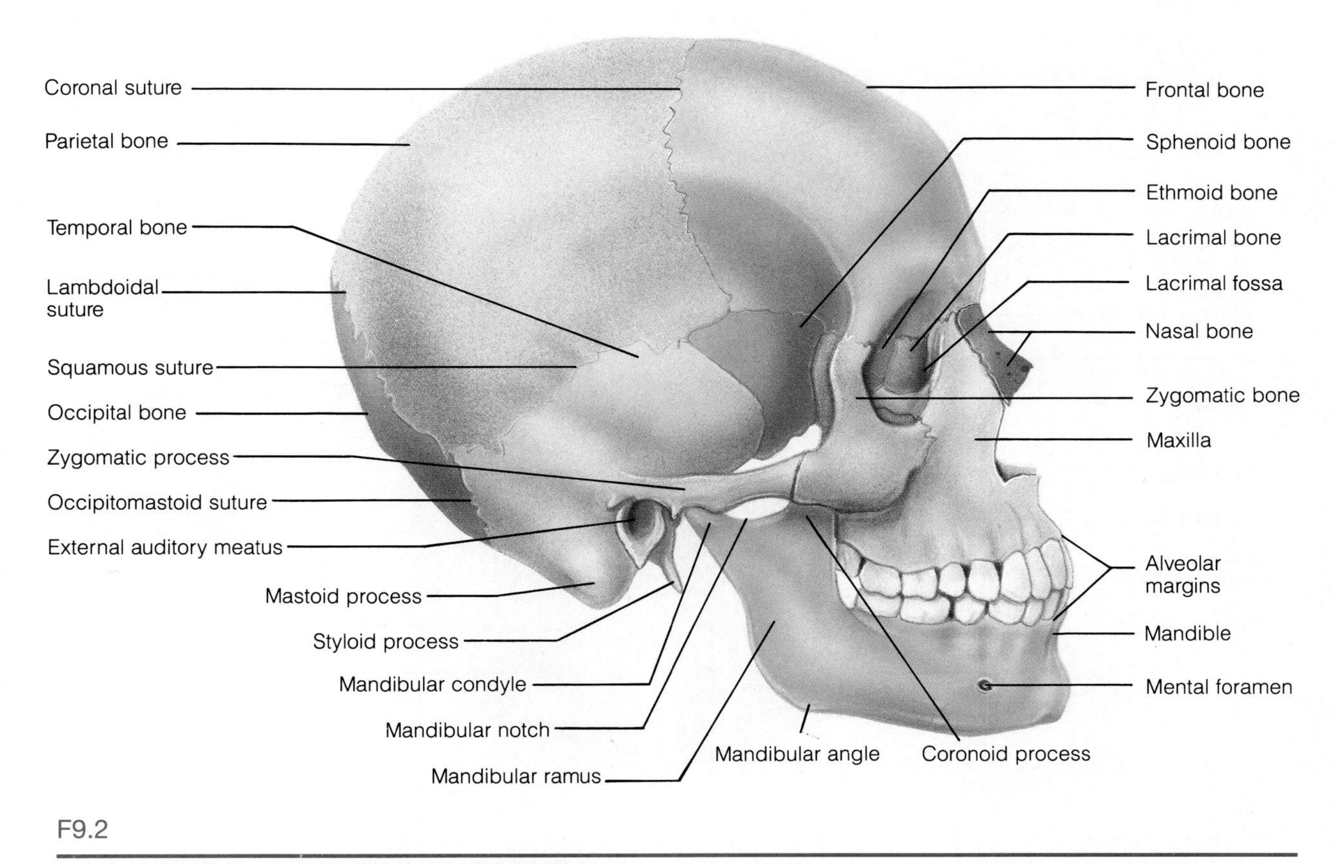

F9.2

External anatomy of the right lateral aspect of the skull.

Occipitomastoid suture: point of articulation between the occipital bone and the mastoid region of the temporal bone on each side.

SPHENOID See Figures 9.1 through 9.4. Bat-shaped bone forming part of the middle cranial fossa across the width of the skull.

Greater wings: portions of the sphenoid seen extending anterior to the temporal and forming a portion of the orbits.

Superior orbital fissures: jagged openings in orbits providing passage for cranial nerves III, IV, V, and VI.

Sella turcica (Turk's saddle): small depression in sphenoid midline in which the pituitary gland rests in the living person.

Lesser wings: bat-shaped portion of sphenoid anterior to sella turcica.

Optic foramina: openings in the base of the lesser wings through which the optic nerves enter the orbits to serve the eyes.

Foramen rotundum: opening lateral to sella turcica providing passage for a branch of the fifth cranial nerve. (This foramen is *not* visible on an inferior view of the skull.)

Foramen ovale: opening posterior to the sella turcica providing passage for a branch of the fifth cranial nerve.

Foramen spinosum: tiny opening posterolateral to the foramen ovale that transmits the middle meningeal artery.

Foramen lacerum: a jagged opening between the temporal bone and the sphenoid providing passage for a number of small nerves, and for the internal carotid artery to enter the middle cranial fossa (after it passes through part of the temporal bone).

ETHMOID See Figures 9.1, 9.2, and 9.4. Irregularly shaped bone anterior to the sphenoid; forms the roof of the nasal cavity, the upper nasal septum, and part of the medial orbit walls.

Crista galli (cock's comb): vertical projection providing a point of attachment for the dura mater.

Cribriform plates: bony plates lateral to the crista galli through which olfactory fibers pass to the brain.

Superior and middle nasal conchae (turbinates): thin plates of bone extending medially from the lateral aspects (masses) of the ethmoid into the nasal cavity.

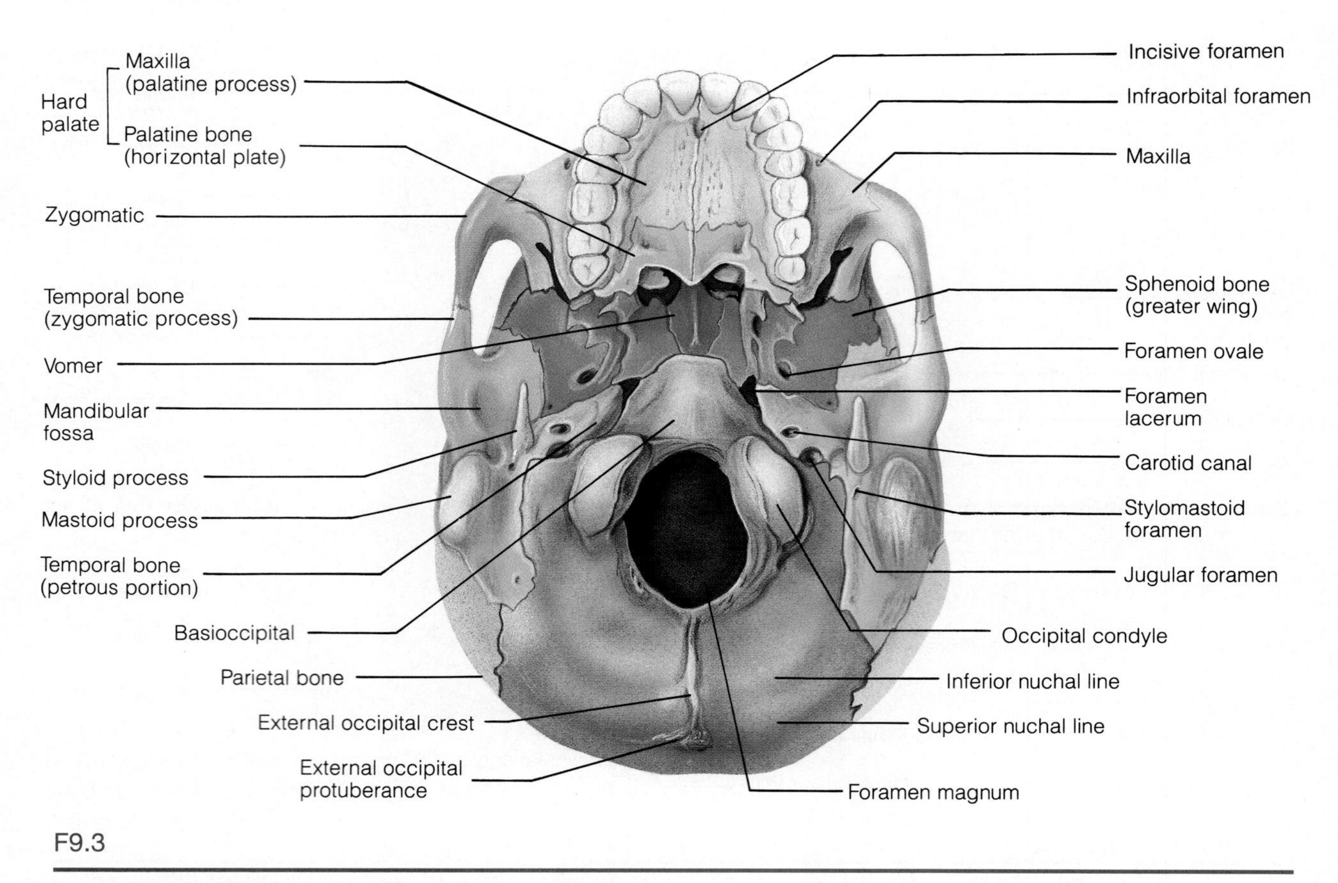

F9.3

Inferior superficial view of the skull, mandible removed.

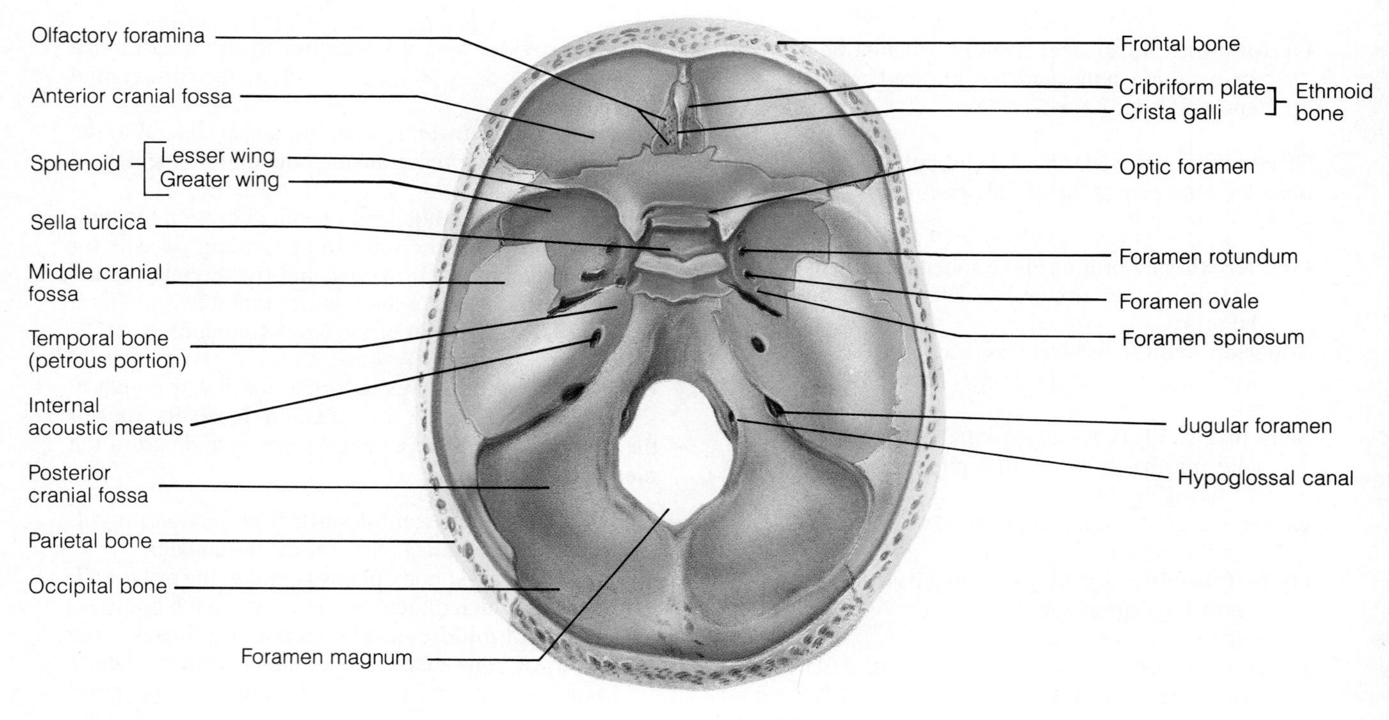

F9.4

Superior view of the floor of the cranial cavity, calvaria removed.

Facial Bones

There are 14 bones composing the face, and 12 of these are paired. *Only the mandible and vomer are single bones.* An additional bone, the hyoid bone, although not a facial bone, is considered here because of its location. Refer to Figures 9.1 through 9.4 to find the structures described below.

MANDIBLE See Figures 9.1 and 9.2. The lower jawbone, which articulates with the temporal bones, providing the only freely movable joints of the skull.

Body: horizontal portion; forms the chin.
Ramus: vertical extension of the body on either side.
Mandibular condyle: point of articulation of the mandible with the mandibular fossa of the temporal bone.
Coronoid process: jutting anterior portion of the ramus; site of muscle attachment.
Angle: posterior point at which ramus meets the body.
Mental foramen: prominent opening on the body (lateral to the midline) that transmits the mental blood vessels and nerve (branch of cranial nerve V) to the lower jaw.
Mandibular foramen: Open the lower jaw of the skull to identify this prominent foramen on the medial aspect of the ramus. This foramen permits passage of the nerve involved with tooth sensation (mandibular branch of cranial nerve V) and is the site where the dentist injects Novocain to prevent pain while working on the lower teeth.
Alveoli: sockets on superior margin of mandible in which the teeth lie.
Mandibular symphysis: anterior median depression indicating point of fusion of the two (right and left) "halves" of the mandibular body.

MAXILLAE See Figures 9.1, 9.2, and 9.3. Two bones fused in a median suture that form the upper jawbone and part of the orbits. Keystone bones of the facial skeleton.

Alveoli: sockets on the inferior margin in which teeth lie.
Palatine processes: form the anterior parts of the hard palate.
Infraorbital foramen: opening under the orbit carrying the infraorbital nerves and blood vessels to the nasal region.
Incisive foramen: large bilateral foramen located posterior to the central incisor tooth of the maxilla, and piercing the hard palate; transmits the nasopalatine blood vessels.

PALATINE See Figure 9.3. Paired bones posterior to the palatine processes; form posterior hard palate and part of the orbit.

ZYGOMATIC See Figures 9.1, 9.2, and 9.3. Lateral to the maxilla; forms the portion of the face commonly called the cheekbone and forms part of the lateral orbit. Its three processes are named for the bones with which they articulate.

LACRIMAL See Figures 9.1 and 9.2. Fingernail-sized bones forming a part of the medial orbit walls between the maxilla and the ethmoid. Each lacrimal bone is pierced by an opening, the **lacrimal fossa,** which serves as a passageway for tears.

NASAL See Figures 9.1 and 9.2. Small rectangular bones forming the bridge of the nose.

VOMER (*vomer* = plow) See Figure 9.1. Irregularly shaped bone in median plane of nasal cavity that forms the posterior and inferior nasal septum.

INFERIOR NASAL CONCHAE (turbinates) See Figure 9.1. Thin curved bones protruding medially from the lateral walls of the nasal cavity.

HYOID BONE See Figure 9.5. Not really considered a skull bone. Located in the throat above the larynx; serves as a point of attachment for many tongue and mouth muscles; does not articulate with any other bone and is thus unique; horseshoe shaped, with a body and two pairs of horns, or **cornua.**

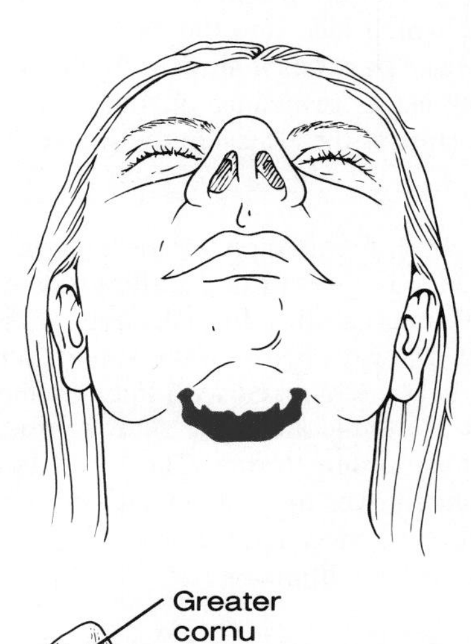

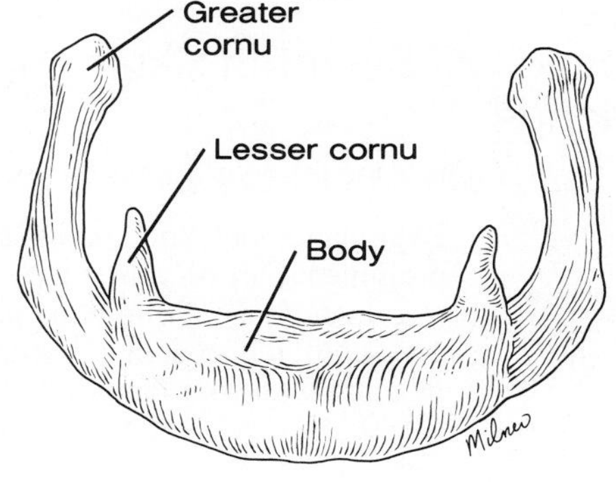

F9.5

Hyoid bone.

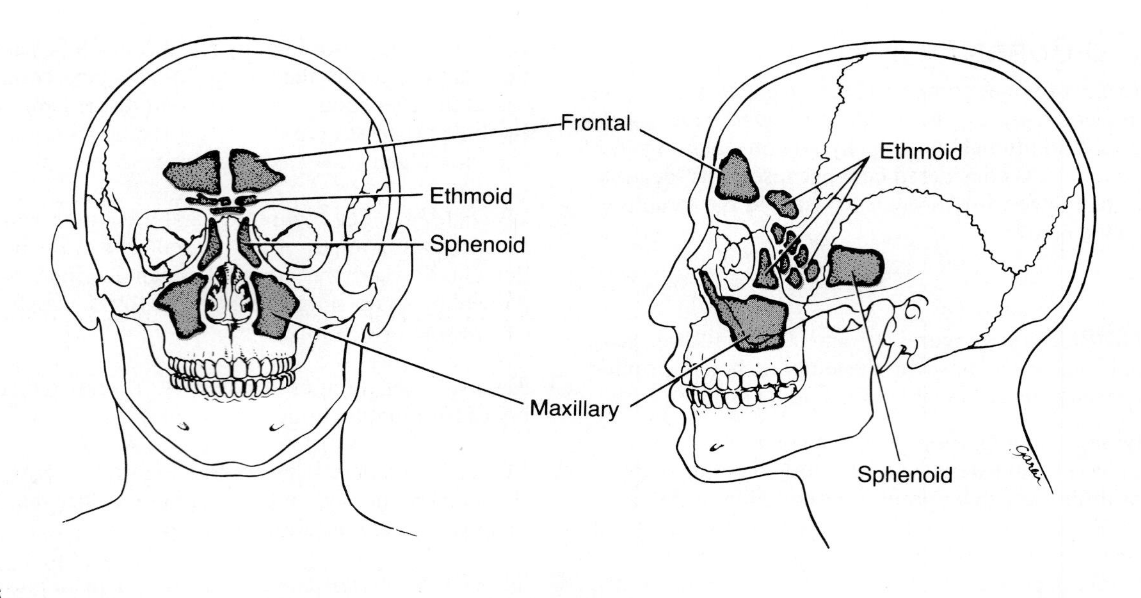

F9.6

Paranasal sinuses.

Paranasal Sinuses

Four skull bones—the frontal, ethmoid, sphenoid, and maxillary bones—contain sinuses (air cavities lined with mucosa), which lead into the nasal passages (see Figure 9.6). These **paranasal sinuses** lighten the facial bones and may act as resonance chambers for speech. The maxillary sinus is the largest of the sinuses found in the skull.

Sinusitis, or inflammation of the sinuses, sometimes occurs as a result of an allergy or bacterial invasion of the sinus cavities. In such cases, some of the connecting passageways between the sinuses and nasal passages may become blocked with thick mucus or infectious material. As the air in the sinus cavities is absorbed, a partial vacuum forms. The result is a sinus headache localized over the inflammed sinus area. Severe sinus infections may require surgical drainage to relieve this painful condition. ■

Palpation of Selected Skull Markings

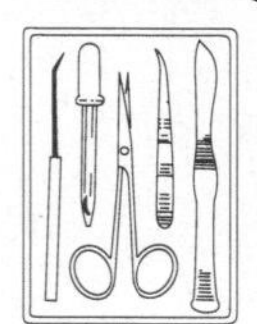

Palpate the following areas on yourself:

- Zygomatic bone and arch. (The most prominent part of your cheek is your zygomatic bone. Follow the posterior course of the zygomatic arch to its junction with your temporal bone.)
- Mastoid process (the rough area just behind your ear).
- Temporomandibular joints. (Open and close your jaws to locate these.)
- Greater wing of sphenoid. (Find the indentation posterior to the orbit and superior to the zygomatic arch on your lateral skull.)
- Superior orbital foramen. (Apply firm pressure along the superior orbital margin to find the indentation resulting from this foramen.)
- Inferior orbital foramen. (Apply firm pressure along the inferomedial border of the orbit to locate this large foramen.)
- Mandibular angle (most inferior and posterior aspect of the mandible).
- Mandibular symphysis (midline of chin).
- Nasal bones. (Run your index finger and thumb along opposite sides of the bridge of your nose until they "slip" medially at the inferior end of the nasal bones.)
- External occipital protuberance. (This midline projection is easily felt by running your fingers up the furrow at the back of your neck to the skull.)
- Hyoid bone. (Place a thumb and finger high behind the lateral edge of the mandible and squeeze medially.)

The Fetal Skull (Optional)

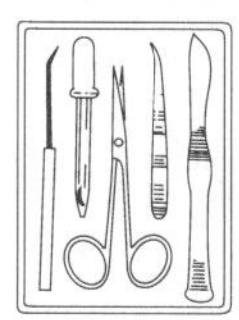

1. Obtain a fetal skull, and study it carefully. Make observations as needed to answer the following questions. Does it have the same bones as the adult skull?

____

What frontal bone difference is *obvious*?

____

____

How does the size of the fetal face relate to the cranium?

____

____

How does this compare to what is seen in the adult skull?

____

____

____

2. Identations between the bones of the fetal skull, called **fontanels,** are fibrous membranes. These areas will become bony (ossify) after birth, completing the process by the age of 20 to 22 months. The fontanels allow the fetal skull to be compressed slightly during birth and also allow for brain growth during late fetal life. Locate the following fontanels on the fetal skull with the aid of Figure 9.7: *anterior fontanel, mastoid fontanel, sphenoidal fontanel,* and *posterior fontanel.*

THE VERTEBRAL COLUMN

To some people, the term *vertebral column* might suggest a rather rigid supporting rod, but this picture is far from the truth. The **vertebral column** consists of 24 single bones (**vertebrae**) and two composite, or fused, bones (the sacrum and coccyx) that are connected in such a way as to provide a flexible curved structure (Figure 9.8). Of the 24 single bones, the 7 vertebrae of the neck are cervical vertebrae, the next 12 are the thoracic vertebrae, and the 5 supporting the lower back are lumbar vertebrae. Remembering common mealtimes, 7 A.M., 12 noon, and 5 P.M., may help you to recall the number of bones in these three regions of the vertebral column. The vertebrae are separated by pads of fibrocartilage—**intervertebral discs**—that cushion the vertebrae and absorb shocks.

As a person ages, the water content of the discs decreases (as it does in other tissues throughout the body) and the discs become less compressible. This situation, along with a weakening of the ligaments and tendons of the vertebral column, predisposes older people to slipped discs. ■

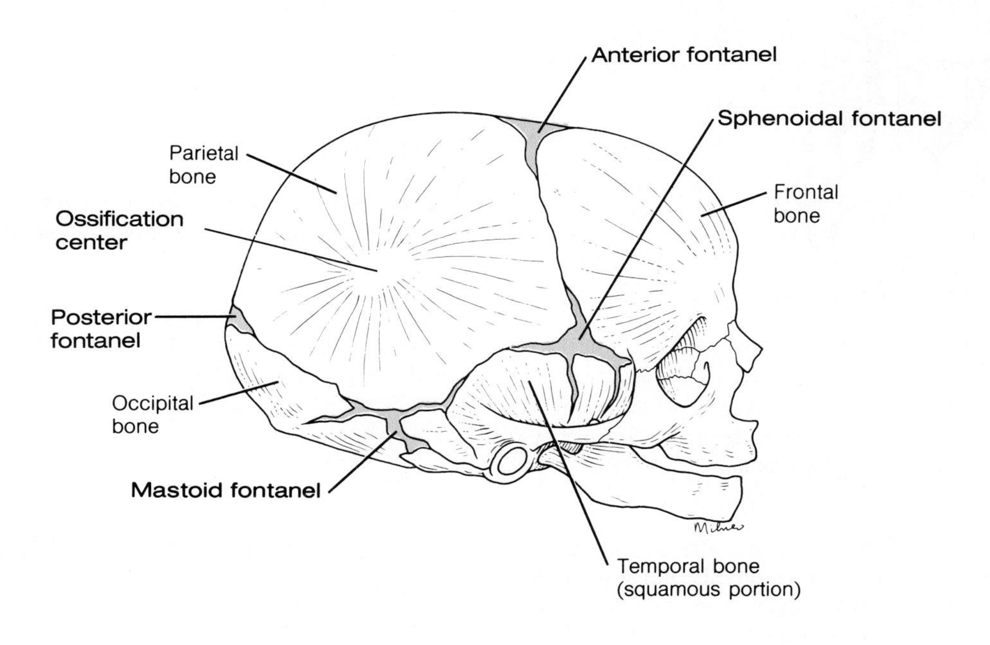

F9.7

The fetal skull, lateral view.

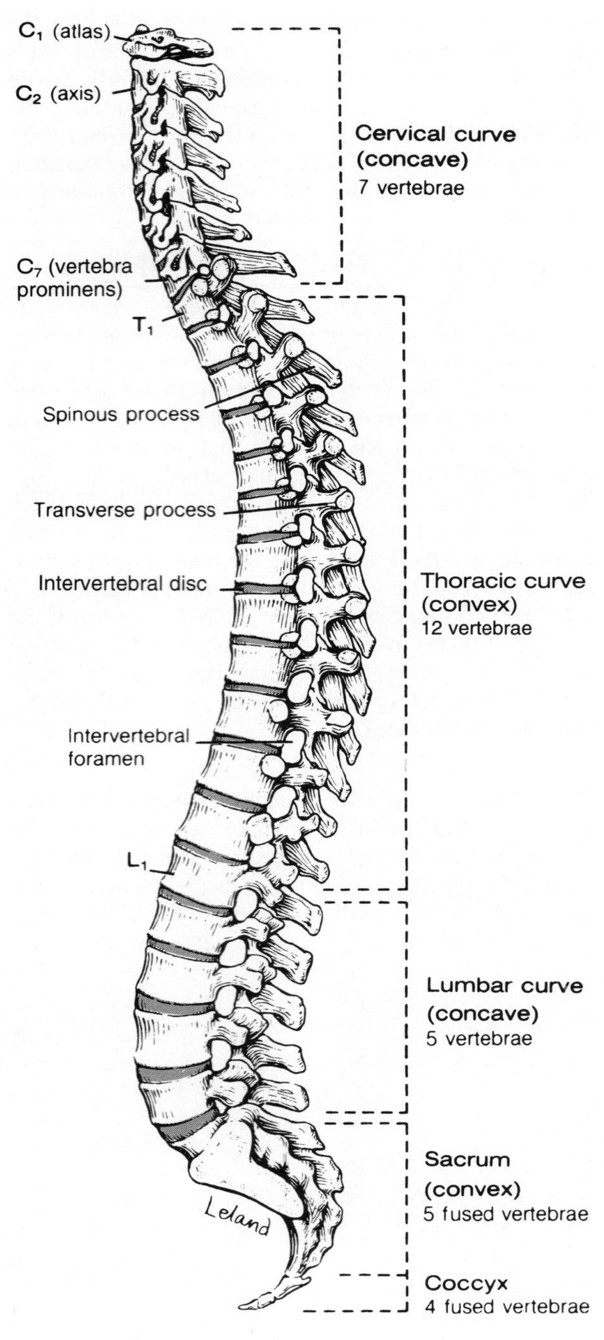

F9.8

The vertebral column. Notice the curvatures in this lateral view. (The terms *convex* and *concave* refer to the curvature of the posterior aspect of the vertebral column.)

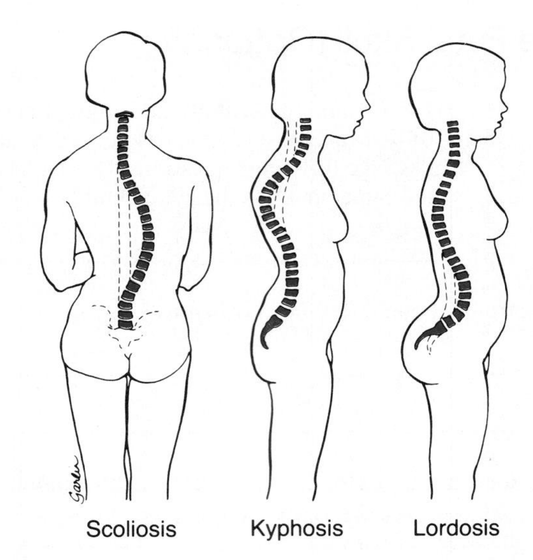

F9.9

Abnormal spinal curvatures.

The presence of the discs and the S-shaped, or springlike, construction of the vertebral column prevent shock to the head in walking and running and allow for flexibility in the movement of the body trunk. The thoracic and sacral curvatures of the spine are referred to as *primary curvatures* because they are present at birth. Later the *secondary curvatures* are formed. The cervical curvature appears when the baby begins to hold its head up independently, and the lumbar curvature develops when the baby begins to walk.

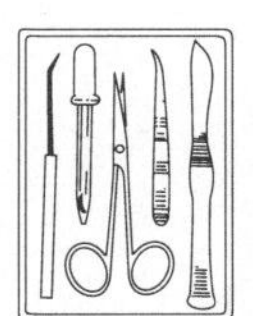

1. Note the normal curvature of the vertebral column in your laboratory specimen, and then examine Figure 9.9, which depicts three abnormal spinal curvatures—scoliosis, kyphosis, and lordosis. These abnormalities may result from disease or poor posture. Also examine X-rays, if they are available, showing the same conditions in a living patient.

2. Using an articulated spinal column (or an articulated skeleton), examine the freedom of movement between two lumbar vertebrae separated by an intervertebral disc.

When the fibrous disc is properly positioned, are the spinal cord or peripheral nerves impaired in any way?

____

What would happen to the spinal nerves in areas of malpositioned or "slipped" discs?

____

____

Structure of a Typical Vertebra

Although they differ in size and specific features, all vertebrae have some structures in common (Figure 9.10).

Body (or centrum): rounded central portion of the vertebra that faces anteriorly in the human vertebral column.

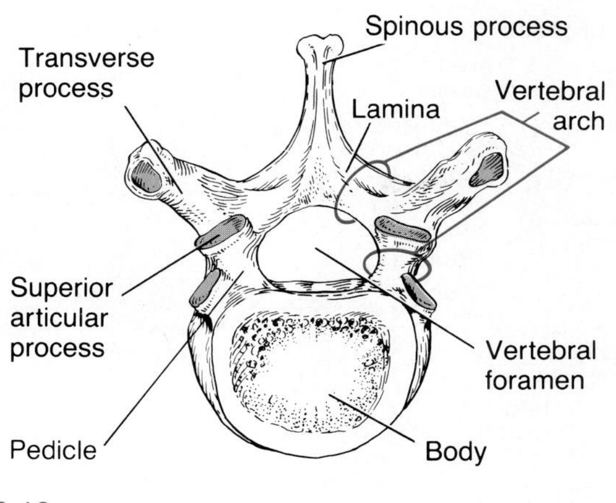

F9.10

A typical vertebra, superior view. (Inferior articulating surfaces are not shown.)

Vertebral arch: composed of pedicles, laminae, and a spinous process; represents the junction of all posterior extensions from the vertebral body.

Vertebral foramen: opening enclosed by the body and vertebral arch; a conduit for the spinal cord.

Transverse processes: two lateral projections from the vertebral body.

Spinous process: single medial and posterior projection.

Superior and inferior articular processes: paired projections lateral to the vertebral foramen that allow articulation with adjacent vertebrae; superior articular processes face toward the spinous process, whereas inferior articular processes face away from the spinous process.

Intervertebral foramina: the right and left pedicles have notches on their inferior and superior surfaces that create openings, the intervertebral foramina, for spinal nerves to leave the spinal cord between adjacent vertebrae.

Figures 9.11 and 9.12 show how specific vertebrae differ; refer to them as you read the following sections.

Cervical Vertebrae

The seven cervical vertebrae (referred to as C_1 through C_7) form the neck portion of the vertebral column. The first two cervical vertebrae (atlas and axis) are highly modified to perform special functions (see Figure 9.11). The **atlas** (C_1) lacks a body, and its lateral processes

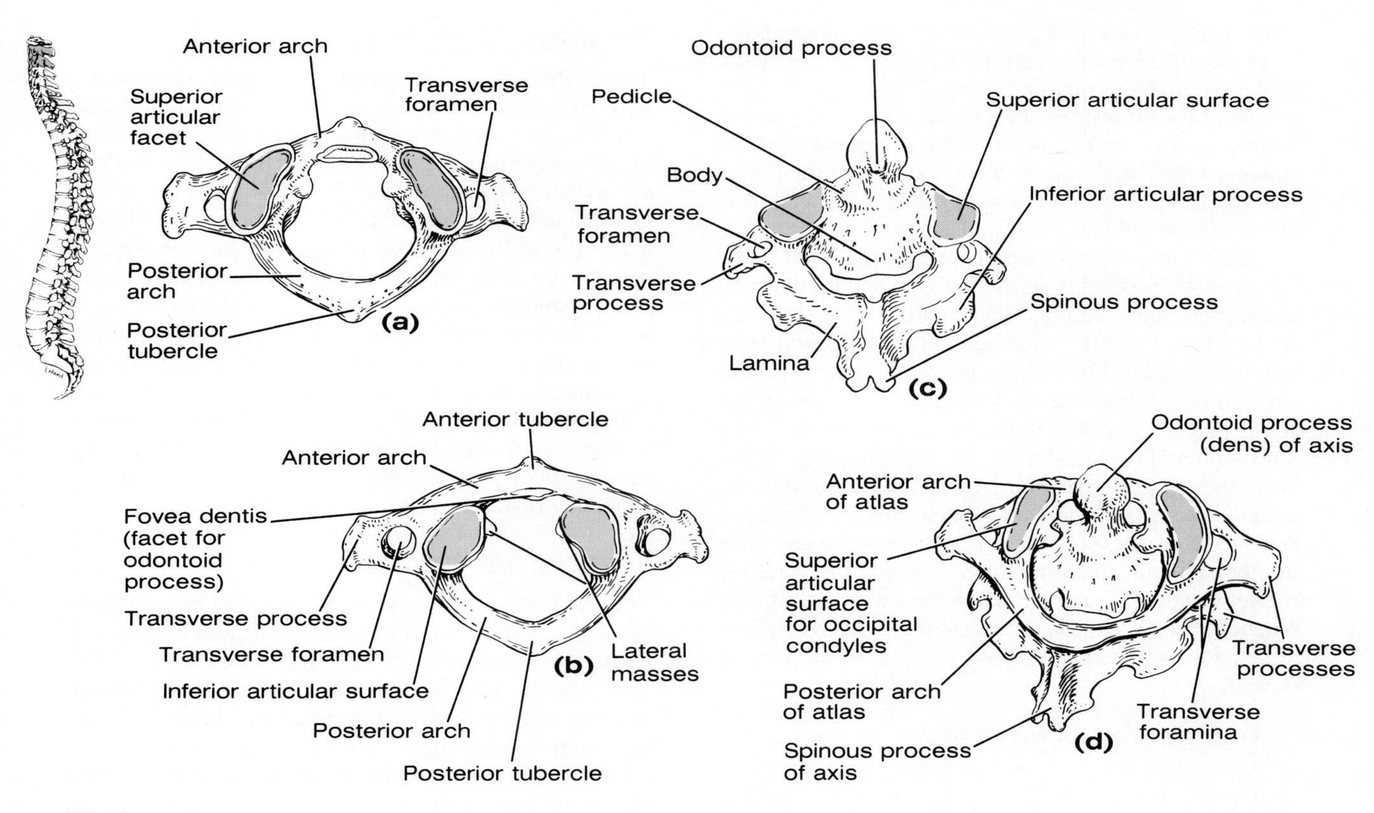

F9.11

Cervical vertebrae C_1 and C_2. (a) Superior view of the atlas (C_1); (b) inferior view of the atlas (C_1); (c) superior view of the axis (C_2); (d) superior lateral view of the articulated atlas and axis.

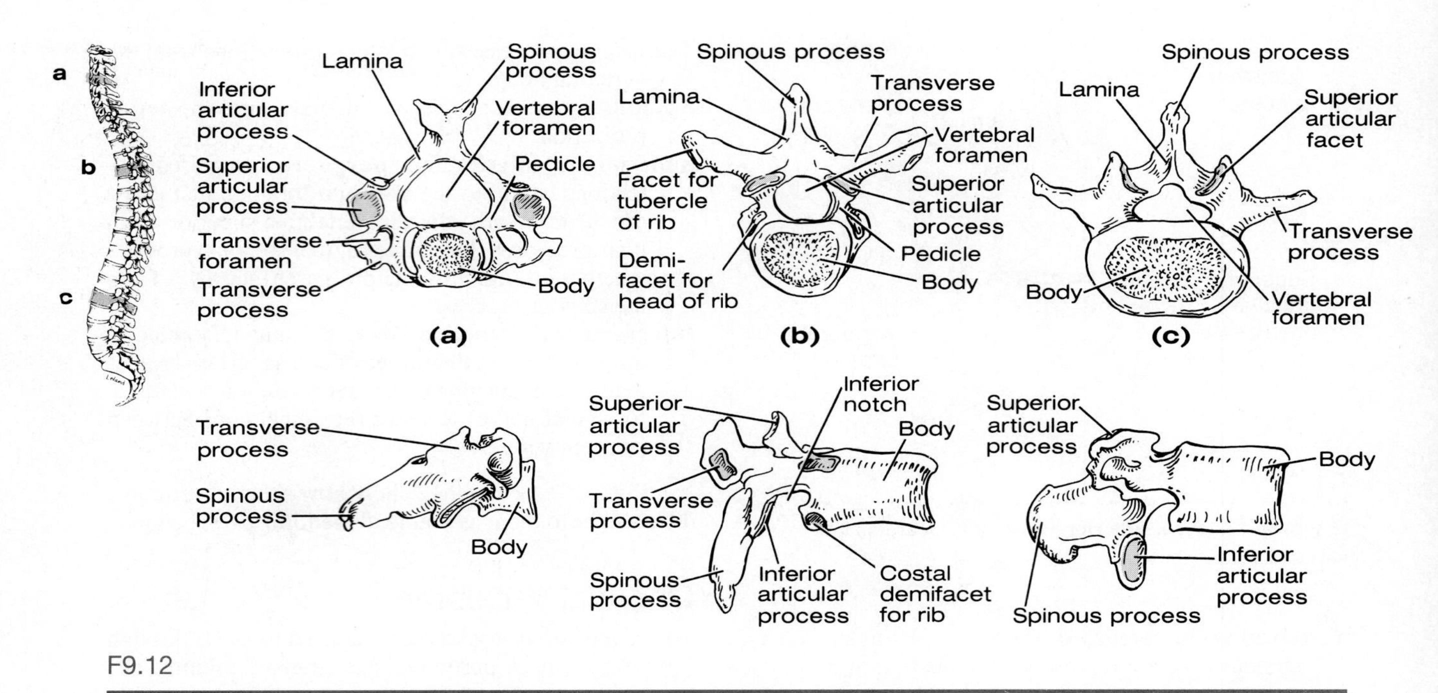

F9.12

Comparison of typical cervical, thoracic, and lumbar vertebrae (superior view above, lateral view below). (a) Cervical vertebra; (b) thoracic vertebra; (c) lumbar vertebra.

contain large concave depressions on their superior surfaces that receive the occipital condyles of the skull. This joint enables you to nod "yes." The **axis** (C_2) acts as a pivot for the rotation of the atlas (and skull) above. It bears a large vertical process, the **odontoid process,** or **dens,** which serves as the pivot point. The articulation between C_1 and C_2 allows you to rotate your head from side to side to indicate "no."

The "more typical" cervical vertebrae (C_3 through C_7) are distinguished from the thoracic and lumbar vertebrae by several features (see Figure 9.12a). They are the smallest, lightest vertebrae, and the vertebral foramen is triangular. The spinous process is short and usually bifurcated, or divided, into two branches; the exception is that of C_7, called the *vertebra prominens,* which is not split and is much longer and more prominent. Because the spinous process of C_7 is visible through the skin, it is used as a landmark for counting the vertebrae. Transverse processes of the cervical vertebrae are wide, and they contain foramina through which the vertebral arteries usually pass superiorly on their way to the brain. Whenever you see these foramina in the transverse processes, you can be sure that it is a cervical vertebra being viewed.

- Palpate your vertebra prominens.

Thoracic Vertebrae

The 12 thoracic vertebrae (referred to as T_1 through T_{12}) may be recognized by the following structural characteristics. As you can see in Figure 9.12b, they have a larger body than the cervical vertebrae. The body is somewhat heart shaped and has two costal demifacets on each side (one superior, the other inferior), close to the origin of the vertebral arch, that articulate with the heads of the ribs. The vertebral foramen is oval or round and the spinous process is long, with a sharp downward hook. The closer the thoracic vertebra is to the lumbar region, the less sharp and shorter is the spinous process. Articular facets on the transverse processes articulate with the tubercles of the ribs. Besides forming the thoracic part of the spine, these vertebrae also form the posterior aspect of the bony thoracic cage (commonly called the rib cage).

Lumbar Vertebrae

The five lumbar vertebrae (L_1 through L_5) have massive blocklike bodies and short, thick, hatchet-shaped spinous processes extending backward (see Figure 9.12c). The superior articular facets are directed posteromedially; the inferior ones are directed anterolaterally. These structural features help to reduce the mobility of the lumbar region of the spine. Because most of the stress of the vertebral column occurs in the lumbar region, these are the sturdiest of the vertebrae.

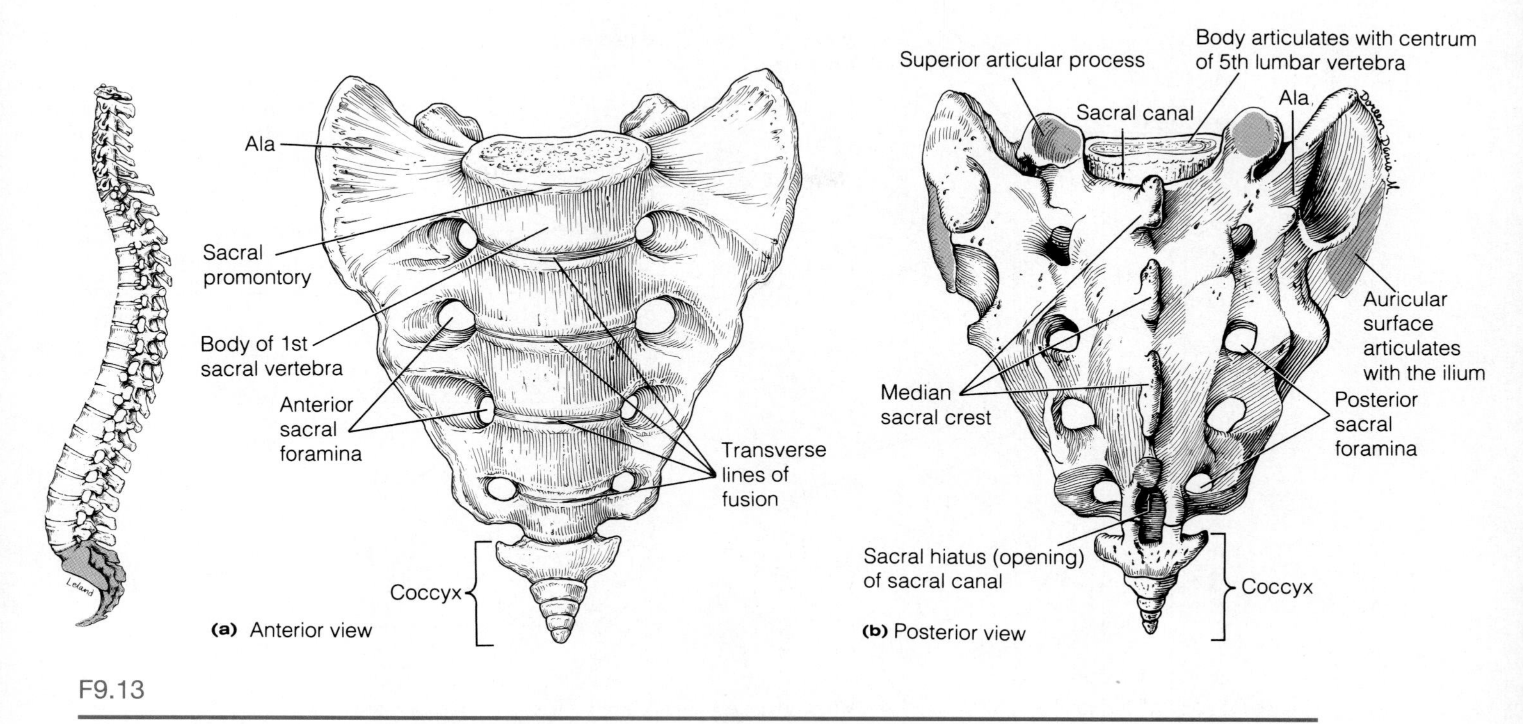

F9.13

Sacrum and coccyx. (a) Anterior view; (b) posterior view.

The Sacrum

The sacrum (Figure 9.13) is a composite bone formed from the fusion of five vertebrae. Superiorly it articulates with L_5, and inferiorly it connects with the coccyx. The **median sacral crest** is a remnant of the spinous processes of the fused vertebrae. The winglike **alae,** formed from the fusion of the transverse processes, articulate laterally with the hip bones. The sacrum is slightly concave anteriorly and forms the posterior border of the pelvis. Four ridges (lines of fusion) cross the anterior part of the sacrum, and **sacral foramina** are located at either end of these ridges. These foramina allow for the passage of blood vessels and nerves. The vertebral canal continues inside the sacrum as the **sacral canal.** The **sacral promontory** (anterior border of the body of S_1) is an important anatomic landmark for obstetricians.

- Attempt to palpate the median sacral crest of your sacrum. (This is more easily done in thin people.)

The Coccyx

The coccyx (see Figure 9.13) is formed from the fusion of three to five small, irregularly shaped vertebrae. It is literally the human tailbone, a vestige of the tail that other vertebrates have. The coccyx is attached to the sacrum by ligaments.

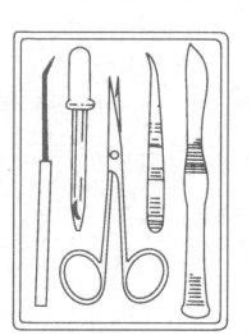

Obtain examples of each type of vertebra and examine them carefully, comparing them to Figures 9.11, 9.12, and 9.13 and to each other.

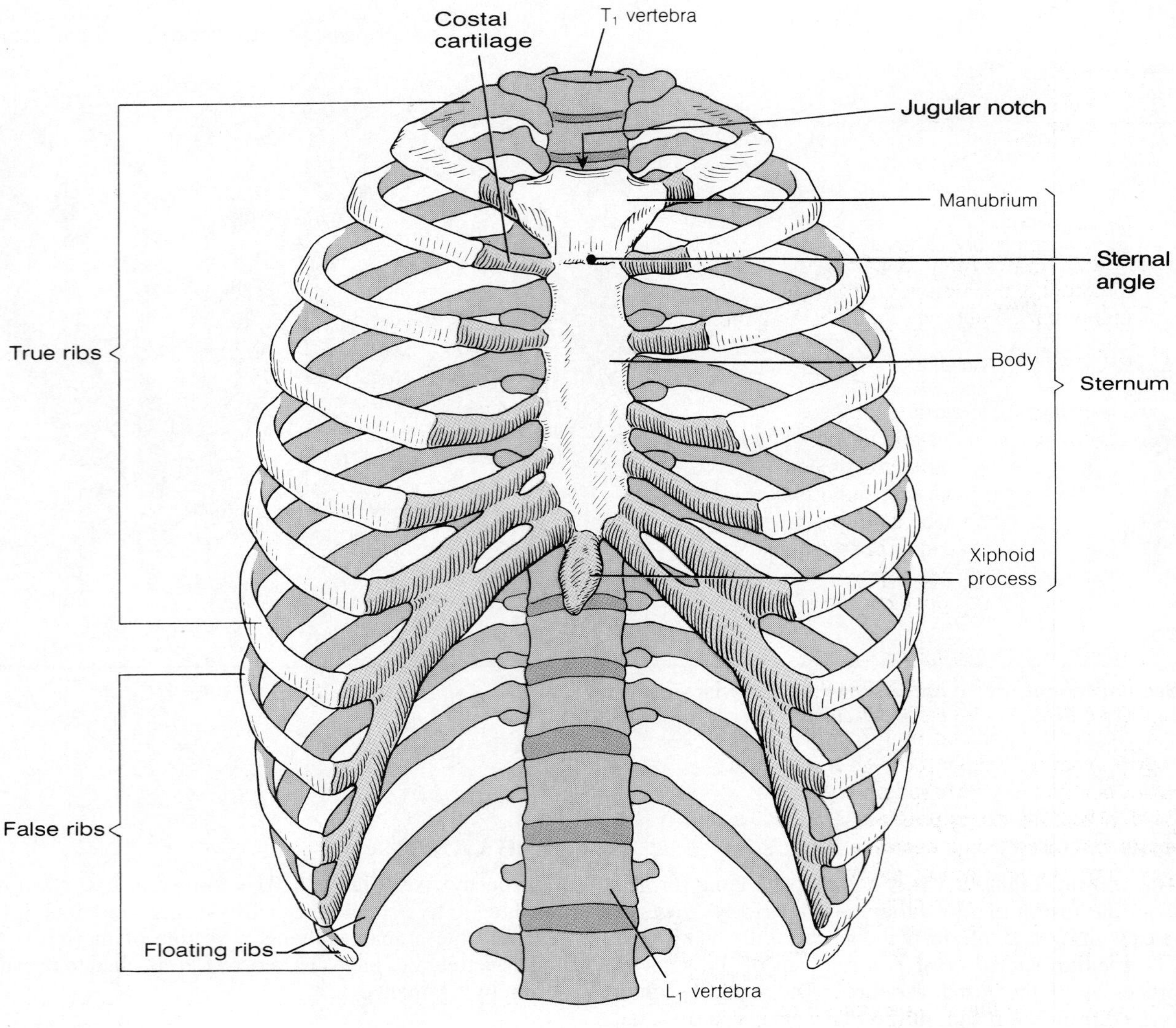

F9.14

Bony thorax, anterior view.

THE BONY THORAX

The **bony thorax** is composed of the sternum, ribs, and thoracic vertebrae (Figure 9.14). It is also referred to as the **thoracic cage** because of its appearance and because it forms a protective cone-shaped enclosure around the organs of the thoracic cavity (heart and lungs, for example).

The Sternum

The sternum (breastbone), a typical flat bone, is attached to the first seven pairs of ribs. The sternum results from the fusion of three bones: the manubrium, body, and xiphoid process. The superiormost **manubrium** looks like the knot of a man's tie; it articulates with the clavicle (collarbone) laterally. The **body** forms the bulk of the sternum. The **xiphoid** process forms the inferior end of the sternum and lies at the level of the fifth intercostal space. Although it is made of hyaline cartilage in children, it is usually ossified in adults.

In some people, the xiphoid process tends to project dorsally. This may present a problem because chest trauma can push such a xiphoid into the heart or liver (both immediately deep to the xiphoid process), causing massive hemorrhage. ■

The sternum has two important bony landmarks—the **jugular notch** and the **sternal angle.** The jugular notch (concave upper border of the manubrium) can be palpated easily; generally it is at the level of the disc

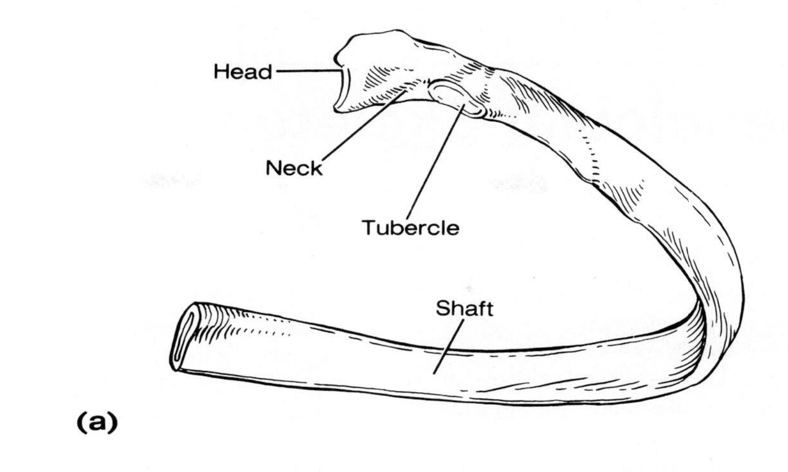

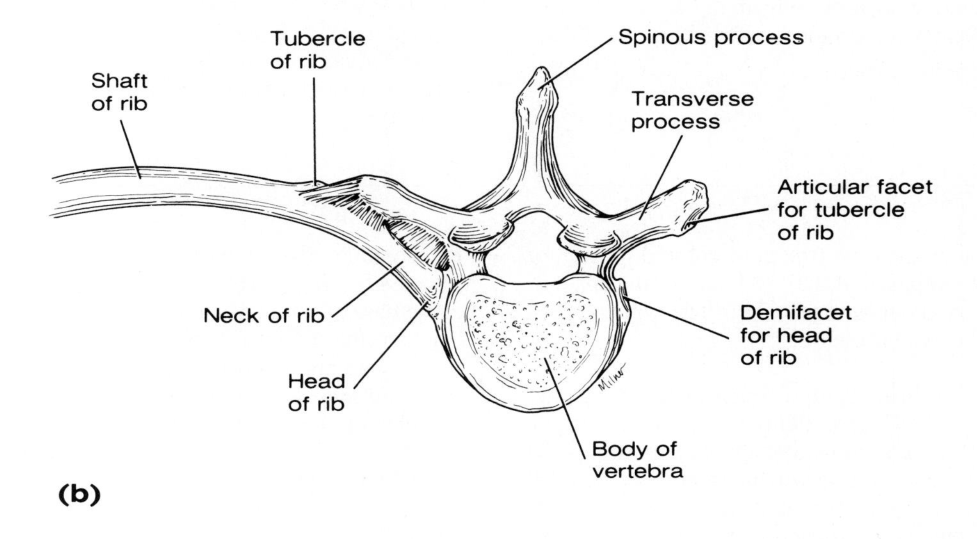

F9.15

Rib anatomy and articulations with the vertebral column. (a) Typical true rib from right side of body (costal cartilage not shown); (b) superior view of the articulation between a rib and a thoracic vertebra.

between the second and third thoracic vertebrae. The sternal angle is a result of the manubrium and body meeting at a slight angle to each other, so that a transverse ridge is formed at the level of the second ribs. It provides a handy reference point for counting ribs to locate the second intercostal space for listening to certain heart valves and is an important anatomical landmark for thoracic surgery.

- Palpate your sternal angle and jugular notch.

The sternum is a major hematopoietic site in adults and, because of its accessibility, it is a favored site for obtaining samples of blood-forming tissue for the diagnosis of suspected blood diseases. A needle is inserted into the marrow of the sternum, and the sample withdrawn (sternal puncture).

The Ribs

The 12 pairs of ribs form the walls of the thoracic cage. (See Figures 9.14 and 9.15.) All of the ribs articulate posteriorly with the vertebral column, and then curve downward and toward the anterior body surface. The "*true ribs,*" the first seven pairs, attach directly to the sternum by individual costal cartilages. "*False ribs,*" the next five pairs, have indirect cartilage attachments to the sternum or no sternal attachment at all, as in the case of the last two pairs (which are also called floating ribs).

- First take a deep breath to expand your chest. Notice how your ribs seem to move outward and how your sternum rises. Then examine an articulated skeleton to observe the relationship between the ribs and the vertebrae.

10 EXERCISE

The Appendicular Skeleton

OBJECTIVES

1. To identify on an articulated skeleton the bones of the shoulder and pelvic girdles, and their attached limbs.
2. To arrange unmarked, disarticulated bones in their proper relative position to form the entire skeleton, and to identify important bone markings.
3. To differentiate between a male and a female pelvis.
4. To discuss the common features of the human appendicular girdles (pectoral and pelvic) and to note how their structure relates to their specialized functions.

MATERIALS

Articulated skeletons
Disarticulated skeletons (complete)
Articulated pelves (male and female for comparative study)

The appendicular skeleton (the gold-colored portion of Figure 8.1) is composed of the 126 bones of the appendages and the shoulder and pelvic girdles, which attach the limbs to the axial skeleton.

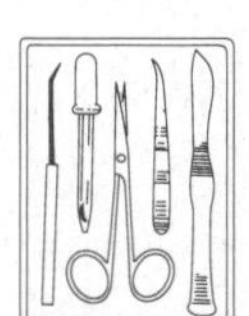

Carefully examine each of the bones described, and identify the characteristic bone markings of each. The markings aid in determining whether a bone is the right or left member of its pair. *This is a very important instruction because before completing this laboratory exercise, you will be constructing your own skeleton.*

BONES OF THE SHOULDER GIRDLE AND UPPER EXTREMITY

The Shoulder Girdle

Each **shoulder girdle** or **pectoral girdle** (Figure 10.1) consists of two bones—the anterior clavicle and the posterior scapula. The paired shoulder girdles (right and left) function to attach the upper limbs to the axial skeleton. In addition, the bones of the shoulder girdles serve as attachment points for many trunk and neck muscles.

The **clavicle,** or collarbone, is a slender, doubly curved bone, rounded on the medial end, which attaches to the sternal manubrium, and flattened on the lateral end, where it articulates with the scapula to form part of the shoulder joint. The clavicle serves as an anterior brace, or strut, to hold the arm away from the top of the thorax.

The **scapulae,** or shoulder blades, are generally triangular. Each scapula has a flattened body and two important processes—the **acromion** (the enlarged end of the spine of the scapula) and the beaklike **coracoid process** (*corac* = crow, raven). The supracapular notch at the base of the coracoid process allows nerves to pass. The acromion connects with the clavicle; the coracoid process points anteriorly over the tip of the shoulder joint and serves as a point of attachment for some of the muscles of the upper limb. The scapula has no direct attachment to the axial skeleton but is loosely held in place by trunk muscles.

The scapula has three angles: superior, inferior, and lateral. The inferior angle provides a landmark for auscultating (listening to) lung sounds. The scapula also has three named borders: the superior, the medial (vertebral), and the lateral (axillary). Several shallow depressions (fossae) appear on both sides of the scapula and are named according to location; i.e., there are the anterior subscapular fossa and the posterior infraspinous and supraspinous fossae. All of these fossae are clothed by muscles, many of which are named for their fossa of origin. The **glenoid cavity,** a shallow socket that receives the head of the arm bone, is located in the lateral angle.

The shoulder girdle is exceptionally light and allows the upper limb a degree of mobility not observed anywhere else in the body. This is due to the following factors:

- The sternoclavicular joints are the *only* sites of attachment of the shoulder girdles to the axial skeleton.
- The relative looseness of the scapular attachment allows it to slide back and forth against the thorax with muscular activity.
- The glenoid cavity is shallow, and does little to stabilize the shoulder joint.

However, this exceptional flexibility exacts a price: the arm bone (humerus) is extremely susceptible to dislocation, and fracture of the clavicle disables the entire upper limb.

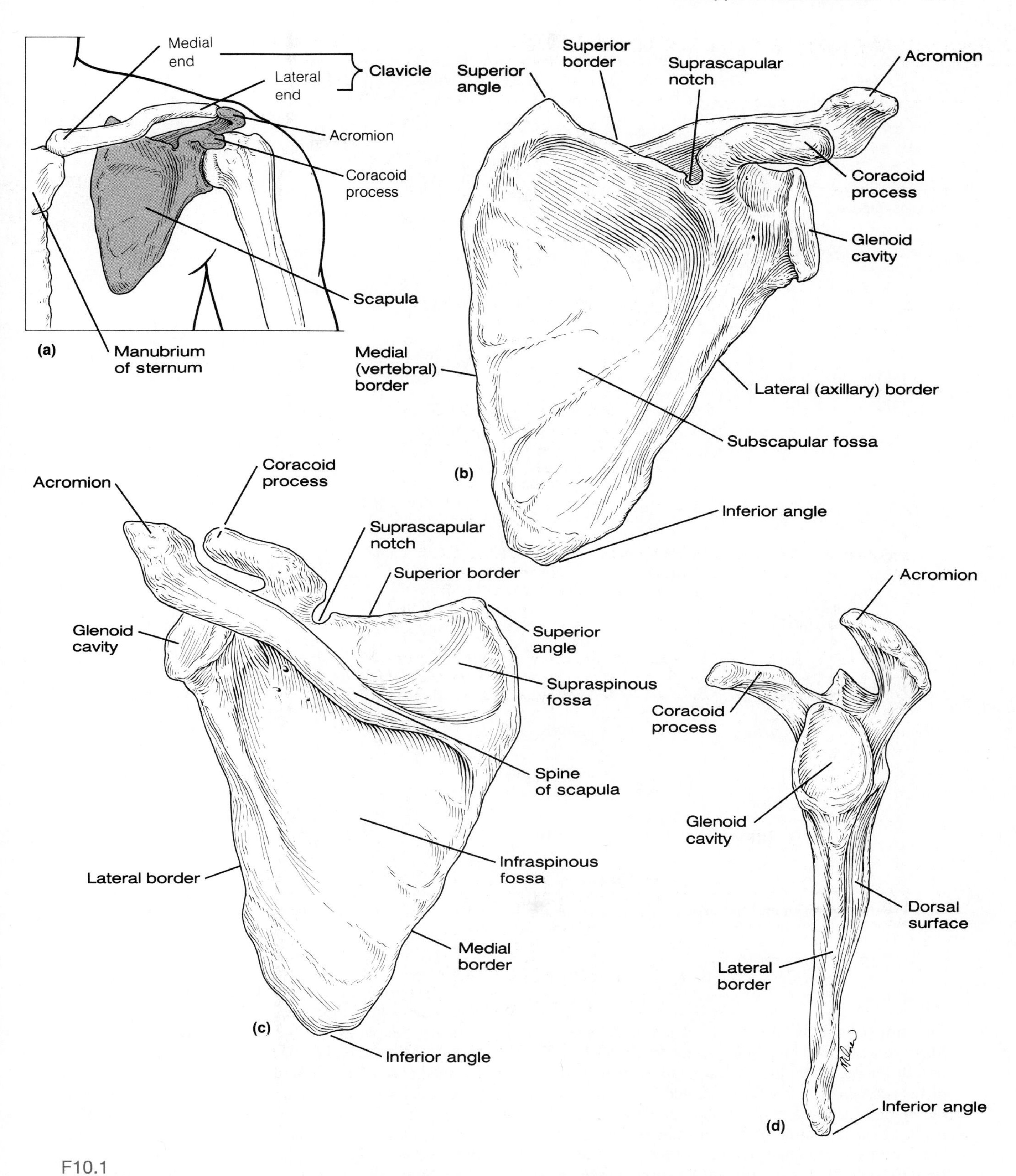

F10.1

Bones of the shoulder girdle. (a) Left shoulder girdle articulated to show the relationship of the girdle to the bones of the thorax and arm; (b) left scapula, anterior view; (c) left scapula, posterior view; (d) left scapula, lateral view.

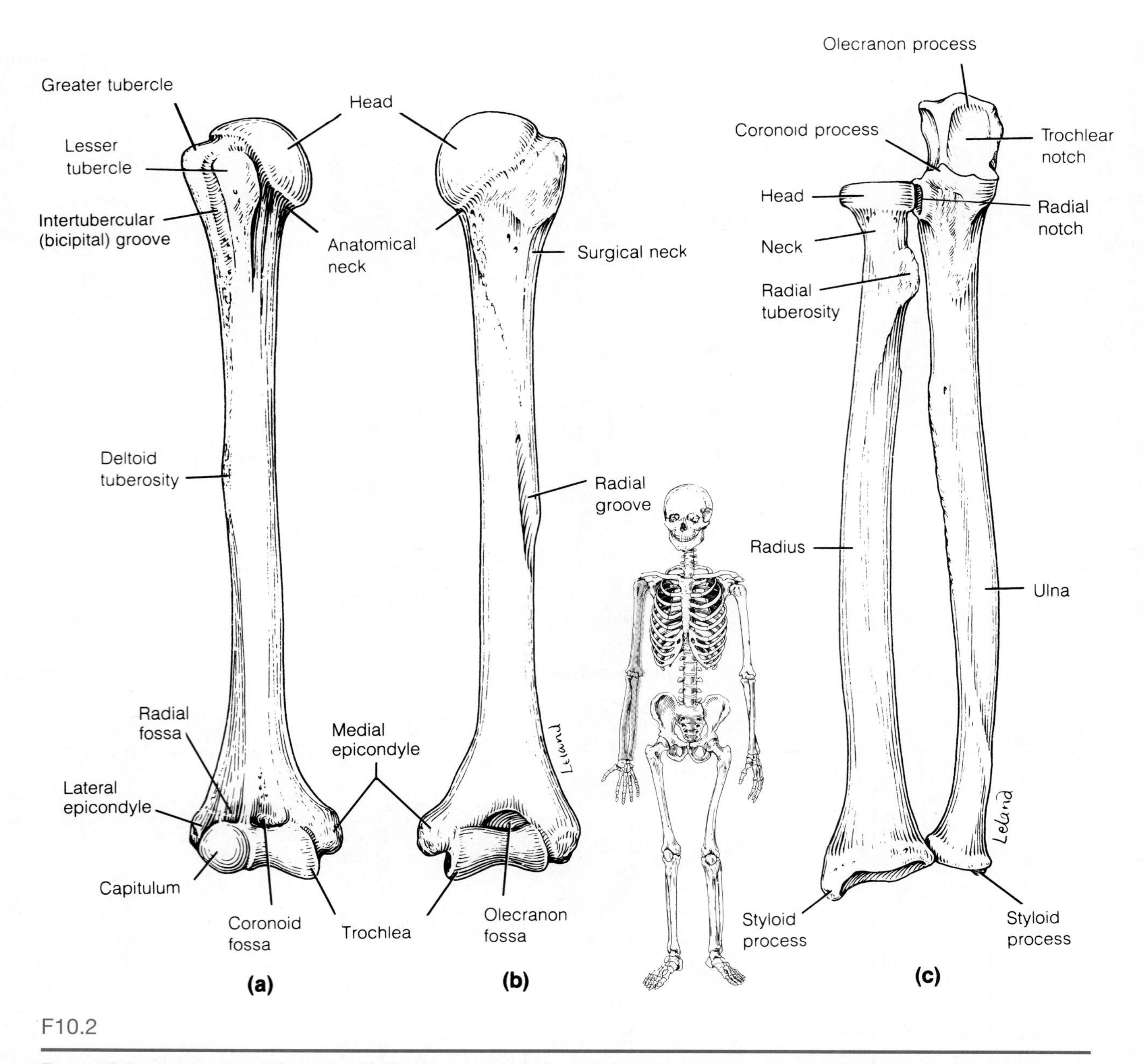

F10.2

Bones of the right arm and forearm. (a) Humerus, anterior view; (b) humerus, posterior view; (c) radius and ulna, anterior view.

The Arm

The arm (Figure 10.2) consists of a single bone—the **humerus,** a typical long bone. At its proximal end is the rounded head, which fits into the shallow glenoid cavity of the scapula. The head is separated from the shaft by the *anatomical neck* and the more constricted *surgical neck,* which is a common site of fracture. Opposite the head are two prominences, the **greater** and **lesser tubercles,** separated by a groove (the **intertubercular** or **bicipital groove**) that guides the tendon of the biceps muscle to its point of attachment (the superior rim of the glenoid cavity). In the midpoint of the shaft is a roughened area called the **deltoid tuberosity,** where the large fleshy shoulder muscle, the deltoid, attaches. Just inferior to the deltoid tuberosity is the **radial groove,** which indicates the pathway of the radial nerve.

At the distal end of the humerus are two condyles—the medial **trochlea** (looking rather like a spool), which articulates with the ulna, and the lateral **capitulum,** which articulates with the radius of the forearm. This condyle pair is flanked medially by the **medial epicondyle** and laterally by the **lateral epicondyle.**

The medial epicondyle is commonly referred to as the "funny bone." The large ulnar nerve runs in a groove beneath the medial epicondyle; and when this region is sharply bumped, we are quite likely to experience a temporary, but excruciatingly painful, tingling

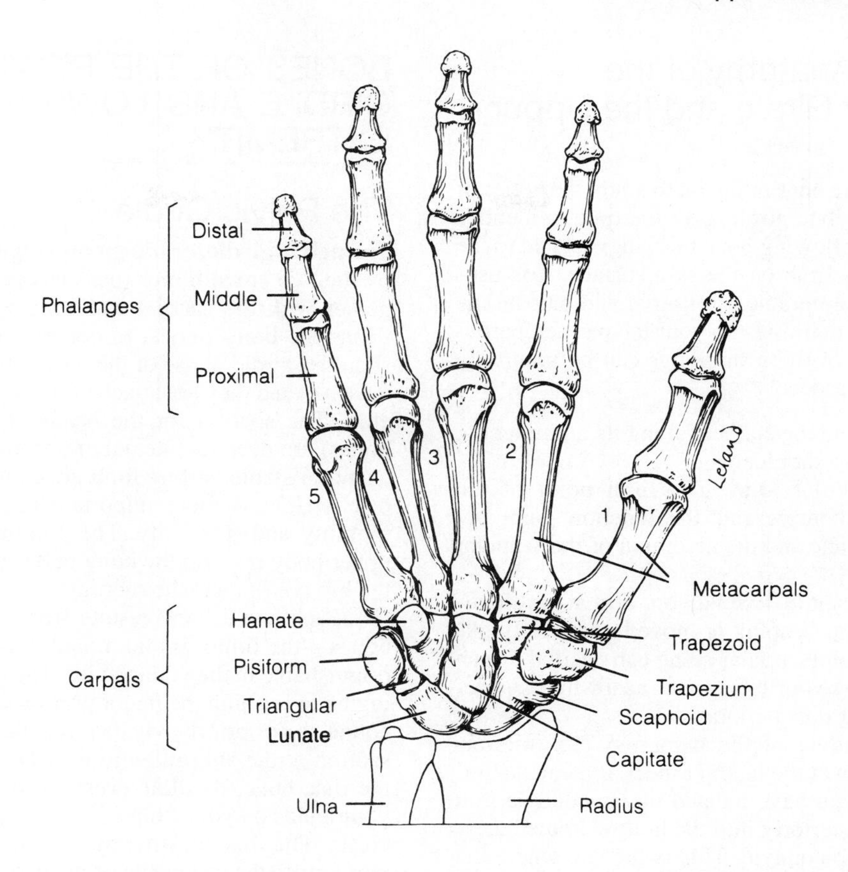

F10.3

Bones of the right wrist and hand, anterior view.

sensation. This event is called "hitting the funny bone," a strange expression, because it is certainly *not* "funny"!

Above the trochlea on the anterior surface is a depression, the **coronoid fossa;** on the posterior surface is the **olecranon fossa.** These two depressions allow the corresponding processes of the ulna to move freely when the elbow is flexed and extended. A small **radial fossa,** lateral to the coronoid fossa, receives the head of the radius when the elbow is flexed.

The Forearm

Two bones, the radius and the ulna, compose the skeleton of the forearm, or antebrachium. (See Figure 10.2.) When the body is in the anatomical position, the radius and ulna are uncrossed. Proximally, the disc-shaped head of the **radius** articulates with the capitulum of the humerus. Just below the head, on the medial aspect of the shaft, is a prominence called the **radial tuberosity,** the point of attachment for the tendon of the biceps muscle of the arm.

The **ulna** is the medial bone of the forearm. Its proximal end bears the anterior **coronoid process** and the posterior **olecranon process,** which are separated by the **trochlear notch.** Together these processes grip the trochlea of the humerus in a plierslike joint. The small **radial notch** on the lateral side of the coronoid process articulates with the head of the radius. The smaller distal end of the ulna bears a small medial **styloid process,** which serves as a point of attachment for the ligaments of the wrist.

The Wrist

The wrist is referred to anatomically as the **carpus,** and the eight bones comprising it are the **carpals.** The carpals are arranged in two irregular rows of four bones each, which are named and illustrated in Figure 10.3. The carpals are bound closely together by ligaments, which restrict movements between them.

The Hand

The hand, or manus (Figure 10.3), consists of two groups of bones: the **metacarpals** (bones of the palm) and the **phalanges** (bones of the fingers). The metacarpals are numbered 1 to 5 from the thumb side of the hand toward the little finger. When the fist is clenched, the heads of the metacarpals become prominent as the knuckles. Each hand contains 14 phalanges. There are three phalanges in each finger except the thumb, which has only proximal and distal phalanges.

Surface Anatomy of the Shoulder Girdle and the Upper Limb

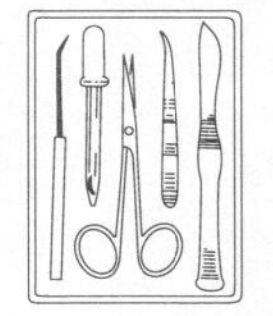

Before continuing on to study the bones of the pelvic girdle, take the time to identify the following bone markings related to the upper limb on the skin surface. It is usually preferable to observe and palpate the bone markings on your lab partner, particularly since many of these markings can be seen only from the posterior aspect.

- Clavicle: Palpate the clavicle along its entire length from sternum to shoulder.
- Acromioclavicular joint: The high point of the shoulder which represents the junction point between the clavicle and the acromion of the scapular spine.
- Spine of the scapula: Extend your arm at the shoulder so that your scapula is moved posteriorly. As you do this, your scapular spine can easily be seen and palpated by your lab partner as a winglike protrusion on your dorsal thorax.
- Lateral epicondyle of the humerus: The inferiormost projection at the lateral aspect of the distal humerus. After you have located the epicondyle, run your finger posteriorly into the hollow immediately dorsal to the epicondyle. This is the site where the extensor muscles of the hand are attached and is a common site of the often excruciating pain of tennis elbow, a condition in which those muscles and their tendons are "abused" physically.
- Medial epicondyle of the humerus: Feel out this medial projection at the distal end of the humerus.
- Olecranon process of the ulna: "Work" your elbow—flexing and extending—as you palpate its dorsal aspect to feel the olecranon process of the ulna moving into and out of the olecranon fossa on the dorsal aspect of the humerus.
- Styloid process of the ulna: With the hand in the anatomical position, feel out this small inferior projection on the medial aspect of the distal end of the ulna.
- Styloid process of the radius: Find this projection from the distal end of the radius (lateral aspect). It is most easily located by moving the hand medially at the wrist. Once you have palpated the styloid process, move your fingers just medially to the process (onto the anterior wrist). Press firmly and then let up slightly on the pressure. You should be able to feel your pulse at this pressure point, which lies over the radial artery (radial pulse).
- Metacarpophalangeal joints (knuckles): Clench your fist and find the first set of "flexed-joint protrusions" beyond the wrist—these are the knuckles.

BONES OF THE PELVIC GIRDLE AND LOWER EXTREMITY

The Pelvic Girdle

The **pelvic girdle,** or **hip girdle** (Figure 10.4), is formed by the two **coxal bones** (**ossa coxae** or *hip bones*). Together with the sacrum and coccyx, the pelvic girdle forms the **bony pelvis.** In contrast to the bones of the shoulder girdle, those of the pelvic girdle are heavy and massive, and they are attached securely to the axial skeleton. The sockets for the heads of the femurs (thigh bones) are deep and heavily reinforced by ligaments to ensure a stable, strong limb attachment. The ability to bear weight is more important here than exceptional mobility and flexibility. The combined weight of the upper body rests on the bony pelvis (specifically, where the hip bones meet the sacrum).

Each coxal bone results from the fusion of three bones—the ilium, ischium, and pubis—which are distinguishable in the young child. The **ilium** is a large flaring bone forming the major portion of the coxal bone. It connects posteriorly, via its **auricular surface,** with the sacrum at the **sacroiliac joint.** The superior margin of the iliac bone, the **iliac crest,** is rough; when you rest your hands on your "hips," you are palpating your iliac crests. The iliac crest terminates in the **anterior superior spine** (the origin site of many thigh flexor muscles) and the **posterior superior spine,** to which many muscles extending the thigh attach. Two inferior spines are located below these. The shallow **iliac fossa** marks its internal surface, and a shallow ridge, the **arcuate line,** outlines the pelvic inlet or pelvic brim.

The **ischium** is the "sit-down" bone, forming the most inferior and posterior portion of the coxal bone. The most outstanding marking on the ischium is the **ischial tuberosity,** which receives the weight of the body when sitting. The **ischial spine,** superior to the ischial tuberosity, is an important anatomical landmark of the pelvic cavity (see Comparison of the Male and Female Pelves, p. 86). Two other anatomical features are the **lesser** and **greater sciatic notches,** which allow passage of nerves and blood vessels to and from the leg.

The **pubis** is the most anterior portion of the coxal bone. The fusion of the **rami** of the pubic bone anteriorly and the ischium posteriorly forms a bar of bone enclosing the **obturator foramen,** through which blood vessels and nerves run from the pelvic cavity into the leg. The pubic bones of each hip bone meet anteriorly at the thickened **pubic crest** to form a cartilaginous joint called the **pubic symphysis.**

The ilium, ischium, and pubis fuse at the deep hemispherical socket called the **acetabulum** (literally, "vinegar cup"), which receives the head of the thigh bone.

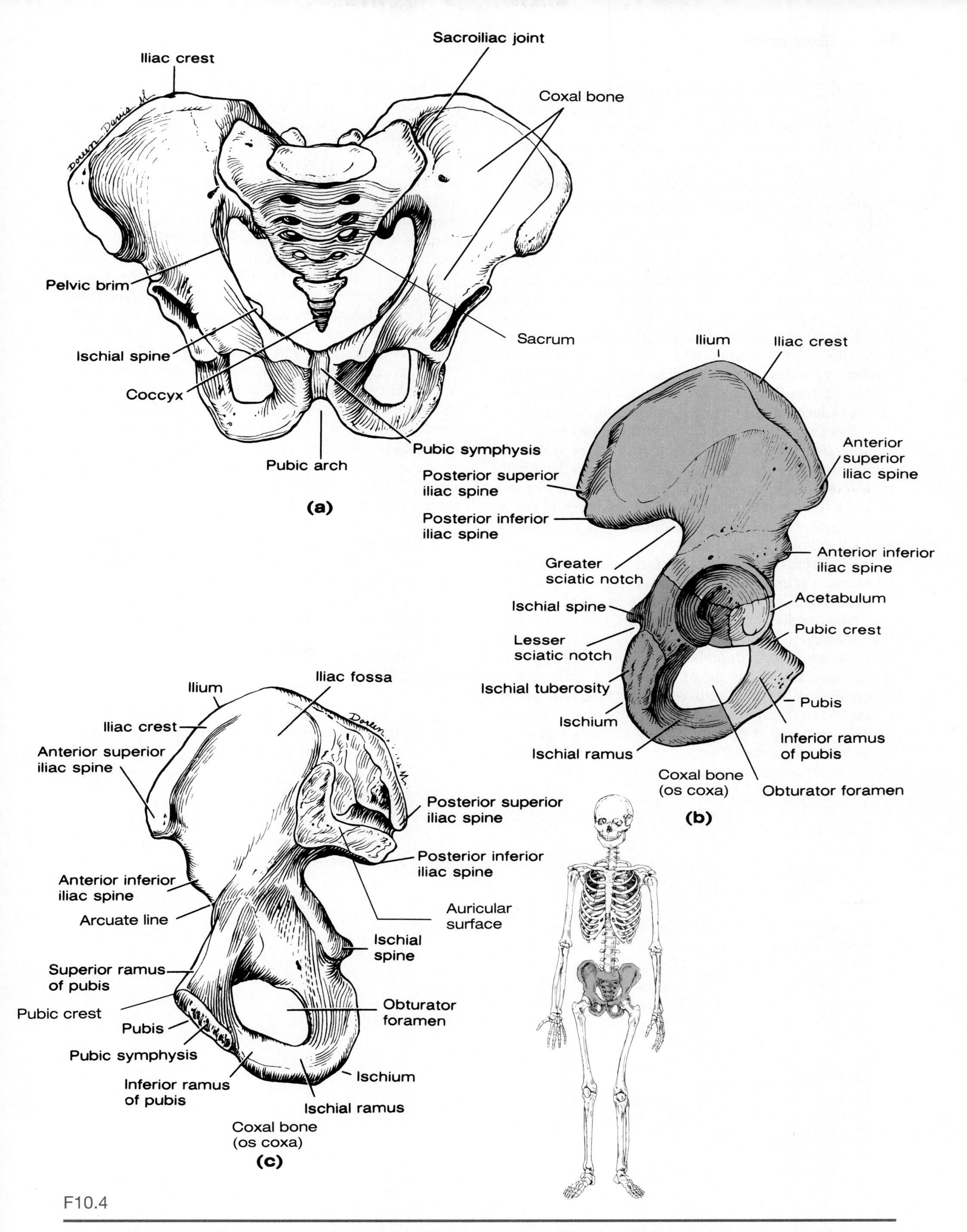

F10.4

Bones of the pelvic girdle. (a) Articulated bony pelvis, showing the two coxal bones, which together comprise the pelvic girdle, and the sacrum; (b) right coxal bone (lateral aspect) showing the point of fusion of the ilium, ischium, and pubic bones; (c) right coxal bone, medial view.

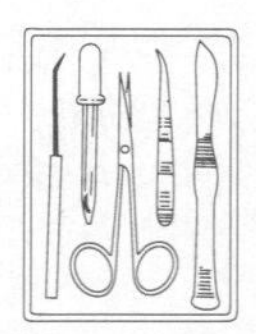

Before continuing with the bones of the lower limbs, take the time to examine an articulated pelvis. Note how each coxal bone articulates with the sacrum posteriorly and how the two coxal bones join at the pubic symphysis. The sacroiliac joint, because of the pressure it must bear, is often a site of lower back problems.

COMPARISON OF THE MALE AND FEMALE PELVES Although bones of males are usually larger and heavier and have more prominent bone markings, the bones of the male and female skeletons are very similar. The outstanding exception to this generalization is pelvic structure (see Table 10.1). So striking are the differences between male and female pelves that a trained anatomist can immediately determine the sex of the skeleton during a casual examination of the pelvis.

The female pelvis reflects modifications for childbearing, and generally speaking the female pelvis is wider, more shallow, lighter, and rounder than that of the male. Not only must her pelvis support the increasing size of a fetus, but it must also be large enough to allow the infant's head (its largest dimension) to descend through the birth canal at birth.

To describe pelvic sex differences, a few more terms must be introduced. Anatomically, the pelvis can be described in terms of a false pelvis and a true pelvis. The **false pelvis** is that portion superior to the arcuate line; it is bounded by the alae of the ilia laterally and the sacral promontory and lumbar vertebrae posteriorly. Although the false pelvis supports the abdominal viscera, it does not restrict childbirth in any way. The **true pelvis** is the region inferior to the arcuate line that is almost entirely surrounded by bone. Its posterior boundary is formed by the sacrum. The ilia, ischia, and pubic bones define its limits laterally and anteriorly.

The dimensions of the true pelvis, particularly its inlet and outlet, are critical if delivery of a baby is to be nonproblematic; and they are carefully measured by the obstetrician. The **pelvic inlet,** or **pelvic brim,** is the opening delineated by the sacral promontory posteriorly and the arcuate lines of the ilia anterolaterally. It is the superiormost margin of the true pelvis. Its widest dimension is from left to right, that is, along the frontal plane. Generally the infant's head enters the inlet with the forehead facing toward one ilium and the occipital region facing the other. The **pelvic outlet** is the inferior margin of the true pelvis. It is bounded anteriorly by the pelvic arch, laterally by the ischia, and posteriorly by the sacrum and coccyx. Since both the coccyx and the ischial spines protrude into the outlet opening, a sharply angled coccyx or large sharp ischial spines can dramatically narrow the outlet. The largest dimension of the outlet is the anterior-posterior diameter. Generally, as the baby's head passes through the inlet and enters the pelvic cavity, it rotates so that the forehead faces posteriorly and the occiput faces anteriorly. Thus, the normal descent of the head during birth mirrors, or follows, the widest dimensions of the pelvic openings.

The diameters of these regions may be determined by palpation through the mother's vaginal wall or, more accurately, by measuring dimensions on pelvic X-rays. When any of the pelvic diameters are too narrow to allow normal descent, a Caesarian delivery is indicated. In most such cases, the outlet dimensions (particularly the distance between the ischial spines) are the limiting factor. ■

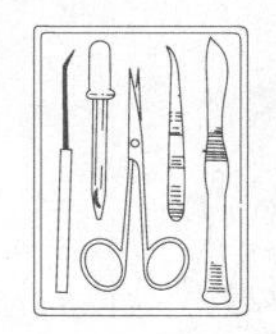

Examine male and female pelves for the differences described and illustrated in Table 10.1.

(Text continues on p. 88.)

TABLE 10.1 Comparison of the Male and Female Pelves

Characteristic	Female	Male
General structure and functional modifications	Tilted forward; adapted for childbearing; true pelvis defines the birth canal; cavity of the true pelvis is broad, shallow, and has a greater capacity	Tilted less far forward; adapted for the male's heavier build and stronger muscles; cavity of the true pelvis is narrow and deep.
Bone thickness	Less; bones lighter, thinner, and smoother	Greater; bones heavier and thicker, and markings are more prominent
Acetabula	Smaller; farther apart	Larger; closer
Pubic angle/arch	Broader (over 90°); more rounded	Angle is more acute (less than 90°)
Anterior view	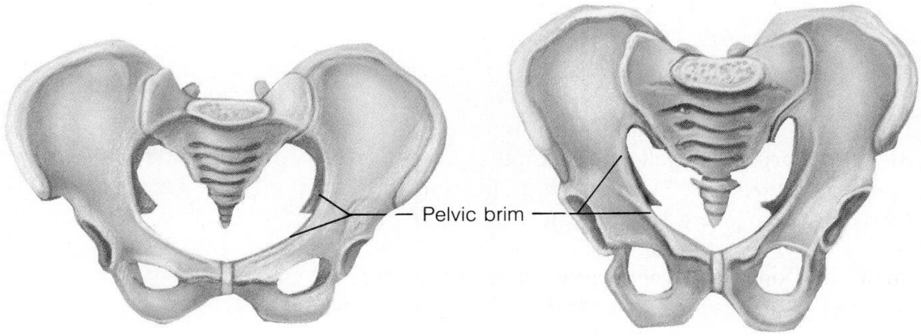	
Sacrum	Wider; shorter; sacral curvature is accentuated	Narrow; longer; sacral promontory more ventral
Coccyx	More movable; straighter	Less movable; curves ventrally
Left lateral view	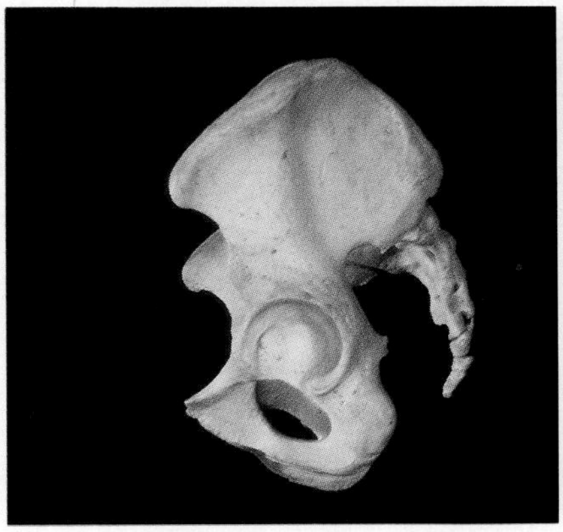	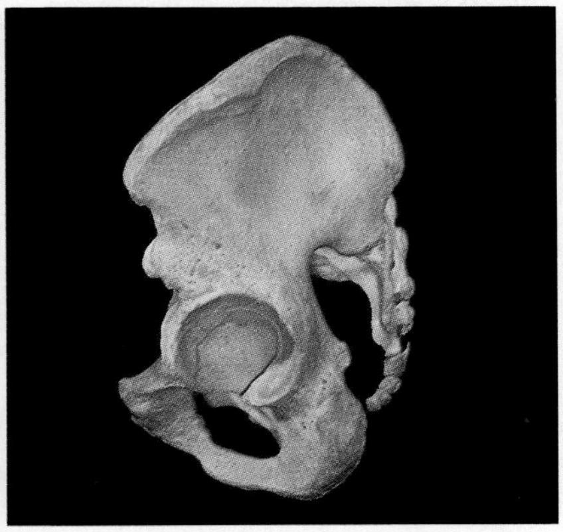
Pelvic inlet (brim)	Wider; oval from side to side	Narrow; basically heart shaped
Pelvic outlet	Wider; ischial tuberosities shorter, farther apart, and everted	Narrower; ischial tuberosities longer, sharper, and point more medially
Posteroinferior view	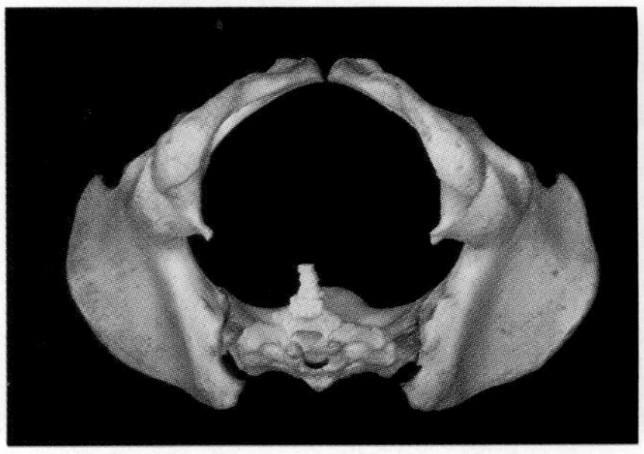	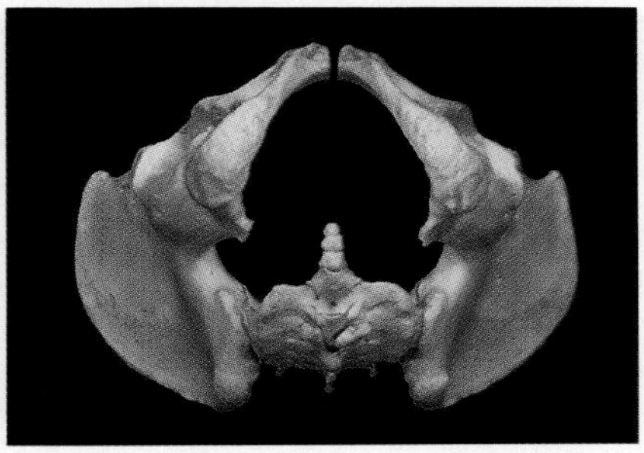

Pelvic brim

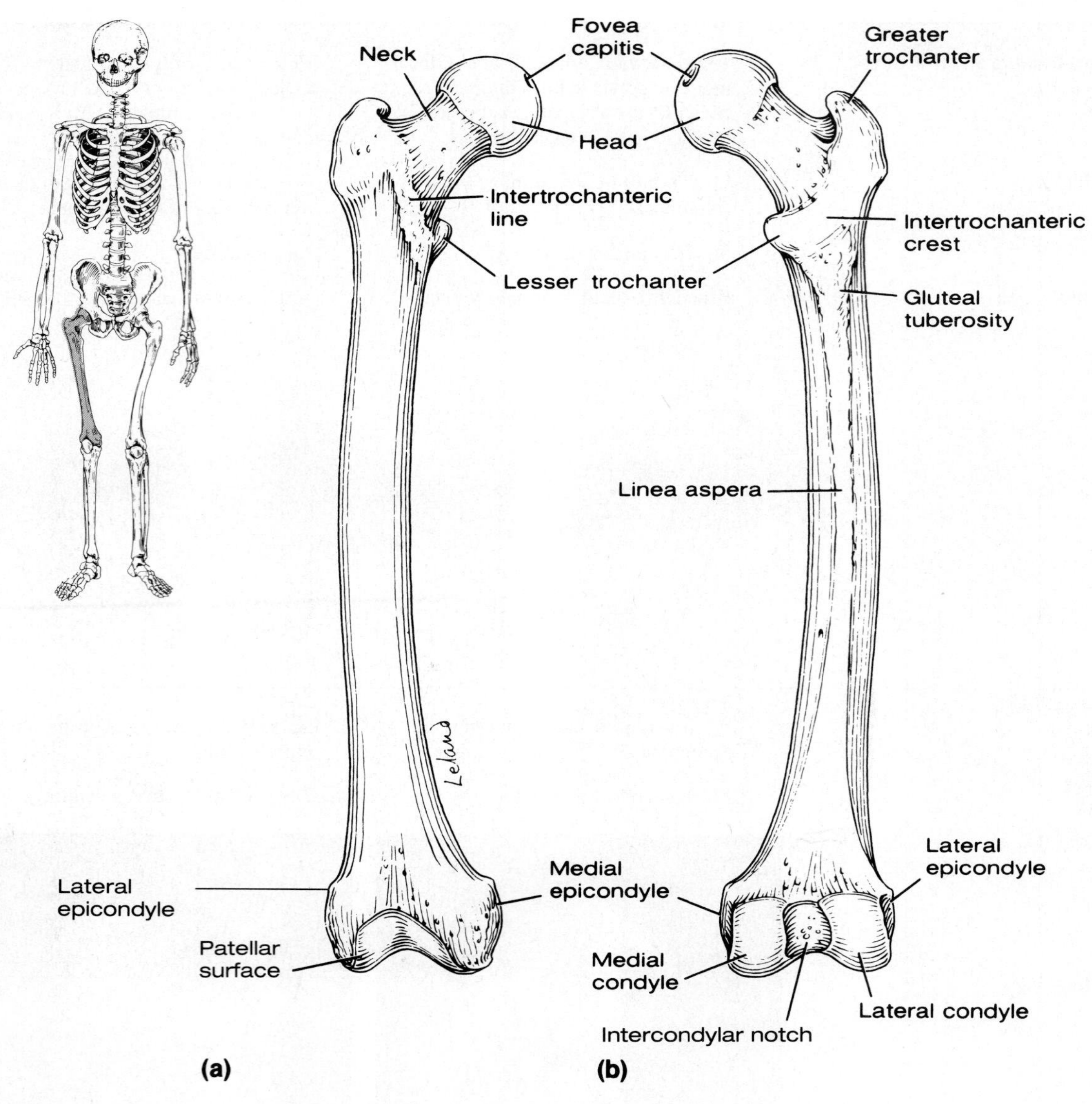

F10.5

Bone of the right thigh: femur. (a) Anterior view; (b) posterior view. Diagrammatic views are shown on the left; photos are on the right.

The Thigh

The **femur,** or thigh bone (Figure 10.5), is the sole bone of the thigh and is the heaviest, strongest bone in the body. Its proximal end bears a ball-like head, a neck, and **greater** and **lesser trochanters** (separated posteriorly by the **intertrochanteric crest** and anteriorly by the **intertrochanteric line**). The head of the femur articulates with the acetabulum of the hip bone.

The femur inclines medially as it runs downward to the lower leg bones; this brings the knees in line with the body's center of gravity or maximum weight.

Distally, the femur terminates in the **lateral** and **medial condyles,** which articulate with the tibia below. The **lateral** and **medial epicondyles,** just superior to these condyles, are separated by the **intercondylar notch.**

The trochanters and trochanteric crest, as well as the **gluteal tuberosity** and the **linea aspera** located on the shaft, serve as sites for muscle attachment.

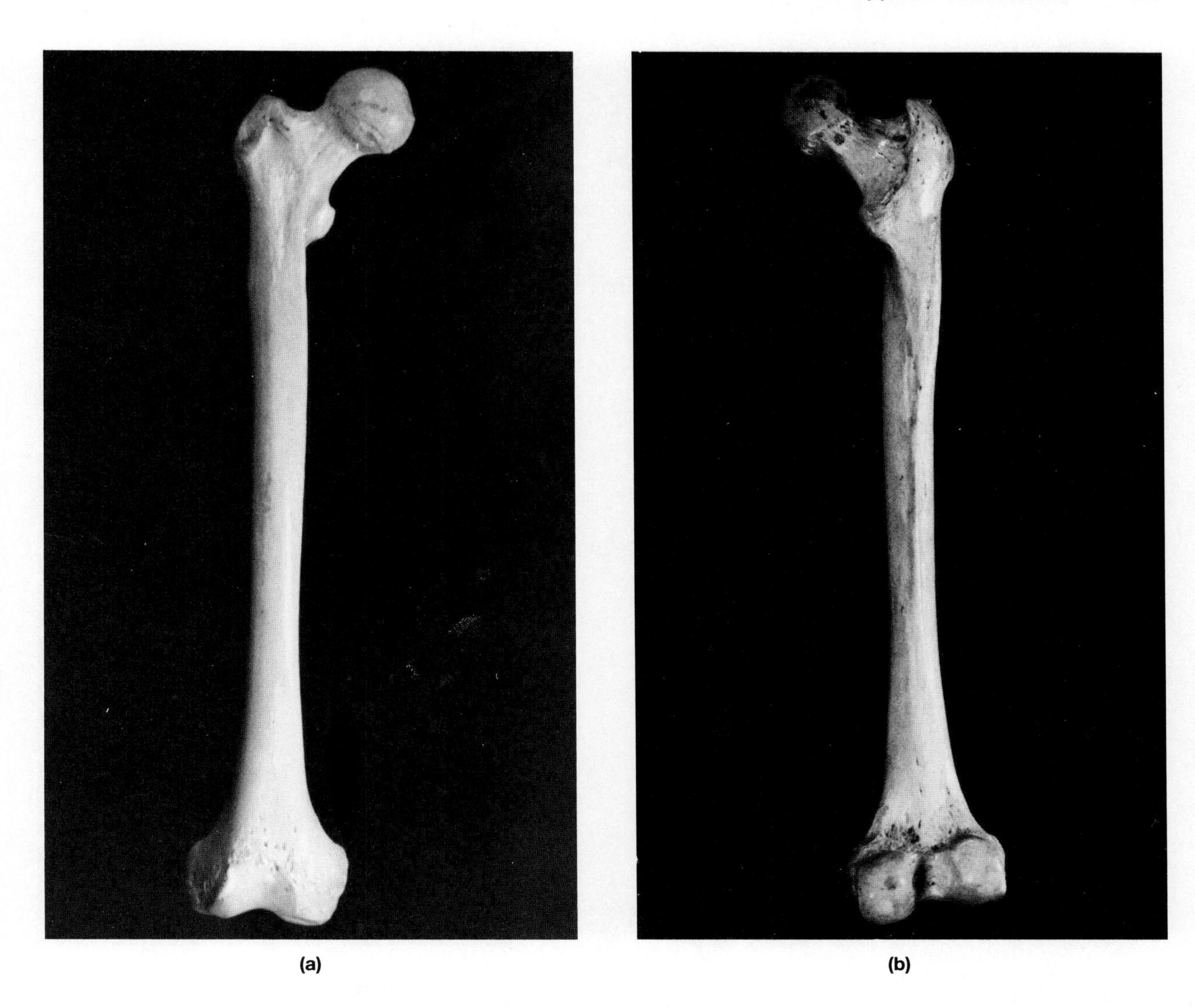

F10.5 (continued)

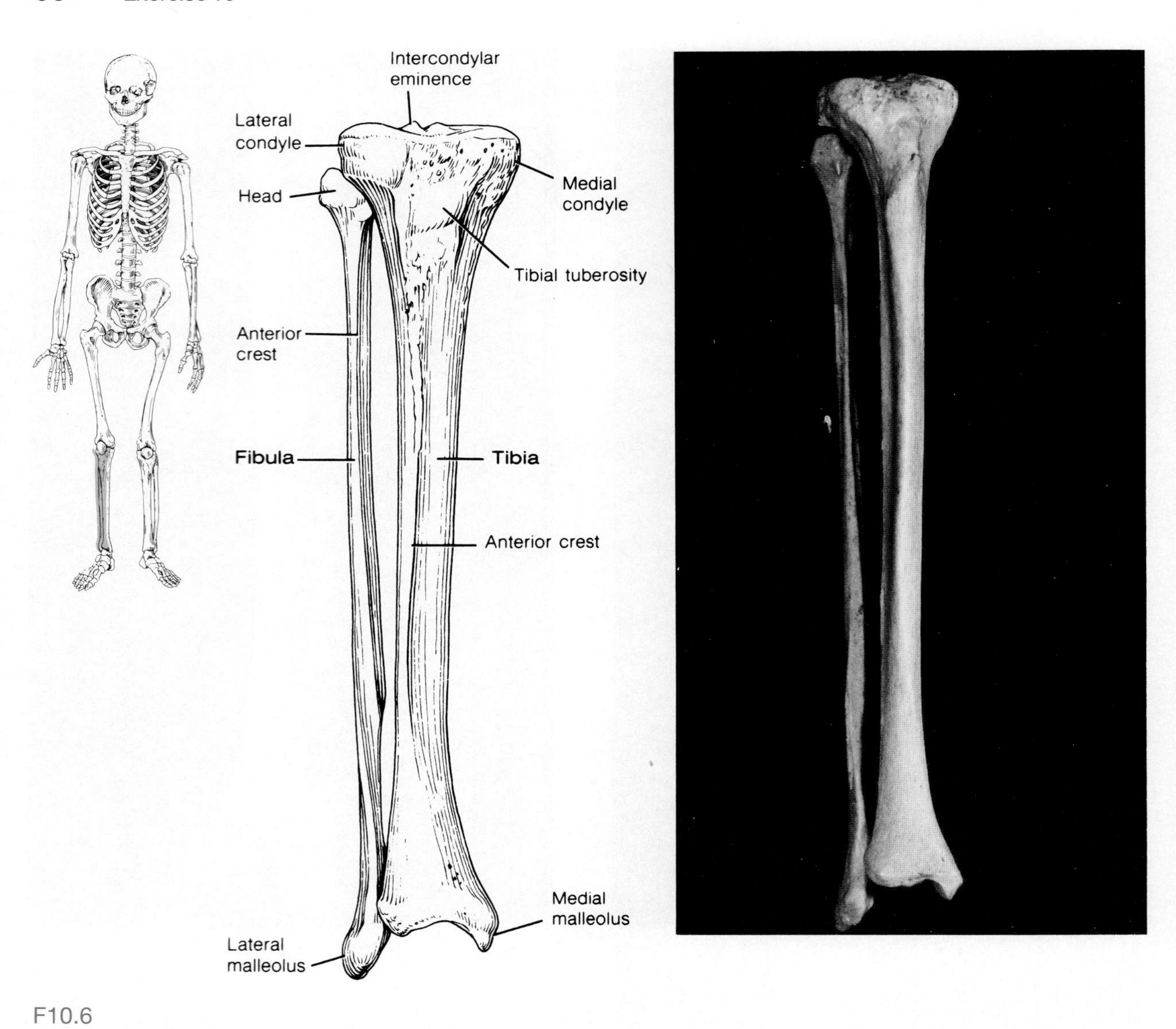

F10.6

Bones of the right leg: tibia and fibula, anterior view. Diagram on the left; photo on the right.

The Leg

Two bones, the tibia and the fibula, form the leg (Figure 10.6). The **tibia,** or *shin bone,* is the larger and more medial of the two leg bones. At the proximal end, the **medial** and **lateral condyles** (separated by the **intercondylar eminence**) receive the distal end of the femur to form the knee joint. The **tibial tuberosity,** a roughened protrusion on the anterior tibial surface (just below the condyles), serves as the site for attachment of the patellar (kneecap) ligament. Small facets on its superior and inferior lateral surface articulate with the fibula. Distally, a process called the **medial malleolus** forms the medial bulge of the ankle, and the smaller distal end articulates with the talus bone of the foot. The anterior surface of the tibia is a sharpened ridge (*anterior crest*) that is relatively unprotected by muscles. It is easily felt beneath the skin.

The **fibula,** which lies parallel to the tibia, takes no part in forming the knee joint. Its proximal head articulates with the lateral condyle of the tibia. The fibula is thin and sticklike, with a sharp anterior crest. It terminates distally in the **lateral malleolus,** which forms the lateral bulge of the ankle.

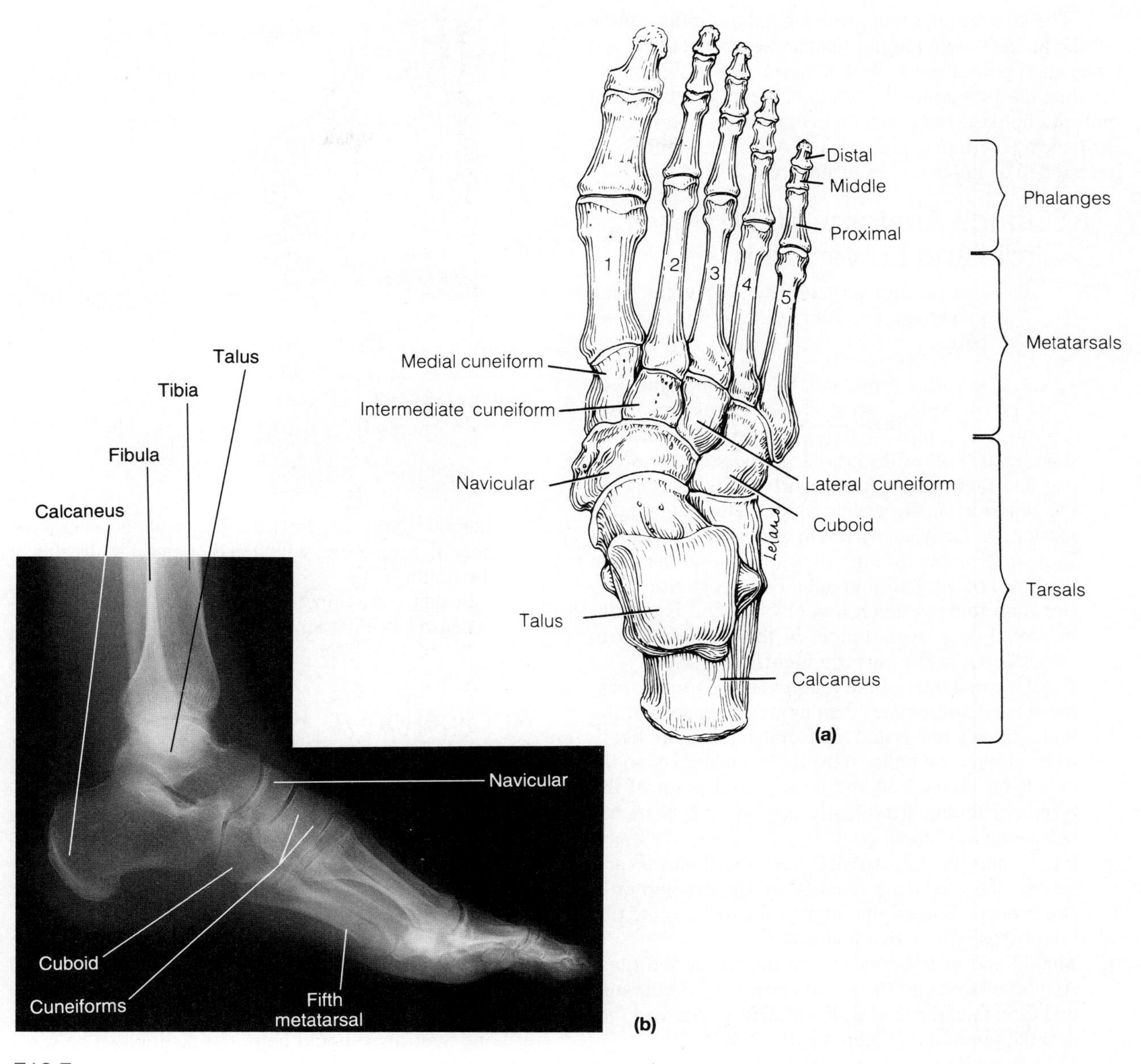

F10.7

Bones of the right ankle and foot. (a) Superior view; (b) X-ray of lateral aspect.

The Foot

The bones of the foot include the 7 **tarsal** bones, 5 **metatarsals,** which form the instep, and 14 **phalanges,** which form the toes. (See Figure 10.7.) Body weight is concentrated on the two largest tarsals which form the posterior aspect of the foot, the *calcaneus* (heel bone) and the *talus,* which lies between the tibia and the calcaneus. (The other tarsals are named and identified in Figure 10.7.) Like the fingers of the hand, each toe has three phalanges, except the great toe, which has two.

The bones in the foot are arranged to produce three strong arches—two longitudinal arches (medial and lateral) and one transverse arch (Figure 10.8). Ligaments, binding the foot bones together, and tendons of the foot muscles hold the bones firmly in the arched position, but still allow a certain degree of give. Weakened arches are referred to as fallen arches or flat feet.

Surface Anatomy of the Pelvic Girdle and Lower Limb

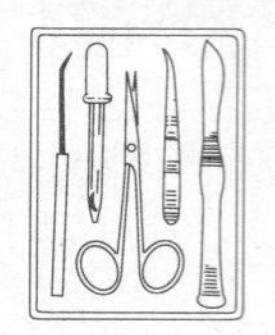

Locate and palpate the following bone markings on yourself and/or your lab partner.

- Iliac crest and anterior superior iliac spine: Rest your hands on your "hips"—they will be overlying the iliac crests. Follow the crest as far posteriorly as you can and then follow it anteriorly to its terminus at the anterior superior iliac spine. This latter bone marking is fairly easily felt in almost everyone, and is clearly visible through the skin (and perhaps the clothing) of very slim people. (The posterior superior iliac spine is much less obvious and is usually indicated only by a dimple in the overlying skin. Check it out in the mirror tonight.)
- Greater trochanter of the femur: This is usually easier to locate in females than in males because of the wider female pelvis and the fact that it is more likely to be clothed by bulky muscles in males. Try to locate it on yourself as the most lateral point of the proximal femur. It typically lies about 6–8 inches below the iliac crest.
- Patella and tibial tuberosity: Feel your kneecap and palpate the ligaments attached to superior and inferior aspects. Follow the inferior ligament to the tibial tuberosity to which it attaches.
- Medial and lateral condyles of the femur and tibia: As you move from the patella inferiorly on the medial (and then the lateral) knee surface, you will feel first the femoral and then the tibial condyle.
- Medial malleolus: Feel the medial protrusion of your ankle, which is the medial malleolus of the distal tibia.

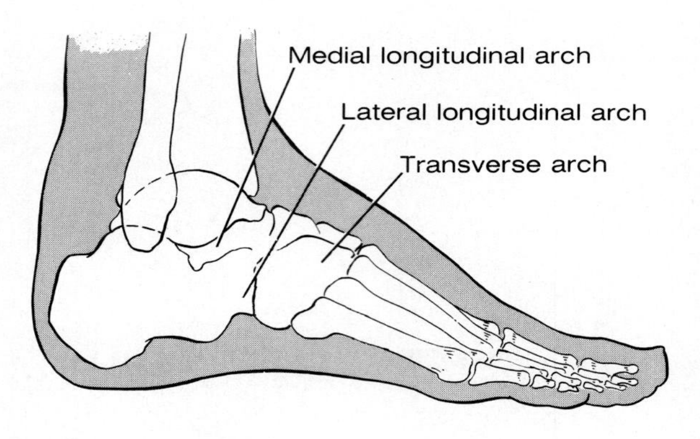

F10.8

Arches of the foot.

- Lateral malleolus: Feel the bulge of the lateral aspect of your ankle, which is the lateral malleolus of the fibula.
- Calcaneus: Attempt to follow the extent of your calcaneus or heel bone.

Application of Knowledge: Constructing a Skeleton

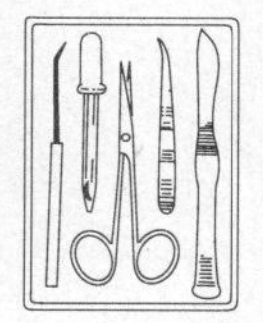

1. When you have finished examining the disarticulated bones of the appendicular skeleton, note their relationships in the articulated skeleton.

2. Work together with your lab partner to arrange the disarticulated bones on the laboratory bench in their proper relative positions to form an entire skeleton. Careful observation of bone markings should help you distinguish between right and left members of bone pairs. Ask the instructor to check your arrangement to ensure that it is correct before you return the bones to their storage areas.

Articulations and Body Movements

OBJECTIVES	MATERIALS
1. To name the three structural categories of joints, and to compare their mobility. 2. To identify the types of movement seen in synovial joints. 3. To define the *origin* and *insertion* of muscles. 4. To demonstrate or identify the various body movements.	Articulated skeleton, skull Diarthrotic beef joint (fresh or preserved), preferably a knee joint Disposable gloves Anatomical chart of joint types (if available) X-rays of normal and arthritic joints (if available)

With rare exceptions, every bone in the body is connected to, or forms a joint with, at least one other bone. Joints, or articulations, perform two functions for the body. They (1) hold the bones together and (2) allow the rigid skeletal system some flexibility so that gross body movements can occur.

TYPES OF JOINTS

Joints may be classified structurally or functionally. The *structural classification* is based on whether there is connective tissue fibers, cartilage, or a joint cavity between the articulating bones. Structurally, there are *fibrous, cartilaginous,* and *synovial joints.*

The *functional classification* focuses on the amount of movement allowed at the joint. On this basis, there are **synarthroses,** or immovable joints; **amphiarthroses,** or slightly movable joints; and **diarthroses,** or freely movable joints. Freely movable joints predominate in the limbs, whereas immovable and slightly movable joints are largely restricted to the axial skeleton, where firm bony attachments and protection of enclosed organs are a priority.

In general, fibrous joints are immovable, and all synovial joints are freely movable. Cartilaginous joints offer both rigid and slightly movable examples. Since the structural categories are more clear-cut, we will use the structural classification here and indicate functional properties where appropriate.

Fibrous Joints

In **fibrous joints,** the bones are joined by fibrous tissue. No joint cavity is present. The amount of movement allowed depends on the length of the fibers uniting the bones. Although a few fibrous joints are slightly movable, most are synarthrotic and permit essentially no movement.

The two major types of fibrous joints are sutures and syndesmoses. In **sutures** (Figure 11.1d) the irregular edges of the bones interlock and are united by very short connective tissue fibers, as in most joints of the skull. In **syndesmoses** the articulating bones are connected by short ligaments of dense fibrous tissue; the bones do not interlock. The joint at the distal end of the tibia and fibula is an example of a syndesmosis (Figure 11.1e). Although this syndesmosis allows some give, it is classed functionally as a synarthrosis.

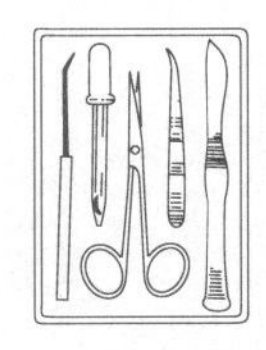

Examine a human skull again. Notice that adjacent bone surfaces do not actually touch but are separated by fibrous connective tissue. Also examine a skeleton and an anatomical chart of joint types for examples of fibrous joints.

Cartilaginous Joints

In **cartilaginous joints,** the articulating bone ends are connected by a plate or pad of cartilage. No joint cavity is present. The two major types of cartilaginous joints are synchondroses and symphyses. Although there is variation, most cartilaginous joints are *slightly movable* (amphiarthroses) functionally. In **symphyses** (*symphysis* means "a growth together") the bones are connected by a broad, flat disc of fibrocartilage. The intervertebral joints and the pubic symphysis of the pelvis are symphyses (see Figure 11.1b and c). In **synchondroses** the bony portions are united by hyaline cartilage. The best examples of synchondroses are the epiphyseal plates seen in the long bones of growing children, and the articulation of the costal cartilage of the first rib with the sternum. The epiphyseal plates are flexible during childhood but they are eventually totally ossified.

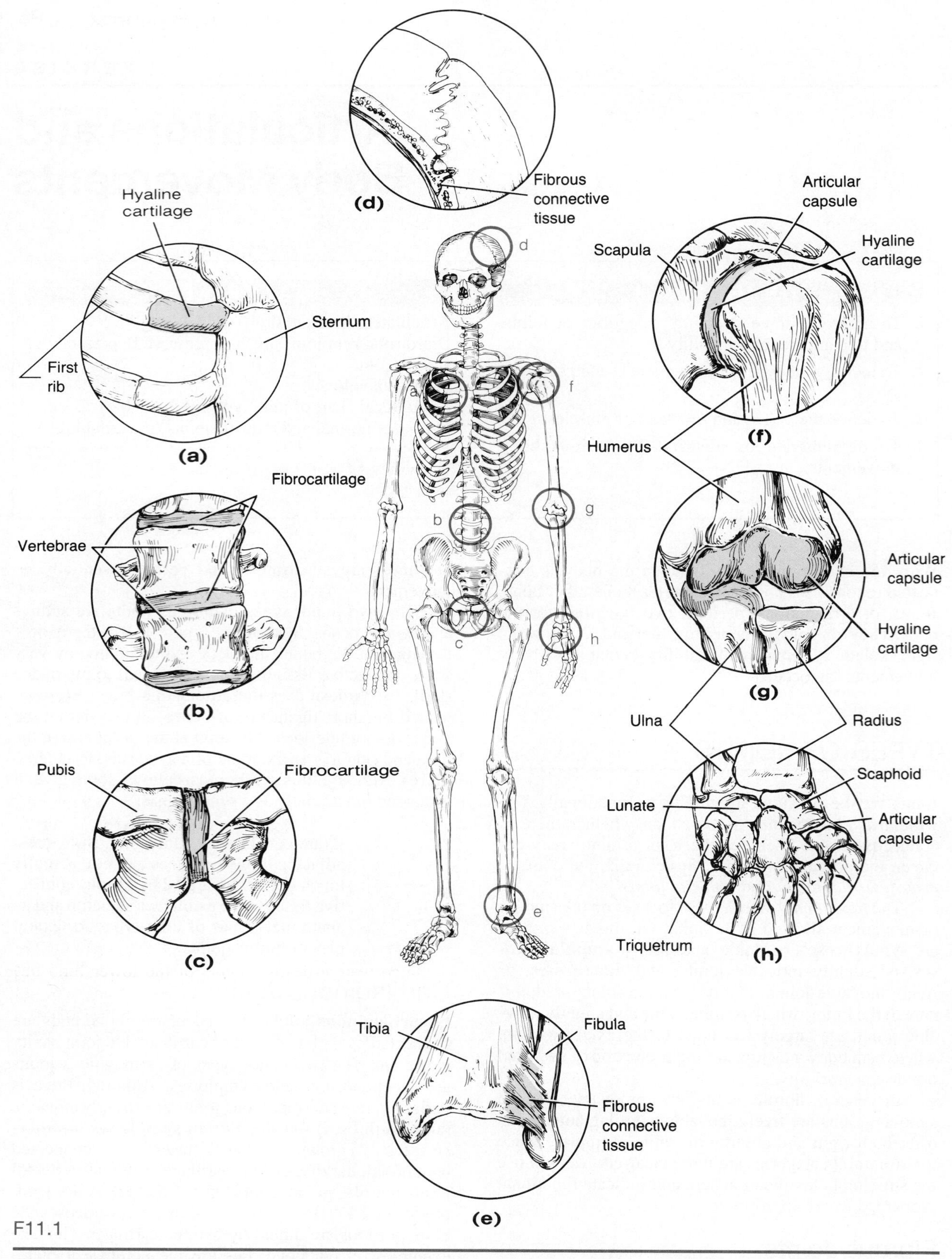

F11.1

Types of joints. Joints to the left of the skeleton are cartilaginous joints; joints above and below the skeleton are fibrous joints; joints to the right of the skeleton are synovial joints. (a) Synchondrosis (joint between costal cartilage of rib 1 and the sternum); (b) symphysis (intervertebral discs of fibrocartilage connecting the vertebrae); (c) symphysis (cartilaginous pubic symphysis connecting the pubic bones anteriorly); (d) suture (fibrous connective tissue connects the interlocking skull bones); (e) syndesmosis (fibrous connective tissue connecting the distal ends of the tibia and fibula); (f) synovial joint (multiaxial shoulder joint); (g) synovial joint (uniaxial elbow joint); (h) synovial joints (biaxial intercarpal joints of the hand).

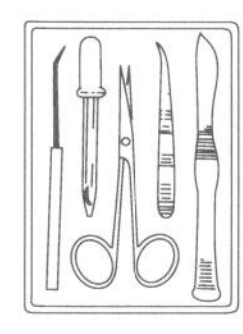

Identify the cartilaginous joints on a human skeleton and on an anatomical chart of joint types.

Synovial Joints

Synovial joints are those in which the articulating bone ends are separated by a joint cavity containing synovial fluid (see Figure 11.1f–h). This arrangement allows a great deal of mobility and all synovial joints are diarthroses, or freely movable joints. This flexibility varies, however; some synovial joints can move in only one plane, and others can move in several directions (multiaxial movement). Most joints in the body are synovial joints.

All synovial joints are characterized by the following structural characteristics (Figure 11.2):

- The joint surfaces are enclosed by an *articular capsule* (a sleeve of fibrous connective tissue).
- The interior of this capsule is lined with a smooth connective tissue membrane, called *synovial membrane,* which produces a lubricating fluid (synovial fluid) that reduces friction.
- Articulating surfaces of the bones forming the joint are covered with hyaline *(articular)* cartilage.
- The articular capsule is typically reinforced with ligaments, and may or may not contain bursae (fluid-filled sacs that reduce friction where tendons cross bone).
- Fibrocartilage pads may or may not be present within the capsule.

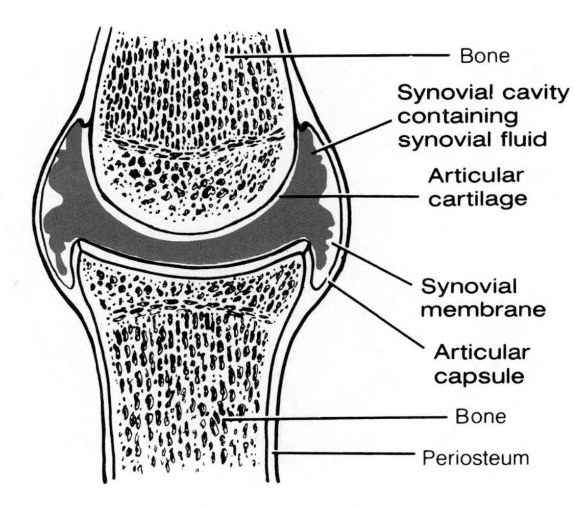

F11.2

Major structural features of a synovial joint.

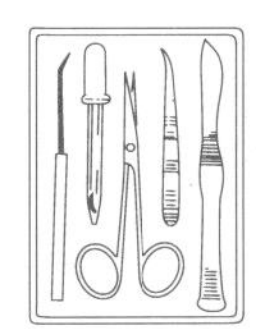

Examine a beef joint to identify the structural features of synovial joints.

⚠ If the joint is freshly obtained from the slaughterhouse and you will be handling it, don plastic gloves before beginning your observations.

Because there are so many types of synovial joints, they have been divided into the following subcategories on the basis of movements allowed:

- **Gliding:** articulating surfaces are flat or slightly curved, allowing slipping movements; examples are the intercarpal and intertarsal joints, and the vertebrocostal joints.
- **Hinge:** the rounded process of one bone fits into the concave surface of another to allow movement in one plane (uniaxial), usually flexion and extension; examples are the elbow and interphalangeal joints.
- **Pivot:** the rounded or conical surface of one bone articulates with a shallow depression or foramen in another bone to allow uniaxial rotation, as in the joint between the atlas and axis (C_1 and C_2).
- **Condyloid:** the oval condyle of one bone fits into an ellipsoidal depression in another bone, allowing biaxial (two-way) movement; the wrist joint and the metacarpal-phalangeal joints (knuckles) are examples.
- **Saddle:** articulating surfaces are saddle shaped; the articulating surface of one bone is convex, and the reciprocal surface is concave; saddle joints, which are biaxial, include the joint between the thumb metacarpal and the trapezium of the wrist.
- **Ball and socket:** the ball-shaped head of one bone fits into a cuplike depression of another; these are multiaxial joints, allowing movement in all directions and pivotal rotation; examples are the shoulder and hip joints.

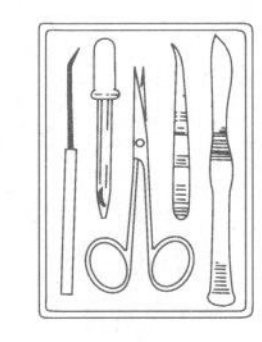

1. Examine the articulated skeleton, anatomic charts, and yourself to identify the subcategories of synovial joints. Make sure you understand the terms *uniaxial, biaxial,* and *multiaxial.*
2. Compare and contrast the structure of the hip and knee joints (Figure 11.3). Both of these joints are large weight-bearing joints of the lower limb but they differ substantially in their security. Read through the questions in the review section that pertain to this exercise before beginning your comparison.

JOINT DISORDERS

Most of us don't think about our joints until something goes wrong with them. Joint pains and malfunctions result from a variety of causes. For example, a hard blow to the knee can cause a painful *bursitis,* known as "water on the knee," due to damage to, or inflammation of, the patellar bursa. Slippage of a fibrocartilage pad or tearing of a ligament may result in a painful condition that persists over a long period, because these poorly vascularized structures heal so slowly.

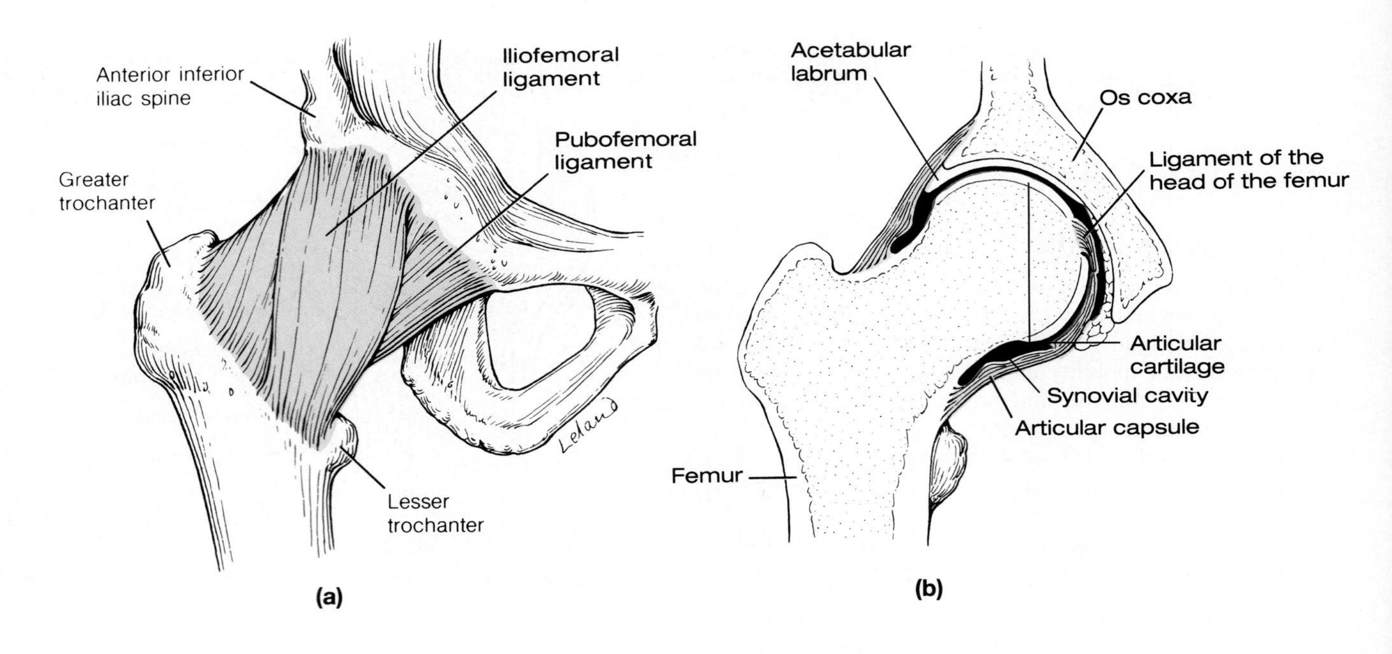

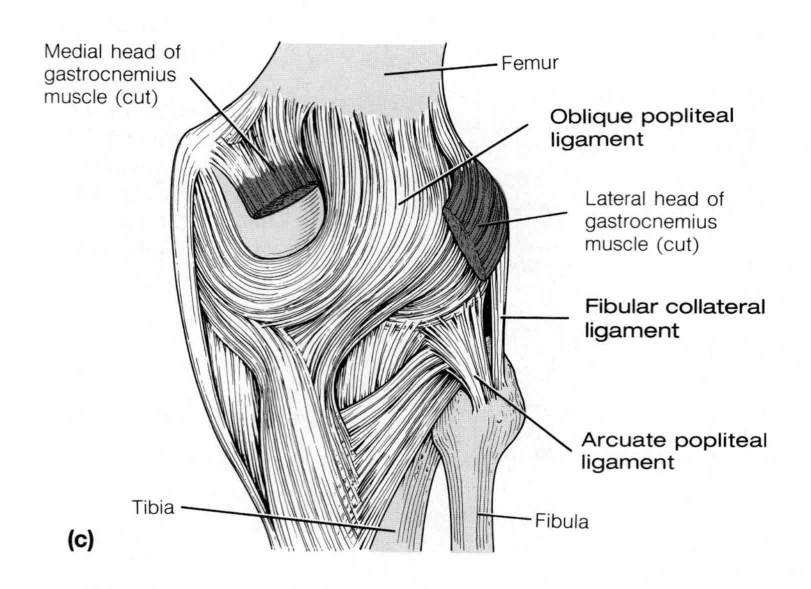

F11.3

Comparative anatomy of the hip and knee joints. (a) Ligaments of the right hip joint, anterior view. (b) Right hip joint, frontal section view. (c) Posterior ligaments of the right knee joint. (d) Midsagittal section of the right knee joint. (e) Anterior view of the flexed right knee. Patella and articular capsule removed to allow ligaments and menisci to be viewed.

Sprains and dislocations are other types of joint problems. In a *sprain,* the ligaments reinforcing a joint are damaged by excessive stretching or are torn away from the bony attachment. Since ligaments are cords of dense connective tissue with a poor blood supply, sprains heal slowly and are quite painful. *Dislocations* occur when bones are forced out of their normal position in the joint cavity. They are normally accompanied by torn or stressed ligaments and considerable inflammation. The process of returning the bone to its proper position, called *reduction,* should be done only by a physician. Attempts by the untrained person to "snap the bone back into its socket" are often more harmful than helpful.

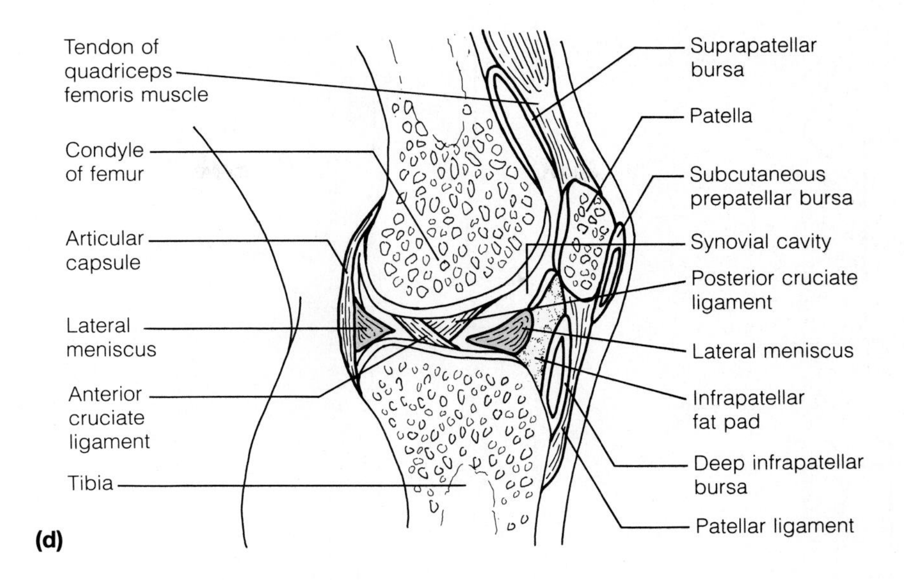

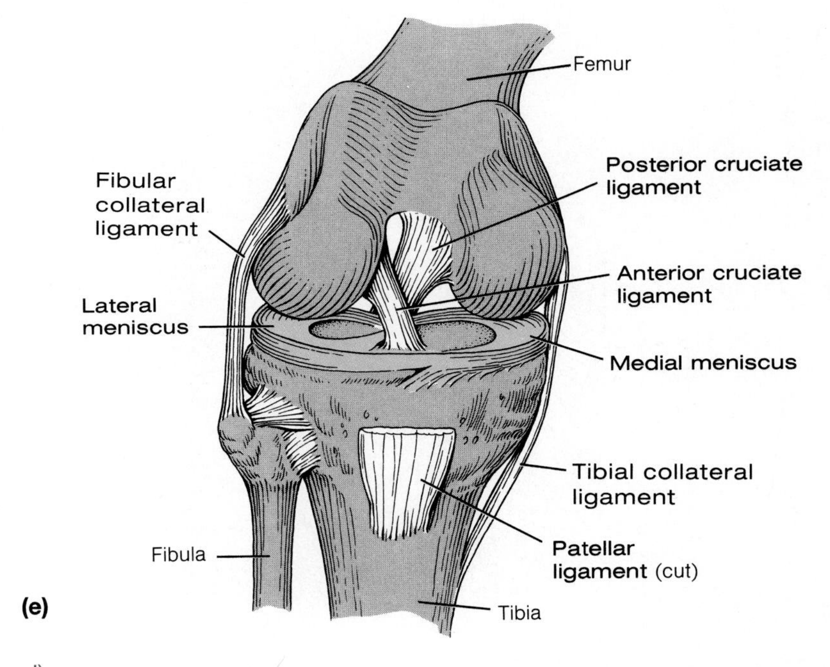

F11.3 (Continued)

Advancing years also take their toll on joints. Weight-bearing joints in particular eventually begin to degenerate. Adhesions (fibrous bands) may form between the surfaces where bones join, and extraneous bone tissue (spurs) may grow along the joint edges. Such degenerative changes lead to the complaint so often heard from the elderly: "My joints are getting so stiff. . . ." ■

- If possible, compare an X-ray of an arthritic joint to one of a normal joint.

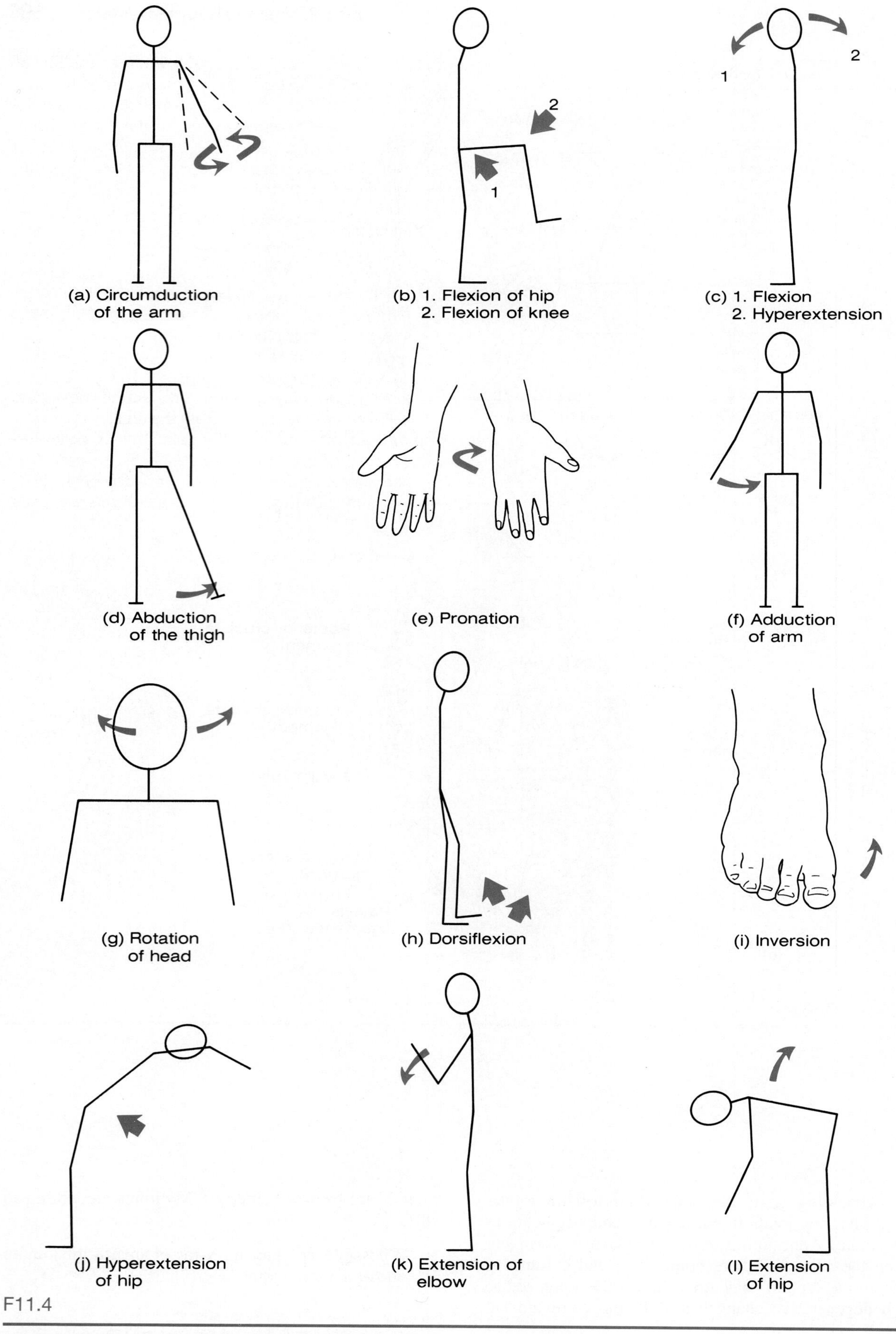

F11.4

Movements occurring at synovial joints of the body.

BODY MOVEMENTS

Every muscle of the body is attached to bone (or other connective tissue structures) at two points—the **origin** (the stationary, immovable, or less movable attachment) and the **insertion** (the movable attachment). Body movement occurs when muscles contract across diarthrotic synovial joints. When the muscle contracts and its fibers shorten, the insertion moves toward the origin. The type of movement depends on the construction of the joint (nonaxial, uniaxial, biaxial, or multiaxial) and on the placement of the muscle relative to the joint. The most common types of body movements are described below and illustrated in Figure 11.4.

Attempt to demonstrate each movement as you read through the following material:

Flexion: a movement, generally in the sagittal plane, that decreases the angle of the joint and lessens the distance between the two bones. Flexion is typical of hinge joints (bending the knee or elbow), but is also common at ball-and-socket joints (bending forward at the hip).

Extension: a movement that increases the angle of a joint, and the distance between two bones or parts of the body (straightening the knee or elbow). Extension is the opposite of flexion. If extension is greater than 180 degrees (bending the trunk backward), it is termed *hyperextension.*

Abduction: movement of a limb away from the midline or median plane of the body, generally on the frontal plane, or the fanning movement of fingers or toes when they are spread apart.

Adduction: movement of a limb toward the midline of the body. Adduction is the opposite of abduction.

Rotation: movement of a bone around its longitudinal axis without lateral or medial displacement. Rotation, a common movement of ball-and-socket joints, also describes the movement of the atlas around the odontoid process of the axis.

Circumduction: a combination of flexion, extension, abduction, and adduction commonly observed in ball-and-socket joints like the shoulder. The proximal end of the limb remains stationary, and the distal end moves in a circle (the limb as a whole outlining a cone and the proximal end of the limb serving as the apex).

Pronation: movement of the palm of the hand from an anterior or upward-facing position to a posterior or downward-facing position. This action moves the distal end of the radius across the ulna.

Supination: movement of the palm from a posterior-facing position to an anterior-facing position (the anatomical position). Supination is the opposite of pronation. During supination, the radius and ulna are parallel.

The last four terms refer to movements of the foot:

Inversion: a movement that results in the medial turning of the sole of the foot.

Eversion: a movement that results in the lateral turning of the sole of the foot; the opposite of inversion.

Dorsiflexion: a movement of the ankle joint in a dorsal direction (standing on one's heels).

Plantar flexion: a movement of the ankle joint in which the foot is flexed downward (standing on one's toes or pointing the toes).

12 EXERCISE

Microscopic Anatomy, Organization, and Classification of Skeletal Muscle

OBJECTIVES

1. To describe the structure of skeletal muscle from gross to microscopic levels.
2. To define and explain the role of the following:

actin	*perimysium*
myosin	*aponeurosis*
fiber	*tendon*
myofibril	*endomysium*
myofilament	*epimysium*

3. To describe the structure of a myoneural junction and to explain its role in muscle function.
4. To define: *prime mover (agonist), antagonist, synergist,* and *fixator.*
5. To cite criteria used in naming skeletal muscles.

MATERIALS

Three-dimensional model of skeletal muscle cells (if available)
Forceps
Dissecting needles
Microscope slides and coverslips
0.9% saline solution in dropper bottles
Chicken breast or thigh muscle (obtained fresh from the meat market)
Compound microscope
Histological slides of skeletal muscle (longitudinal and cross-sectional) and skeletal muscle showing myoneural junctions
Three-dimensional model of skeletal muscle showing myoneural junction (if available)

The bulk of the body's muscle is called **skeletal muscle** because it is attached to the skeleton (or underlying connective tissue). Skeletal muscle influences body contours and shape, allows you to smile and frown, provides a means of locomotion, and enables you to manipulate the environment. The balance of the body's muscle—smooth and cardiac muscle—is the major component of the walls of hollow organs and the heart, where it is involved with the transport of materials within the body.

Each of the three muscle types has a structure and function uniquely suited to its task in the body. However, because the term *muscular system* applies specifically to skeletal muscle, the primary objective of this unit is to investigate the structure and function of skeletal muscle.

Skeletal muscle is also known as *voluntary muscle* (because it can be consciously controlled) and as *striated muscle* (because it appears to be striped). As you might guess from both of these alternative names, skeletal muscle has some very special characteristics. Thus our investigation of skeletal muscle begins at the cellular level.

THE CELLS OF SKELETAL MUSCLE

Most skeletal muscle is composed of relatively large, long cylindrical cells ranging from 10 to 100 μm in diameter and typically 4 cm in length. However, the cells of large, hard-working muscles like the antigravity muscles of the hip are extremely coarse, ranging up to 30 cm (more than a foot!) in length, and can be seen with the naked eye.

Skeletal muscle cells (Figure 12.1a) are multinucleate: Multiple oval nuclei can be seen just beneath the plasma membrane (called the *sarcolemma* in these cells). The nuclei are pushed peripherally by the longitudinally arranged **myofibrils,** which nearly fill the sarcoplasm (Figure 12.1b). Alternating light (I) and dark (A) bands along the length of the perfectly aligned myofibrils give the muscle fiber as a whole its striped appearance.

Electron microscope studies have revealed that the myofibrils are made up of even smaller threadlike structures called **myofilaments** (Figure 12.1b and d). The myofilaments are composed largely of two varieties of contractile proteins—**actin** and **myosin**—which slide past each other during muscle activity to bring about shortening or contraction of the muscle cells. It is the highly specific arrangement of the myofilaments within the myofibrils that is responsible for the banding pattern in skeletal muscle. The actual contractile units of muscle, called **sarcomeres,** extend from the middle of one I band (its Z line) to the middle of the next along the length of the myofibrils. (See Figure 12.1c and d.)

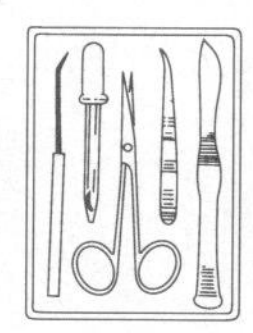

1. Look at the three-dimensional model of skeletal muscle cells, and examine the relative shape and size of the cells. Identify the nuclei, myofibrils, and the light I bands and dark A bands.

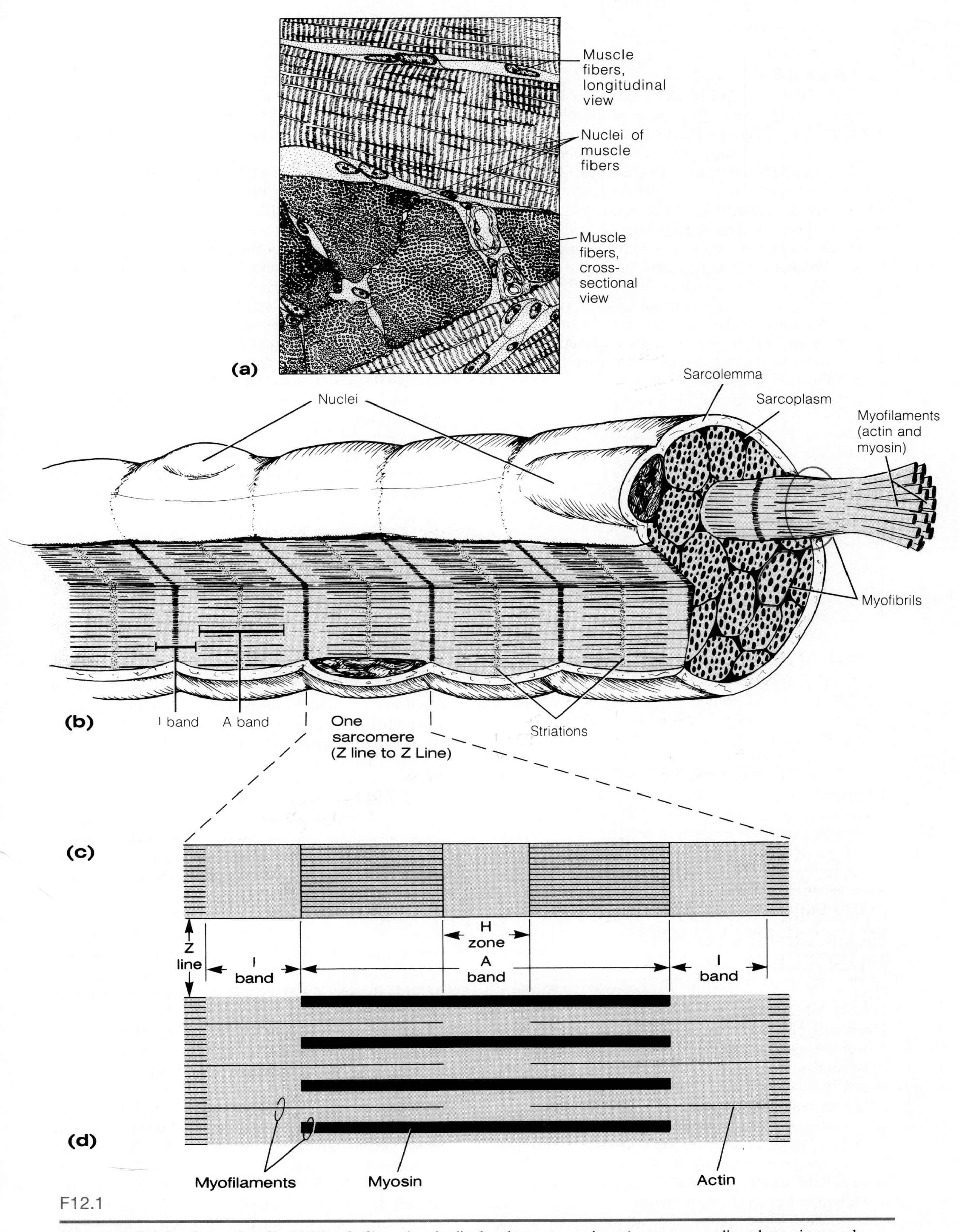

F12.1

Structure of skeletal muscle cells. (a) Muscle fibers, longitudinal and transverse views (see corresponding photomicrograph, plate 2 of the histology atlas). (b) Portion of a skeletal muscle cell. One myofibril has been extended and disrupted to indicate its myofilament composition. (c) One sarcomere of the myofibril. (d) Banding pattern in the sarcomere. (From H. E. Huxley, "The Contraction of Muscle." Copyright November 1958 by Scientific American, Inc. All rights reserved.)

2. Obtain forceps, two dissecting needles, slide and coverslip, and a dropper bottle of saline solution. With forceps, remove a very small piece of muscle from the chicken breast (or thigh). Place the tissue on a clean microscope slide, and add a drop of the saline solution. Pull the muscle fibers apart with the dissecting needles (tease them) until you have a fluffy-looking mass of tissue. Cover the teased tissue with a coverslip, and observe it under the high-power lens of a microscope. Look for the banding pattern. Regulate the light carefully to obtain the highest possible contrast.

3. Now compare your observations with what can be seen with professionally prepared muscle tissue. Obtain a slide of skeletal muscle (longitudinal section), and view it under high power. From your observations, draw a small section of a muscle fiber in the space provided here. ***Label*** the nuclei, sarcolemma, and A and I bands.

What structural details are more apparent in the prepared slide?

____

____

ORGANIZATION OF SKELETAL MUSCLE CELLS INTO MUSCLES

Muscle fibers are soft and surprisingly fragile. Thousands of muscle fibers are bundled together with connective tissue to form the organs we refer to as skeletal muscles (Figure 12.2). Each muscle fiber is enclosed in a delicate, (areolar) connective tissue sheath called **endomysium.** Several sheathed muscle fibers are wrapped by a collagenic membrane called **perimysium,** forming a bundle of fibers called a **fascicle,** or **fasciculus.** A large number of fascicles are bound together by a substantially coarser "overcoat" of dense connective tissue wrappings called **epimysium** which sheathes the entire muscle. These epimysia blend into the **deep fascia** surrounding groups of muscles and into the strong cordlike **tendons** or sheetlike **aponeuroses,** which attach muscles to each other or indirectly to bones.

The tendons perform several functions, two of the most important being to provide durability and to conserve space. Because tendons are tough collagenic connective tissue, they can span rough bony prominences that would destroy the more delicate muscle tissues. And because of their relatively small size, more tendons than fleshy muscles can pass over a joint.

In addition to supporting and binding the muscle fibers, and providing strength to the muscle as a whole, the connective tissue wrappings provide a route for the entry and exit of nerves and blood vessels that serve the muscle fibers. The larger, more powerful muscles have more connective tissue than the muscles involved in fine or delicate movements. As we age, the amount of muscle fiber decreases, and the amount of connective tissue increases; thus the skeletal muscles gradually become more sinewy, or "stringier."

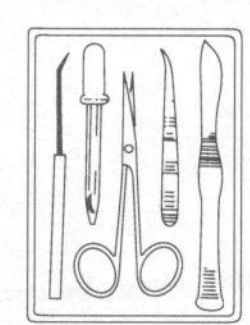

Obtain a slide showing a cross section of skeletal muscle tissue. Using Figure 12.2 as a guide, identify the muscle fibers, endomysium, perimysium, and epimysium (if visible).

THE NEUROMUSCULAR JUNCTION

Voluntary muscle cells are always stimulated by nerve impulses via motor neurons. The junction between a nerve fiber (axon) and a muscle cell is called a **neuromuscular,** or **myoneural, junction** (Figure 12.3).

Each motor axon breaks up into many branches, called *axonal terminals,* as it approaches the muscle, and each of these branches stimulates a different muscle cell. Thus a single neuron may stimulate many muscle fibers. Together, a neuron and all the muscle cells it stimulates make up the functional structure called the **motor unit.** Part of a motor unit is shown in Figure 12.4.

Each axonal terminal (also called a *motor end plate*) has numerous projections called **sole feet.** The neuron and muscle fiber membranes, close as they are, do not actually touch. They are separated by a small fluid-filled gap of 300 to 500 Å, called the **synaptic cleft.**

Within the sole foot are many mitochondria and vesicles containing a neurotransmitter chemical called acetylcholine. When a nerve impulse reaches the end plate, some of these vesicles liberate their contents into the synaptic cleft. The acetylcholine rapidly diffuses across the junction and combines with the receptors on the sarcolemma. If sufficient acetylcholine has been released, a transient change in the permeability of the sarcolemma briefly allows more sodium ions to diffuse into the muscle fiber, resulting in the depolarization of the sarcolemma and subsequent contraction of the muscle fiber.

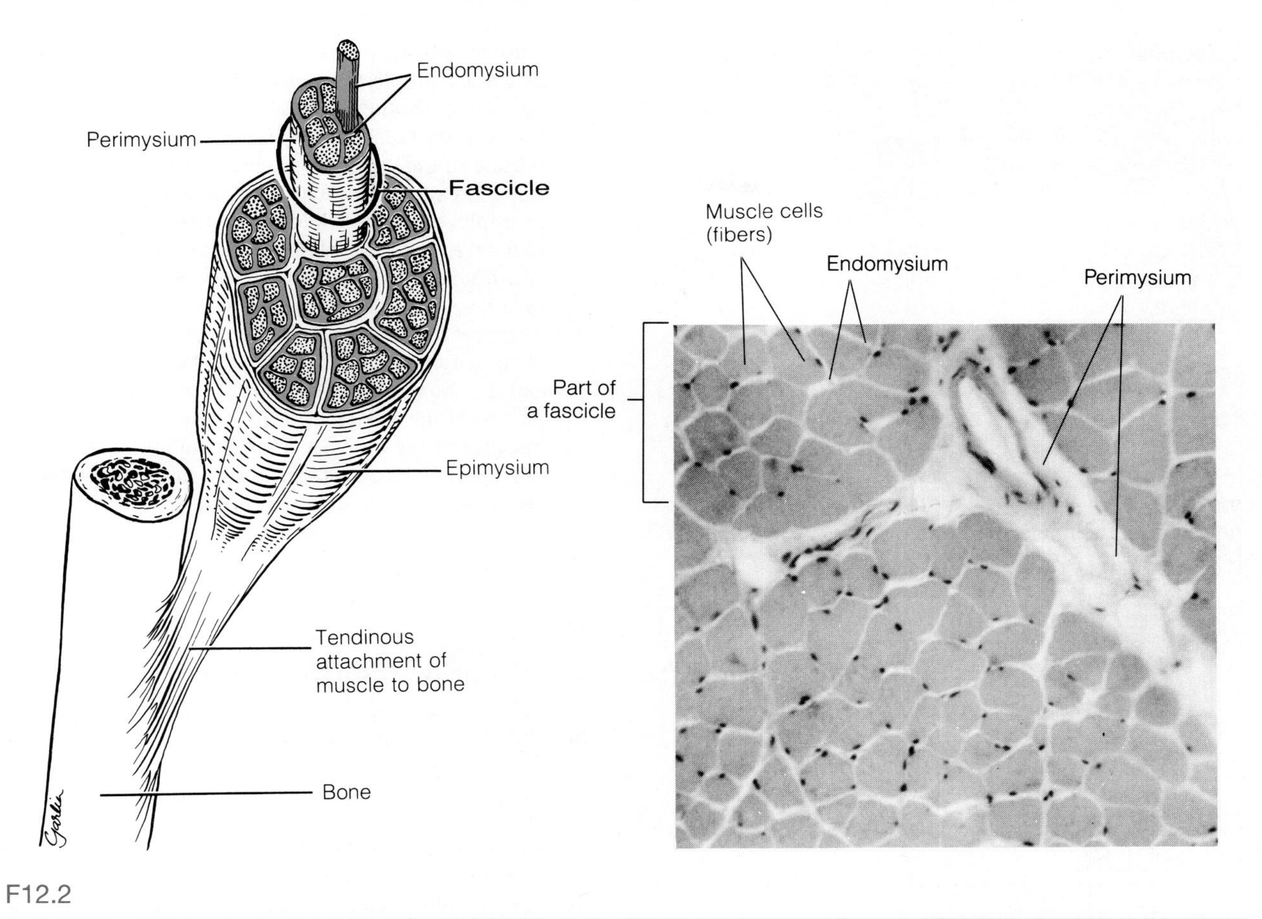

Connective tissue coverings of skeletal muscle (64×).

1. If possible, examine a three-dimensional model of skeletal muscle cells that illustrates the neuromuscular junction. Identify the structures described above.

2. Obtain a slide of skeletal muscle stained to show part of a motor unit. Examine the slide under high power and compare your observations to Figure 12.4. Sketch a small section in the space provided, labeling the motor axon, its terminal branches, motor end plates, and muscle fibers.

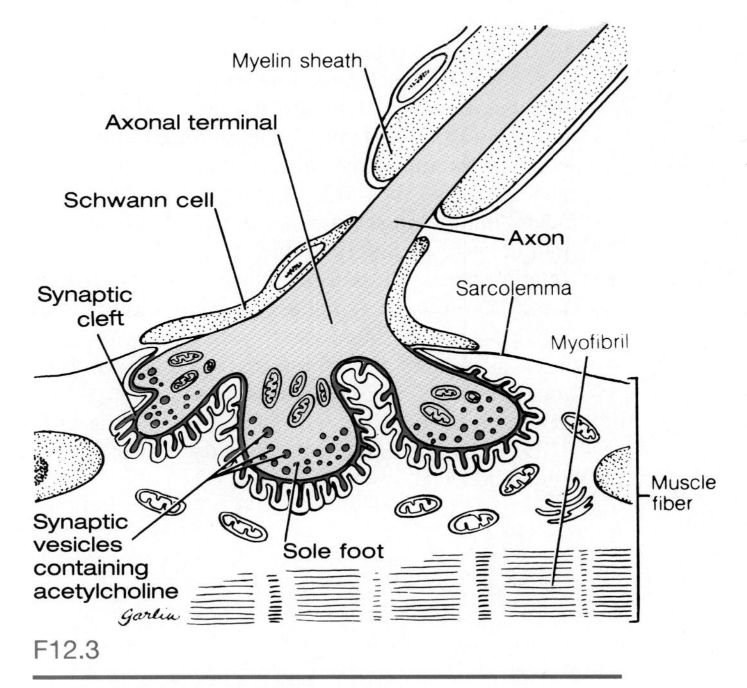

The neuromuscular junction.

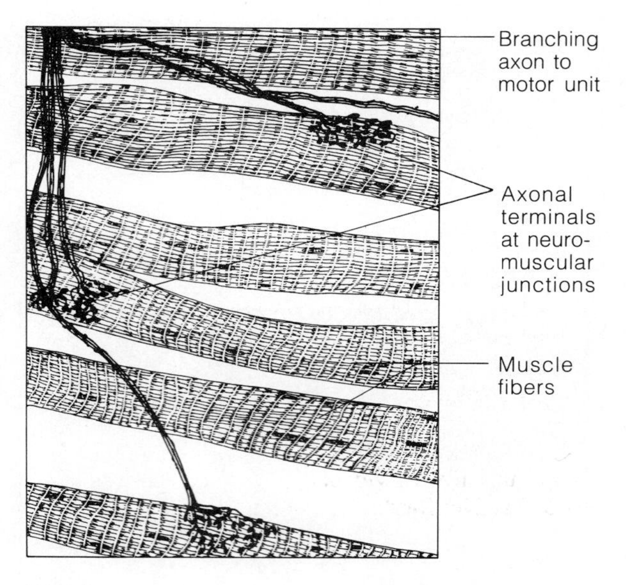

F12.4

A portion of a motor unit. Corresponding photomicrograph is plate 3 of the histology atlas.

CLASSIFICATION OF SKELETAL MUSCLES

Naming Skeletal Muscles

Remembering the names of the skeletal muscles is a monumental task, but certain clues help. Muscles are named on the basis of the following criteria:

- Direction of muscle fibers: Some muscles are named in reference to some imaginary line, usually the midline of the body or the longitudinal axis of a limb bone. A muscle with fibers (and fascicles) running parallel to that imaginary line will have the term *rectus* (straight) in its name. For example, the rectus abdominis is the straight muscle of the abdomen. Likewise, the terms *transverse* and *oblique* indicate that the muscle fibers run at right angles and obliquely (respectively) to the imaginary line.
- Relative size of the muscle: Terms such as *maximus* (largest), *minimus* (smallest), *longus* (long), and *brevis* (short) are often used in the names of muscles—as in gluteus maximus and gluteus minimus.
- Location of the muscle: Some muscles are named according to the bone with which they are associated. For example, the frontalis muscle overlies the frontal bone.
- Number of origins: When the term *biceps, triceps,* or *quadriceps* forms part of a muscle name, you can generally assume that the muscle has two, three, or four origins (respectively). For example, the biceps muscle of the arm has two heads, or origins.
- Location of the muscle's origin and insertion: For example, the sternocleidomastoid muscle has its origin on the sternum (*sterno*) and clavicle (*cleido*), and inserts on the *mastoid* process of the temporal bone.
- Shape of the muscle: For example, the deltoid muscle is roughly triangular (*deltoid* means "triangle"), and the trapezius muscle resembles a trapezoid.
- Action of the muscle: For example, all the adductor muscles of the anterior thigh bring about its adduction, and all the extensor muscles of the wrist extend the wrist.

Types of Muscles

Most often, body movements are not a result of the contraction of a single muscle but are the coordinated action of several muscles acting together. Muscles that are primarily responsible for producing a particular movement are called **prime movers,** or **agonists.**

Muscles that oppose or reverse a movement are called **antagonists.** When a prime mover is active, the fibers of the antagonist are stretched and in the relaxed state. The antagonist can also regulate the prime mover by providing some resistance, to prevent overshoot or to stop its action.

It should be noted that antagonists can be prime movers in their own right; for example, the biceps muscle of the arm (a prime mover of elbow flexion) is antagonized by the triceps (a prime mover of elbow extension).

Synergists contribute substantially to the action of prime movers by reducing undesirable or unnecessary movement. Contraction of a muscle crossing two or more joints would cause movement at all joints spanned if synergists were not there to stabilize them. For example, you can make a fist without bending your wrist only because synergist muscles stabilize the wrist joint and allow the prime mover to exert its force at the finger joints.

Fixators, or fixation muscles, are specialized synergists. They immobilize the origin of a prime mover so that all the tension is exerted at the insertion. Muscles that help maintain posture are fixators; so too are muscles of the back that stabilize or "fix" the scapula during arm movements.

EXERCISE 13

Identification of Human Muscles

OBJECTIVES

1. To name and locate the major muscles of the human body (on a torso model, a human cadaver, laboratory chart, or diagram), and to state the action of each.
2. To explain how muscle actions are related to their location.
3. To identify antagonists of the major prime movers.
4. To name origins and insertions as required by the instructor.

MATERIALS

Human torso model
Large anatomical charts showing anterior and posterior human musculature
Colored pencils
Human musculature videotape (if available)
Human cadaver for demonstration (if available)

MUSCLES OF THE HEAD AND NECK

The muscles of the head serve many specific functions. For instance, the muscles of facial expression differ from most skeletal muscles because they insert into the skin (or other muscles) rather than into bone. As a result, they move the facial skin, allowing a wide range of emotions to be shown on the face. Other muscles of the head are the muscles of mastication, which manipulate the mandible during chewing, and the six extrinsic eye muscles located within the orbit, which aim the eye. (The extrinsic eye muscles are studied in conjunction with the anatomy of the eye in Exercise 21.) Neck muscles are primarily concerned with the movement of the head and shoulder girdle. Figures 13.1 and 13.2 are summary figures illustrating the superficial musculature of the body as a whole. The head and neck muscles are discussed in Tables 13.1 and 13.2 and shown in Figures 13.3 and 13.4.

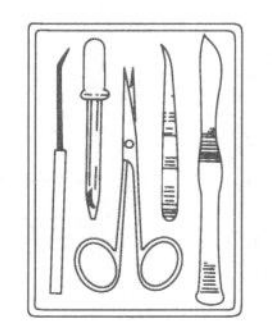

Carefully read the description of each muscle and attempt to visualize what happens when the muscle contracts. Once you have read through the tables and have identified the head and neck muscles in Figures 13.3 and 13.4, use a torso model or an anatomic chart to again identify as many of these muscles as possible. (If a human cadaver is available for observation, specific instructions for muscle examination will be provided by your instructor.) Then carry out the following palpations on yourself:

- To demonstrate how the *temporalis* works, clench your teeth. The *masseter* can also be palpated at this time at the angle of the jaw.

MUSCLES OF THE TRUNK

The trunk musculature includes muscles that move the vertebral column; anterior thorax muscles that act to move ribs, head, and arms; and muscles of the abdominal wall that play a role in the movement of the vertebral column but more importantly form the "natural girdle," or the major portion, of the abdominal body wall.

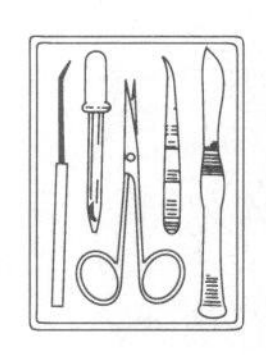

The trunk muscles are described in Tables 13.3 and 13.4 and shown in Figures 13.5 and 13.6. As before, identify the muscles in the figure as you read the tabular descriptions and then identify them on the torso or laboratory chart.

When you have completed this study, work with a partner to demonstrate the operation of the following muscles. One of you can demonstrate the movement (the following steps are addressed to this partner); the other can supply the necessary resistance and palpate the muscle being tested.

1. Start by fully abducting the arm and extending the elbow. Now try to adduct the arm against resistance. You are exercising the *latissimus dorsi.*
2. To observe the *deltoid,* attempt to abduct your shoulder against resistance. Now attempt to elevate your shoulder against resistance; you are contracting the upper portion of the *trapezius.*
3. The *pectoralis major* comes into play when you press your hands together at chest level with your elbows widely abducted.

(*Text continues on p. 116*)

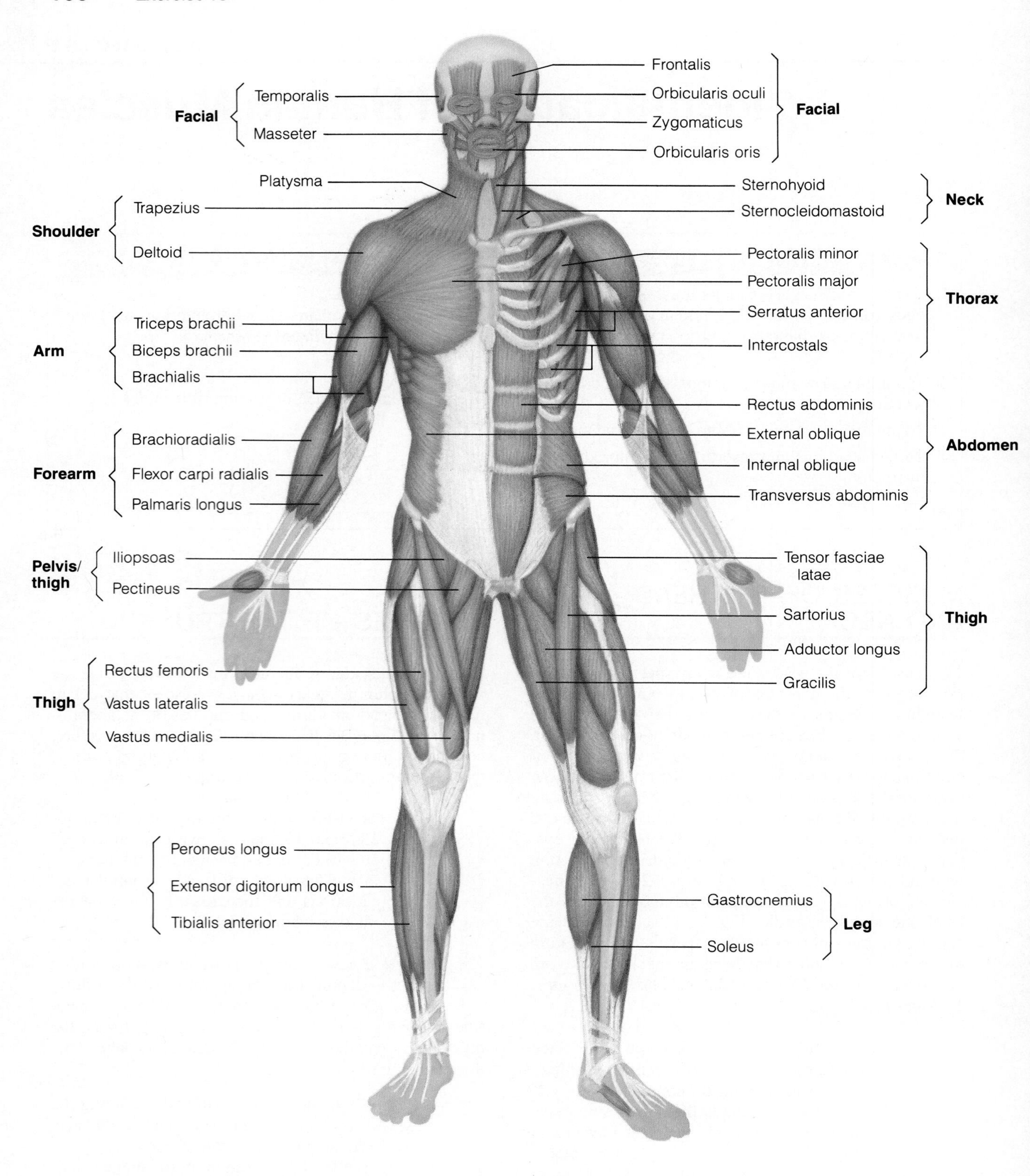

F13.1

Anterior view of superficial muscles of the body. The abdominal surface has been partly dissected on the right side of the diagram to show somewhat deeper muscles.

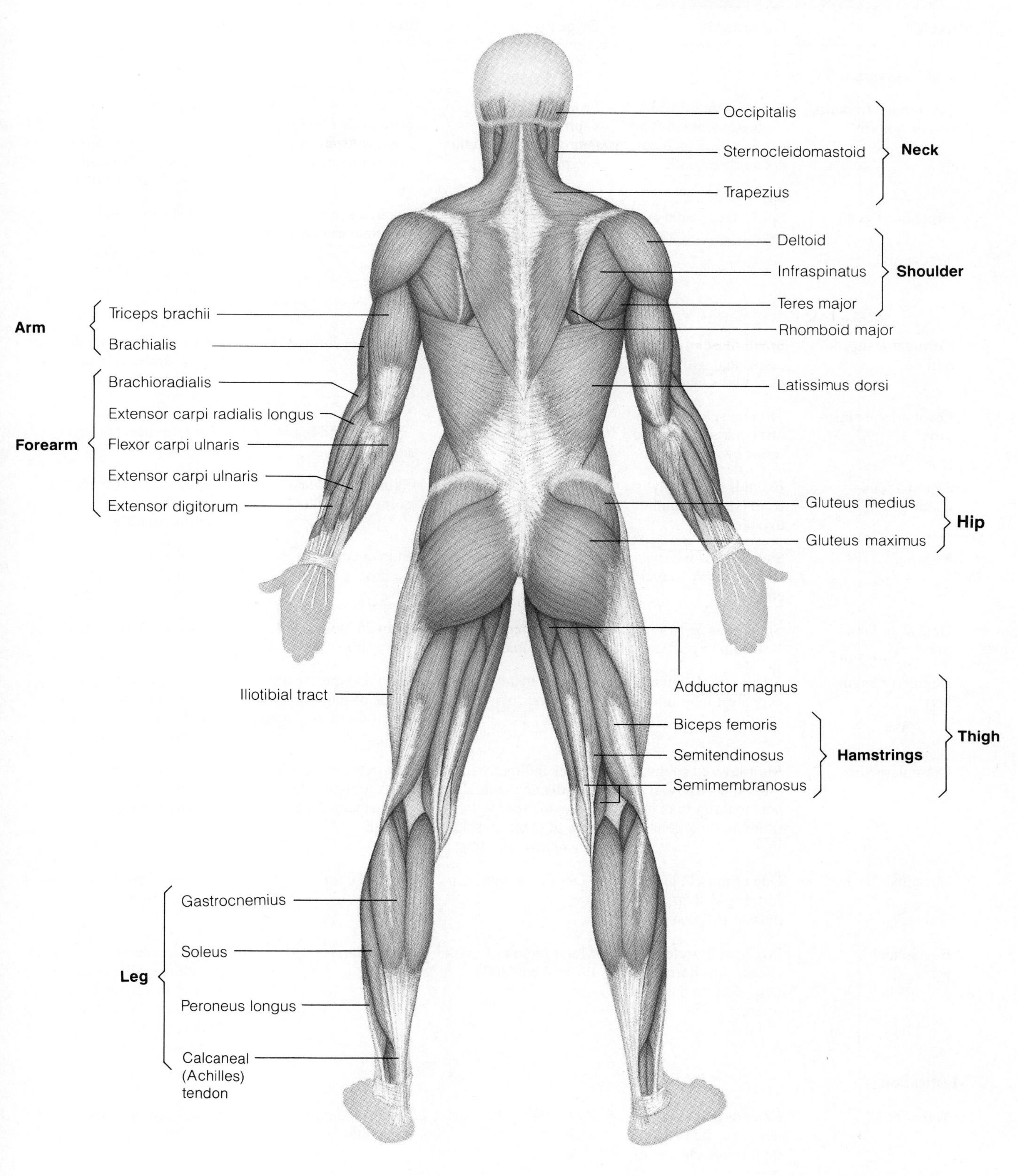

F13.2

Posterior view of superficial muscles of the body, diagrammatic view.

TABLE 13.1 Major Muscles of Human Head (see Figure 13.3)

Muscle	Comments	Origin	Insertion	Action
Facial Expression (Figure 13.3a)				
Epicranius: frontalis and occipitalis	Bipartite muscle consisting of frontalis and occipitalis, which covers dome of skull	Frontalis: cranial aponeurosis (galea aponeurotica); occipitalis: occipital bone	Frontalis: skin of eyebrows and root of nose; occipitalis: cranial aponeurosis	With aponeurosis fixed, frontalis raises eyebrows; occipitalis fixes aponeurosis and pulls scalp posteriorly
Orbicularis oculi	Sphincter muscle of eyelids	Frontal and maxillary bones and ligaments around orbit	Encircles orbit and inserts in tissue of eyelid	Various parts can be activated individually; closes eyes, produces blinking, squinting, and draws eyebrows downward
Corrugator supercilii	Small muscle; activity associated with that of orbicularis oculi	Arch of frontal bone above nasal bone	Skin of eyebrow	Draws eyebrows medially; wrinkles skin of forehead vertically
Levator labii superioris	Thin muscle between orbicularis oris and inferior eye margin	Zygomatic bone and infraorbital margin of maxilla	Skin and muscle of upper lip and border of nostril	Raises and furrows upper lip; flares nostril (as in disgust)
Zygomaticus—major and minor	Extends diagonally from corner of mouth to cheekbone	Zygomatic bone	Skin and muscle at corner of mouth	Raises lateral corners of mouth upward (smiling muscle)
Risorius	Slender muscle; runs laterally to zygomaticus	Fascia of masseter muscle	Skin at corner of mouth	Draws corner of lip laterally; tenses lip; zygomaticus synergist
Depressor labii inferioris	Small muscle from lower lip to jawbone	Body of mandible lateral to its midline	Skin and muscle of lower lip	Draws lower lip downward
Depressor anguli oris	Small muscle lateral to depressor labii inferioris	Body of mandible below incisors	Skin and muscle at angle of mouth below insertion of zygomaticus	Zygomaticus antagonist; draws corners of mouth downward and laterally
Orbicularis oris	Multilayered sphincter muscle of lips with fibers that run in many different directions	Arises indirectly from maxilla and mandible; fibers blended with fibers of other muscles associated with lips	Encircles mouth; inserts into muscle and skin at angles of mouth	Closes mouth; purses and protrudes lips (kissing muscle)
Mentalis	One of muscle pair forming V-shaped muscle mass on chin	Mandible below incisors	Skin of chin	Protrudes lower lip; wrinkles chin
Buccinator	Principal muscle of cheek; runs horizontally, deep to the masseter	Molar region of maxilla and mandible	Orbicularis oris	Draws corner of mouth laterally; compresses cheek (as in whistling); holds food between teeth during chewing
Mastication				
Masseter	Extends across jawbone; can be palpated on forcible closure of jaws	Zygomatic process and arch	Angle and ramus of mandible	Closes jaw and elevates mandible
Temporalis	Fan-shaped muscle over temporal bone	Temporal fossa	Coronoid process of mandible	Closes jaw; elevates and retracts mandible
Buccinator	(See muscles of facial expression.)			

(*Table continues on p. 110*)

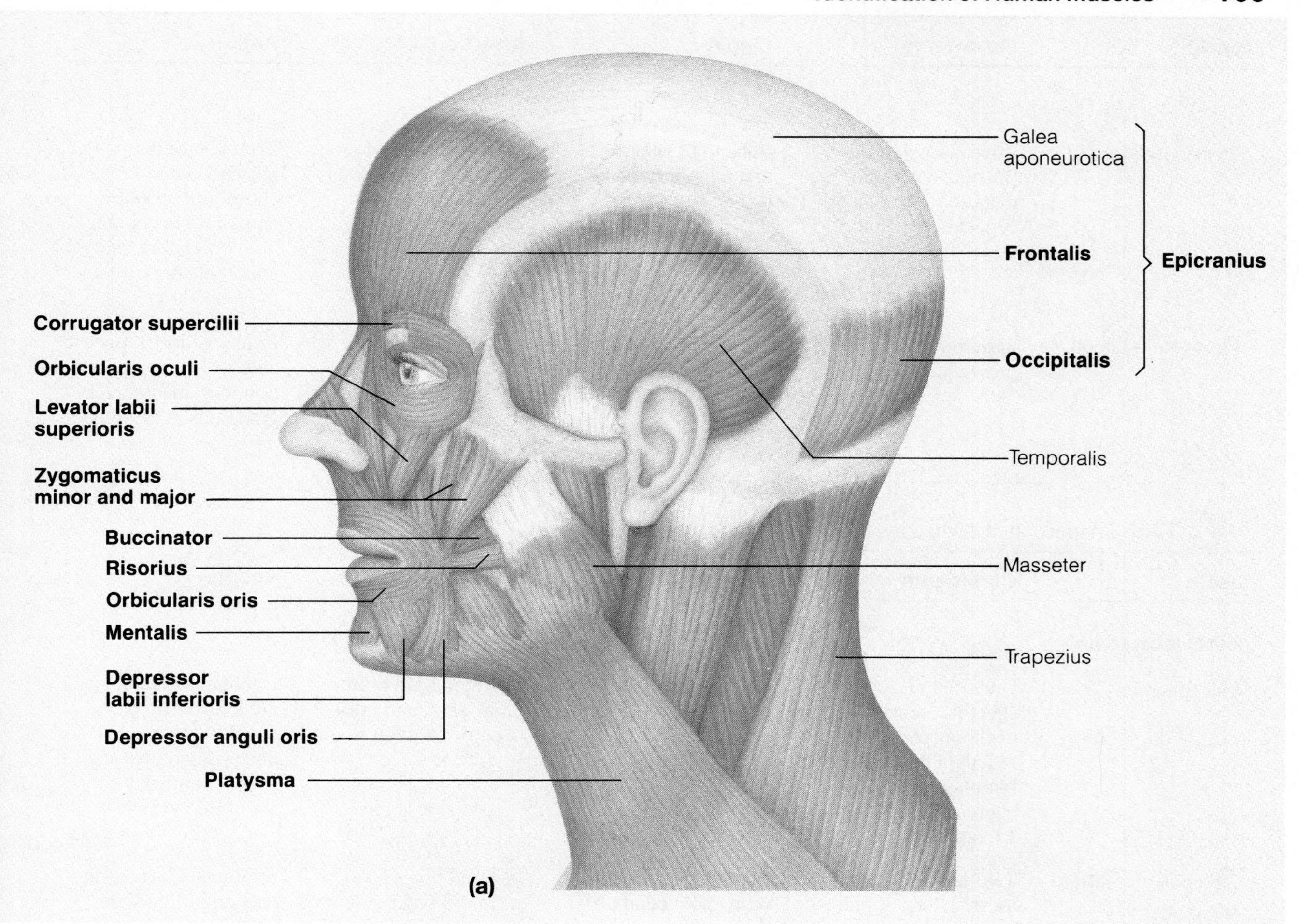

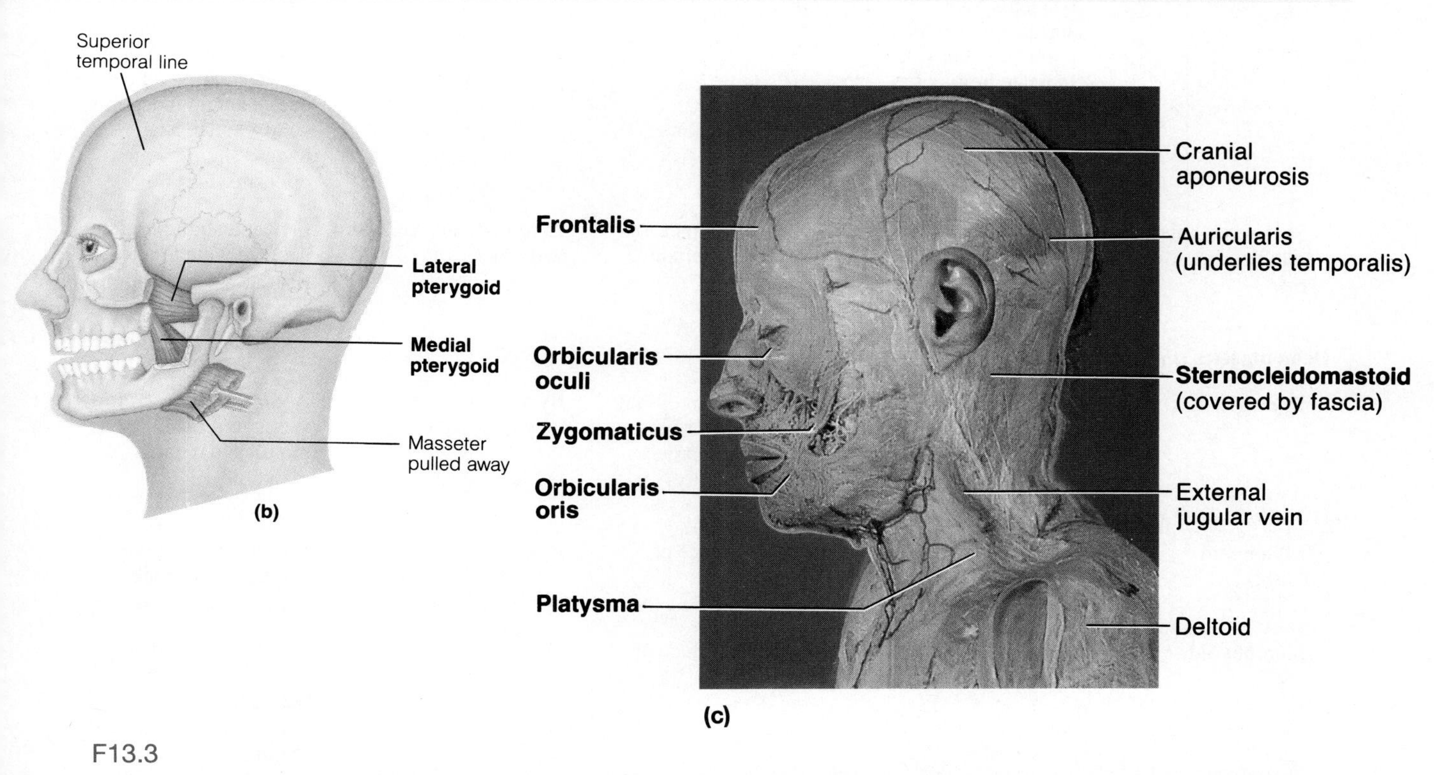

F13.3

Muscles of the scalp, face, and neck. (a) Lateral view; (b) lateral view of deep chewing muscles, the medial and lateral pterygoid muscles; (c) photo of superficial structures of the head and neck, lateral view.

TABLE 13.1 *(Continued)*

Muscle	Comments	Origin	Insertion	Action
Figure 13.3b				
Pterygoid—medial	Runs along internal (medial) surface of mandible (thus largely concealed by that bone)	Sphenoid, palatine, and maxillary bones	Medial surface of mandibular ramus and angle	Synergist of temporalis and masseter; closes and elevates mandible; in conjunction with lateral pterygoid, aids in grinding movements
Pterygoid—lateral	Superior to medial pterygoid	Greater wing of sphenoid bone	Mandibular condyle	Protracts jaw (moves it anteriorly); in conjunction with medial pterygoid, aids in grinding movements of teeth

TABLE 13.2 Anterolateral Muscles of Human Neck (see Figure 13.4)

Muscle	Comments	Origin	Insertion	Action
Superficial muscles				
Platysma	Unpaired muscle: thin, sheetlike superficial neck muscle, not strictly a head muscle but plays role in facial expression (see Fig. 13.3a)	Fascia of chest (over pectoral muscles and deltoid)	Lower margin of mandible, skin, and muscle at corner of mouth	Depresses mandible; pulls lower lip back and down; i.e., produces downward sag of the mouth
Sternocleidomastoid	Two-headed muscle located deep to platysma on anterolateral surface of neck; fleshy parts on either side indicate limits of anterior and posterior triangles of neck	Manubrium of sternum and medial portion of clavicle	Mastoid process of temporal bone	Simultaneous contraction of both muscles of pair causes flexion of neck forward, generally against resistance (as when lying on the back); acting independently, rotate head toward shoulder on opposite side
Scalenes—anterior, middle, and posterior	Located more on lateral than anterior neck; deep to platysma (see Fig. 13.4b)	Transverse processes of cervical vertebrae	Anterolaterally on first two ribs	Flex and slightly rotate neck; elevate first two ribs (aid in inspiration)
Deep muscles (Figure 13.4a)				
Digastric	Consists of two bellies united by an intermediate tendon; assumes a V-shaped configuration under chin	Lower margin of mandible (anterior belly) and mastoid process (posterior belly)	By a connective tissue loop to hyoid bone	Acting in concert, elevate hyoid bone; open mouth and depress mandible
Mylohyoid	Just deep to digastric; forms floor of mouth	Medial surface of mandible	Hyoid bone	Elevates hyoid bone and base of tongue during swallowing
Sternohyoid	Runs most medially along neck; straplike	Posterior surface of manubrium	Lower margin of body of hyoid bone	Acting with sternothyroid and omohyoid (all below hyoid bone), depresses larynx and hyoid bone if mandible is fixed; may also flex skull

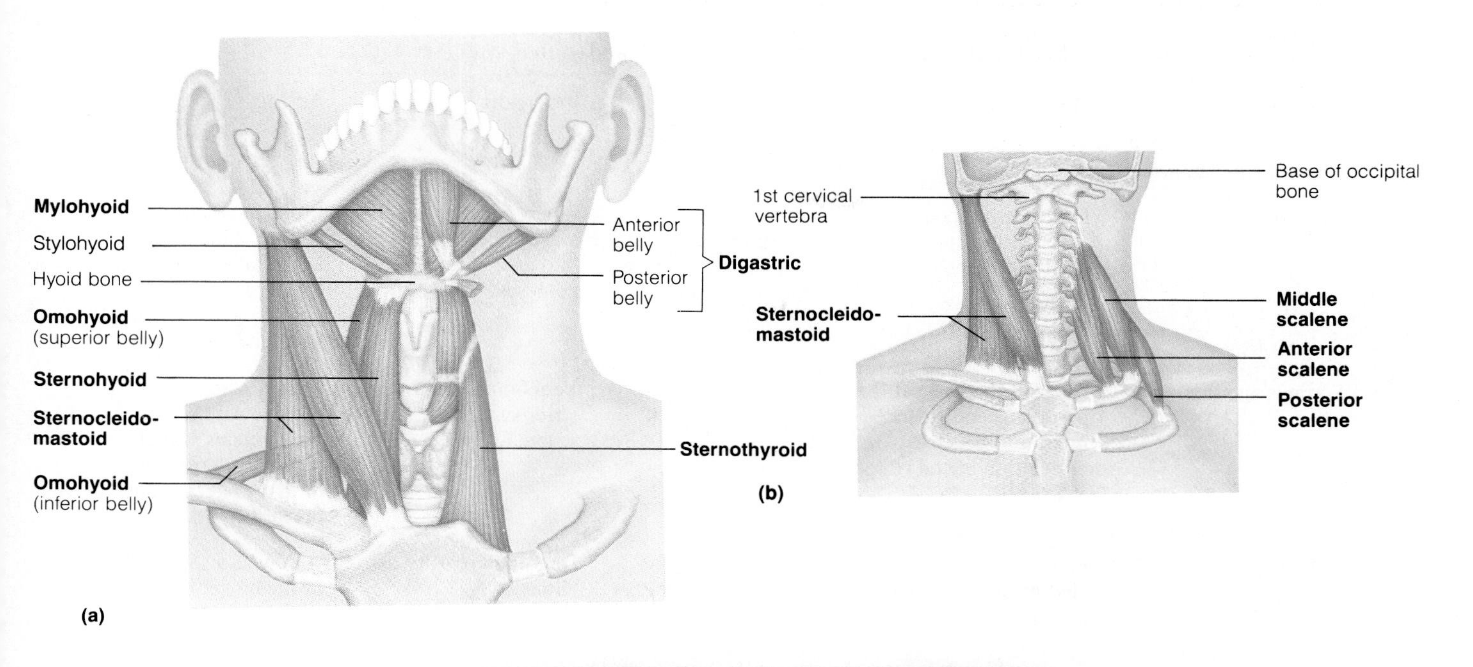

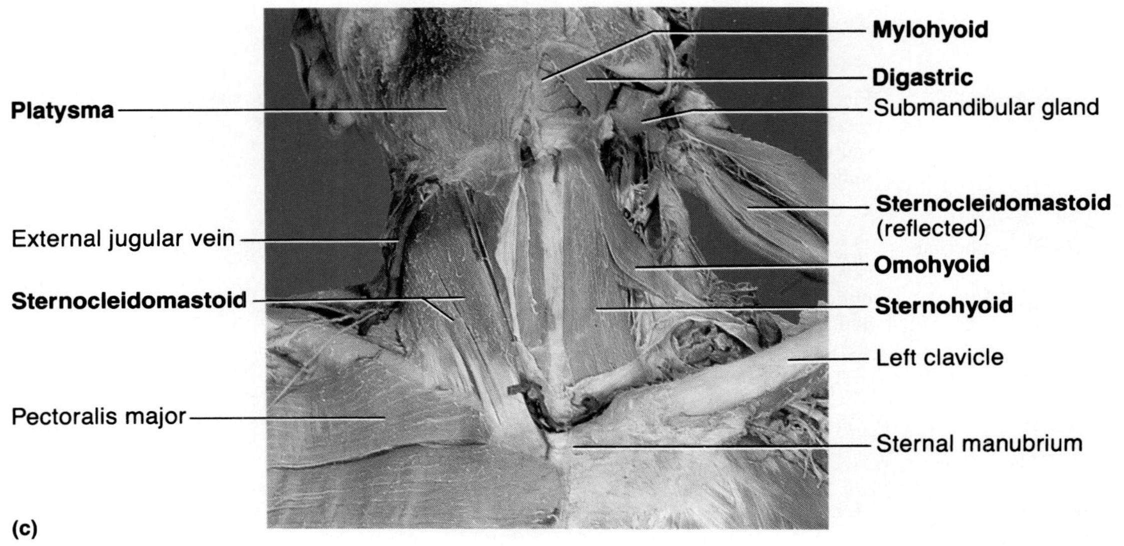

F13.4

Muscles of the neck and throat. (a) Anterior view of the deep neck (suprahyoid and infrahyoid) muscles. (b) Muscles of the anterolateral neck. The superficial platysma muscle and the deeper neck muscles have been removed to show the origins and insertions of the sternocleidomastoid and scalene muscles clearly. (c) Photo of the anterior and lateral regions of the neck. The fascia has been partially removed (left side of photo) to expose the sternocleidomastoid muscle. On the right side of the photo, the sternocleidomastoid muscle is reflected to expose the sternohyoid and omohyoid muscles.

TABLE 13.2 *(Continued)*

Muscle	Comments	Origin	Insertion	Action
Sternothyroid	Lateral to sternohyoid; straplike	Manubrium and medial end of clavicle	Thyroid cartilage of larynx	(See Sternohyoid, above)
Omohyoid	Straplike with two bellies; lateral to sternohyoid	Superior surface of scapula	Hyoid bone	(See Sternohyoid, above)

TABLE 13.3 Anterior Muscles of Human Thorax, Shoulder, and Abdominal Wall (see Figure 13.5)

Muscle	Comments	Origin	Insertion	Action
Thorax and Shoulder (Figure 13.5a):				
Pectoralis major	Large fan-shaped muscle covering upper portion of chest	Clavicle, sternum, cartilage of first six ribs, and aponeurosis of external oblique muscle	Fibers converge to insert by short tendon into greater tubercle of humerus	Prime mover of arm flexion; adducts, medially rotates arm; with arm fixed, pulls chest upward (thus also acts in forced inspiration)
Serratus anterior	Deep and superficial portions; beneath and inferior to pectoral muscles on lateral rib cage	Lateral aspect of first to eighth (or ninth) ribs	Vertebral border of anterior surface of scapula	Moves scapula forward toward chest wall; rotates scapula causing inferior angle to move laterally and upward
Deltoid	Fleshy triangular muscle forming shoulder muscle mass	Lateral third of clavicle; acromion process and spine of scapula	Deltoid tuberosity of humerus	Acting as a whole, prime mover of arm abduction; when only specific fibers are active, can aid in flexion, extension, and rotation of humerus
Pectoralis minor	Flat, thin muscle directly beneath and obscured by pectoralis major	Anterior surface of third, fourth, and fifth ribs, near their costal cartilages	Coracoid process of scapula	With ribs fixed, draws scapula forward and inferiorly; with scapula fixed, draws rib cage superiorly
Intercostals—external	11 pairs lie between ribs; fibers run obliquely downward and forward toward sternum	Inferior border of rib above (not shown in figure)	Superior border of rib below	Pulls ribs toward one another to elevate rib cage; aids in inspiration

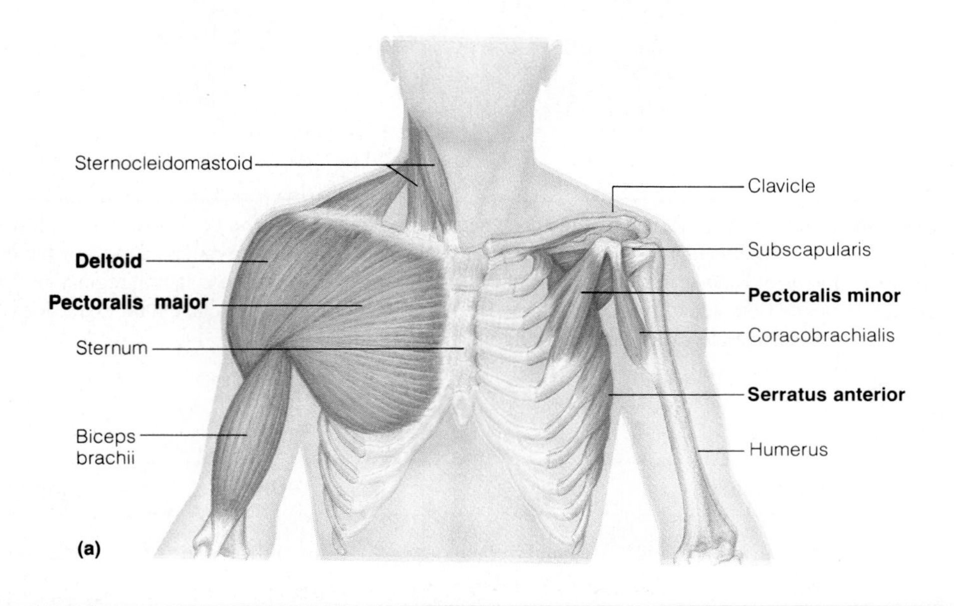

F13.5

Anterior muscles of the thorax, shoulder, and abdominal wall. (a) Anterior thorax. The superficial pectoralis major and deltoid muscles that effect arm movements are illustrated on the left. These muscles have been removed on the right side of the figure to illustrate (1) the pectoralis minor and serratus anterior muscles and (2) the subscapularis muscle. (b) Anterior view of the muscles forming the anterolateral abdominal wall. The superficial muscles have been partially cut away on the left side of the diagram to reveal the deeper internal oblique and transversus abdominis muscles.

TABLE 13.3 (*Continued*)

Muscle	Comments	Origin	Insertion	Action
Thorax and Shoulder *(continued)*				
Intercostals—internal	11 pairs lie between ribs; fibers run deep and at right angles to those of external intercostals	Superior border of rib below	Inferior border of rib above (not shown in figure)	Draws ribs together to depress rib cage; aids in forced expiration; antagonistic to external intercostals
Abdominal Wall (Figure 13.5b and c):				
Rectus abdominis	Medial superficial muscle, extends from pubis to rib cage; ensheathed by aponeuroses of oblique muscles; segmented	Pubic crest and symphysis	Xiphoid process and costal cartilages of fifth through seventh ribs	Flexes vertebral column; increases abdominal pressure; fixes and depresses ribs; stabilizes pelvis during walking
External oblique	Most superficial lateral muscle; fibers run downward and medially; ensheathed by an aponeurosis	Anterior surface of last eight ribs	Linea alba,* pubic tubercles, and iliac crest	See Rectus abdominis, above; also aids muscles of back in trunk rotation and lateral flexion
Internal oblique	Fibers run at right angles to those of external oblique, which it underlies	Lumbodorsal fascia, iliac crest, and inguinal ligament	Linea alba, pubic crest, and costal cartilages of last three ribs	As for External oblique
Transversus abdominis	Deepest muscle of abdominal wall; fibers run horizontally	Inguinal ligament, iliac crest, and cartilages of last five or six ribs	Linea alba and pubic crest	Compresses abdominal contents

*The linea alba ("white line") is a narrow, tendinous sheath that runs along the middle of the abdomen from the sternum to the pubic symphysis. It is formed by the fusion of the aponeurosis of the external oblique and transversus muscles.

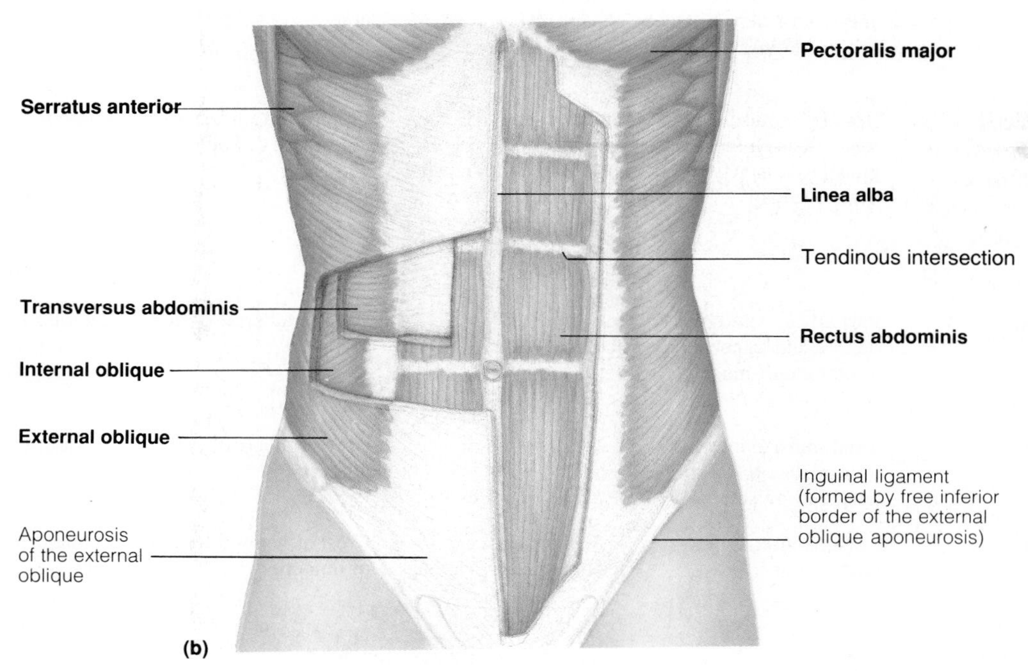

(*F13.5 continues*)

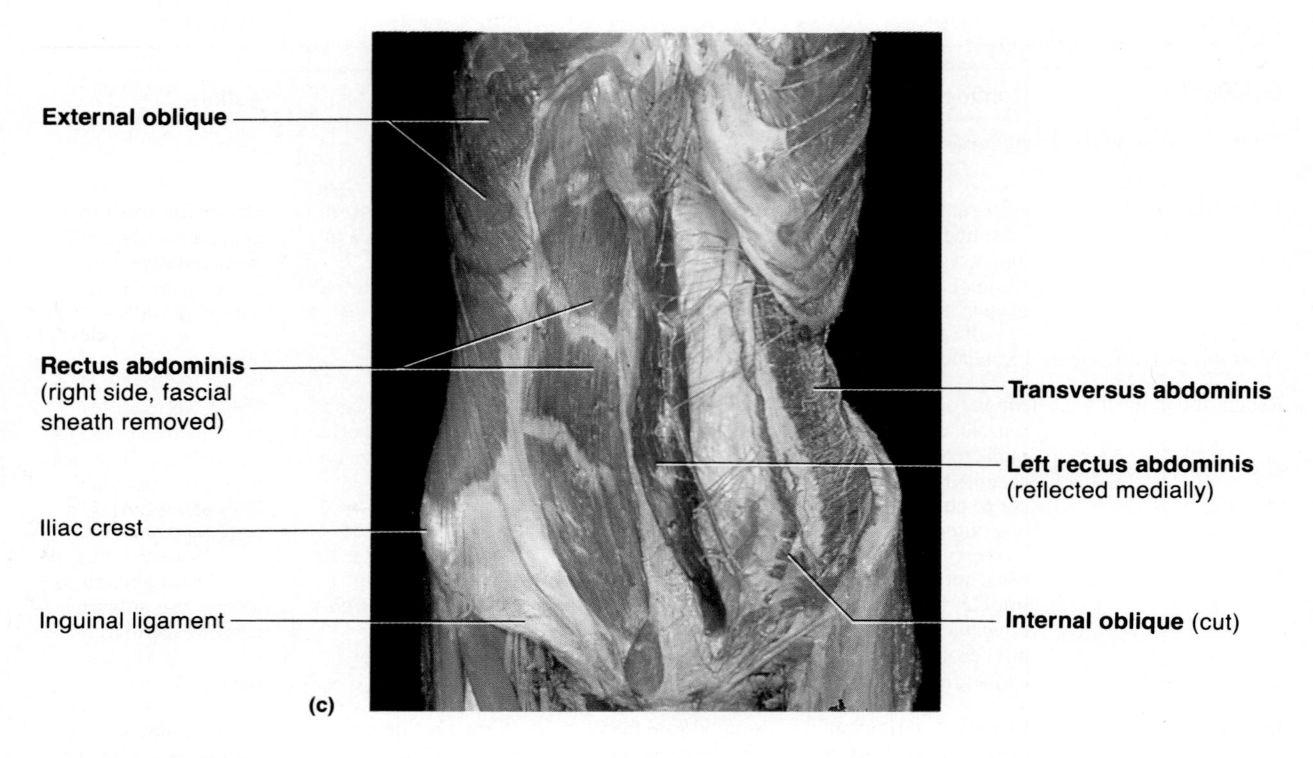

F13.5 *(continued)*

Anterior muscles of the thorax, shoulder, and abdominal wall. (c) Photo of the anterolateral abdominal wall.

TABLE 13.4 Posterior Muscles of Human Trunk (see Figure 13.6)

Muscle	Comments	Origin	Insertion	Action
Muscles of the Neck, Shoulder, and Thorax (Figure 13.6a):				
Trapezius	Most superficial muscle of posterior neck and thorax; very broad origin and insertion	Occipital bone; ligamentum nuchae; spines of C_7 and all thoracic vertebrae	Acromion and spinous process of scapula; lateral third of clavicle	Extends head; retracts (adducts) scapula and stabilizes it; upper fibers elevate scapula; lower fibers depress it
Latissimus dorsi	Broad flat muscle of lower back (lumbar region); extensive superficial origins	Indirect attachment to spinous processes of lower six thoracic vertebrae, lumbar vertebrae, lower 3 to 4 ribs, and iliac crest	Floor of intertubercular groove of humerus	Prime mover of arm extension; adducts and medially rotates arm; depresses scapula; brings arm down in power stroke, as in striking a blow
Infraspinatus	Partially covered by deltoid and trapezius; a rotator cuff muscle	Infraspinous fossa of scapula	Greater tubercle of humerus	Lateral rotation of humerus; helps hold head of humerus in glenoid cavity
Teres minor	Small muscle inferior to infraspinatus; a rotator cuff muscle	Lateral margin of scapula	Greater tuberosity of humerus	As for infraspinatus
Teres major	Located inferiorly to teres minor	Posterior surface at inferior angle of scapula	Crest of lesser tubercle of humerus	Extends, medially rotates, and adducts humerus; synergist of latissimus dorsi

Muscle	Comments	Origin	Insertion	Action
Supraspinatus	Obscured by trapezius and deltoid; a rotator cuff muscle	Supraspinous fossa of scapula	Greater tubercle of humerus	Assists abduction of humerus; stabilizes shoulder joint
Levator scapulae	Located at back and side of neck, deep to trapezius	Transverse processes of C_1 through C_4	Superior vertebral border of scapula	Raises and adducts scapula; with fixed scapula, flexes neck to the same side
Rhomboids—major and minor	Beneath trapezius and inferior to levator scapulae; run from vertebral column to scapula	Spinous processes of C_7 and T_1 through T_5	Vertebral border of scapula	Pull scapula medially (retraction) and elevate it
Muscles Associated with the Vertebral Column (Figure 13.6b):				
Semispinalis	Deep composite muscle of the back—thoracis, cervicis, and capitis portions	Transverse processes of C_7–T_{12}	Occipital bone and spinous processes of cervical vertebrae and T_1–T_4	Acting together, extend head and vertebral column; acting independently (right vs. left) causes rotation toward the opposite side
Erector spinae	A long tripartite muscle composed of iliocostalis (lateral), longissimus, and spinalis (medial) muscle columns; superficial to semispinalis muscles; extends from pelvis to head	Sacrum, iliac crest, transverse processes of lumbar, thoracic and cervical vertebrae, and/or ribs 3–12 depending on specific part	Ribs and transverse processes of vertebrae about six segments above origin. Longissimus also inserts into mastoid process	All act to extend and abduct the vertebral column; fibers of the longissimus also extend head

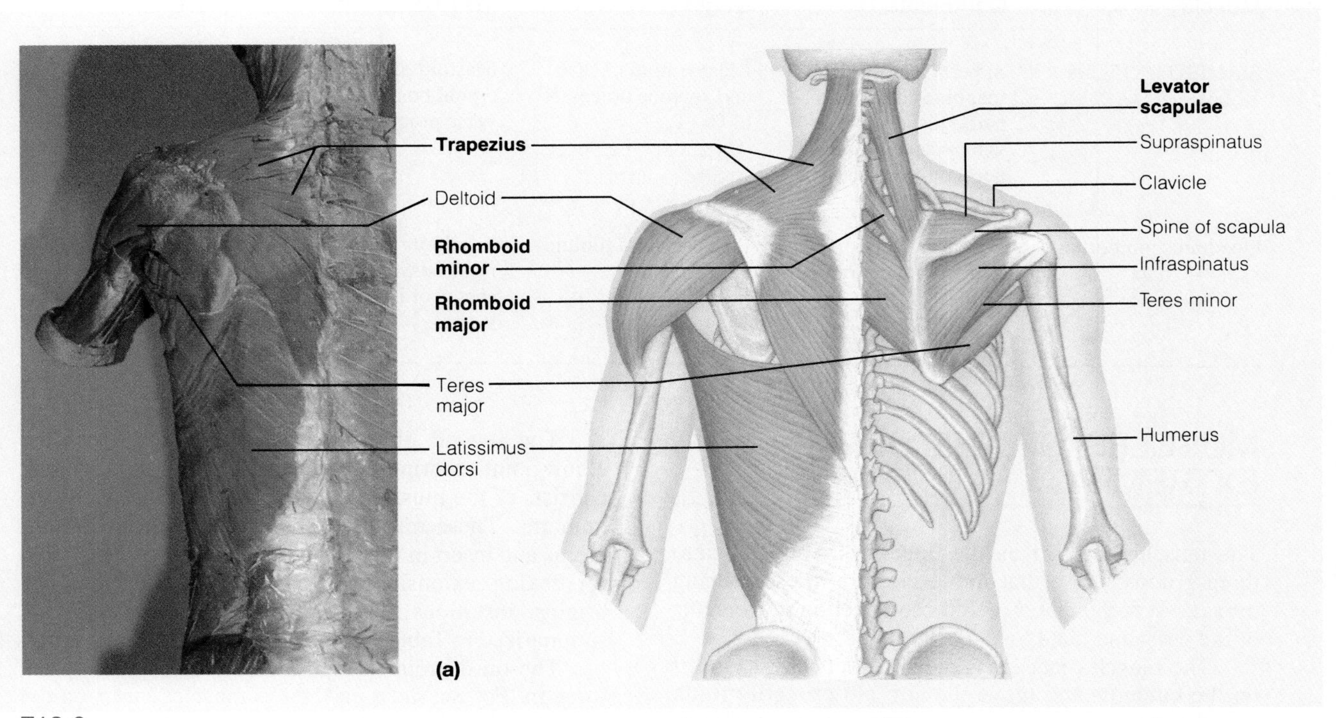

F13.6

Muscles of the neck, shoulder, and thorax, posterior view. (a) The extensive superficial muscles of the back are shown on the left side of the illustration and corresponding photo. The superficial muscles are removed on the right side of the illustration to reveal the deeper muscles acting on the scapula and the rotator cuff muscles that help stabilize the shoulder joint.

(*F13.6 continues*)

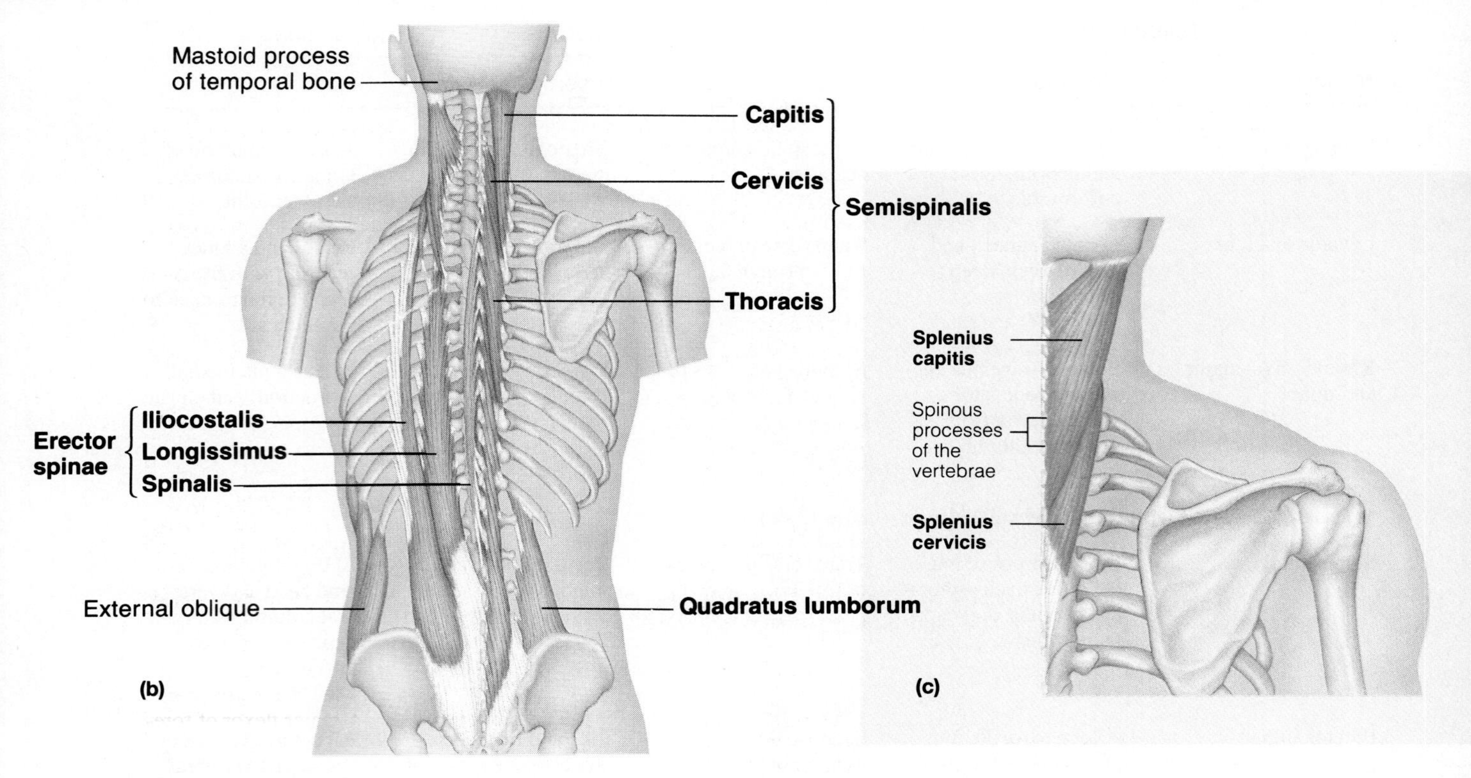

F13.6 (*continued*)

Muscles of the neck, shoulder, and thorax, posterior view. (b) The composite erector spinae and semispinalis muscles, which respectively form the intermediate and deep muscle layers of the back associated with the vertebral column. (c) The splenius muscle (capitis and cervicis parts), which lies superficial to the erector spinae.

TABLE 13.4 *(continued)*

Muscle	Comments	Origin	Insertion	Action
Splenius (see Figure 13.6c)	Superficial muscle (capitis and cervicis parts) just deep to levator scapulae and superficial to erector spinae	Ligamentum nuchae and spinous processes of C_7–T_6	Mastoid process, occipital bone, and transverse processes of C_2–C_4	As a group, extend or hyperextend head; when only one side is active, head is rotated and bent toward the same side
Quadratus lumborum	Forms greater portion of posterior abdominal wall	Iliac crest and iliolumbar fascia	Inferior border twelfth rib; transverse processes of lumbar vertebrae	Each flexes vertebral column laterally; together extend lumbar spine and fix 12th rib

MUSCLES OF THE UPPER EXTREMITY

The muscles that act on the upper extremity fall into three groups: those that move the arm, those causing movement at the elbow, and those effecting movements of the wrist and hand.

The muscles that cross the shoulder joint to insert on the humerus and move the arm are primarily trunk muscles (subscapularis, supraspinatus and infraspinatus, deltoid, and so on) that originate on the axial skeleton or shoulder girdle. These muscles are included with the trunk muscles.

The second group of muscles, which cross the elbow joint to bring about movement of the forearm, consists of the muscles forming the musculature of the humerus. These muscles arise primarily from the humerus and insert in forearm bones. They are responsible for flexion, extension, pronation, and supination. The origins, insertions, and actions of these muscles are summarized in Table 13.5 and are shown in Figure 13.7.

The third group composes the musculature of the forearm. For the most part, these muscles insert on the digits, producing movements at the wrist and fingers. In general, muscles acting on the wrist and hand can more easily be identified if their insertion tendons are located first. These muscles are described in Table 13.6 and illustrated in Figure 13.8.

(*Text continues on p. 120*)

TABLE 13.5 Muscles of Human Humerus That Act on the Forearm (see Figure 13.7)

Muscle	Comments	Origin	Insertion	Action
Triceps brachii	Sole, large fleshy muscle of posterior humerus; three-headed origin	Long head: inferior margin of glenoid cavity; lateral head: posterior humerus; medial head: distal radial groove on posterior humerus	Olecranon process of ulna	Powerful forearm extensor; antagonist of forearm flexors (brachialis and and biceps brachii)
Biceps brachii	Most familiar muscle of anterior humerus because this two-headed muscle bulges when forearm is flexed	Short head: coracoid process; tendon of long head runs in intertubercular groove and within capsule of shoulder joint	Radial tuberosity	Flexion (powerful) of elbow and supination of forearm; "it turns the corkscrew and pulls the cork"; weak arm flexor
Brachioradialis	Superficial muscle of lateral forearm; forms lateral boundary of antecubital fossa	Lateral ridge at distal end of humerus	Base of styloid process of radius	Forearm flexor (weak)
Brachialis	Immediately deep to biceps brachii	Distal portion of anterior humerus	Coronoid process of ulna	A major flexor of forearm

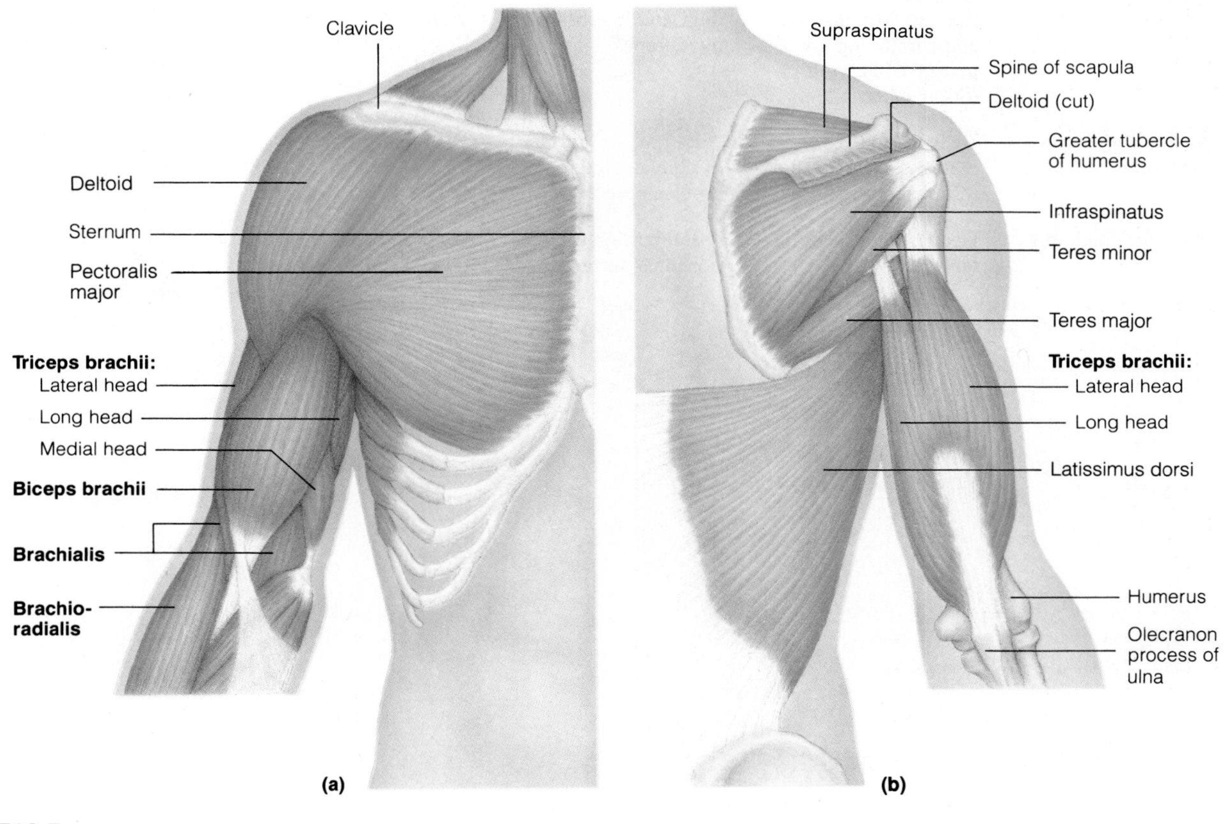

F13.7

Muscles causing movements of the forearm. (a) Superficial muscles of the anterior thorax, shoulder, and arm, anterior view; (b) posterior aspect of the arm showing the lateral and long heads of the triceps brachii muscle.

TABLE 13.6 Muscles of Human Forearm That Act on Hand and Fingers (see Figure 13.8)

Muscle	Comments	Origin	Insertion	Action
Anterior Compartment (Figure 13.8a,b,c,d)				
Superficial muscles				
Pronator teres	Seen in a superficial view between proximal margins of brachioradialis and flexor carpi ulnaris	Medial epicondyle of humerus and coronoid process of ulna	Midshaft of radius	Acts synergistically with pronator quadratus to pronate forearm; weak forearm flexor
Flexor carpi radialis	Superficial; runs diagonally across forearm	Medial epicondyle of humerus	Base of second and third metacarpals	Powerful flexor of wrist; abducts hand
Palmaris longus	Small fleshy muscle with a long tendon; medial to flexor carpi radialis	Medial epicondyle of humerus	Palmar aponeurosis	Flexes wrist (weak)
Flexor carpi ulnaris	Superficial; medial to palmaris longus	Medial epicondyle of humerus and olecranon process of ulna	Base of fifth metacarpal	Powerful flexor of wrist; adducts hand
Flexor digitorum superficialis	Deeper muscle; overlain by muscles named above; visible at distal end of forearm	Medial epicondyle of humerus, medial surface of ulna, and anterior border of radius	Middle phalanges of second through fifth fingers	Flexes wrist and middle phalanges of second through fifth fingers
Deep muscles				
Flexor pollicis longus	Deep muscle of anterior forearm; distal to and paralleling lower margin of flexor digitorum superficialis	Anterior surface of radius, and interosseous membrane	Distal phalanx of thumb	Flexes thumb (*pollix* is Latin for "thumb"); weak flexor of wrist
Flexor digitorum profundus	Deep muscle; overlain entirely by flexor digitorum superficialis	Anteromedial surface of ulna and interosseous membrane	Distal phalanges of second through fifth fingers	Sole muscle that flexes distal phalanges; assists in wrist flexion
Pronator quadratus	Deepest muscle of distal forearm	Distal portion of anterior ulnar surface	Anterior surface of radius, distal end	Pronates forearm

(*Table continues on p. 120*)

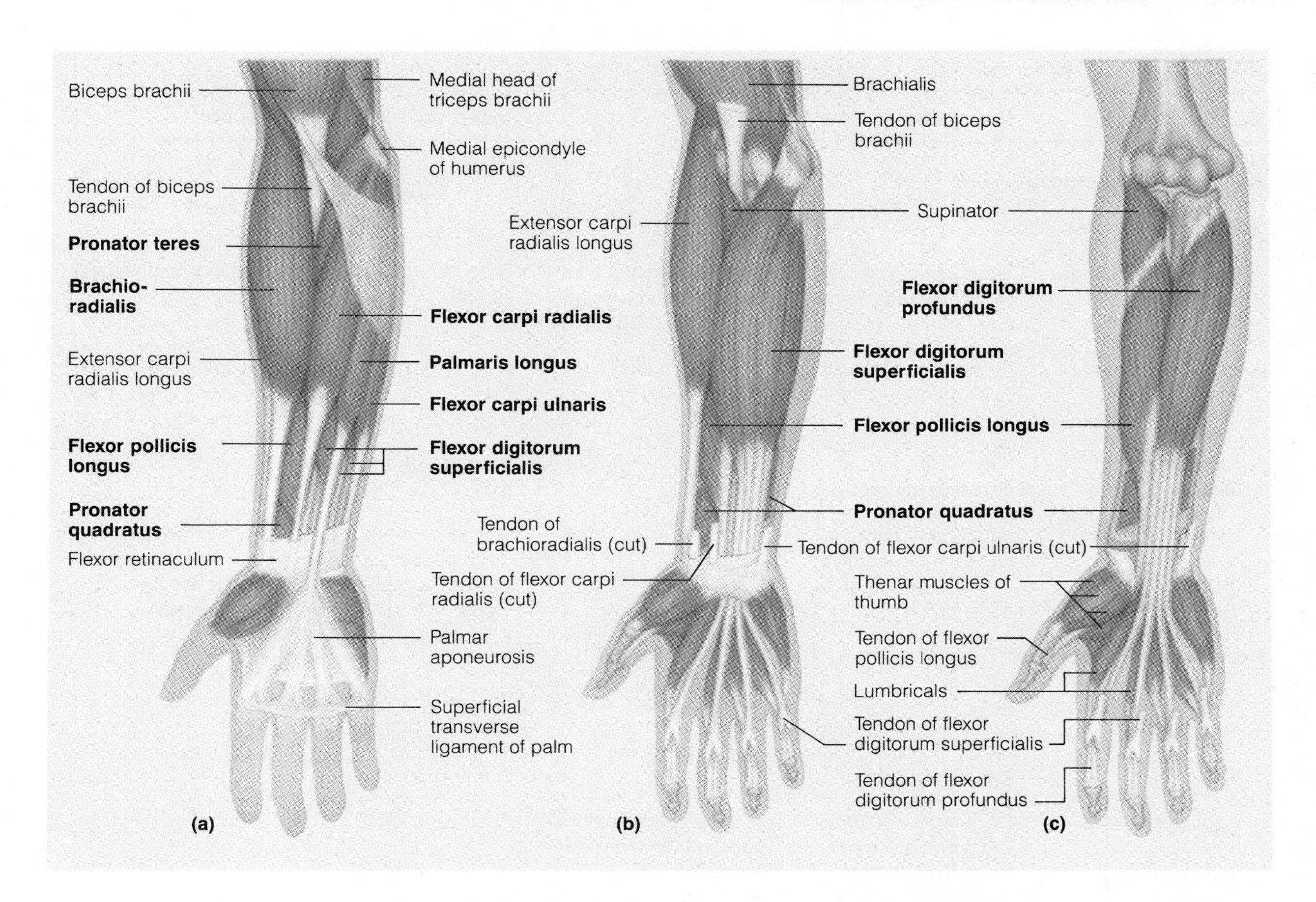

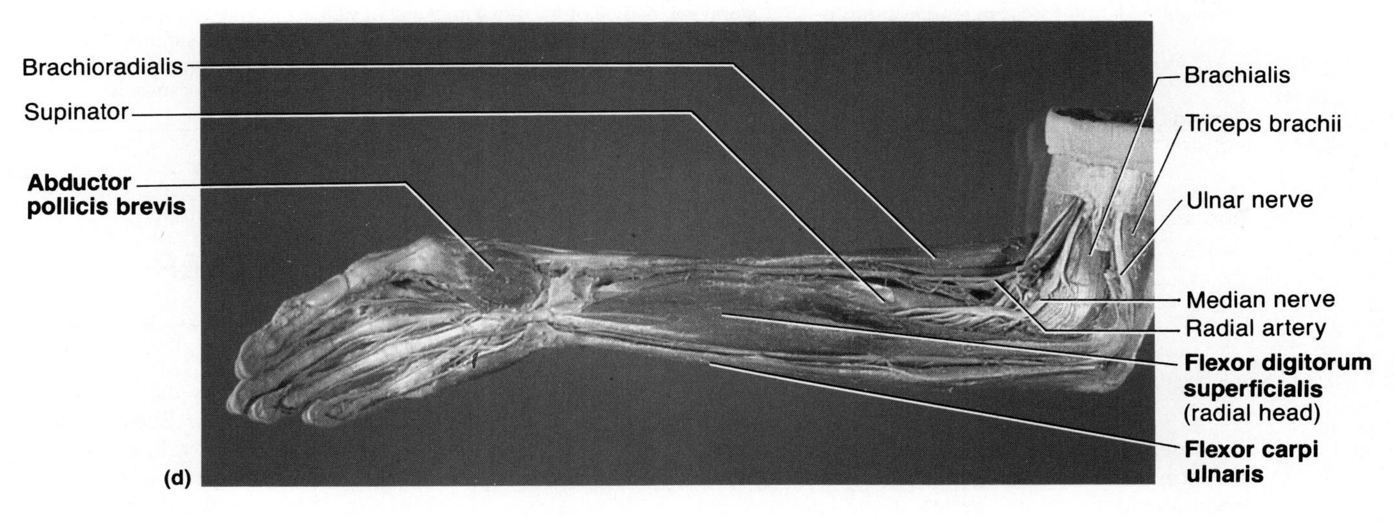

F13.8

Muscles of the forearm and wrist. (a) Superficial anterior view of muscles of the right forearm and hand. (b) The brachioradialis, flexors carpi radialis and ulnaris, and palmaris longus muscles have been removed to reveal the position of the somewhat deeper flexor digitorum superficialis. (c) Deep muscles of the anterior compartment; superficial muscles have been removed. (d) Photo of muscles of the anteromedial aspect of the forearm. The pronator teres and flexor carpi radialis muscles have been removed.

(*F13.8 continues*)

TABLE 13.6 (*Continued*)

Muscle	Comments	Origin	Insertion	Action
Posterior Compartment (Figure 13.8e,f)				
Superficial muscles				
Extensor carpi radialis longus	Superficial; parallels brachioradialis on lateral forearm	Lateral supracondylar ridge of humerus	Base of second metacarpal	Extends and abducts wrist
Extensor carpi radialis brevis	Posterior to extensor carpi radialis longus	Lateral epicondyle of humerus	Base of third metacarpal	Extends and abducts wrist; steadies wrist during finger flexion
Extensor carpi ulnaris	Superficial; medial posterior forearm	Lateral epicondyle of humerus	Base of fifth metacarpal	Extends and adducts wrist
Extensor digitorum	Superficial; between extensor carpi ulnaris and extensor carpi radialis brevis	Lateral epicondyle of humerus	By four tendons into distal phalanges of second through fifth fingers	Prime mover of finger extension; extends wrist; can flare (abduct) fingers
Deep muscles				
Extensor pollicis longus and brevis	Deep muscle pair with a common origin and action; overlain by extensor carpi ulnaris	Dorsal shaft of ulna and radius, interosseous membrane	Base of distal phalanx of thumb (longus) and proximal phalanx of thumb (brevis)	Extends thumb
Abductor pollicis longus	Deep muscle; lateral and parallel to extensor pollicis longus	Posterior surface of radius and ulna; interosseous membrane	First metacarpal	Abducts and extends thumb
Supinator	Deep muscle at posterior aspect of elbow	Lateral epicondyle of humerus	Proximal end of radius	Acts with biceps brachii to supinate forearm; antagonistic to pronator muscles

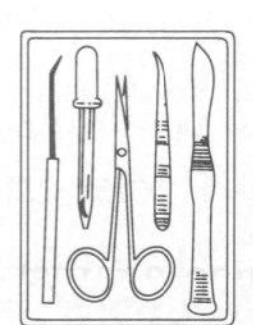

First study the tables and figures, then see if you can identify these muscles on a torso model or anatomic chart.

Complete this portion of the exercise with palpation demonstrations as outlined next.

- To observe the *biceps brachii,* attempt to flex your forearm (hand supinated) against resistance. The insertion tendon can also be felt in the lateral aspect of the antecubital fossa (where it runs toward the radius to attach).
- If you acutely flex your elbow and then try to extend it against resistance, you can demonstrate the action of your *triceps brachii.*
- Strongly flex your wrist and make a fist. Palpate your contracting wrist flexor muscles (which originate from the medial epicondyle of the humerus) and their insertion tendons, which can be easily felt at the anterior aspect of the wrist.
- Flare your fingers to identify the tendons of the *extensor digitorum* muscle on the dorsum of your hand.

MUSCLES OF THE LOWER EXTREMITY

Muscles that act on the lower extremity cause movement at the hip, knee, and foot joints. Because the human pelvic girdle is composed of heavy, fused bones that allow little movement, no special group of muscles is necessary to stabilize it. This is unlike the shoulder girdle, where many muscles (mainly trunk muscles) are necessary to stabilize the scapulae.

Muscles acting on the thigh (femur) cause various movements at the multiaxial hip joint (flexion, extension, rotation, abduction, and adduction). These include the iliopsoas, the adductor group, and other muscles summarized in Tables 13.7 and 13.8 and illustrated in Figures 13.9 and 13.10.

Muscles acting on the leg form the major musculature of the thigh. (Anatomically the term *leg* refers only to that portion between the knee and the ankle.) The thigh muscles cross the knee to allow its flexion and ex-

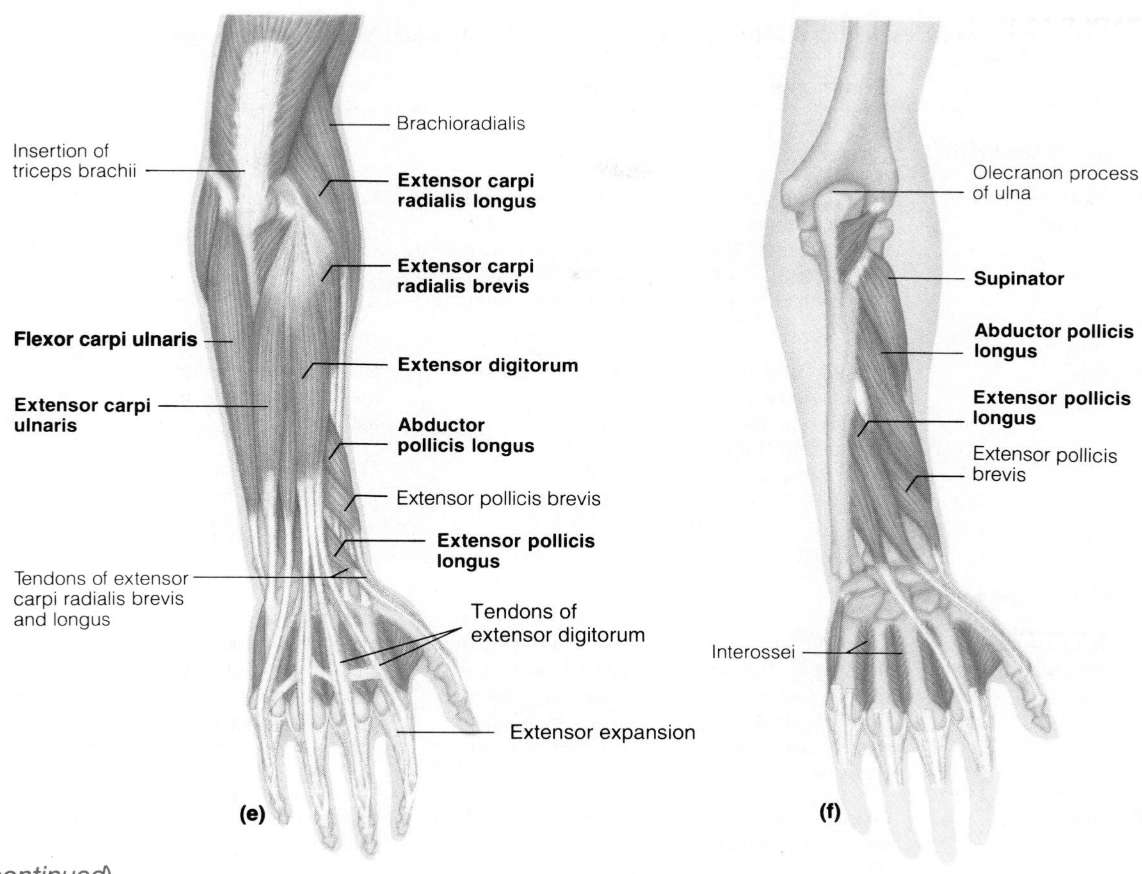

F13.8 (*continued*)

Muscles of the forearm and wrist. (e) Superficial muscles, posterior view. (f) Deep posterior muscles; superficial muscles have been removed. The interossei, the deepest layer of intrinsic hand muscles, are also illustrated. Note: The thenar muscles of the thumb and the lumbricals that help move the fingers illustrated here are not described in Table 13.6.

tension. They include the hamstrings and the quadriceps and, along with the muscles acting on the thigh, are described in Tables 13.7 and 13.8 and illustrated in Figures 13.9 and 13.10. Since some of these muscles also have attachments on the pelvic girdle, they can cause movement at the hip joint.

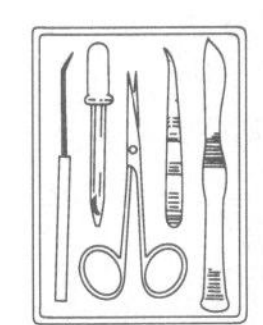

The muscles originating on the leg and acting on the foot and toes are described in Table 13.9 and shown in Figures 13.11 and 13.12. Identify the muscles as instructed previously.

Complete this exercise by performing the following palpation demonstrations.

- Go into a deep knee bend and palpate your own *gluteus maximus* muscle as you extend your hip to resume the upright posture.
- You can demonstrate the contraction of the *quadriceps femoris* by trying to extend the knee against resistance. Do this while seated and note how the patellar ligament reacts. The *biceps femoris* comes into play when you flex your knee against resistance.
- Now stand on your toes. Have your partner palpate the lateral and medial heads of the *gastrocnemius* and follow it to its insertion in the calcaneal tendon.
- Dorsiflex and invert your foot while palpating your *tibialis anterior* muscle (which parallels the sharp anterior crest of the tibia laterally).

TABLE 13.7 Muscles Acting on Human Thigh and Leg, Anterior and Medial Aspects (see Figure 13.9)

Muscle	Comments	Origin	Insertion	Action
Origin on the Pelvis				
Iliopsoas—iliacus and psoas major	Two closely related muscles; fibers pass under inguinal ligament to insert into femur via a common tendon	Iliacus: iliac fossa; psoas major: transverse processes, bodies, and discs of T_{12} and lumbar vertebrae	Lesser trochanter of femur	Flex trunk on thigh; major flexor of hip (or thigh on pelvis when pelvis is fixed)
Sartorius	Straplike superficial muscle running obliquely across anterior surface of thigh to knee	Anterior superior iliac spine	By an aponeurosis into medial aspect of proximal tibia	Flexes and laterally rotates thigh; flexes knee; known as "tailor's muscle" because it helps bring about cross-legged position in which tailors are often depicted
Medial Compartment				
Adductors—magnus, longus, and brevis	Large muscle mass forming medial aspect of thigh; arise from front of pelvis and insert at various levels on femur	Magnus: ischial and pubic rami; longus: pubis near pubic symphysis; brevis: body and inferior ramus of pubis	Magnus: linea aspera and adductor tubercle of femur; longus and brevis: linea aspera	Adduct and laterally rotate and flex thigh; posterior part of magnus is also a synergist in thigh extension
Pectineus	Overlies adductor brevis on proximal thigh	Pectineal line of pubis	Inferior to lesser trochanter of femur	Adducts, flexes, and laterally rotates thigh
Gracilis	Straplike superficial muscle of medial thigh	Inferior ramus and body of pubis	Medial surface of head of tibia	Adducts thigh; flexes and medially rotates leg, especially during walking
Anterior Compartment				
Quadriceps*				
Rectus femoris	Superficial muscle of thigh; runs straight down thigh; only muscle of group to cross hip joint; arises from two heads	Anterior inferior iliac spine and superior margin of acetabulum	Tibial tuberosity	Extends knee and flexes thigh at hip
Vastus lateralis	Forms lateral aspect of thigh	Greater trochanter and linea aspera	Tibial tuberosity	Extends knee
Vastus medialis	Forms medial aspect of thigh	Linea aspera	Tibial tuberosity	Extends knee
Vastus intermedius	Obscured by rectus femoris; lies between vastus lateralis and vastus medialis on anterior thigh	Anterior and lateral surface of femur (not shown in figure)	Tibial tuberosity	Extends knee
Tensor fasciae latae	Enclosed between fascia layers of thigh	Anterior aspect of iliac crest and anterior superior iliac spine	Iliotibial band of fascia lata	Flexes, abducts, and medially rotates thigh

*The quadriceps form the flesh of the anterior thigh and have a common insertion in the tibial tuberosity via the patellar tendon. They are powerful leg extensors, enabling humans to kick a football, for example.

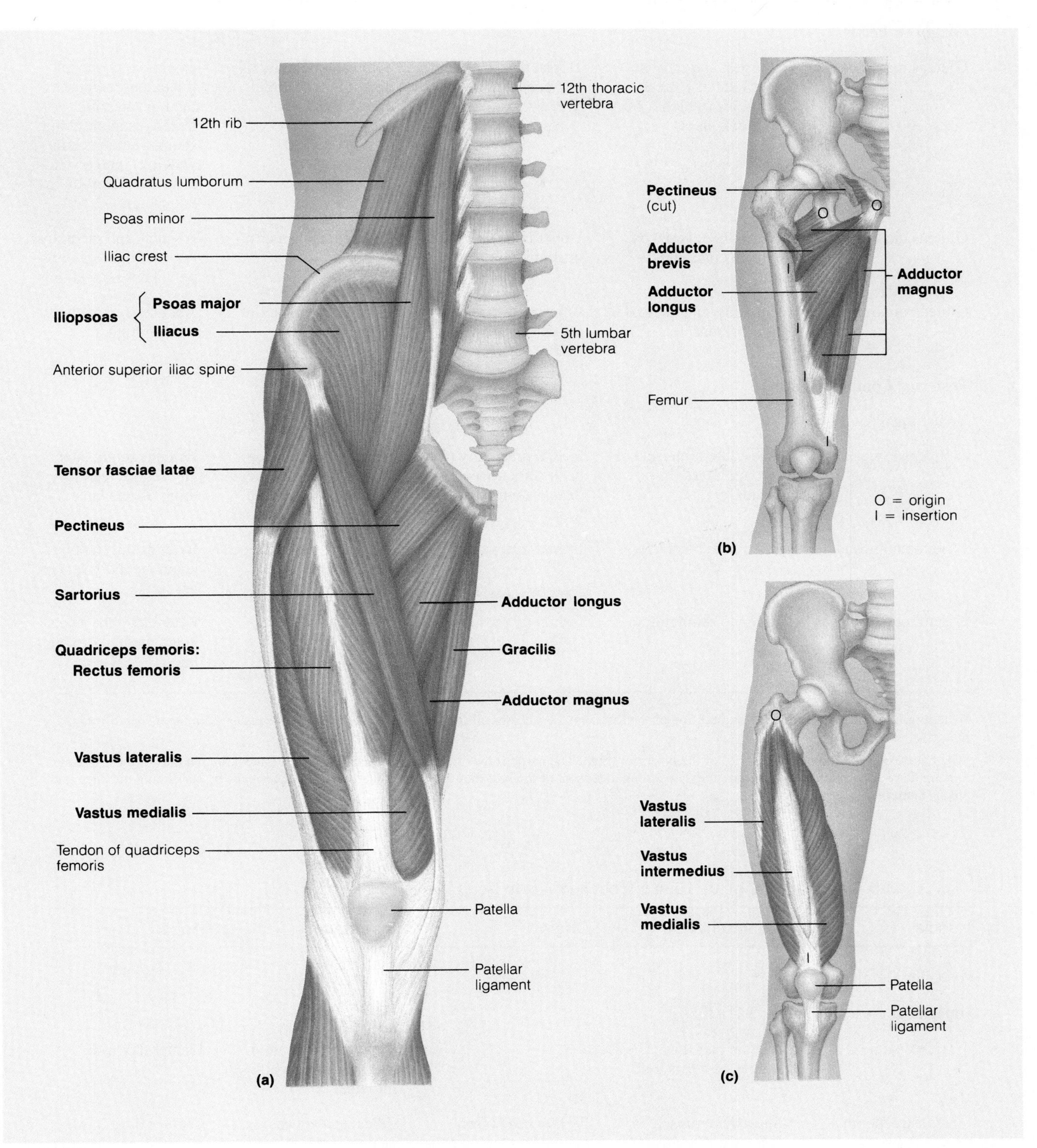

F13.9

Anterior and medial muscles promoting movements of the thigh and leg. (a) Anterior view of the deep muscles of the pelvis and superficial muscles of the right thigh. (b) Adductor muscles of the medial compartment of the thigh. Other muscles have been removed so that the origins and insertions of the adductor muscles can be seen. (c) The vastus muscles of the quadriceps group. The rectus femoris muscle of the quadriceps and surrounding muscles have been removed to reveal the attachments and extent of the vastus muscles.

TABLE 13.8 Muscles Acting on Human Thigh and Leg, Posterior Aspect (see Figure 13.10)

Muscle	Comments	Origin	Insertion	Action
Origin on Pelvis				
Gluteus maximus	Largest and most superficial of gluteal muscles (which form buttock mass)	Dorsal ilium, sacrum, and coccyx	Gluteal tuberosity of femur and iliotibial tract*	Complex, powerful hip extensor (most effective when hip is flexed, as in climbing stairs—but not as in walking); antagonist of iliopsoas; laterally rotates thigh
Gluteus medius	Partially covered by gluteus maximus	Upper lateral surface of ilium	Greater trochanter of femur	Abducts and medially rotates thigh; steadies pelvis during walking
Gluteus minimus	Smallest and deepest gluteal muscle	Inferior surface of ilium (not shown in figure)	Greater trochanter of femur	Abducts and medially rotates thigh
Posterior Compartment				
Hamstrings†				
Biceps femoris	Most lateral muscle of group; arises from two heads	Ischial tuberosity (long head); linea aspera and distal femur (short head)	Tendon passes laterally to insert into head of fibula and lateral condyle of tibia	Extends thigh; laterally rotates leg on thigh; flexes knee
Semitendinosus	Medial to biceps femoris	Ischial tuberosity	Medial aspect of upper tibial shaft	Extends thigh; flexes knee; medially rotates leg
Semimembranosus	Deep to semitendinosus	Ischial tuberosity	Medial condyle of tibia	Extends thigh; flexes knee; medially rotates leg

*The iliotibial tract, a thickened lateral portion of the fascia lata, ensheaths all the muscles of the thigh. It extends as a tendinous band from the iliac crest to the knee.

†The hamstrings are the fleshy muscles of the posterior thigh. The name comes from the butchers' practice of using the tendons of these muscles to hang hams for smoking. As a group, they are strong extensors of the hip; they counteract the powerful quadriceps by stabilizing the knee joint when standing.

TABLE 13.9 Muscles Acting on Human Foot and Ankle (see Figures 13.11 and 13.12)

Muscle	Comments	Origin	Insertion	Action
Posterior Compartment				
Superficial muscles (Figure 13.11a,b)				
Triceps surae	Muscle pair that shapes posterior calf		Via common tendon (calcaneal or Achilles) into heel	Plantar flex foot
Gastrocnemius	Superficial muscle of pair; two prominent bellies	By two heads from medial and lateral condyles of femur	Calcaneus via calcaneal tendon	Crosses knee joint; thus also can flex knee (when foot is dorsiflexed)
Soleus	Deep to gastrocnemius	Proximal portion of tibia and fibula	Calcaneus via calcaneal tendon	Plantar flexion; is an important muscle for locomotion

(Table continues on p. 126)

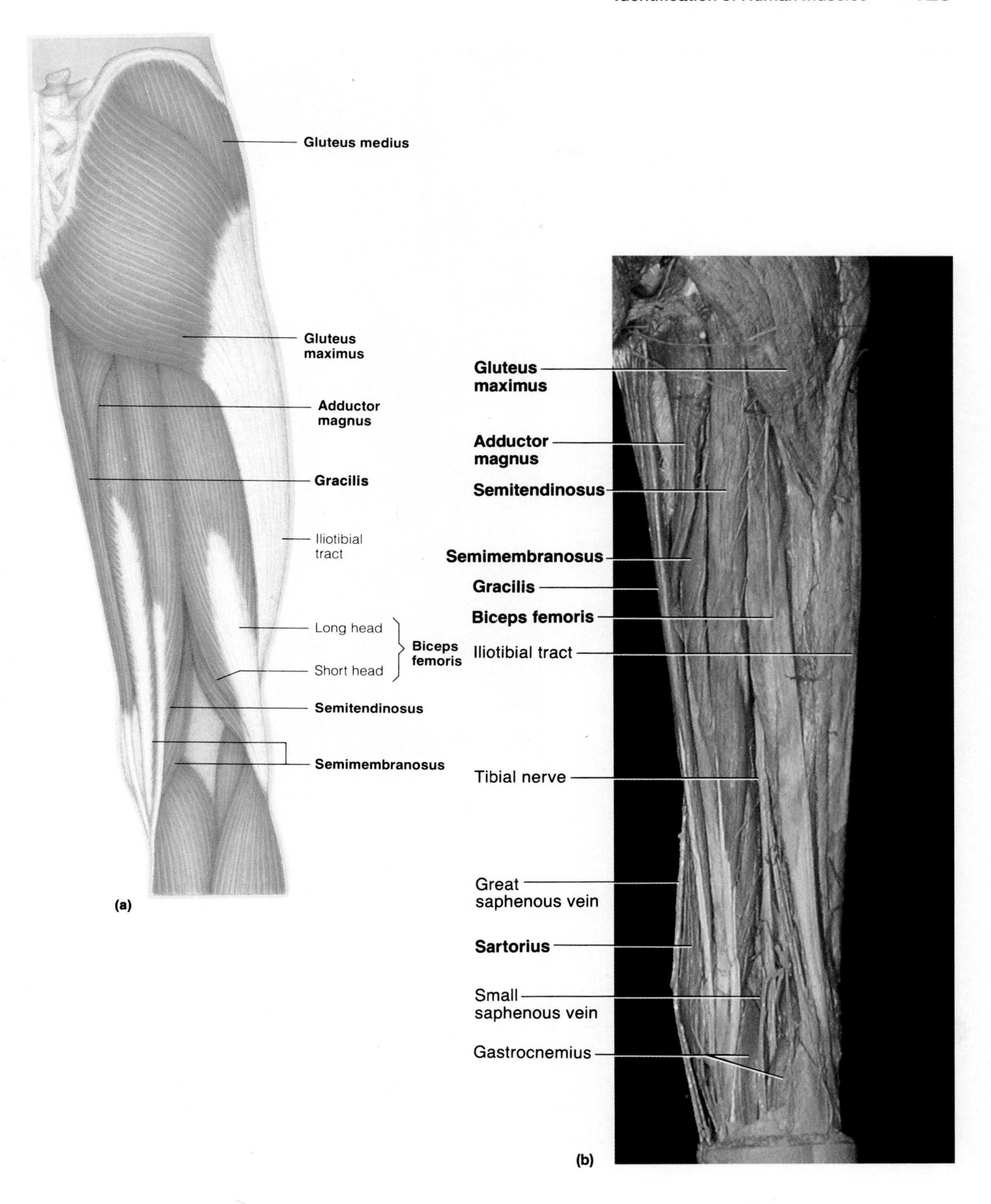

F13.10

Posterior muscles of the right hip and thigh. (a) Superficial view showing the gluteus muscles of the buttock and hamstring muscles of the thigh. (b) Photo of muscles of the posterior thigh.

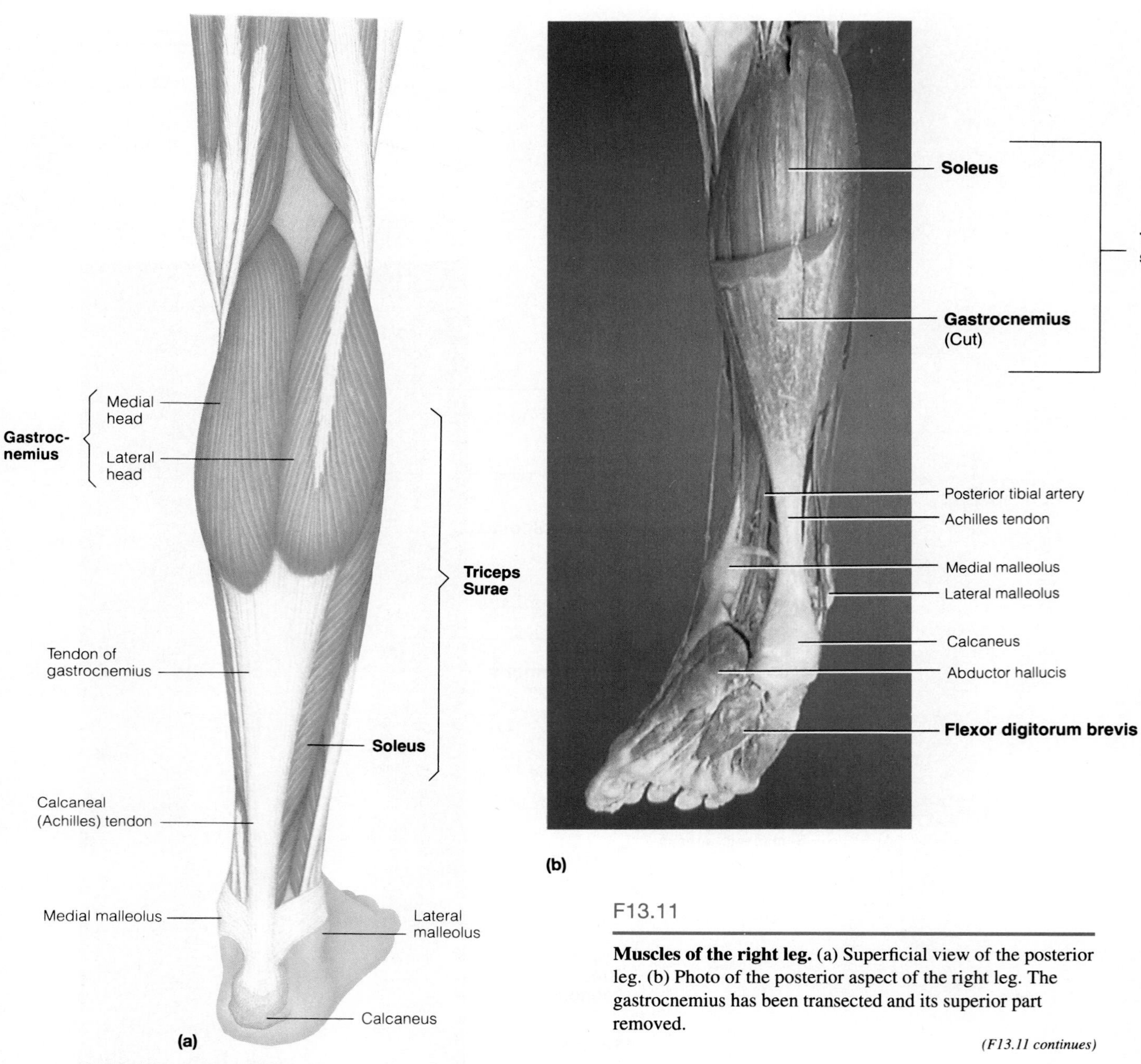

F13.11

Muscles of the right leg. (a) Superficial view of the posterior leg. (b) Photo of the posterior aspect of the right leg. The gastrocnemius has been transected and its superior part removed.

(F13.11 continues)

TABLE 13.9 *(Continued)*

Muscle	Comments	Origin	Insertion	Action
Deep muscles (Figure 13.11c,d)				
Popliteus	Thin muscle at posterior aspect of knee	Lateral condyle of femur	Proximal tibia	Flexes and rotates leg medially to "unlock" extended knee when flexion begins
Tibialis posterior	Thick muscle deep to soleus	Superior portion of tibia and fibula and interosseous membrane	Tendon passes obliquely behind medial malleolus and under arch of foot; inserts into several tarsals and metatarsals 2–4	Prime mover of foot inversion; plantar flexes foot

TABLE 13.9 (*Continued*)

Muscle	Comments	Origin	Insertion	Action
Flexor digitorum longus	Runs medial to and partially overlies tibialis posterior	Posterior surface of tibia	Distal phalanges of second through fifth toes	Flexes toes; plantar flexes and inverts foot
Flexor hallucis longus (see also Figure 13.12)	Lies lateral to inferior aspect of tibialis posterior	Middle portion of fibula shaft	Tendon runs under foot to insert on distal phalanx of great toe	Flexes great toe; plantar flexes and inverts foot; the "push-off muscle" during walking

(*Table continues on p. 129*)

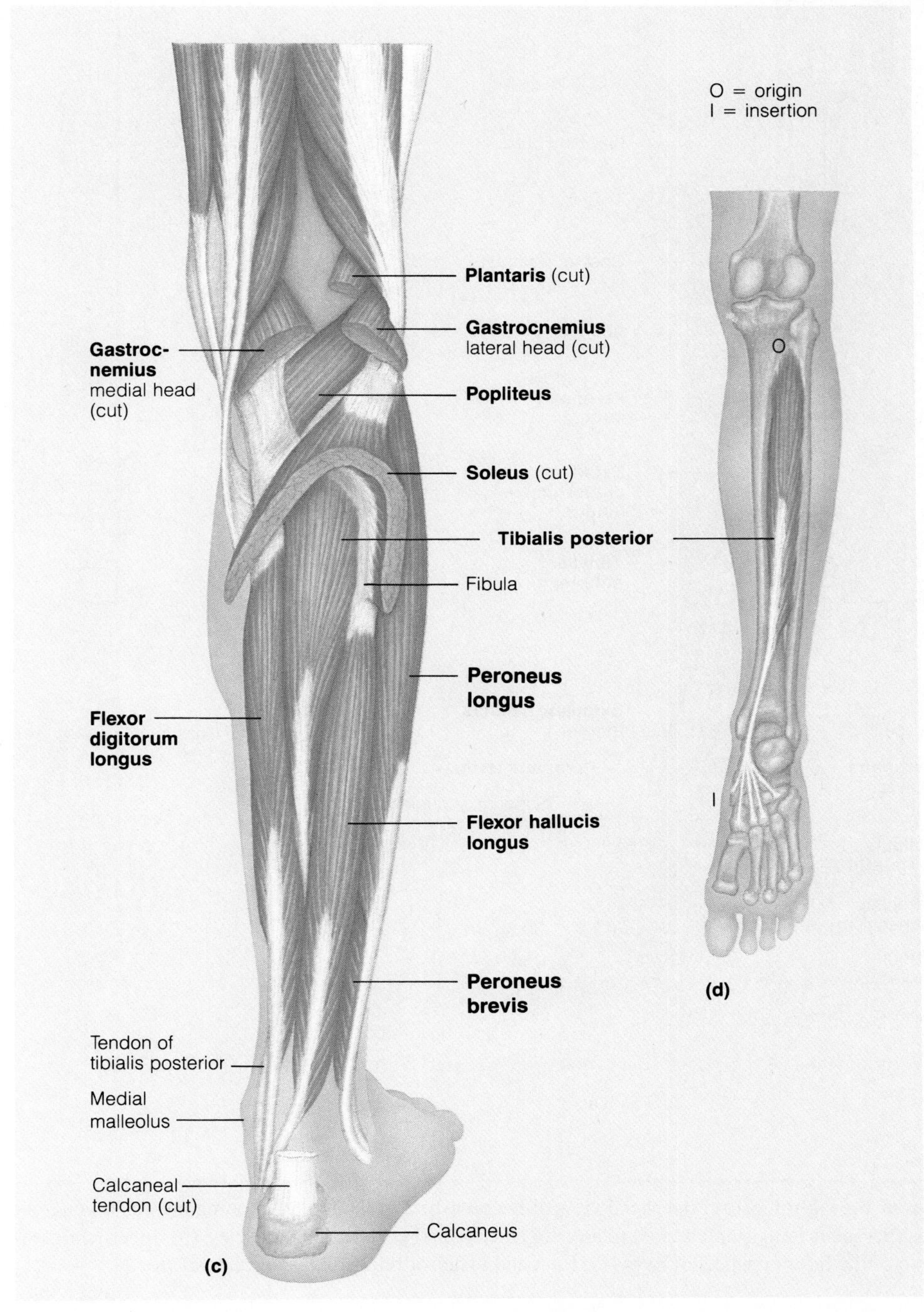

F13.11 (*continued*)

Muscles of the right leg. (c) The triceps surae has been removed to show the deep muscles of the posterior compartment. (d) Tibialis posterior shown in isolation so that its origin and insertion may be visualized.

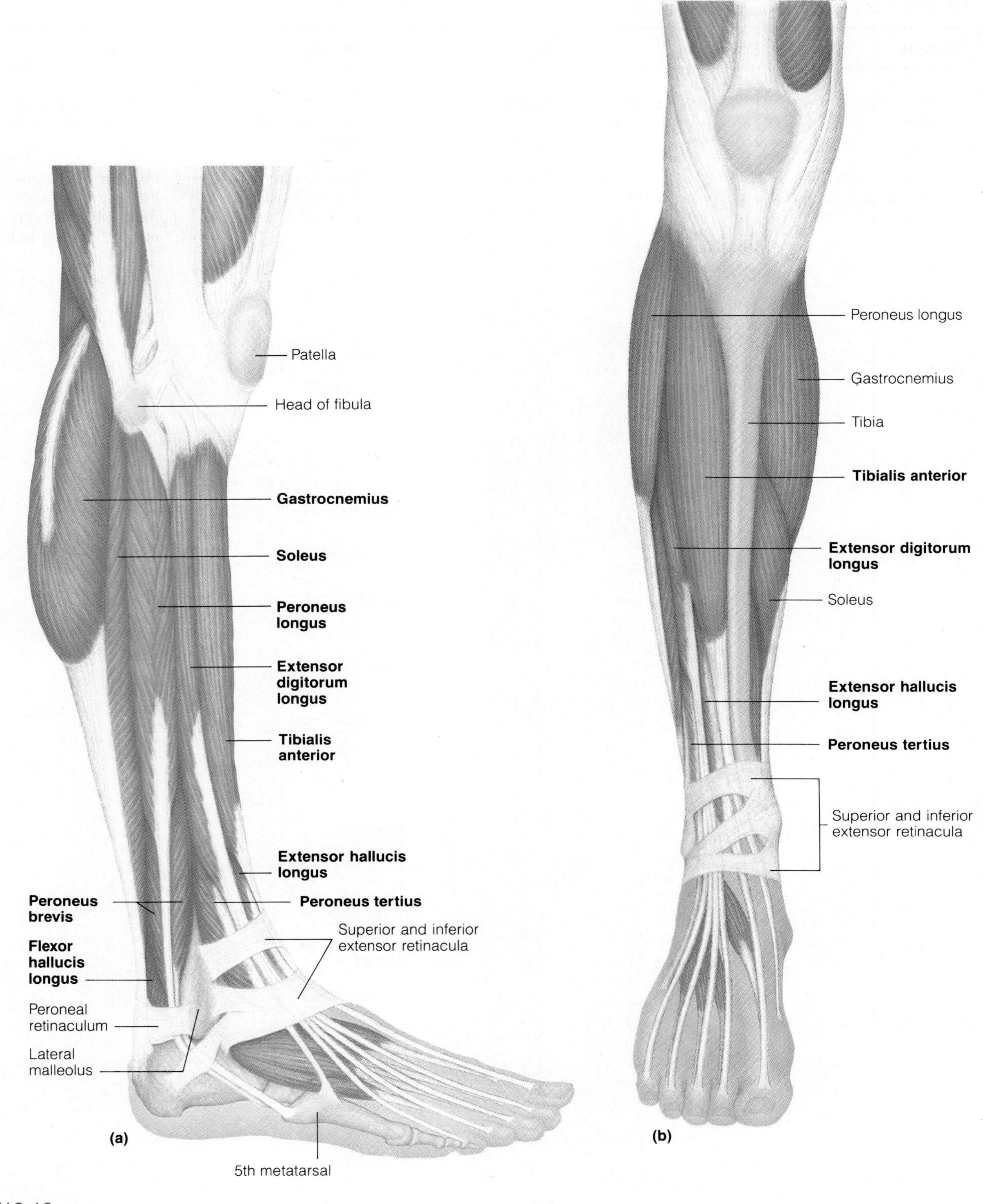

F13.12

Muscles of the right leg. (a) Superficial view of lateral aspect of the leg, illustrating the positioning of the lateral compartment muscles (peroneus longus and brevis) relative to anterior and posterior leg muscles. (b) Superficial view of anterior leg muscles. (c) Photo of the inferior aspect of right leg, foot, and extensor retinacula, anterolateral view.

TABLE 13.9 *(Continued)*

Muscle	Comments	Origin	Insertion	Action
Lateral Compartment (Figure 13.11c and Figure 13.12a,b,c)				
Peroneus longus	Superficial lateral muscle; overlies fibula	Head and upper portion of fibula	By long tendon under foot to first metatarsal and medial cuneiform	Plantar flexes and everts foot; helps keep foot flat on ground
Peroneus brevis	Smaller muscle; deep to peroneus longus	Distal portion of fibula shaft	By tendon running behind lateral malleolus to insert on proximal end of fifth metatarsal	Plantar flexes and everts foot, as part of peronei group
Anterior Compartment (Figure 13.12a,b,c)				
Tibialis anterior	Superficial muscle of anterior leg; parallels sharp anterior margin of tibia	Lateral condyle and upper 2/3 of tibia; interosseous membrane	By tendon into inferior surface of first cuneiform and metatarsal 1	Prime mover of dorsiflexion; inverts foot
Extensor digitorum longus	Anterolateral surface of leg; lateral to tibialis anterior	Lateral condyle of tibia; proximal 3/4 of fibula; interosseous membrane	Tendon divides into four parts; insert into middle and distal phalanges of toes 2–5	Prime mover of toe extension; dorsiflexes and everts foot
Peroneus tertius	Small muscle; often fused to distal part of extensor digitorum longus	Distal anterior surface of fibula	Tendon passes anterior to lateral malleolus and inserts on dorsum of fifth metatarsal	Dorsiflexes and everts foot
Extensor hallucis longus	Deep to extensor digitorum longus and tibialis anterior	Anteromedial shaft of fibula and interosseous membrane	Tendon inserts on distal phalanx of great toe	Extends great toe; dorsiflexes foot

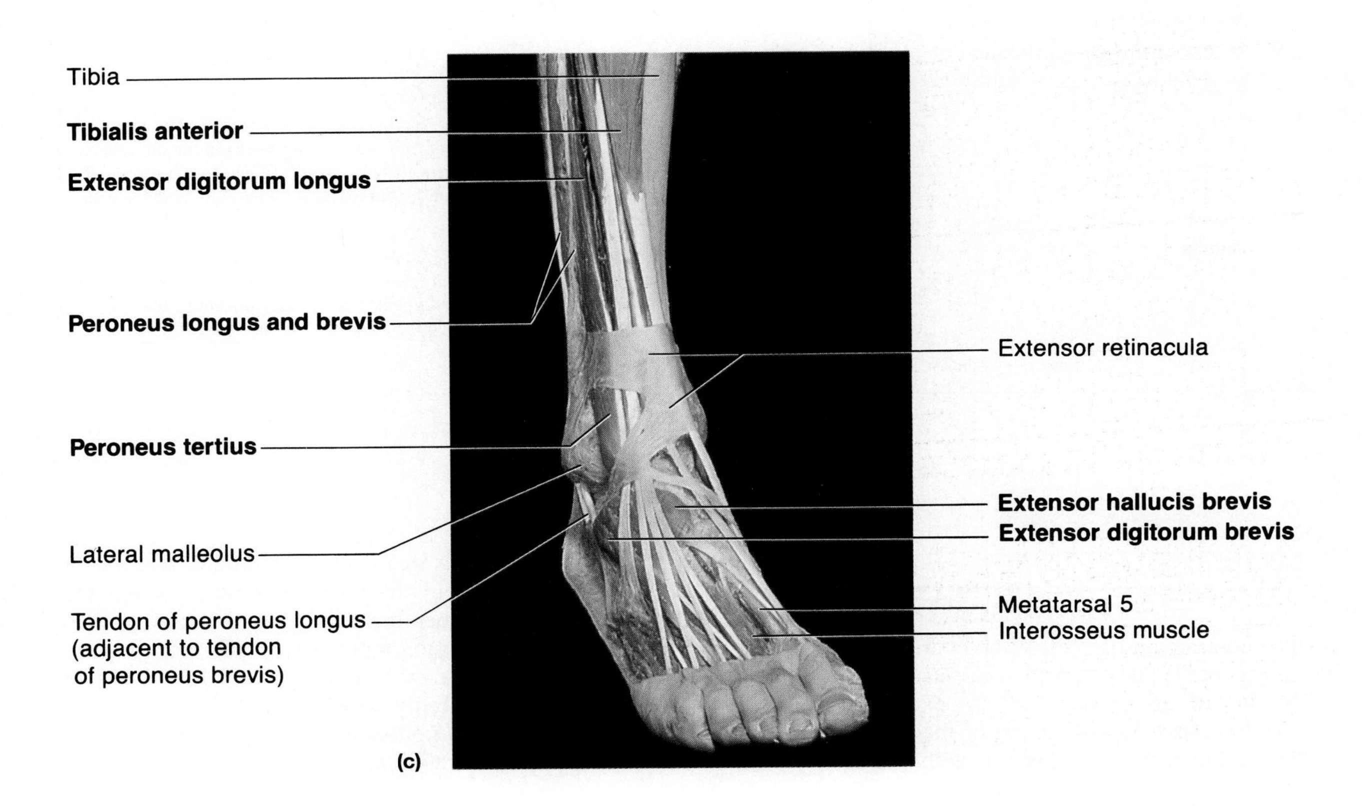

(c)

14A EXERCISE

Muscle Physiology (Frog Experimentation)

OBJECTIVES

1. To observe muscle contraction on the microscopic level, and to describe the role of ATP and various ions in muscle contraction.
2. To define and explain the physiological basis for the events of muscle contraction.
3. To trace the events that result from the electrical stimulation of a muscle.
4. To explain why the "all or none" law is reflected in the activity of a single muscle cell but not in an intact skeletal muscle.
5. To recognize that the graded response of skeletal muscle is a function of the number of muscle fibers stimulated and the frequency of stimulation.
6. To name and describe the phases of a muscle twitch.
7. To distinguish between a muscle twitch and a sustained (tetanic) contraction, and to describe their importance in normal muscle activity.
8. To demonstrate how the kymograph or polygraph can be used to obtain pertinent and representative recordings of various physiologic events of skeletal muscle activity.
9. To explain the significance of muscle tracings obtained during experimentation.

MATERIALS

ATP muscle kits (glycerinated rabbit psoas muscle*; ATP and salt solutions obtainable from Carolina Biological Supply)

Petri dishes
Microscope slides
Cover glasses
Millimeter ruler
Compound microscope
Dissecting microscope
Small beaker (50 ml)
Frog Ringer's solution
Disposable plastic gloves
Apparatus A or B
Scissors
Metal needle probes
Pointed glass probes (teasing needles)
Medicine dropper
Cotton thread
Forceps
Glass or porcelain plate
Pithed bullfrog†

A. Physiograph (polygraph), polygraph paper and ink, myograph, pin and clip electrodes, stimulator output extension cable, transducer cable, straight pins, frog board, laboratory stand, clamp

B. Kymograph, kymograph paper (smoking stand, burner, and glazing fluid if using a smoke-writing apparatus), skeletal muscle lever, signal magnet, laboratory stand, clamp, electronic stimulators

Notes to Instructor:

* At the beginning of the lab, the muscle bundle should be removed from the test tube and cut into ~2-cm lengths. Both the cut muscle segments and the entubed glycerol should be put into a petri dish. One muscle *segment* is sufficient for each two to four students making observations.

† Bullfrogs to be pithed by lab instructor as needed for student experimentation. If the instructor would prefer that students pith their own specimens, an instructional sheet on that procedure suitable for copying for student handouts is provided in the Instructor's Guide.

MUSCLE ACTIVITY

The contraction of skeletal and cardiac muscle can be considered in terms of three events—the electrical excitation of the muscle cell, the excitation-contraction coupling, and the shortening of the muscle cell due to the sliding of the myofilaments within it.

At rest, all cells maintain a potential difference, or voltage, across their plasma membrane in which the inner face of the membrane is approximately −60 to −90 millivolts (mv) compared with the cell exterior. This potential difference is a result of differences in membrane permeability to cations, most importantly sodium (Na^+) and potassium (K^+) ions. Intracellular potassium concentration is much greater than its extracellular concentration, and intracellular sodium concentration is considerably less than its extracellular concentration. Hence, steep concentration gradients across the membrane exist for both cations. However, because the plasma membrane is slightly more permeable to K^+ than Na^+, Na^+ influx into the cell is inadequate to balance K^+ outflow. The result of this unequal Na^+-K^+ diffusion across the membrane establishes the cell's **resting membrane potential.** The resting membrane potential is of particular interest in excitable cells like muscle cells or neurons, because changes in that voltage underlie their ability to do work (contract or signal respectively in muscle cells and neurons).

Action Potential

When a muscle cell is stimulated, the sarcolemma becomes temporarily permeable to sodium, which rushes into the cell. This sudden influx of sodium ions alters the membrane potential. That is, the cell interior becomes less negatively charged at that point, an event called **depolarization.** When depolarization reaches a certain level and the sarcolemma momentarily changes its polarity, a depolarization wave travels along the sarcolemma. Even as the influx of sodium ions occurs, the sarcolemma becomes impermeable to sodium and permeable to potassium ions. Consequently, potassium ions leak out of the cell, restoring the resting membrane potential (but not the original ionic conditions), an event called **repolarization.** The repolarization wave follows the depolarization wave across the cell membrane. This rapid depolarization and repolarization of the membrane that is propagated along the entire membrane from the point of stimulation is called the **action potential.**

Until repolarization of the membrane has been completed, the muscle cell cannot be stimulated to contract again. In this condition it is said to be in the **absolute refractory period.** Repolarization restores the muscle cell's irritability. Temporarily, the sodium-potassium pump, which actively transports potassium ions into the cell and sodium ions out of the cell, need not be activated. But if the cell is stimulated to contract again and again, the loss of K^+ and gain of Na^+ occurring during action potential generation begins to hamper its ability to respond. And so, eventually the sodium-potassium pump must be activated to reestablish the ionic concentrations of the resting state.

Contraction

The propagation of the action potential along the sarcolemma causes the release of calcium ions (Ca^{2+}) from storage depots (tubules of the sarcoplasmic reticulum) within the muscle cell. When the calcium ions reach the myofilaments and bind to regulatory proteins on the actin filaments, they act as an ionic trigger that initiates contraction, and the actin and myosin filaments slide past each other. Once the action potential ends, the calcium ions are almost immediately reabsorbed into the tubules of the sarcoplasmic reticulum. Instantly the muscle cell relaxes.

The events of the contraction process can most simply be summarized as follows: muscle cell contraction is initiated by generation and transmission of an action potential along the sarcolemma. This electrical event is coupled with the sliding of the myofilaments—contraction—by the release of Ca^{2+}. Keep in mind this sequence of events as you conduct the experiments.

OBSERVATION OF MUSCLE FIBER CONTRACTION

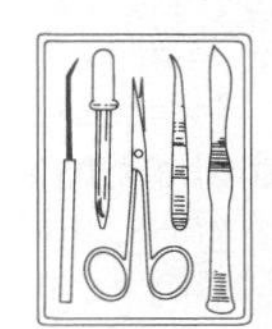

In this simple observational experiment, you will have the opportunity to review your understanding of muscle cell anatomy and to watch the fiber contracting (or not contracting) in response to the presence of certain chemicals (ATP, and potassium and magnesium ions).

1. Obtain the following materials from the supply area: 2 glass teasing needles; 3 glass microscope slides and cover glasses; millimeter ruler; dropper vials containing the following solutions: a. 0.25% ATP in triply distilled water, b. 0.25% ATP plus 0.05 *M* KCl plus 0.001 *M* $MgCl_2$ in distilled water, and c. 0.05 *M* KCl plus 0.001 *M* $MgCl_2$ in distilled water; a petri dish; and a previously cut muscle bundle segment. While you are at the supply area, place the muscle bundle in the petri dish and pour a small amount of glycerol (the fluid in the supply petri dish) over your muscle segment. Also obtain both a compound and a dissecting microscope, and bring them to your laboratory bench.

2. Using the fine glass needles, tease the muscle segment to separate its fibers. The objective is to isolate *single* muscle cells or fibers for observation. Be patient and work carefully so that the fibers do not get torn during this isolation procedure.

3. Transfer one or more of the fibers (or the thinnest strands you have obtained) onto a clean microscope slide with a glass needle, and cover them with a cover glass. Examine the fibers under low- and then high-power magnifications to observe the striations and the smoothness of the fibers when they are in the relaxed state.

4. Using a glass needle, transfer three or four fibers to a second clean microscope slide. Using the needle as a prod, carefully position the fibers so that they are parallel to one another and are as straight as possible. Place this slide under a dissecting microscope, and measure the length of each fiber by holding a millimeter ruler adjacent to it. Alternatively, you can rest the microscope slide *on* the millimeter ruler to make your length determinations. Record the fiber lengths on the chart that follows on page 132.

5. Flood the fibers (situated under the dissecting microscope) with several drops of the solution containing ATP + potassium ions and magnesium ions. Watch the reaction of the fibers after adding the solution. After 30 seconds (or slightly longer), remeasure each fiber and

record the observed lengths on the chart. Also, observe the fibers to see if any width changes have occurred. Calculate the degree (or percentage) of contraction by using the following simple formula, and record this information on the chart also:

$$\text{Initial length (mm)} - \text{contracted length (mm)} = \text{degree of contraction (mm)}$$

Then:

$$\frac{\text{degree of contraction (mm)}}{\text{initial length (mm)}} \times 100 = \%\ \text{contraction}$$

6. Carefully transfer one of the contracted fibers to a clean microscope slide, cover with a cover glass, and observe with the compound microscope. Mentally compare your initial observations with the view you are observing now. What differences do you see?

(Be specific.) ______________________________

____

____

What zones (or bands) have disappeared?

____

7. Repeat steps 3 through 6, using clean slides and fresh muscle cells. First use the solution of ATP in distilled water (no salts), and then use the solution containing only salts (no ATP) for the third series.

What degree of contraction was observed when ATP was applied in the absence of potassium and magnesium ions?

____

What degree of contraction was observed when the muscle fibers were flooded with a K^+ and Mg^{2+}-containing solution that lacked ATP?

____

What conclusions can you draw about the importance of ATP and potassium and magnesium ions to the muscle contractile process?

____

____

____

INDUCTION OF CONTRACTION IN THE FROG GASTROCNEMIUS MUSCLE

Physiologists have learned a great deal about the way muscles function by isolating muscles from laboratory animals and then stimulating these muscles to observe their responses. Various stimuli—electrical shock, temperature changes, extremes of pH, certain chemicals—elicit muscle activity, but laboratory experiments of this type typically use electrical shock. This is because it is easier to control the onset and cessation of electrical shock, as well as the strength of the stimulus.

Preparing a Muscle for Experimentation

The preparatory work that precedes the recording of muscle activity tends to be quite time-consuming. If you work in teams of two or three, the work can be divided. While one of you is setting up the recording apparatus (physiograph or kymograph), one or two students can dissect the frog leg. Experimentation should begin as soon as the dissection has been completed.

Various types of apparatus are used to record muscle contraction. All include a way to mark time intervals, a way to indicate exactly when the stimulus was applied, and a way to measure the magnitude of the contraction response. Instructions are provided here for setting up physiograph (Figure 14A.1) and kymograph (Figure 14A.2) apparatus. Specific instructions for use of the recording apparatus during recording will be provided by your instructor.

	Muscle Fiber 1	**Muscle Fiber 2**	**Muscle Fiber 3**
Initial length (mm)			
Contracted length (mm)			
% contraction			

(*Text continues on p. 135*)

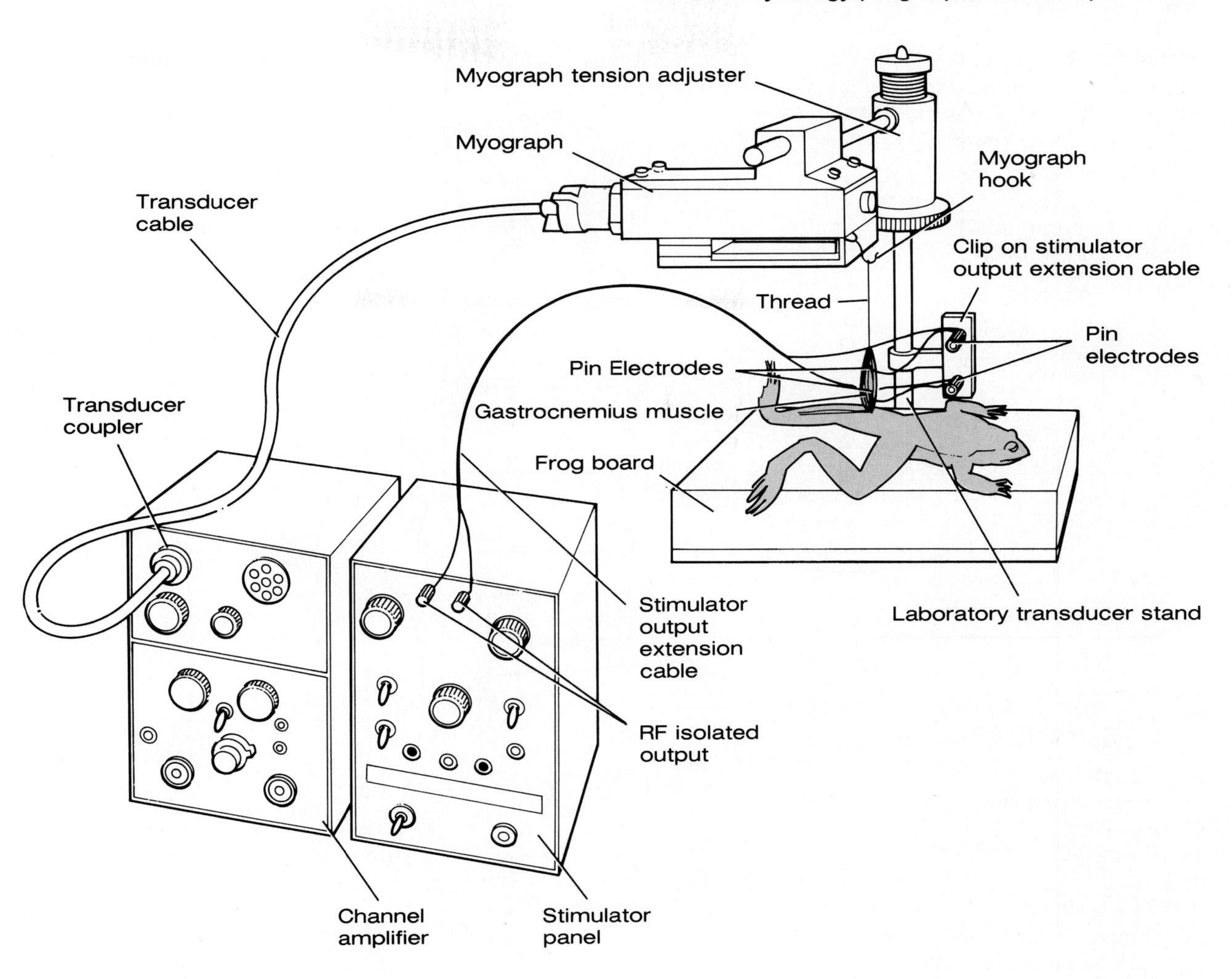

Materials:
Channel amplifier and stimulator transducer cable
Stimulator panel and stimulator output extension cable
Myograph
Myograph tension adjuster
Transducer stand
Two pin electrodes
Frog board and straight pins
Prepared frog (gastrocnemius muscle freed and
 Achilles tendon ligated with thread)
Frog Ringer's solution

1. Connect myograph to transducer stand and attach frog board to stand.

2. Attach transducer cable to myograph and to input connection on amplifier channel.

3. Attach stimulator output extension cable to output on stimulator panel (red to red, black to black).

4. Using clip at opposite end of extension cable, attach cable to bottom of transducer stand adjacent to frog board.

5. Attach two pin electrodes securely to electrodes on clip.

6. Place knee of prepared frog in clip on frog board and secure by inserting a straight pin through tissues of frog. Keep frog muscle moistened with Ringer's solution.

7. Attach thread ligating the Achilles tendon of frog to myograph leaf-spring hook.

8. Adjust position of myograph on stand to produce a constant tension on thread attached to muscle (taut but not tight). Gastrocnemius muscle should hang vertically directly below myograph hook.

9. Insert free ends of pin electrodes into muscle, one at proximal end and other at distal end.

F14A.1

Physiograph setup for frog gastrocnemius experiments.

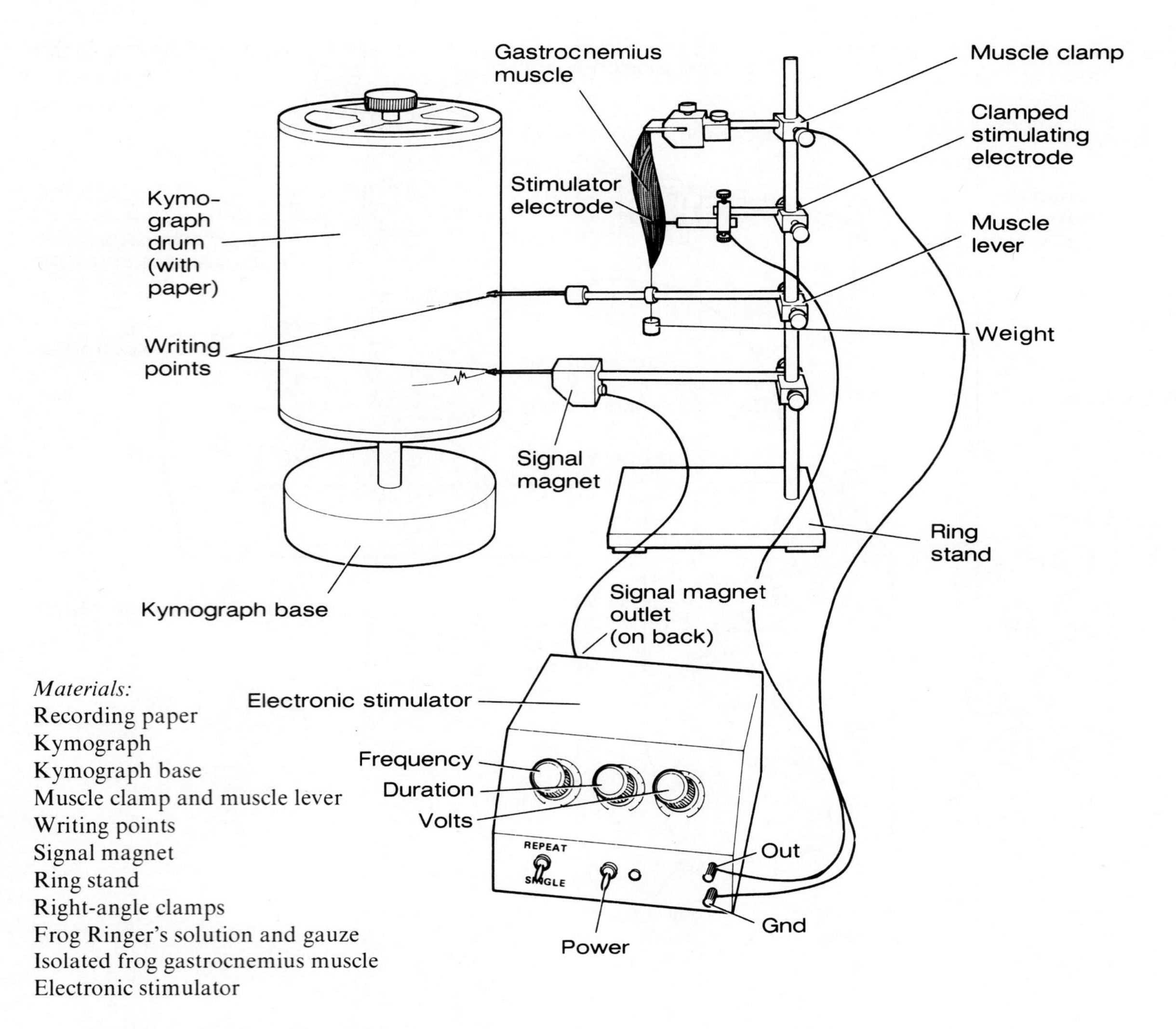

Materials:
Recording paper
Kymograph
Kymograph base
Muscle clamp and muscle lever
Writing points
Signal magnet
Ring stand
Right-angle clamps
Frog Ringer's solution and gauze
Isolated frog gastrocnemius muscle
Electronic stimulator

1. Attach muscle clamp to supporting rod and secure cut femur in clamp.

2. Position and clamp muscle lever to supporting rod beneath muscle tissue. Tie thread ligating the Achilles tendon to hook on the muscle lever so that muscle hangs vertically. Keep muscle moist with Ringer's solution from this point until completion of preparations and experimentation.

3. Hang a 10-g weight on muscle lever just below thread attachment to provide proper tension on lever. Muscle lever should rest horizontally. Writing tip should lightly touch paper surface on drum. Test its position by pulling lightly and directly upward on thread. Writing tip should remain in contact with drum surface.

4. Clamp signal magnet to supporting rod. Tip of its writing stylus should line up directly beneath that of muscle lever. *Precise alignment is important* so that a correlation can be seen between the point of stimulus and onset of muscle contractions.

5. Position stimulator to one side so that area immediately in front of recording apparatus is free for manipulations during and after experimentation. Turn all its switches to OFF and all dials to left. Plug the stimulator into the power outlet.

6. Connect two gray wires from signal magnet outlet (back of stimulator) to binding posts on signal magnet. To operate signal magnet, toggle switch (upper left on front of stimulator) must be ON (flipped up). Signal magnet will indicate up to 30 pulses per second when it is operating. If you use frequencies greater than 30 pulses per second during experimentation, turn signal magnet toggle switch to OFF (down) position.

7. Wash platinum tips of stimulator electrode in distilled water and dry. Connect electrode to white and black output terminals of stimulator and clamp electrode to supporting rod so that its tips rest on muscle tissue.

F14A.2

Kymograph setup for frog gastrocnemius experiments.

1. Before beginning the frog dissection, have the following supplies ready at your laboratory bench: a small beaker containing 20 to 30 ml of frog Ringer's solution, scissors, a metal needle probe, a glass probe with a pointed tip, a medicine dropper, cotton thread, forceps, and a glass or porcelain plate. While these supplies are being accumulated, one member of your team should notify the instructor that you are ready to begin experimentation so that the frog can be prepared (pithed). Preparation of the frog in this manner renders it unable to feel pain and prevents reflex movements (like hopping) that would interfere with the experiments.

2. All students who will be handling the frog should obtain and don disposable plastic gloves. Obtain the pithed frog and place it, ventral surface down, on the glass plate. Make an incision into the skin approximately midthigh (Figure 14A.3), and then con-

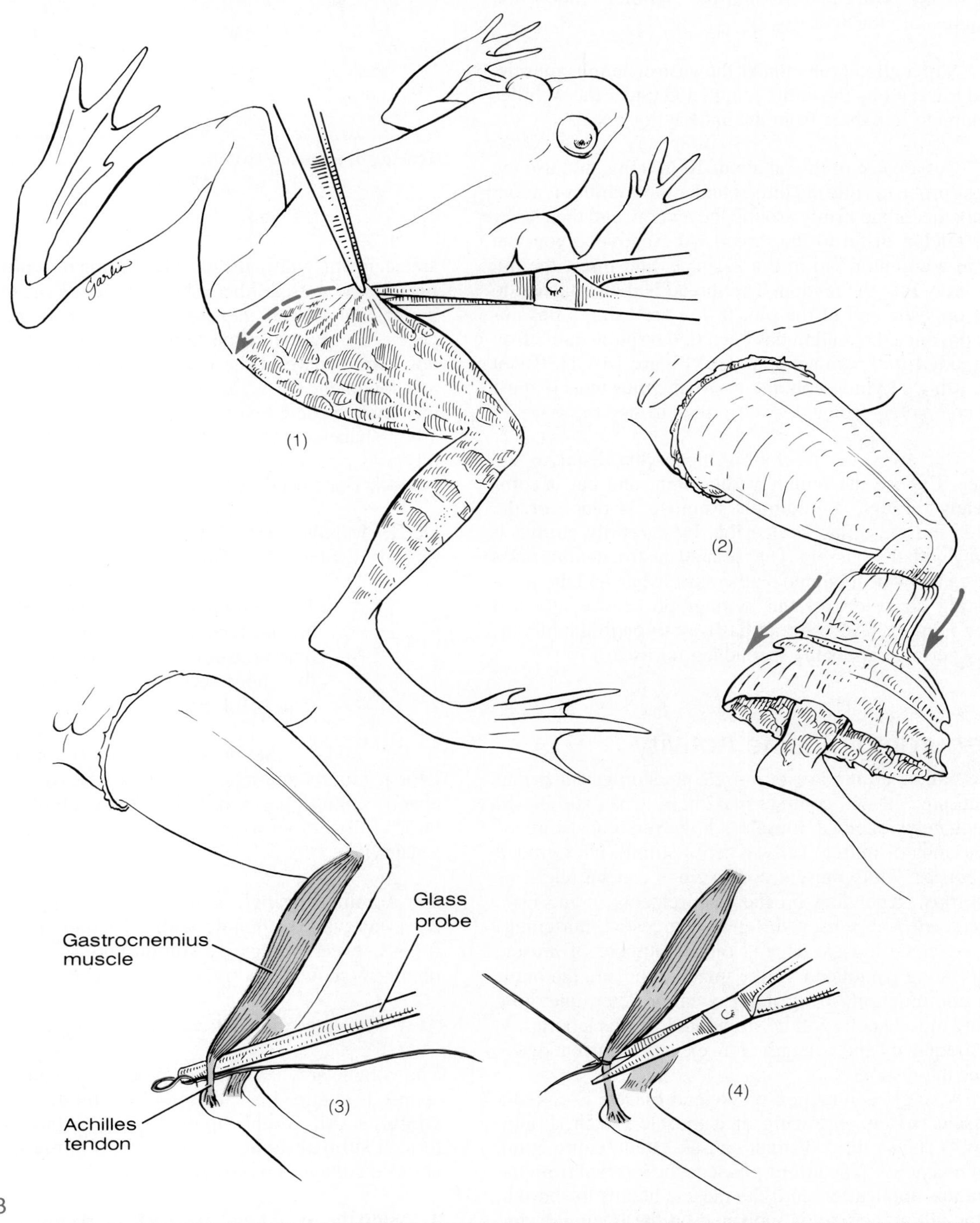

F14A.3

Preparation of the frog gastrocnemius muscle. Numbers indicate sequence of manipulations.

tinue the cut completely around the thigh. Grasp the skin with the forceps and strip it from the leg and hind foot. The skin tends to adhere more at the joints, but a careful, persistent pulling motion—somewhat like pulling off a nylon stocking—will enable you to remove it in one piece. *From this point on, the exposed muscle tissue should be routinely moistened* with the Ringer's solution to prevent spontaneous twitches.

3. Identify the gastrocnemius muscle (the fleshy muscle of the posterior calf) and the Achilles tendon that secures it to the heel.

4. Slip a glass probe under the gastrocnemius muscle, and run it along the entire length and under the Achilles tendon to free them from the underlying tissues.

5. Cut a piece of thread about 10 in. long, and use the glass probe to slide the thread under the Achilles tendon. Knot the thread firmly around the tendon, and then sever the tendon distal to the thread. Alternatively, you can bend a common pin into a Z-shape, and insert the pin securely into the tendon. The thread is then attached to the opposite end of the pin. If you are using a physiograph, once the tendon has been tied or pinned, the frog is ready for experimentation (see Figure 14A.1). If you are using a kymograph, the gastrocnemius muscle must be completely isolated, as described in step 6.

6. Cut away the fibulotibial bone just distal to the knee. Expose the femur of the thigh, and cut it completely through at midthigh. Remove as much of the thigh muscle tissue as possible by carefully cutting it away with the scissors. The isolated gastrocnemius muscle can now be mounted on the muscle bar, and the stimulating electrodes of the kymograph can be attached (see Figure 14A.2). About halfway through the laboratory period, dissect the second leg for use.

Recording Muscle Activity

The **"all or none" law** of muscle physiology states that a muscle cell will contract maximally when stimulated adequately. Skeletal muscles, however, consisting of thousands of muscle cells, react to stimuli with graded responses. Thus muscle contractions can be slight or vigorous, depending on the requirements of the task. The *graded responses* (different degrees of shortening) of a skeletal muscle depend on the number of muscle cells being stimulated. In the intact organism, the number of motor units firing at any one time determines how many muscle cells will be stimulated; in this laboratory, the frequency and strength of an electrical current determine the response.

A single contraction of skeletal muscle is called a **muscle twitch.** A tracing of a muscle twitch (Figure 14A.4) shows three distinct phases: latent, contraction, and relaxation. The **latent phase** is the interval from the stimulus application until the muscle begins to shorten. Although no activity is indicated on the tracing during this phase, important electrical and chemical changes

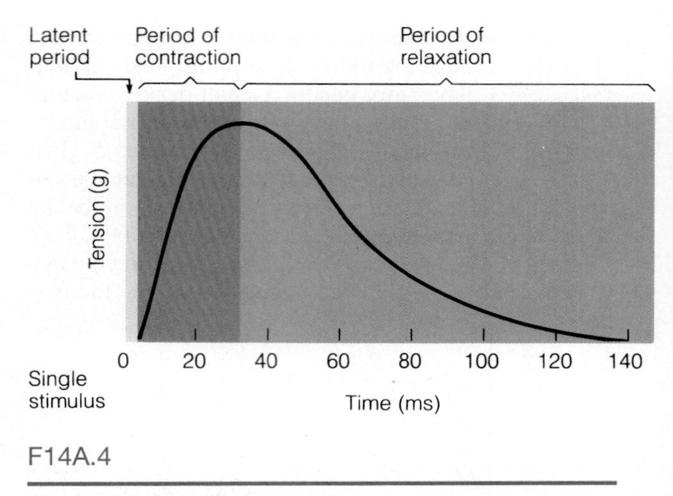

F14A.4

Tracing of a muscle twitch.

are occurring within the muscle. During the **contraction period,** the muscle fibers shorten; the tracing shows an increasingly higher needle deflection, and the tracing peaks. During the **relaxation period,** represented by a downward curve of the tracing, the muscle fibers relax and lengthen. On a slowly moving recording surface, the single muscle twitch appears as a spike (rather than a bell-shaped curve, as in Figure 14A.4); but on a rapidly moving recording surface, the three distinct phases just described become recognizable.

DETERMINING THE MINIMAL, OR THRESHOLD, STIMULUS

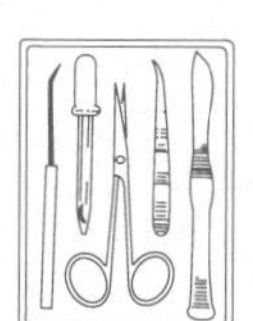

1. Assuming that you have already set up the recording apparatus, set the time marker to deliver one pulse per second and the paper speed at a slow rate, approximately 0.1 cm per second.

2. Set the duration control on the stimulator at 7 msec, multiplier × one, and the voltage control at zero v, multiplier × one. Turn the knob of the stimulator fully clockwise (to lowest value, greatest sensitivity).

3. Administer single stimuli to the muscle at 1- to 2-sec intervals, beginning with 0.1 v and increasing by 0.1 v between successive stimuli until a contraction is obtained (shown by a spike on the paper).

At what voltage did contraction occur? ____________ v

The voltage at which the first perceptible contractile response is obtained is called the **threshold,** or **minimal, stimulus.** All stimuli applied prior to this point are termed **subthreshold,** or **subliminal, stimuli,** because at those voltages no response was elicited.

4. Stop the recording, and mark the record to indicate the threshold stimulus and voltage and time. *Do not re-*

move the record from the recording surface; continue with the next experiment. *Remember:* keep the muscle preparation moistened with Ringer's solution at all times.

GRADED MUSCLE RESPONSE TO INCREASED STIMULUS INTENSITY

1. Follow the previous setup instructions, but set the voltage control at the threshold voltage (as determined in the first experiment).
2. Deliver single stimuli at 1- or 2-sec intervals. Initially increase the voltage between shocks by 0.5 v; then increase the voltage by 1 to 2 v between shocks as the experiment continues, until contraction height increases no further. Stop the recording apparatus.

   What voltage produced the highest spike (thus the maximal strength of contraction)?

   ______________ v

   This voltage, called the **maximal stimulus** (for *your* muscle specimen), is the weakest stimulus at which all muscle cells are being stimulated. As the voltage was increased to this point, more and more muscle cells (motor units) were activated, resulting in stronger and stronger contractions. Past this point, an increase in the intensity of the stimulus will not produce any greater contractile response. **Multiple motor unit summation** is the process by which the increased contractile strength reflects the relative number of muscle cells stimulated.
3. Mark the record *multiple motor unit summation,* and record the maximal stimulus voltage and the time you completed the experiment. Continue on to the next experiment.

TIMING THE MUSCLE TWITCH

1. Follow the previous setup direction; but set the voltage for the maximal stimulus (as determined in the preceding experiment), and set the paper advance or recording speed at maximum. Record the paper speed setting:

   ______________ mm/sec

2. Determine the time required for the paper to advance 1 mm by using the formula:

$$\frac{1 \text{ mm}}{\text{mm/sec (paper speed)}}$$

   (Thus, if your paper speed is 25 mm/sec, each mm on the chart equals 0.04 sec.) Record the computed value:

   1 mm = ______________ sec

3. Deliver single stimuli at 2- to 3-sec intervals to obtain several "twitch" curves. Stop the recording.

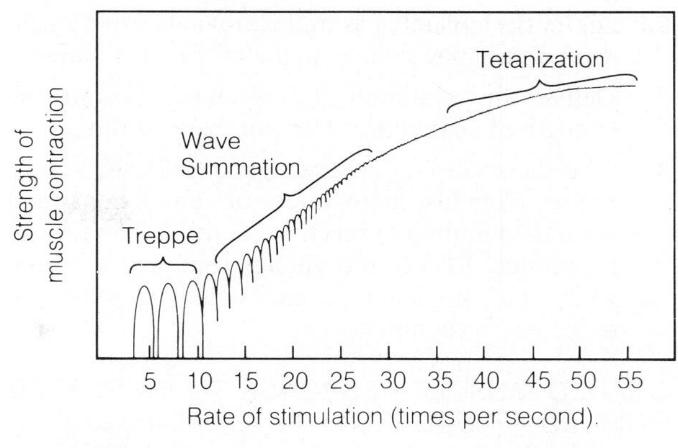

F14A.5

Treppe, wave summation, and tetanization. Treppe occurs during the first few contractions in response to a constant intensity stimulus (when the rate of stimulation allows complete relaxation between stimuli). Progressive summation of successive contractions occurs as the rate of stimulation is increased. Tetanization occurs when the rate of stimulation reaches approximately 35 per second, and maximum contraction force occurs at a stimulation rate of approximately 50/second.

4. Determine the duration of the latent, contraction, and relaxation phases of the twitches; and record here:

   Duration of latent period: ______________ sec

   Duration of contraction period: ______________ sec

   Duration of relaxation period: ______________ sec

5. Label the record to indicate the point of stimulus, the beginning of contraction, the end of contraction, and the end of relaxation.
6. Allow the muscle to rest (but keep it moistened) before continuing with the next experiment.

THE TREPPE, OR STAIRCASE, PHENOMENON

As a muscle is stimulated to contract, a curious phenomenon is observed in the tracing pattern of the first few twitches. Even though the stimulus intensity is unchanged, the height of the individual spikes increases in a stepwise manner—producing a sort of staircase pattern (Figure 14A.5).

This phenomenon is not well understood, but the following explanation has been offered: In the muscle cell's resting state, there is less Ca^{2+} in the sarcoplasmic reticulum, and the enzyme systems in the muscle cell are less efficient than after it has contracted a few times. As the muscle tissue begins to contract, Ca^{2+} moves into the cells from the extracellular fluid. The heat generated by muscle activity increases the efficiency of the enzyme systems, and the muscle becomes more efficient and contracts more vigorously. This is the physiologic basis of the *warm-up period* prior to competition in sports events.

1. Set up the apparatus as in the previous experiment, again setting the voltage to the maximal stimulus.
2. Deliver single stimuli at 1-sec intervals until the strength of contraction does not increase further.
3. Stop the recording apparatus and mark the record *treppe*. Note also the number of contractions (and seconds) required to reach the constant contraction magnitude. Record the voltage used and the time when you completed this experiment. Continue on to the next experiment.

GRADED MUSCLE RESPONSE TO INCREASED STIMULUS FREQUENCY Muscles subjected to frequent stimulation, without a chance to relax, exhibit two kinds of responses—wave summation and tetanus—depending on the level of stimulus frequency (Figure 14A.5).

Wave Summation: If a muscle is stimulated with a rapid series of stimuli of the same intensity before it has had a chance to relax completely, the response to the second and subsequent stimuli will be greater than to the first stimulus. This phenomenon, called wave summation, occurs because the muscle is already in a partially contracted state when the subsequent stimuli are delivered.

1. With the recorder running at slow speed and the stimulus intensity set at the maximal stimulus, stimulate the muscle at a rate of 15 to 25 stimuli per second.
2. Shut off the recorder and label the record as *wave summation*. Note also the time, the voltage, and the frequency.

Tetanus: Stimulation of a muscle at an even higher frequency will produce a "fusion" of the summated twitches (*tetanization*). In effect, a single sustained contraction is achieved in which *no* evidence of relaxation can be seen (Figure 14A.5). Tetanus is a feature of normal skeletal muscle functioning; the single muscle twitch is primarily a laboratory phenomenon.

1. To demonstrate tetanus, maintain the conditions you used for wave summation, except for the frequency of stimulation. Set the stimulator to deliver 60 stimuli per second.
2. As soon as you obtain a single smooth, sustained contraction (with no evidence of relaxation), discontinue stimulation and shut off the recorder.
3. Label the tracing with the conditions of experimentation, the time, and the area of tetanus.

THE EFFECT OF LOAD ON SKELETAL MUSCLE When the fibers of a skeletal muscle are slightly stretched by a weight or tension, the muscle responds by contracting more forcibly and thus is capable of doing more work. If the load is increased beyond the optimum, the latent period becomes longer, contraction height decreases, and relaxation (fatigue) occurs more quickly. With excessive stretching, the muscle is unable to develop any tension, and no contraction occurs. Since the myofilaments no longer overlap at all with this degree of stretching, the sliding force cannot be generated.

If your equipment allows you to add more weights to the muscle specimen or to increase the tension on the muscle, perform the following experiment to determine the effect of loading on skeletal muscle and to develop a work curve for the frog's gastrocnemius muscle.

1. Set the stimulator to deliver the maximal voltage as previously determined.
2. Stimulate the unweighted muscle with single shocks at 1- to 2-sec intervals to achieve three or four muscle twitches.
3. Stop the recording apparatus, and add 10 g of weight or tension to the muscle. Restart and advance the recording about 1 cm, and then stimulate again to obtain three or four spikes.
4. Repeat the previous step seven more times increasing the weight by 10 g each time, until the total load on the muscle is 80 g or the muscle fails to respond. If the Achilles tendon tears, the weight will drop, thus ending the trial. In such cases, another frog leg will need to be prepared for continuing the experiments and the maximal stimulus will have to be determined for the new muscle preparation.
5. Discontinue recording and remove the tracing. Mark the curves on the record to indicate the load (in grams).
6. Measure the height of contraction (in millimeters) for each of the sequence of twitches obtained with each load, and insert this information into the chart below.
7. Compute the work done by the muscle for each twitch (load) sequence.

$$\text{Weight of load (g)} \times \frac{\text{distance load}}{\text{lifted (mm)}} = \text{work done}$$

Enter these calculations into the chart in the columns labeled "Trial 1."

Load (g)	Distance Load Lifted (mm)		Work Done	
	Trial 1	Trial 2	Trial 1	Trial 2
0				
10				
20				
30				
40				
50				
60				
70				
80				

8. Allow the muscle to rest for 5 min. Then conduct a second trial in the same manner (i.e., repeat steps 2 through 7), but record your calculated data in the columns labeled "Trial 2." Be sure to keep the muscle well moistened with Ringer's solution during the resting interval.
9. Using two different colors, plot a line graph of work done against the weight on the accompanying grid for each trial. Label each plot appropriately.

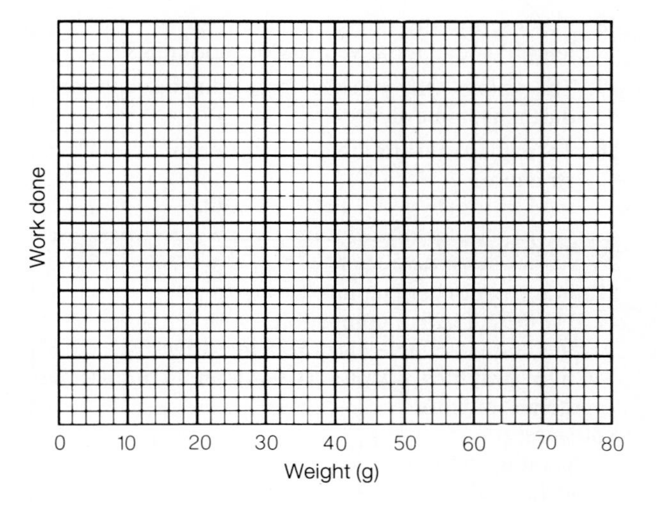

MUSCLE FATIGUE Muscle fatigue, the loss of the ability to contract, is believed to be a result of the oxygen debt that occurs in the tissue after prolonged activity (through the accumulation of such waste products as lactic acid, as well as the depletion of ATP). True muscle fatigue rarely occurs in the body, because it is most often preceded by a subjective feeling of fatigue. Furthermore, fatigue of the neuromuscular junctions typically precedes fatigue of the muscle.

1. To demonstrate muscle fatigue, set up an experiment like the tetanus experiment; but allow the stimulus to be delivered until the muscle completely relaxes and the contraction curve returns to the base line.
2. Measure the time interval between the beginning of complete tetanus and the beginning of fatigue (when the tracing begins its downward curve). Mark the record appropriately.
3. Determine the time required for complete fatigue to occur (the time interval from the beginning of fatigue until the return of the curve to the base line). Mark the record appropriately.
4. Allow the muscle to rest (keeping it moistened with Ringer's solution) for 10 min, and then repeat the experiment.

   What was the effect of the rest period on the fatigued muscle?

   ____

   ____

   What is the physiological basis for this reaction?

   ____

   ____

   ____

5. ⚠ Dispose of the frog remains in the appropriate organic debris container. Dispose of the gloves and wash and dry your hands.
6. Dismantle all apparatus and prepare the equipment for storage. Inspect your records of the experiments; and make sure each is fully labeled with the experimental conditions, the date, and the names of those who conducted the experiments. (If you used a smoked drum kymograph recording system, "fix" the records before leaving the laboratory.) For future reference, attach to this page a tracing (or a photocopy of the tracing) for each experiment.

14B EXERCISE

Muscle Physiology (Computerized Simulations)

OBJECTIVES

1. To define important terms used in describing muscle physiology.
2. To identify the two ways that the mode of stimulation can affect muscle force production.
3. To draw a graph relating stimulus strength and twitch force to illustrate graded muscle response.
4. To explain how slow, smooth, sustained contraction is possible in a skeletal muscle.
5. To graphically understand the relationships between passive, active, and total forces.
6. To identify the conditions under which muscle contraction is isometric or isotonic.
7. To describe in terms of length and force the transitions between isometric and isotonic conditions during a single muscle twitch.
8. To describe the effects of afterload and starting length on the initial velocity of shortening.
9. To explain why muscle force remains constant during isotonic shortening.
10. To explain results obtained in terms of muscle structure.

MATERIALS

Minimum equipment required:
- IBM PC/XT/AT or compatible
- 256K RAM—single disk (5.25 in.) drive
- Color graphics adapter (CGA) and compatible graphics monitor

Software:
- Operating system
- Mechanical Properties of Active Muscle

Many important physiological concepts of skeletal muscle contraction can be demonstrated using this set of computer simulations, which investigates the mechanical properties of active skeletal muscle. The programs graphically present all the equipment and materials necessary for you, the investigator, to set up experimental conditions and observe the results. In student-conducted laboratory investigations there are many ways to approach a problem and the same is true of these simulations. The instructions provided are intended to guide you in your investigation, but you should also attempt alternate approaches to gain insight into the logical methods employed in scientific experimentation.

Try the following approach: The first time through the programs, follow the instructions closely and answer the questions it poses as you go along. Then try your own ideas, asking "What if . . . ?" questions to test the validity of your theories. Major advantages of these computer simulations are that the muscle cannot be damaged accidentally, lab equipment will not break down at the worst possible time, and you will have ample time to think critically about the processes in question.

Various types of apparatus are used in research settings, and you will find minor differences in the operation of the equipment used in these simulations. This requires that you think about what is happening in each situation because it is important to understand how you are experimentally manipulating the muscle for a complete understanding of the actual results. To aid you in this endeavor, each of the exercises has the following elements:

(1) Introduction to the System (explains equipment used);
(2) Some Definitions (briefly defines terms); and
(3) Experiments (designed to manipulate skeletal muscle).

Work your way through each section in turn so that you are familiar with the simulated equipment you will be using and the terminology involved.

GETTING STARTED

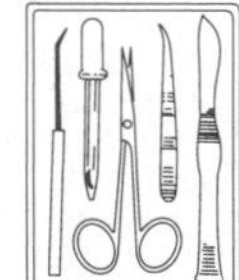

Begin by making sure you have the computer equipment listed in the materials section. If your computer is already running, proceed to step 2.

1. Insert the DOS (operating system) diskette into drive A. Turn the computer on and answer any questions that appear on the screen. When you see "A>" (the A prompt) remove the DOS diskette.
2. Insert the muscle program diskette into drive A, type the word "GO" (but without quotes), and press the ENTER (return) key. You must keep the program diskette in drive A for the entire time you are using the simulation.

The title screen will appear after a few seconds. Press any key to move on to the next information screen and again to advance to the main Index of Programs. This is a list of all the choices available to you. Select Option #1 to see an overview of all programs available. This will display the opening screen of each program in turn and let you examine it in more detail if you wish. Instructions will appear in boxes on the screen. If you get lost, pressing the ESC key will get you back to the Index of Programs.

ELECTRICAL STIMULATION

The contraction of skeletal muscle will produce force and/or shortening when nervous or electrical stimulation is applied. Unlike single cells or motor units, which follow the all-or-none law of muscle physiology, a whole muscle responds to stimuli with a graded response. A *motor unit* consists of a motor neuron and all the muscle cells it innervates. Hence, activation of the neuron innervating a single motor unit will cause all muscle cells in that unit to fire simultaneously in an all-or-none fashion. The graded contractile response of a whole muscle reflects the number of motor units firing at a given time. Strong muscle contraction implies many motor units are activated (and each unit has maximally contracted); weak contraction means few motor units are active (however, the activated units are maximally contracted). By increasing the number of motor units firing, we can produce a slow, steady increase in muscle force, a process called *recruitment* or *motor unit summation.*

Regardless of the number of motor units activated, a single contraction of skeletal muscle is called a *muscle twitch.* A recorded tracing of a twitch is divided into three phases: latent, contraction, and relaxation (Figure 14B.1). The **latent phase** is a short period between the time of stimulation and the beginning of contraction. Although no force is generated during this interval, many chemical changes are taking place intracellularly in preparation for contraction. During **contraction,** the myofilaments are sliding and the muscle shortens. **Relaxation** takes place when contraction has ended and the muscle returns to its normal resting state (and length).

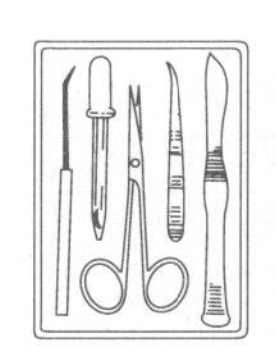

This selection simulates an **isometric** (fixed-length) **contraction** of an isolated skeletal muscle and allows you to investigate how the strength and frequency of an electrical stimulus affect whole muscle function. Note that these simulations involve *indirect* stimulation by an electrode placed on the surface of the muscle. This differs from the situation *in vivo,* where each fiber in the muscle receives *direct* stimulation via a nerve ending. Now spend a few moments on the introduction to the system and definitions of terms used.

Single Stimulus

Select option 3: *Single Stimulus.* The opening screen will appear in a few seconds (Figure 14B.2).

The oscilloscope display is the most important part of the screen, because it is where all contraction data are graphically presented for analysis. *Time* is displayed on the horizontal axis, where a full sweep is 1 second. It is marked off in 0.1-second intervals. *Force* is displayed on the vertical axis on an arbitrary scale from 0 to 5. Familiarize yourself with the control keys (Figure 14B.3) before proceeding.

1. Press the "S" key once. Since the VOLTAGE is set to ZERO, no muscle activity will result. However, notice that a yellow line moves across the bottom of the tracing—this will indicate the muscle force later. The red line moving across the top of the screen indicates time. A small vertical line marks the exact time of stimulus.

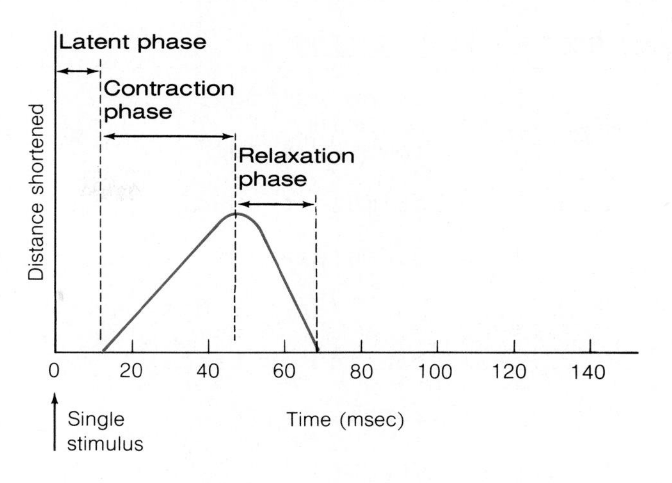

F14B.1

Tracing of a single muscle twitch. (From RA Meiss, *Mechanical Properties of Active Muscle* © 1987, COMPress, a division of Queue, Inc.)

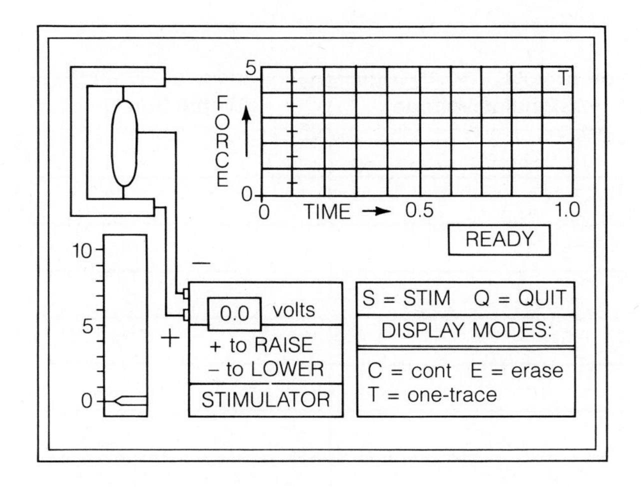

F14B.2

Opening screen of the electrical stimulation simulation. (From RA Meiss, *Mechanical Properties of Active Muscle* © 1987, COMPress, a division of Queue, Inc.)

```
KEY.......FUNCTION

+.........raise voltage

-.........lower voltage

S.........stimulate muscle

P.........pause display

H.........halt (abort) display

C.........continuous trace

T.........single trace

E.........erase all tracings

Q.........quit (return to menu)
```

F14B.3

Summary of single stimulus control keys.

2. Press the "+" key until the voltmeter reads 5. Then press the "S" key once and release it. You will see the muscle react, and a TWITCH FORCE will appear on the screen. Note that the latent phase is not represented on the screen, allowing a cleaner tracing for experimental purposes. You can use the ERASE key "E" as needed to clear up the screen. The CONTINUOUS key "C" allows the tracing to wrap around the screen continuously until force = 0; its companion, the ONE-TRACE key "T," will stop the tracing when it has reached the end of a single sweep. The HALT key "H" aborts the tracing, and the PAUSE key "P" causes the tracing to pause momentarily until you press it again to resume. The QUIT key "Q" returns you to the program menu.

Stimulus Voltage	Twitch Force

3. Try changing the voltage and notice how force of contraction also changes. Observe the contraction and relaxation phases in the tracings. What happens if you stimulate again before the muscle has a chance to fully relax?

____

4. Feel free to experiment with anything that comes to mind, as this will give you a sense of how a whole muscle will respond to an electrical stimulus.

DETERMINING THE THRESHOLD (MINIMAL) STIMULUS

1. Erase the oscilloscope display and set the voltage control to zero. Set the display mode to single trace.

2. Using single stimuli, increase the voltage in 0.1-volt increments until the first trace of contraction is seen. This is the **threshold** voltage, below which no contraction occurs. What is the threshold voltage?

_______________ v

GRADED MUSCLE RESPONSE TO INCREASED STIMULUS INTENSITY Examine the tracings just recorded above. Note that as voltage is increased, contractile force also increases. Also notice the amount of the muscle mass being stimulated (it turns red as stimulus is applied). Try using higher voltages. As more voltage is delivered to the whole muscle, greater numbers of muscle fibers are activated, thereby increasing the total force produced by the muscle. This result is similar to that occurring *in vivo,* where the recruitment of additional motor units increases total force production. This phenomenon is called **multiple motor unit summation.**

1. Is there a stimulus voltage beyond which there appears to be no further increase in muscle contraction?

____

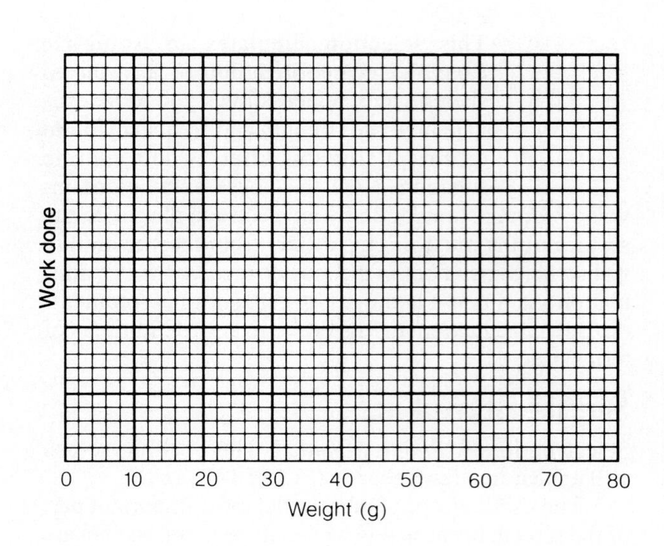

Why is this so? ______________________________

____

This voltage is the lowest stimulus intensity (strength) necessary to activate all cells within the muscle and is called the *maximal stimulus.* The maximum contraction thus produced is called the *maximal response.*

2. Do a systematic study of force produced over the entire range of stimulus voltages in 1.0-volt increments and record your data in the chart on p. 142. Then plot the results in the accompanying grid.

From the graph, estimate the threshold voltage. ______ v

The maximal stimulus. ______________ v

Is a muscle twitch an all-or-none response? __________

Justify your answer.

____

____

GRADED MUSCLE RESPONSE TO INCREASED STIMULUS FREQUENCY In addition to multiple motor unit summation, another way to increase the amount of force produced by muscle is wave (temporal) summation. **Wave summation** is achieved by increasing the stimulus frequency (rate of stimulus delivery to the muscle), unlike motor unit summation which relies on increased stimulus intensity. Wave summation occurs because the muscle is already in a partially contracted state when subsequent stimuli are delivered.

Tetanus can be considered an extreme form of wave summation that results in a steady, sustained contraction. In effect, the muscle does not have any chance to relax due to the very high frequency of stimulation. This "fuses" the force peaks so that we observe a smooth tracing.

WAVE SUMMATION

1. Erase all tracings, set voltage at maximal stimulus, and use the continuous tracing mode. Stimulate once and then again when the muscle has relaxed about halfway. Is the peak force produced greater than that produced by the first stimulus?

____

2. Try stimulating again at greater frequencies. Is the contraction more forceful?

____

3. Try manipulating the stimulus frequency to produce a constant sustained contraction (tetanus) at Force = 2. Is the force produced steady, or does it rise and fall periodically?

____

4. Try again at Force = 3. Is the tracing smoother this time?

____

5. So far, you have been using maximum stimulus voltage. What do you think would happen if you used a lower voltage?

____

____

Try it to prove (or disprove) your hypothesis.

6. Try adjusting both stimulus voltage and frequency to achieve sustained contraction at Force = 2 and Force = 3. Is it possible?

____

7. Use the concepts of stimulus intensity and frequency to explain how human skeletal muscles work to produce smooth, steady contractions at all desired levels of force.

____

____

____

8. When finished experimenting in this area, press "Q" to quit and return to the menu.

Multiple Stimulus

Select option 4: *Multiple Stimulus.* The screen will appear much like that for the Single Stimulus equipment. The first thing you should notice is that the equipment works differently from that in Single Stimulus. The reason for this is that you can encounter many different types of physiological equipment in the laboratory, and this set of exercises reflects that idea. The equipment in this section functions in the same way as that of Single Stimulus, but you now have the additional ability to select exact stimulus rates. Basic function keys are the same as for Single Stimulus with a few additions (Figure 14B.4). As before, familiarize yourself with all control keys before beginning the experiment.

Press the "R" key and notice that a chart comes up allowing you to select stimulus rate. The "I" key INCREASES stimulation rate, the "D" DECREASES it. As before, the "+" and "−" keys increase or decrease VOLTAGE. Pressing the "K" key will start stimulation, pressing it again will stop it, and pressing it once more will stimulate again. Pressing the "C" key puts the oscilloscope in the continuous mode: the tracing will continue running until you stop it or the force is decreased to zero.

```
KEY........FUNCTION
K..........stimulate on/off
R..........selection box to set rate
I..........increase stimulus rate
D..........decrease stimulus rate
V..........return to experiment
C..........continuous trace
T..........single trace
E..........erase all tracings
Q..........quit (return to menu)
```

F14B.4

Summary of multiple stimulus control keys.

FUSION FREQUENCY

1. Set the stimulator at the maximal stimulus.
2. Stimulate at this intensity for all available rates, keeping all tracings on the screen.

   At what stimulation rate is tetany produced?

   ______________ stimuli per second

   This rate is the *fusion frequency.* As before, try producing smooth, sustained contraction (tetany) at Force = 2 and Force = 3 by adjusting stimulus rate. Do these present results support your earlier ideas?

   ____

MUSCLE FATIGUE A prolonged period of sustained contraction will result in **muscle fatigue,** a condition in which the tissue loses its ability to contract. Fatigue is easily demonstrated using the *Multiple Stimulus program.*

1. Set voltage control at the maximal stimulus, stimulus rate at 20/sec, and screen on continuous sweep. Stimulate once and observe the results.

   Does the force begin to fall (muscle fatigue)?

   ____

2. Erase the screen, stimulate as before, but when fatigue has developed, turn off the stimulator (press "K" again) for a second or two, then turn back on.

   Do you see any evidence of recovery? __________

3. When finished with this section, return to the Index of Programs by pressing the ESC key.

ISOMETRIC CONTRACTION

Isometric contraction is the condition in which muscle length does not change regardless of the amount of force generated by the muscle (*iso* = same, *metric* = length). This is accomplished experimentally by holding both ends of the muscle in a fixed position while stimulating it electrically. *Resting length* (length of the muscle before contraction) is an important factor in determining the amount of force that a muscle can develop. *Passive force* is generated by stretching the muscle and is due to the elastic properties of the tissue itself. *Active force* is generated by the physiological contraction of the muscle. *Total force* is the sum of passive and active forces.

This program allows you to set the resting length of the experimental muscle and stimulate it with a single maximum stimulus shock. You can then construct a graph relating the forces generated to length of the muscle. These principles can then be applied to human muscles in order to understand how optimum resting length results in maximum force production. In order to understand why muscle tissue behaves as it does, it is necessary to comprehend *how* contraction works at the cellular level. If you have conceptual difficulty with the results of this exercise, review the sliding filament model of muscle contraction.

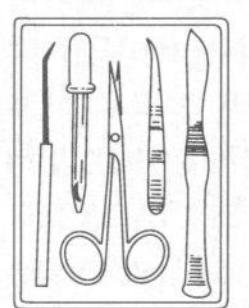

Spend a few moments on the "Introduction to the System" and "Some Definitions" sections to gain an understanding of the essential elements of the system. Select option 3: *Isometric Contraction program.* The opening screen will appear in a few seconds (Figure 14B.5).

The tip of the length-changing device indicates muscle length on the length axis below and is set at approximately 21 units. The Force-Time graph displays the single muscle twitch. Notice a small horizontal line next to the force scale; this will display resting force of the muscle. Familiarize yourself with the control keys (Figure 14B.6) before proceeding.

1. Shorten the muscle until a length of 14 units is achieved.

2. Press the "M" key once to mark PASSIVE (resting) FORCE (a small dot should appear just above the length axis on the Force-Length graph).

3. Stimulate once by pressing the "S" key. You should see a single force tracing on the Force-Time graph.

4. Press "M" once again to mark PEAK TWITCH (active) FORCE.

5. Increasing the muscle length by 1 unit each time and repeating the MARK-STIMULATE-MARK pattern, complete the tracings for the entire length range.

What is happening to the resting and peak twitch forces as muscle length is increased?

____

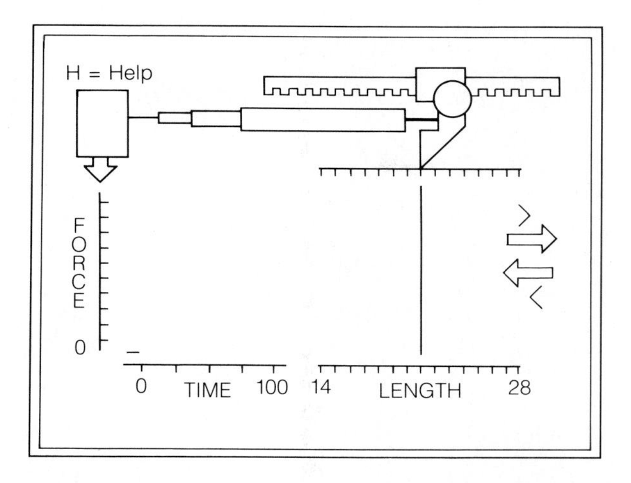

F14B.5

Opening screen of isometric contraction simulation. (From RA Meiss, *Mechanical Properties of Active Muscle* © 1987, COMPress, a division of Queue, Inc.)

6. When finished with plotting all data, connect points by pressing the "L" key.

Can you explain the dip in the upper curve (keep in mind you are measuring total muscle force)?

____

____

7. Display the ideal curves by pressing "D." Now is the reason for the dip evident?

8. Press the "X" key to label the graph and note that what we have been calling peak force is actually total force and is the sum of the resting and active forces.

9. When you understand these concepts, return to the Index of Programs (ESC).

ISOTONIC CONTRACTION

During **isotonic contraction,** muscle length changes, but the force produced does not (*iso* = same, *tonic* = force). Unlike the isometric exercise, where both ends of the muscle were held in a fixed position, the setup for isotonic contraction requires one end of the muscle to be free. Variable weights can then be attached to the free end, while the other end of the muscle is held fixed. If the weight is not too great, the muscle will be able to lift it with a certain velocity. You can think of lifting an object from the floor as an example: If the object is light, it can be lifted quickly (high velocity), whereas a heavier weight will be lifted with a lower velocity. Try to transfer the idea of what is happening in the simulation to the muscles of your arm when you are attempting to lift a weight. The two important variables in this exercise are *starting length* of the muscle and the *afterload* (weight) applied. As before, some background

KEY........FUNCTION

>..........lengthen muscle

<..........shorten muscle

S..........stimulate muscle

M..........mark data point

F..........data values

L..........draw line graph

D..........display ideal curves

C..........clear screen

E..........erase tracings only

R..........resume work

H..........help screen

Q..........quit (return to menu)

F14B.6

Summary of isometric contraction control keys.

in the sliding filament model of contraction will be helpful.

This program allows you to change muscle length and afterload so that you can investigate these effects on the speed of skeletal muscle shortening. Both variables can be independently altered and results are graphically presented directly on the screen.

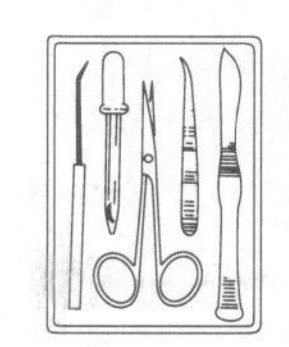

Choose Isotonic Contraction from the Index of Programs and, as before, spend some time on the introductory sections before proceeding. Select option 3: *Contraction and Relaxation.* The opening screen will appear in a few seconds (Figure 14B.7).

The length scale displays 5 to 9 arbitrary units and the force scale indicates a range from 0 to 5 units. Initially, muscle length is set at 9 units and afterload weight at 3 units. Familiarize yourself with the equipment layout and control keys (Figure 14B.8) before proceeding.

Stimulate the muscle once and observe the tracings. Notice the READY box while the tracing is being produced. It will indicate ISOMETRIC conditions (length is constant but force may be changing) and ISOTONIC conditions (force is constant but length is changing). To change weight or length, press the appropriate key ("W" or "L") and select the variable you wish. The upper tracing displays Force as a function of Time.

Does the flat part of the tracings correspond to isotonic conditions or isometric conditions?

____

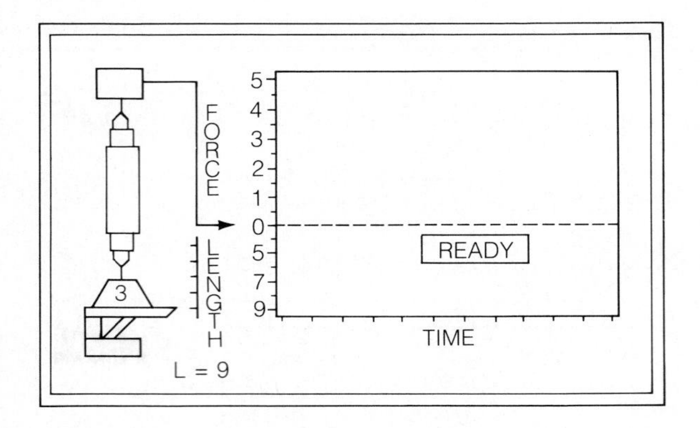

F14B.7

Opening screen of isotonic contraction simulation. (From RA Meiss, *Mechanical Properties of Active Muscle* © 1987, COMPress, a division of Queue, Inc.)

```
KEY........FUNCTION
S..........stimulate muscle
V..........view screen
W..........change afterload
L..........change length
G..........graph data
E..........erase screen
H..........help screen
Q..........quit (return to menu)
```

F14B.8

Summary of isotonic contraction control keys.

The lower tracing indicates Length as a function of Time. Analyze length tracings carefully because they indicate initial velocity of shortening. Think about what the curve means and the following concept should become clear: Using a given weight, the length of the muscle will change in a certain amount of time (heavier afterloads require more time). This means that the *steepness* of the tracing is an indication of initial velocity or speed of shortening; the steeper the tracing, the higher the initial velocity.

THE EFFECT OF LOAD ON SKELETAL MUSCLE

1. Keeping the resting length constant at 9 units, run through the entire range of afterload weights. After each weight tracing has been completed, press "G" to activate graph, then press "P" to plot each data point. When the entire set of afterload weights has been plotted, press "D" while in the graph option to draw the plot.

What is the relationship between afterload weight (Fa) and the initial velocity of shortening?

____

____

What about afterload and maximum length change?

____

2. Now keep afterload weight constant (Wt = #2) and run through the entire range of starting lengths.

What is the relationship between starting length and initial velocity of shortening?

____

____

3. Try the same experiment with a different weight.

Is there a similar pattern? ______________

4. Can you set up a contraction that is entirely isometric?

____

One that is entirely isotonic? ______________

Explain why or why not.

____

____

____

____

5. When you are finished with this exercise, press ESC to return to the Index of Programs.

6. If time allows, try the BIKER selection. This program simulates a bike rider trying to finish a race before exhaustion sets in. You control the gears of the bicycle, and the object of the exercise is to shift effectively so that the biker achieves his or her goal. Bar graphs indicating conditions and a Force-Velocity curve are provided to aid you in your task. Apply what you have learned in this exercise along with a logical interpretation of the Force-Velocity curve and it will be possible to finish the race.

7. When finished with all exercises, escape to the Index of Programs and select option 8: *Quit*. Remove the simulation program diskette and return the computer to the original starting condition.

EXERCISE 15

Neuron Anatomy and the Nerve Impulse

OBJECTIVES

1. To differentiate between the functions of neurons and neuroglia.
2. To identify the important anatomical characteristics of a neuron on an appropriate diagram or projected slide.
3. To state the functions of axons, dendrites, axonal terminals, and myelin sheaths.
4. To classify neurons according to structure and function.
5. To list the two major physiological properties of neurons.
6. To describe the polarized and depolarized states of the neuronal plasma membrane and to describe the events that lead to generation and conduction of a nerve impulse.
7. To explain how a nerve impulse is transmitted from one neuron to another.
8. To define *depolarization, repolarization,* and *refractory period.*
9. To list various substances and factors that can stimulate neurons.

MATERIALS

Supply area 1:

Model of a "typical" neuron (if available)
Compound microscope
Histological slides of an ox spinal cord smear, teased myelinated nerve fibers

Supply area 2:

*Rana pipiens**
Dissecting instruments and tray
Ringer's solution (frog) in dropper bottles, some at room temperature and some in an ice bath
Thread
Glass rods or probes
Glass plates or slides
Ring stand and clamp
Stimulator; platinum electrodes
Oscilloscope
Nerve chamber
Filter paper
0.01% hydrochloric acid (HCl) solution
Sodium chloride (NaCl) crystals
Heat-resistant mitts
Bunsen burner
Disposable plastic gloves

* Instructor to provide freshly pithed frogs (*Rana pipiens*) for student experimentation.

The nervous system is the master integrating and coordinating system, continuously monitoring and processing sensory information both from the external environment and from within the body. Every thought, action, and sensation is a reflection of its activity. Like a computer, it processes and integrates new "inputs" with information previously fed into it ("programmed"), to produce an appropriate response ("readout"). However, no man-made computer can possibly compare in complexity and scope to the human nervous system.

Despite its complexity, nervous tissue is made up of just two principal cell populations: the **neurons** and the **neuroglia** (glial cells). The neuroglia, literally "nerve glue," include *astrocytes, oligodendrocytes, microglia,* and *ependymal cells.* These cells serve the needs of the neurons by acting as supportive and protective cells, myelinating cells, and phagocytes. In addition, they probably serve some nutritive function by acting as part of the blood-brain barrier. Although neuroglia resemble neurons in some ways (they have fibrous cellular extensions), they are not capable of generating and transmitting nerve impulses, a capability that is highly developed in neurons. Our focus in this exercise is the highly irritable neurons.

NEURON ANATOMY

The delicate **neurons** are highly specialized to transmit messages (nerve impulses) from one part of the body to another. Although neurons differ structurally, they have many identifiable features in common (Figure 15.1). All have a **cell body** from which slender **processes** or fibers extend. Neuron cell bodies, which are found only in the CNS (brain or spinal cord), typically in clusters called **nuclei,** or in **ganglia** (collections of neuron cell bodies outside the CNS), make up the gray matter of the nervous system. Neuron processes running through the CNS form **tracts** of white matter; outside the CNS they form the peripheral **nerves.**

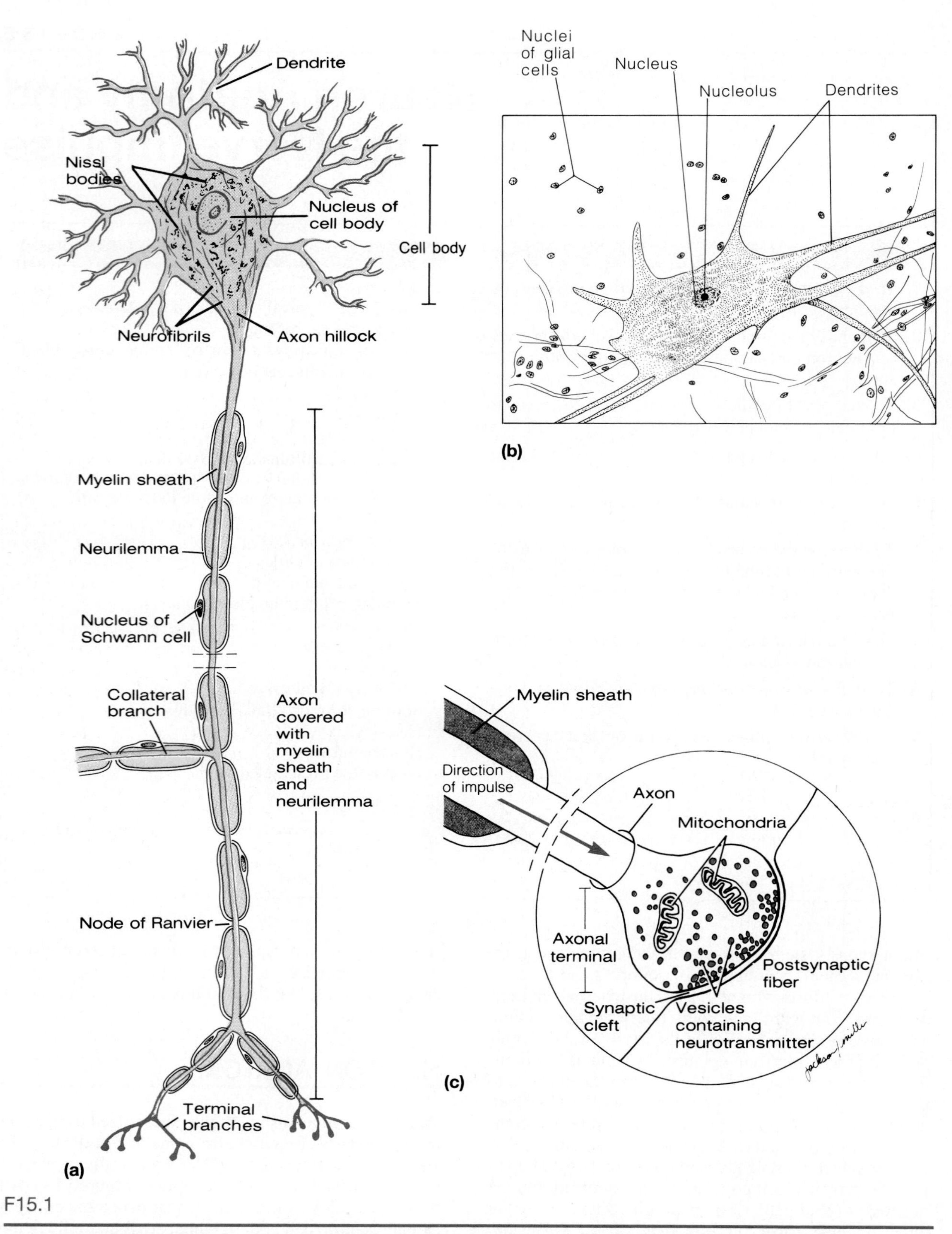

F15.1

Structure of a typical motor neuron. (a) Diagrammatic view; (b) illustration reproducing photomicrograph of multipolar neuron (plate 4 of histology atlas); (c) enlarged view of a synapse.

The neuron cell body contains a large round nucleus surrounded by cytoplasm (neuroplasm). The cytoplasm is riddled with neurofibrils and with darkly staining structures called Nissl bodies. **Neurofibrils,** the cytoskeletal elements of the neuron, have a support and intracellular transport function. **Nissl bodies,** an elaborate type of rough endoplasmic reticulum, are involved in the metabolic activities of the cell.

According to the "older," traditional scheme, neuron processes that conduct electrical currents *toward* the cell body are called **dendrites,** and those that carry impulses *away from* the cell body are called **axons.** However, when it was later discovered that this anatomic scheme had pitfalls (some axons carry impulses *both* toward and away from the cell body), a newer physiological definition of neuron processes was adopted. According to this scheme, dendrites are *receptive* regions (bear receptors for neurotransmitters released by other neurons), whereas axons are nerve impulse generators and transmitters. As a rule, neurons have only one axon (which may branch into **collaterals**) but may have many dendrites, depending on the neuron type. Note that the term *nerve fiber* is a synonym for axon and is, thus, quite specific.

In general, a neuron is excited by other neurons when their axons release neurotransmitters close to its dendrites or cell body. The electrical current travels across the cell body and down the axon. As Figure 15.1 shows, the axon fiber (in motor neurons) begins at a slightly enlarged cell body structure called the **axon hillock** and ends in many small structures called **axonal terminals,** or synaptic knobs. These terminals store the neurotransmitter chemical in tiny vesicles. The enlarged view of the axonal terminal reveals that it is separated from the cell body or dendrites of the next (postsynaptic) neuron by a tiny gap called the **synaptic cleft.** Although they are close, there is no actual physical contact between neurons. When an impulse reaches the axonal terminals, the synaptic vesicles rupture and release the neurotransmitter into the synaptic cleft. The neurotransmitter then diffuses across the synaptic cleft to bind to membrane receptors on the next neuron, initiating the action potential.*

Many drugs can influence the transmission of impulses at synapses. Some, like caffeine, are stimulants which decrease the receptor neuron's threshold and make it more irritable. Others block transmission by binding competitively with the receptor sites or by interfering with the release of neurotransmitter by the axonal terminals. As might be anticipated, some of these drugs are used as painkillers or tranquilizers. ■

Most long nerve fibers are covered with a fatty material called myelin, and such fibers are referred to as **myelinated fibers.** Axons in the PNS are typically heavily myelinated by special cells called **Schwann cells,** which wrap themselves tightly around the axon, jelly-roll fashion (Figure 15.2). During the wrapping process, the cytoplasm is squeezed from between adjacent layers of the Schwann cell membranes—so that when the process is completed, a tight core of plasma membrane material encompasses the axon. This wrapping is the **myelin sheath.** The Schwann cell nucleus and the bulk of its cytoplasm ends up just beneath the outermost portion of the Schwann cell plasma membrane; this, plus the peripherally exposed part of the Schwann cell plasma membrane, is referred to as the **neurilemma.** Since the myelin sheath is formed by many individual Schwann cells, it is a discontinuous sheath; the gaps or indentations in the sheath are called **nodes of Ranvier** (see Figure 15.1).

Within the CNS, myelination is accomplished by glial cells called oligodendrocytes. The CNS sheaths do not exhibit the neurilemma seen in fibers myelinated by Schwann cells. Because of its chemical composition, myelin acts to insulate the fibers and greatly increases the speed of neurotransmission by neuron fibers.

1. Study the typical motor neuron shown in Figure 15.1, noting the structural details described above, and then identify these structures on a neuron model.

2. Obtain a prepared slide of the ox spinal cord smear, which has large, easily identifiable neurons. Study one representative neuron under oil immersion; and identify the cell body, the nucleus, the large prominent "owl's-eye" nucleolus, and the granular Nissl bodies. If possible, distinguish the axon from the many dendrites. Sketch the cell in the space below and label the important anatomical details you have observed. Compare your sketch to plates 4 and 5 of the histology atlas.

* Specialized synapses in skeletal muscle are called neuromuscular junctions. They are discussed in Exercise 12, along with excitation of the muscle cells.

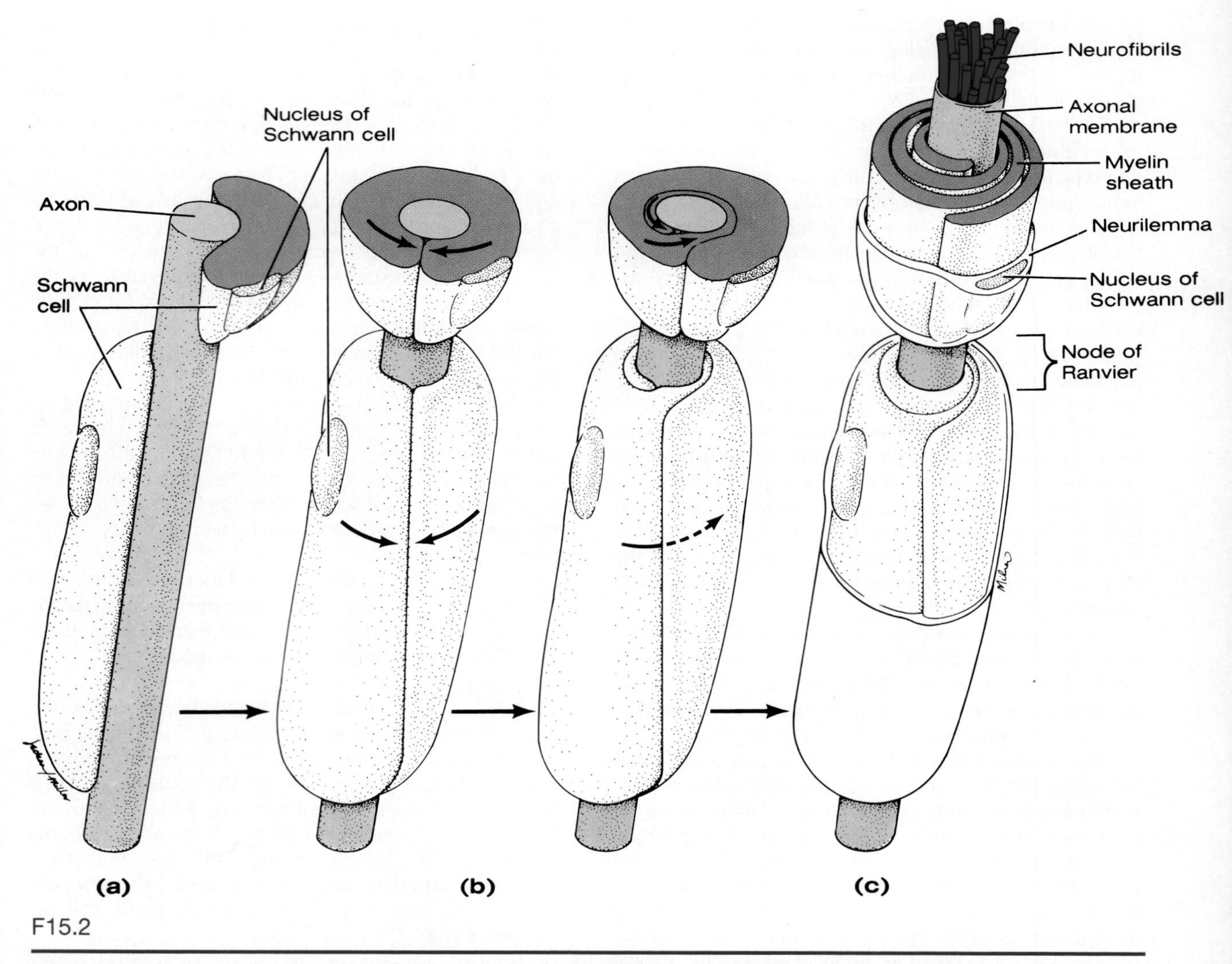

F15.2

Myelination of axons in the peripheral nervous system by individual Schwann cells. (a) A Schwann cell becomes apposed to the axon. (b) The Schwann cell coils around the axon. (c) The process is completed and a myelinated fiber is formed.

3. Obtain a prepared slide of teased myelinated nerve fibers. Using plate 6 of the histology atlas as a guide, identify the following: nodes of Ranvier, neurilemma, axis cylinder (the axon itself), Schwann cell nuclei, and myelin sheath.

Do the nodes seem to occur at consistent intervals, or are they irregularly distributed?

____

Explain the significance of this finding: ____________

____

____

____

____

Sketch a portion of a myelinated nerve fiber in the space provided here, illustrating two or three nodes of Ranvier. Label the axon, myelin sheath, nodes, and neurilemma.

NEURON CLASSIFICATION

Neurons may be classified on the basis of structure or of function. Figure 15.3 depicts both schemes.

Basis of Structure

Structurally neurons may be differentiated according to the number of processes attached to the cell body. In **unipolar neurons,** one very short process, which divides into *peripheral* and *central processes,* extends from the cell body. Physiologically, only the most distal portions of the peripheral process act as dendrites; the rest acts as an axon. Nearly all neurons that conduct impulses toward the CNS are unipolar (Figure 15.3a).

Bipolar neurons have two processes—one axon and one dendrite—attached to the cell body. This neuron type is quite rare, typically found only as part of the receptor apparatus of the eye, ear, and olfactory mucosa.

Many processes issue from the cell body of **multipolar neurons,** all classified as dendrites, except for a single axon. Most neurons in the brain and spinal cord (CNS neurons) and those with axons carrying impulses away from the CNS fall into this last category.

Basis of Function

In general, neurons carrying impulses from the sensory receptors in the internal organs (viscera) or in the skin are termed **sensory,** or **afferent, neurons** (see Figure 15.3b, p. 152). The dendritic endings of sensory neurons are generally equipped with specialized receptors that are stimulated by specific changes in their immediate environment. The cell bodies of sensory neurons are always found in a ganglion outside the CNS and these neurons are typically unipolar.

Neurons carrying impulses from the CNS to the viscera and/or body muscles and glands are termed **motor,** or **efferent, neurons.** Motor neurons are most often multipolar and their cell bodies are almost always located in the CNS.

The third functional category of neurons is the **interneurons,** or **association neurons,** which are situated between and contribute to pathways that connect sensory and motor neurons. Their cell bodies are always located within the CNS and they are multipolar neurons structurally.

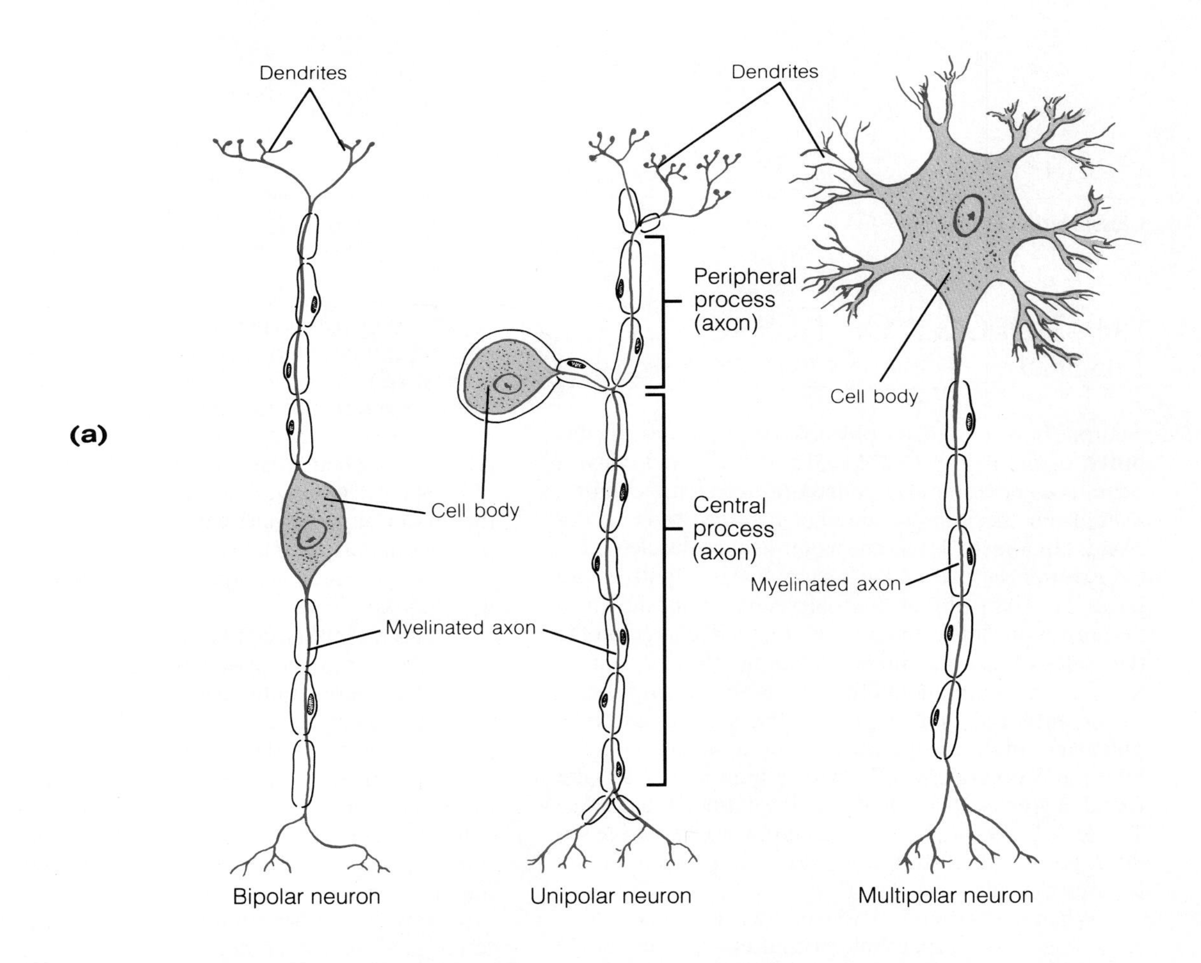

F15.3

Classification of neurons. (a) On the basis of structure. *(continues on p. 152)*

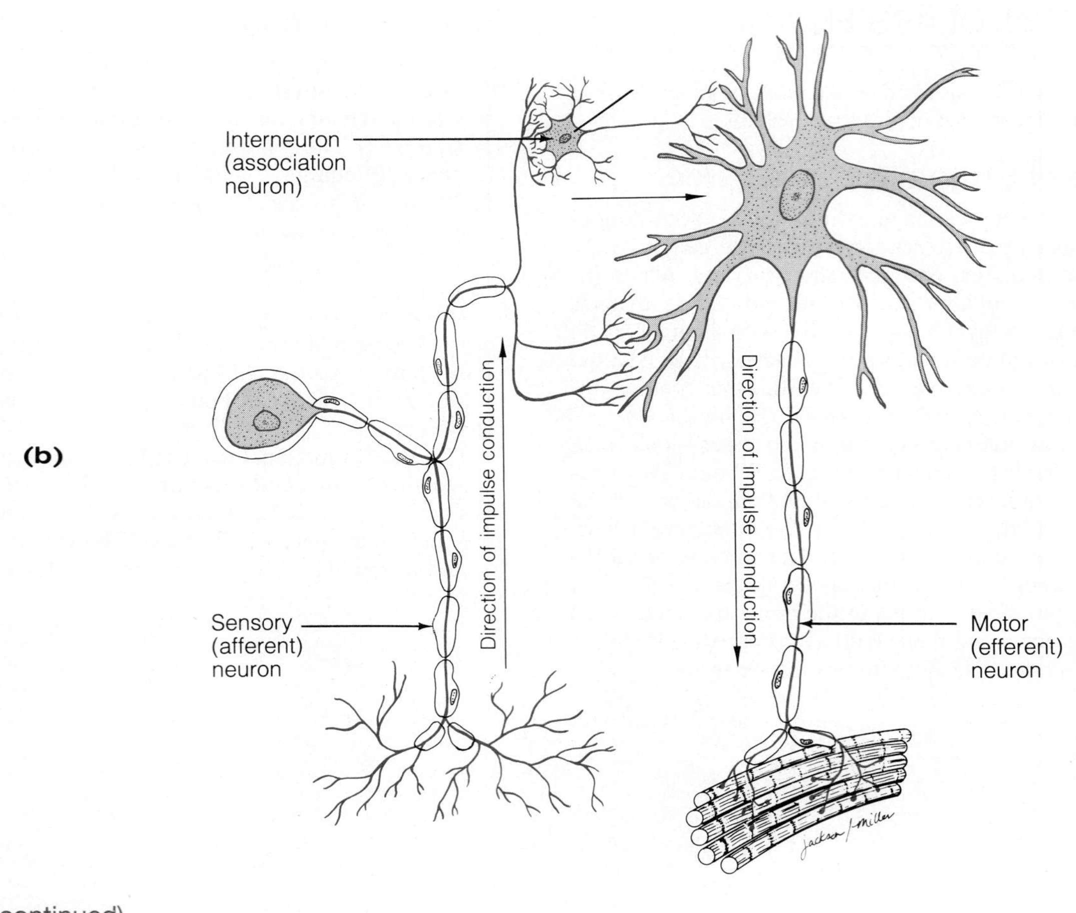

F15.3 (continued)

Classification of neurons. (b) On the basis of function.

PHYSIOLOGY OF NERVE FIBERS

Neurons have two major physiologic properties: **irritability,** or the ability to respond to stimuli and convert them into nerve impulses, and **conductivity,** the ability to transmit the impulse to other neurons, muscles, or glands. In a resting neuron (as in resting muscle cells), the exterior surface of the membrane is slightly more positively charged than the inner surface, as shown in Figure 15.4a. This difference in electrical charge on the two sides of the membrane results in a voltage across the plasma membrane referred to as the **resting membrane potential,** and a neuron in this state is said to be **polarized.** In the resting state, the predominant intracellular ion is potassium (K^+), and sodium ions (Na^+) are found in greater concentration in the extracellular fluids. The resting potential is maintained by a very active sodium pump, which transports Na^+ out of the cell and K^+ into the cell.

When the neuron is activated by a stimulus of adequate intensity—a **threshold stimulus**—the membrane at its *trigger zone,* typically the axon hillock (or the most peripheral part of a sensory neuron's axon), briefly becomes more permeable to sodium (sodium gates are opened). Sodium ions rush into the cell, increasing the number of positive ions inside the cell (Figure 15.4b). Thus the interior of the membrane becomes less negative at that point and the exterior surface becomes less positive—a phenomenon called **depolarization.**

When depolarization reaches a certain point such that the local membrane polarity changes (momentarily the external face becomes negative and the internal face becomes positive), it initiates an **action potential** (Figure 15.4c).*

Within a millisecond after the inward influx of sodium, the membrane permeability is again altered. As a result, Na^+ permeability decreases, K^+ permeability increases, and K^+ rushes out of the cell. Since K^+ ions are positively charged, their movement out of the cell reverses the membrane potential again, so that the external membrane surface is again positive relative to the internal membrane face (Figure 15.4d). This event, called **repolarization,** reestablishes the resting membrane potential. During the time of repolarization, the neuron is insensitive to further stimulation; thus this period is referred to as the **refractory period.**

* If the stimulus is of less than threshold intensity, depolarization is limited to a small area of the membrane, and no action potential is generated.

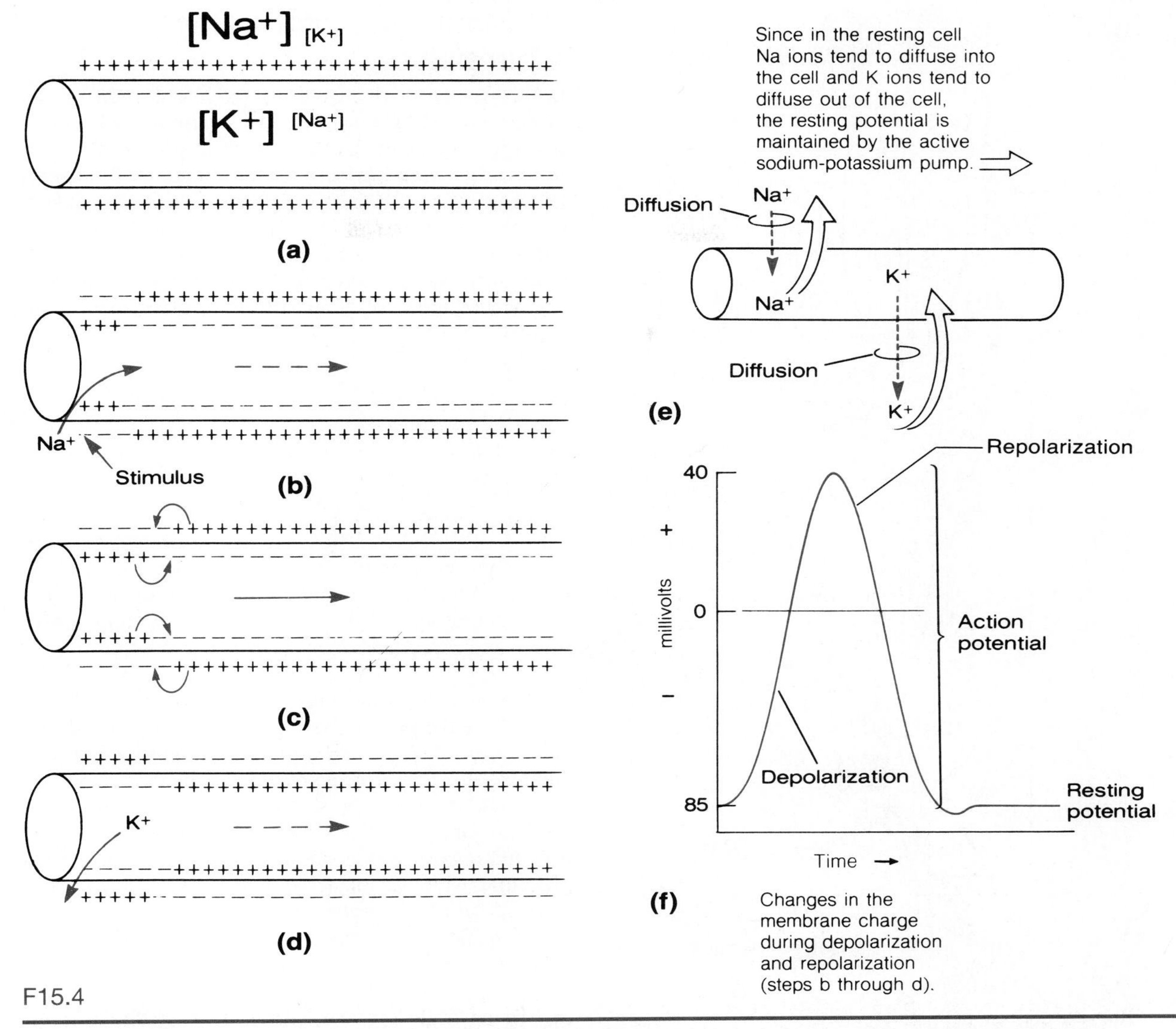

F15.4

The nerve impulse. Events depicted are occurring at the neuron's trigger zone (e.g., axon hillock). (a) Resting membrane potential (−85 mv). There is an imbalance of sodium and potassium ions on the two sides of the plasma membrane, with Na^+ the predominant extracellular ion and K^+ the predominant intracellular ion. The plasma membrane has a lower permeability to sodium ions. (b) Depolarization: reversal of the resting membrane potential. Application of a stimulus changes membrane permeability and Na^+ is allowed to diffuse rapidly into the cell. (c) Generation of an action potential or nerve impulse. If the stimulus is of adequate intensity, the depolarization wave spreads rapidly across the membrane. (d) Repolarization: reestablishment of the resting potential. The negative charge on the internal plasma membrane surface and the positive charge on its external surface are reestablished by diffusion of K^+ out of the cell. Repolarization proceeds in the same direction as depolarization. (e) The original ionic concentrations of the resting state are restored by the sodium-potassium pump. (f) A tracing of an action potential.

Once generated, the action potential is a self-propagating phenomenon that spreads rapidly along the entire length of the neuron. It is never partially transmitted; that is, it is an all-or-none response. This propagation of the action potential in neurons is also called the **nerve impulse.** When the nerve impulse reaches the axonal terminals, they release a neurotransmitter that acts either to stimulate or to inhibit the next neuron in the transmission chain. (Note that only stimulatory transmitters are considered here.)

Because only minute amounts of sodium and potassium ions have changed places, once repolarization has been completed the neuron can quickly respond again to a stimulus. In fact, thousands of impulses can be generated before ionic imbalances prevent the neuron from transmitting impulses. Eventually, however, it is necessary to restore the original ionic concentrations on the two sides of the membrane. This is accomplished by the activity of the Na^+-K^+ pump (Figure 15.4e).

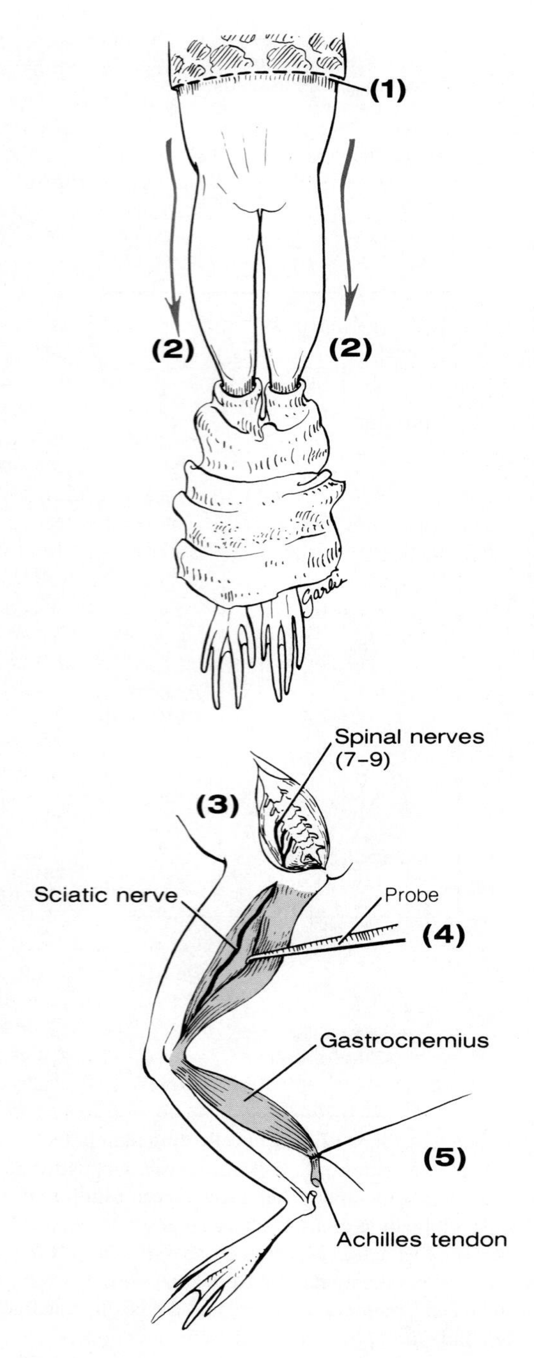

F15.5

Removal of the sciatic nerve and gastrocnemius muscle. (1) Cut through the frog's skin around the circumference of the trunk. (2) Pull the skin down over the trunk and legs. (3) Make a longitudinal cut through the abdominal musculature and expose the roots of the sciatic nerve (arising from spinal nerves 7–9). Ligate the nerve and cut the roots proximal to the ligature. (4) Use a glass probe to expose the sciatic nerve beneath the posterior thigh muscles. (5) Ligate the Achilles tendon and cut it free distal to the ligature. (6) Release the gastrocnemius muscle from the connective tissue of the knee region.

Stimulation of the Nerve Impulse

In this experiment you will investigate the functioning of nerve fibers by subjecting the sciatic nerve of a frog to various types of stimuli. (Work in groups of two to four to lighten the work load.) Stimulation of the nerve and generation of the action potential will be indicated by the contraction of the gastrocnemius muscle. Because you will make no mechanical recording (unless your instructor asks you to), you must keep complete and accurate records of all experimental procedures and results.

1. Don gloves to protect yourself from any parasites the frogs might have. Request and obtain a pithed frog from your instructor and bring it to your laboratory bench. Also obtain dissecting instruments and a tray from the supply area.

2. Prepare the sciatic nerve as illustrated in Figure 15.5. Place the pithed frog on the dissecting tray, dorsal side down. Make a cut through the skin around the circumference of the frog approximately halfway down the trunk, and then pull the skin down over the muscles of the legs. Open the abdominal cavity; and push the abdominal organs to one side to expose the origin of the glistening white sciatic nerve, which arises from the last three spinal nerves. *Once the sciatic nerve has been exposed, it should be kept continually moist with Ringer's solution.*

3. Using a glass probe, slip a piece of thread moistened with Ringer's solution under the sciatic nerve, close to its origin at the vertebral column. Make a single ligature (tie it firmly with the thread), and then cut through the nerve roots to free the proximal end of the sciatic nerve from its attachments. Using a glass rod or probe, carefully separate the posterior thigh muscles to locate and free the sciatic nerve, which runs down the posterior aspect of the thigh.

4. Tie a piece of thread around the Achilles tendon of the gastrocnemius muscle, and then cut through the tendon distal to the ligature to free the gastrocnemius muscle from the heel. Using a scalpel, very carefully release the gastrocnemius muscle from the connective tissue in the knee region. At this point, you should have completely freed both the gastrocnemius muscle and the sciatic nerve, which innervates it.

5. With glass rods, transfer the muscle to a glass plate or slide, and then attach the slide to a ring stand with a clamp. Allow the end of the sciatic nerve to hang free over the edge of the glass slide so that it is easily stimulated. *Remember to keep the nerve moist at all times.*

6. You are now ready to investigate the response of the sciatic nerve to various stimuli, beginning with electrical stimulation. Using the stimulator and platinum electrodes, stimulate the sciatic nerve with single shocks, gradually increasing the intensity of the stimulus until the threshold stimulus is determined. (The muscle as a whole will just barely contract at this stimulus.) Record the voltage of this stimulus:

_______________ v

Continue to increase the voltage until you find the point beyond which no further increase in the strength of muscle contraction occurs—that is, the point at which the maximal contraction of the muscle is obtained. Record this voltage below.

_______________ v

Delivering multiple or repeated shocks to the sciatic nerve causes volleys of impulses in the nerve. Shock the muscle with multiple stimuli. Observe the response of the muscle. How does this response compare with the response to the single electrical shocks?

____

____

____

7. To investigate mechanical stimulation, pinch the free end of the nerve by firmly pressing it between two glass rods or by pinching it with a forceps. What is the result?

____

8. Chemical stimulation can be tested by applying a small piece of filter paper saturated with 0.01% hydrochloric acid (HCl) solution to the free end of the nerve. What is the result?

____

Drop a few grains of salt (NaCl) on the free end of the nerve. What is the result?

____

9. Now test thermal stimulation. Wearing the heat-resistant mitts, heat a glass rod for a few moments over a Bunsen burner. Then touch the rod to the free end of the nerve. What is the result? What do these muscle reactions say about the irritability and conductivity of neurons?

____

____

____

Although most neurons within the body are stimulated to the greatest degree by a particular stimulus (in many cases, a chemical neurotransmitter), a variety of other stimuli may trigger nerve impulses, as illustrated in the experimental series just conducted. Generally, no matter what type of stimulus is present, if the part affected responds by becoming activated, it will always react in the same way. The most striking illustration of this is that if the optic nerve could be attached to the ear, and the auditory nerve to the eye, then you would *see* thunder and *hear* lightning! More familiar examples are the well-known phenomena of "seeing stars" when you receive a blow to the head or press on your eyeball (try it), both of which trigger impulses in your optic nerves.

Likewise, although certain afferent neurons are specialized to respond to pressure stimuli, deep continued pressure on all body nerves may block neural transmission. This phenomenon is frequently seen when you sit on your leg or doze with your head on your bent arm. When the pressure is relieved from the body part (which we say has "gone to sleep"), there is an initial heaviness and lack of response. This is quickly followed by an excruciatingly painful tingling sensation as neural transmission resumes.

Visualization of the Action Potential with an Oscilloscope

The *oscilloscope* is an instrument that visually displays the rapid but extremely minute changes in voltage that occur during the production of an action potential. The oscilloscope is similar to a TV set in that the screen display is produced by a stream of electrons generated by an electron gun (cathode) at the rear of a tube. The electrons pass through the tube and between two sets of plates that lie alongside the beam pathway. When the electrons reach the fluorescent screen, they create a tiny glowing spot. The plates determine the placement of the glowing spot by controlling the vertical or horizontal sweep of the beam. Vertical movement represents the voltage of the input signal, and horizontal movement represents the time base. When there is no electrical input signal to the oscilloscope, the electron beam sweeps horizontally (left to right) across the screen, but when the plates are electrically stimulated, the path of electrons is deflected vertically.

In this exercise, a frog's sciatic nerve will be electrically stimulated, and the action potentials generated will be observed on the oscilloscope. The dissected nerve will be placed in contact with two pairs of electrodes—*stimulating* and *recording*. As experimentation begins, the stimulating electrodes will be used to deliver a pulse of electricity to a point on the sciatic nerve. At another point on the nerve, a pair of recording electrodes connected to the oscilloscope will deliver the current to the plates inside the tube, and the electrical pulse will be recorded on the screen as a vertical deflection, or a *stimulus artifact*. As the nerve is stimulated with increasingly higher voltage, the stimulus artifact increases in amplitude as well. When the stimulus voltage reaches a high enough level (threshold), an action potential will

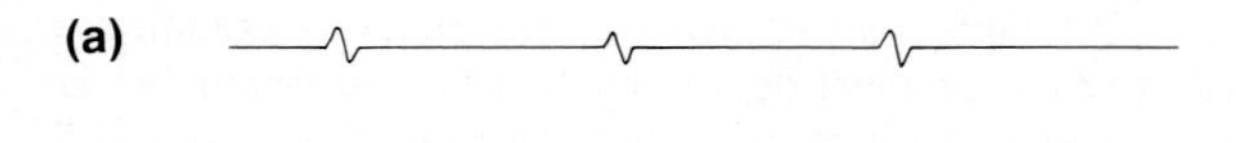

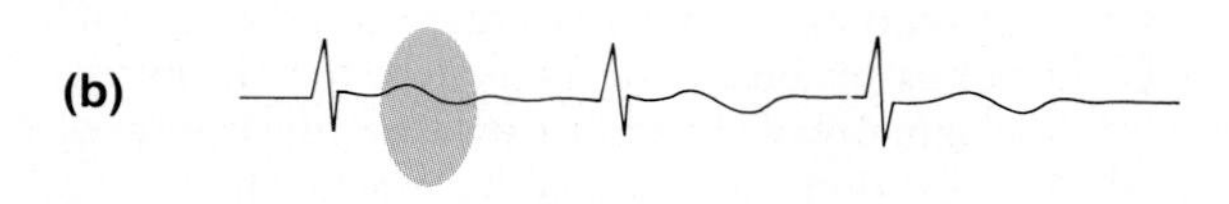

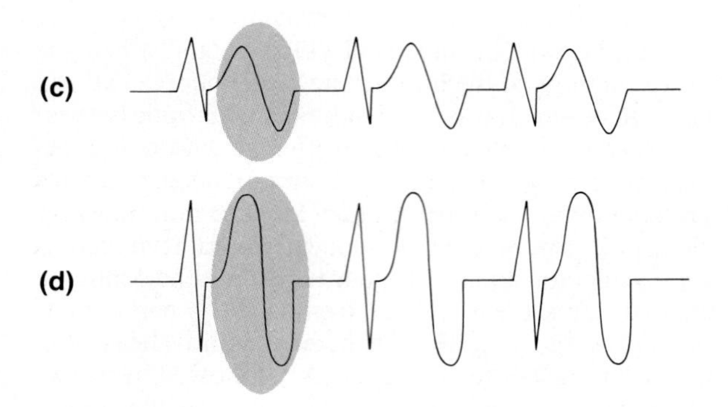

F15.6

Oscilloscope scans of nerve stimulation using stimuli with increasing intensities. The first action potential in each scan is circled. (a) Stimulus artifacts only; no action potential produced. Subthreshold stimulation. (b) Threshold stimulation. (c) Submaximal stimulation. (d) Maximal stimulus.

be generated by the nerve, and a second vertical deflection will appear on the screen, approximately 2 milliseconds after the stimulus artifact (Figure 15.6b–d). This second deflection reports the potential difference between the two recording electrodes—that is, between the first recording electrode (which has already depolarized and is in the process of repolarizing as the action potential travels along the nerve) and the second recording electrode.

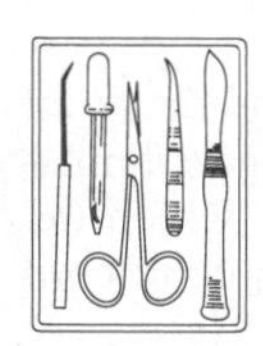

1. Obtain a nerve chamber, an oscilloscope, a stimulator, frog Ringer's solution (room temperature), a dissecting needle, and glass probes. Set up the experimental apparatus as illustrated in Figure 15.7. Connect the two stimulating electrodes to the output terminals of the stimulator and the two recording electrodes to the preamplifier of the oscilloscope.

2. Check with your instructor to find out if the frog sciatic nerve used in the previous experiment may be used for this experimental series as well. If not, prepare your frog's second sciatic nerve for experimentation as indicated on page 154 in steps 2 and 3 under Stimulation of the Nerve Impulse. While working, be careful not to touch the nerve with your fingers, and do not allow the nerve to touch the frog's skin.

3. When you have freed the sciatic nerve to the knee region with the glass probe, slip another thread length beneath that end of the nerve and make a ligature. Cut the nerve distal to this tied thread and then carefully lift the cut nerve away from the thigh of the frog by holding the threads at its proximal and distal ends. Place the nerve in the nerve chamber so that it rests across all four electrodes (the two stimulating and two recording electrodes) as shown in Figure 15.7. Flush the nerve with room temperature frog Ringer's solution.

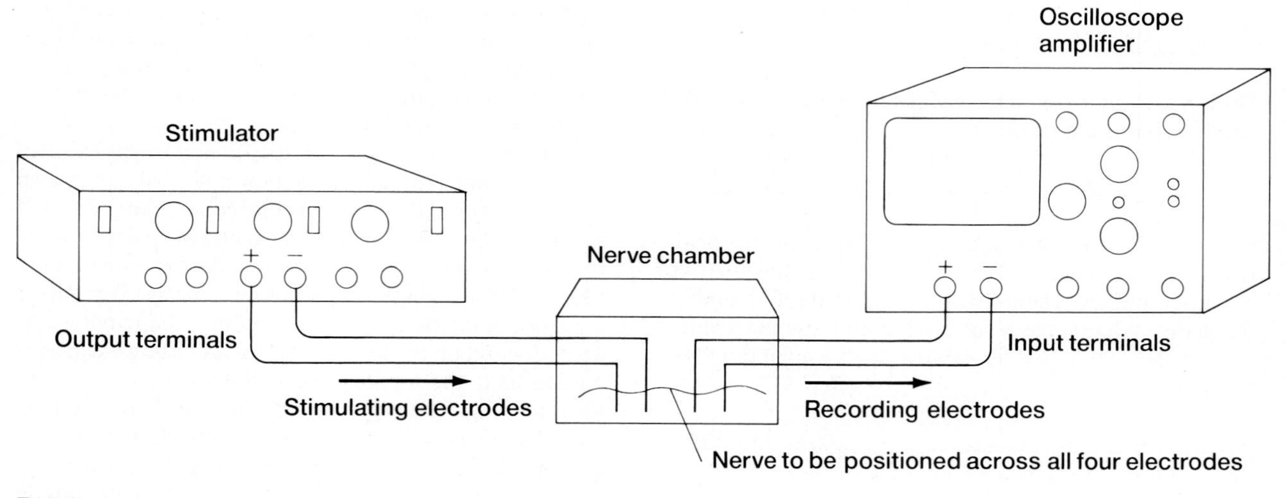

F15.7

Setup for oscilloscope visualization of action potentials in a nerve.

4. Adjust the horizontal sweep according to the instructions given in the manual or by your instructor, and set the stimulator duration, frequency, and amplitude to their lowest settings.

5. Begin to stimulate the nerve with single stimuli, slowly increasing the stimulus voltage until a threshold stimulus is achieved. The action potential will appear as a small rounded "hump" immediately following the stimulus artifact. Record the voltage of the threshold stimulus.

_______________ v

6. Flush the nerve with the Ringer's solution and continue to increase the voltage, watching as the vertical deflections produced by the action potentials become diphasic (show both upward and downward vertical deflections). Record the voltage at which the action potential reaches its maximal amplitude; this is the maximal stimulus.

_______________ v

7. Set the stimulus voltage at a level just slightly lower than the maximal stimulus and gradually increase the frequency of stimulation. What is the effect on the size (amplitude) of the action potential?

____

8. Flush the nerve with the saline once again, and allow it to sit for a few minutes while you obtain a bottle of Ringer's solution from the ice bath. Repeat steps 5 and 6 while your partner continues to flush the nerve preparation with the cold saline. Record the threshold and maximal stimulus and watch the oscilloscope pattern carefully to detect any differences in the velocity or speed of conduction from what was seen previously.

Threshold stimulus _______________ v

Maximal stimulus _______________ v

9. Flush the nerve preparation with room temperature Ringer's solution again and then gently lift the nerve by its attached threads and turn it around so that the end formerly resting on the stimulating electrodes now rests on the recording electrodes and vice versa. Stimulate the nerve. Is the impulse conducted in the opposite direction?

____

10. Dispose of the frog remains and gloves in the appropriate containers, and return your equipment to the proper supply area.

16 EXERCISE

Gross Anatomy of the Brain and Cranial Nerves

OBJECTIVES

1. To identify or locate the following brain structures on a dissected specimen, human brain model (or slices), or appropriate diagram, and to state their functions:
   - *cerebral hemisphere structures:* lobes, important sulci, lateral ventricles, basal nuclei, corpus callosum, fornix, septum pellucidum
   - *diencephalon structures:* thalamus, intermediate mass, hypothalamus, optic chiasma, pituitary gland, mammillary bodies, pineal body, choroid plexus of the third ventricle, interventricular foramen of Monro
   - *brain stem structures:* corpora quadrigemina, cerebral aqueduct, cerebral peduncles; pons, medulla, fourth ventricle
   - *cerebellum structures:* cerebellar hemispheres, vermis, arbor vitae
2. To describe the composition of the gray and white matter.
3. To locate the well-recognized functional areas of the human cerebral hemispheres.
4. To identify the three meningeal layers and state their function.
5. To state the function of the arachnoid villi and dural sinuses.
6. To discuss the formation, circulation, and drainage of cerebrospinal fluid.
7. To identify the cranial nerves by number and name on an appropriate model or diagram, stating the function of each.

MATERIALS

Human brain model (dissectible)
Preserved human brain (if available)
Coronally sectioned human brain slice (if available)
3-D model of ventricles
Preserved sheep brain (meninges and cranial nerves intact)
Dissecting tray and instruments
Protective skin cream or disposable gloves
Materials as needed for cranial nerve testing

When viewed alongside all nature's animals, humans are indeed unique and the key to their uniqueness is found in the brain. Only in humans has the brain region called the cerebrum become so elaborated and grown so out of proportion that it overshadows other brain areas. Other animals are primarily concerned with informational input and response for the sake of survival and preservation of the species, but human beings devote considerable time to nonsurvival ends. They are the only animals who manipulate abstract ideas and search for knowledge for its own sake, who are capable of emotional response and artistic creativity, or who can anticipate the future and guide their lives according to ethical and moral values. For all this, humans can thank their overgrown cerebrum (cerebral hemispheres).

We can be considered composite reflections of our brain's experience. If all past sensory input could mysteriously and suddenly be "erased," we would be unable to walk, talk, or communicate in any manner. Spontaneous movement would occur, as in a fetus, but no voluntary integrated function of any type would be possible. Clearly we would cease to be the same individuals.

Because of the complexity of the nervous system, its anatomic structures are usually considered in terms of two principal divisions: the **central nervous system (CNS)** and the **peripheral nervous system (PNS).** The central nervous system consists of the brain and spinal cord, which primarily interpret incoming sensory information and issue instructions based on past experience. The peripheral nervous system consists of the cranial and spinal nerves, ganglia, and sensory receptors. These structures serve as communication lines as they carry impulses from the sensory receptors to the CNS and from the CNS to the appropriate glands or muscles.

In this exercise both CNS (brain) and PNS (cranial nerves) structures will be studied, because of their close anatomic relationship.

THE HUMAN BRAIN

During embryonic development of all vertebrates, the CNS first makes its appearance as a simple tubelike structure, the **neural tube,** that extends down the dorsal medial plane. By the fourth week, the human brain begins to form as an expansion of the anterior, or rostral, end of the neural tube (the end toward the head). Shortly thereafter, constrictions appear, dividing the developing brain into three major regions—the **forebrain, midbrain,** and **hindbrain** (Figure 16.1). The remainder of the neural tube becomes the spinal cord.

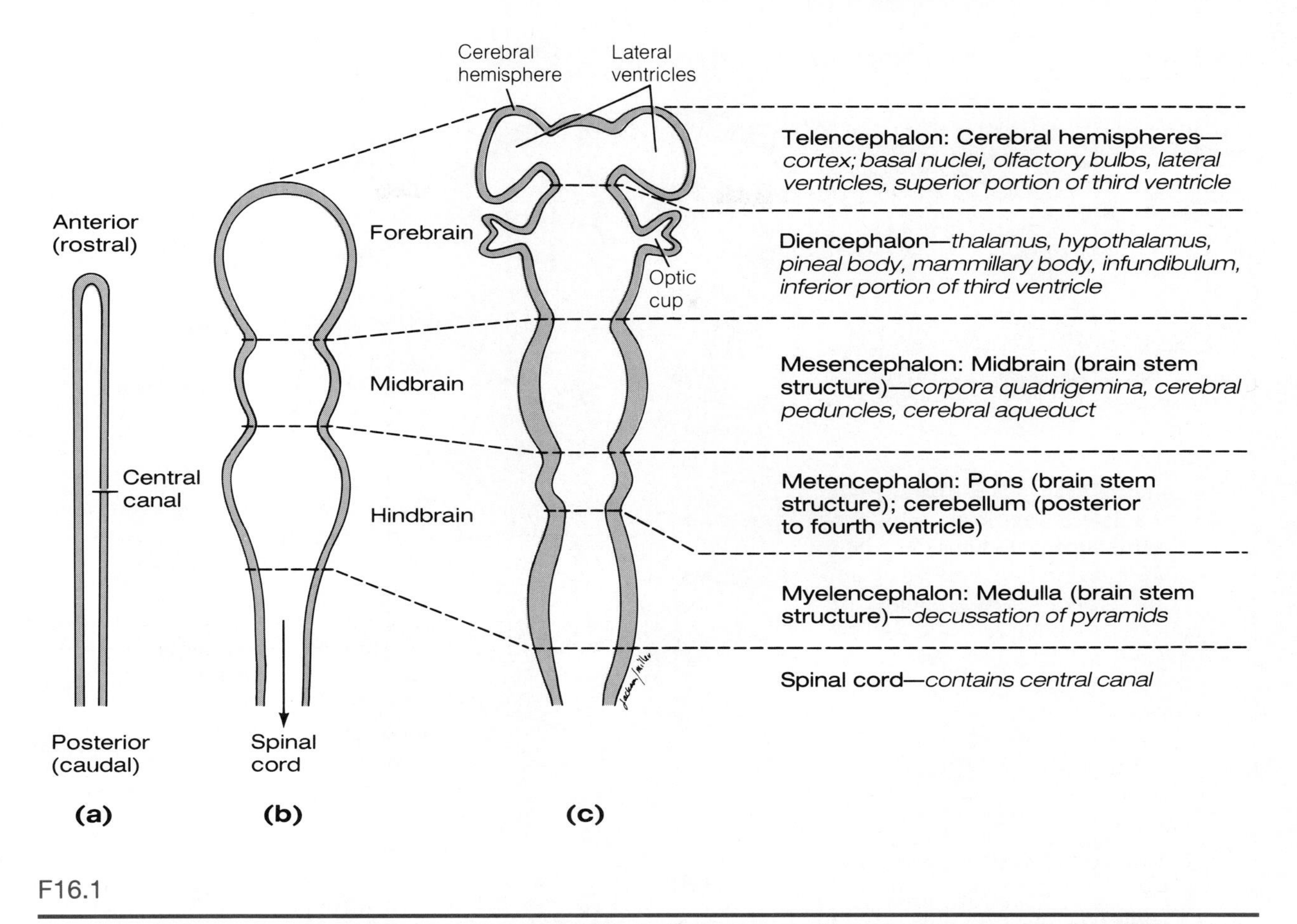

F16.1

Embryonic development of the brain. Adult brain structures developing from forebrain, midbrain, and hindbrain subdivisions. (a) Neural tube precedes (b) subdivisions of the neural tube, which precede (c) the adult brain structures.

During fetal development, two anterior outpocketings extend from the forebrain and grow rapidly to form the cerebral hemispheres. Because of space restrictions imposed by the skull, the cerebral hemispheres are forced to grow posteriorly and inferiorly, and finally end up enveloping and obscuring the rest of the forebrain and midbrain structures. Somewhat later in development, the dorsal portion of the hindbrain also enlarges, producing the cerebellum. The central canal of the neural tube, which remains continuous throughout the brain and cord, becomes enlarged in four regions of the brain, forming chambers called **ventricles.**

External Anatomy

Generally, in studying the major brain areas in the laboratory, the brain is considered in terms of four major regions: the cerebral hemispheres, diencephalon, brain stem, and cerebellum. The correlation between these anatomic regions and the structures of the forebrain, midbrain, and hindbrain is also outlined in Figure 16.1.

CEREBRAL HEMISPHERES The **cerebral hemispheres** are the most superior portion of the brain (Figure 16.2). Their entire surface is thrown into elevated ridges of tissue called **gyri** that are separated by depressed areas called **fissures,** or **sulci.** Of the two types of depressions, the fissures are deeper. Many of the fissures and gyri are important anatomical landmarks.

The cerebral hemispheres are divided by a single deep fissure, the **longitudinal fissure.** The **central sulcus** divides the **frontal lobe** from the **parietal lobe,** and the **lateral sulcus** separates the **temporal lobe** from the parietal lobe. The **parieto-occipital sulcus,** which divides the **occipital lobe** from the parietal lobe, is not visible externally. Notice that the cerebral lobes are named for the cranial bones that lie over them.

The functional areas of the cerebral hemispheres have also been located. The **primary** (somatic) **sensory area** is located in the **postcentral gyrus** of the parietal lobe. Impulses traveling from the body's sensory receptors (such as those for pressure, pain, and temperature) are localized in this area of the brain. ("This information is from my big toe.") Immediately posterior to the primary sensory area is the **somatosensory association area,** where the meaning of incoming stimuli is analyzed. ("Ouch! I have a *pain* there.") Thus, the somatosensory association area allows you to become aware of pain, coldness, a light touch, and the like.

Impulses from the special sense organs are interpreted in other specific areas also noted in Figure 16.2b.

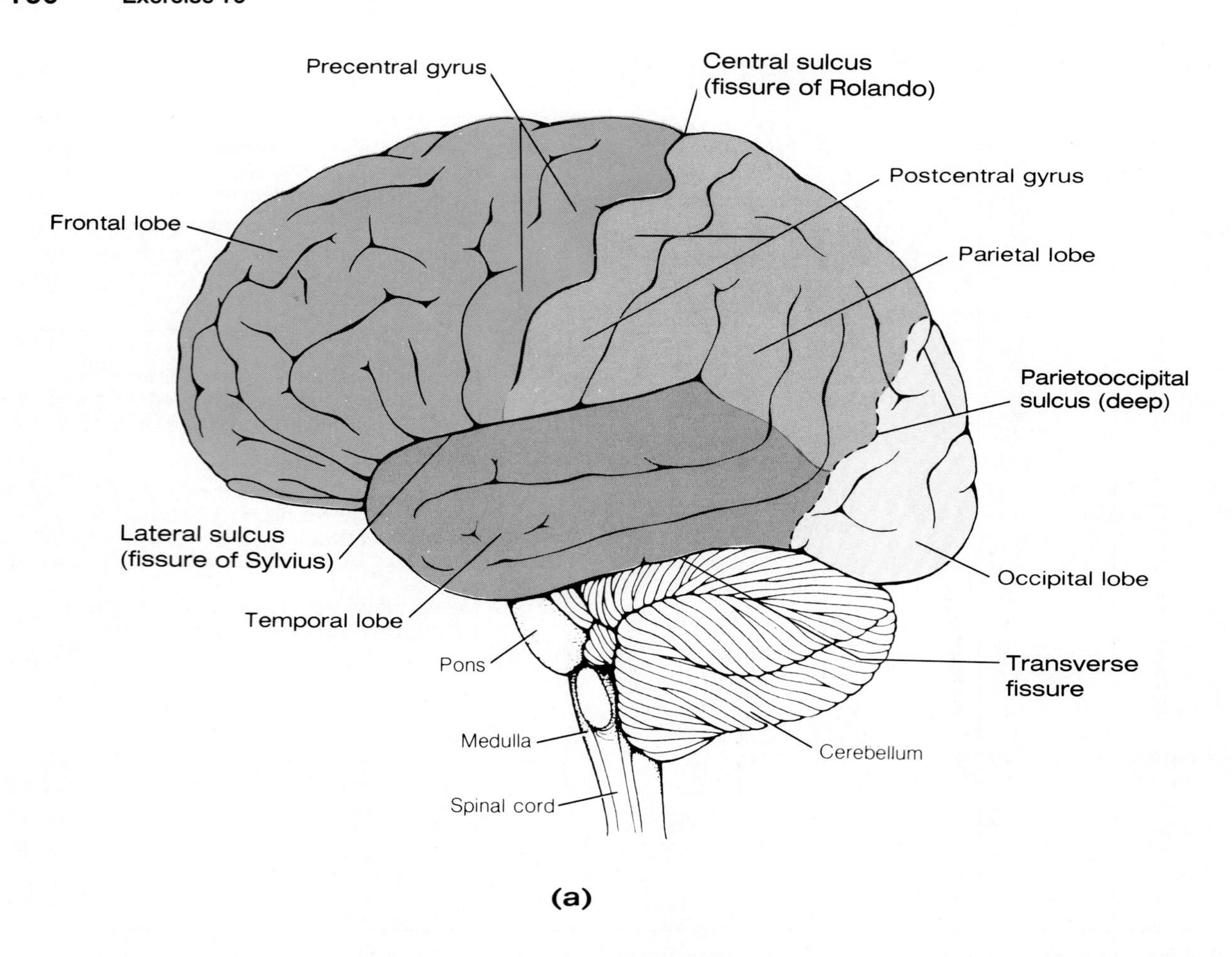

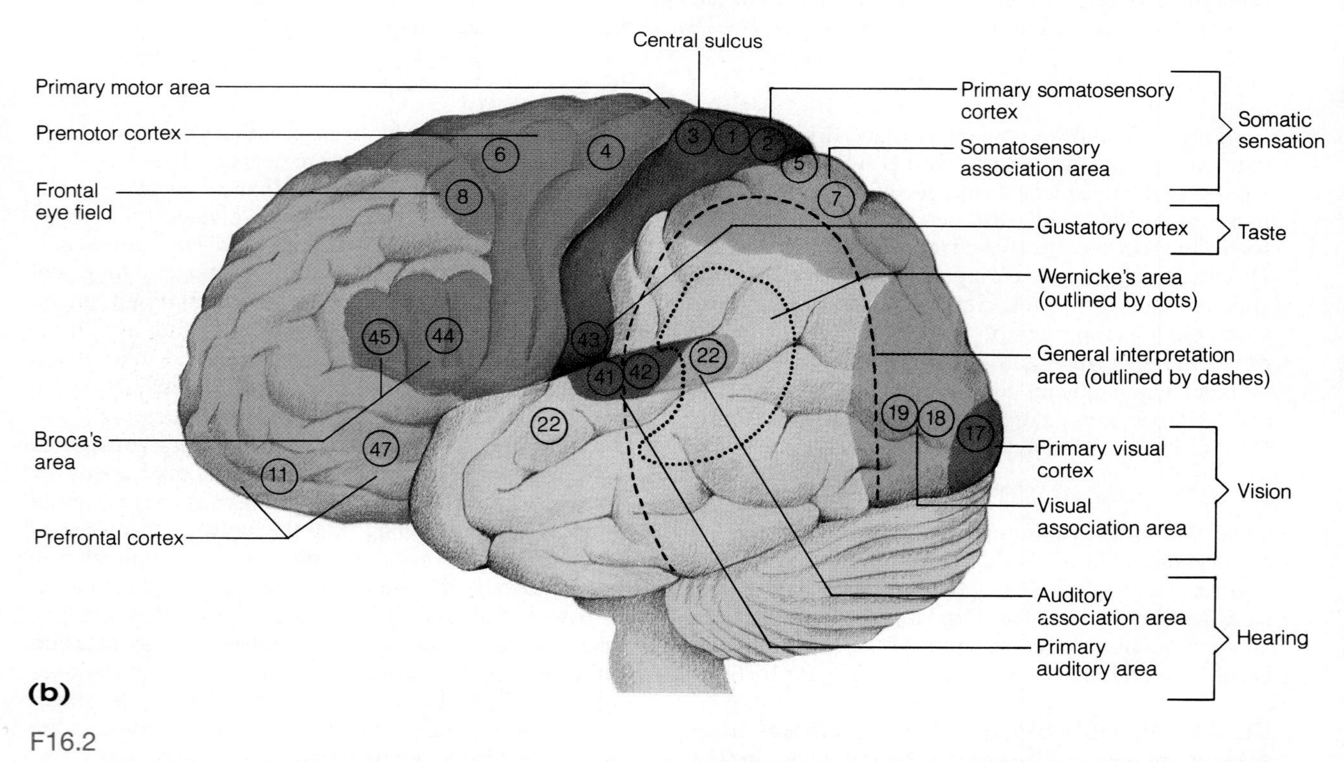

F16.2

Cerebral hemispheres of the human brain. Major structural areas: (a) left lateral view and (b) Major functional areas of the left cerebral hemisphere.

For example, the visual areas are located in the occipital lobe and the auditory area is in the temporal lobe, bordering the lateral sulcus.

The **primary motor area,** which is responsible for conscious or voluntary movement of the skeletal muscles, is located in the **precentral gyrus** of the frontal lobe. A specialized motor speech area called **Broca's area** is found at the base of the precentral gyrus, just above the lateral sulcus. Damage to this area (which is located only in one cerebral hemisphere, usually the left) reduces or eliminates the ability to articulate words.

Areas involved in higher intellectual reasoning are believed to lie in the anterior portions of the frontal lobes. A rather poorly defined region at the junction of the temporal, parietal, and occipital lobes is a **speech area** (*Wernicke's area*), in which unfamiliar words are sounded out. Like Broca's area, Wernicke's area is located in one cerebral hemisphere only, typically the left.

Although there are many similar functional areas in both cerebral hemispheres, for example, motor and sensory areas, each hemisphere is also a "specialist" in certain ways. For example, the left hemisphere is our "language brain," because it houses centers associated with language skills and speech. The right hemisphere is more specifically concerned with abstract, conceptual, or spatial processes—skills associated with artistic or inventive pursuits.

Generally, girls tend to excel in left-brain activities, whereas boys are more adept at tasks that require spatial skills. Research on this observation seems to indicate that this dichotomy largely reflects our society's concept of what is feminine or masculine. Little girls are talked to more, have more stories read to them, and receive dolls and books for gifts. Boys are more likely to receive erector or building-block sets, tools, and puzzles of various types that challenge their spatial abilities. However, there *are* biological differences between the male and female brain which cannot be discounted.

The cell bodies of cerebral neurons involved in these functions are found only in the outermost gray matter of the cerebrum, the area called the **cerebral cortex.** Most of the balance of cerebral tissue—the deeper **cerebral white matter**—is composed of fiber tracts carrying impulses to or from the cortex.

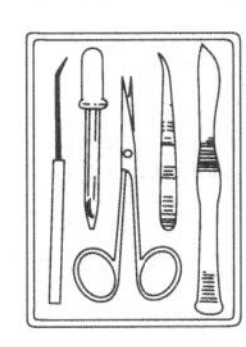

Using a model of the human brain (and a preserved human brain, if available), identify the areas and structures of the cerebral hemispheres described above.

DIENCEPHALON The **diencephalon,** sometimes considered the most superior portion of the brain stem, is embryologically part of the forebrain, along with the cerebral hemispheres.

1. Turn the brain model so the ventral surface of the brain can be viewed. Using Figure 16.3 as a guide, start superiorly and identify the externally visible structures that mark the position of the floor of the diencephalon. These are the **olfactory bulbs** and **tracts, optic nerves, optic chiasma** (where the fibers of the optic nerves partially cross over), **optic tracts, pituitary gland,** and **mammillary bodies.**

BRAIN STEM

1. Continue inferiorly to identify the **brain stem** structures—the **cerebral peduncles** (fiber tracts in the **midbrain** connecting the pons below with cerebrum above), the pons, and the medulla oblongata. *Pons* means "bridge," and the **pons** consists primarily of motor and sensory fiber tracts connecting the brain with lower CNS centers. The lowest brain stem region, the **medulla,** is also composed primarily of fiber tracts. You can see the **decussation of pyramids,** a crossover point for the major motor tract (pyramidal tract) descending from the motor areas of the cerebrum to the cord, on the medulla's anterior surface. The medulla also houses many vital autonomic centers involved in the control of heart rate, respiratory rhythm, and blood pressure as well as involuntary centers involved in the initiation of vomiting, swallowing, and so on.

CEREBELLUM

1. Turn the brain model so you can see the dorsal aspect. Identify the large cauliflowerlike **cerebellum,** which projects dorsally from under the occipital lobe of the cerebrum. Note that, like the cerebrum, the cerebellum has two hemispheres and a convoluted surface. It also has an outer cortex made up of gray matter with an inner region of white matter.

2. Remove the cerebellum to view the **corpora quadrigemina,** a brain stem structure located on the posterior aspect of the midbrain. The two superior prominences are the **superior colliculi** (visual reflex centers); the two smaller inferior prominences are the **inferior colliculi** (auditory reflex centers).

Internal Anatomy

The deeper structures of the brain have also been well mapped. Like the external structures, these can be studied in terms of the four major regions.

CEREBRAL HEMISPHERES

1. Take the brain model apart so you can see a median sagittal view of the internal brain structures (Figure 16.4). Observe the model closely to see the extent of the outer cortex (gray matter), which contains the cell bodies of cerebral neurons.

2. Now observe the deeper area of white matter, which is composed of fiber tracts. The fiber tracts found in the cerebral hemisphere white matter are named *association tracts* if they connect two portions of the same hemisphere, *projection tracts* if they run between the cerebral cortex and the lower brain or spinal cord, and *commissures* if they run from one hemisphere to another. Observe the large **corpus callosum,** the major commissure

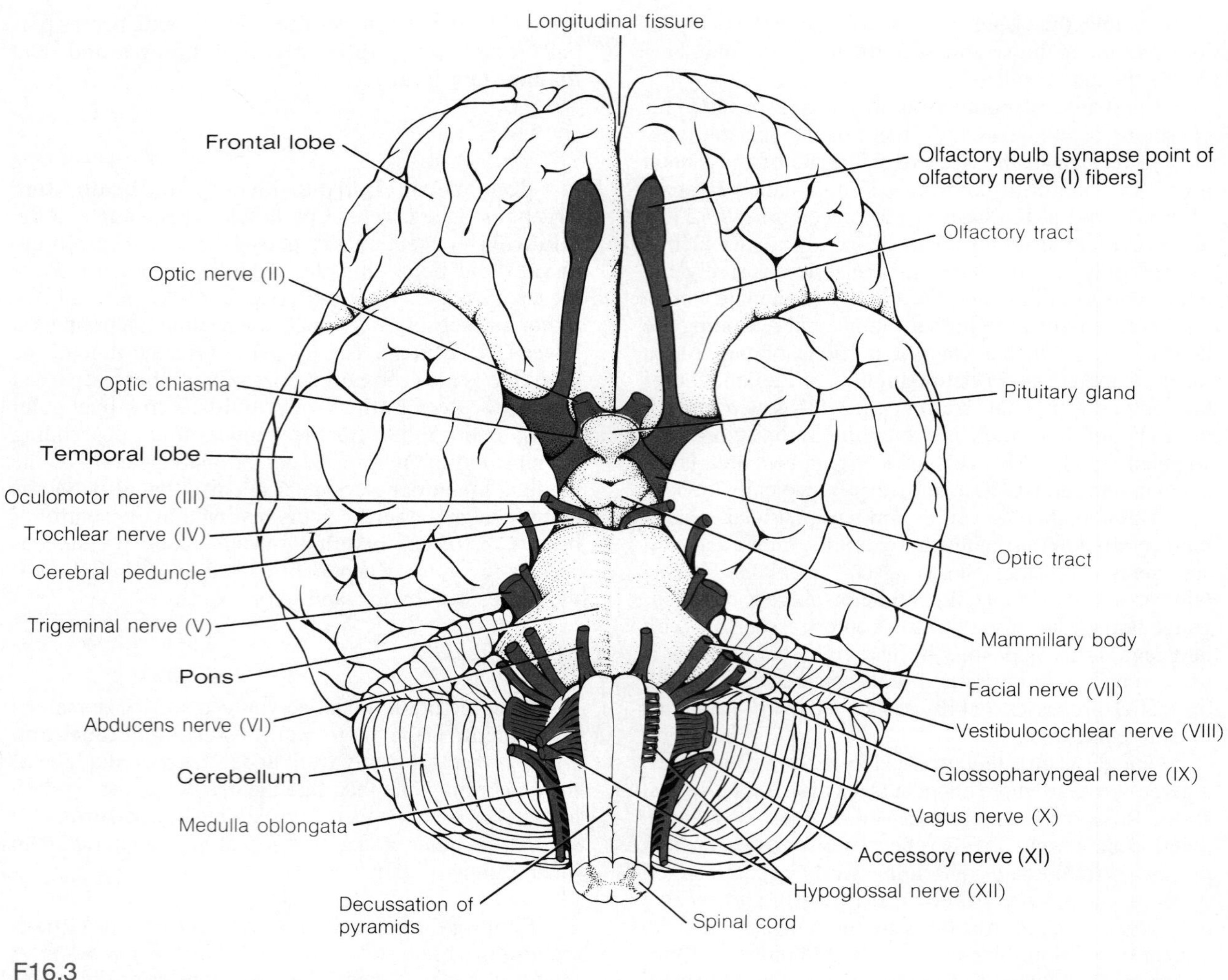

F16.3

Ventral surface of the human brain.

connecting the cerebral hemispheres. The corpus callosum arches above the structures of the diencephalon and roofs over the lateral ventricles. Note also the **fornix,** a bandlike fiber tract concerned with olfaction, and the membranous **septum pellucidum,** which separates the lateral ventricles of the cerebral hemispheres.

3. In addition to the gray matter of the cerebral cortex, there are several "islands" of gray matter (clusters of neuron cell bodies) called **nuclei** buried deep within the white matter of the cerebral hemispheres. One important group of cerebral nuclei, called the **basal nuclei,** flank the lateral and third ventricles. You can see the basal nuclei if you have an appropriate dissectible model or a coronally sectioned human brain slice. Otherwise, Figure 16.5 will suffice.

The basal nuclei, which are important subcortical motor nuclei (and part of the so-called extrapyramidal system), are involved in the regulation of voluntary motor activities. The most important of them are the **caudate nucleus,** the **claustrum,** the **amygdaloid nucleus** (located at the tip of the caudate nucleus), and the **lentiform nucleus,** which is composed of the **putamen** and **globus pallidus nuclei.** The **corona radiata,** a spray of projection fibers coursing down from the precentral (motor) gyrus, combines with sensory fibers traveling to the sensory cortex to form a broad band of fibrous material called the **internal capsule** that passes between the diencephalon and the basal nuclei.

4. Examine the relationship of the lateral ventricles and corpus callosum to the hypothalamus, thalamus, and third ventricle—from the coronal viewpoint (see Figure 16.5b).

DIENCEPHALON

1. The major internal structures of the diencephalon are the thalamus, hypothalamus, and epithalamus (see Figure 16.4). The **thalamus** consists of two large lobes of gray matter that laterally enclose the shallow third ventricle of the brain. A slender stalk of thalamic tissue, the **intermediate mass,** or **massa intermedia,** connects the two thalamic lobes and bridges the ventricle. The thalamus is a major integrating and relay station for sen-

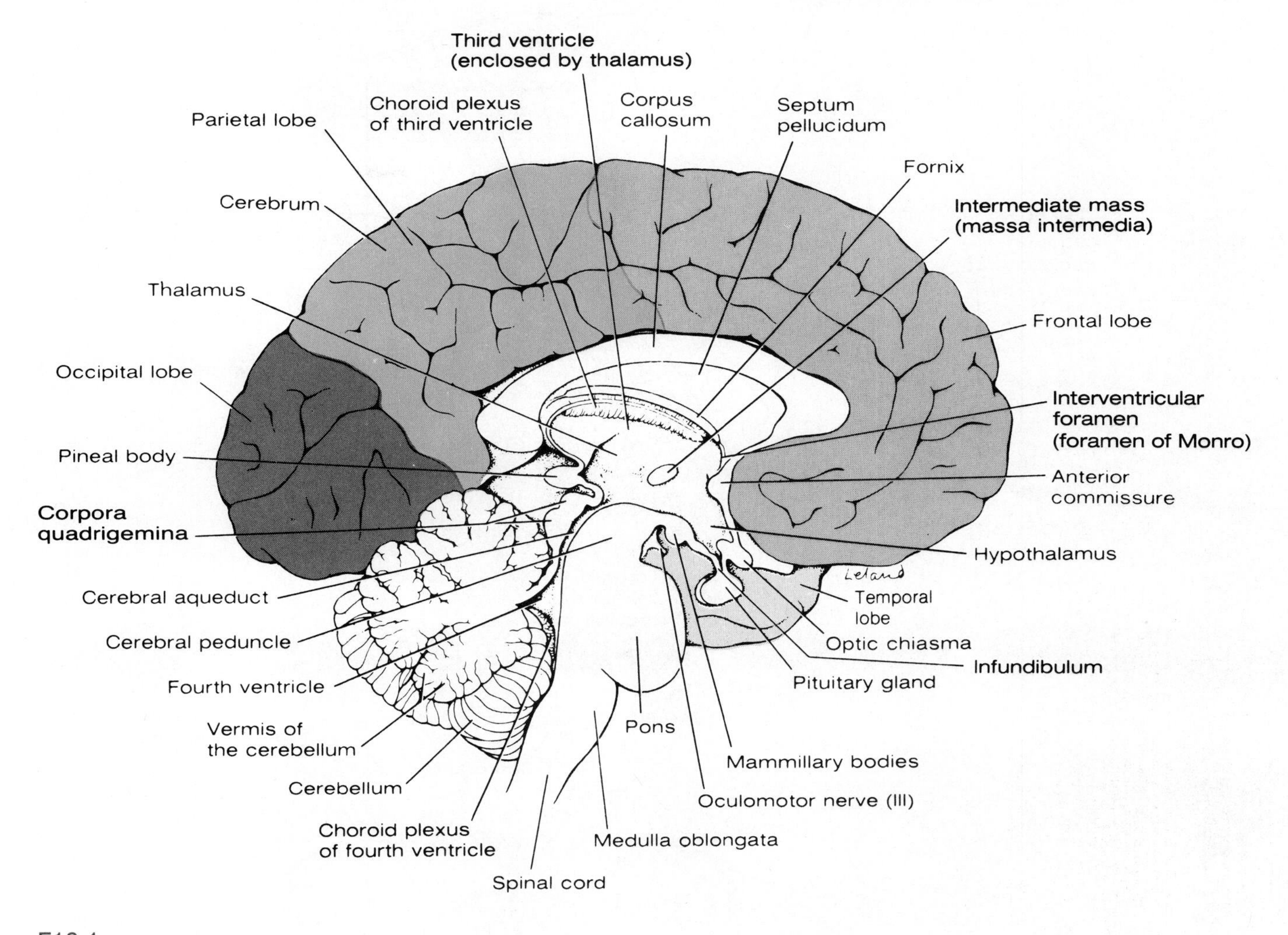

F16.4

Midsagittal section of the human brain.

sory impulses passing upward to the cortical sensory areas for localization and interpretation. Locate also the **interventricular foramen (foramen of Monro),** a tiny orifice connecting the third ventricle with the lateral ventricle on the same side.

2. The **hypothalamus** makes up the floor and a part of the inferior lateral walls of the third ventricle. It is an important autonomic center involved in the regulation of body temperature, water balance, and fat and carbohydrate metabolism, as well as in many other activities and drives (sex, hunger, thirst). If possible, also locate the pituitary gland, or **hypophysis,** which hangs from the anterior floor of the hypothalamus by a slender stalk, the **infundibulum.** (The pituitary gland is usually not present in preserved brain specimens.) In life, the pituitary rests in the fossa of the sella turcica portion of the sphenoid bone. Its function is discussed in Exercise 24.

Anterior to the pituitary, the optic chiasma portion of the optic pathway to the brain can also be identified. The **mammillary bodies,** reflex centers for olfaction, bulge exteriorly from the floor of the hypothalamus, posterior to the pituitary gland.

3. The **epithalamus** forms the roof of the third ventricle. Important structures in the epithalamus are the **pineal body** or **gland** (a neuroendocrine structure), and the **choroid plexus** of the third ventricle. The choroid plexuses, knotlike collections of capillaries within each ventricle, form the cerebrospinal fluid.

BRAIN STEM

1. Now trace the short midbrain from the mammillary bodies to the rounded pons below. Continue to refer to Figure 16.4. The **cerebral aqueduct** is a slender canal traveling through the midbrain; it connects the third ventricle to the fourth ventricle in the hindbrain below. The cerebral peduncles and the rounded corpora quadrigemina make up the midbrain tissue anterior and posterior (respectively) to the cerebral aqueduct.

2. Locate the hindbrain structures. Trace the rounded pons to the medulla oblongata below, and identify the fourth ventricle posterior to these structures.

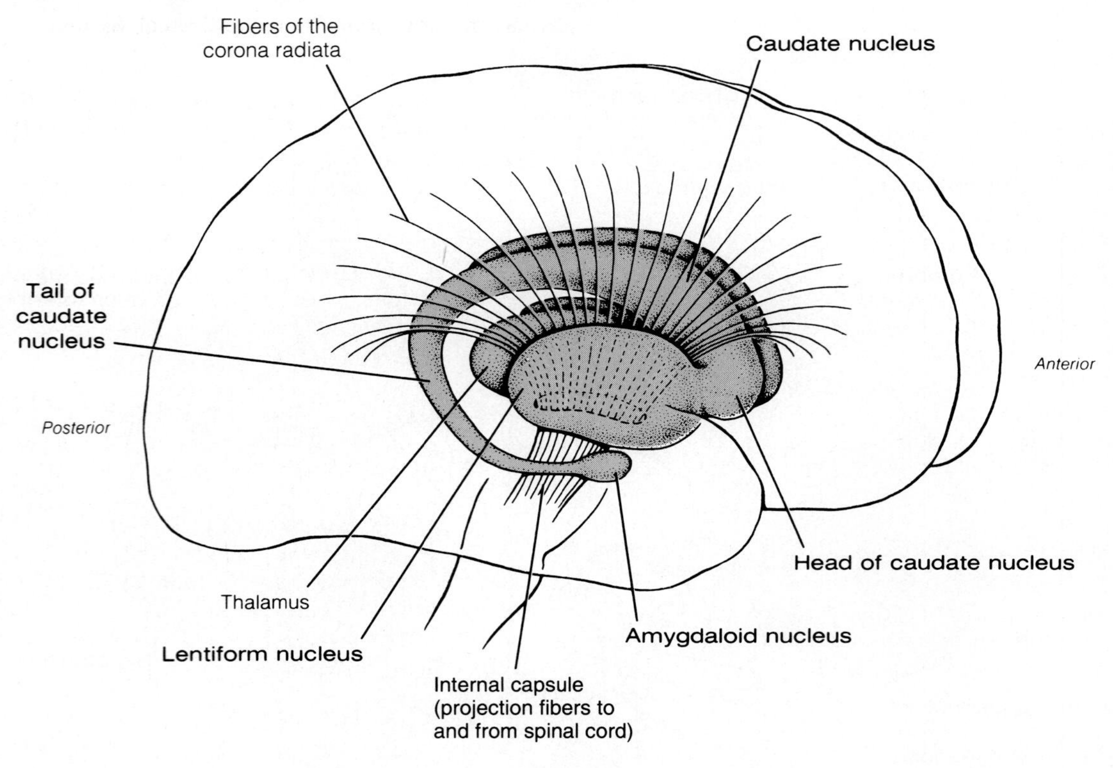

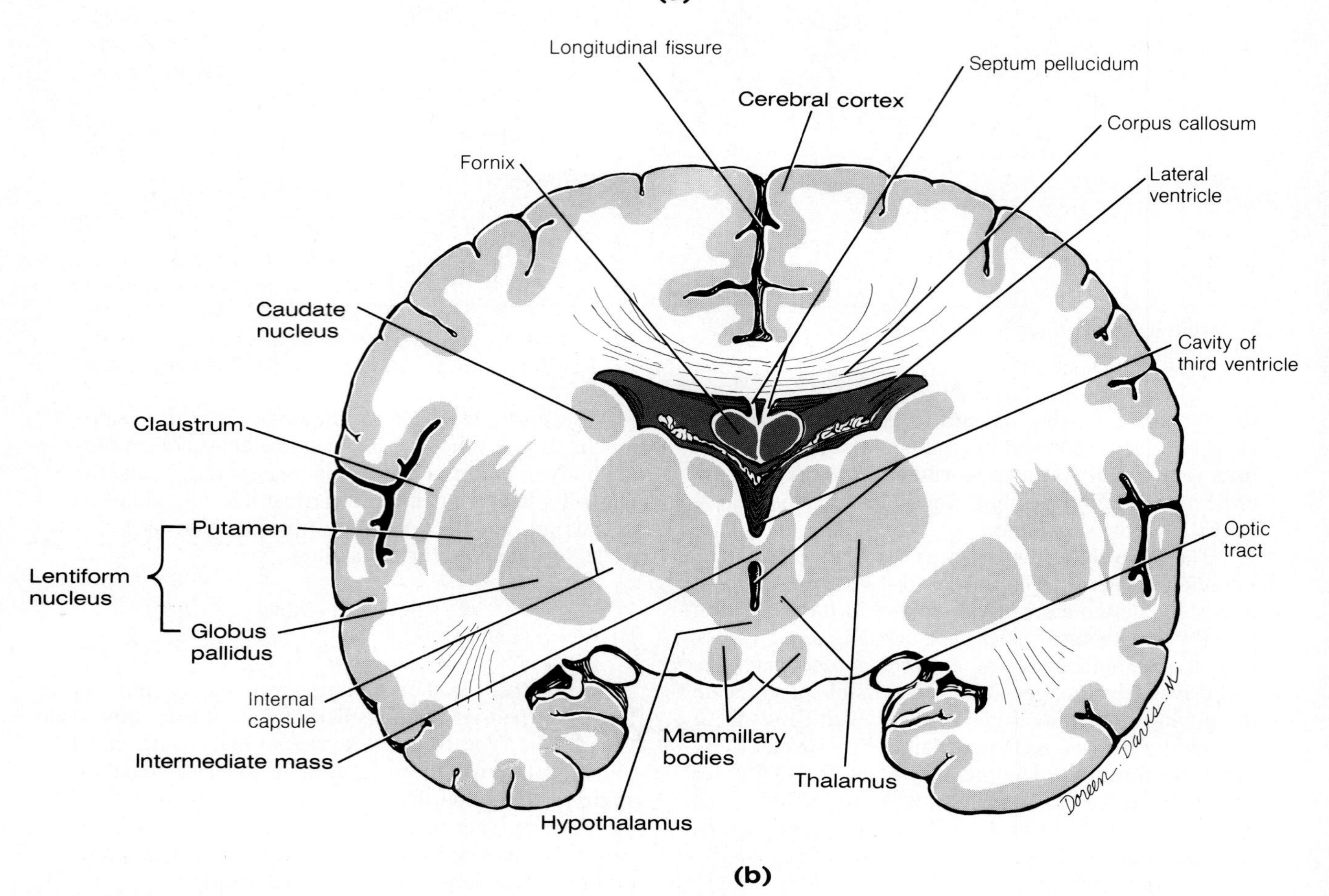

F16.5

Location of the basal nuclei. (a) Three-dimensional view of the basal nuclei showing their position within the cerebrum; (b) frontal section of the human brain.

CEREBELLUM Examine the cerebellum. Notice that it is composed of two lateral hemispheres connected by a midline lobe called the **vermis.** As in the cerebrum, the cerebellum has an outer cortical area of gray matter and an inner area of white matter. The treelike branching of the cerebellar white matter is referred to as the **arbor vitae,** or tree of life. The cerebellum is concerned with the unconscious coordination of skeletal muscle activity and the control of balance and equilibrium. Fibers converge on the cerebellum from the equilibrium apparatus of the inner ear, visual pathways, proprioceptors of the tendons and skeletal muscles, and many other areas. Thus the cerebellum remains constantly aware of the position and state of tension of the various body parts.

CRANIAL NERVES

The cranial nerves are part of the peripheral nervous system and not part of the brain proper, but they are most appropriately identified in conjunction with the study of brain anatomy. The 12 pairs of cranial nerves primarily serve the head and neck. Only one pair (the vagus nerves) extends to the thoracic and abdominal cavities. All but the first two pairs (olfactory and optic nerves) arise from the brain stem and pass through foramina in the base of the skull to reach their destination.

The cranial nerves are numbered consecutively, and in most cases their names reflect the major structures they control. The cranial nerves are described by name, number (Roman numeral), origin, course, and function in Table 16.1. This information should be committed to memory. A mnemonic device that might be helpful for remembering the cranial nerves in order is, "*O*h, *o*h, *o*h, *t*o *t*ouch *a*nd *f*eel *v*ery *g*ood *v*elvet, *Ah*." The first letter of each word, and both letters of the final "Ah," will remind you of the first letter of the cranial nerve name.

Most cranial nerves are mixed nerves (containing both motor and sensory fibers). However, close scrutiny of Table 16.1 will reveal that three pairs of cranial nerves (optic, olfactory, and vestibulocochlear) are purely sensory in function.

You may recall that the cell bodies of neurons are always located within the central nervous system (cortex or nuclei) or in specialized collections of cell bodies (ganglia) outside the CNS. Neuron cell bodies of the sensory cranial nerves are located in ganglia; those of the mixed cranial nerves are found both within the brain and in peripheral ganglia.

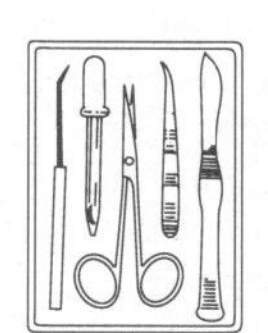

1. Observe the anterior surface of the brain model to identify the cranial nerves. Figure 16.3 may also aid you in this study. Note that the first (olfactory) cranial nerves are not visible on the model because they consist only of those short axons that run from the nasal mucosa through the cribriform plate of the ethmoid bone. (However, the synapse points of the first cranial nerves, the *olfactory bulbs,* are visible on the model.)

2. The last column of Table 16.1 describes techniques for testing cranial nerves, which is an important part of any neurologic examination. This information may help you understand cranial nerve function, especially as it pertains to some aspects of brain function. Conduct tests of cranial nerve function following directions given in the "testing" column of the table.

MENINGES OF THE BRAIN

The brain (and spinal cord) are covered and protected by three connective tissue membranes called **meninges** (Figure 16.6). The outermost meninx is the leathery **dura mater,** a double-layered membrane. One of its layers (the *periosteal layer*) is attached to the inner surface

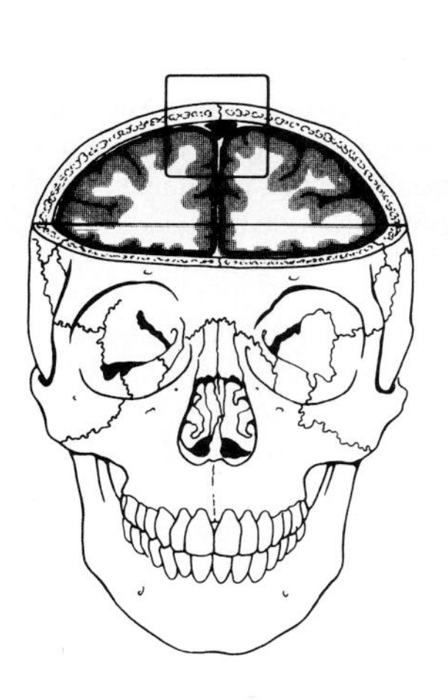

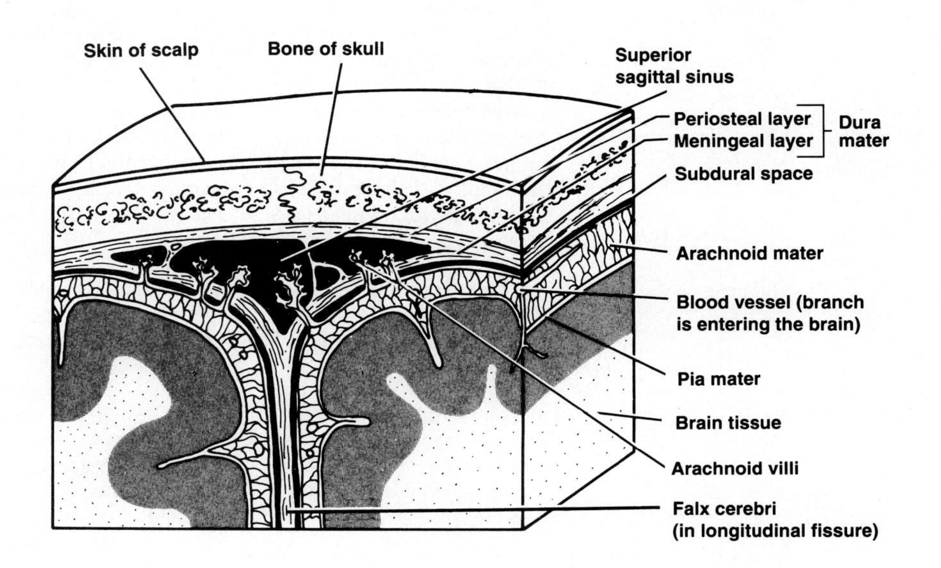

F16.6

Meninges of the brain.

TABLE 16.1 The Cranial Nerves (see Figure 16.3)

Number and name	Origin and course	Function	Testing
I. Olfactory	Fibers arise from olfactory mucosa and run through cribriform plate of ethmoid bone to synapse with olfactory bulbs.	Purely sensory—carries impulses associated with sense of smell.	Person is asked to sniff aromatic substances, such as oil of cloves and vanilla, and to identify each.
II. Optic	Fibers arise from retina of eye and pass through optic foramen in sphenoid bone. Fibers of two optic nerves then take part in forming optic chiasma (with partial crossover of fibers), after which they continue on to thalamus as the optic tracts. Final fibers of this pathway travel from the thalamus to the optic cortex as the optic radiation.	Purely sensory—carries impulses associated with vision.	Vision and visual field are determined with eye chart and by testing the point at which the person first sees an object (finger) moving into the visual field. Fundus of eye viewed with ophthalmoscope to detect papilledema (swelling of optic disc), or point at which optic nerve leaves the eye and to observe blood vessels.
III. Oculomotor	Fibers emerge from midbrain and exit from skull via superior orbital fissure to run to eye.	Somatic motor fibers to inferior oblique and superior, inferior, and medial rectus muscles, which direct eyeball, and to levator palpebrae muscles of eyelid; parasympathetic fibers to iris and smooth muscle controlling lens shape (control reflex responses to varying light intensity and focusing of eye for near vision); contains proprioceptive sensory fibers carrying impulses from extrinsic eye muscles.	Pupils are examined for size, shape, and equality. Pupillary reflex is tested with penlight (pupils should constrict when illuminated). Convergence for near vision is tested, as is subject's ability to follow objects up, down, side to side, and diagonally.
IV. Trochlear	Fibers emerge from midbrain and exit from skull via superior orbital fissure to run to eye.	Provides somatic motor fibers to superior oblique muscle (an extrinsic eye muscle); conveys proprioceptive impulses from same muscle to brain.	Tested in common with cranial nerve III.
V. Trigeminal	Fibers emerge from pons and form three divisions, which exit separately from skull: mandibular division through foramen ovale in sphenoid bone, maxillary division via foramen rotundum in sphenoid bone, and ophthalmic division through superior orbital fissure of eye socket.	Major sensory nerve of face; conducts sensory impulses from skin of face and anterior scalp, from mucosae of mouth and nose, and from surface of eyes; mandibular division also contains motor fibers that innervate muscles of mastication and muscles of floor of mouth.	Sensations of pain, touch, and temperature are tested with safety pin and hot and cold objects. Corneal reflex tested with wisp of cotton. Motor branch assessed by asking person to clench his teeth, open mouth against resistance, and move jaw side to side.
VI. Abducens	Fibers leave inferior region of pons and exit from skull via superior orbital fissure to run to eye.	Carries motor fibers to lateral rectus muscle of eye and proprioceptive fibers from same muscle to brain.	Tested in common with cranial nerve III.
VII. Facial	Fibers leave pons and travel through temporal bone via internal acoustic meatus, exiting via stylomastoid foramen to reach the face.	Mixed—supplies somatic motor fibers to muscles of facial expression and parasympathetic motor fibers to lacrimal and salivary glands; carries sensory fibers from taste receptors of anterior portion of tongue.	Anterior two-thirds of tongue is tested for ability to taste sweet (sugar), salty, sour (vinegar), and bitter (quinine) substances. Symmetry of face is checked. Subject is asked to close eyes, smile, whistle, and so on. Tearing is assessed with ammonia fumes.

TABLE 16.1 Continued

Number and name	Origin and course	Function	Testing
VIII. Vestibulocochlear	Fibers run from inner ear equilibrium and hearing apparatus, housed in temporal bone, through internal acoustic meatus to enter pons.	Purely sensory—vestibular branch transmits impulses associated with sense of equilibrium from vestibular apparatus and semicircular canals; cochlear branch transmits impulses associated with hearing from cochlea.	Hearing is checked by air and bone conduction using tuning fork.
IX. Glossopharyngeal	Fibers emerge from medulla and leave skull via jugular foramen to run to throat.	Mixed—somatic motor fibers serve pharyngeal muscles, and parasympathetic motor fibers serve salivary glands; sensory fibers carry impulses from pharynx, tonsils, posterior tongue (taste buds), and pressure receptors of carotid artery.	Position of the uvula is checked. Gag and swallowing reflexes are checked. Subject is asked to speak and cough. Posterior third of tongue may be tested for taste.
X. Vagus	Fibers emerge from medulla and pass through jugular foramen and descend through neck region into thorax and abdomen.	Fibers carry somatic motor impulses to pharynx and larynx and sensory fibers from same structures; very large portion is composed of parasympathetic motor fibers, which supply heart and smooth muscles of abdominal visceral organs; transmits sensory impulses from viscera.	As for cranial nerve IX (IX and X are tested in common, since they both innervate muscles of throat and mouth).
XI. Accessory	Fibers arise from medulla and superior aspect of spinal cord and travel through jugular foramen to reach muscles of neck and back.	Provides somatic motor fibers to sternocleidomastoid and trapezius muscles and to muscles of soft palate, pharynx, and larynx (spinal and medullary fibers respectively); proprioceptive impulses are conducted from these muscles to brain.	Sternocleidomastoid and trapezius muscles are checked for strength by asking person to rotate head and shrug shoulders against resistance.
XII. Hypoglossal	Fibers arise from medulla and exit from skull via hypoglossal canal to travel to tongue.	Carries somatic motor fibers to muscles of tongue and proprioceptive impulses from tongue to brain.	Person is asked to protrude and retract tongue. Any deviations in position are noted.

of the skull, forming the periosteum; the other (the *meningeal layer*) forms the outermost brain covering and is continuous with the dura mater of the spinal cord.

The dural layers are fused together, except in three areas where the inner membrane extends inward to form a septum that secures the brain to structures inside the cranial cavity. One such extension, the **falx cerebri,** dips into the longitudinal fissure between the cerebral hemispheres to attach to the crista galli of the ethmoid bone of the skull. The cavity created at this point is the large **superior sagittal sinus,** which collects blood draining from the brain tissue. The **falx cerebelli,** separating the two cerebellar hemispheres, and the **tentorium cerebelli,** separating the cerebrum from the cerebellum below, are two other important inward folds of the inner dural membrane.

The middle meninx, the weblike **arachnoid mater,** underlies the dura mater and is partially separated from it by the **subdural space.** Threadlike projections bridge the **subarachnoid space** to attach the arachnoid mater to the innermost meninx, the **pia mater.** The delicate pia mater is extensively vascularized and clings tenaciously to the surface of the brain, following its convolutions.

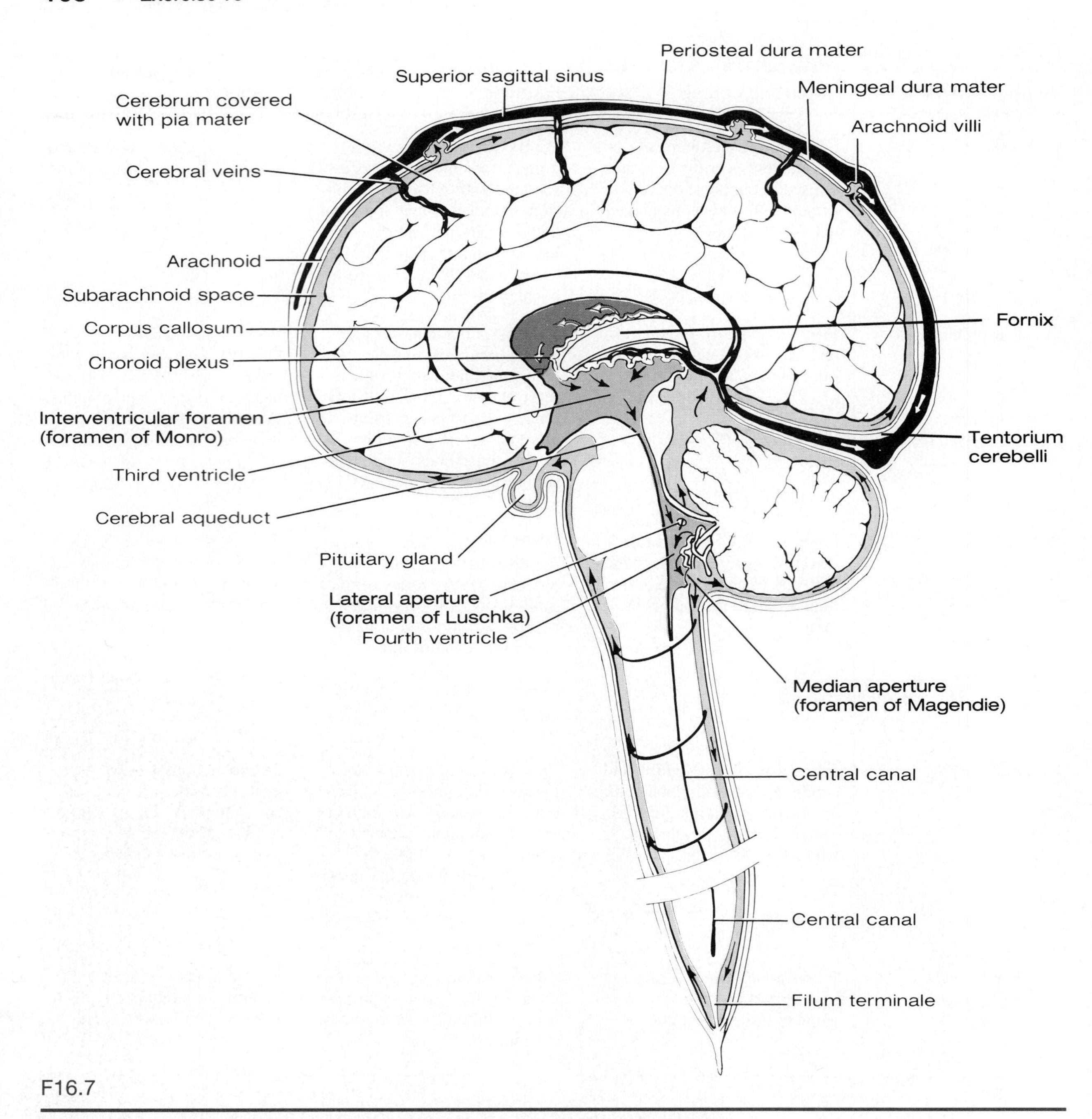

F16.7

Sagittal view of the brain showing the circulatory pathway of cerebrospinal fluid.

In life, the subarachnoid space is filled with cerebrospinal fluid. Specialized projections of the arachnoid tissue called **arachnoid villi** protrude through the dura mater to allow the cerebrospinal fluid to drain back into the venous circulation via the dural sinuses.

Meningitis, an inflammation of the meninges, is a serious threat to the brain because of the intimate association between the brain and the meninges. Should the infection spread to the neural tissue of the brain itself, life-threatening *encephalitis* may occur. ■

CEREBROSPINAL FLUID

The cerebrospinal fluid, much like plasma in composition, is continually formed by the choroid plexuses, small capillary knots hanging from the roof of the ventricles of the brain. The cerebrospinal fluid in and around the brain forms a watery cushion that protects the delicate brain tissue against blows to the head.

Within the brain, the cerebrospinal fluid circulates from the two lateral ventricles (in the cerebral hemispheres) into the third ventricle via the interventricular foramina, and then through the cerebral aqueduct to the midbrain into the fourth ventricle in the hindbrain (Figure 16.7). A portion of the fluid reaching the fourth ventricle continues down the central canal of the spinal cord, but the bulk of it circulates into the subarachnoid space, exiting through three openings in the walls of the fourth ventricle, the two *lateral apertures* (foramina of Luschka) and the single *median aperture* (foramen of Magendie). The fluid returns to the blood in the dural sinuses through the arachnoid villi.

Ordinarily, cerebrospinal fluid forms and drains at a constant rate. However, under certain conditions—for example, obstructed drainage or circulation resulting from tumors or anatomical deviations—the cerebrospinal fluid begins to accumulate and exerts increasing pressure on the brain, which, uncorrected, causes neurological damage in adults. In infants, *hydrocephalus* (literally, water on the brain) is indicated by a gradually enlarging skull. Since the infant's skull is still flexible and contains fontanels, it can expand to accommodate the increasing size of the brain. This condition is usually corrected by installing a shunt that drains the excess cerebrospinal fluid and returns it to the venous circulation. ■

DISSECTION OF THE SHEEP BRAIN

The brain of any mammal is enough like the human brain to warrant comparison. Obtain a sheep brain, protective skin cream or disposable gloves, dissecting pan, and instruments, and bring them to your laboratory bench.

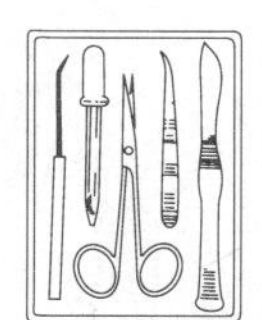

1. Place the intact sheep brain ventral surface down on the dissecting pan, and observe the dura mater. Feel its consistency and note its toughness. Cut through the dura mater along the line of the longitudinal fissure (which separates the cerebral hemispheres) to enter the superior sagittal sinus. Gently force the cerebral hemispheres apart laterally to expose the corpus callosum deep to the longitudinal fissure.

2. Carefully remove the dura mater, and examine the superior surface of the brain. Note that its surface is thrown into convolutions (fissures and gyri), just as the

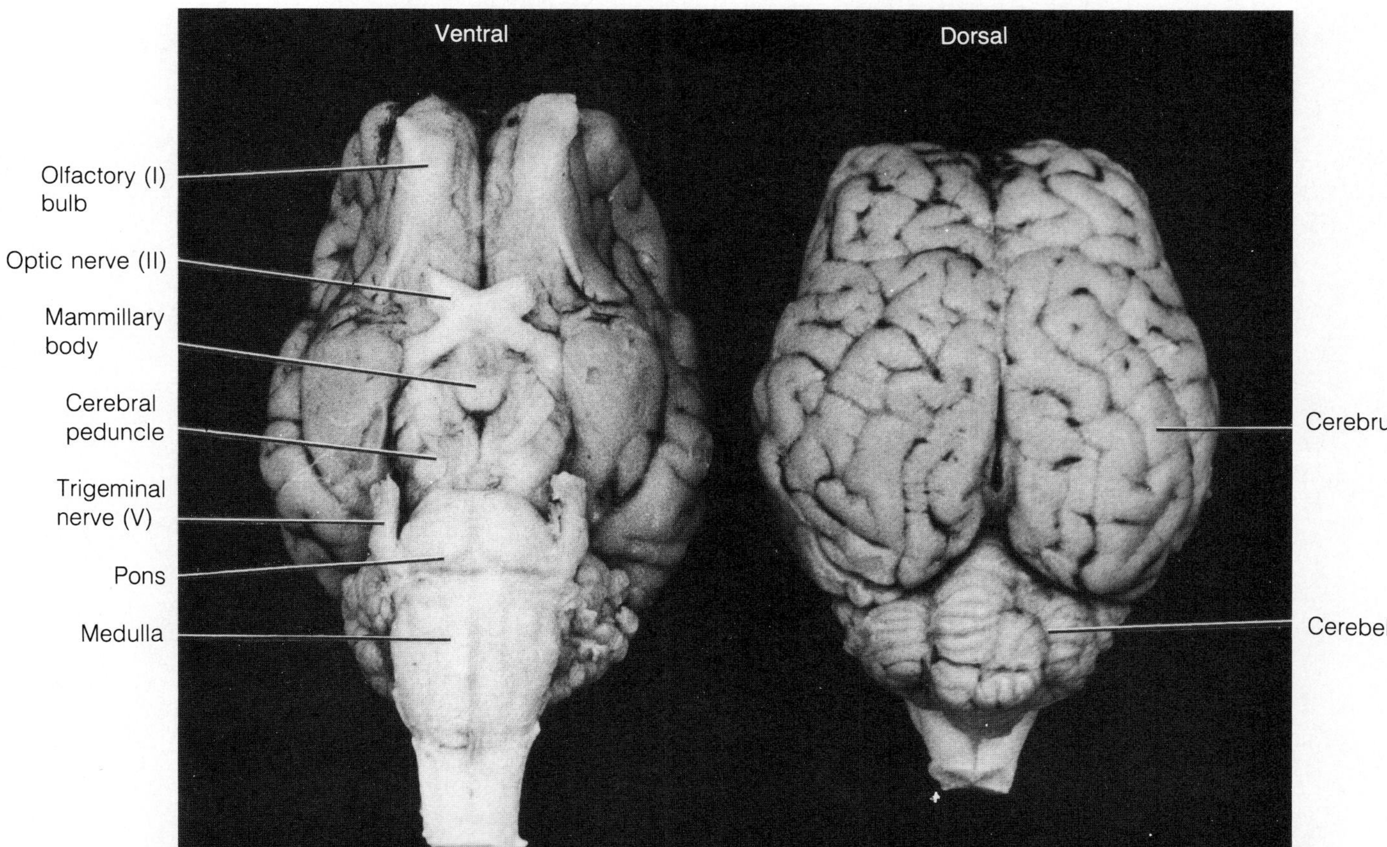

F16.8

Intact sheep brain. (a) Photograph showing ventral and dorsal views *(continues)*.

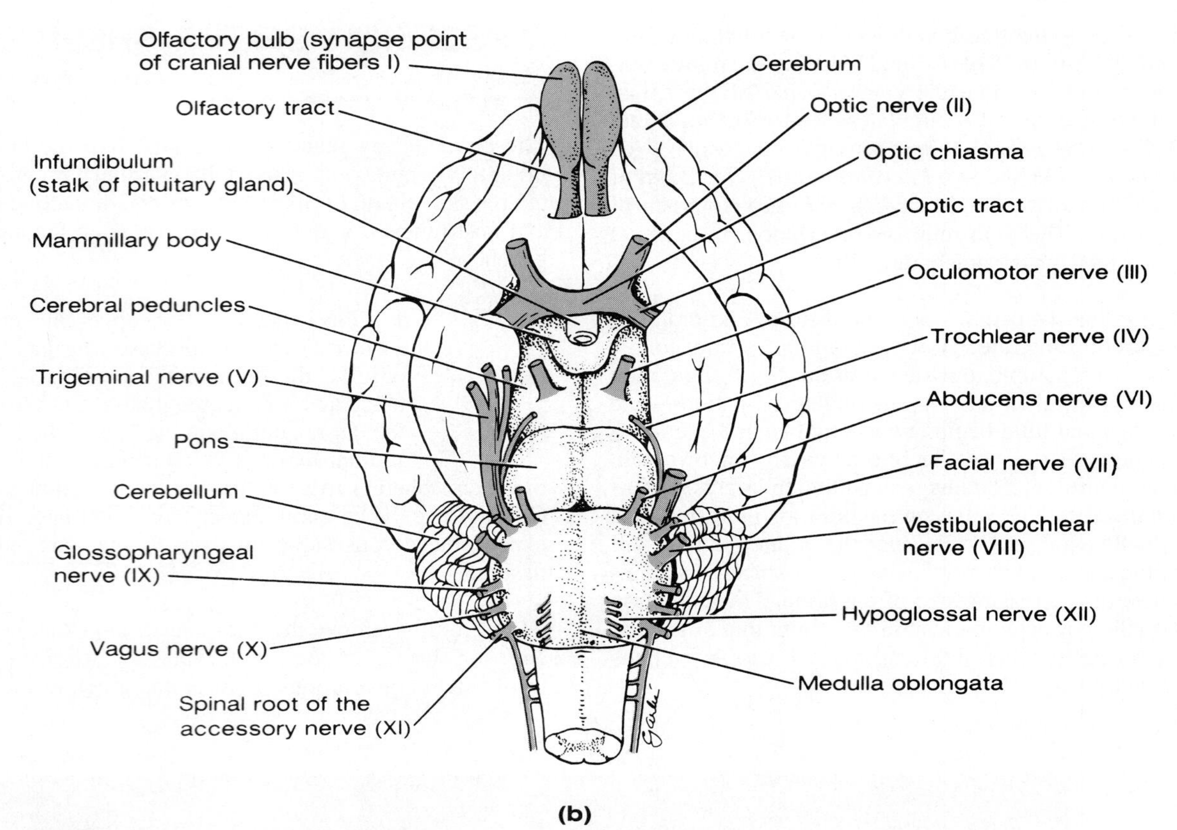

F16.8 (continued)

Intact sheep brain. (b) ventral view; (c) dorsal view.

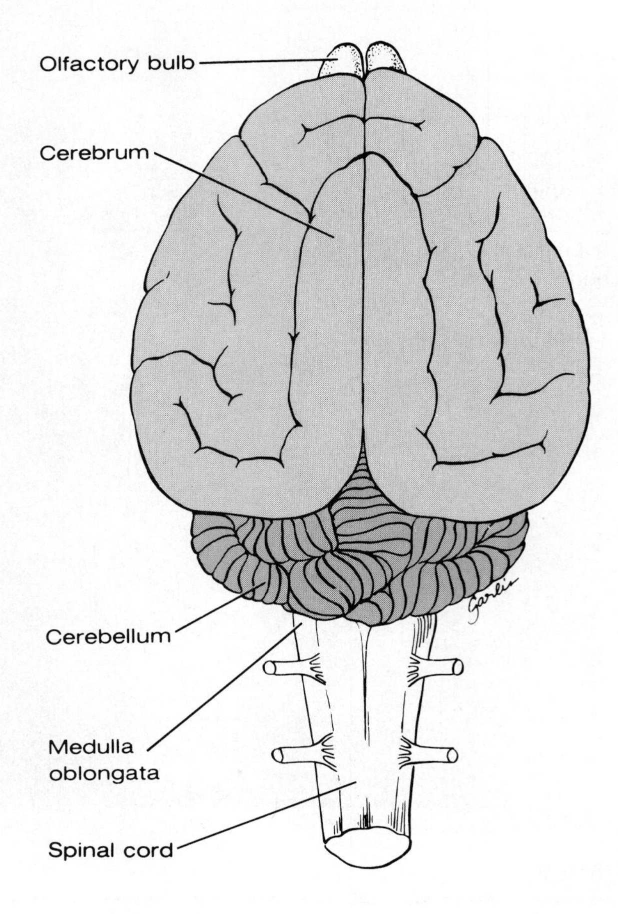

human brain is (Figure 16.8). Locate the arachnoid meninx, which appears on the brain surface as a delicate "cottony" material *spanning* the fissures. In contrast, the innermost meninx, the pia mater, closely follows the cerebral contours.

Dorsal Structures

1. Refer to Figure 16.8a and c as a guide in identifying the following structures. The cerebral hemispheres should be easy to locate. How do the size of the sheep's cerebral hemispheres and the depth of the fissures compare to those in the human brain?

____

____

____

2. Carefully examine the cerebellum. Note that it is not divided longitudinally, in contrast to the human cerebellum, and that its fissures are oriented differently.

What dural falx is missing that is present in humans?

____

____

3. Locate the three pairs of cerebellar peduncles, fiber tracts that connect the cerebellum to other brain structures, by lifting the cerebellum dorsally away from the brain stem. The most posterior pair, the inferior cerebellar peduncles, connect the cerebellum to the medulla. The middle cerebellar peduncles attach the cerebellum to the pons, and the superior cerebellar peduncles run from the cerebellum to the midbrain.

4. To expose the dorsal surface of the midbrain, gently spread the cerebrum and cerebellum apart, as shown in Figure 16.9. Identify the corpora quadrigemina, which appear as four rounded prominences on the dorsal midbrain surface.

What is the function of the corpora quadrigemina?

___

___

___

Also locate the pineal body, a structure of the epithalamus which appears in the midline just anterior to the corpora quadrigemina.

Ventral Structures

Figures 16.8 a and b show the important features of the ventral surface of the brain.

1. Look for the clublike olfactory bulbs, anteriorly on the inferior surface of the frontal lobes of the cerebral hemispheres.

How does the size of these olfactory bulbs compare with those of humans?

___

Is the sense of smell more important as a protective or food-getting sense in sheep *or* in humans?

___

___

2. The optic nerve (II) carries sensory impulses from photoreceptor cells of the retina of the eye. Thus this cranial nerve is involved in the sense of vision. Identify the optic nerves, optic chiasma, and optic tracts.

3. Posterior to the optic chiasma, two structures protrude from the ventral aspect of the hypothalamus—the

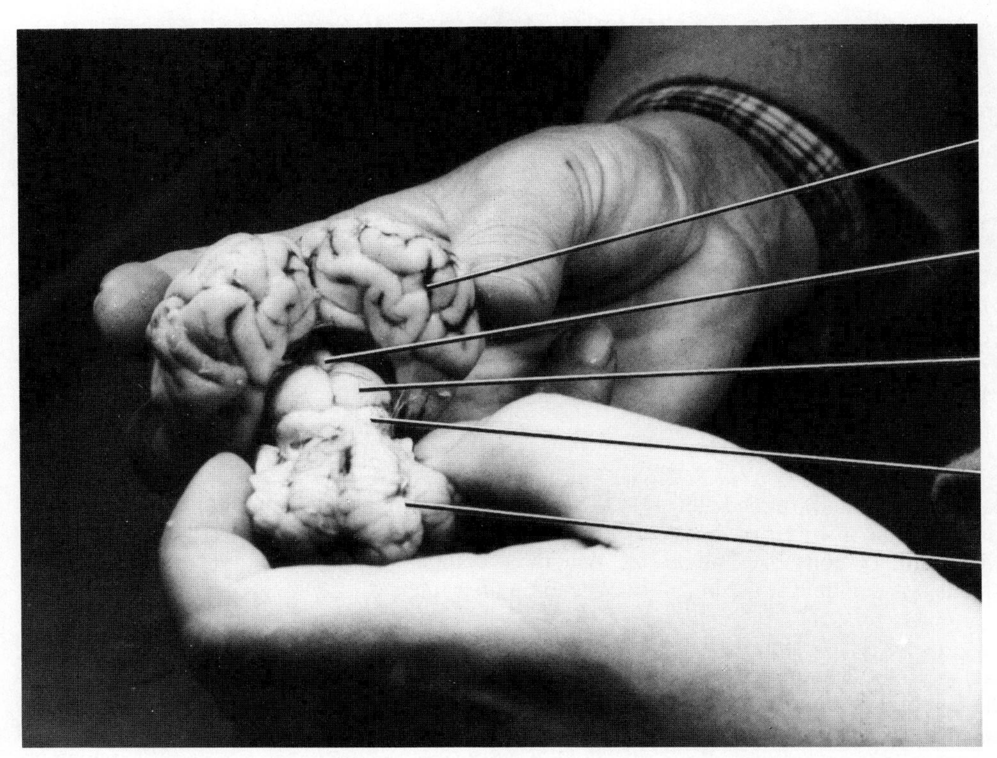

F16.9

Means of exposing the dorsal midbrain structures of the sheep brain.

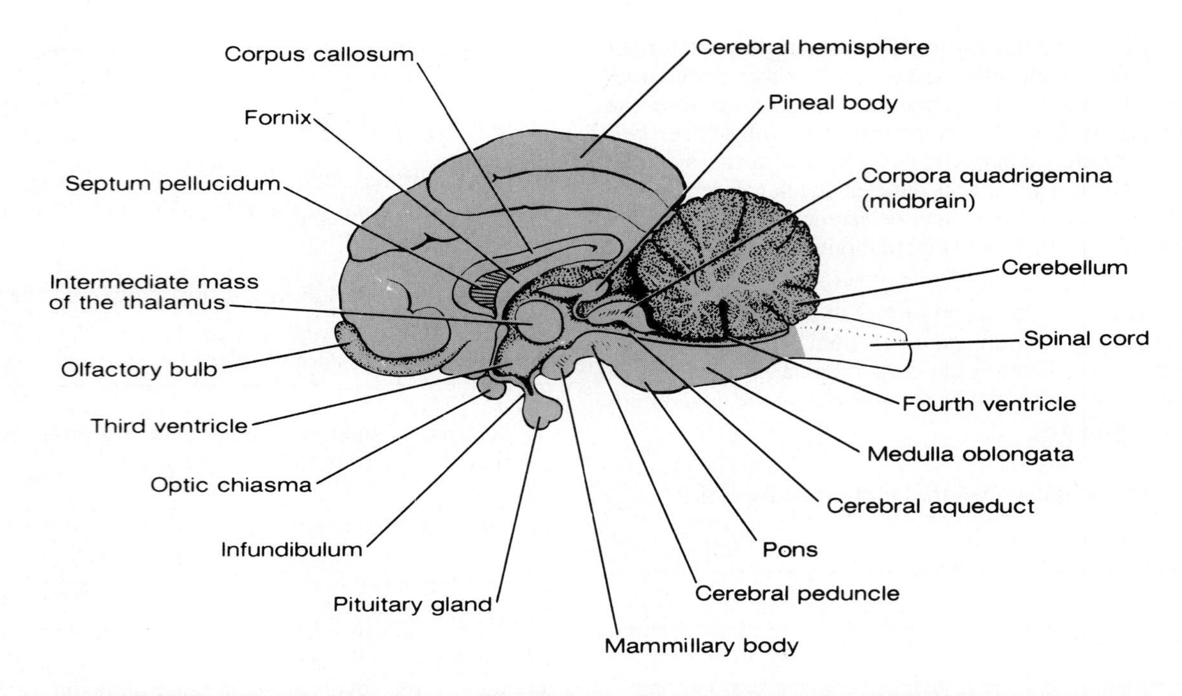

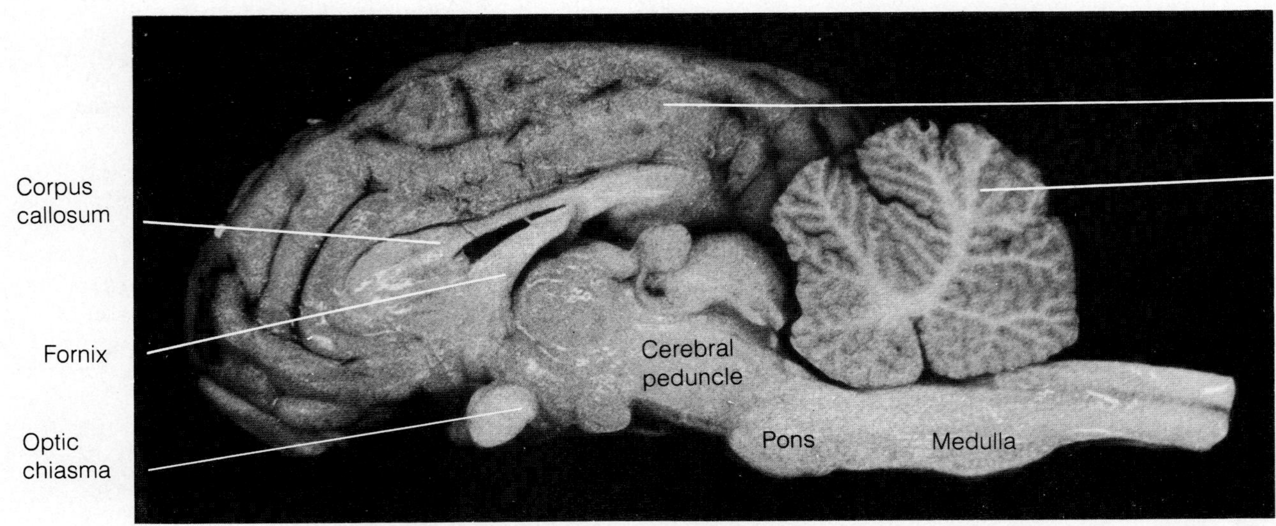

F16.10

Sagittal section of the sheep brain showing internal structures. (a) Diagrammatic view; (b) photograph.

infundibulum (stalk of the pituitary gland) immediately posterior to the optic chiasma and the mammillary body. Notice that the sheep's mammillary body is a single rounded eminence; in humans it is a double structure.

4. Identify the cerebral peduncles on the ventral aspect of the midbrain, just posterior to the mammillary body of the hypothalamus. The cerebral peduncles are fiber tracts containing ascending and descending fibers. Identify the large oculomotor nerves (III), which arise from the ventral midbrain surface, and the tiny trochlear nerves (IV), which can be seen at the junction of the midbrain and pons. Both of these cranial nerves provide motor fibers to extrinsic muscles of the eyeball.

5. Move posteriorly from the midbrain to identify first the pons and then the medulla oblongata, both hindbrain structures composed primarily of ascending and descending fiber tracts.

6. Return to the junction of the pons and midbrain, and proceed posteriorly to identify the following cranial nerves, all arising from the pons: the trigeminal nerves (V), which are involved in chewing and sensations of the head and face; the abducens nerves (VI), which abduct the eye (and thus work in conjunction with cranial nerves III and IV); and the large facial nerves (VII), which are involved in taste sensation, gland function (salivary and lacrimal glands), and facial expression.

7. Continue posteriorly to identify the purely sensory vestibulocochlear nerves (VIII), which are involved in the sensations of hearing and equilibrium; the glossopharyngeal nerves (IX), which contain motor fibers that innervate throat structures and sensory fibers that transmit taste stimuli (in conjunction with cranial nerve VII); the vagus nerves (X), often called "wanderers," which serve many organs of the head, thorax, and abdominal cavity; the accessory nerves (XI), which serve muscles of the neck and shoulder; and the hypoglossal nerves (XII), which stimulate tongue and neck muscles. Note that the accessory nerves arise from both the medulla and the spinal cord.

Internal Structures

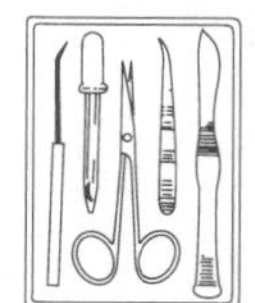

1. The internal structure of the brain can only be examined after further dissection. Place the brain ventral side down on the dissecting pan, and make a cut completely through it in a superior to inferior direction. Cut through the longitudinal fissure, corpus callosum, and midline of the cerebellum. Refer to Figure 16.10 as you work.

2. The thin nervous tissue membrane immediately ventral to the corpus callosum that separates the lateral ventricles is the septum pellucidum. Pierce this membrane, and probe the lateral ventricle cavity. The fiber tract ventral to the septum pellucidum and anterior to the third ventricle is the fornix.

How does the size of the fornix in this brain compare with the human fornix?

___

___

Why do you suppose this is so? (Hint: What is the function of this band of fibers?)

___

___

3. Identify the thalamus, which forms the walls of the third ventricle and is located posterior and ventral to the fornix. The intermediate mass spanning the ventricular cavity appears as an oval protrusion of the thalamic wall. Anterior to the intermediate mass, locate the interventricular foramen, a canal connecting the lateral ventricle on the same side with the third ventricle.

4. The hypothalamus forms the floor of the third ventricle. Identify the optic chiasma, infundibulum, and mammillary body on its exterior surface. You can see the pineal body at the superior-posterior end of the third ventricle, just beneath the junction of the corpus callosum and fornix.

5. Locate the midbrain by identifying the corpora quadrigemina that form its dorsal roof. Follow the cerebral aqueduct (the narrow canal connecting the third and fourth ventricles) through the midbrain tissue to the fourth ventricle. Identify the cerebral peduncles, which form its anterior walls.

6. The pons and medulla can be found anterior to the fourth ventricle. The medulla continues into the spinal cord without any obvious anatomical change, but the point at which the fourth ventricle narrows to a small canal is generally accepted as the beginning of the spinal cord.

7. The cerebellum can be seen posterior to the fourth ventricle. Note its internal treelike arrangement of white matter, the arbor vitae.

8. If time allows, obtain another sheep brain and section it along the frontal plane so that the cut passes through the infundibulum. Compare your specimen to the diagrammatic view in Figure 16.11, and attempt to identify all the structures shown in the figure.

9. When you have completed your study of the sheep brain, dispose of all organic debris, and wash the dissecting instruments and pan.

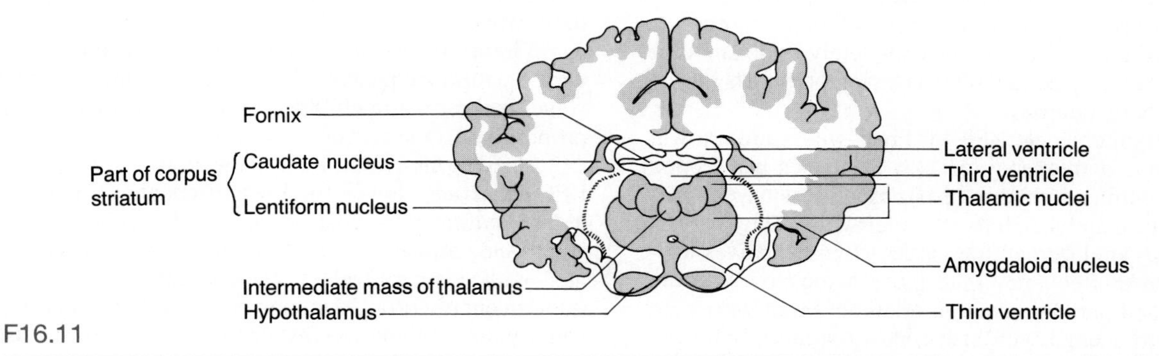

F16.11

Frontal section of a sheep brain. Major structures revealed are the location of major basal nuclei deep in the interior, the thalamus, hypothalamus, and the lateral and third ventricles.

17 EXERCISE

Electroencephalography

OBJECTIVES

1. To define *electroencephalogram* and to discuss its clinical significance.
2. To describe or recognize typical tracings of the most common brain wave patterns (alpha, beta, theta, and delta waves), and to note the conditions under which each is most likely to be predominant.
3. To state the source of brain waves.
4. To define *alpha block.*
5. To monitor electroencephalography and recognize alpha rhythm.
6. To describe the effects of a sudden sound, mental concentration, and alkalosis on brain wave patterns.

MATERIALS

Oscilloscope and EEG lead-selector box, or polygraph and high-gain preamplifier
Cot
Electrode gel
EEG electrodes and leads
Collodion gel or long elastic EEG straps

BRAIN WAVE PATTERNS AND THE ELECTROENCEPHALOGRAM

Any physiologic investigation of the brain can emphasize and expose only a very minute portion of its function. Higher brain functions, such as consciousness and logical reasoning, are extremely difficult to investigate. It is obviously much easier to do experiments on the brain's input-output functions, some of which can be detected with appropriate recording equipment. Still, the ability to record brain activity does not necessarily guarantee an understanding of the brain.

The **electroencephalogram (EEG),** a record of the electrical activity of the brain, can be obtained through electrodes placed at various points on the skin or scalp of the head. This electrical activity, which is recorded as waves (Figure 17.1), is not completely understood at present but may be regarded as action potentials generated by brain neurons.

Certain characteristics of brain waves are known. They have a frequency of 1 to 30 cycles per second (cps); a dominant rhythm of 10 cps; and an average amplitude (voltage) of 20 to 100 microvolts. They vary in frequency in different brain areas, occipital waves having a lower frequency than those associated with the frontal and parietal lobes. In addition, brain waves are known to change with age, sensory stimuli, brain pathology or disease, and the physiochemical state of the body. (Glucose deprivation, oxygen poisoning, and sedatives all interfere with the rhythmic activity of brain output by disturbing the metabolism of the neurons.)

The first of the brain waves to be described by scientists were the **alpha waves** (or alpha rhythm). Alpha waves have an average frequency range of 8 to 13 cps and are produced when the individual is in a relaxed state with the eyes closed. **Alpha block,** suppression of the alpha rhythm, occurs if the eyes are opened or if the individual begins to concentrate on some mental problem or visual stimulus. Under these conditions, the waves decrease in amplitude but increase in frequency. Under conditions of fright or excitement, the frequency increases still more.

Beta waves are faster (14 to 25 cps) than alpha waves and have a lower amplitude. They are typical of the attentive or alert state.

Very large (high-amplitude) waves with a frequency of 4 cps or less that are seen in deep sleep are **delta waves.**

Theta waves are large, abnormally contoured waves with a frequency of 4 to 7 cps. Although theta waves are normal in children, they represent emotional problems or some sort of neural imbalance in adults.

Sleeping individuals and patients in a stupor have EEGs that are slower (or lower frequency) than the alpha rhythm of normal adults. Fright, epileptic seizures, and various types of drug intoxication are associated with comparatively faster cortical activity. Thus impairment of cortical function is indicated by neuronal activity that is either too fast or too slow; unconsciousness occurs at both extremes of the frequency range. In young infants, the frequency of the major wave pattern

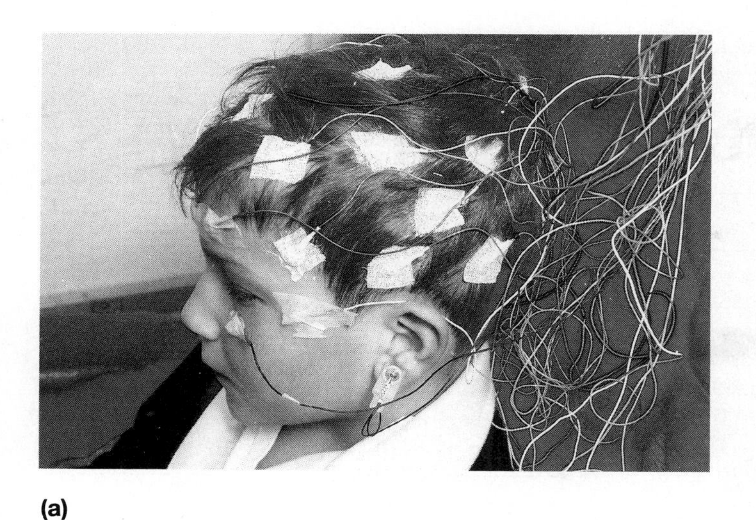

(a)

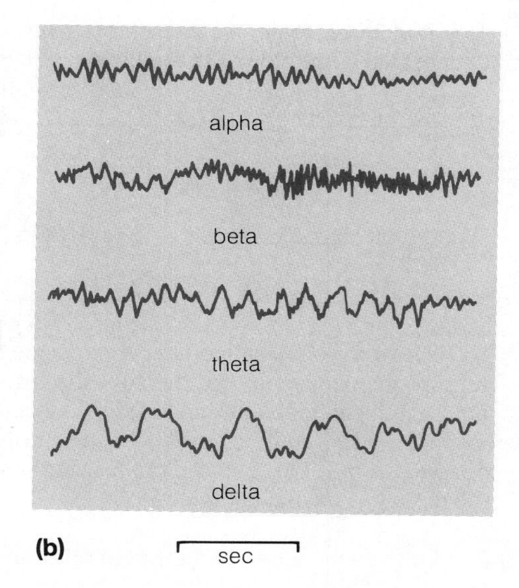

(b)

F17.1

Electroencephalography and brain waves. (a) To obtain a recording of brain wave activity (an EEG), electrodes are positioned on the subject's scalp and attached to a recording device called an electroencephalograph. (b) Typical EEGs. Alpha waves are typical of the awake, relaxed state; beta waves occur in the awake, alert state; theta waves are common in children, but not in normal adults; delta waves occur during deep sleep.

is quite slow (0.5 to 2 cps), but the frequency of the major waves increases with age, and at about age 12 the EEG recording is similar to that of an adult.

Since spontaneous brain waves are always present, even during unconsciousness, the absence of brain waves (a "flat" EEG) is taken as evidence of clinical death. The EEG is used clinically to aid in the diagnosis and localization of a number of brain lesions, including epileptic lesions, infections, abscesses, and brain tumors. ■

 OBSERVING BRAIN WAVE PATTERNS

If one electrode (the *active electrode*) is placed over a particular cortical area and another (the *indifferent electrode*) is placed over an inactive part of the head, such as the earlobe, all of the activity of the cortex underlying the active electrode will, theoretically, be recorded. The inactive area provides a zero reference point, or a baseline; and the EEG represents the difference between "activities" occurring under the two electrodes.

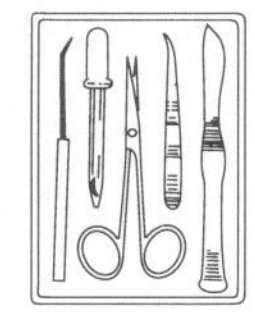

1. Connect the EEG selector box to the oscilloscope preamplifier, or connect the high-gain preamplifier to the polygraph channel amplifier. Adjust the horizontal sweep and sensitivity according to the directions given in the manual or by your instructor.

2. Prepare the subject. The subject should lie undisturbed, with eyes closed, in a quiet, dimly lit area. (Someone who is able to relax easily makes a good subject.) Apply a small amount of electrode gel to the subject's forehead above the left eye and to the left earlobe. Press an electrode to each prepared area, and secure them by (1) applying a film of collodion gel to the electrode surface and the adjacent skin or (2) with a long elastic EEG strap (knot tied at the back of the head). If the collodion gel is used, allow it to dry before continuing.

3. Connect the active frontal lead (forehead) to the EEG selector box outlet marked "L Frontal" and the lead from the indifferent electrode (earlobe) to the ground outlet (or to the appropriate input terminal on the high-gain preamplifier).

4. Turn on the oscilloscope or polygraph, and observe the EEG pattern of the relaxed subject for a period of 5 min. If the subject is truly relaxed, you should see a typical alpha-wave pattern. (If the subject is unable to relax and the alpha-wave pattern does not appear in this time interval, test another subject.) Discourage all muscle movement during the monitoring period.*

5. Abruptly and loudly clap your hands. The subject's eyes should open and alpha block should occur. Observe

* Note that 60 cycle "noise" (appearing as fast, regular low-amplitude waves superimposed on the more irregular brain waves) may interfere with the tracings being made, particularly if the laboratory has a lot of electrical equipment.

the immediate brain wave pattern. How do the frequency and amplitude of the brain waves change?

____

____

Would you characterize this as beta rhythm? ________

Why? ______________________________________

6. Allow the subject about 5 min to achieve complete relaxation once again, and then ask him or her to compute a number problem that requires concentration (for example, add 3 and 36, subtract 7, multiply by 2, add 50, and so on). Observe the brain wave pattern during the period of mental computation.

7. Once again allow the subject to relax until the alpha rhythm is observed, and then ask him or her to hyperventilate for 3 min. (Be sure to tell the subject when to stop.) Hyperventilation rapidly flushes carbon dioxide out of the lungs, decreasing carbon dioxide levels in the blood and producing respiratory alkalosis. Record the changes noted in the rhythm and amplitude of the brain waves during the period of hyperventilation. Record your observations below.

____

____

____

____

____

____

____

____

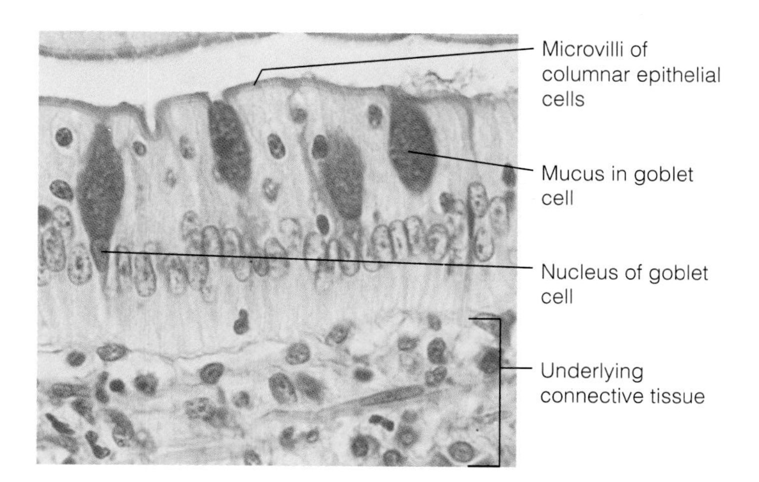

PLATE 1 Simple columnar epithelium. Mucus in goblet cells stains pink in this view (543X)

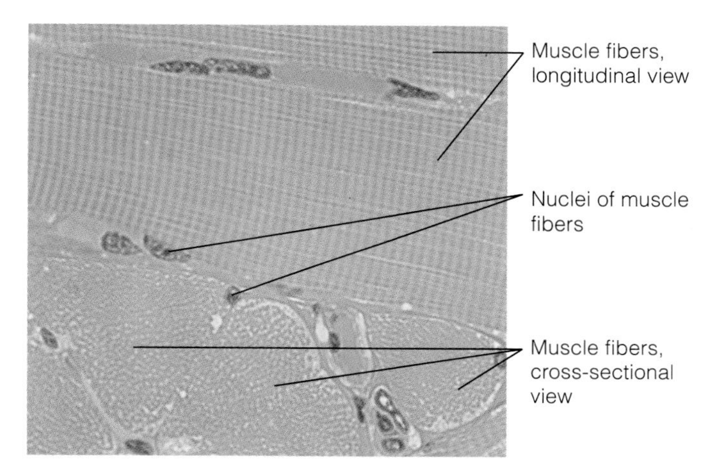

PLATE 2 Skeletal muscle, transverse and longitudinal views shown (543X)

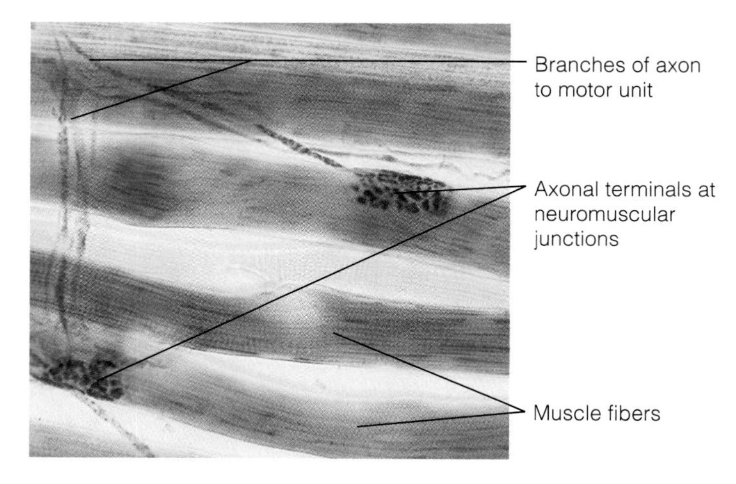

PLATE 3 Part of a motor unit (265X)

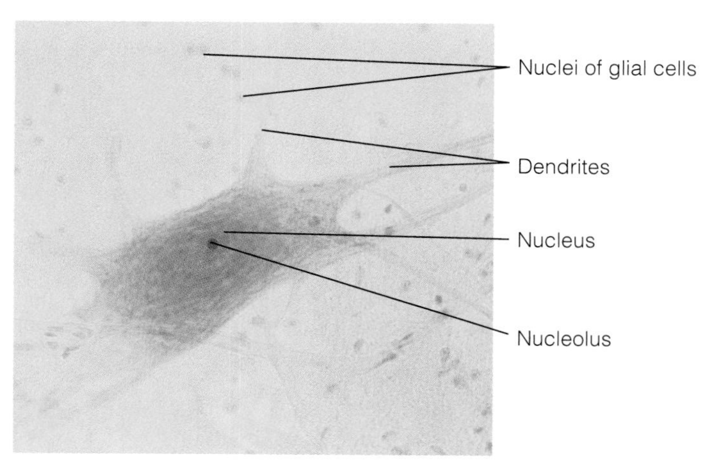

PLATE 4 Multipolar neuron in spinal cord smear (212X)

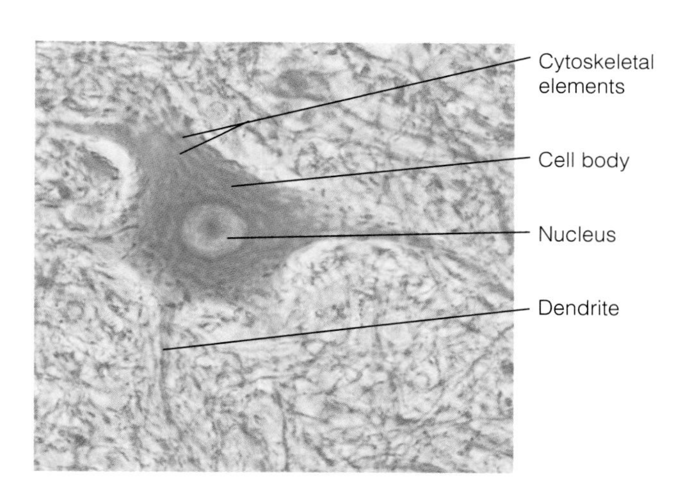

PLATE 5 Neuron stained to allow cytoskeletal elements in their processes to be seen (265X)

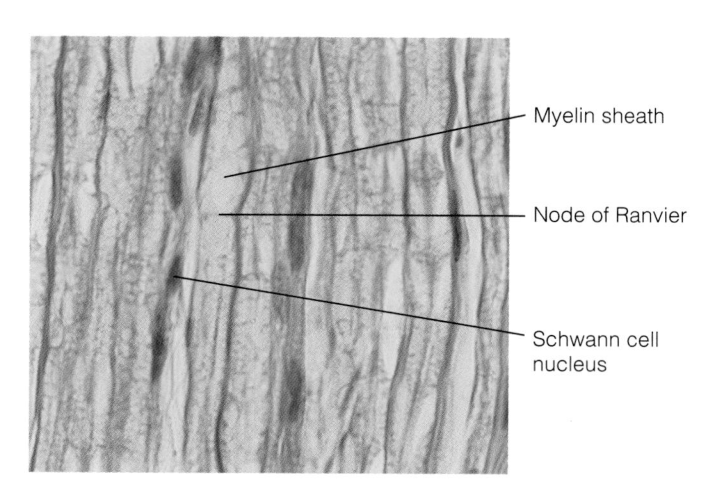

PLATE 6 Teased myelinated axons (543X). Myelin sheaths appear "bubbly" because most of the fatty myelin is dissolved during slide preparation

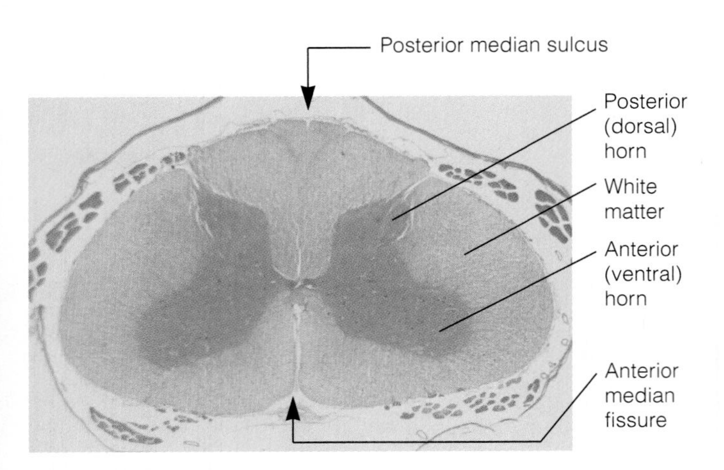

PLATE 7 Adult spinal cord, cross-sectional view (12X)

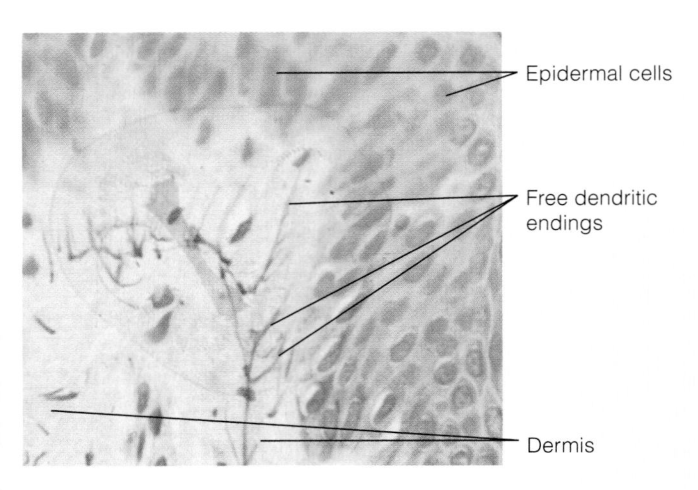

PLATE 10 Free dendritic endings at the dermal–epidermal junction (480X)

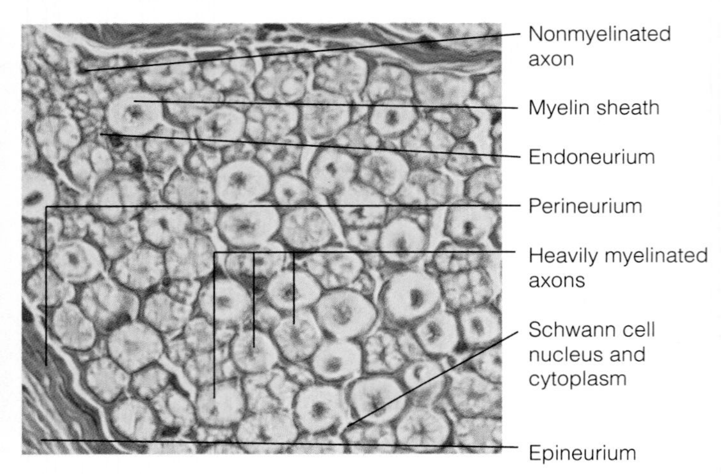

PLATE 8 Cross section of a portion of a peripheral nerve (543X). Heavily myelinated fibers are identified by a centrally-located axon, surrounded by an unstained ring of myelin, and then a peripheral rim of pink-staining Schwann cell

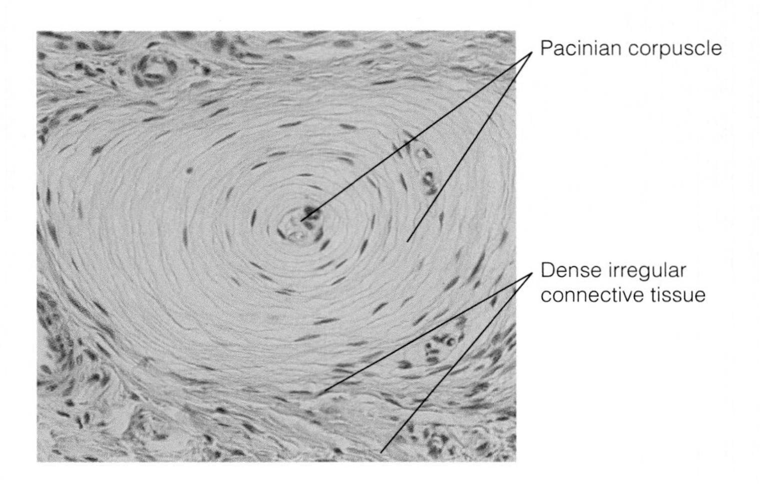

PLATE 11 Pacinian corpuscle in the hypodermis (212X)

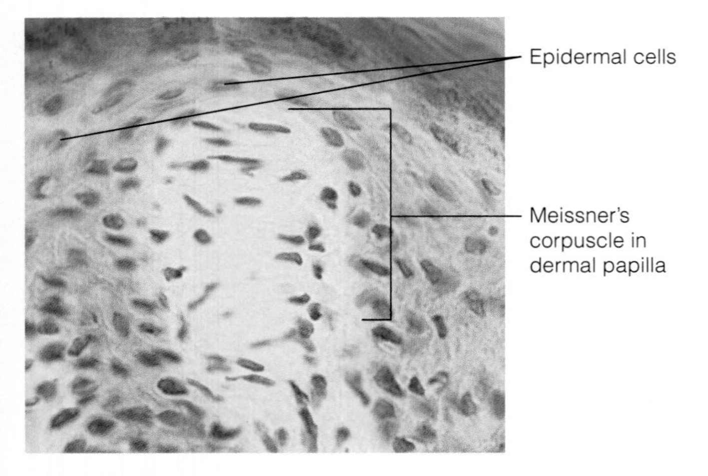

PLATE 9 Meissner's corpuscle in a dermal papilla (559X)

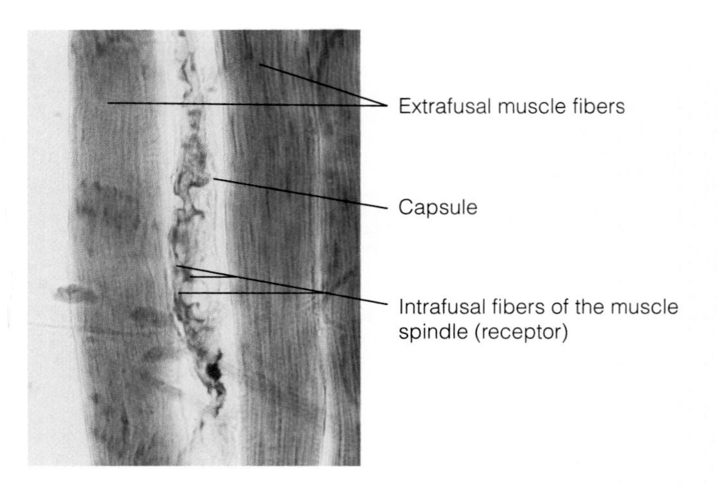

PLATE 12 Longitudinal section of a muscle spindle (305X)

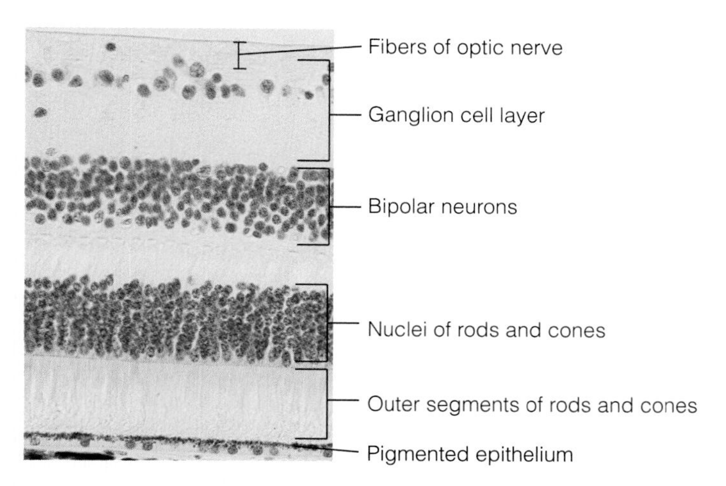

PLATE 13 Structure of the retina of the eye (353X)

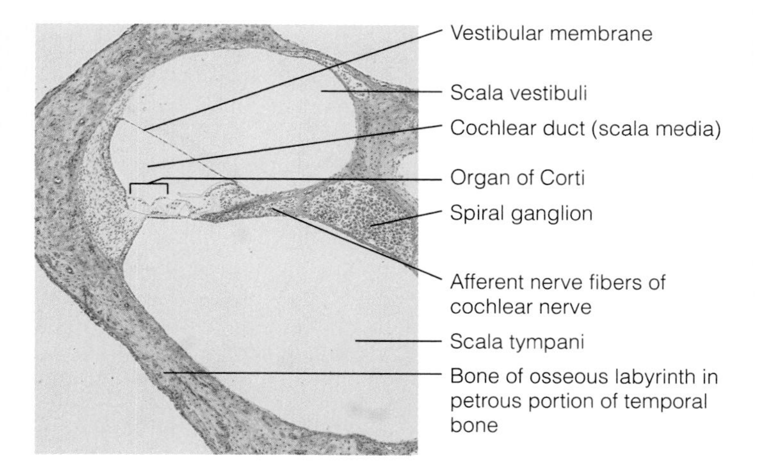

PLATE 14 Cross-sectional view of one turn of the cochlea showing the three scalas, and the location of the organ of Corti (36X)

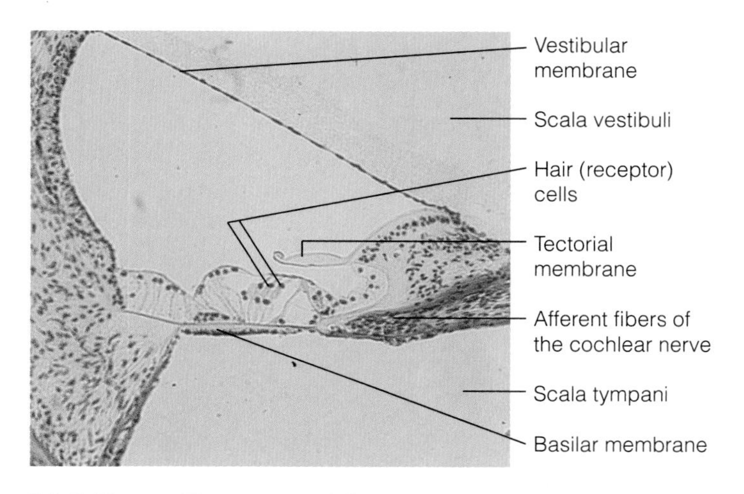

PLATE 15 The organ of Corti (106X)

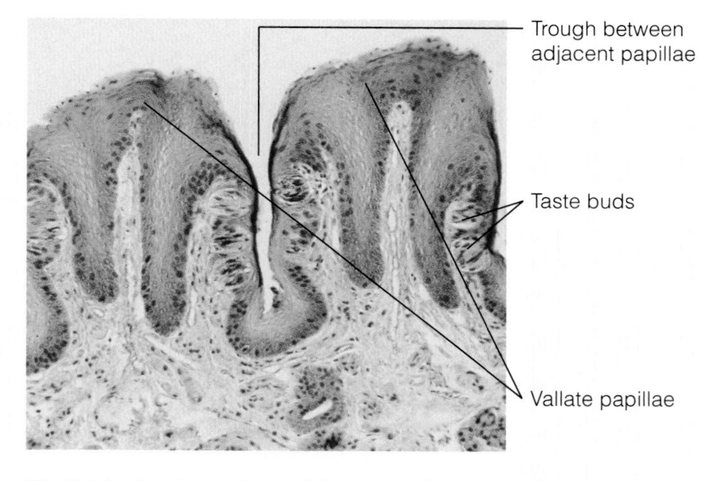

PLATE 16 Location of the taste buds on the lateral aspects of vallate papillae of the tongue (133X)

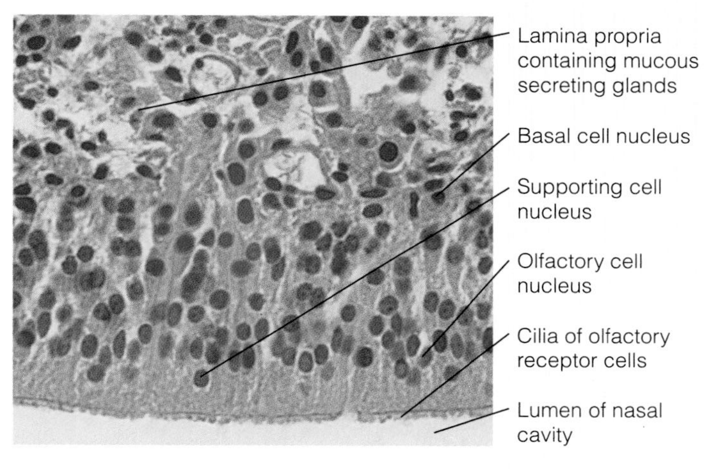

PLATE 17 Olfactory epithelium. From the lamina propria to the nasal cavity, the general arrangement of cells in this pseudostratified epithelium is basal cells, olfactory receptor cells, and supporting cells (543X)

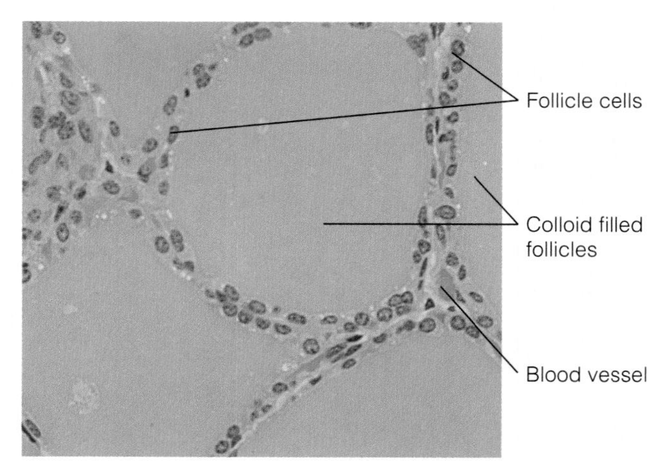

PLATE 18 The thyroid gland (345X)

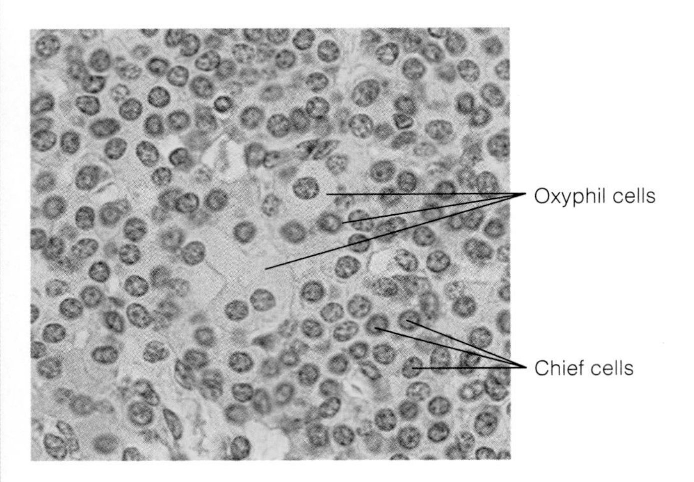

PLATE 19 Parathyroid gland tissue (543X)

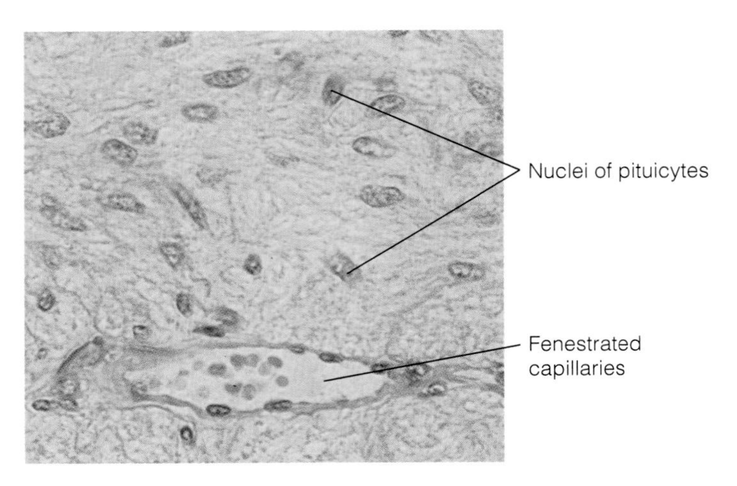

PLATE 22 Posterior pituitary (543X). The axons of the neurosecretory cells are indistinguishable from the cytoplasm of the pituicytes.

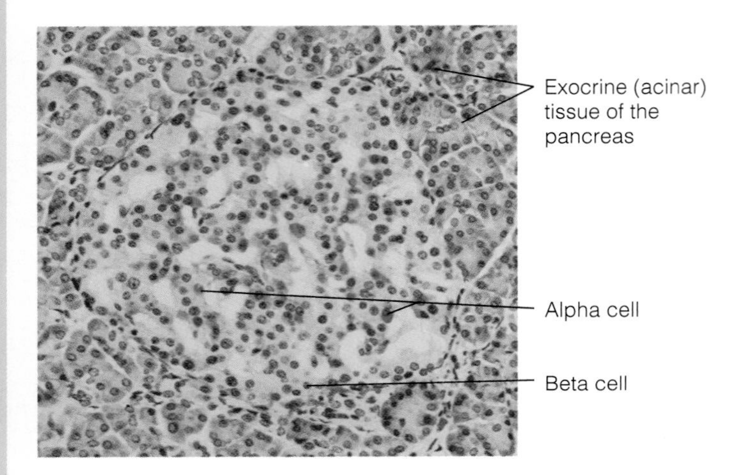

PLATE 20 Pancreatic islet stained differentially to allow identification of the glucagon-secreting alpha cells and the insulin-secreting beta cells (199X)

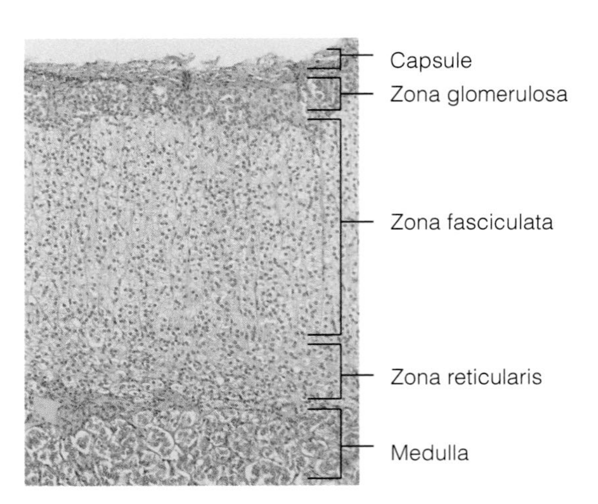

PLATE 23 Histologically distinct regions of the adrenal gland (56X)

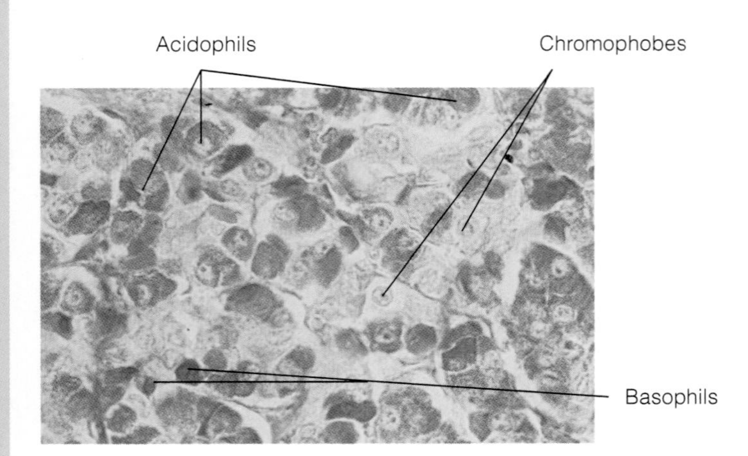

PLATE 21 Anterior pituitary gland. Differentially staining allows the acidophils, basophils, and chromophobes to be distinguished (434X)

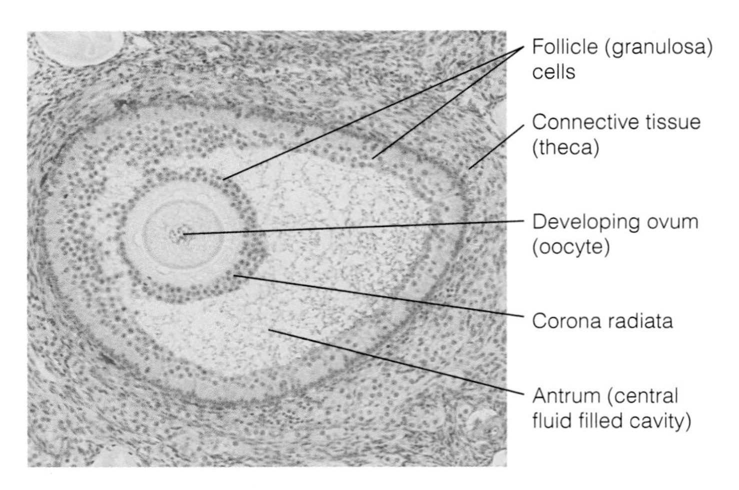

PLATE 24 A vesicular follicle of the ovary (106X)

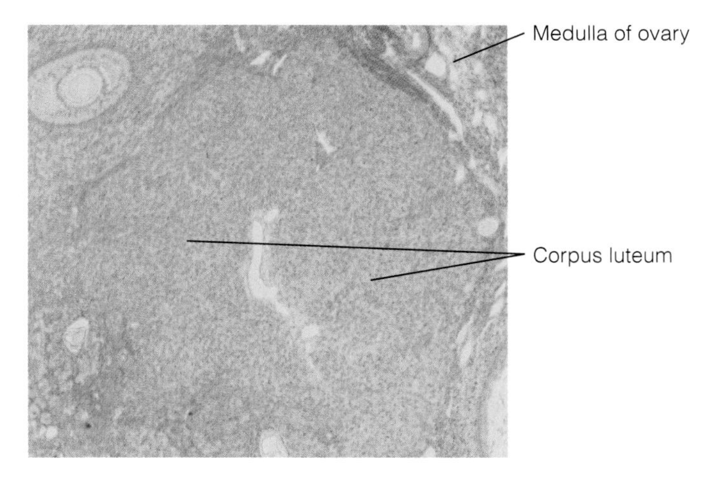

PLATE 25 The glandular corpus luteum of an ovary (45X)

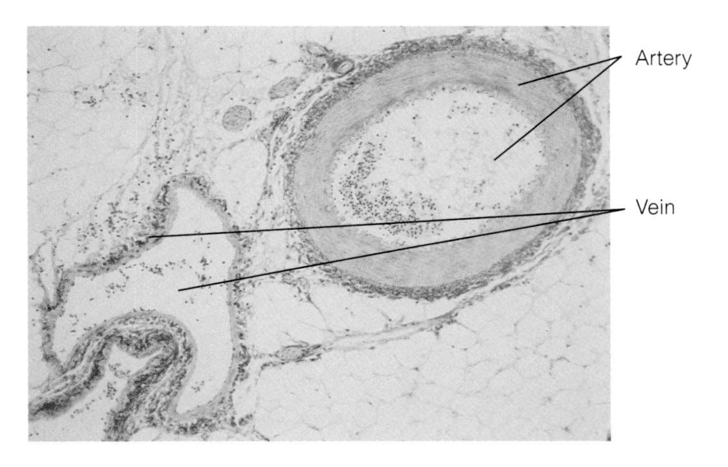

PLATE 26 Cross-sectional view of a small artery and vein (158X)

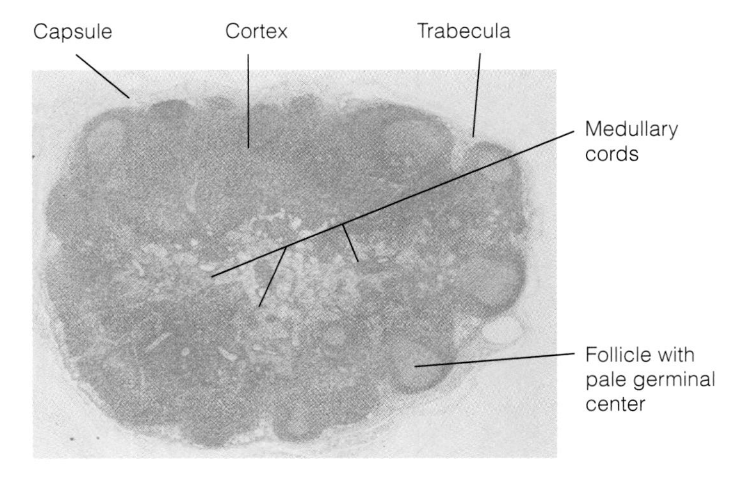

PLATE 27 Main structural features of a lymph node (18X)

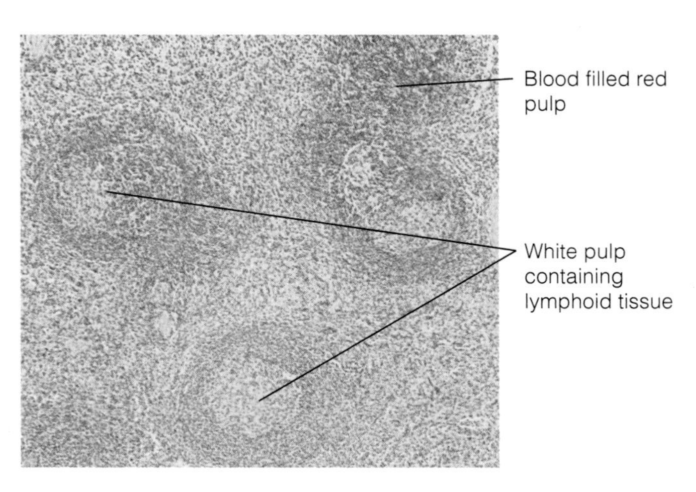

PLATE 28 Microscopic anatomy of a portion of the spleen showing the red and white pulp regions (56X)

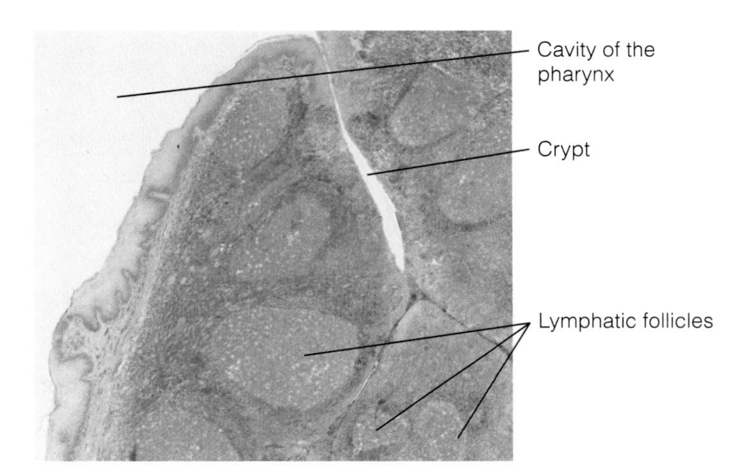

PLATE 29 Histology of a palatine tonsil. The luminal surface is covered with epithelium which invaginates deeply to form crypts (27X)

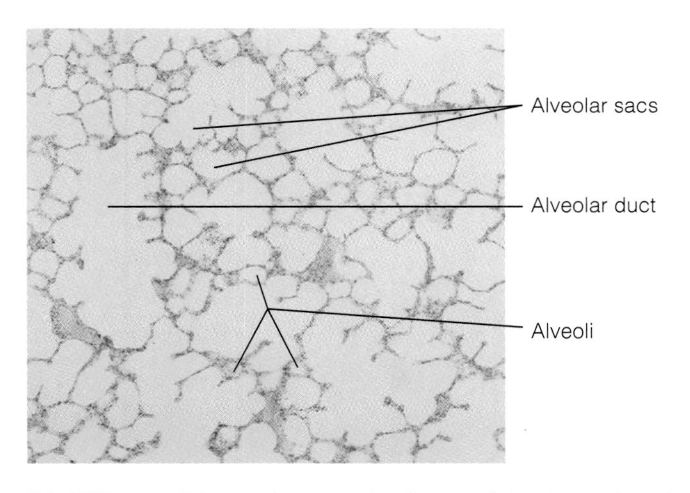

PLATE 30 Photomicrograph of part of the lung showing alveoli and alveolar ducts and sacs (34X)

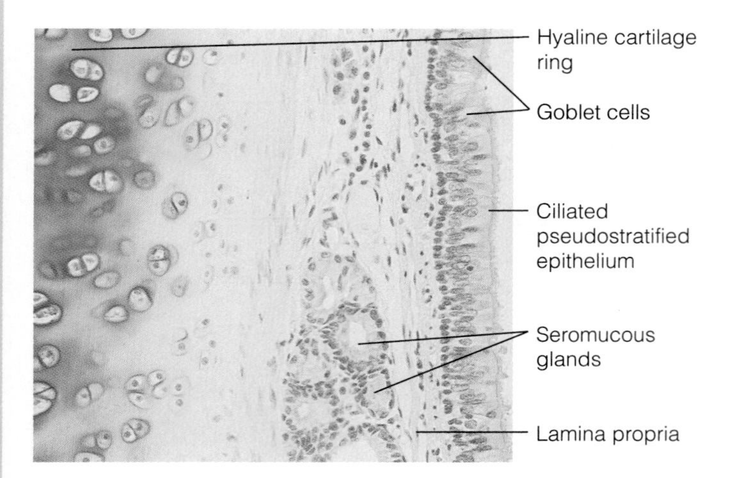

PLATE 31 Cross-section through the trachea showing the pseudostratified ciliated epithelium, glands, and part of the supporting ring of hyaline cartilage (159X)

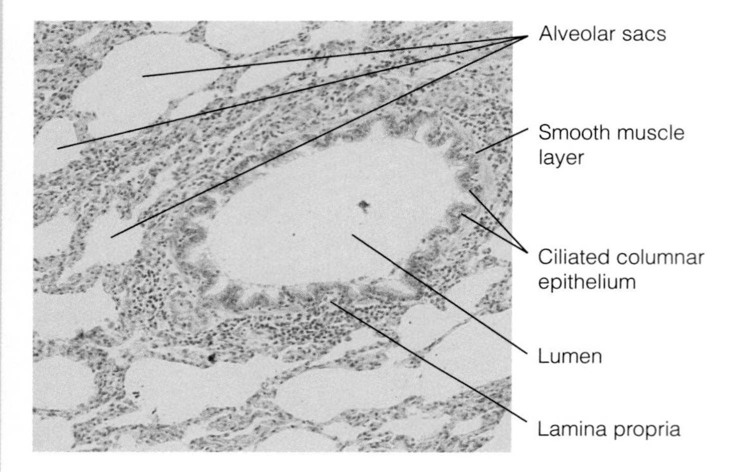

PLATE 32 Bronchiole, cross-sectional view (106X)

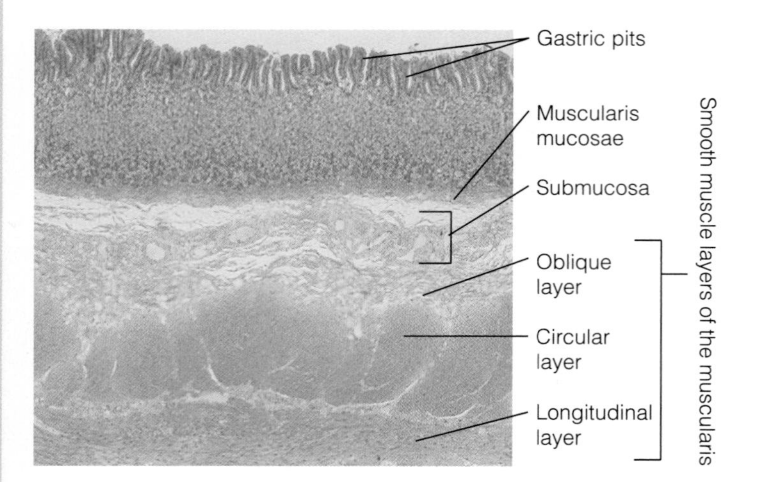

PLATE 33 Stomach. Low-power cross-sectional view through its wall showing three tunics (34X)

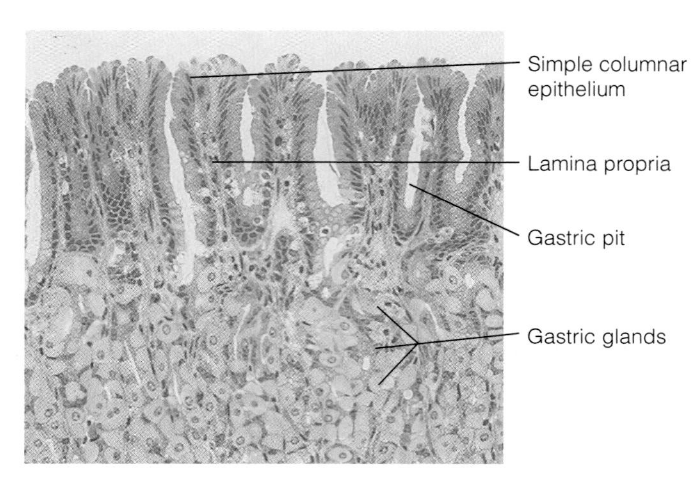

PLATE 34 Detailed structure of the gastric glands and pits (212X)

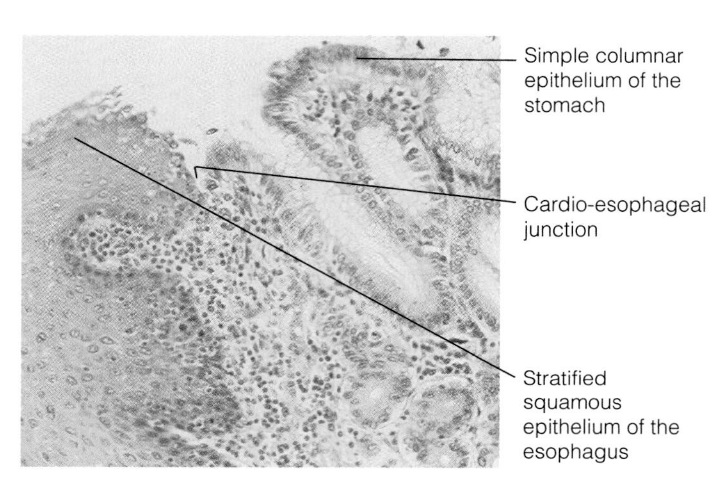

PLATE 35 Gastroesophageal junction showing the meeting of the simple columnar epithelium of the stomach and the stratified squamous epithelium of the esophagus (133X)

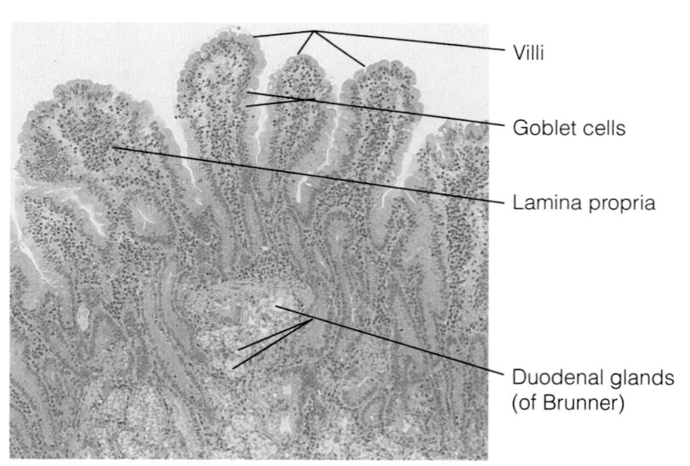

PLATE 36 Cross-sectional view of the duodenum showing villi and duodenal glands (66X)

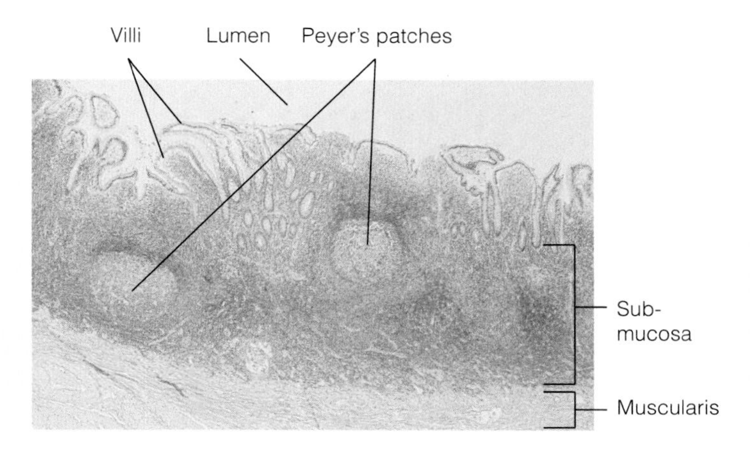

PLATE 37 Ileum, showing Peyer's patches (36X)

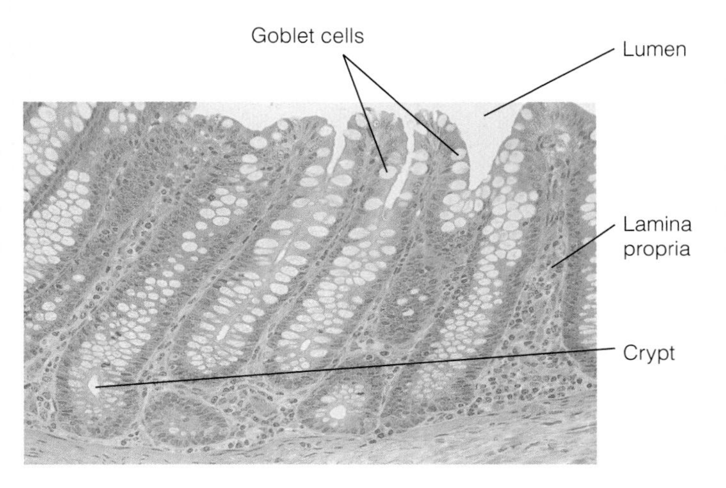

PLATE 38 Large intestine. Cross-sectional view showing the abundant goblet cells of the mucosa (127X)

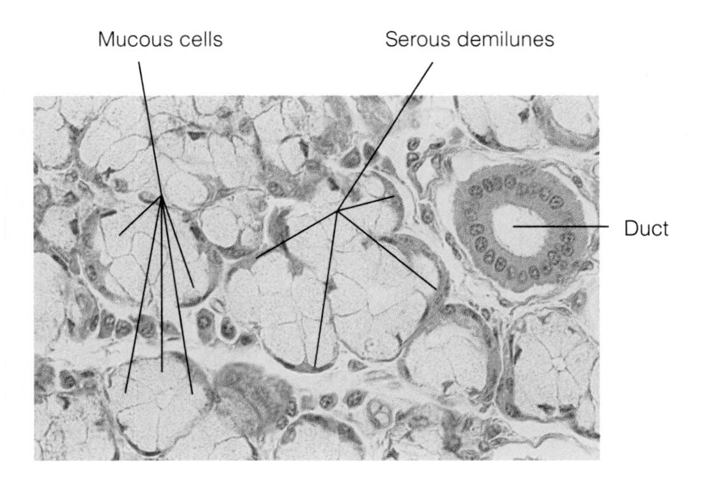

PLATE 39 Sublingual salivary glands (350X)

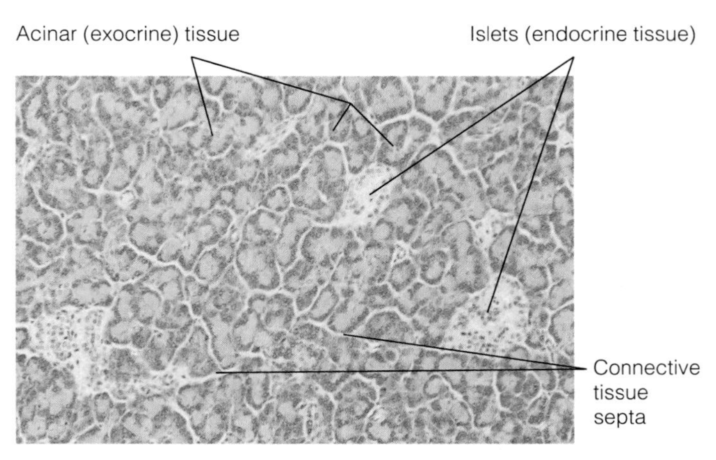

PLATE 40 Pancreas tissue. Exocrine and endocrine (islets of Langerhans) areas clearly visible (106X)

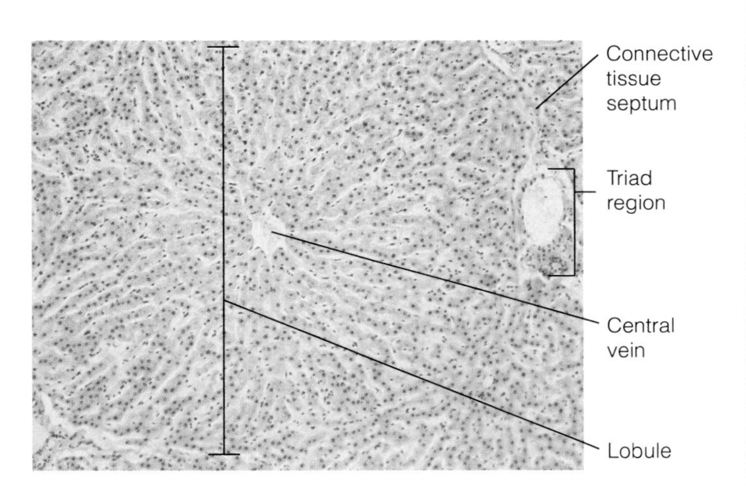

PLATE 41 Pig liver. Structure of the liver lobules (66X)

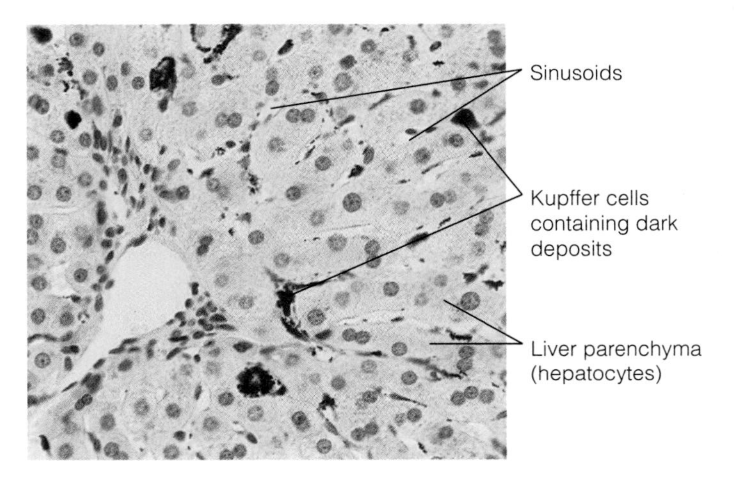

PLATE 42 Liver stained to show the location of the phagocytic cells (Kupffer cells) lining the sinusoids (265X)

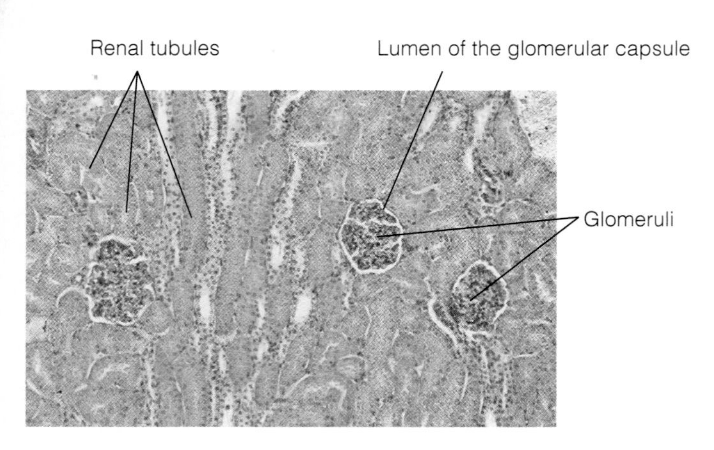

PLATE 43 Renal cortex of the kidney (85X)

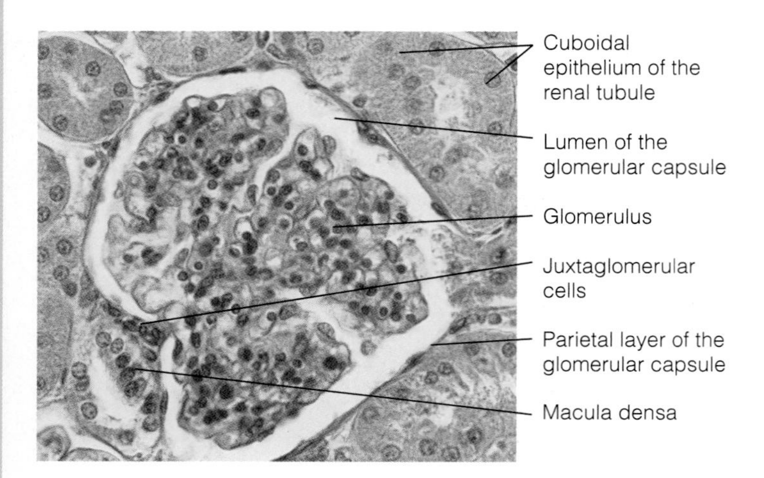

PLATE 44 Detailed structure of a glomerulus (345X)

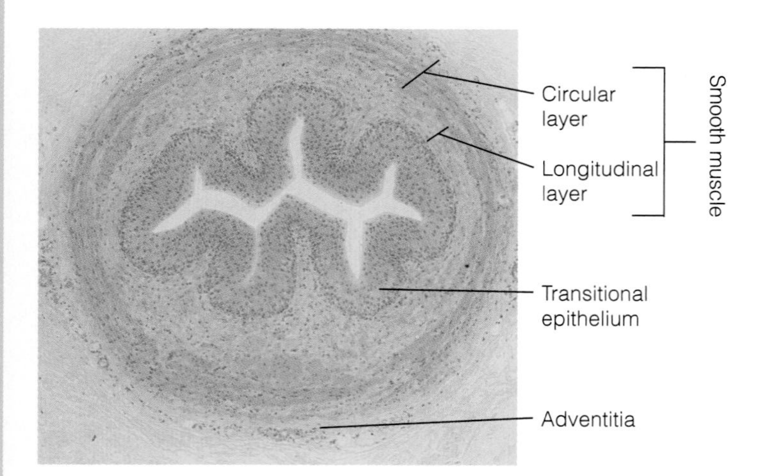

PLATE 45 Cross-section of the ureter (45X)

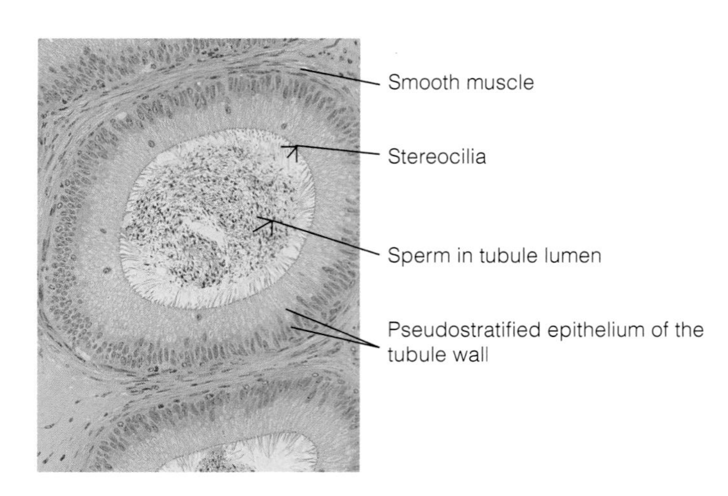

PLATE 46 Epididymis (148X)

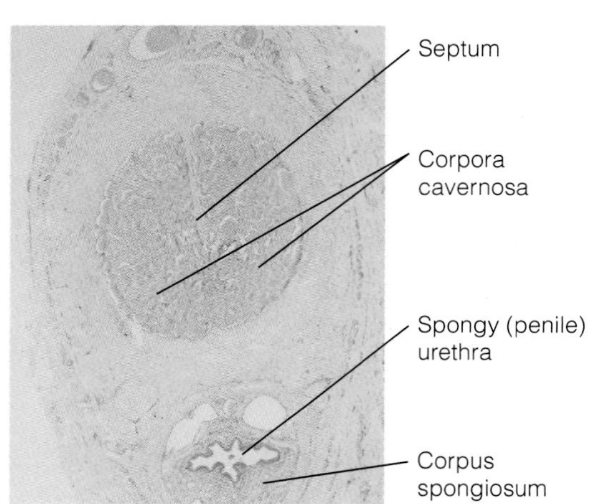

PLATE 47 Penis, transverse section (13X)

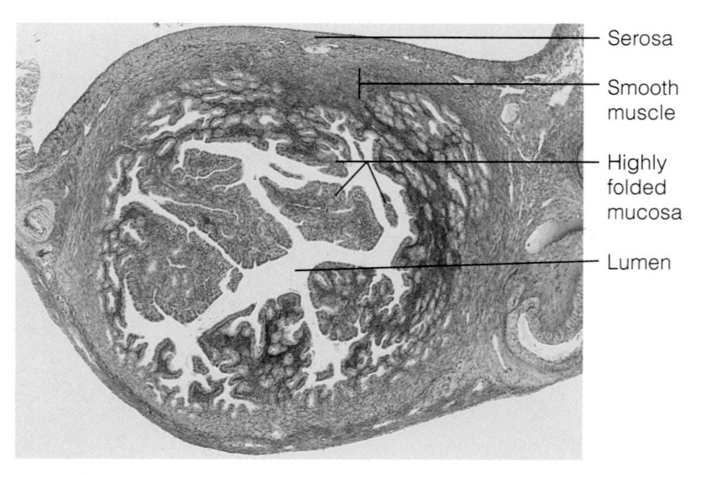

PLATE 48 Cross-sectional view of the uterine tube (34X)

Spinal Cord, Spinal Nerves, and the Autonomic Nervous System

OBJECTIVES

1. To identify important anatomical areas on a spinal cord model or appropriate diagram of the spinal cord.
2. To define *conus medullaris, cauda equina,* and *filum terminale.*
3. To list two major functions of the spinal cord.
4. To describe the origin, fiber composition, and distribution of the spinal nerves, differentiating between *roots,* the *spinal nerve proper,* and *rami,* and to discuss the result of transecting these structures.
5. To discuss the distribution of the dorsal rami and ventral rami of the spinal nerves.
6. To identify the four major nerve plexuses, the major nerves of each, and their distribution.
7. To describe the structure of a nerve, identifying the connective tissue coverings and citing their functions.
8. To identify the site of origin and the function of the sympathetic and parasympathetic divisions of the autonomic nervous system, and to state how the autonomic nervous system differs from the somatic nervous system.

MATERIALS

Spinal cord model (cross section)
Laboratory charts of the spinal cord, spinal nerves, and sympathetic chain
Preserved cow spinal cord sections with meninges and nerve roots intact
Prepared slide of a nerve (cross section)
Prepared slide of spinal cord (cross section)
Compound microscope
Dissecting microscope
Dissecting tray and instruments
Protective skin cream or disposable gloves

ANATOMY OF THE SPINAL CORD

The cylindrical spinal cord, a continuation of the brain stem, is an association and communication center. It plays a major role in spinal reflex activity, and provides neural pathways to and from higher nervous centers. Enclosed within the vertebral canal of the spinal column, the spinal cord extends from the foramen magnum of the skull to the first or second lumbar vertebra, where it terminates in the cone-shaped **conus medullaris** (Figure 18.1). Like the brain, it is cushioned and protected by meninges. The dura mater and arachnoid meningeal coverings extend beyond the conus medullaris, approximately to the level of S_2, and a fibrous extension of the pia mater extends even farther (into the coccygeal canal) as the **filum terminale.**

The fact that the meninges, filled with cerebrospinal fluid, extend well beyond the end of the spinal cord provides an excellent site for removing cerebrospinal fluid for analysis (as when bacterial or viral infections of the spinal cord or its meningeal coverings are suspected) without endangering the delicate spinal cord. This procedure, called a *lumbar tap,* is usually performed below L_3. Additionally, the administration of "saddle block" or caudal anesthesia for childbirth is normally done between L_3 and L_5.

In humans, 31 pairs of spinal nerves arise from the spinal cord and pass through intervertebral foramina to serve the body area at their approximate level of emergence. The cord is about the size of a thumb in circumference for most of its length, but there are obvious enlargements in the cervical and lumbar areas where the nerves serving the upper and lower limbs issue from the cord.

Because the cord does not extend to the end of the spinal column, the spinal nerves emerging from the inferior end of the cord must travel through the vertebral canal for some distance before exiting at the appropriate intervertebral foramina. This collection of spinal nerves transversing the inferior end of the vertebral canal is called the **cauda equina** because of its similarity to a horse's tail (the literal translation of *cauda equina*).

Obtain a model of a cross section of a spinal cord, and identify its structures as they are described next.

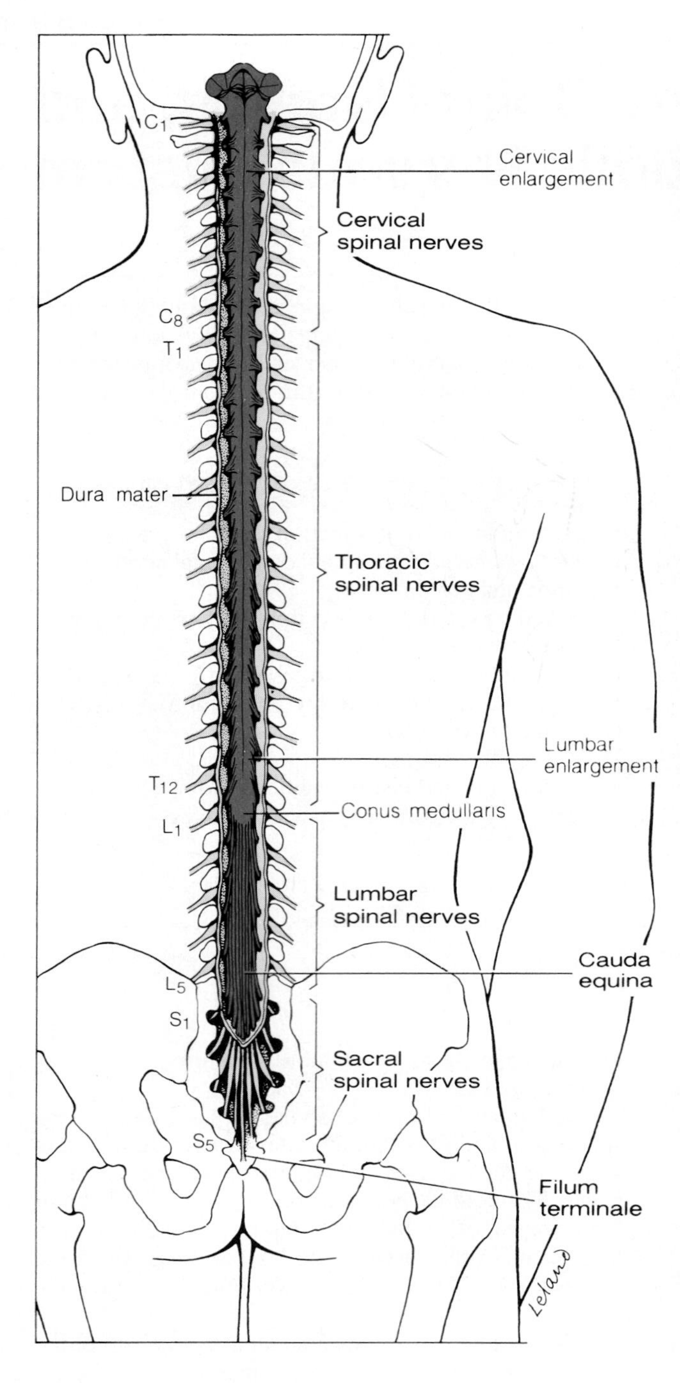

F18.1

Anatomy of the human spinal cord, dorsal view, showing the extent of the filum terminale. The tiny coccygeal nerves (C_0) are not illustrated.

Gray Matter

In cross section, the **gray matter** of the spinal cord looks like a butterfly or the letter *H* (Figure 18.2). The two posterior projections are called the **dorsal,** or **posterior, horns;** the two anterior projections are the **ventral,** or **anterior, horns.** The tips of the ventral horns are broader and less tapered than those of the dorsal horns. In the thoracic and lumbar regions of the cord, there is also a lateral outpocketing of gray matter on each side referred to as the **lateral horn.** The central area of gray matter connecting the two vertical regions is the **gray commissure.** The gray commissure surrounds the **central canal** of the cord, which contains cerebrospinal fluid.

Neurons with specific functions can be localized in the gray matter. The dorsal horns, for instance, contain interneurons and sensory fibers that enter the cord from the body periphery via the **dorsal root.** The cell bodies of these sensory neurons are found in an enlarged area of the dorsal root called the **dorsal root ganglion.** The ventral horns contain cell bodies of motor neurons of the somatic nervous system (voluntary system), which send their axons out via the **ventral root** of the cord to enter the adjacent spinal nerve. The **spinal nerves** are formed from the fusion of the dorsal and ventral roots. The lateral horns, where present, contain cell bodies of sympathetic motor neurons of the autonomic nervous system. Their axons also leave the cord via the ventral roots, along with those of the motor neurons of the ventral horns.

White Matter

The **white matter** of the spinal cord is nearly bisected by fissures. The more open anterior fissure is the **anterior median fissure,** and the posterior one is the **posterior median sulcus.** The white matter is composed of myelinated fibers—some running to higher centers, some traveling from the brain to the cord, and some conducting impulses from one side of the cord to the other.

Because of the irregular shape of the gray matter, the white matter on each side of the cord can be divided into three primary regions or *white columns: the* **posterior, lateral,** and **anterior funiculi.** Each funiculus contains a number of fiber **tracts** composed of axons with the same origin, terminus, and function. Tracts conducting sensory impulses to the brain are called ascending, or sensory, tracts; those carrying impulses from the brain to the skeletal muscles are descending, or motor, tracts.

Because it serves as the transmission pathway between the brain and the body periphery, the spinal cord is an extremely important functional area. Even though it is protected by meninges and cerebrospinal fluid in the vertebral canal, it is highly vulnerable to traumatic injuries, such as might occur in an automobile accident.

When the cord is transected (or severely traumatized), both motor and sensory functions are lost in body areas normally served by that (and lower) regions of the spinal cord. Injury to certain cord areas may even result

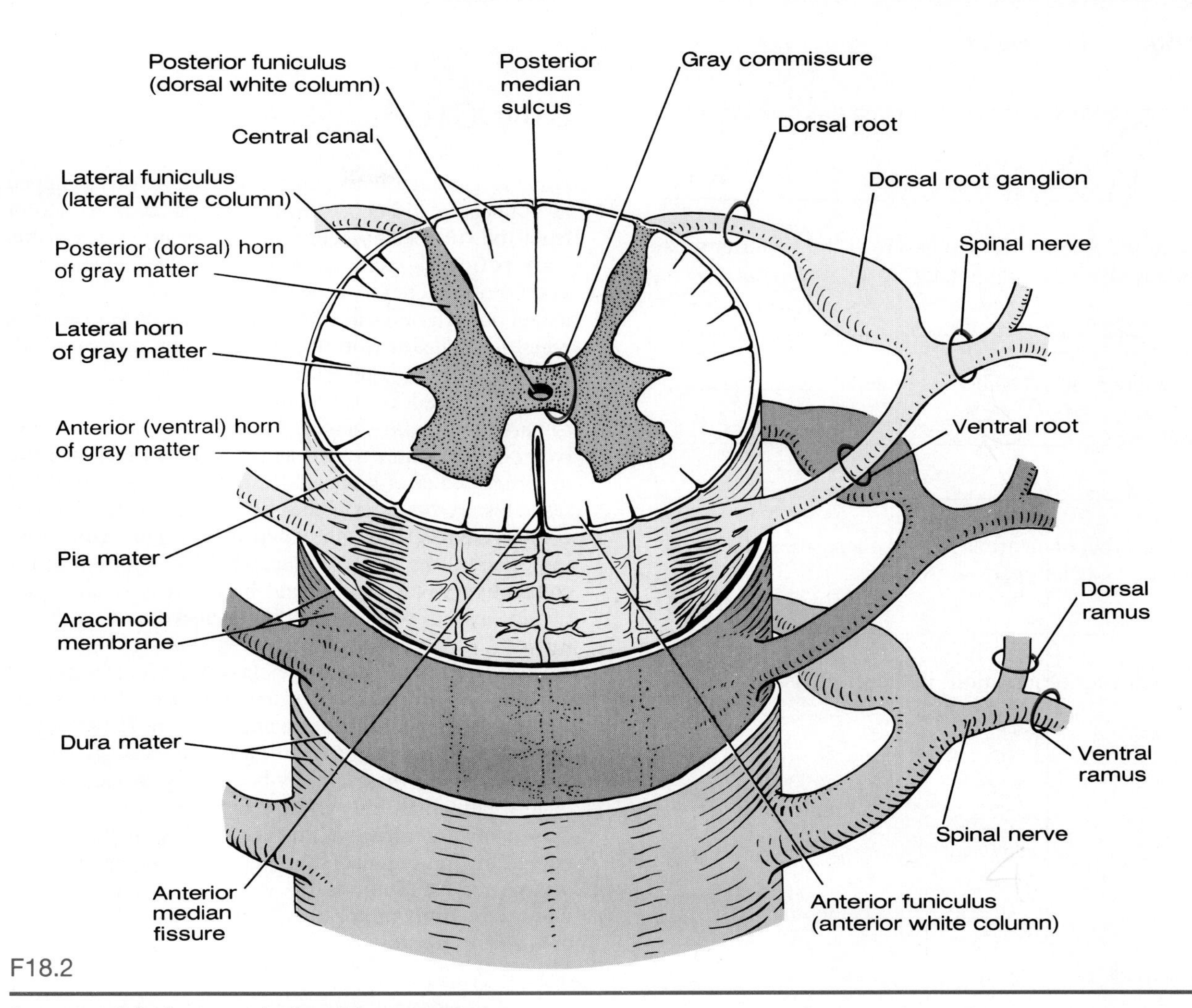

F18.2

Anatomy of the human spinal cord (cross section).

in a permanent flaccid paralysis of both legs (paraplegia) or of all four limbs (quadraplegia).

Paralysis may also be caused by viral infections. For example, *poliomyelitis* is a viral disease that attacks the anterior horns of the spinal cord. Depending on the localization of the viral lesion, the resulting paralysis may affect skeletal muscles or respiratory muscles, or both. ■

Spinal Cord Dissection

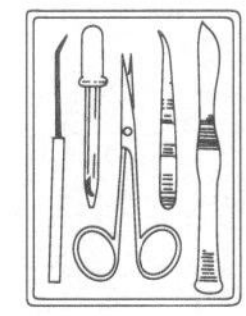

1. Obtain a dissecting tray and instruments, and a segment of preserved spinal cord. Identify the tough outer meninx (dura mater) and the weblike arachnoid membrane.

What name is given to the third meninx, and where is it found?

____

Peel back the dura mater, and observe the fibers making up the dorsal and ventral roots. If possible, identify a dorsal root ganglion.

2. Cut a thin cross section of the cord and, with the naked eye or with the aid of a dissecting microscope, identify the dorsal and ventral horns of the gray matter.

How can you be certain that you are correctly identifying the dorsal and ventral horns?

____

____

____

Also identify the central canal; white matter; anterior median fissure; posterior median sulcus; and posterior, anterior, and lateral funiculi.

3. Obtain a prepared slide of the spinal cord (cross section) and a compound microscope. Refer to plate 7 in the histology atlas as you examine the slide carefully under low power. Observe the shape of the central canal.

Is it basically circular or oval? ____________________

Name the glial cell type that lines this canal. ________

What would you expect to find in this canal in a living animal?

___

Does the anterior median fissure or the posterior median sulcus touch the gray matter? If so, which does?

___

Can any neuron cell bodies be seen? ___

Where? ___

___

What type of neurons would these most likely be—motor, association, or sensory?

___

Sketch your observations in the space below.

STRUCTURE OF A NERVE

A nerve is a bundle of neuron fibers or axons wrapped in connective tissue coverings that extends to and/or from the CNS and visceral organs or structures of the body periphery (such as skeletal muscles, glands, and skin). Within a nerve, each fiber is surrounded by a delicate connective tissue sheath called an **endoneurium,** which insulates it from the other neuron processes adjacent to it. (The endoneurium is often mistaken for the myelin sheath; it is instead an additional sheath that surrounds the myelin sheath.) Groups of fibers are bound by a coarser connective tissue, called the **perineurium,** to form bundles of fibers called **fascicles.** Finally, all the fascicles are bound together by a tough, white, fibrous connective tissue sheath called the **epineurium,** forming the cordlike nerve (Figure 18.3). In addition to the connective tissue wrappings, blood vessels and lymphatic vessels serving the fibers also travel within a nerve.

Like neurons, nerves are classified according to the direction in which they transmit impulses. Nerves carrying both sensory (afferent) and motor (efferent) fibers are called **mixed nerves;** all spinal nerves are mixed nerves. Nerves that carry only sensory processes and conduct impulses only toward the CNS are referred to as **sensory,** or **afferent, nerves.** A few of the cranial nerves are pure sensory nerves, but the majority are mixed nerves. The ventral roots of the spinal cord, which carry only motor fibers, can be considered **motor,** or **efferent, nerves.**

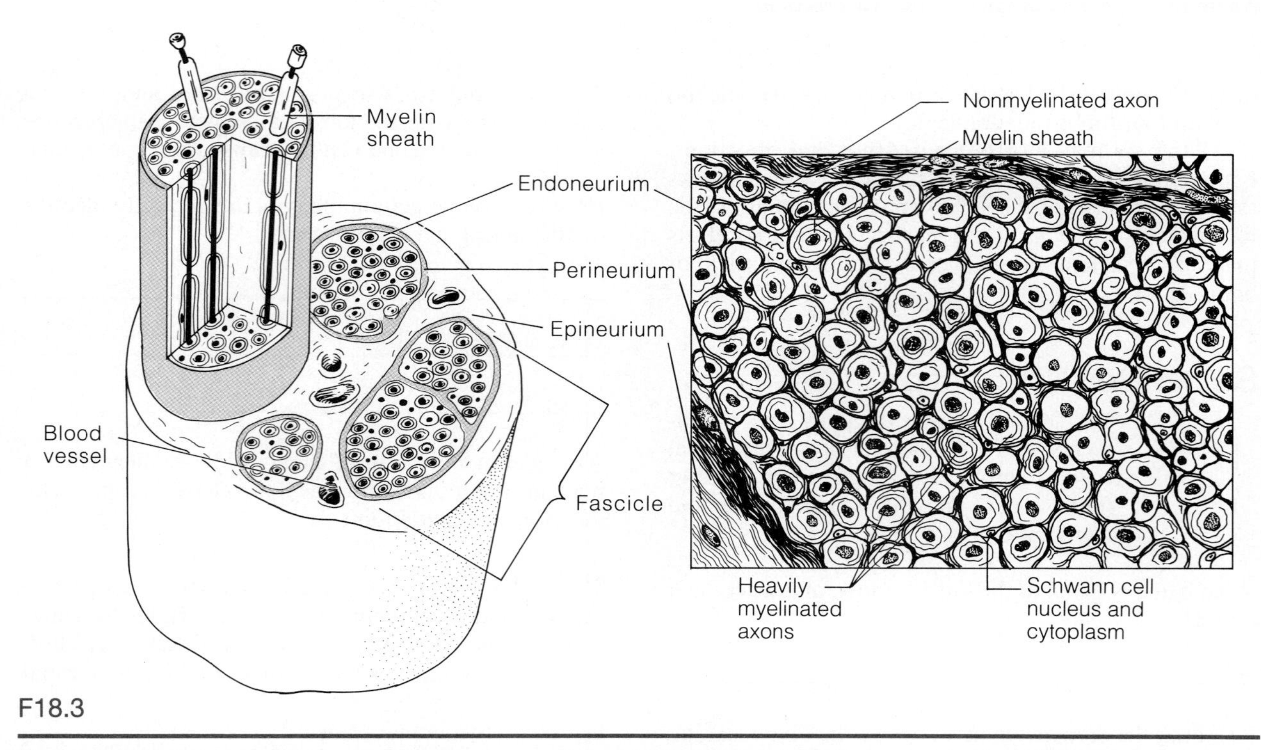

F18.3

Structure of a nerve, showing connective tissue wrappings. (a) Three-dimensional diagrammatic view; (b) Cross-sectional view corresponding to plate 8 of the histology atlas.

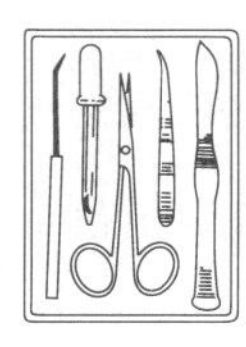

Examine under the compound microscope a prepared cross section of a peripheral nerve. Identify nerve fibers, myelin sheaths, fascicles, and the endoneurium, perineurium, and epineurium sheaths.

SPINAL NERVES AND NERVE PLEXUSES

The 31 pairs of human spinal nerves arise from the fusions of the ventral and dorsal roots of the spinal cord. (Figure 18.4 shows how the nerves are named according to their point of issue.) Because the ventral roots contain the myelinated axons of motor neurons located in the cord and the dorsal roots carry sensory fibers entering the cord, all spinal nerves are mixed nerves. The first pair of spinal nerves leaves the vertebral canal between the base of the occiput and the atlas, but all the rest exit via the intervertebral foramina. The second through seventh pairs of cervical nerves emerge *above* the vertebra for which they are named. C_8 emerges between C_7 and T_1. (Notice that there are *7* cervical vertebrae but *8* pairs of cervical nerves.) The remaining pairs emerge *below* the vertebrae for which they are named.

Almost immediately after emerging, each nerve divides into **dorsal** and **ventral rami** (Figure 18.4b). (Thus each spinal nerve is only about 1 or 2 cm long.) The rami, like the spinal nerves, contain both motor and sensory fibers. The smaller dorsal rami serve the skin and musculature of the posterior body trunk at their approximate level of emergence. The ventral rami of spinal nerves T_2 through T_{12} pass anteriorly as the **intercostal nerves,** supplying the muscles of intercostal spaces, and the skin and muscles of the anterior and lateral

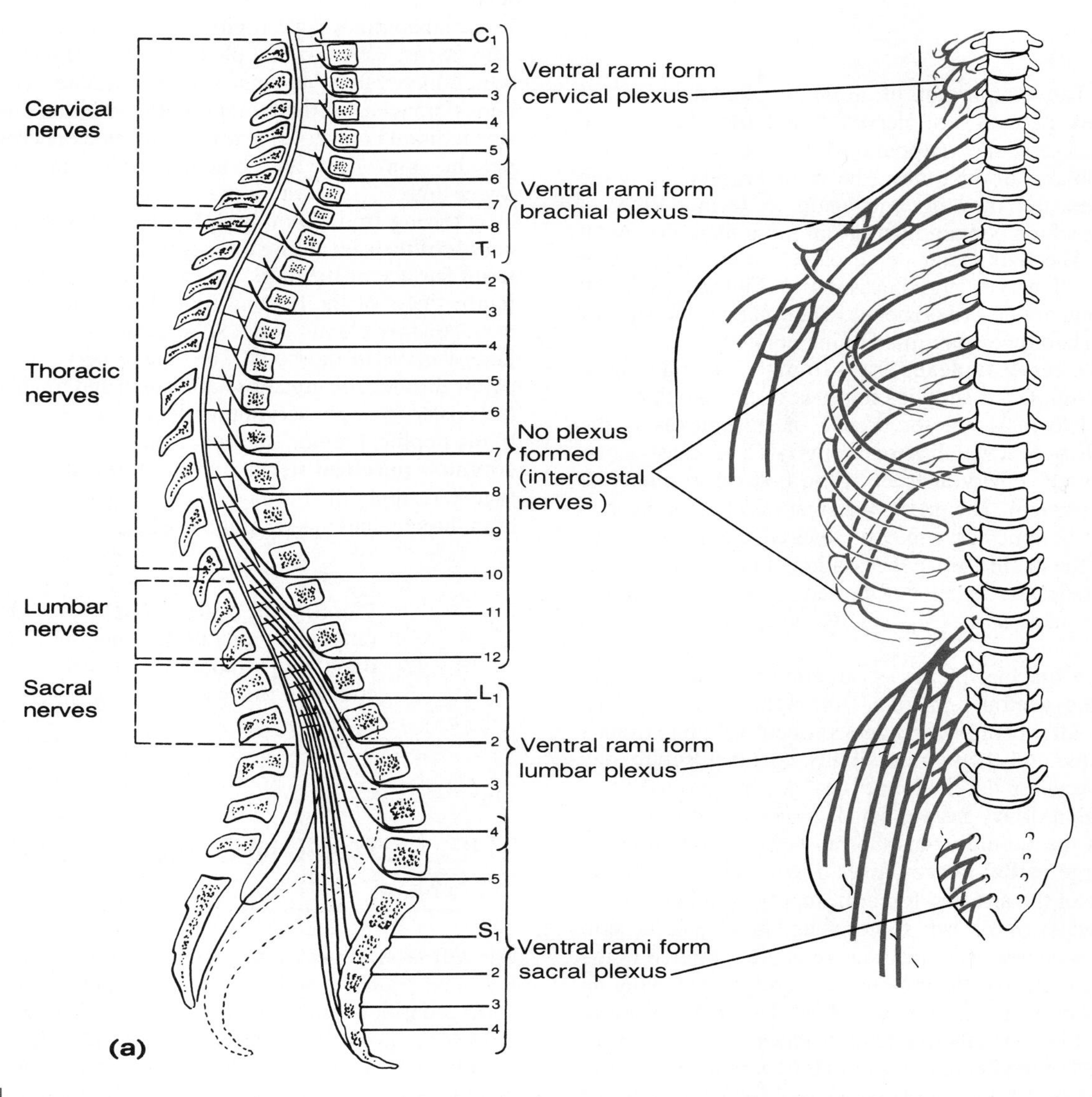

F18.4

Human spinal nerves. (a) Relationship of spinal nerves to vertebrae (areas of plexuses formed by the anterior rami are indicated) (*continues*).

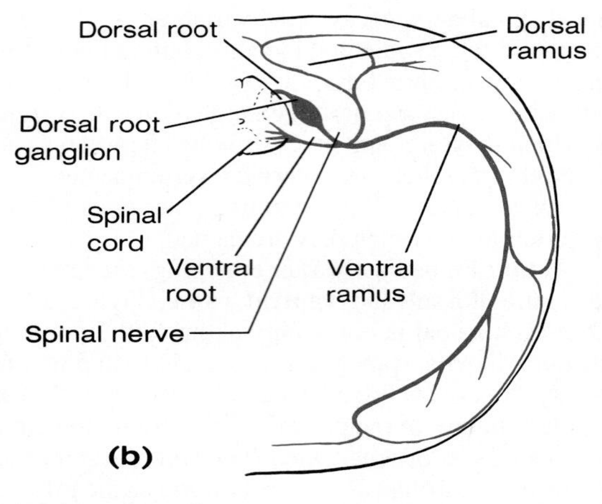

F18.4 (continued)

Human spinal nerves. (b) Relative distribution of the ventral and dorsal rami of a spinal nerve (cross section of left trunk).

trunk. The ventral rami of all other spinal nerves form complex networks of nerves called **plexuses.** These plexuses serve the motor and sensory needs of the muscles and skin of the limbs. The ventral rami unite in the plexuses and then diverge again to form peripheral nerves, which contain fibers from more than one spinal nerve. The four major nerve plexuses and their major peripheral nerves are illustrated in Figures 18.4 and 18.5, and are described below. Their names and sites of origin should be committed to memory.

The **cervical plexus** arises from the ventral rami of C_1 through C_5; it supplies muscles of the shoulder and neck. The major motor branch of this plexus is the **phrenic nerve,** which arises from C_3–C_5 and passes into the thoracic cavity in front of the first rib to innervate the diaphragm. The primary danger of a broken neck is that the phrenic nerve may be severed, leading to paralysis of the diaphragm and cessation of breathing. (A jingle to help you remember the rami [roots] forming the phrenic nerves is: "C_3, C_4, C_5 keep the diaphragm alive.")

The brachial plexus is large and complex, arising from the ventral rami of C_5 through C_8 and T_1. The plexus, after being rearranged consecutively into trunks, divisions, and then cords, finally becomes subdivided into five major peripheral nerves.

The **axillary nerve,** which serves the muscles and skin of the shoulder, has the most limited distribution. The large **radial nerve** passes down the posterolateral surface of the arm and forearm, supplying all the extensor muscles of the arm, forearm, and hand, and the skin along its course. The radial nerve is often injured in the axillary region by the pressure of a crutch or by hanging one's arm over the back of a chair. The **median nerve** passes down the anteromedial surface of the arm to supply most of the flexor muscles in the forearm and several muscles in the hand (plus the skin of the lateral surface of the palm of the hand).

- Hyperextend your wrist to identify the long obvious tendon of your palmaris longus muscle, which crosses the exact midline of the anterior wrist. Your median nerve lies immediately deep to that tendon and the radial nerve lies just *lateral* to it.

The **musculocutaneous nerve** supplies the arm muscles that flex the forearm and the skin of the lateral surface of the forearm. The **ulnar nerve** travels down the posteromedial surface of the arm. It courses around the medial epicondyle of the humerus to supply the flexor carpi ulnaris, the ulnar head of the flexor digitorum profundus of the forearm, and all intrinsic muscles of the hand not served by the median nerve. It supplies the skin of the medial third of the hand, both the anterior and posterior surfaces.

The *lumbrosacral plexus,* which serves the pelvic region of the trunk and the lower limbs, is actually a complex of two plexuses, the lumbar plexus and the sacral plexus (see Figure 18.5). The **lumbar plexus** arises from ventral rami of L_1 through L_4. Its nerves serve the lower abdominopelvic region and the anterior thighs. The largest nerve of this plexus is the **femoral nerve,** which innervates the anterior thigh muscles. The cutaneous branches of the femoral nerve (median and anterior femoral cutaneous, and the saphenous nerves) supply the skin of the anteromedial surface of the entire lower limb.

Arising from L_4 through L_5 and S_1 through S_4, the nerves of the **sacral plexus** supply the buttock, the posterior surface of the thigh, and virtually all sensory and motor fibers of the leg and foot. The major peripheral nerve of this plexus is the **sciatic nerve,** which is the largest nerve in the body. The sciatic nerve leaves the pelvis through the greater sciatic notch and travels down the posterior thigh, serving its flexor muscles and skin. In the popliteal region, the sciatic nerve divides into the **common peroneal nerve** and the **tibial nerve,** which together supply the balance of the leg muscles and skin, both directly and via several branches.

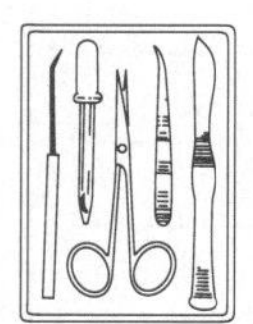

Identify the four major nerve plexuses (and their major nerves) in Figure 18.5 and on a large laboratory chart. Trace the course of the nerves.

THE AUTONOMIC NERVOUS SYSTEM

The **autonomic nervous system** is the division of the PNS that regulates body activities that are generally not under conscious control. It is composed of a special group of motor neurons serving cardiac muscle, smooth muscle, and internal glands. Because these structures function without conscious control, this system is often referred to as the *involuntary nervous system.*

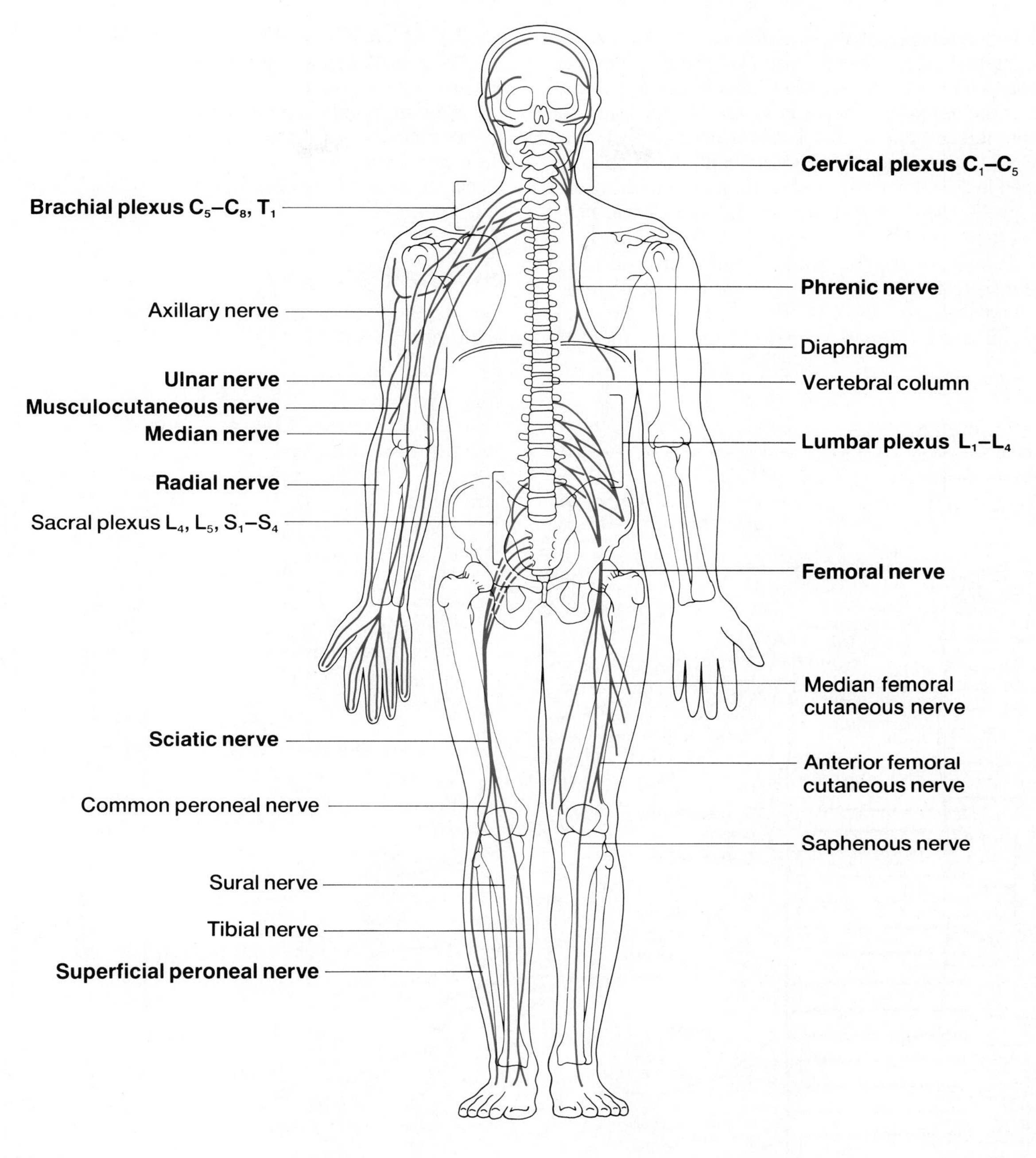

F18.5

Nerve plexuses and major nerves arising from each. For clarity, each plexus is illustrated only on one side of the body.

There is a basic anatomical difference between the motor pathways of the **somatic** (voluntary) **nervous system,** which innervates the skeletal muscles, and those of the autonomic nervous system. In the somatic division, the cell bodies of the motor neurons reside in the CNS (spinal cord or brain); and their axons, sheathed in spinal nerves, extend all the way to the skeletal muscles they serve. However, the autonomic nervous system consists of chains of two motor neurons. The cell body of the first motor neuron of each pair, called the *preganglionic neuron,* resides in the brain or cord. Its axon leaves the CNS to synapse with the second motor neuron (the *postganglionic neuron*), the cell body of which is located in a ganglion outside the CNS. The axon of the postganglionic neuron then extends to the organ it serves.

The autonomic nervous system has two major functional subdivisions (Figure 18.6). These subdivisions, the sympathetic and parasympathetic divisions, serve the same organs, but generally cause opposing or antagonistic effects.

Sympathetic Division

The preganglionic neurons of the **sympathetic,** or **thoracolumbar, division** are in the lateral horns of the gray

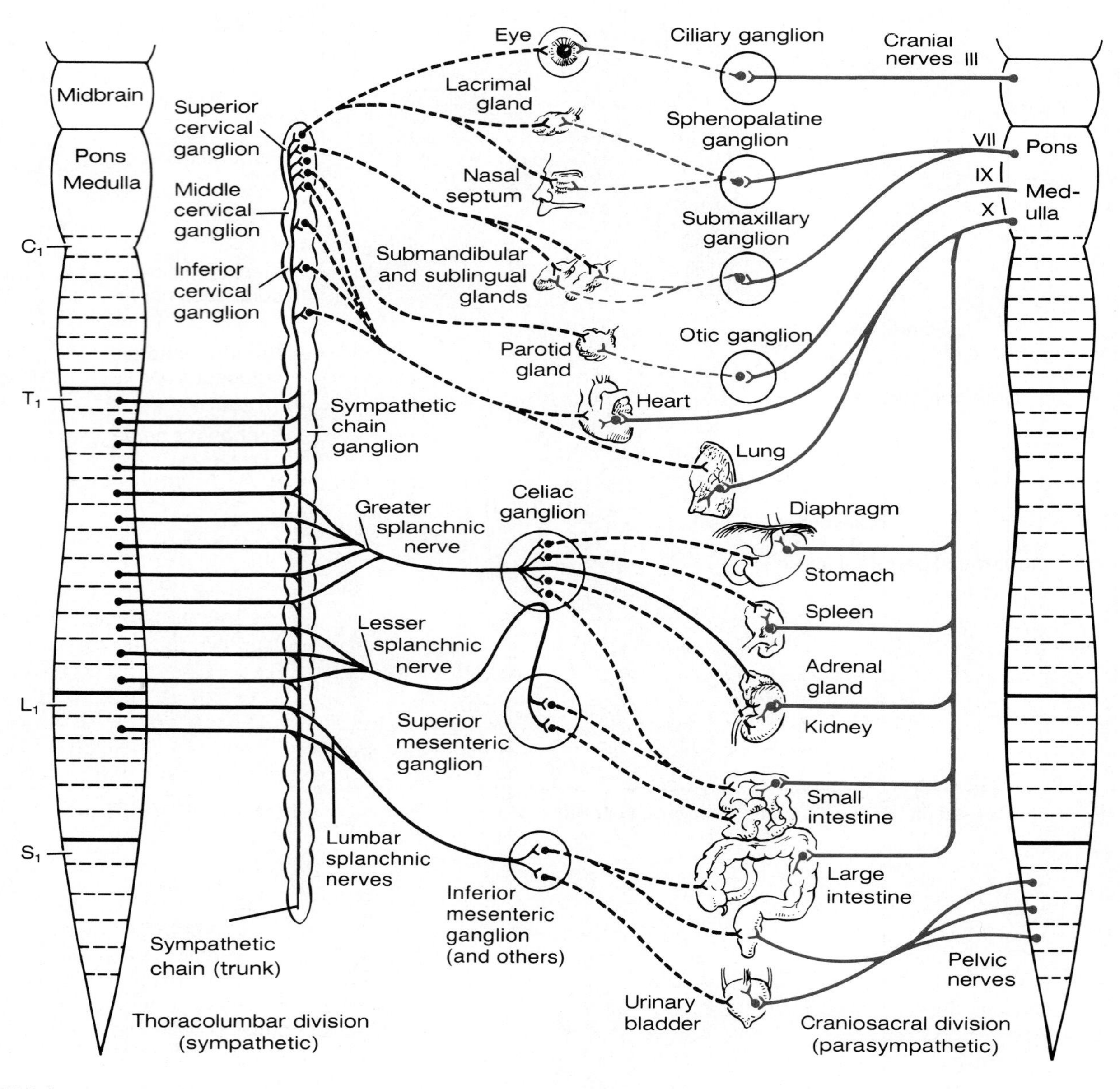

F18.6

The autonomic nervous system. The solid lines indicate preganglionic fibers. The dashed lines denote postganglionic nerve fibers.

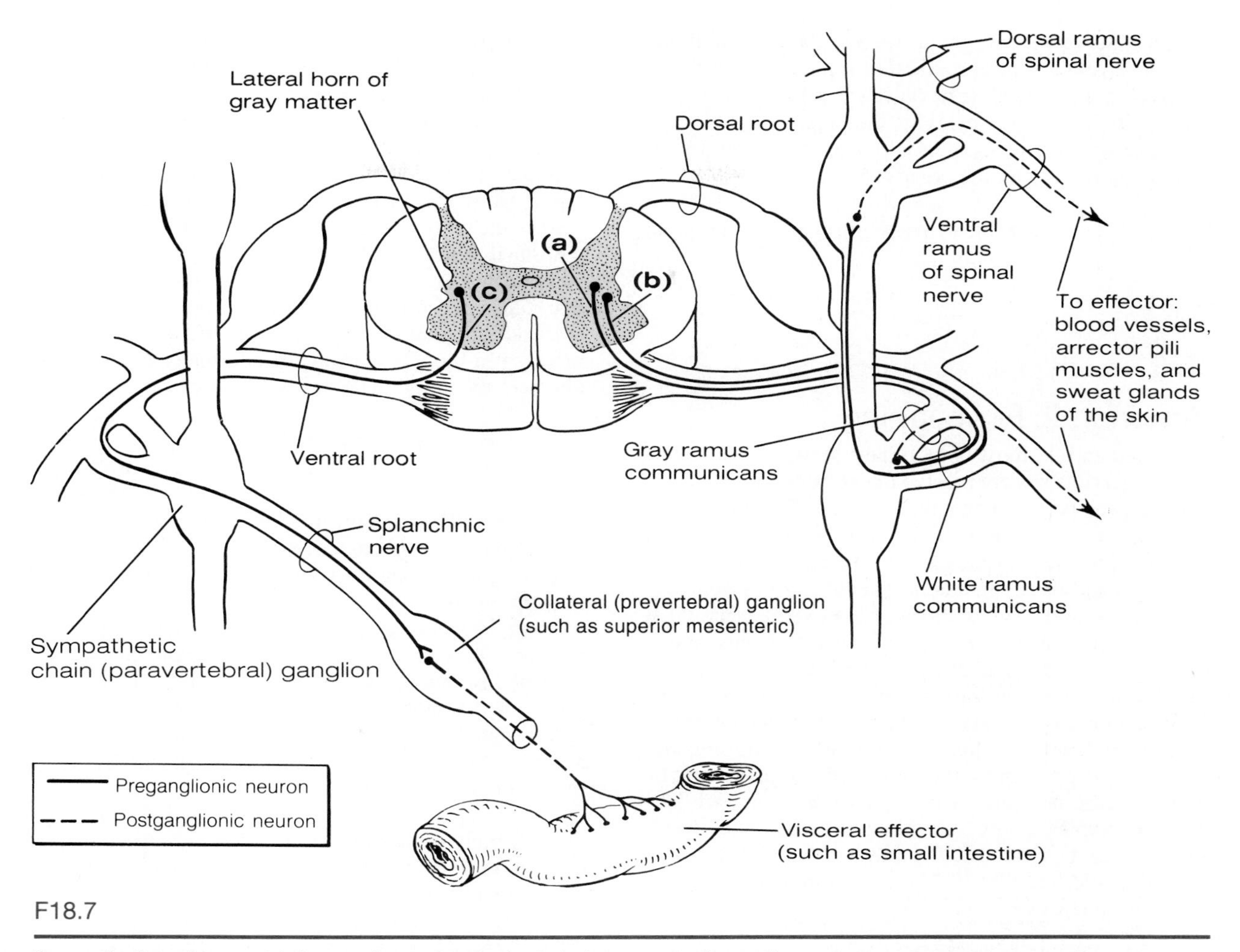

F18.7

Sympathetic pathways. (a) Synapse in a sympathetic chain (paravertebral) ganglion at the same level; (b) synapse in a sympathetic chain ganglion at a different level; (c) synapse in a collateral (prevertebral) ganglion.

matter of the spinal cord from T_1 through L_2. The preganglionic axons leave the cord via the ventral root (in conjunction with the axons of the somatic motor neurons), enter the spinal nerve, and then travel briefly in the ventral ramus (Figure 18.7). From the ventral ramus, they pass through a small branch called the **white ramus communicans** to enter a **paravertebral ganglion** in the **sympathetic chain,** or **trunk,** which lies alongside the vertebral column (the literal meaning of paravertebral).

Having reached the ganglion, an axon may take one of three courses (see Figure 18.7). First, it may synapse with a postganglionic neuron in the sympathetic chain at *that* level. Second, the axon may travel upward or downward through the sympathetic chain to synapse with a postganglionic neuron in a paravertebral ganglion at another level. In either of these two instances, the postganglionic axons then reenter the ventral or dorsal ramus of a spinal nerve via a **gray ramus communicans** and travel in the ramus to innervate skin structures (sweat glands, arrector pili muscles attached to hair follicles, and the smooth muscles of blood vessel walls). Third, the axon may pass through the ganglion without synapsing and form part of the **splanchnic nerves,** which travel to the viscera to synapse with a postganglionic neuron in a **collateral,** or **prevertebral, ganglion.** (The major collateral ganglia—the celiac, superior mesenteric, and inferior mesenteric ganglia—supply the abdominal and pelvic visceral organs.) The postganglionic axon then leaves the collateral ganglion and travels to a nearby visceral organ which it innervates.

Parasympathetic Division

The preganglionic neurons of the **parasympathetic,** or **craniosacral, division** are located in brain nuclei of cranial nerves III, VII, IX, and X, and in the S_2 through S_4 level of the spinal cord. The axons of the preganglionic neurons of the cranial region travel in their respective cranial nerves to the *immediate area* of the head and neck organs to be stimulated. There they synapse with the postganglionic neuron in a **terminal** or **intramural** (literally "within the walls") **ganglion.** The postganglionic neuron then sends out a very short axon to the organ it serves. In the sacral region, the preganglionic axons leave the ventral roots of the spinal cord and collectively form the **pelvic nerves,** which travel to the pelvic cavity. In the pelvic cavity, the preganglionic axons synapse with the postganglionic neurons in ganglia located on, or close to, the organs served.

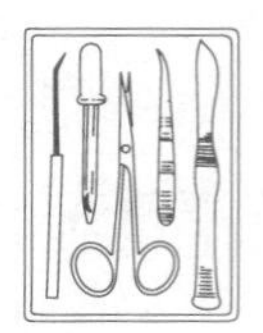

Locate the sympathetic chain on the spinal nerve chart.

Autonomic Functioning

As noted earlier, most body organs served by the autonomic nervous system receive fibers from both the sympathetic and parasympathetic divisions. The only exceptions are the structures of the skin (sweat glands and arrector pili muscles attached to the hair follicles), the pancreas and liver, the adrenal medulla, and essentially all blood vessels except those of the external genitalia, all of which receive sympathetic innervation only. When both divisions serve an organ, they have antagonistic effects. This is because their postganglionic axons release different neurotransmitters. The parasympathetic fibers, called **cholinergic fibers,** release acetylcholine; the sympathetic postganglionic fibers, called **adrenergic fibers,** release norepinephrine. (However, there are isolated examples of postganglionic sympathetic fibers, such as those serving blood vessels in the skeletal muscles, that release ACh.) The preganglionic fibers of both divisions release acetylcholine.

The parasympathetic division is often referred to as the *housekeeping,* or *"resting and digesting," system* because it maintains the visceral organs in a state most suitable for normal functions and internal homeostasis; that is, it promotes normal digestion and elimination. In contrast, activation of the sympathetic division is referred to as the "fight or flight" response because it readies the body to cope with situations that threaten homeostasis. Under such emergency conditions, the sympathetic nervous system induces an increase in heart rate, an increase in blood pressure (by inducing vasoconstriction), dilation of the bronchioles of the lungs, an increase in blood sugar levels, and many other effects that help the individual cope with a stressor.

As we grow older, our sympathetic nervous system gradually becomes less and less efficient, particularly in causing vasoconstriction of blood vessels. When elderly people stand up quickly after sitting or lying down, they often become light-headed or faint because the sympathetic nervous system is not able to react quickly enough to counteract the pull of gravity by activating the vasoconstrictor fibers; and so blood pools in the feet. This condition, *orthostatic hypotension,* is a type of low blood pressure resulting from changes in body position as described. Orthostatic hypotension can be prevented to some degree if *slow* changes in position are made. This gives the sympathetic nervous system a little more time to react and adjust. ■

Several body organs are listed below. Using your textbook as a reference, list the effect of the sympathetic and parasympathetic divisions on each.

Organ or Function	Parasympathetic Effect	Sympathetic Effect
Heart		
Bronchioles of lungs		
Digestive tract activity		
Urinary bladder		
Iris of the eye		
Blood vessels (most)		
Sweat glands		
Penis/clitoris		
Adrenal medulla		

EXERCISE 19

General Sensation

OBJECTIVES

1. To recognize various types of general sensory receptors as studied in the laboratory, and to describe the function and location of each type.
2. To define *exteroceptor, interoceptor,* and *proprioceptor.*
3. To demonstrate and relate differences in relative density and distribution of tactile and thermoreceptors in the skin.
4. To define *tactile localization* and to describe how this ability varies in different areas of the body.
5. To explain the tactile two-point discrimination test and to state its anatomical basis.
6. To define *referred pain* and attempt to explain it.
7. To define *adaptation, negative afterimage,* and *projection.*

MATERIALS

Compound microscope
Histological slides of Pacinian corpuscles (longitudinal section), Meissner's corpuscles, Golgi tendon organs, and muscle spindles
Mall probes (obtainable from Carolina Biological Supply, code #627400), or pointed metal rods of approximately 1-in. diameter and 2- to 3-in. length, with insulated blunt end
Large beaker of ice water; chipped ice
Hot water bath set at 45°C
Thermometer
Fine-point, felt-tipped markers (black, red, and blue)
Small metric ruler
Von Frey's hairs (or 1-in. lengths of horsehair glued to a matchstick) or sharp pencils
Towel
Caliper or esthesiometer
Four coins (nickels or quarters)
Three large finger bowls or 1000-ml beakers

It can be said without reservation that people are irritable creatures. Hold a sizzling steak before them and their mouths water. Flash your high beams in their eyes on the highway and they cuss. Stroke their arms gently and they smile. These "irritants" (the steak, the light, and the soft touch) and many others are stimuli that continually assault us.

The body's sense organs, which include its sensory receptors, react to stimuli or changes within the body and in the external environment. The tiny sensory receptors of the general senses react to touch, pressure, pain, heat, cold, and changes in position, and are distributed throughout the body. In contrast to the widely distributed general receptors, the special senses are large, complex sensory organs or small, localized groups of receptors. The special senses include sight, hearing, equilibrium, smell, and taste.

Sensory receptors may be classified according to the source of their stimulus. **Exteroceptors** react to stimuli in the external environment, and typically they are found close to the body surface. Exteroceptors include the cutaneous receptors in the skin and the highly specialized receptor structures of the special senses (the vision apparatus of the eye, and the hearing and equilibrium receptors of the ear, for example). **Interoceptors** respond to stimuli arising within the body. Interoceptors are found in the internal visceral organs, and include stretch receptors (found in walls of hollow organs), chemoreceptors, and others. A subdivision of the interoceptors, the **proprioceptors,** are located in the skeletal muscles and their tendons. They provide information on the position and degree of stretch of the skeletal muscles and tendons.

The receptors of the special sense organs are complex and deserve considerable study. Thus the special senses (vision, hearing, equilibrium, taste, and smell) are covered separately in Exercises 21 through 23. Only the anatomically simpler cutaneous sensory receptors and proprioceptors will be studied in this exercise.

STRUCTURE OF SENSORY RECEPTORS

You cannot become aware of changes in the environment unless your sensory neurons and their receptors are operating properly. Sensory receptors are modified dendritic endings (or specialized cells associated with the dendrites) that are sensitive to certain environmental stimuli. They react to such stimuli by initiating a nerve impulse. Although there is a good deal of functional overlap, several histologically distinct types of receptors in the skin have been identified; their structures are de-

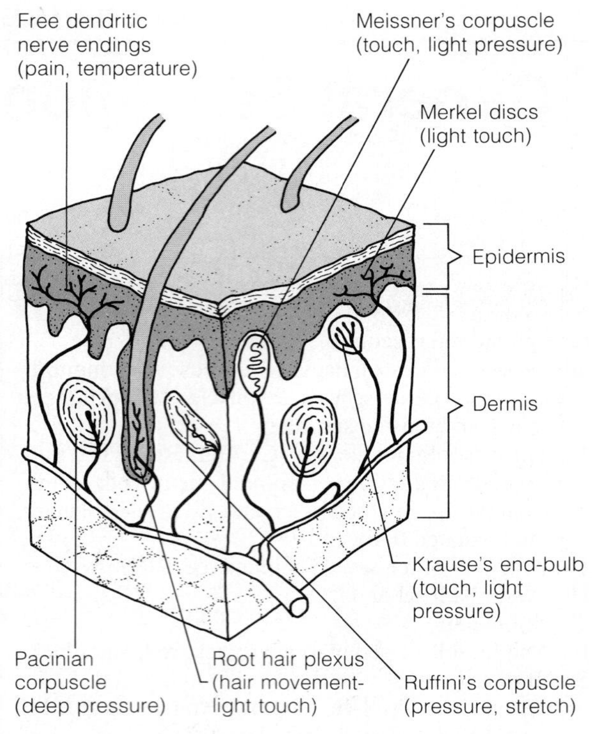

F19.1

Cutaneous receptors (free dendritic endings, Merkel discs, root hair plexus, Meissner's corpuscle, Pacinian corpuscle, Krause's end bulb, and Ruffini's corpuscle). See also plates 9, 10, and 11 of the histology atlas.

picted in Figure 19.1. The least specialized of the cutaneous receptors are the **free dendritic endings** of sensory neurons (Figure 19.1 and plate 10), which respond chiefly to pain and temperature. Certain free dendritic endings associate with specific epidermal cells to form **Merkel discs,** or entwine in hair follicles to form **root hair plexuses.** Both Merkel discs and root hair plexuses function as light touch receptors.

The other cutaneous receptors are more complex, with the dendritic endings *encapsulated* by connective tissue cells. **Meissner's corpuscles,** commonly referred to as *tactile receptors* because they respond to light touch, are located in the dermal papillae of the skin. Although **Krause's end bulbs** and **Ruffini's corpuscles** were presumed to be thermoreceptors, now most authorities classify Krause's end bulbs with Meissner's corpuscles as a special kind of touch, or light pressure, receptor. Ruffini's corpuscles appear to respond to pressure and stretch stimuli. As you inspect Figure 19.1, notice that all of the encapsulated receptors are quite similar, with the possible exception of the **Pacinian corpuscles** (deep pressure receptors), which are anatomically more distinctive and lie deepest in the dermis.

1. Obtain histological slides of Pacinian and Meissner's corpuscles. Locate, under low power, a Meissner's corpuscle in the dermal layer of the skin. (Most often, Meissner's corpuscles are found immediately beneath the epidermis, in the dermal papillae.) Then switch to the oil immersion lens for a detailed study. Notice that the naked dendritic fibers within the capsule are aligned parallel to the skin surface. Compare your observations to Figure 19.1 and plate 9 of the histology atlas.

2. Next observe the Pacinian corpuscles located deeper in the dermis. If possible, note the slender, naked dendrite ending in the center of the receptor and the heavy capsule of connective tissue surrounding it (which looks rather like an onion cut lengthwise). Also, note how much larger the Pacinian corpuscles are than the Meissner's corpuscles. Compare your observations to the views shown in Figure 19.1 and plate 11 of the histology atlas.

3. Now obtain slides of **muscle spindles** and **Golgi tendon organs,** the two major types of proprioceptors (see Figure 19.2). In the slide of muscle spindles, note that minute extensions of the dendrites of the sensory neuron coil around specialized slender skeletal muscle cells called **intrafusal cells,** or **fibers.** The Golgi tendon organs are composed of dendrites that ramify through the tendon tissue close to the muscle tendon attachment. Stretching of the muscle or tendon excites both types of receptors, which then transmit impulses that ultimately reach the cerebellum for interpretation. Compare your observations to Figure 19.2.

RECEPTOR PHYSIOLOGY

Sensory receptors act as transducers, changing the environmental stimulus into afferent nerve impulses. Because the action potential generated in all nerves is essentially identical, the stimulus is identified by the area of the brain's sensory cortex that is stimulated (which, of course, differs for the various afferent nerves).

Four qualities of cutaneous sensations have traditionally been recognized: tactile (touch), heat, cold, and pain. Mapping these sensations on the skin has revealed that the sensory receptors for these qualities have discrete locations and are characterized by clustering at certain points—**punctate distribution**—rather than by uniform distribution.

The simple pain receptors, extremely important in protecting the body, are the most numerous. Touch receptors tend to be clustered where greater sensitivity is desirable, as on the hands and face. It is surprising to learn that large areas of the skin are quite insensitive to touch because of a relative lack of touch receptors.

There are several simple experiments you can conduct to investigate the location and physiology of the cutaneous receptors. In each of the following activities, work in pairs with one person as the subject and the other as the experimenter. After you have completed an

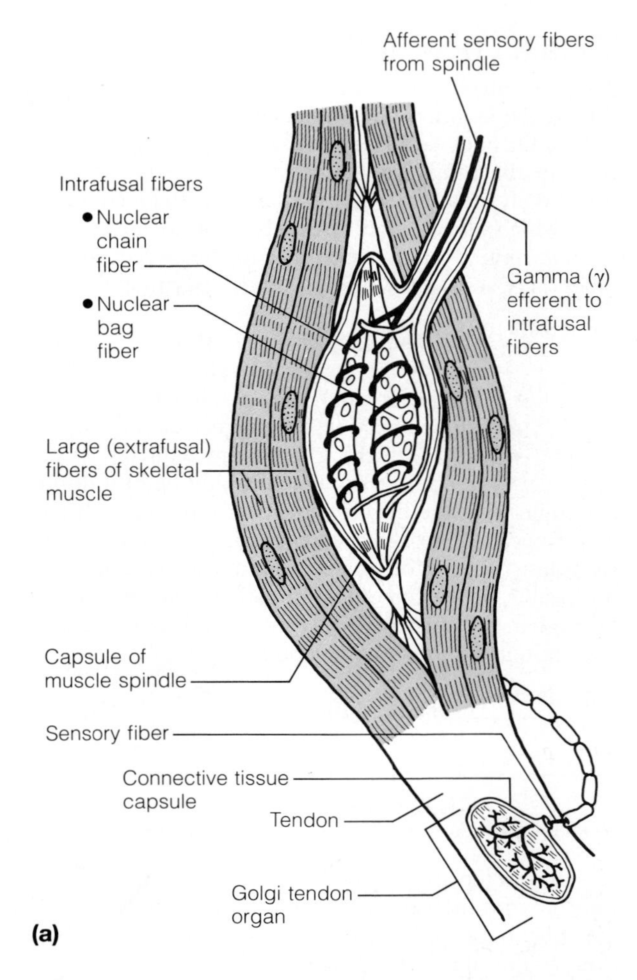

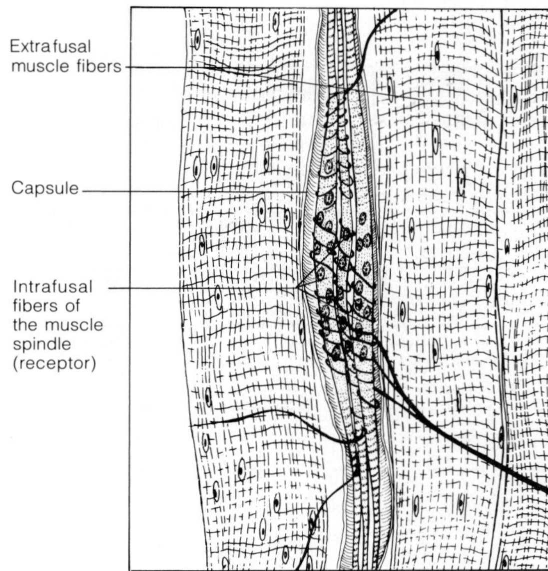

F19.2

Proprioceptors (a) Diagrammatic view of a muscle spindle and Golgi tendon organ; (b) drawing of a photomicrograph of a muscle spindle. See corresponding plate 12 in the histology atlas.

experiment, switch roles and go through the procedures again so that all class members obtain individual results. Keep an accurate account of each test that you perform.

Density and Location of Temperature and Touch Receptors

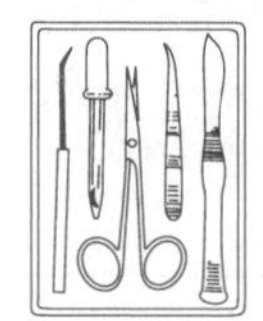

1. Place one Mall probe (or temperature cylinder) in a beaker of ice water and the other in a water bath controlled at 45°C. On the ventral surface of the subject's forearm, draw a square (2 cm on each side) with a felt marker. During the following tests, the subject's eyes are to remain closed; the subject should tell the examiner when a stimulus is detected.

2. Working in a systematic manner from one side of the marked square to the other, gently touch the Von Frey's hairs or a sharp pencil tip to different points within the square. The *hairs* should be applied with a pressure that just causes them to bend; use the same pressure for each contact. Do not apply deep pressure; the goal is to stimulate only the more superficially located Meissner's corpuscles (as opposed to the Pacinian corpuscles located in the subcutaneous tissue). Mark with a *black dot* all points at which the touch is perceived.

3. Remove the Mall probe or temperature rod from the ice water, and quickly wipe it dry. Repeat the procedure outlined above, marking all points of cold perception with a *blue dot.** Perception should be for temperature, not simply touch.

4. Remove the second temperature rod from the 45°C water bath; and repeat the procedure once again, marking all points of heat perception with a *red dot.**

5. After each student has acted as the subject, produce a "map" of receptor areas in the squares below.

2cm

Student 1 ____________________

2cm

Student 2 ____________________

* The temperature rods will have to be returned to the water baths approximately every 2 minutes to maintain the desired testing temperatures.

How does the density of the heat receptors correspond to that of the touch receptors?

____

To that of the cold receptors? ______________

____

On the basis of your observations, which type of receptor appears to be most abundant (in the area tested)?

____

Two-Point Discrimination Test

The density of the touch receptors varies significantly in different areas of the body. In general, areas that have the greatest density of tactile receptors have a heightened ability to "feel." These areas correspond to areas that receive the greatest motor innervation; thus they are also typically areas of fine motor control.

On the basis of this information, which areas of the body do you *predict* will have the greatest density of touch receptors?

____

____

Using a caliper or esthesiometer and a metric ruler, test the ability of the subject to differentiate two distinct sensations when the skin is touched simultaneously at two points. Beginning with the face, start with the caliper arms completely together. Gradually increase the distance between the arms, testing the subject's skin after each adjustment. Continue with this testing procedure until the subject reports that two points of contact can be felt. This measurement, the smallest distance at which two points of contact can be felt, is the **two-point threshold.** Repeat this procedure on the back and palm of the hand, fingertips, lips, back of the neck, and back of the calf. Record your results in the chart.

Body Area Tested	Two-Point Threshold (millimeters)
Face	
Back of hand	
Palm of hand	
Fingertips	
Lips	
Back of neck	
Back of calf	

Tactile Localization

Tactile localization is the ability to determine which portion of the skin has been touched. The tactile receptor field of the body periphery has a corresponding "touch" field in the brain's somatosensory association area. Some body areas are well represented with touch receptors, which allows tactile stimuli to be localized with great accuracy; but the density of the touch receptors in other body areas allows only a crude discrimination.

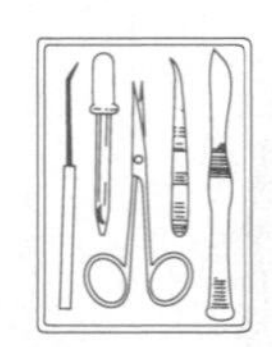

1. The subject's eyes should be closed during the testing. Touch the palm of the subject's hand with a pointed, black felt-tipped marker. The subject should then try to touch the exact point with his or her own marker, which should be of a different color. Measure the error of localization in millimeters. Repeat the test in the same spot two more times, recording the error of localization for each test. Average the results of the three determinations, and record the average in the chart below. Does the ability to localize the stimulus improve the second time?

______________ The third time? ______________ Explain.

____

____

2. Repeat the above procedure on a fingertip; the ventral forearm; the ventral surface of the upper arm; and the upper back, between the shoulder blades. Record the averaged results in the chart below.

Body Area	Average Error of Localization (millimeters)
Palm of hand	
Fingertip	
Ventral forearm	
Ventral surface of upper arm	
Upper back	

Adaptation of Touch Receptors

The number of impulses transmitted by sensory receptors often changes both with the intensity of the stimulus and with the length of time the stimulus is applied. In many cases, when a stimulus is applied for a prolonged period, the rate of receptor discharge slows; and conscious awareness of the stimulus is decreased or lost until some type of stimulus change occurs. This phenomenon is referred to as **adaptation.** The touch receptors adapt particularly rapidly, which is highly desirable. Who, for instance, would want to be continually aware of the pressure of clothing on their skin? The simple experiments to be conducted next allow you to investigate this phenomenon of adaptation.

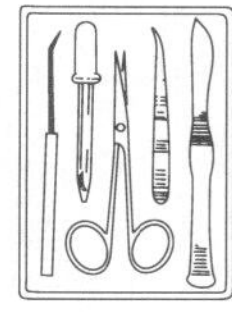

1. The subject's eyes should be closed. Place a coin on the anterior surface of the subject's forearm, and determine how long the sensation persists for the subject. Record the duration of the sensation:

_______________ sec

2. Repeat the test, placing the coin at a different forearm location. How long does the sensation persist at the second location?

_______________ sec

After awareness of the sensation has been lost at the second site, stack three more coins atop the first one. Does the pressure sensation return?

____

If so, for how long is the subject aware of the pressure in this instance?

_______________ sec

Are the same receptors being stimulated when the four coins, rather than the one coin, are used?

____________ Explain. ___________________________

____

3. To further illustrate the adaptation of touch receptors—in this case, the root hair plexuses of the hair follicles—gently and slowly bend one hair shaft with a pen or pencil until it springs back (away from the pencil) to its original position. Is the tactile sensation greater when the hair is being slowly bent or when it springs back?

____

Why is the adaptation of the touch receptors in the hair follicles particularly important to a woman who wears her hair in a ponytail? If the answer is not immediately apparent, consider the opposite phenomenon: What would happen, in terms of sensory input from her hair follicles, if these receptors did not exhibit adaptation?

____

____

____

Adaptation of Temperature Receptors

Adaptation of the temperature receptors can be tested through very unsophisticated methods.

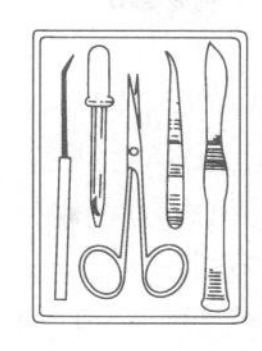

1. Obtain three large finger bowls or 1000-ml beakers and fill the first with 45°C water. Have the subject immerse her or his left hand in the water and report the sensation. Keep the left hand immersed for 1 min and then also immerse the right hand in the same bowl.

What is the sensation of the left hand when it is first immersed?

____

____

What is the sensation of the left hand after 1 min as compared to the sensation in the right hand just immersed?

____

Had adaptation occurred in the left hand? ____________

2. Rinse both hands in tap water, dry them, and wait 5 min before conducting the next test. Just before beginning the test, refill the finger bowl with fresh 45°C water, fill a second with ice water, and fill a third with water at room temperature.

3. Place the left hand in the ice water and the right hand in the 45°C water. What is the sensation in each hand after 2 min as compared to the sensation perceived when the hands were first immersed?

____

____

Which hand seemed to adapt more quickly?

____

4. After reporting these observations, the subject should then place both hands simultaneously into the finger bowl containing the water at room temperature. Record the sensation in the left hand:

____

The right hand: ___________________________________

The sensations that the subject experienced when both hands were put into room-temperature water are called **negative afterimages.** They are explained by the fact that sensations of heat and cold depend on the rapidity of heat loss or gain by the skin and differences in the temperature gradient.

Referred Pain

Experiments on pain receptor localization and adaptation are commonly conducted in the laboratory. However, there are certain problems in conducting such experiments. The pain receptors are densely distributed in the skin, and they adapt very little, if at all. (This lack of adaptability is due to the protective function of the receptors. The sensation of pain often indicates tissue damage or trauma to body structures.) Thus no attempt will be made in this exercise to localize the pain receptors or to prove their nonadaptability, since both would cause needless discomfort to those of you acting as subjects and would not add any additional insight.

However, the phenomenon of referred pain is easily demonstrated in the laboratory, and such experiments provide information that may be useful in explaining common examples of this phenomenon. **Referred pain** is a sensory experience in which pain is perceived as arising in one area of the body when in fact another, often quite remote, area is receiving the painful stimulus. Thus the pain is said to be "referred" to a different area. The phenomenon of **projection,** the process by which the brain refers sensations to their *usual* point of stimulation, provides the most simple explanation of such experiences. Many of us have experienced referred pain as a radiating pain in the forehead after quickly swallowing an ice-cold drink.

Referred pain is important in many types of clinical diagnosis; for example, inadequate oxygenation of the heart muscle often results in pain being referred to the chest wall and shoulder (*angina pectoris*), and reflux of gastric juice into the esophagus causes a sensation of intense discomfort in the thorax known as *heartburn.* In addition, amputees often report *phantom limb pain*—feelings of pain that appear to be coming from a part of the body that is no longer there.

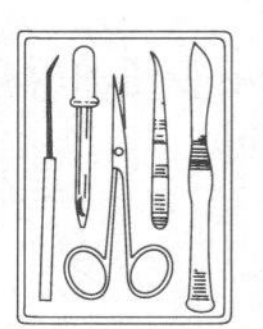

Immerse the subject's elbow in a finger bowl containing ice water. Record the quality (such as discomfort, tingling, or pain) and the quality progression of the sensation for 2 minutes. Also record the location of the perceived sensations.

The ulnar nerve, which serves the medial third of the hand, is involved in the phenomenon of referred pain experienced during this test. How does the localization of this referred pain correspond to the areas served by the ulnar nerve?

____

____

____

____

Time of Observation	Quality of Sensation	Localization of Sensation
On immersion		
After 1 min		
After 2 min		

EXERCISE 20

Human Reflex Physiology

OBJECTIVES

1. To define *reflex.*
2. To name, identify, and describe the function of each element of a reflex arc.
3. To state why reflex testing is an important part of every physical examination.
4. To describe and discuss several types of reflex activities as observed in the laboratory, to note the functional or clinical importance of each, and to categorize each as a somatic or autonomic reflex action.
5. To explain why cord-mediated reflexes are generally much faster than those involving input from the higher brain centers.

MATERIALS

Reflex hammer
Sharp pencils
Cot (if available)
Absorbent cotton
Tongue depressor
Metric and 12-in. ruler
Flashlight
100- or 250-ml beaker
10- or 25-ml graduated cylinder
Lemon juice
Wide-range pH paper
Large bucket with freshly prepared 10% bleach solution
Disposable autoclave bag
Wash bottle containing 10% bleach solution

THE REFLEX ARC

Reflexes are rapid, predictable, involuntary motor responses to stimuli; they are mediated over neural pathways called **reflex arcs.**

There are five essential components of all reflex arcs (Figure 20.1a):

1. The *receptor,* that reacts to a stimulus.
2. The *sensory neuron,* which conducts the afferent impulses to the CNS.
3. The *integration center,* which consists of one to several synapses in the CNS.
4. The *motor neuron,* which conducts the efferent impulses from the integration center to an effector organ.
5. The *effector,* the muscle fibers or glands that respond to the efferent impulses by contracting or secreting a product (respectively).

The simple patellar, or knee-jerk, reflex shown in Figure 20.1b is an example of a simple, two-neuron, monosynaptic reflex arc; and it will be demonstrated in the laboratory. However, most reflexes are more complex, involving the participation of one or more association neurons in the reflex arc pathway. A three-neuron reflex arc (flexor reflex) is diagramed in Figure 20.1c. Because delay or inhibition of the reflex may occur at the synapses, the larger the number of synapses encountered in a reflex pathway, the greater the time required to effect the reflex.

Reflexes of many types may be considered programed into the neural anatomy. For example, many *spinal reflexes,* reflexes that are initiated and completed at the spinal cord level such as the flexor reflex, occur without the involvement of higher brain centers. These reflexes work equally well in decerebrate animals (those in which the brain has been destroyed), as long as the spinal cord is functional. Conversely, other reflexes require the involvement of functional brain tissue, because many different inputs must be evaluated before the appropriate reflex is determined. The superficial cord reflexes and pupillary responses to light are in this category. In addition, although many spinal reflexes do not require the involvement of higher centers, the brain is frequently "advised" of spinal cord reflex activity, and may alter it by facilitating or inhibiting the reflexes.

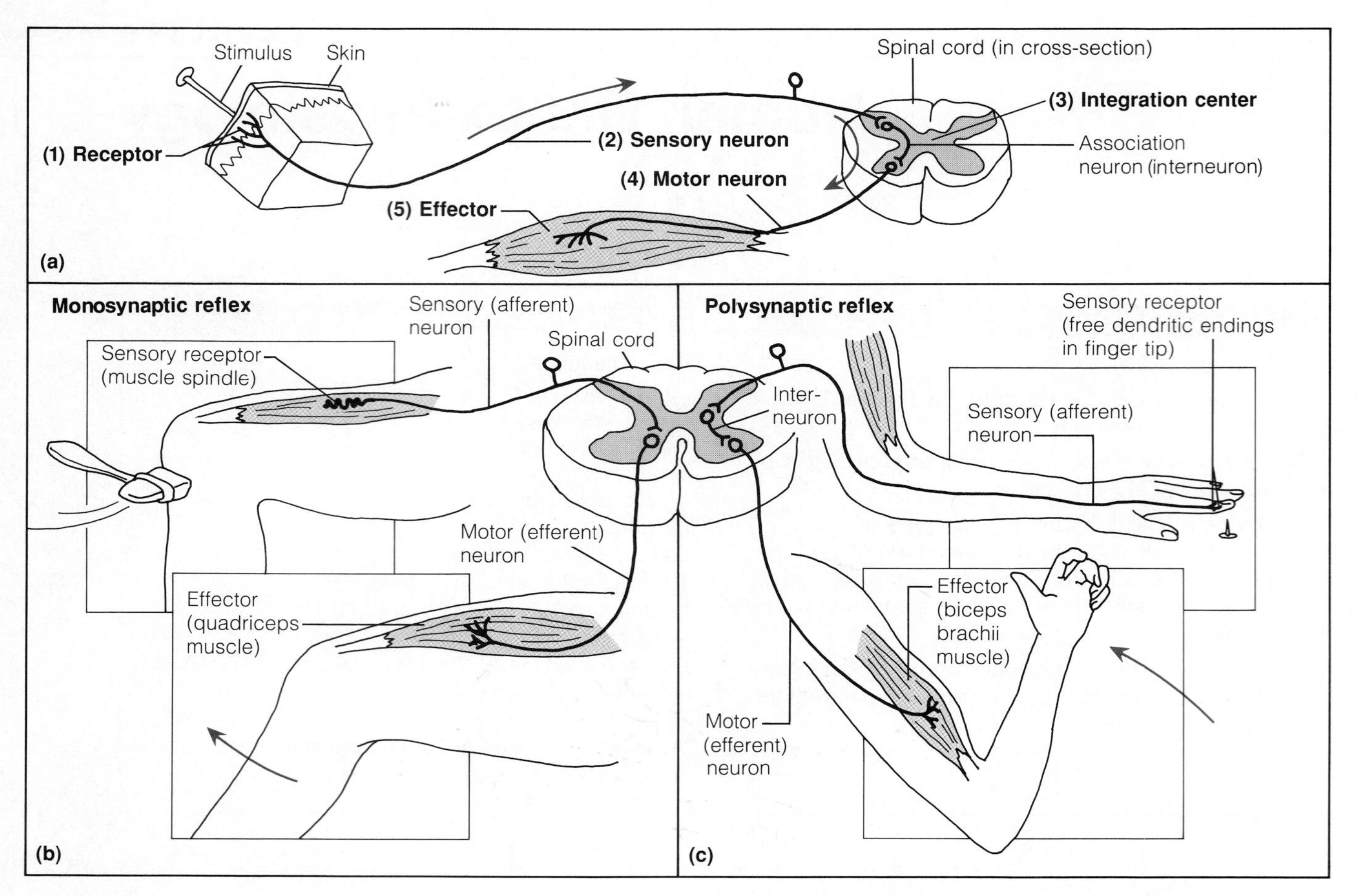

F20.1

Simple reflex arcs. (a) Components of all human reflex arcs: receptor, sensory neuron, integration center (one or more synapses in the CNS), motor neuron, and effector. (b) Monosynaptic reflex arc. (c) Polysynaptic reflex arc. The integration center is the spinal cord, and in each example the receptor and effector are in the same limb.

Reflex testing is an important diagnostic tool to assess the condition of the nervous system. Exaggerated, distorted, or absent reflex responses may indicate degeneration or pathology of portions of the nervous system, often before other signs are apparent.

If the spinal cord is damaged, the easily performed reflex tests can help to pinpoint the area (level) of spinal cord injury. Motor nerves above the injured area may be unaffected, whereas those at, or below, the lesion site may be unable to participate in normal reflex activity. ■

Reflexes can be categorized into one of two large groups: the somatic reflexes and the autonomic reflexes. **Autonomic** (or visceral) **reflexes** are mediated through the autonomic nervous system and are not subject to conscious control. These reflexes result in the activation of smooth muscles, cardiac muscle, and the glands of the body; they involve the regulation of such body functions as digestion, elimination, blood pressure, salivation, and sweating. **Somatic reflexes** include all those reflexes that involve the stimulation of skeletal muscles by the somatic division of the nervous system. An example of such a reflex is the rapid withdrawal of a hand from a hot object.

SOMATIC REFLEXES

There are several types of somatic reflexes, including the stretch, crossed extensor, superficial cord, corneal, and gag reflexes. Some (spinal reflexes) require only spinal cord activity, whereas others also require brain involvement.

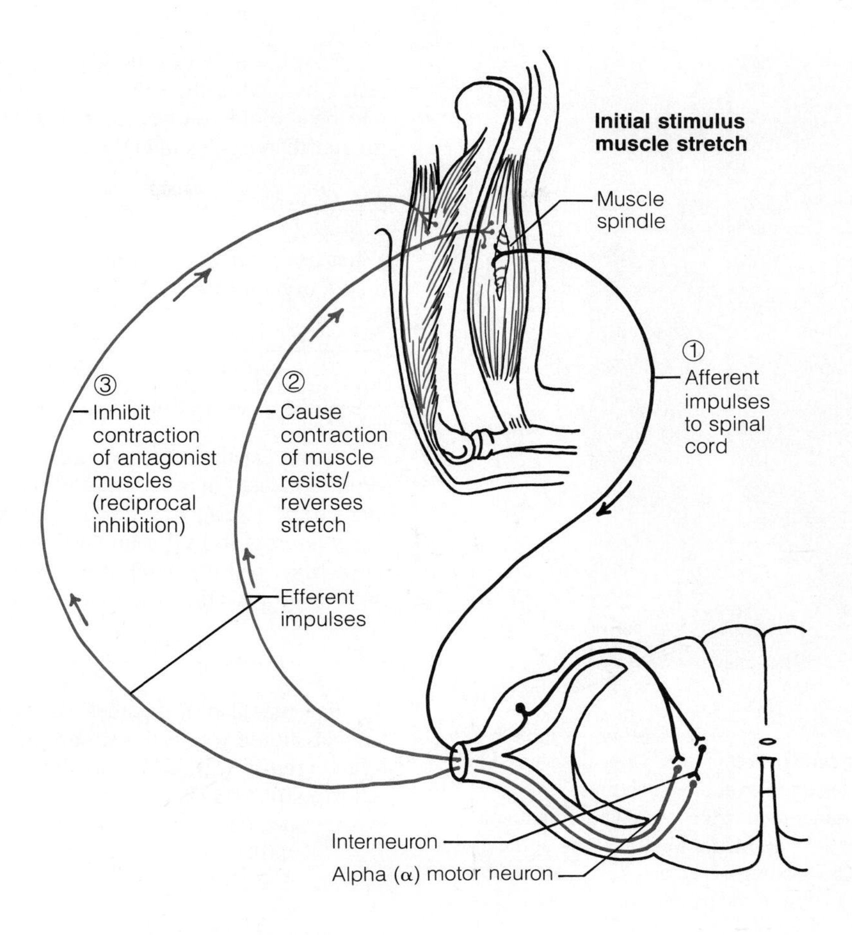

F20.2

Events of the stretch reflex by which muscle stretch is damped. The events are shown in circular fashion. (1) Stretching of the muscle activates a muscle spindle. (2) Impulses transmitted by afferent fibers from the muscle spindle to alpha motor neurons in the spinal cord result in activation of the stretched muscle, causing it to contract. (3) Impulses transmitted by afferent fibers from muscle spindle to interneurons in the spinal cord result in reciprocal inhibition of the antagonist muscle.

Spinal Reflexes

STRETCH REFLEXES **Stretch reflexes** are important postural reflexes, normally acting to maintain posture, balance, and locomotion. Stretch reflexes are initiated by tapping a tendon, which stretches the muscle to which the tendon is attached. This stimulates the muscle spindles and results in reflex contraction of the stretched muscle or muscles, which resist further stretching. Even as the primary stretch reflex is occurring, impulses are being relayed to other destinations as well. For example, branches of the afferent fibers (from the muscle spindles) also synapse with interneurons (association neurons) controlling the antagonist muscles (Figure 20.2). The inhibition of the antagonistic muscles that ensues, called *reciprocal inhibition,* causes the antagonists to relax and prevents them from resisting (or reversing) the contraction of the stretched muscle caused by the main reflex arc. Additionally, impulses are relayed to higher brain centers (largely via the dorsal white columns) to advise of muscle length, speed of shortening, and the like—information needed to maintain muscle tone and posture. Stretch reflexes tend to be hypoactive or absent in cases of peripheral nerve damage or ventral horn disease, and hyperactive in corticospinal tract lesions. They are absent in deep sedation and coma.

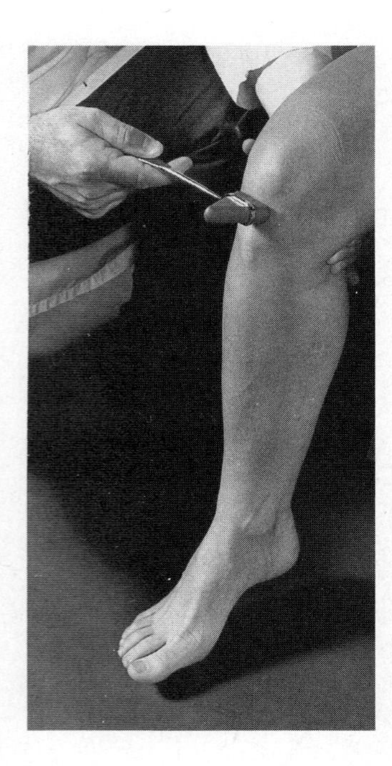

F20.3

Testing the patellar reflex. The subject's muscles should be relaxed (supporting the knee promotes such muscle relaxation). The examiner then strikes the patellar ligament position indicated by the side-facing straight arrow at the base of the patella. The location of the patellar ligament may be ascertained by palpation of the patella curved arrow.

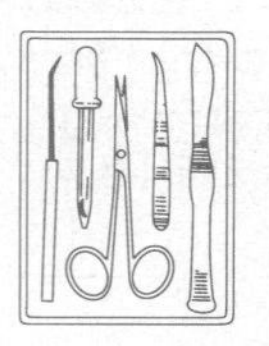

1. Test the **patellar,** or knee-jerk, **reflex** by seating a subject on the laboratory bench, with legs hanging free (or with knees crossed). With the reflex hammer, tap the patellar ligament *sharply* just below the knee to elicit the knee-jerk response (Figure 20.3). Test both knees and record your observations.

____

____

____

____

Which muscles contracted? ____________________

____

What nerve is carrying the afferent and efferent impulses?

____

What muscles are undergoing reciprocal inhibition?

____

2. Test the effect of mental distraction on the patellar reflex by having the subject add a column of three-digit numbers while you test the reflex again. Is the response greater than or less than the first response?

____

What are your conclusions about the effect of mental distraction on reflex activity?

____

____

3. Now test the effect of muscular activity occurring simultaneously in other areas of the body. Have the subject clasp the edge of the laboratory bench and vigorously attempt to pull it upward with both hands. At the same time, test the patellar reflex. Is the response more or less vigorous than the first response?

____

4. Fatigue also influences the reflex response. The subject should jog in position until she or he is very fatigued (**really fatigued**—no slackers). Test the patellar reflex again, and record the results (that is, is it more or less vigorous?).

____

Would you say that nervous system activity or muscle function is responsible for the changes you have just observed?

____

Explain your reasoning. ______________________

____

____

____

____

5. To demonstrate the **Achilles,** or ankle-jerk, **reflex,** your partner should kneel on a chair with his or her feet dangling (relaxed) over the seat edge. Sharply tap his or her Achilles (calcaneal) tendon with the reflex hammer. What is the result?

____

____

Does the contraction of the gastrocnemius normally result in the activity you have observed?

____

CROSSED EXTENSOR REFLEX The **crossed extensor reflex** is more complex than the stretch reflex, consisting of a flexor, or withdrawal, reflex followed by extension of the opposite limb.

This reflex is quite obvious when, for example, a stranger suddenly and strongly grips one's arm. The immediate response is to withdraw the clutched arm and push the intruder away with the other arm. The reflex is more difficult to demonstrate in a laboratory because it is anticipated, and under these conditions the extensor part of the reflex may be inhibited.

The subject should be seated with eyes closed and with the dorsum of one hand resting on the laboratory bench. Obtain a sharp pencil and suddenly prick the subject's index finger. What are the results?

____

____

____

Did the extensor part of this reflex seem slow compared to the other reflexes you have observed?

____

What are the reasons for this? ______________

____

____

The reflexes that have been demonstrated so far—the stretch and crossed extensor reflexes—are examples of reflexes in which the reflex pathway is initiated and completed at the cord level.

Superficial Cord Reflexes

The **superficial cord reflexes** (abdominal, cremaster, and plantar reflexes) result from pain and temperature changes. They are initiated by stimulation of receptors in the skin and mucosae. The superficial cord reflexes depend *both* on functional upper motor pathways and on the cord-level reflex arc.

ABDOMINAL REFLEX To test the **abdominal reflex,** have the subject lie in the supine position on a cot or on the laboratory bench. Since the skin of the abdominal wall will have to be exposed, the subject should wear a blouse or shirt that can be pulled up and aside to expose the umbilical area. Gently stroke the skin on one side of the abdomen; stroke slightly above the umbilicus, and toward the umbilicus, and record the response.

____

____

Then stroke the skin on the same side, but below and to the side of the umbilicus. What is the result?

____

In each case, the skin of the umbilicus should have moved toward the point of stimulation. In cases of upper-motor-neuron lesions involving the pyramidal or corticospinal tract, the reflex is absent.

How does the speed of this superficial type of reflex compare with that of the stretch reflex investigated earlier?

____

PLANTAR REFLEX The **plantar reflex,** an important neurological test, is elicited by stimulating the cutaneous receptors in the sole of the foot. In adults, stimulation of these receptors causes the toes to flex and move closer together; however, damage to the pyramidal (or corticospinal) tract produces *Babinski's sign,* in which the toes flare and the great toe moves in an upward direction. In newborn infants, Babinski's sign is seen and is due to incomplete myelination of the nervous system. Babinski's sign in adults is an abnormal finding.

Have the subject remove a shoe and lie on the cot or laboratory bench with knees slightly bent and thighs rotated so that the lateral side of the foot is resting on the cot. Alternatively, the subject may sit up and rest the lateral surface of the foot on a chair. Draw the handle of the reflex hammer firmly down the lateral side of the exposed sole from the heel to the base of the toes. What is the response?

____

Is this a normal plantar reflex or a Babinski response?

____

Cranial Nerve Reflex Tests

In these experiments, you will be working with your lab partner to illustrate two somatic reflexes mediated by cranial nerves.

CORNEAL REFLEX The **corneal reflex** is mediated through the trigeminal nerve (cranial nerve V). The absence of this reflex is an ominous sign, as it often indicates brain stem damage resulting from compression of the brain or other trauma.

Stand to one side of the subject; the subject should look away from you, toward the opposite wall. Wait a few seconds, and then quickly, *but gently,* touch the subject's cornea (on the side toward you) with a wisp of absorbent cotton. What is the reaction?

____

What is the function of this reflex?

____

Was the sensation experienced that of touch *or* of pain?

____

Why? _______________________________________

____

____

GAG REFLEX The **gag reflex** tests the somatic motor responses of cranial nerves IX and X. When the oral mucosa on the side of the uvula is stroked, each side of the mucosa should rise, and the amount of elevation should be equal.*

For this experiment, select a subject who does not have a "queasy" stomach, because regurgitation is a possibility. Stroke the subject's oral mucosa on each side of the uvula with a tongue depressor. What are the results?

____

____

Discard the used tongue depressor in the disposable autoclave bag before continuing. Do *not* lay it on the laboratory bench.

AUTONOMIC REFLEXES

The autonomic reflexes include the pupillary, ciliospinal, and salivary reflexes, as well as a multitude of other reflexes that are difficult to observe in a laboratory situation. Work with your lab partner to demonstrate the four automic reflexes described next.

Pupillary Reflexes

There are several types of pupillary reflexes. The **pupillary light reflex** and the **consensual reflex** will be examined here. In both of these pupillary reflexes, the retina of the eye is the receptor, the optic nerve (cranial nerve II) contains the afferent fibers, cranial nerve III is responsible for conducting efferent impulses to the eye, and the smooth muscle of the iris is the effector. Many central nervous system centers are involved in the integration of these responses. Absence of the normal pupillary reflexes is generally a late indication of severe trauma or deterioration of the vital brain stem tissue due to metabolic imbalance.

1. Conduct the reflex testing in an area where the lighting is relatively dim. Before beginning, measure and record the size of the subject's pupils.

Right pupil: ________ mm Left pupil: ________ mm.

* The uvula is the fleshy tab hanging from the roof of the mouth, just above the root of the tongue.

2. Stand to the left of the subject while conducting the testing. The subject should shield his or her right eye by holding one hand vertically between the eyes, along the right side of the nose.

3. Shine a flashlight into the subject's left eye. What is the pupillary response?

____

Measure the size of the left pupil: ____________ mm

4. Observe the right pupil. Has the same type of change (called a *consensual response*) occurred in the right eye?

____

Measure the size of the right pupil: ____________ mm

The consensual response, or any reflex observed on one side of the body when the other side has been stimulated, is called a **contralateral response.** The pupillary light response, or any reflex occurring on the same side stimulated, is referred to as an **ipsilateral response.**

When a contralateral response occurs, what does this indicate about the pathways involved?

____

____

____

Was it the sympathetic *or* the parasympathetic division of the autonomic nervous system that was active during the testing of these reflexes?

____

What is the function of these pupillary responses?

____

____

Ciliospinal Reflex

The **ciliospinal reflex** is another example of reflex activity in which pupillary responses can be observed. This response may initially seem a little bizarre, especially in view of the consensual reflex just demonstrated.

While observing the subject's eyes, gently stroke the skin (or just the hairs) on the left side of the back of the subject's neck, close to the hairline.

What is the reaction of the left pupil? ____________

The reaction of the right pupil? ______________

If you see no reaction, repeat the test using a gentle pinch in the same area.

The response you should have noted—pupillary dilation—is consistent with the pupillary changes occurring when the sympathetic nervous system is stimulated. Such a response may also be elicited in a single pupil when, for any reason, more impulses from the sympathetic nervous system reach it. For example, when the left side of the subject's neck was stimulated, sympathetic impulses to the left iris increased, resulting in the ipsilateral reaction of the left pupil.

On the basis of your observations, would you say that the sympathetic innervation of the two irises is closely integrated?

Why or why not? ______________________________

____

Salivary Reflex

Unlike the other reflexes, in which the effectors were smooth or skeletal muscles, the effectors of the **salivary reflex** are glands. Salivary gland secretion varies in amount according to reflex stimulation.

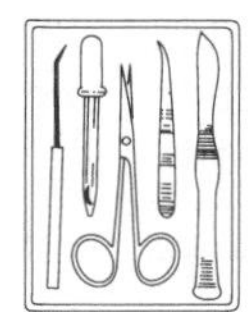

1. Obtain a small beaker, a graduated cylinder, lemon juice, and wide-range pH paper. After refraining from swallowing for 2 minutes, the subject is to expectorate (spit) the accumulated saliva into a small beaker. Using the graduated cylinder, measure the volume of the expectorated saliva and determine its pH.

Volume: ______________ cc; pH: ______________

2. Now place 2 or 3 drops of lemon juice on the subject's tongue. Allow the lemon juice to mix with the saliva for 5 to 10 seconds, and then determine the pH of the subject's saliva by touching a piece of pH paper to the tip of his or her tongue.

pH: ______________________________

As before, the subject is to refrain from swallowing for 2 minutes. After the 2 minutes is up, again collect and measure the volume of the saliva, and determine its pH.

Volume: ______________ cc; pH: ______________

3. How does the volume of saliva collected after the application of the lemon juice compare with the volume of the first saliva sample?

____

How does the final saliva pH reading compare to the initial reading?

____

To that obtained 10 seconds after the application of lemon juice?

____

What division of the autonomic nervous system mediates the reflex release of saliva?

____

Dispose of the saliva-containing beakers and the graduated cylinders in the laboratory bucket containing bleach, and put the used pH paper in the disposable autoclave bag. Wash the bench down with 10% bleach solution before continuing.

Other Autonomic Reflexes

Very few autonomic reflexes were investigated in this exercise, because they are difficult to illustrate in a laboratory situation. To rectify this omission, several autonomic reflexes are listed below. Name the organ involved, receptors stimulated, and resulting action involved in each case. (Use an appropriate reference as necessary.)

Micturition (urination):

Organ/receptors: ______________________________

Result: ______________________________

____

Hering-Breuer:

Organ/receptors: ______________________________

Result: ______________________________

____

Defecation:

Organ/receptors: ______________________________

Result: ______________________________

____

Carotid sinus:

Organ/receptors: ______________________________

Result: ______________________________

____

REACTION TIME OF UNLEARNED RESPONSES

The time required for reaction to a stimulus depends on many factors—the sensitivity of the receptors, the velocity of nerve conduction, the number of neurons and synapses involved, and the speed of effector activation, to name just a few. The type of response to be elicited is also important. If the response involves the reflex arc, the synapses are facilitated; and response time will be short. If, on the other hand, the response can be categorized as an unlearned response, then a far larger number of neural pathways and many types of higher intellectual activities—including choice and decision making—will be involved; and the time for response will be considerably lengthened.

There are various ways of testing the reaction time of unlearned responses. The tests range in difficulty from simple to ultrasophisticated. Because the objective here is to demonstrate the major time difference between reflexes and unlearned responses, the simple approach will suffice.

1. Using a reflex hammer, elicit the patellar reflex in your partner. Note the relative reaction time needed for this reflex to occur.

2. Now test the reaction time for unlearned responses. The subject should hold his hand out, with the thumb and index finger extended. Hold a 12-in. ruler so that its end is exactly 1 in. above the subject's outstretched hand. The ruler should be in the vertical position, with the numbers reading from the bottom up. When the ruler is dropped, the subject should be able to grasp it between thumb and index finger as it passes, without having to change position. Have the subject catch the ruler five times, varying the time between each trial. The relative speed of reaction can be determined by reading the number on the ruler at the point of the subject's fingertips. (Thus if the number at the fingertips is 6, the subject was unable to catch the ruler until 7 in. of length had passed through his fingers—6 in. of ruler length, plus 1 in. to account for the initial distance of the ruler above the hand.) Record the number of inches that pass through the subject's fingertips for each trial:

Trial 1: __________ in. Trial 2: __________ in.

Trial 3: __________ in. Trial 4: __________ in.

Trial 5: __________ in.

3. Perform the test again, but this time say a simple word each time you release the ruler. Designate a specific word as a signal for the subject to catch the ruler; on all other words, the subject is to allow the ruler to pass through his fingers. Trials in which the subject erroneously catches the ruler are to be disregarded. Record the distance the ruler travels for five successful trials:

Trial 1: __________ in. Trial 2: __________ in.

Trial 3: __________ in. Trial 4: __________ in.

Trial 5: __________ in.

Did the addition of a specific word to the stimulus increase or decrease the reaction time?

____

4. Perform the testing once again, this time to investigate the subject's reaction to word association. As you drop the ruler, say a word—for example, *hot*. The subject is to respond with a word he associates with the stimulus word—for example, *cold*—catching the ruler as he responds. If he is unable to make the word association, he must allow the ruler to pass through his fingers. Record the distance the ruler travels for five successful trials, as well as the number of times the ruler is not caught by the subject:

Trial 1: __________ in. Trial 2: __________ in.

Trial 3: __________ in. Trial 4: __________ in.

Trial 5: __________ in.

The number of times the subject was unable to catch the ruler:

____

You should have noted quite a large variation in reaction time in this series of trials. Why is this so?

____

____

Special Senses: Vision

OBJECTIVES

1. To identify the structural components of the eye when provided with a model, an appropriate diagram, or a preserved sheep or cow eye, and to list the function(s) of each.
2. To describe the cellular makeup of the retina.
3. To discuss the mechanism of image formation on the retina.
4. To define the following terms:

refraction	*myopia*
accommodation	*hyperopia*
convergence	*cataract*
astigmatism	*glaucoma*
emmetropia	*conjunctivitis*

5. To explain the difference between the rods and cones with respect to visual perception and retinal localization.
6. To discuss the importance of the pupillary and convergence reflexes.
7. To state the importance of an ophthalmoscopic examination.

MATERIALS

Dissectible eye model
Chart of eye anatomy
Preserved cow or sheep eye
Dissecting pan and instruments
Protective skin cream or disposable gloves

Demonstration: microscope focused on histological section of the eye, showing retinal layers

Snellen eye chart (floor marked with chalk to indicate 20-ft. distance from posted Snellen chart)
Ishihara's color-blindness plates
Metric ruler; meter stick
Common straight pins
Test tubes
Laboratory lamp or penlight
Ophthalmoscope

ANATOMY OF THE EYE

External Anatomy and Accessory Structures

The adult human eye is a sphere measuring about 1 inch (2.5 cm) in diameter. Only about one-sixth of the eye's anterior surface is observable; the remainder is enclosed by a cushion of fat and the walls of the bony orbit.

Six **external,** or **extrinsic, eye muscles** attached to the exterior surface of each eyeball control eye movement and make it possible for the eye to follow a moving object. The names, positioning, and actions of these extrinsic muscles are noted in Figure 21.1.

The anterior surface of each eye is protected by the **eyelids,** or **palpebrae.** (See Figure 21.2.) The medial and lateral junctions of the upper and lower eyelids are referred to as the *medial* and *lateral canthus,* respectively. A mucous membrane, the **conjunctiva,** lines the internal surface of the eyelids (as the *palpebral conjunctiva*) and continues over the anterior surface of the eyeball to its junction with the corneal epithelium (as the *ocular,* or *bulbar, conjunctiva*). The conjunctiva secretes mucus, which aids in lubricating the eyeball. Inflammation of the conjunctiva, often accompanied by redness of the eye, is called **conjunctivitis.**

Projecting from the border of each eyelid is a row of short hairs, the **eyelashes.** The **ciliary glands,** a type of sweat gland, lie between the eyelash hair follicles and help lubricate the eyeball. An inflammation of one of these glands is called a **sty.** Small sebaceous glands associated with the hair follicles and the larger **meibomian glands,** located posterior to the eyelashes, secrete an oily substance.

The **lacrimal apparatus** consists of the **lacrimal gland, lacrimal canals, lacrimal sac,** and **nasolacrimal duct.** The lacrimal glands are situated superior to the lateral aspect of each eye. They continually liberate a dilute salt solution (tears) that flows onto the anterior surface of the eyeball through several small ducts. The tears flush across the eyeball into the lacrimal canals medially, then into the lacrimal sac, and finally into the nasolacrimal duct, which empties into the nasal cavity. The lacrimal secretion also contains **lysozyme,** an antibacterial enzyme. Because it constantly flushes the eyeball, the lacrimal fluid cleanses and protects the eye surface as it moistens and lubricates it. As we age, our eyes tend to become dry due to decreased lacrimation, and

Name	Innervation (cranial nerve)	Action
Lateral rectus	VI	Moves eye horizontally (laterally)
Medial rectus	III	Moves eye horizontally (medially)
Superior rectus	III	Elevates eye
Inferior oblique	III	Elevates eye and turns it laterally
Inferior rectus	III	Depresses eye
Superior oblique	IV	Depresses eye and turns it laterally

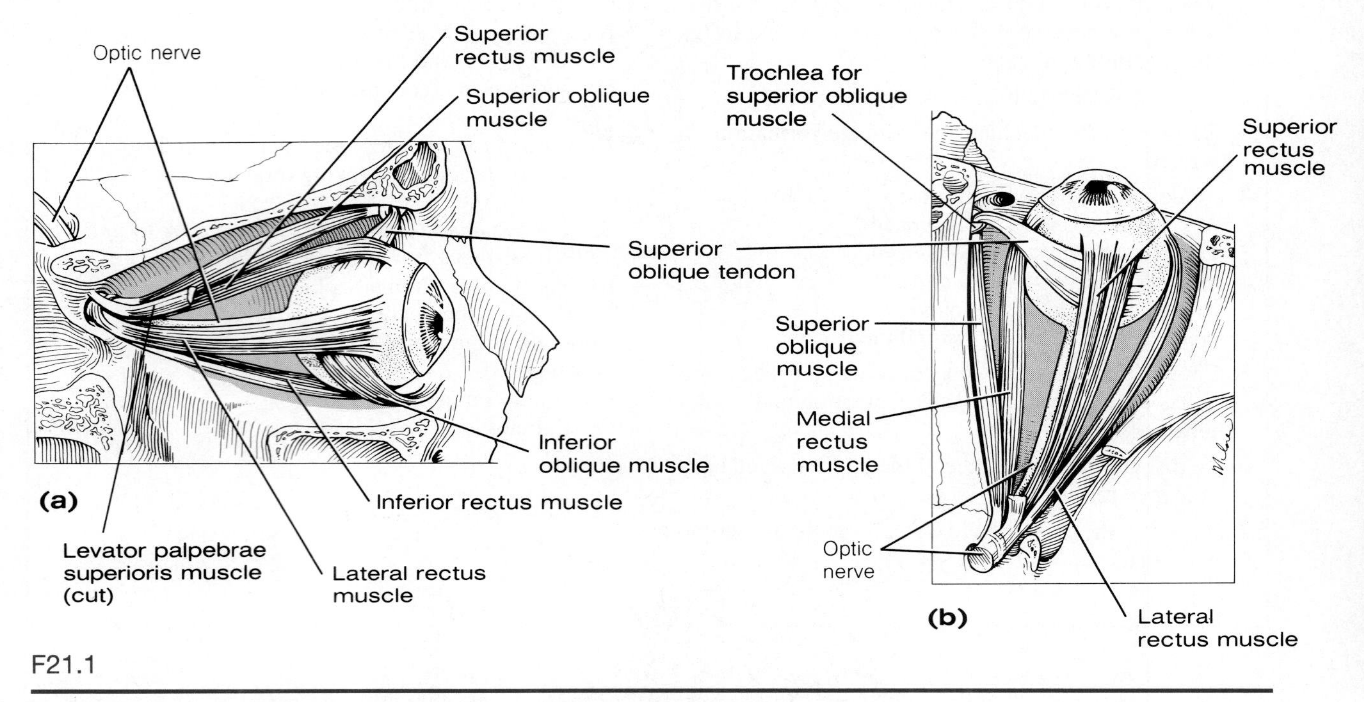

F21.1

Extrinsic muscles of the eye. (a) Lateral view of right eye; (b) superior view of right eye.

thus are more vulnerable to bacterial invasion and irritation.

Observe the eyes of another student, and identify as many of the accessory structures as possible. Ask the student to look to the left. What extrinsic eye muscles are responsible for this action?

Right eye ____________________

Left eye ____________________

Internal Anatomy of the Eye

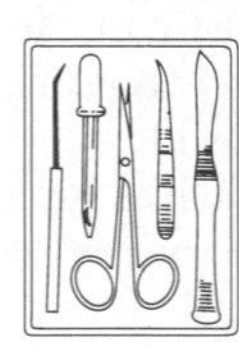

Obtain a dissectible eye model, and identify its internal structures as they are described below. As you work, also refer to Figure 21.3.

Anatomically, the wall of the eye is constructed of three tunics, or coats. The outermost **fibrous tunic** is a protective layer composed of dense avascular connective tissue. It has two obviously different regions: The opaque white **sclera** forms the bulk of the fibrous tunic and is observable anteriorly as the "white of the eye." Its anteriormost portion is modified structurally to form the transparent **cornea,** through which light enters the eye.

The middle tunic, called the **uvea,** is the **vascular tunic.** Its posteriormost part, the **choroid,** is a richly vascular nutritive layer that contains a dark pigment that prevents light scattering within the eye. Anteriorly, the choroid is modified to form the **ciliary body,** to which the lens is attached, and then the pigmented **iris.** The iris is incomplete, resulting in a rounded opening, the **pupil,** through which light passes.

The iris is composed of circularly and radially arranged smooth muscle fibers, and acts as a reflexively activated diaphragm in regulating the amount of light

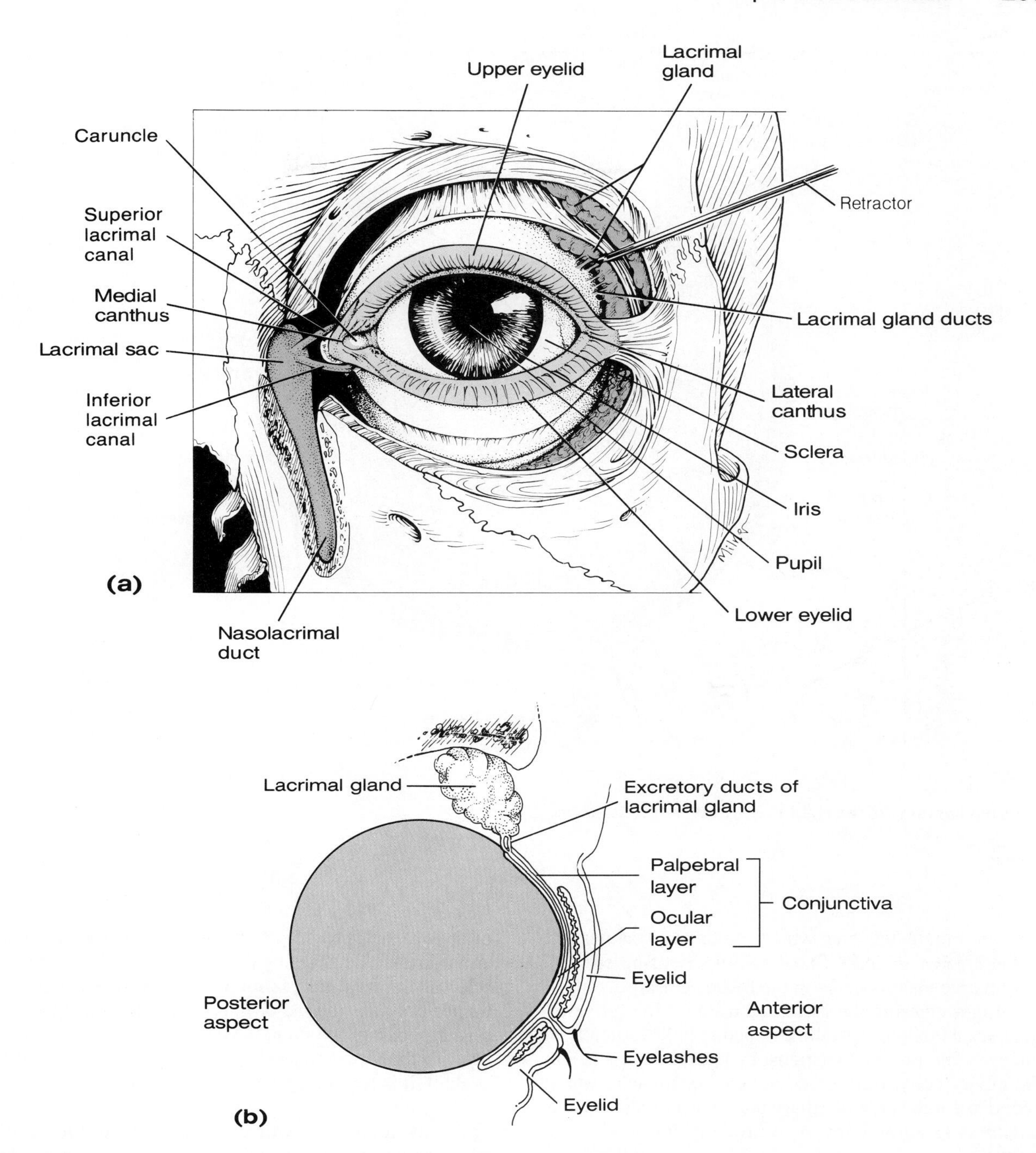

F21.2

External anatomy of the eye and accessory structures. (a) Anterior view; (b) sagittal section.

entering the eye. In close vision and in bright light, the circular muscles of the iris contract, and the pupil constricts. In distant vision and in dim light, the radial fibers contract, enlarging (dilating) the pupil, and allowing more light to enter the eye.

The innermost **sensory tunic** of the eye is the delicate two-layered **retina.** The outer **pigmented layer** abuts and lines the entire uvea. The transparent inner **neural (nervous) layer** extends anteriorly only to the ciliary body. It contains the photoreceptor cells, the **rods** and **cones,** which begin the chain of electrical events that ultimately result in the transduction of light energy into nerve impulses that are transmitted to the optic cortex of the brain. Vision is the result. The photoreceptor cells are distributed over the entire neural retina, except where the optic nerve leaves the eyeball. This site is called the **optic disc,** or blind spot. Lateral to each blind spot, and directly posterior to the lens, is an area called the **macula lutea** (yellow spot), an area of high cone density. In its center is the **fovea centralis,** a minute pit

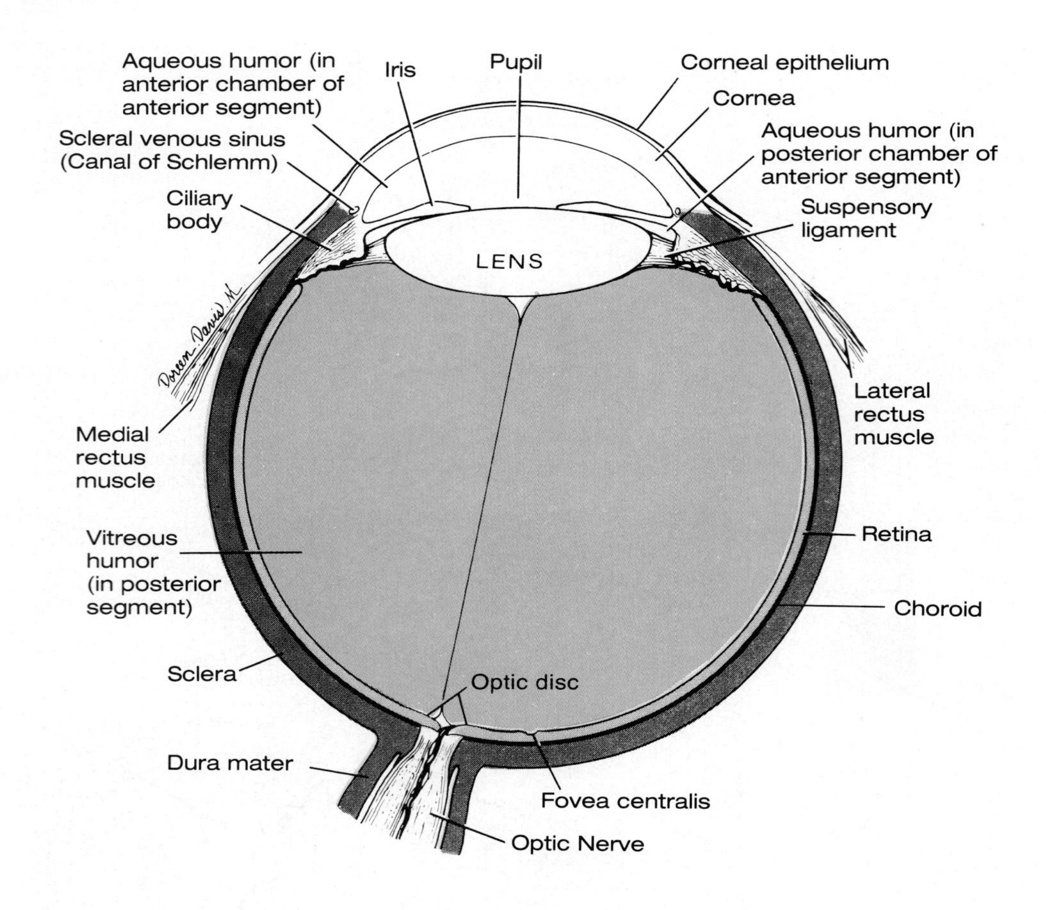

F21.3

Internal anatomy of the right eye (transverse section).

about ½ mm in diameter, which contains only cones and is the area of greatest visual acuity. Focusing for discriminative vision occurs in the fovea centralis.

Light entering the eye is focused on the retina by the **lens,** a flexible crystalline structure held vertically in the eye's interior by a **suspensory ligament** attached to the ciliary body. Activity of the **ciliary muscle,** which accounts for the bulk of ciliary body tissue, changes lens thickness to allow light to be properly focused on the retina.

In the elderly the lens becomes increasingly hard and opaque. *Cataracts,* which are often the end result of this process, cause vision to become hazy or entirely obstructed. ■

The lens divides the eye into two segments: the **anterior segment** anterior to the lens, which contains a clear watery fluid called the **aqueous humor;** and the **posterior segment** posterior to the lens, filled with a gel-like substance, the **vitreous humor,** or **vitreous body.** The anterior segment is further divided into **anterior** and **posterior chambers,** located before and after the iris, respectively. The aqueous humor is continually formed by the capillaries of the **ciliary processes** of the ciliary body. It helps maintain the intraocular pressure of the eye, and provides nutrients for the avascular lens and cornea. The aqueous humor is reabsorbed into the **scleral venous sinus (canal of Schlemm).** The vitreous humor provides the major internal reinforcement of the posterior part of the eyeball, and helps to keep the neural layer of the retina pressed firmly against the wall of the eyeball. It is formed *only* before birth.

Any inference with aqueous fluid drainage increases intraocular pressure. When the intraocular pressure reaches dangerously high levels the retina and optic nerve are compressed, resulting in pain and possible blindness, a condition called *glaucoma.* ■

DISSECTION OF THE COW (SHEEP) EYE

1. Obtain a preserved cow or sheep eye, dissecting instruments, a dissecting pan and, if desired, disposable gloves or protective skin cream.

2. Examine the external surface of the eye, noting the thick cushion of adipose tissue. Identify the optic nerve (cranial nerve II) as it leaves the eyeball,

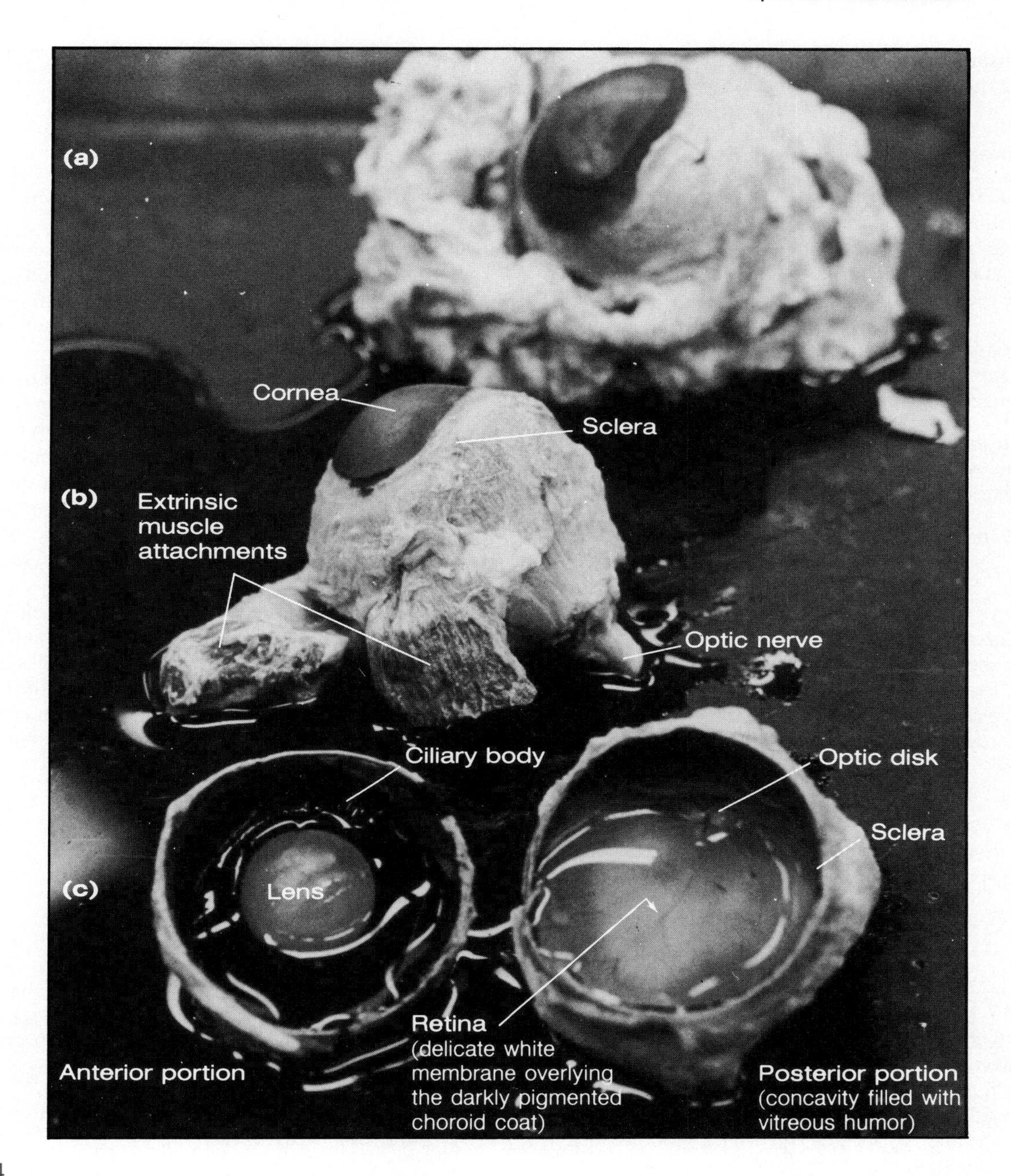

F21.4

Anatomy of the cow eye. (a) Cow eye (entire) removed from orbit (note the large amount of fat cushioning the eyeball); (b) cow eye (entire) with fat removed to show the extrinsic muscle attachments and optic nerve; (c) cow eye cut along the coronal plane to reveal internal structures.

the remnants of the extrinsic eye muscles, the conjunctiva, the sclera, and the cornea. The normally transparent cornea is opalescent or opaque if the eye has been preserved. Refer to Figure 21.4 as you work.

3. Trim away most of the fat and connective tissue, but leave the optic nerve intact. Holding the eye with the cornea facing downward, make an incision with a sharp scalpel into the sclera about ¼ inch above the cornea. Complete the incision around the circumference of the eyeball, parallelling the corneal edge.

4. Carefully lift the anterior part of the eyeball away from the posterior portion. (Conditions being proper, the vitreous body should remain with the posterior part of the eyeball.)

5. Examine the anterior part of the eye and identify the following structures:

Ciliary body: black pigmented body that appears to be a halo encircling the lens.

Lens: biconvex structure that appears opaque in preserved specimens.

Suspensory ligament: a halo of delicate fibers attaching the lens to the ciliary body.

Carefully remove the lens and identify the adjacent structures:

Iris: anterior continuation of the ciliary body, penetrated by the pupil.

Cornea: more convex, anteriormost portion of the sclera; normally transparent, but cloudy in preserved specimens.

6. Examine the posterior portion of the eyeball. Remove the vitreous humor and identify the following structures:

Retina: the neural layer of the retina appears as a delicate white, probably crumpled membrane that separates easily from the pigmented choroid.

Note its point of attachment. What is this point called?

____

Pigmented choroid coat: appears iridescent in the cow or sheep eye owing to a special reflecting surface called the **tapetum lucidum.** This specialized surface reflects the light within the eye and is found in the eyes of animals that live under conditions of low-intensity light. It is not found in humans.

MICROSCOPIC ANATOMY OF THE RETINA (OPTIONAL)

As described above, the retina consists of two main types of cells: a pigmented *epithelial* layer, which abuts the choroid, and an inner cell layer composed of *neurons,* which is in contact with the vitreous humor (Figure 21.5). The inner nervous layer is composed of three major neuronal populations. These are, from outer to inner aspect, the **photoreceptor layer** (rods and cones), the **bipolar cells,** and the **ganglion cell layer.**

The **rods** are the specialized receptors for dim light. Visual interpretation of their activity is in gray tones. The **cones** are color receptors that permit high levels of visual acuity, but they function only under conditions of high light intensity. Only cones are found in the fovea centralis, and their number decreases as the retinal periphery is approached. Conversely, the rods are most dense in the periphery, and their number decreases as the macula is approached.

Light must pass through the ganglion cell neuron layer and the bipolar neuron layer to reach and excite the rods and cones, which then undergo changes in their membrane potential that ultimately influence the bipolar neurons. These in turn stimulate the ganglion cells, the axons of which leave the retina in the tight bundle of fibers known as the optic nerve.

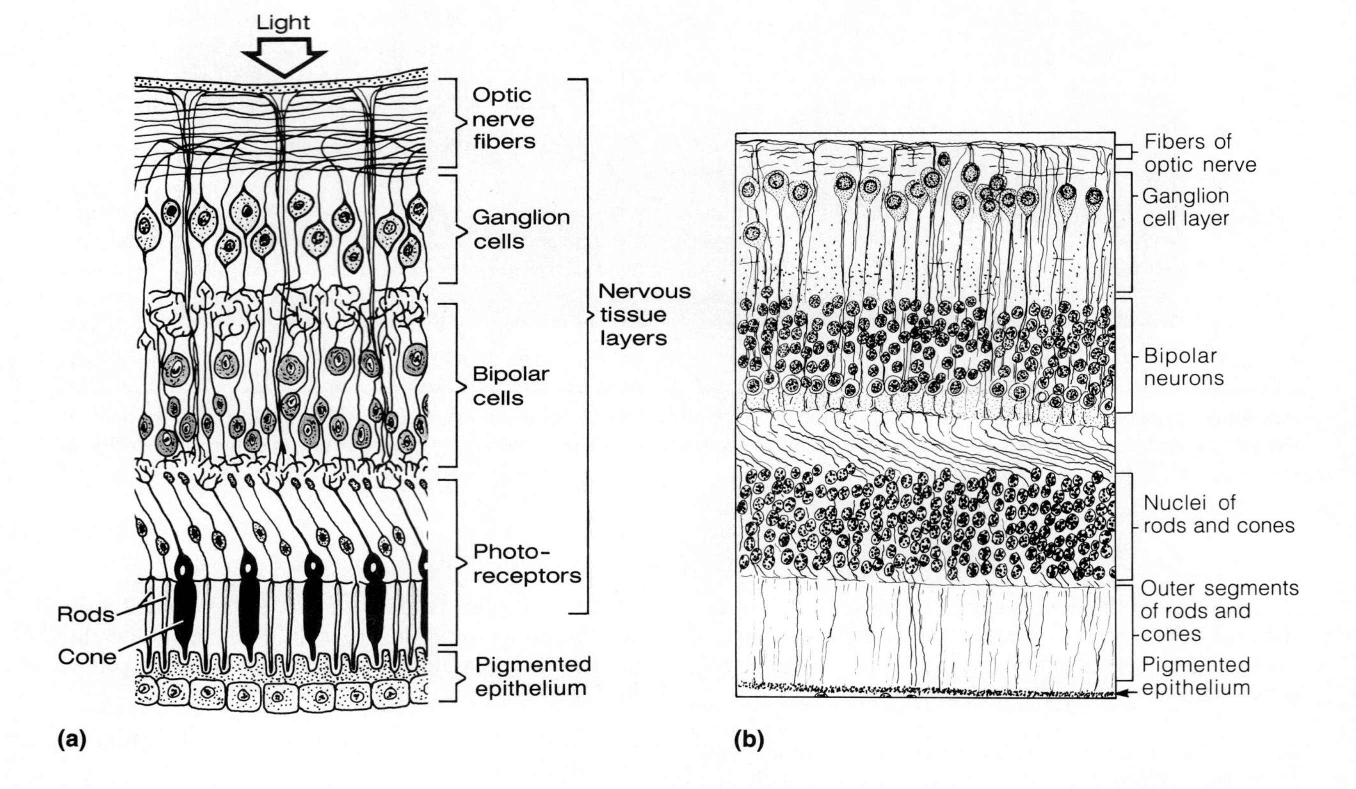

Microscopic anatomy of the cellular layers of the retina. (a) Diagrammatic view; (b) line drawing of the photomicrograph provided in plate 13 of the histology atlas.

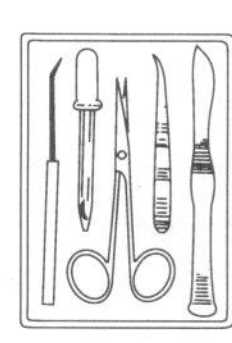

Go to the demonstration area, and examine the histological slide of a longitudinal section of the eye. Identify the retinal layers by comparing it to Figure 21.5.

VISUAL TESTS AND EXPERIMENTS

Demonstration of the Blind Spot

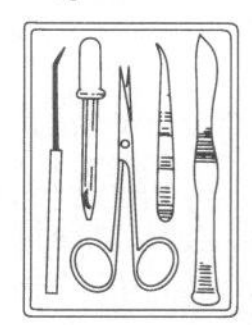

1. Hold Figure 21.6 about 18 inches from your eyes. Close your left eye, and focus your right eye on the X, which should be positioned so that it is directly in line with your right eye. Move the figure slowly toward your face, keeping your right eye focused on the X. When the dot focuses on the blind spot, which lacks photoreceptors, it will disappear.

2. Have your laboratory partner record in metric units the distance at which this occurs. (The dot will reappear as the figure is moved closer.) Distance at which the dot disappears:

Right eye ______________________________

Repeat the test for the left eye, this time closing the right eye and focusing the left eye on the dot. Record the distance at which the X disappears:

Left eye ______________________________

Afterimages

When light from an object strikes **rhodopsin,** the purple pigment contained in the rods of the retina, it triggers a photochemical reaction that causes the rhodopsin to be split into its colorless precursor molecules (vitamin A and a protein called opsin). This event, called *bleaching of the pigment,* initiates a chain of events leading to impulse transmission along fibers of the optic nerve. Once bleaching has occurred in a rod, its photoreceptor pigment must be resynthesized before the rod can be restimulated. This takes a certain period of time. Both phenomena—that is, the stimulation of the photoreceptor cells and their subsequent inactive period—can be demonstrated indirectly in terms of positive and negative afterimages.

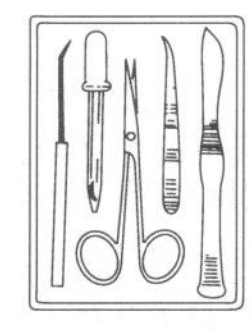

1. Stare at a bright light bulb for a few seconds, and then gently close your eyes for approximately one minute.

2. Record, in the sequence of occurrence, what you "saw" after closing your eyes:

F21.6

Blind-spot test figure.

____

____

____

The bright image of the light bulb initially seen was a **positive afterimage** caused by the continued firing of the rods. The dark image of the light bulb that subsequently appeared against a lighter background was the **negative afterimage,** an indication that the rhodopsin in the affected photoreceptor cells had been bleached.

Refraction, Tests for Visual Acuity, and Astigmatism

When light rays pass from one medium to another, their velocity, or speed of transmission, changes, and the rays are bent or refracted.

Although the cornea is the major refractory apparatus of the eye, its refractive index (bending power) is constant. The refractive index or strength of the lens, on the other hand, can be varied by changing its shape—that is, by making it more or less convex. The greater the lens convexity, or bulge, the more the light will be bent and the stronger the lens. Conversely, the less the lens convexity (the flatter it is), the less it bends the light.

In general, light from a distant source (over 20 feet) approaches the eye as parallel rays, and no change in lens convexity is necessary. However, light from a close source tends to diverge, and the convexity of the lens must increase to make close vision possible. To achieve this, the ciliary muscle contracts, decreasing the tension on the suspensory ligament attached to the lens, allowing the elastic lens to "round up." The ability of the eye to focus differentially for objects of near vision (less than 20 feet) is called **accommodation.** It should be noted that the image formed on the retina is a **real image** (reversed from left to right, inverted, and smaller than the object) (Figure 21.7).

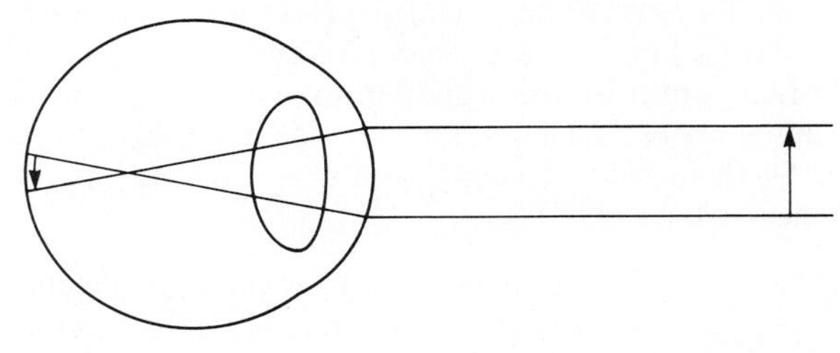

F21.7

Refraction of light by the lens resulting in the production of a real image on the retina.

Thus, a lens capable of bringing a *close* object into a sharp focus is stronger than a lens focusing a more distant object. For example, a lens that brings an object 0.3 m away into sharp focus is stronger (more convex) than one focusing light from an object 1 meter away.

The strength, or refractive power, of a lens is expressed in **diopters.** The distance beyond a lens at which incident light is brought to a focus is referred to as the **focal length** of the lens. The relationship between lens strength and focal length is expressed by the equation

$$\text{Lens strength (diopters)} = \frac{1}{\text{focal length (meters)}}$$

In normal eyes, lens convexity can be adjusted to provide a range of power from 67 diopters (least convex state, for distant vision) to 79 diopters (most convex state, for near vision).

The normal, or **emmetropic,** eye is able to accommodate properly. However, problems of visual acuity may result from lenses that are too strong or too "lazy" (overconverging and underconverging, respectively) or from structural problems such as an eyeball that is too long or too short to provide for proper focusing by the lens, or a cornea or lens with improper curvatures.

Individuals in whom the image normally focuses in front of the retina are said to have **myopia,** or "nearsightedness;" they can see close objects without difficulty, but distant objects are blurred or seen indistinctly. Correction requires a concave lens, which causes the light reaching the eye to diverge. The strength of the corrective lens is given in diopters, preceded by a minus (−) sign.

If the image focuses behind the retina, the individual is said to have **hyperopia,** or "farsightedness." Such persons have no problems with distant vision but need glasses with convex lenses to augment the converging power of the lens for close vision. The lens prescription is given in diopters, preceded by a plus (+) sign.

Irregularities in the curvatures of the lens and/or the cornea lead to a blurred-vision problem called **astigmatism.** Cylindrically ground lenses, which compensate for inequalities in the curvatures of the refracting surfaces, are prescribed to correct the condition. Both the strength of the lens and the axis of the defect are indicated in the prescription.

TEST FOR NEAR-POINT ACCOMMODATION The elasticity of the lens decreases dramatically with age, resulting in difficulty in focusing for near or close vision. This condition is called **presbyopia**—literally, old vision. Lens elasticity can be tested by measuring the **near point of accommodation.** The near point of vision is about 7.5 cm (or 3 inches) at age 10, 20–25 cm (or 8–10 inches) in young adults, and 83 cm (or 33 inches) at the age of 60.

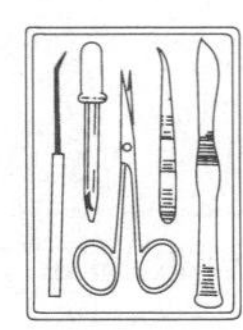

To determine your near point of accommodation, hold a common straight pin at arm's length in front of one eye. Slowly move the pin toward that eye until the pin image becomes distorted. Have your lab partner measure the distance from your eye to the pin at this point, and record the distance below. Repeat the procedure for the other eye.

Near point for right eye ______________

Near point for left eye ______________

TEST FOR VISUAL ACUITY Visual acuity, or sharpness of vision, is generally tested with a Snellen eye chart, which consists of letters of various sizes printed on a white card. This test is based on the fact that letters of a certain size can be seen clearly by eyes with normal vision at a specific distance. The distance at which the normal, or emmetropic, eye can read each line of letters is printed at the end of that line.

1. Have your partner stand 20 feet from the posted Snellen eye chart, with one eye covered by a card or hand. As your partner reads each consecutive line aloud, check for accuracy. (If this individual wears glasses, the test should be taken twice—first with glasses off and then with glasses on.)

2. Record the number of the line with the smallest-sized letters read. If it is 20/20, the person's vision for that eye is normal. If it is 20/40 (or any ratio with a value less than one), his or her vision is less than the normal acuity. (Such an individual is myopic.) If the visual acuity is 20/15, vision is better than normal, because this person can stand 20 feet from the chart and read letters that are only discernible by the normal eye at 15 feet.

3. Have your partner test and record your visual acuity. If you wear glasses, the test results *without* glasses should be recorded first.

Visual acuity right eye ______________

Visual acuity left eye ______________

TEST FOR ASTIGMATISM The astigmatism chart (Figure 21.8) is designed to test for defects in the refracting surface of the lens and/or cornea.

View the chart first with one eye and then with the other, focusing on the center of the chart. If all the radiating lines appear equally dark and distinct, there is no distortion of your refracting surfaces. If some of the lines are blurred or appear less dark than others, at least some degree of astigmatism is present.

Is astigmatism present in your left eye? ______________

Right eye? ______________

Test for Color Blindness

Ishihara's color plates are designed to test for deficiencies in the cones, or color photoreceptor cells. Studies suggest that there are three cone types, each containing a different photoreceptor pigment. One type primarily absorbs the red wavelengths of the visible light spectrum, another the blue wavelengths, and a third the green wavelengths. Nervous impulses reaching the brain from these different photoreceptor types are then interpreted (seen) as red, blue, and green, respectively.

The interpretation of the intermediate colors of the visible light spectrum is a result of overlapping input from more than one cone type.

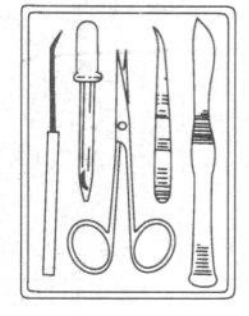

1. View the various color plates in bright light or sunlight while holding them about 30 inches away and at right angles to your line of vision. Report to your laboratory partner what you see in each plate. (Take no more than 3 seconds for your decision.)

2. Your partner is to write down your responses and then check their accuracy with the correct answers given at the front of the color plate book. Is there any indication that you have some degree of color blindness?

If so, what type? ______________________________

Repeat the procedure to test your partner's color vision.

Tests for Binocular Vision

Humans, cats, predatory birds, and most primates are endowed with **binocular,** or two-eyed, vision. Although both eyes look in approximately the same direction, they see slightly different views. Their visual fields, each about 170 degrees, overlap to a considerable extent; thus there is two-eyed vision at the overlap area (Figure 21.9).

In contrast, the eyes of many animals (rabbits, pigeons, and others) are more on the sides of their head. Such animals see in two different directions and thus have a panoramic field of view and **panoramic vision.**

Although both types of vision have their good points, we now know that binocular vision provides three-dimensional vision and an accurate means of locating objects in space. The slight differences between the views seen by the two eyes are fused by the higher centers of the visual cortex to give us *depth perception.* Because of the manner in which the visual cortex resolves these two different views into a single image, it is often referred to as the "cyclopean eye of the binocular animal."

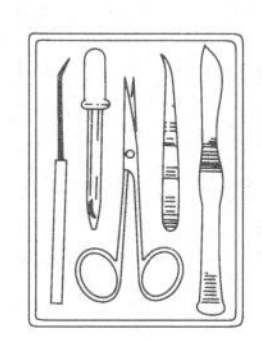

1. To demonstrate that a slightly different view is seen by each eye, perform the following simple experiment: Close your left eye. Hold a pencil at arm's length directly in front of your right eye. Hold another pencil directly beneath it and move it about half the distance toward you. (As you move the lower pencil make sure it remains in the *same plane* as the stationary pencil, so that the two pencils continually form a straight line.) Then, without moving the pencils, close your right eye and open your left eye. Note that with only the right eye open, the moving pencil stays in the same plane as the fixed pencil, but that when viewed with the left eye, the moving pencil is displaced laterally away from the plane of the fixed pencil.

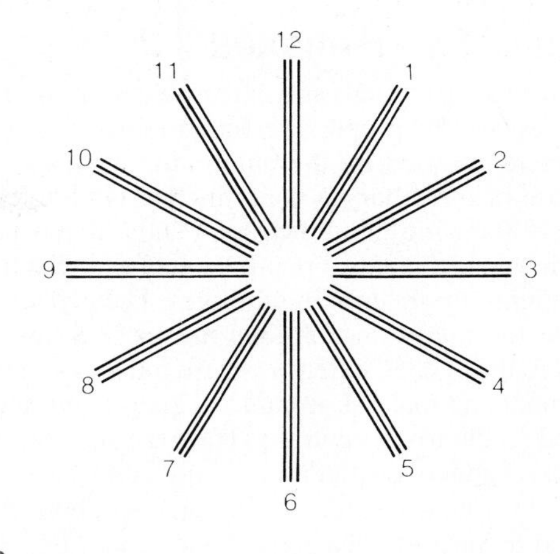

F21.8

Astigmatism testing chart.

2. To demonstrate the importance of two-eyed binocular vision for depth perception, perform this second simple experiment.

Have your laboratory partner hold a test tube erect about an arm's length in front of you. With both eyes open, *quickly* insert a pencil into the test tube. Remove the pencil, bring it back close to your body, close one eye, and quickly and without hesitation reinsert the pencil into the test tube. (Do not "feel for" the test tube with the pencil!) Repeat with the other eye closed.

Was it as easy to dunk the pencil with one eye closed as with both eyes open?

____

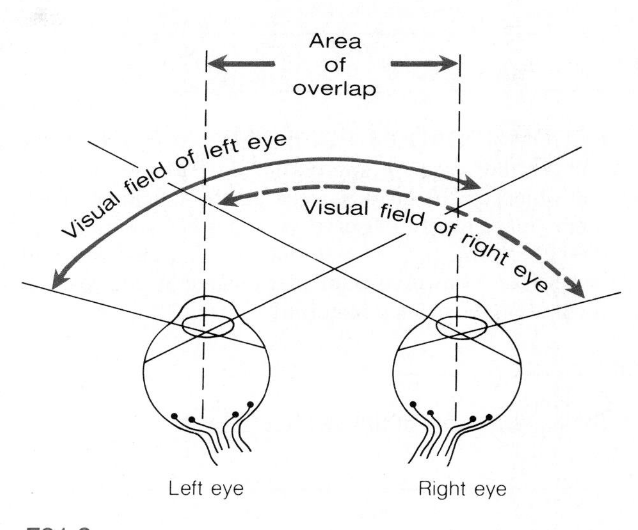

F21.9

Overlapping of the visual fields.

Tests of Eye Reflexes

Both intrinsic (internal) and extrinsic (external) muscles are necessary for proper eye functioning. The *intrinsic muscles,* controlled by the autonomic nervous system, are those of the ciliary body (which alters the lens curvature in focusing), and the radial and circular muscles of the iris (which control pupillary size and thus regulate the amount of light entering the eye). The *extrinsic muscles* are the rectus and oblique muscles, which are attached to the eyeball exterior. These muscles control eye movement and make it possible to keep moving objects focused on the fovea centralis. They are also responsible for **convergence,** or medial eye movements, which are essential for near vision. (In convergence, both eyes are directed toward the near object viewed.) The extrinsic eye muscles are controlled by the somatic nervous system.

Involuntary activity of both of these muscle types is brought about by reflex actions that can be observed in the following experiments.

PHOTOPUPILLARY REFLEX A sudden illumination of the retina by a bright light causes the pupil to contract reflexly in direct proportion to the light intensity. This protective response prevents damage to the delicate photoreceptor cells.

Obtain a laboratory lamp or penlight. Have your laboratory partner sit with eyes closed and hands over eyes. Turn on the light, and position it so that it shines on the subject's right hand. After 1 minute, ask your partner to uncover and open his or her right eye. Quickly observe the pupil of that eye.

What happens to the pupil? ______________________

____

Shut off the light, and ask your partner to uncover and open the left eye. What are your observations?

____

____

____

ACCOMMODATION PUPILLARY REFLEX Have your partner gaze for approximately 1 minute at a distant object in the lab—*not* toward the windows or another light source. Observe your partner's pupils, and then hold some printed material 6 to 10 inches from his or her face. How does pupil size change as your partner focuses on the printed material?

____

Explain the value of this reflex. ___________________

____

____

____

CONVERGENCE REFLEX Repeat the previous experiment, this time using a pen or pencil as the close object to be focused on. Note the position of your partner's eyeballs, both while he or she is gazing at the distant object and while gazing at the close object. Do they change position as the object of focus is changed?

_____ In what way? ___________________________

____

____

Explain the importance of the convergence reflex.

____

____

____

Ophthalmoscopic Examination of the Eye (Optional)

The ophthalmoscope is an instrument used to examine the *fundus,* or eyeball interior, to determine visually the condition of the retina, optic disc, and internal blood vessels. Certain pathologic conditions such as diabetes, arteriosclerosis, and degenerative changes of the optic nerve and retina can be detected by such an examination. The ophthalmoscope consists of a set of lenses mounted on a rotating disc (the **lens selection disc**), a light source regulated by a **rheostat control,** and a mirror that reflects the light so that the eye interior can be illuminated (Figure 21.10).

The lens selection disc is positioned in a small slit in the mirror, and the examiner views the eye interior through this slit, appropriately called the *viewing window.* The focal length of each lens is indicated in diopters preceded by a + sign if the lens is convex and by a − sign if the lens is concave. When the zero (0) is seen in the *diopter window,* there is no lens in position in the slit. The depth of focus for viewing the eye interior is changed by changing the lens.

The light is turned on by depressing the red *rheostat lock* button and then rotating the rheostat control in the clockwise direction. The aperture selection disc on the front of the instrument allows the nature of the light beam to be altered. (Generally, green light allows for clearest viewing of the blood vessels in the eye interior and is most comfortable for the subject.) Now that you are familiar with the ophthalmoscope, you are ready to conduct an eye examination.

1. Conduct the examination in a dimly lit or darkened room, with the subject comfortably seated and gazing straight ahead. To examine the right eye, sit facing the subject, and hold the instrument in your right hand. Use your right eye to view the eye interior. To view the left eye, hold the instrument in your left hand, and use your left eye. When

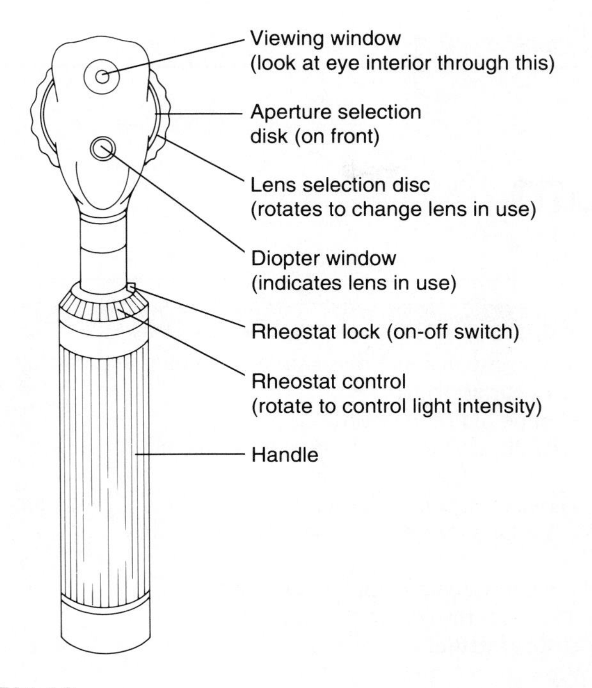

F21.10

Structure of an ophthalmoscope.

the ophthalmoscope is correctly set, the fundus of the right eye should appear as shown in Figure 21.11.

2. Begin the examination with the 0 (no lens) in position. Hold the instrument so that the lens disc may be rotated with the index finger. Hold the ophthalmoscope about 6 inches from the subject's eye, and direct the light into the pupil at a slight angle—through the pupil edge rather than directly through its center. You will see a red circular area that is the illuminated eye interior.

3. Move in as close as possible to the subject's cornea as you continue to observe the area. Steady your instrument-holding hand on the subject's cheek if necessary. If both your eye and that of the subject are normal, the fundus can be viewed clearly without further adjustment of the ophthalmoscope. If the fundus cannot be focused, slowly rotate the lens disc counterclockwise until the fundus is clearly seen. (Note: If a positive [convex] lens is required and your eyes are normal, the subject has hyperopia. If a negative [concave] lens is necessary to view the fundus and your eyes are normal, the subject is myopic.)

When the examination is proceeding correctly, the subject can often see images of retinal vessels in his own eye that appear rather like cracked glass. If you are unable to achieve a sharp focus or to see the optic disc, move medially or laterally and begin again.

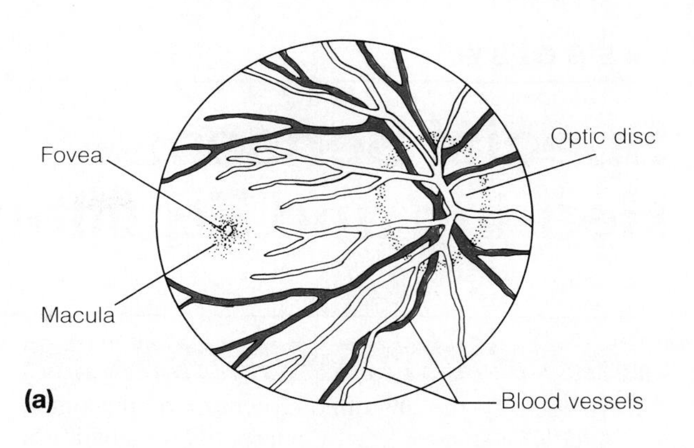

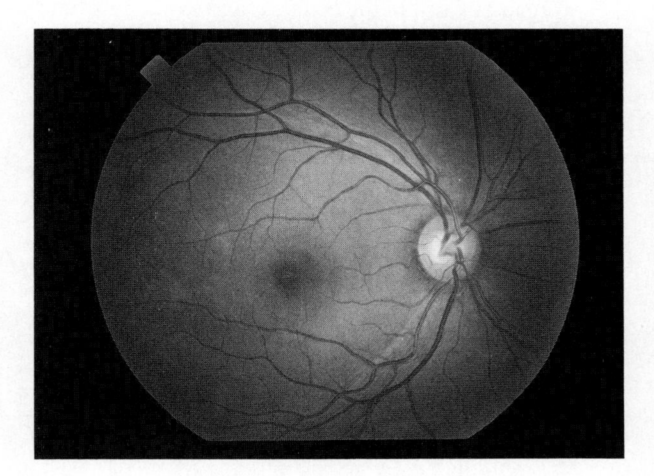

(b)

F21.11

Posterior portion of the right retina. (a) Diagrammatic view; (b) photograph taken with slit-lamp camera.

4. Examine the optic disc for color, elevation, and sharpness of outline, and observe the blood vessels radiating from near its center. Locate the macula, which is lateral to the optic disc. It is a darker area in which blood vessels are absent, and the fovea appears to be a slightly lighter area in its center. The macula is most easily seen when the subject looks directly into the light of the ophthalmoscope.

22 EXERCISE

Special Senses: Hearing and Equilibrium

OBJECTIVES

1. To identify the anatomic structures of the outer, middle, and inner ear by appropriately labeling a diagram.
2. To describe the anatomy of the organ of hearing (organ of Corti in the cochlea) and explain its function in sound reception.
3. To describe the anatomy of the equilibrium organs of the inner ear (cristae ampullares and maculae) and to explain their relative function in maintaining equilibrium.
4. To define or explain *central deafness, conduction deafness,* and *nystagmus.*
5. To state the purpose of the Rinne and Romberg tests.
6. To explain how one is able to localize the source of sounds.

MATERIALS

Three-dimensional dissectible ear model and/or chart of ear anatomy
Compound microscope
Histologic slides of the cochlea of the ear

Demonstration: Microscope focused on a crista ampullaris receptor of a semicircular canal

Pocket watch or small clock that ticks
Tuning forks (range of frequencies)
Rubber mallet
Absorbent cotton
Otoscope (if available)
Alcohol swabs
12-inch ruler

ANATOMY OF THE EAR

Gross Anatomy

The ear is a complex structure containing sensory receptors for hearing and equilibrium. It is divided into three major areas: the **outer** or **external, ear;** the **middle ear;** and the **inner ear** (Figure 22.1). The outer and middle ear structures serve the needs of the sense of hearing *only,* whereas the inner ear structures function both in equilibrium and in sound reception.

Obtain a dissectible ear model, and identify its structures described below. Refer to Figure 22.1 as you work.

The outer ear is composed primarily of the external **pinna,** or **auricle,** and the **external auditory canal.** The pinna is the skin-covered cartilaginous structure encircling the auditory canal opening. In many animals, it collects and directs sound waves into the auditory canal. In humans this function of the pinna is largely lost.

The external auditory canal is a short, narrow chamber (about 1 inch long by ¼ inch wide) carved into the temporal bone. In its skin-lined walls are wax-secreting glands called **ceruminous glands.** The sound waves that enter the external auditory canal eventually encounter the **tympanic membrane,** or **eardrum,** which vibrates at exactly the same frequency as the sound wave(s) hitting it. The membranous eardrum separates the outer from the middle ear.

The middle ear is essentially a small chamber—the **tympanic cavity**—found within the temporal bone. The cavity is spanned by three small bones, collectively called the **ossicles** (hammer, anvil, and stirrup),* which articulate to form a lever system that transmits the vibratory motion of the eardrum to the fluids of the inner ear via the **oval window.**

Connecting the middle ear chamber with the nasopharynx is the **auditory (eustachian) tube.** Normally this tube is flattened and closed, but swallowing or yawning can cause it to open temporarily to equalize the pressure of the middle ear cavity with external air pressure. This is an important function, because the eardrum does not vibrate properly unless the pressure on both of its surfaces is the same.

Because the mucosal membranes of the middle ear cavity and the nasopharynx are continuous through the eustachian tube, *otitis media,* or inflammation of the middle ear, is a fairly common condition, especially

* The ossicles are commonly referred to by their Latin names—that is, **malleus, incus,** and **stapes,** respectively.

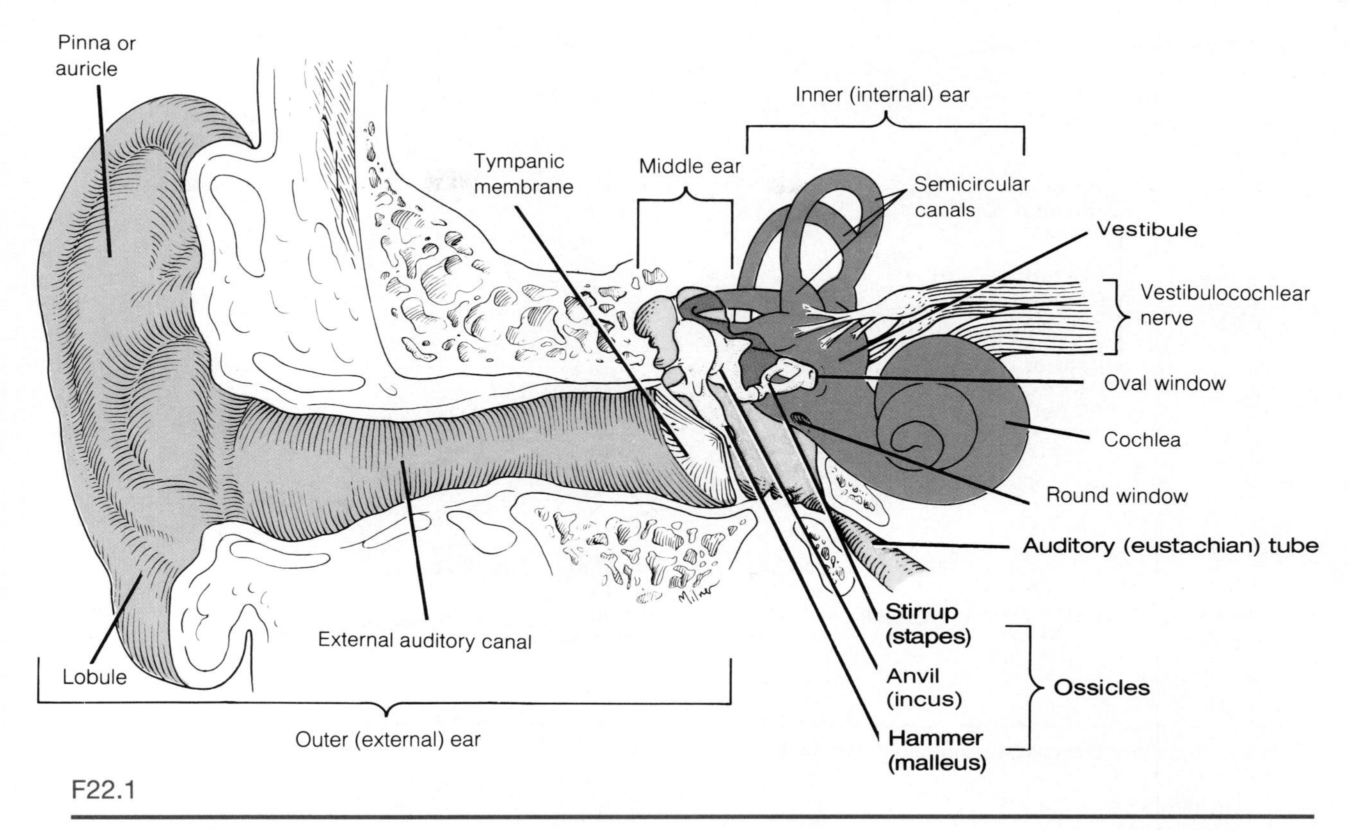

F22.1

Anatomy of the ear.

among youngsters prone to sore throats. In cases in which large amounts of fluid or pus accumulate in the middle ear cavity, an emergency myringotomy (lancing of the eardrum) may be necessary to relieve the pressure. Frequently, tiny ventilating tubes are put in during the procedure. ■

The inner ear consists of a system of bony and rather tortuous chambers called the **osseous,** or **bony, labyrinth,** which is filled with an aqueous fluid called **perilymph** (Figure 22.2). Suspended in the perilymph is the **membranous labyrinth,** a system that pretty much follows the contours of the osseous labyrinth. The interior of the membranous labyrinth is filled with a more viscous fluid called **endolymph.** The three subdivisions of the bony labyrinth are the **cochlea,** the **vestibule,** and the **semicircular canals,** with the vestibule situated between the cochlea and semicircular canals.

The snail-like cochlea (see Figures 22.2 and 22.3) contains the sensory receptors for hearing. The cochlear membranous labyrinth, the **cochlear duct,** is a soft wormlike tube about 1½ inches long. It winds through the full two and three-quarter turns of the cochlea, and separates the perilymph-containing cochlear cavity into upper and lower chambers, the **scala vestibuli** and **scala tympani,** respectively. The scala vestibuli terminates at the oval window, which "seats" the foot plate of the stirrup located laterally in the tympanic cavity. The scala tympani is bounded by a membranous area called the **round window.** The cochlear duct, itself filled with endolymph, supports the **organ of Corti,** which contains the receptors for hearing—the sensory hair cells and nerve endings of the cochlear division of the vestibulocochlear nerve (VIII).

Otoscopic Examination of the Ear (Optional)

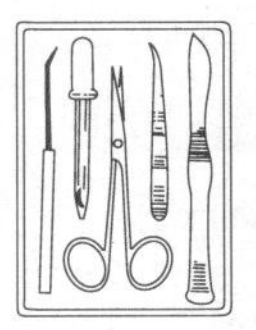

1. Obtain an otoscope and two alcohol swabs. Inspect your partner's ear canal and then select the largest-*diameter* (not length!) speculum that will fit comfortably in his or her ear to permit full visualization. Clean the speculum chosen thoroughly with an alcohol swab, and then attach it to the battery-containing otoscope handle. Check the otoscope light before beginning to make sure that the light beam is strong. (If not, obtain another otoscope or new batteries.)

2. Hold the lighted otoscope securely between your thumb and forefinger (like a pencil) and rest the little finger of the otoscope-holding hand against your partner's head when you are ready to begin the examination. This maneuver forms a brace that allows the speculum to move whenever your partner moves and prevents the speculum from penetrating too deeply into his or her ear canal during unexpected movements.

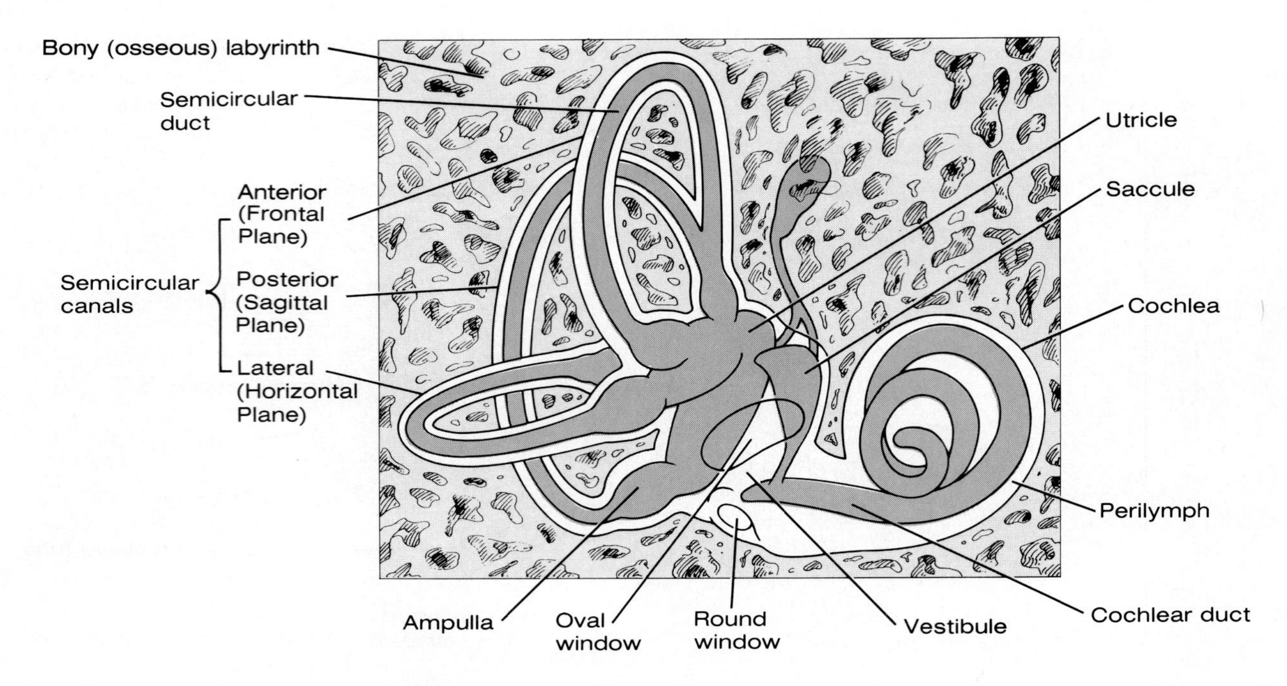

F22.2

Inner ear. Right membranous labyrinth shown within the bony labyrinth.

3. Grasp the ear pinna firmly and pull it up, back, and slightly laterally. If your partner experiences pain or discomfort when the pinna is manipulated, an inflammation or infection of the external ear may be present. If this occurs, do not attempt to examine the ear canal.

4. Carefully insert the speculum of the otoscope into the external auditory canal in a downward and forward direction only far enough to permit examination of the tympanic membrane, or eardrum. Note its shape, color, and vascular network. The healthy tympanic membrane is pearly white. During the examination, note whether there is any discharge or redness in the canal; and identify earwax.

5. After the examination, thoroughly clean the speculum with the second alcohol swab before returning it to the supply area.

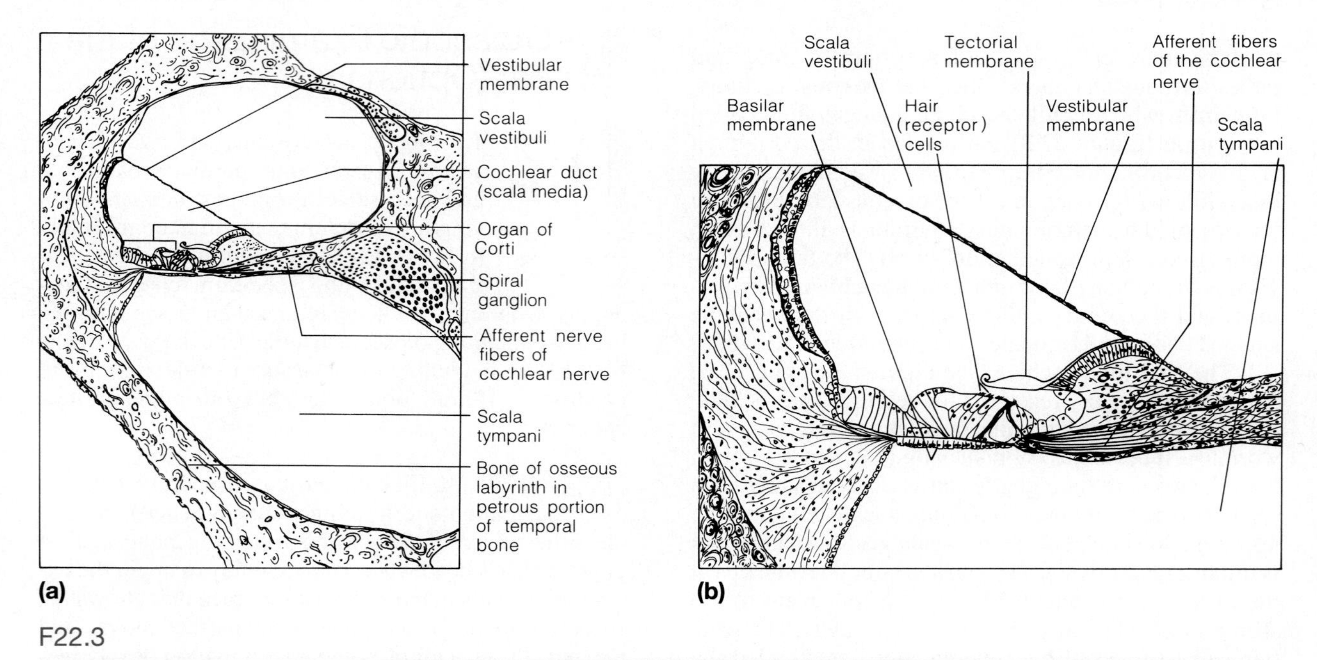

F22.3

Organ of Corti. (a) Cross section through one turn of the cochlea showing the position of the organ of Corti. (b) Enlarged view of the organ of Corti. See corresponding views in plates 14 and 15 of the histology atlas.

Microscopic Anatomy of the Receptors

ORGAN OF CORTI AND THE MECHANISM OF HEARING The anatomical details of the organ of Corti are shown in Figure 22.3. The hair (auditory receptor) cells rest on the **basilar membrane,** which forms the floor of the cochlear duct, and their "hairs" (stereocilia) project into a gelatinous membrane, the **tectorial membrane,** which overlies them. The roof of the cochlear duct is called the **vestibular membrane.** The endolymph-filled chamber of the cochlear duct is called the **scala media.**

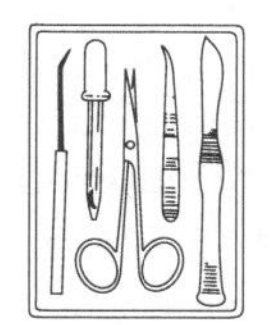

Obtain a compound microscope and a microscope slide of the cochlea, and identify the areas noted in Figure 22.3a and b.

The mechanism of hearing begins as sound waves pass through the external auditory canal and the middle ear, and into the inner ear, where the vibration eventually reaches the organ of Corti, which contains the receptors for hearing. Many theories have attempted to explain how the organ of Corti actually responds to sound.

The popular "traveling wave" hypothesis of Von Békésy suggests that vibration of the stirrup at the oval window initiates traveling waves that cause maximal displacements of the basilar membrane where they peak and stimulate the hair cells of the organ of Corti in that region. Since the area at which the traveling waves peak is a high-pressure area, the vestibular membrane is compressed at this point and, in turn, compresses the endolymph and the basilar membrane of the cochlear duct. The resulting pressure on the perilymph in the scala tympani causes the membrane of the round window to bulge outward into the middle ear chamber, thus acting as a relief valve for the compressional wave (Figure 22.4).

Von Békésy found that high-frequency waves (high-pitched sounds) peaked close to the oval window and that low-frequency waves (low-pitched sounds) peaked farther up the basilar membrane, near the apex of the cochlea. Although the mechanism of sound reception by the organ of Corti is not completely understood, we do know that the hair cells are uniquely stimulated by sounds of various frequencies (interpreted as differences in pitch) and amplitude (interpreted as differences in loudness), and that once stimulated, they depolarize and begin the chain of nervous impulses to the auditory centers of the temporal lobe cortex. This series of events results in the phenomenon we call hearing.

EQUILIBRIUM APPARATUS AND MECHANISMS OF EQUILIBRIUM The equilibrium apparatus of the inner ear is in the vestibular and semicircular canal portions of the body labyrinth. Their chambers are filled with perilymph, in which membranous labyrinth structures are suspended. (The vestibule contains the saclike **utricle** and **saccule,** and the semicircular canals contain the membranous **semicircular ducts.**) Like the cochlear duct, these membranes are filled with endolymph and contain receptor cells that are activated by the disturbance of their cilia.

By the time most people are in their 60s, a gradual deterioration and atrophy of the organ of Corti begins, and leads to a loss in the ability to hear high tones and speech sounds. This condition, *presbycusis,* is a type of sensorineural deafness. Because many elderly people refuse to accept their hearing loss and resist using hearing aids, they begin to rely more and more on their vision for clues as to what is going on around them, and may be accused of ignoring people.

Although presbycusis is considered to be a disability of old age, it is becoming much more common in

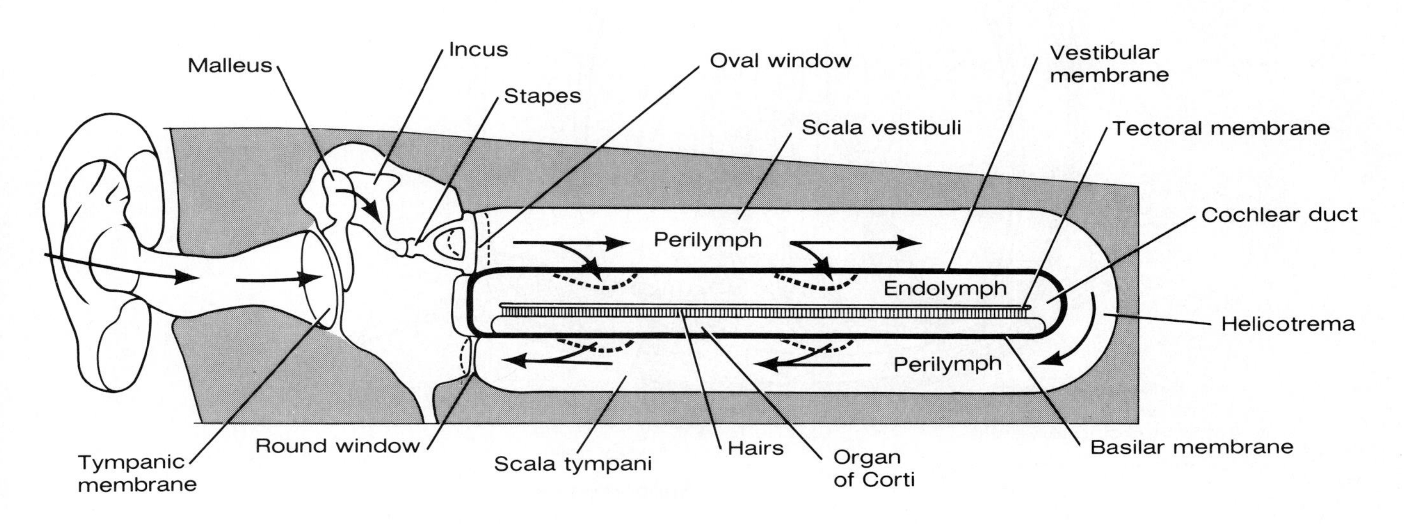

F22.4

Fluid movement in the cochlea following stirrup thrust on the oval window. Compressional wave created causes the round window to bulge into the middle ear.

younger people as our world grows noisier day by day. Noise pollution has become a major health problem, and the damage (breakage of the "hairs" of the hair cells) caused by excessively loud sounds is progressive and cumulative. Each assault causes a bit more damage. Rock music played and listened to at deafening levels is definitely a contributing factor to the deterioration of the hearing receptors. ■

The semicircular canals are centrally involved in the **mechanism of dynamic equilibrium.** They are about ½ inch in circumference and are oriented in three planes—horizontal, frontal, and sagittal. At the base of each semicircular duct is an enlarged region, the **ampulla,** which communicates with the utricle of the vestibule. Within each ampulla is a receptor region called a **crista ampullaris,** which consists of a tuft of hair cells covered with a gelatinous cap, or **cupula** (Figure 22.5). When your head position changes in an angular direction, as when twirling on the dance floor or when taking a rough boat ride, the endolymph in the canal lags behind, pushing the cupula—like a swinging door—in a direction opposite to that of the angular motion. This movement initiates the action potential of the hair cells. These impulses are then transmitted up the vestibular division of the eighth cranial nerve to the brain. Then, when the angular motion stops suddenly, the inertia of the endolymph causes it to continue to move, pushing the cupula in the same direction as the previous motion. This movement again initiates electrical changes in the hair cells. (This phenomenon accounts for the reversed motion sensation you feel when you stop suddenly after twirling.) If you begin to move at a constant rate of motion, the cupula gradually returns to its original position. The hair cells, no longer bent, send no new signals; and you lose the sensation of spinning. Thus the response of these dynamic equilibrium receptors is a reaction to *changes* in angular motion rather than to motion itself.

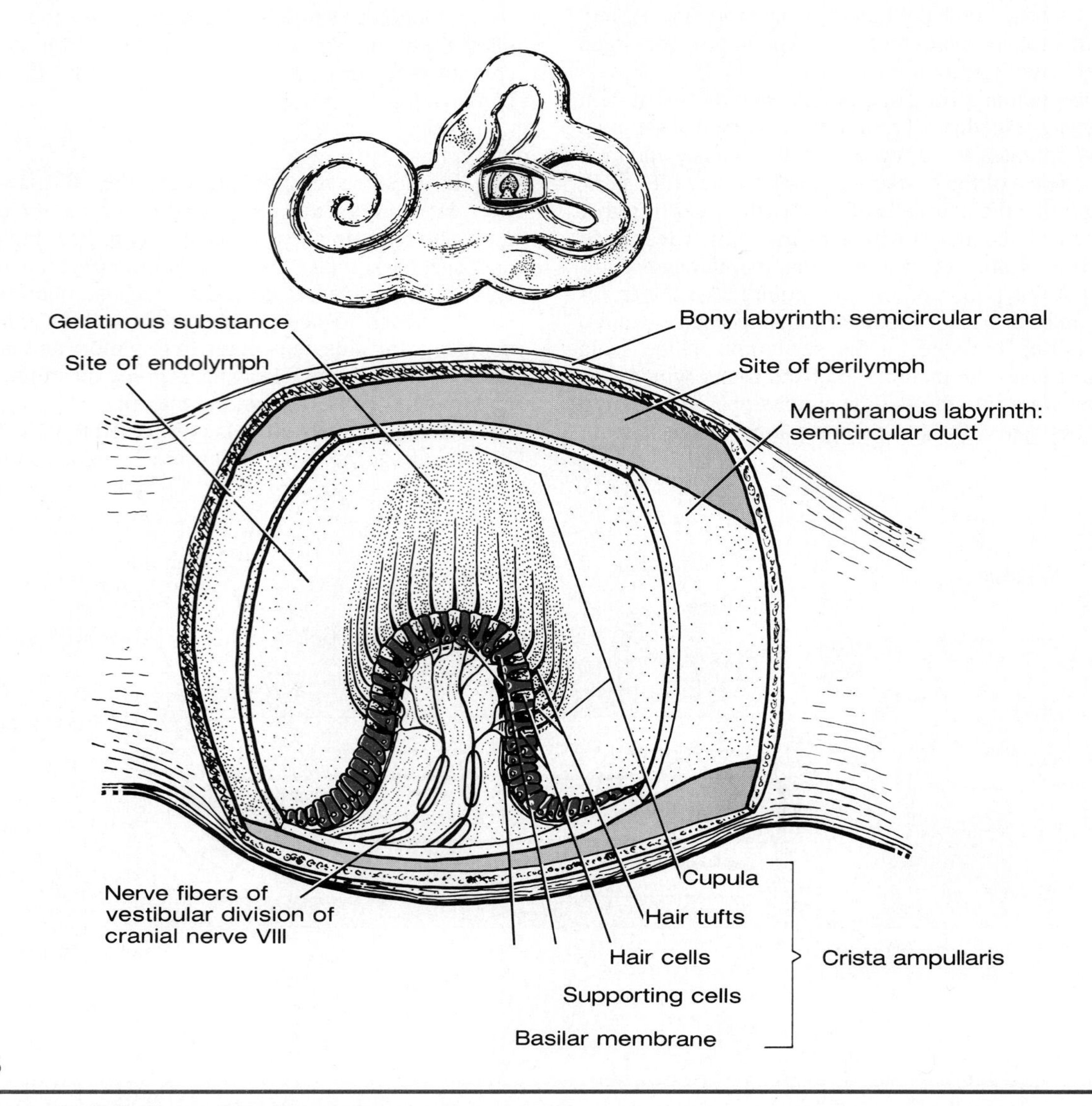

F22.5

Crista ampullaris in the semicircular canal.

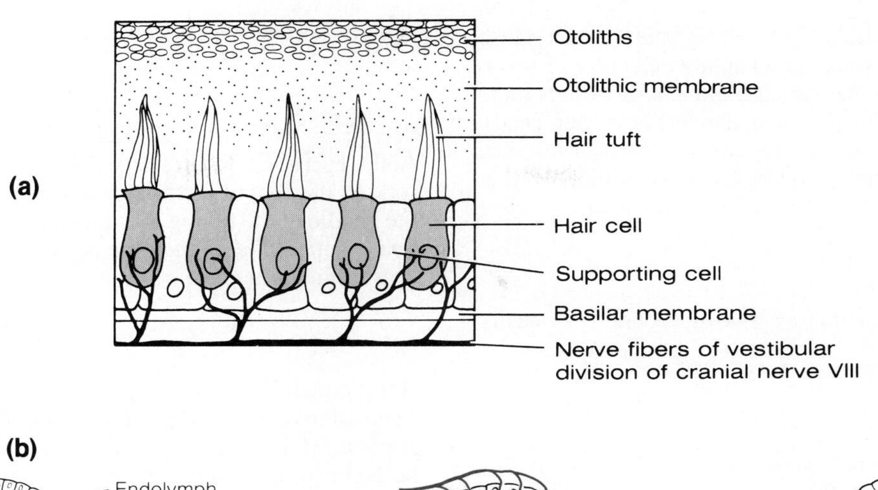

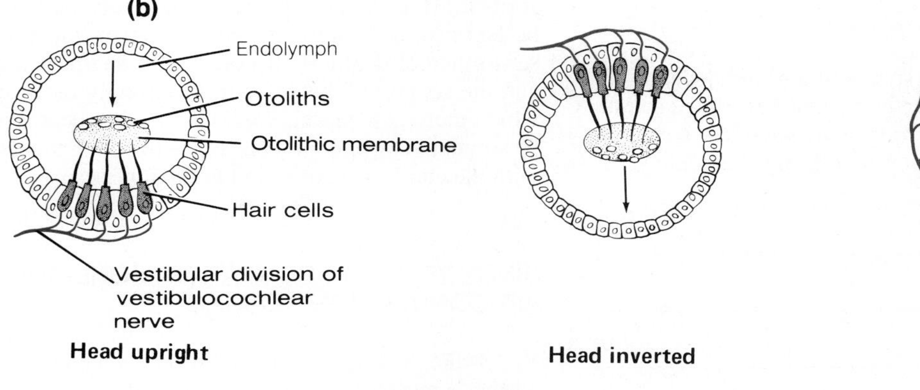

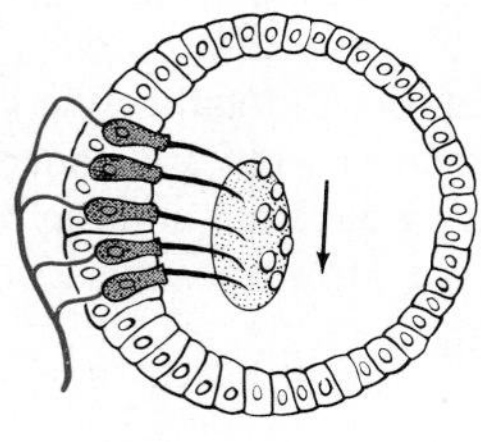

F22.6

Structure and function of static equilibrium receptors (maculae). (a) Diagrammatic view of a portion of a macula; (b) stimulation of the maculae by movement of the otoliths in the gelatinous otolithic membrane that creates a pull on the hair cells. Arrows indicate direction of gravitational pull.

Go to the demonstration area and examine the slide of the crista ampullaris. Identify the areas depicted in Figure 22.5.

The vestibule contains receptors (**maculae**) that are essential to the **mechanism of static equilibrium.** The maculae respond to gravitational pull, thus providing information on which way is up or down, and to linear or straightforward changes in speed. They are located on the walls of the saccule and utricle. The **otolithic membrane,** gelatinous material containing small grains of calcium carbonate (**otoliths**), overrides the hair cells in each macula. As the head moves, the otoliths move in response to variations in gravitational pull and deflect different hair cells, thus triggering nerve impulses along the vestibular nerve (Figure 22.6).

Although the receptors of the semicircular canals and the vestibule are responsible for dynamic and static equilibrium respectively, they rarely act individually. Complex interaction of many of the receptors is often the rule. Also, the information these equilibrium, or balance, senses provide is significantly augmented by the proprioceptors and by sight, as some of the following laboratory experiments demonstrate.

LABORATORY TESTS

Hearing Tests

Perform the following hearing tests in a quiet area.

ACUITY TEST Have your lab partner pack one ear with cotton and sit quietly with eyes closed. Obtain a ticking pocket watch and hold it very close to his or her unpacked ear. Then slowly move it away from the ear until your partner signals that the ticking is no longer audible. Record the distance (in inches) at which ticking is inaudible.

Right ear ______________ Left ear ______________

Is the threshold of audibility sharp or indefinite?

____

SOUND LOCALIZATION Have your partner close both eyes. Hold a watch at an audible distance (about 6 inches) from his or her ear, and move it to various locations (front, back, sides, and above his or her head). Have your partner locate the position in each instance by pointing. Can the sound be localized equally well at all positions?

___

If not, at what position(s) was the sound less easily located?

___

___

The ability to localize the source of a sound depends on two factors—the difference in the loudness of the sound reaching each ear and the time of arrival of the sound at each ear. How does this information help to explain your findings?

___

___

___

FREQUENCY RANGE OF HEARING Obtain three tuning forks: one with a low frequency (75 to 100 cps), one with a frequency of approximately 1000 cps, and one with a frequency of 4000 to 5000 cps. Strike the lowest-frequency fork with a rubber mallet, and hold it close to your partner's ear. Repeat with the other two forks. Which fork was heard most clearly and comfortably?

___ cps

Which was heard least well? ___ cps

WEBER TEST TO DETERMINE CONDUCTIVE AND PERCEPTIVE DEAFNESS Strike a tuning fork on the heel of your hand or with a mallet and place the handle of the tuning fork medially on your forehead. See Figure 22.7a. Is the tone equally loud in both ears, or is it louder in one ear?

___

If it is equally loud in both ears, you have equal hearing, or equal loss of hearing, in both ears. If perceptive (sensorineural) deafness is present in one ear, the tone will be heard in the unaffected ear but not in the ear with sensorineural deafness. If conduction deafness is present, the sound will be heard more strongly in the ear in which there is a hearing loss. Conduction deafness can be simulated by plugging one ear with cotton to interfere with the conduction of sound to the inner ear.

RINNE TEST FOR COMPARING BONE-AND AIR-CONDUCTION HEARING

1. Strike the tuning fork, and place its handle on your partner's mastoid process (Figure 22.7b).

2. When the sound is no longer audible to your partner, hold the still-vibrating prongs close to his auditory canal (Figure 22.7c). If your partner hears the fork again when it is moved to that position (by air conduction), hearing is not impaired, and the test result is recorded as positive (+). (Record below.)

3. Repeat the test, but this time test air-conduction hearing first.

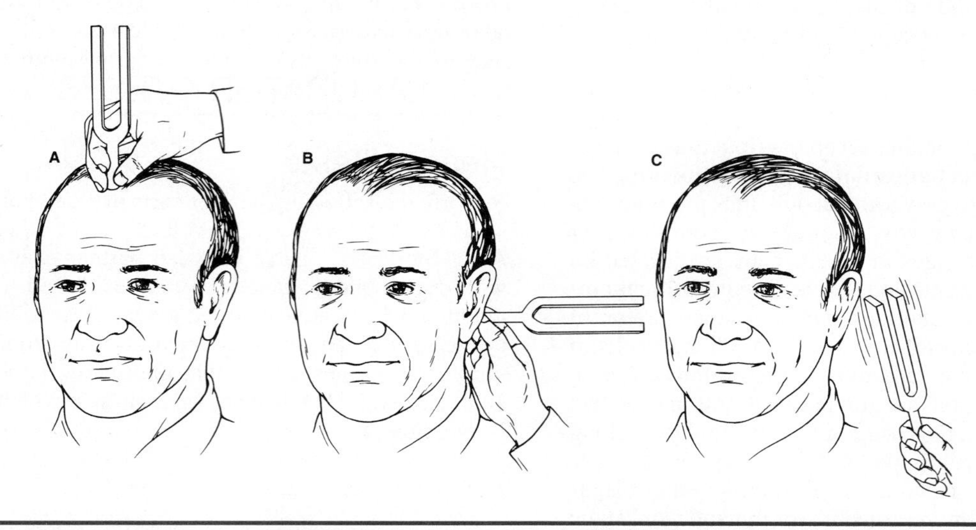

F22.7

The Weber and Rinne tuning fork tests. (a) The Weber test to evaluate whether the sound remains centralized (normal) or lateralizes to one side or the other (indicative of some degree of conductive or sensorineural deafness). (b and c) The Rinne test to compare bone and air conduction.

4. After the tone is no longer heard by air conduction, hold the handle of the tuning fork on the bony mastoid process. If the subject hears the tone again by bone conduction after hearing by air conduction is lost, there is some conductive deafness; and the result is recorded as negative (−).

5. Repeat the sequence for the other ear.

Right ear ______________ Left ear ______________

Does the subject hear better by bone or air conduction?

____

Equilibrium Experiments

The function of the semicircular canals and vestibule is not routinely tested in the laboratory, but the following simple tests should serve to illustrate normal equilibrium-apparatus functioning.

BALANCE TEST Have your partner walk a straight line, placing one foot directly in front of the other. Is he or she able to walk without undue wobbling from side to side?

____

Did he or she experience any dizziness? ____________

The ability to walk with balance and without dizziness, unless subject to rotational forces, indicates normal function of the equilibrium apparatus.

Was nystagmus* present? ______________________

BARANY TEST (INDUCTION OF NYSTAGMUS AND VERTIGO†) This experiment evaluates the semicircular canals and should be conducted as a group effort to protect the test subject(s) from possible injury.

⚠ The following precautionary notes should be read before beginning:

- The subject(s) chosen should not be easily inclined to dizziness on rotational or turning movements.
- Rotation should be stopped immediately if the subject complains of feeling nauseous.
- Because the subject(s) will experience vertigo and loss of balance as a result of the rotation, several classmates should be prepared to catch, hold, or support the subject(s) as necessary until the symptoms pass.

1. Instruct the subject to sit on a rotating chair or stool, and to hold on to the arms or seat of the chair, feet on stool rungs. The subject's head should be tilted forward approximately 30 degrees (almost touching the chest). The horizontal (lateral) semicircular canal will be stimulated when the head is in this position. The subject's eyes are to *remain open* during the test.

2. Four classmates should position themselves so that the subject is surrounded on all sides. The classmate posterior to the subject will rotate the chair.

3. Rotate the chair to the subject's right approximately 10 revolutions in 10 seconds, and then suddenly stop the rotation.

4. Immediately note the direction of the subject's resultant nystagmus; and ask him or her to describe the feelings of movement, indicating speed and direction sensation. Record below.

____

____

____

____

If the semicircular canals are operating normally, the subject will experience a sensation that the stool is still rotating immediately after it has stopped and *will* demonstrate nystagmus.

When the subject is rotated to the right, the cupula will be bent to the left, causing nystagmus during rotation in which the eyes initially move slowly to the left and then quickly to the right. Nystagmus will continue until the cupula has returned to its initial position. Then, when rotation is stopped abruptly, the cupula will be bent to the right, producing nystagmus with its slow phase to the right and its rapid phase to the left. In many subjects, this will be accompanied by a feeling of vertigo and a tendency to fall to the right.

ROMBERG TEST The Romberg test determines the integrity of the dorsal white column of the spinal cord, which transmits impulses to the brain from the proprioceptors involved with posture.

1. Have your partner stand with his or her back to the blackboard.

2. Draw one line parallel to each side of your partner's body. He or she should stand erect, with eyes open and staring straight ahead for 2 minutes while you observe any movements. Did you see any gross swaying movements?

____

* **Nystagmus** is the involuntary rolling of the eyes in any direction or the trailing of the eyes slowly in one direction, followed by their rapid movement in the opposite direction. It is normal after rotation, abnormal otherwise. The direction of nystagmus is that of its quick phase on acceleration.

† **Vertigo** is a sensation of dizziness and rotational movement when such movement is not occurring or has ceased.

3. Repeat the test. This time the subject's eyes should be closed. Note and record the degree of side-to-side movement.

____

____

4. Repeat the test with the subject's eyes closed. This time, however, the subject should be positioned with his or her left shoulder toward, but not touching, the board so that you may observe and record the degree of front-to-back swaying.

____

____

Do you think the equilibrium apparatus of the inner ear was operating equally well in all these tests?

____

The proprioceptors? ______________________________

Why was the observed degree of swaying greater when the eyes were closed?

____

____

____

What conclusions can you draw regarding the factors necessary for maintaining body equilibrium and balance?

____

____

____

____

____

ROLE OF VISION IN MAINTAINING EQUILIBRIUM To further demonstrate the role of vision in maintaining equilibrium, perform the following experiment. (Ask your lab partner to record observations and act as a "spotter.") Stand erect, with your eyes open. Raise your left foot approximately 1 foot off the floor, and hold it there for 1 minute.

Record the observations: ______________________________

____

____

Rest for 1 or 2 minutes; and then repeat the experiment with your other foot raised, but with your eyes closed.

Record the observations: ______________________________

____

____

____

____

____

EXERCISE 23

Special Senses: Taste and Olfaction

OBJECTIVES

1. To describe the structure and function of the taste receptors.
2. To describe the location and cellular composition of the olfactory epithelium.
3. To name the four basic types of taste sensation and list the chemical substances that elicit these sensations.
4. To point out, on a diagram of the tongue, the predominant location of the basic types of taste receptors (salty, sweet, sour, and bitter).
5. To explain the interdependence between the senses of smell and taste.
6. To name two factors other than olfaction that influence taste appreciation of foods.
7. To define *olfactory adaptation.*

MATERIALS

Prepared histologic slides: the tongue, showing taste buds; the nasal epithelium (longitudinal section)
Compound microscope
Paper towels
Small mirror
Granulated sugar
Cotton-tipped swabs
Paper cups; paper plates
Prepared vials of 10% NaCl, 0.1% quinine or Epsom salt solution, 5% sucrose solution, and 1% acetic acid
Disposable autoclave bag
Prepared dropper bottles of oil of cloves, oil of peppermint, and oil of wintergreen
Equal-size food cubes of cheese, apple, raw potato, dried prunes, banana, raw carrot, and hard-cooked egg white (These prepared foods should be in an opaque container; a foil-lined egg carton would work fine.)
Chipped ice

The receptors for taste and olfaction are classified as **chemoreceptors** because they respond to chemicals or volatile substances in solution. The chemical senses, taste and smell, are sharpest at birth. This explains why infants seem to relish food adults consider to be bland or tasteless. Some researchers claim that the sense of smell is just as important as the sense of touch in guiding newborn babies to their mother's breast.

Beginning in the mid-40s, our ability to taste and smell diminishes, which reflects the gradual decrease in the number of these receptor cells. This helps to explain why older adults often prefer highly seasoned (although not necessarily spicy) foods and why many elderly people do not seem to notice many odors that younger people find very disagreeable.

Although four relatively specific types of taste receptors have been identified, the olfactory receptors are considered sensitive to a much wider range of chemical sensations. The sense of smell is one of the least understood of the special senses.

LOCALIZATION AND ANATOMY OF TASTE BUDS

The **taste buds,** specific receptors for the sense of taste, are widely, but not uniformly, distributed in the oral cavity. Most are located on the dorsal surface of the tongue (as described below). A few are found on the soft palate, epiglottis, and inner surface of the cheeks.

The dorsal tongue surface is covered with small projections, or **papillae,** of three major types: sharp *filiform* papillae, and the rounded *fungiform* and *circumvallate* papillae. The taste buds are located primarily on

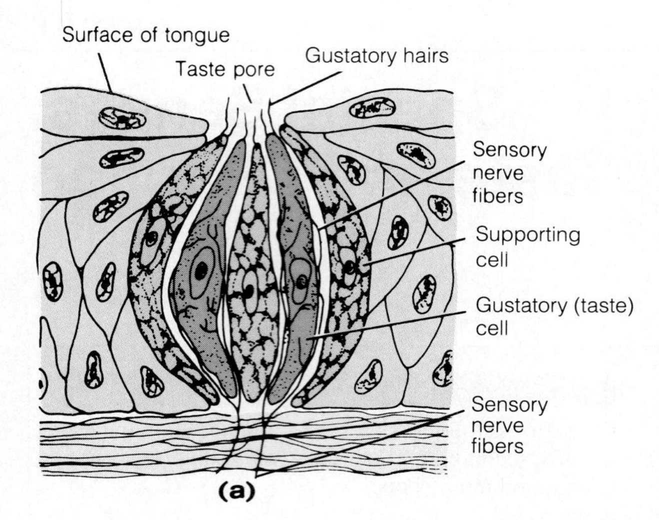

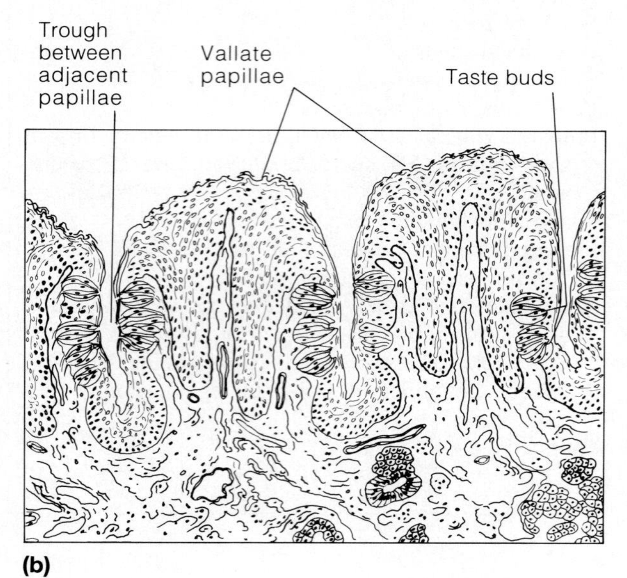

F23.1

Taste bud anatomy and localization. (a) Diagrammatic view of a taste bud; (b) enlarged view of tongue papillae showing positioning of taste buds. (See corresponding photomicrograph, plate 16 of histology atlas.)

the sides of the circumvallate papillae (arranged in a V-formation on the posterior surface of the tongue) and on the more numerous fungiform papillae, which look rather like minute mushrooms and are widely distributed on the tongue.

- Use a mirror to examine your tongue. Can you pick out the various papillae types?

  _______ If so, which? ______________________

  ____

  ____

  ____

Each taste bud consists largely of a globular arrangement of two types of modified epithelial cells: the **gustatory,** or **taste, cells** which are the actual receptor cells, and **supporting cells.** Several nerve fibers enter each taste bud and supply sensory nerve endings to each of the taste cells. The long microvilli of the taste cells penetrate the epithelial surface through an opening called the **taste pore.** When these microvilli (called *gustatory hairs*) contact specific chemicals in solution, the taste cells are excited. The afferent fibers from the taste buds to the sensory cortex in the postcentral gyrus of the brain are carried in three cranial nerves: The *facial nerve* (*VII*) serves the anterior two-thirds of the tongue; the *glossopharyngeal nerve* (*IX*) serves the posterior third of the tongue; and the *vagus nerve* (*X*) carries a few fibers from the pharyngeal region.

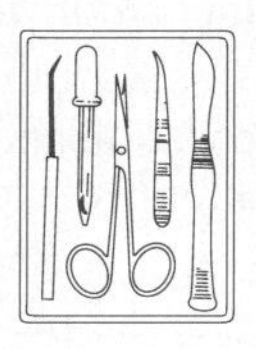

Obtain a prepared slide of a tongue cross section, and use Figure 23.1 as a guide to aid you in locating the taste buds on the papillae. Make a detailed drawing below of one taste bud. Label the taste pore and gustatory hairs if observed.

LOCALIZATION AND ANATOMY OF THE OLFACTORY RECEPTORS

The **olfactory epithelium** (organ of smell) occupies an area of about 2.5 cm in the roof of each nasal cavity. Because the air entering the human nasal cavity must make a hairpin turn to enter the respiratory passages below, the nasal epithelium is in a rather poor position for performing its function. Sniffing, which brings a greater airflow into contact with the receptors, intensifies the sense of smell.

The specialized receptor cells in the olfactory epithelium are surrounded by **supporting cells,** nonsensory

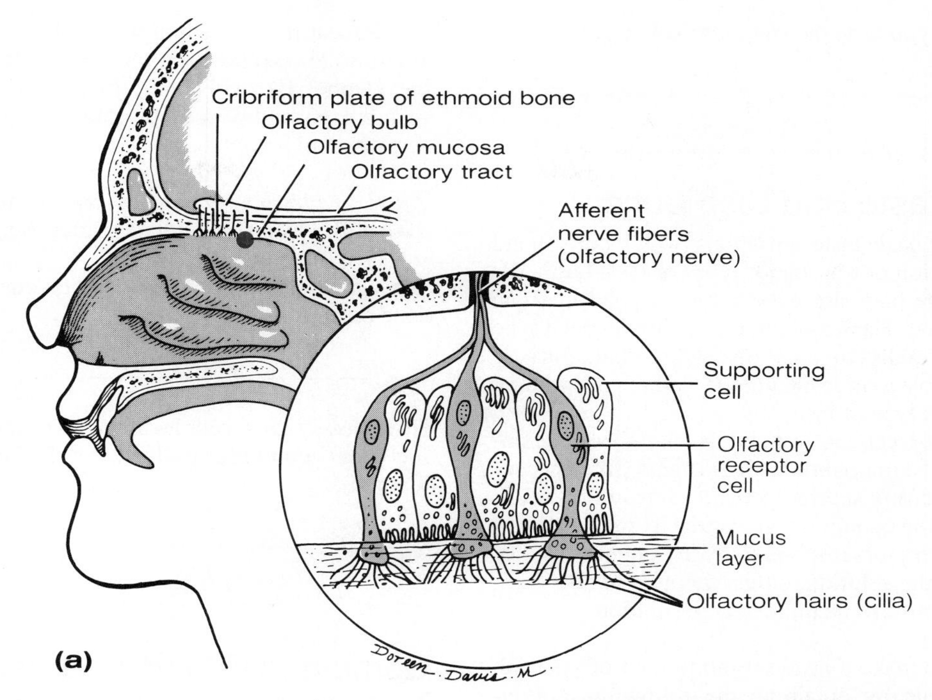

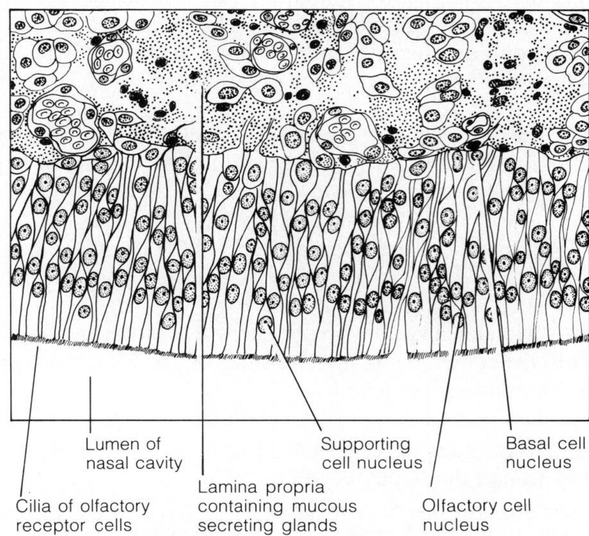

F23.2

Cellular composition of the olfactory epithelium. (a) Diagrammatic representation; (b) line drawing corresponding to photomicrograph in plate 17 of the histology atlas.

epithelial cells. The **olfactory receptor cells** are bipolar neurons, with *olfactory hairs* (actually cilia) extending outward from the epithelium. Emerging from their basal ends are axonal nerve fibers that penetrate the cribriform plate of the ethmoid bone and proceed as the olfactory nerves to synapse in the olfactory bulbs lying on either side of the crista galli of the ethmoid bone. Impulses from neurons of the olfactory bulbs are then conveyed to the olfactory portion of the cortex (uncus).

Obtain a longitudinal section of olfactory epithelium. Examine it closely, comparing it to Figure 23.2.

LABORATORY EXPERIMENTS

Stimulation of Taste Buds

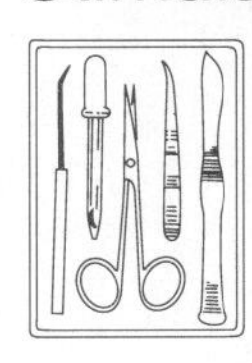

Obtain several paper towels and a disposable autoclave bag and bring them to your bench. With a paper towel, dry the dorsal surface of your tongue.

Immediately dispose of the paper towel in the autoclave bag. Place a few sugar crystals on your dried tongue. Do *not* close your mouth. Time how long it takes to taste the sugar.

______________ sec

Why couldn't you taste the sugar immediately?

Plotting Taste Bud Distribution

There are four basic taste sensations, which correspond to the stimulation of four major types of taste buds. Although all taste buds are believed to respond in some degree to all four classes of chemical stimuli, each type responds optimally to only one. This characteristic makes it possible to map the tongue to show the relative density of each type of taste bud.

The sweet receptors respond to a number of seemingly unrelated compounds such as sugars (fructose, sucrose, and glucose), saccharine, and some amino acids. Some believe the common factor is the hydroxyl (OH^-) group. Sour receptors respond to hydrogen ions (H^+) or the acidity of the solution, bitter receptors to alkaloids, and salty receptors to metallic ions in solution.

1. Prepare to make a taste sensation map of your lab partner's tongue by obtaining the following: cotton-tipped swabs; one vial each of NaCl, quinine or Epsom salt solution, sucrose solution, and acetic acid; paper cups; and a flask of distilled or tap water.

2. Before each test, the subject should rinse his or her mouth thoroughly with water and lightly dry his or her tongue with a paper towel.

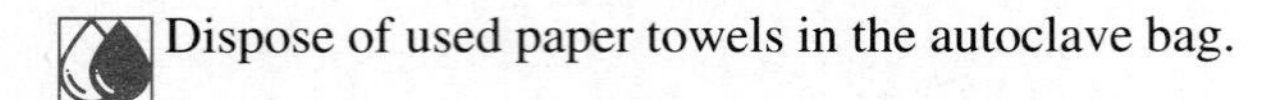
Dispose of used paper towels in the autoclave bag.

3. Moisten a swab with 5% sucrose solution and touch it to the center, back, tip, and sides of the dorsal surface of the subject's tongue.

4. Map, with an O on the tongue outline below, the location of the sweet receptors.

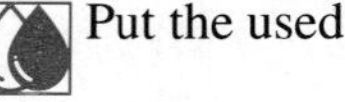
Put the used swab in the autoclave bag.

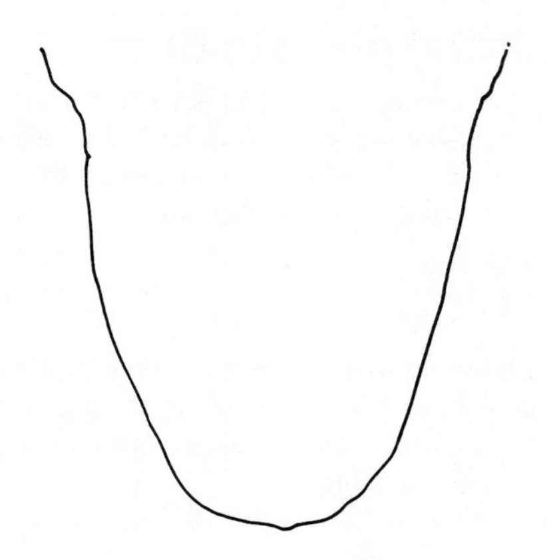

5. Repeat the procedure with quinine (or Epsom salt solution) to map the location of the bitter receptors (use the symbol B), with NaCl to map the salt receptors (symbol +), and with acetic acid to map the sour receptors (symbol −).

Use a fresh swab for each test and properly dispose of the swabs immediately after use.

What area of the dorsum of the tongue seems to lack taste receptors?

How closely does your localization of the different taste receptors coincide with the information in your textbook?

Combined Effects of Smell, Texture, and Temperature on Taste

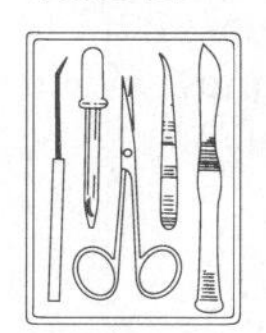

1. Ask the subject to sit with eyes closed and to pinch his or her nostrils shut.

2. Obtain samples of the food items listed in the chart below. At no time should the subject be allowed to see the foods being tested.

3. Use an out-of-sequence order of food testing. For each test, place a cube of food in the subject's mouth and ask him or her to identify the food by using the following sequence of activities:

- First, by manipulating the food with the tongue.
- Second, by chewing the food.
- Third, if a positive identification is not made with these techniques along with the taste sense, ask the subject to release his or her "pinched" nostrils and to continue chewing with nostrils open to determine if a positive identification can be made.

Record the results on the chart by checking the appropriate column.

Was the sense of smell equally important in all cases?

Where did it seem to be important, and why?

Method of Identification

Food	Texture Only	Chewing with Nostrils Pinched	Chewing with Nostrils Open	Identification Not Made
Cheese	______	______	______	______
Apple	______	______	______	______
Raw potato	______	______	______	______
Banana	______	______	______	______
Dried prunes	______	______	______	______
Raw carrot	______	______	______	______
Hard-cooked egg white	______	______	______	______

Effect of Olfactory Stimulation on Taste

There is no question that what is commonly referred to as the sense of taste depends heavily on stimulation of the olfactory receptors, particularly in the case of strongly odoriferous substances. The following experiment should illustrate this fact.

1. Obtain vials of oil of wintergreen, peppermint, and cloves, and some fresh cotton-tipped swabs. Ask the subject to sit so that he or she cannot see which vial is being used, and to dry his or her tongue and close his or her nostrils.

2. Apply a drop of one of the oils to his or her tongue. Can he or she distinguish the flavor?

____

3. Have the subject open his or her nostrils, and record the change in sensation he or she reports.

____

4. Have the subject rinse his or her mouth well and dry his or her tongue.

5. Prepare two swabs, each with one of the two remaining oils.

6. Hold one swab under the subject's open nostrils, and simultaneously touch the second swab to his or her tongue.

Record the reported sensations. ______________________

____

____

7. Appropriately dispose of the used swabs in the autoclave bag before continuing.

Which sense, taste or smell, appears to be more important in the proper identification of a strongly flavored volatile substance?

____

Olfactory Adaptation

Obtain some absorbent cotton and two of the following oils (oil of wintergreen, peppermint, or cloves). Press one nostril shut. Hold the bottle of oil under the open nostril, and exhale through your mouth. Record the time required for the odor to disappear (for olfactory adaptation to occur).

______________ sec

Repeat the procedure with the other nostril.

______________ sec

Immediately test the effect of another oil on the nostril that has just experienced olfactory adaptation. What are the results?

____

____

What conclusions can you draw? ______________________

____

____

____

____

24 EXERCISE

Anatomy and Basic Function of the Endocrine Glands

OBJECTIVES

1. To identify and name the major endocrine glands and tissues of the body when provided with an appropriate diagram.
2. To list the hormones produced by the endocrine glands and discuss the general function of each.
3. To indicate the means by which hormones contribute to body homeostasis by giving appropriate examples of hormonal actions.
4. To describe the structural and functional relationship between the hypothalamus and the pituitary gland.
5. To describe a major pathologic consequence of hypersecretion and hyposecretion of each hormone considered.
6. To correctly identify the histologic structure of the anterior and posterior pituitary, thyroid, parathyroid, adrenal cortex and medulla, pancreas, testis, and ovary by microscopic inspection, or when presented with an appropriate photomicrograph or diagram. (Optional)
7. To name and point out the specialized hormone-secreting cells in the above tissues as studied in the laboratory. (Optional)

MATERIALS

Human torso model
Anatomical chart of the human endocrine system
Compound microscopes
Colored pencils
Histologic slides of the anterior-posterior pituitary (differential staining), thyroid gland, parathyroid glands, adrenal gland, pancreas (differential staining if possible), ovary, and testis tissue

The endocrine system is the second major controlling system of the body. Acting with the nervous system, it helps to coordinate and integrate the activity of the body's cells. However, the nervous system employs electrochemical impulses to bring about rapid control, whereas the more slowly acting endocrine system employs chemical "messengers," or **hormones,** which are released into the blood to be transported throughout the body.

The term *hormone* comes from a Greek word meaning to arouse. The body's hormones, which are steroids or amino acid–based molecules, arouse the body's tissues and cells by stimulating changes in their metabolic activity. These changes lead to growth and development, and to the physiological homeostasis of many body systems. Hormones affect body cells primarily by altering (increasing or decreasing) a metabolic process rather than by initiating a new one. Although all hormones are blood-borne, a given hormone affects only the biochemical activity of a specific organ (or organs). Organs that respond to a particular hormone are referred to as the *target organs* of that hormone. The ability of the target tissue to respond seems to depend on the ability of the hormone to bind with specific receptors (proteins) on the plasma membranes or within the cells.

Although the function of some hormone-producing glands (the anterior pituitary, thyroid, adrenals, parathyroids) is purely endocrine, the function of others (the pancreas and gonads) is mixed—both endocrine and exocrine. Both types of glands are derived from epithelium, but the endocrine or ductless glands release their product (always hormonal) directly into the blood. The exocrine glands release their products at the body's surface or outside an epithelial membrane via ducts. In addition, there are varied numbers of hormone-producing cells within the intestine, stomach, kidney, and placenta, organs whose functions are primarily nonendocrine. Only the major endocrine organs are considered here.

GROSS ANATOMY AND BASIC FUNCTION OF THE ENDOCRINE GLANDS

As the endocrine organs are described, locate and ***identify them by name*** in Figure 24.1. When you have completed the descriptive material, also locate the organs on the anatomical charts or torso.

Pituitary Gland (Hypophysis)

The pituitary gland, or hypophysis, is located in the concavity of the sella turcica of the sphenoid bone. It consists largely of two functional areas—the **adenohy-**

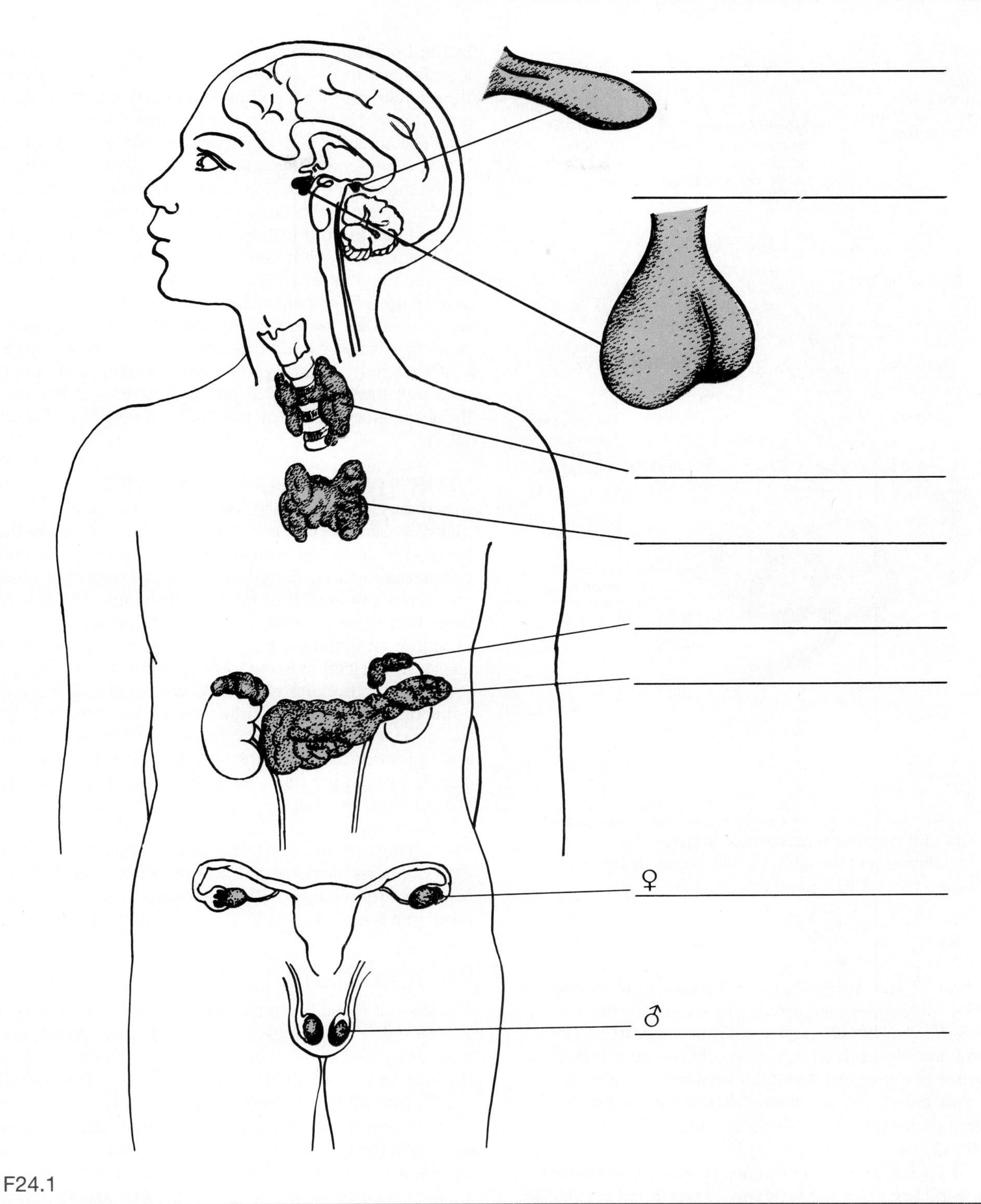

F24.1

Human endocrine glands.

pophysis, or **anterior pituitary,** and the **neurohypophysis,** or **posterior pituitary**—and is attached to the hypothalamus by a stalk called the **infundibulum.**

ADENOHYPOPHYSEAL HORMONES The adenohypophyseal hormones consist of the following tropic hormones or hormone groups: the **gonadotropins**—**follicle-stimulating hormone (FSH)** and **luteinizing hormone (LH)**—regulate gamete production and hormonal activity of the gonads (ovaries and testes). The precise roles of the gonadotropins are covered in Exercise 41, along with other considerations of reproductive system physiology. **Adrenocorticotropic hormone (ACTH)** regulates the endocrine activity of the cortex

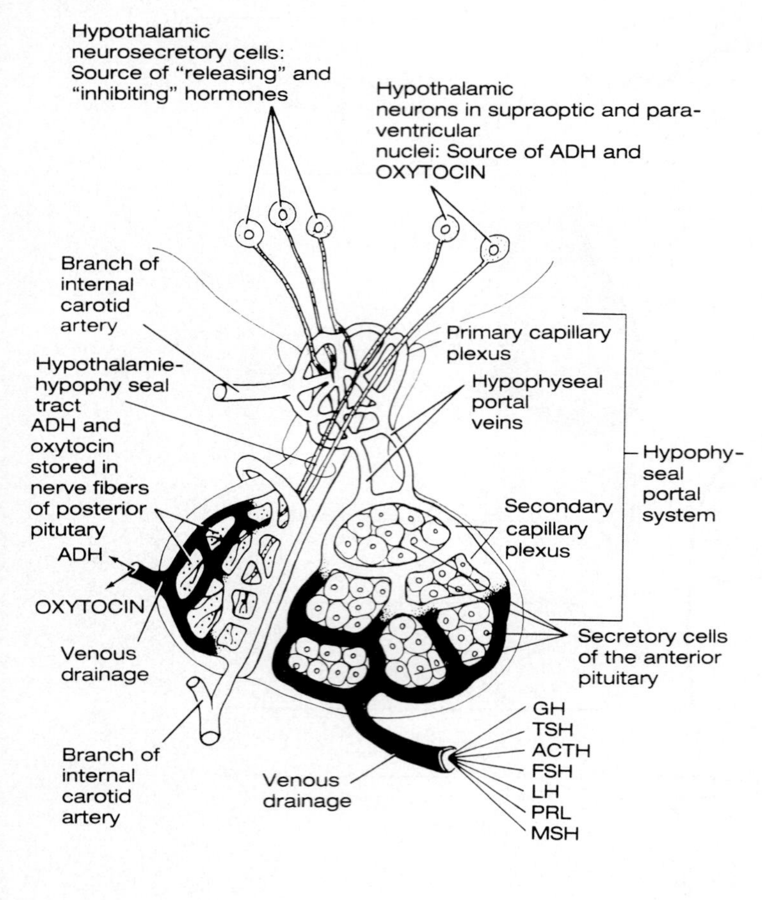

F24.2

Neural and vascular relationships between the hypothalamus and the anterior and posterior lobes of the pituitary. (Note: PRL = prolactin.)

portion of the adrenal gland. **Thyrotropic hormone (TSH)** influences the growth and activity of the thyroid gland. These adenohypophyseal hormones are all **tropic** hormones. In each case, a tropic hormone released by the anterior pituitary stimulates its target organ, which is also an endocrine gland, to secrete its hormones, which then exert their effects on other body organs and tissues.

Two other major hormones produced by the anterior pituitary are not directly involved in the regulation of other endocrine glands of the body. **Growth hormone (GH)** is a general metabolic hormone that plays an important role in determining body size. It is essential for normal retention of body protein and affects many tissues of the body. Its major effects, however, are exerted on the growth of muscle and the long bones of the body. **Prolactin (PRL),** acting synergistically with the female sex hormones (estrogens and progesterone), stimulates breast development and promotes and maintains lactation by the mammary glands after childbirth. Its function in males is unknown. **Melanocyte-stimulating hormone (MSH)** does not appear to be of major significance in humans; but it darkens the skin of reptiles and amphibians, and may similarly affect the melanocytes that produce pigment in human skin.

The anterior pituitary controls the activity of so many other endocrine glands that it has often been called the *master endocrine gland.* Its removal dramatically interferes with body metabolism: the gonads and adrenal and thyroid glands atrophy, and changes resulting from subsequent inadequate hormone production become obvious. However, the anterior pituitary is not autonomous in its control because the release of the anterior pituitary hormones is controlled by neurosecretions, releasing or inhibiting hormones, produced by the hypothalamus. These hypothalamic hormones are liberated into the hypophyseal portal system, which serves the circulatory needs of the anterior pituitary (Figure 24.2).

NEUROHYPOPHYSEAL HORMONES The neurohypophysis, or posterior pituitary, is not an endocrine gland in a strict sense, because it does not synthesize the hormones it releases. Instead, it acts as a storage area for two hormones transported to it from the paraventricular and supraoptic nuclei of the hypothalamus. The first of these hormones is **oxytocin,** which stimulates powerful uterine contractions during birth and coitus, and also causes milk ejection in the lactating mother. The second, **antidiuretic hormone (ADH),** causes the distal and collecting tubules of the kidneys to reabsorb water from the urinary filtrate, thereby reducing urine production and conserving body water. It also plays a role in increasing blood pressure, because of its vasoconstrictor effect on the arterioles.

Hyposecretion of ADH results in dehydration from excessive urine output, a condition called *diabetes insipidus.* Individuals with this condition experience an insatiable thirst. ■

Thyroid Gland

The thyroid gland is composed of two lobes joined by a central mass, or isthmus. It is located in the throat, just inferior to the larynx. It produces two major hormones: **thyroid hormone (TH),** collectively **T_4 (thyroxine)** and **T_3 (triiodothyronine),** and **calcitonin.**

Because the primary function of thyroid hormone is to control the rate of body metabolism and cellular oxidation, it affects every cell in the body. In addition, it is an important regulator of tissue growth and development, especially in the reproductive and nervous systems.

Hyposecretion of thyroid hormone leads to a condition of mental and physical sluggishness, which is called *myxedema* in the adult. ■

Calcitonin, also called *thyrocalcitonin,* decreases blood calcium levels by causing calcium to be deposited in the bones. It acts antagonistically to parathyroid hormone, the hormonal product of the parathyroid glands.

- Try to palpate your thyroid gland by placing your fingers against your windpipe. As you swallow, the thyroid gland will move up and down on the sides and front of the windpipe.

Parathyroid Glands

The parathyroid glands are found embedded in the posterior surface of the thyroid gland. Typically, there are two small oval glands on each lobe. They secrete **parathyroid hormone (PTH),** or **parathormone,** the most important regulator of calcium-phosphate ion homeostasis of the blood. When blood calcium levels decrease below a certain critical level, the parathyroids release PTH, which causes bone to release calcium from the matrix and causes the kidney to decrease reabsorption of phosphate from the filtrate. If blood calcium levels fall too low, *tetany* results and may be fatal.

Adrenal Glands

The two bean-shaped adrenal, or suprarenal, glands are located atop, or close to, a kidney. Anatomically, the **adrenal medulla** develops from neural crest tissue and is directly controlled by sympathetic nervous system neurons. The cells of the medulla respond to this stimulation by releasing **epinephrine** (80%) or **norepinephrine** (20%), both of which act in conjunction with the sympathetic nervous system to elicit the "flight or fight" response to stressors.

The **adrenal cortex** produces three major groups of steroid hormones—the mineralocorticoids, the glucocorticoids, and the gonadocorticoids—collectively called **corticosteroids.** The **mineralocorticoids,** chiefly **aldosterone,** regulate water and electrolyte balance in the extracellular fluids primarily by regulating sodium ion reabsorption by kidney tubules. The **glucocorticoids** (cortisone, hydrocortisone, and corticosterone) enable the body to resist long-term stress, primarily by increasing blood glucose levels. Because of their anti-inflammatory properties at pharmacologic levels, they are often administered to decrease tissue edema and to counteract vascular dilation. The **gonadocorticoids,** or **sex hormones,** produced by the adrenal cortex are chiefly androgens (male sex hormones); but some estrogens (female sex hormones) are also formed. The gonadocorticoids are produced throughout life in relatively insignificant amounts; however, hypersecretion of these hormones produces abnormal hairiness (*hirsutism*), and masculinization occurs.

Pancreas

The pancreas is found close to the stomach. It functions as both an endocrine and an exocrine gland, producing digestive enzymes as well as **insulin** and **glucagon,** important hormones concerned with the regulation of blood sugar levels.

Elevated blood glucose levels stimulate release of insulin, which decreases blood sugar levels, primarily by accelerating the transport of glucose into the body cells, where it is oxidized for energy or converted to glycogen or fat for storage.

Hyposecretion of insulin or some deficiency in insulin receptors leads to *diabetes mellitus,* which is characterized by inability of body cells to utilize glucose and subsequent loss of glucose in the urine. Alterations of protein and fat metabolism also occur, but these are probably secondary to derangements in carbohydrate metabolism. ■

Glucagon acts antagonistically to insulin. Its release is stimulated by low blood glucose levels, and its action is basically hyperglycemic. Its primary target organ is the liver. It stimulates the liver to break down its glycogen stores to glucose and subsequently to release the glucose to the blood.

The Gonads

The female gonads, or ovaries, are paired, almond-sized organs located in the pelvic cavity. In addition to producing the female sex cells (ova), the ovaries produce two steroid hormone groups, the **estrogens** and **progesterone.** The endocrine and exocrine functions of the ovaries do not begin until the onset of puberty, when the anterior pituitary gonadotropic hormones prod the ovary into action that produces rhythmic ovarian cycles in which ova develop, and hormone levels rise and fall. The estrogens are responsible for the development of the secondary sex characteristics of the female at puberty (primarily maturation of the reproductive organs and development of breasts) and act with progesterone to bring about the cyclic changes of the uterine lining that occur during the menstrual cycle. The estrogens also help prepare the mammary glands for lactation.

Progesterone, as already noted, acts with estrogen to bring about the menstrual cycle. During pregnancy it maintains the uterine musculature in a quiescent state and helps to prepare the breast tissue for lactation.

The paired oval testes of the male are suspended in a pouchlike sac, the scrotum, outside the pelvic cavity. In addition to producing the male sex cells, sperm, the testes produce the male sex hormone, **testosterone.** Testosterone promotes the maturation of the male reproductive system accessory structures, brings about the development of the secondary sex characteristics, and is responsible for the sexual drive, or libido. Both the endocrine and exocrine functions of the testes begin at puberty, under the influence of the anterior pituitary gonadotropins.

Two glands not mentioned earlier as major endocrine glands should also be briefly considered here, the thymus and the pineal gland.

Thymus

The thymus is a bilobed gland situated in the upper thorax, posterior to the sternum and overlying the heart and lungs. Conspicuous in the infant, it begins to atrophy at puberty; and by old age it is no longer functional as an endocrine-producing gland. Researchers now believe that the thymus produces a hormone called **thymosin,** and that during early life the thymus acts as an incubator for the maturation and specialization of a unique population of white blood cells called T lymphocytes (T cells). The T lymphocytes are responsible for the cellular immunity aspect of body defense—that is, rejection of foreign grafts and tumors, and destruction of virus-infected cells.

Pineal Body

The pineal body, or **epiphysis cerebri,** is a small cone-shaped gland located in the roof of the third ventricle of the brain. Its major hormone product is **melatonin.**

The endocrine role of the pineal body is still controversial, but it appears to play a role in biological rhythms (particularly mating and migratory behavior of animals). In humans, melatonin seems to exert some inhibitory effect on the reproductive system (especially on the ovaries) that prevents precocious sexual maturation. Serotonin, a neurotransmitter produced by many regions of brain tissue, is known to be a chemical precursor of melatonin in the pineal body.

Once you are satisfied that you can name and locate the endocrine organs, and you have labeled Figure 24.1, use an appropriate reference to define the following pathologic conditions (which have not been described here) resulting from hypersecretion or hyposecretion of the various hormones.

acromegaly ______________________________

____

____

Addison's disease ______________________________

____

____

cretinism ______________________________

____

____

Cushing's syndrome ______________________________

____

____

pituitary dwarfism ______________________________

____

____

eunuchism ______________________________

____

____

exophthalmic goiter ______________________________

____

____

gigantism ______________________________

____

____

Simmonds' disease ______________________________

____

____

MICROSCOPIC ANATOMY OF SELECTED ENDOCRINE GLANDS (OPTIONAL)

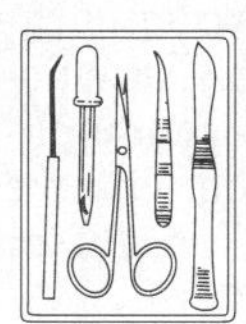

To prepare for the histological study of the endocrine glands, obtain a microscope, one each of the slides listed in the list of materials, and colored pencils. We will study only organs in which it is possible to identify the endocrine-producing cells. Compare your observations with the line drawings of endocrine tissue photomicrographs seen in Figures 24.3 and 24.4.

Thyroid Gland

1. Scan the thyroid under low power, noting the spherical sacs (follicles) containing a pink-stained material (colloid). Stored T_3 and T_4 is attached to the protein colloidal material in the follicles as **thyroglobulin** and is released gradually to the blood. Compare the tissue viewed to Figure 24.3a and plate 18 of the histology atlas.

2. Observe the tissue under high power. Note that the walls of the follicles are formed by simple cuboidal epithelial cells that synthesize the follicular products. The **parafollicular,** or **C, cells** you see between the follicles are responsible for calcitonin production.

3. Color appropriately two or three follicles in Figure 24.3a. Label the colloid, cuboidal epithelial cells (follicular cells), and parafollicular cells.

When the thyroid gland is actively secreting, the follicles appear small, and the colloidal material has a ruffled border. When the thyroid is hypoactive or inactive, the follicles are large and plump. What is the physiological state of the tissue you have been viewing?

____

____

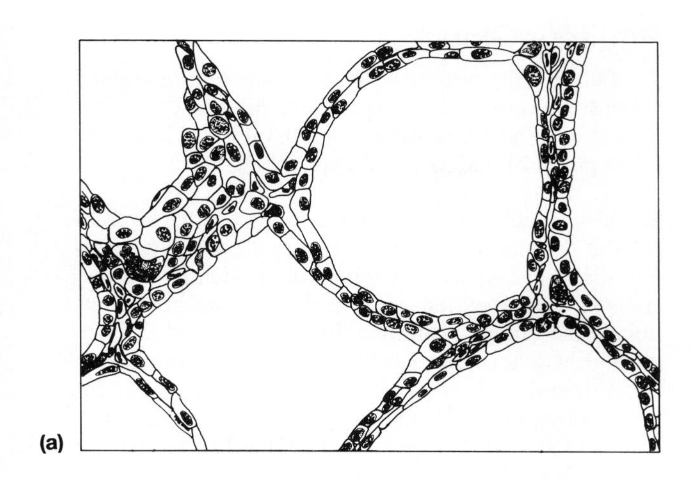

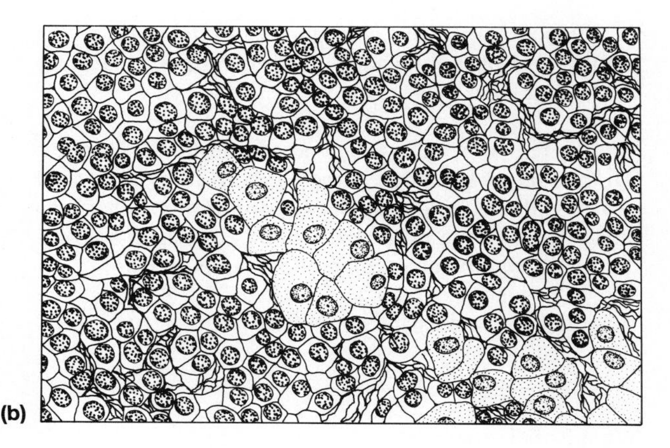

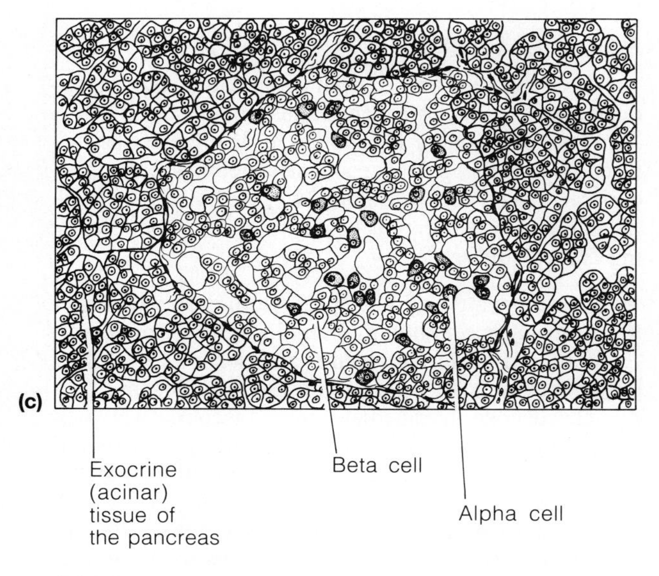

F24.3

Line drawings of photomicrographs of selected endocrine organs. (a) Thyroid, see plate 18 of the histology atlas; (b) parathyroid, see plate 19; (c) pancreas islets of Langerhans and acinar cells, see plate 20.

Parathyroid Glands

1. Observe the parathyroid tissue under low power to view its two major cell types, **chief cells** and **oxyphil cells.** (See Figure 24.3b and plate 19.) The chief cells, which are thought to bear the major responsibility for the synthesis of PTH, are small and abundant and arranged in thick branching cords. The function of the scattered, much larger oxyphil cells is unknown.

2. Color a small portion of the parathyroid tissue in Figure 24.3b. Label the chief cells, the oxyphil cells, and the connective tissue matrix.

Pancreas

1. Observe pancreas tissue under low power to identify the roughly circular **islets of Langerhans,** the endocrine portions of the pancreas. The islets are scattered amid the more numerous acinar cells and stain differentially (usually lighter), which makes their identification possible (see Figure 35.12 on p. 323 and plate 40 of the histology atlas).

2. Focus on an islet and examine its cells under high power. Note that the islet cells are densely packed and have no definite arrangement. In contrast, the cuboidal acinar cells are arranged around secretory ducts. Unless special stains are used, it will not be possible to distinguish the **alpha cells,** which tend to cluster at the periphery of the islets and produce glycogen, from the **beta cells,** which synthesize insulin. With these specific stains, the beta cells are larger and stain gray-blue; and the alpha cells are smaller and appear bright pink (see Figure 24.3c and plate 20 of the histology atlas). What is the product of the acinar cells?

____

____

3. Draw a section of the pancreas below. Label the islets and the acinar cells. If possible, differentiate the alpha and the beta cells by color.

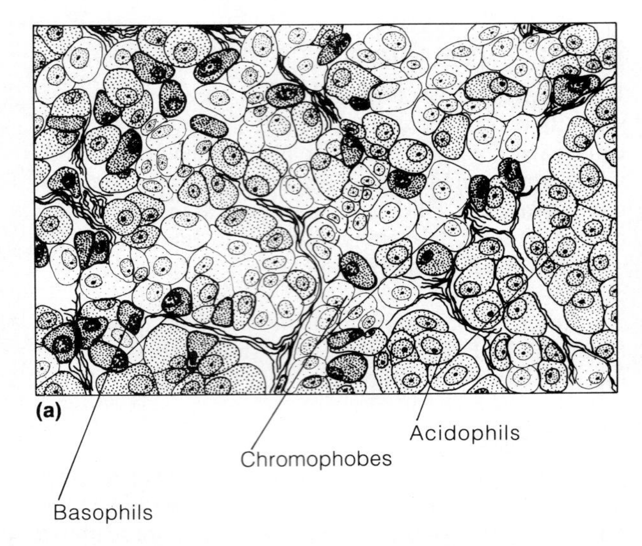

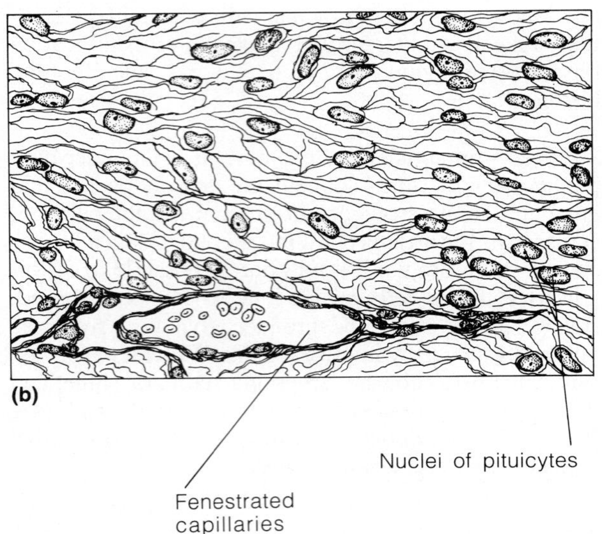

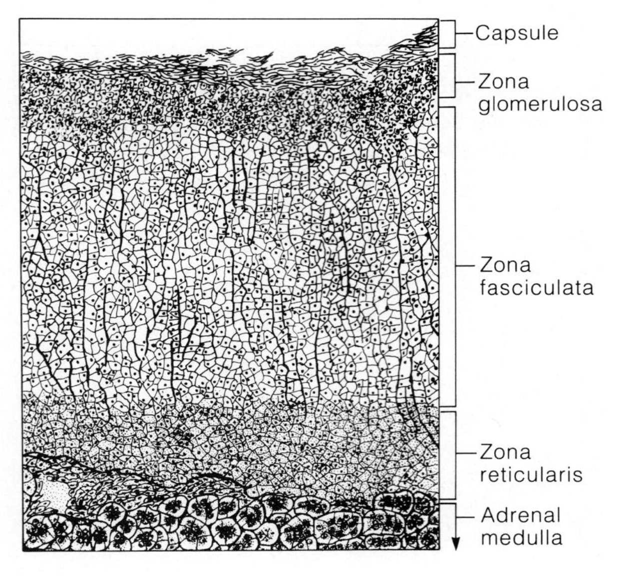

F24.4

Line drawings of photomicrographs. (a) Anterior pituitary, see plate 21; (b) posterior pituitary, see plate 22; (c) adrenal gland, see plate 23.

Pituitary Gland

1. Observe the general structure of the pituitary gland under low power to identify the glandular anterior pituitary and the neural posterior pituitary. Figure 24.4a and b and plates 21 and 22 should help you get started.

2. Using the high-power lens, focus on the nests of cells of the anterior pituitary (Figure 24.4a). When differential stains are used, it is possible to identify the specialized cell types that secrete the specific hormones. Locate the orange-staining **acidophil cells,** which produce growth hormone and PRL, and the **basophil cells,** containing deep-blue granules, that are responsible for the production of the other four anterior pituitary hormones (TSH, ACTH, FSH, and LH). **Chromophobes,** the third cellular population, do not take up the stain, and appear rather dull and colorless. The role of the chromophobes is controversial, but they are apparently not involved in hormone production.

3. Use appropriately colored pencils to indicate the acidophils, basophils, and chromophobes in Figure 24.4a; and label these cells.

4. Switch your focus to the posterior pituitary (Figure 24.4b). Observe the nerve fibers (axons of hypophyseal neurons) that compose most of this portion of the pituitary. Also note the **pituicytes,** glial cells which are randomly distributed among the nerve fibers.

What two hormones are stored here? ____________

____

What is their source? ______________________

Adrenal Gland

1. Hold the slide of the adrenal gland up to the light to distinguish outer cortical and inner medulla areas. Then scan the cortex under low power to distinguish the differences in cell appearance and arrangement in the three cortical areas. Refer to Figure 24.4c and histology plate 23 and to the descriptions below to identify the following cortical areas:

- The connective tissue capsule of the adrenal gland.
- The outermost **zona glomerulosa,** where mineralocorticoid secretion occurs and where the tightly packed cells are arranged in spherical clusters.
- The deeper intermediate **zona fasciculata,** which produces glucocorticoids. This is the thickest part of the cortex, and its cells are arranged in parallel cords.
- The innermost cortical zone (abutting the medulla), the **zona reticularis,** which produces sex hormones and some glucocorticoids. The cells here stain intensely and form a branching network.

2. Switch focus to view the lightly stained cells of the adrenal medulla under high power. Note their clumped arrangement and their relatively large ovoid shape.

What hormonal products are produced by the medulla?

___

3. Draw a representative area of each of the adrenal regions, indicating in your sketch the differences in relative cell size and arrangement.

Zona glomerulosa

Zona fasciculata

Zona reticularis

Adrenal medulla

Ovary

Because you will consider the ovary in greater histologic detail when you study the reproductive system, the objective in this laboratory exercise is just to identify the endocrine-producing parts of the ovary.

1. Scan an ovary slide under low power, and look for a **vesicular (Graafian) follicle,** a circular arrangement of cells enclosing a central cavity (see Figure 40.4, p. 355 and plate 24 of the histology atlas). This structure synthesizes estrogens.

2. Examine the vesicular follicle under high power, identifying the follicular cells that produce estrogen, the antrum (fluid-filled cavity), and developing ovum (if present). The ovum will be the largest cell in the follicle.

3. Draw and color a vesicular follicle below, labeling the antrum, follicle cells, and developing ovum.

4. Switch to low power, and scan the slide to find a **corpus luteum,** a large amorphous-looking area that produces progesterone. A corpus luteum is shown in plate 25 of the histology atlas.

Testis

1. Examine a section of a testis under low power. Identify the seminiferous tubules, which produce sperm, and the **interstitial cells,** which produce testosterone. The interstitial cells are scattered between the seminiferous tubules in the connective tissue matrix. The photomicrograph of seminiferous tubules in plate 49 of the histology atlas will be helpful here.

2. Draw a representative area of the testis in the space provided. Label the seminiferous tubules and area of the interstitial cells.

25 EXERCISE

Experiments on Hormonal Action

OBJECTIVES

1. To describe and explain the effects of hyperinsulinism.
2. To describe and explain the effect of epinephrine on the heart.
3. To understand the physiologic (and clinical) importance of metabolic rate measurement.
4. To investigate the effect of hypo-, hyper-, and euthyroid conditions on oxygen consumption and metabolic rate.
5. To assemble the necessary apparatus and properly use a manometer to obtain experimental results.
6. To calculate metabolic rate in terms of ml O_2/kg/hr.

MATERIALS

500- or 600-ml beakers
10% glucose solution
Commercial insulin in dropper bottle (insulin, zinc—100 units/ml)
Small (1½–2-inch) freshwater fish (bluegill, sunfish, or goldfish—listed in order of preference)

Frog (*Rana pipiens*)
1:10,000 epinephrine (Adrenalin) solution in dropper bottles
Plastic disposable gloves
Dissecting pan, instruments, and pins
Frog Ringer's solution in dropper bottle

Glass desiccator, manometer, 20-ml glass syringe, two-hole cork, and T-valve (1 for every 3 to 4 students)
Soda lime (desiccant)
Hardware cloth squares
Petrolatum
Rubber tubing; tubing clamps; scissors
3-in. pieces of glass tubing
Animal balances
Gloves for handling animal
Young rats (5 weeks) of the same sex, obtained 2 weeks prior to the laboratory session and treated as follows for 14 days:

Group 1: control group—fed normal rat chow and water
Group 2: experimental group A—fed normal rat chow, and drinking water containing 0.02% 6-*n*-propylthiouracil*
Group 3: experimental group B—fed rat chow containing desiccated thyroid (2% by weight) and normal drinking water

Chart set up on chalkboard so that the student groups can record their computed metabolic rate figures under the appropriate headings

* Note to instructor: 6-*n*-propylthiouracil (PTU) is degraded by light, and should be stored in light-resistant containers or in the dark.

The endocrine system exerts many complex and interrelated effects on the body as a whole, as well as on specific organs and tissues. Most scientific knowledge about this system is contemporary, and new information is constantly being presented. Most experiments on the endocrine system require relatively large laboratory animals; are time consuming (requiring days to weeks of observation); and involve technically difficult surgical procedures to remove the glands or parts of them, all of which makes it difficult to conduct more general types of laboratory experiments. Nevertheless, the three technically unsophisticated experiments presented here should illustrate how dramatically hormones affect body functioning.

To conserve laboratory specimens, the experiments can be conducted by groups of four students. The use of larger working groups should not detract from benefits gained, because the major value of these experiments lies in observation.

EFFECTS OF HYPERINSULINISM

Many people with diabetes mellitus need injections of insulin to maintain a proper physiologic balance between the glycogen stored in liver and muscle tissue and the glucose in the blood. Adequate levels of blood glucose are essential for the proper functioning of the nervous system; thus, the administration of insulin must be carefully controlled. If blood glucose levels fall precipitously, the patient will go into insulin shock.

A small fish will be used to demonstrate the effects of hyperinsulinism. Because the action of insulin on the fish parallels that in the human, this experiment should provide valid information concerning its administration to humans.

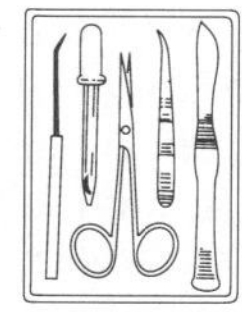

1. Prepare two beakers. Using a wax marker, mark one A and the other B. To beaker A, add 200 ml of water and 10 to 15 drops of commercial insulin. To beaker B, add 200 ml of 10% glucose solution.

2. Place a small fish in beaker A, and carefully observe its actions as the insulin diffuses into its bloodstream through the capillary circulation of its gills.

Approximately how long did it take for the fish to become unconscious?

____

____

What types of activity did you observe in the fish before it lost consciousness?

____

____

3. When the fish is unconscious, carefully transfer it to beaker B, and observe its actions. What happens to the fish after it is transferred to beaker B?

____

____

Approximately how long did it take for this recovery?

____

4. After all observations have been made and recorded, carefully return the fish to the aquarium.

EFFECT OF EPINEPHRINE ON THE HEART

As noted in Exercise 24, the adrenal medulla and the sympathetic nervous system are closely interrelated, specifically because the cells of the adrenal medulla and the postganglionic axons of the sympathetic nervous system both release catecholamines. This experiment demonstrates the effects of epinephrine on the frog heart.

⚠ 1. Obtain a frog, dissecting instruments, pan, and disposable gloves and bring them to your laboratory bench. Don the gloves before beginning Step 2.

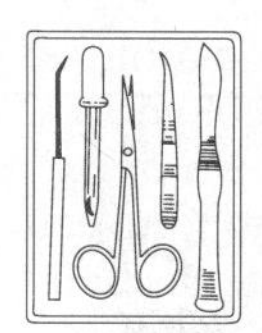

2. Destroy the nervous system of the frog. Insert one blade of a scissors into its mouth as far as possible; and quickly cut off the top of its head, posterior to the eyes. Then identify the spinal cavity, and insert a dissecting needle into it to destroy the spinal cord.

3. Place the frog dorsal side down on a dissecting pan, and carefully open its ventral body cavity by making a vertical incision with the scissors.

4. Identify the beating heart, and carefully cut through the saclike pericardium to expose the heart tissue.

5. Visually count the heart rate for 1 minute and record it below. Keep the heart moistened with frog Ringer's solution during this interval.

Beats per minute: ______________________________

6. Flush the heart with epinephrine solution. Record the heart beat rate per minute for 5 consecutive minutes.

minute 1 ______________

minute 2 ______________

minute 3 ______________

minute 4 ______________

minute 5 ______________

What was the effect of epinephrine on heart rate?

____

Was the effect long-lived? ______________________________

7. Dispose of the frog in an appropriate container, and clean the dissecting pan and instruments before returning them to the supply area.

EFFECT OF THYROID HORMONE ON METABOLIC RATE

Metabolism is a broad term referring to all chemical reactions that are necessary to maintain life. It involves enzymatically controlled processes in which substances are broken down to simpler substances, *catabolism,* and processes in which larger molecules or structures are built from smaller ones, *anabolism.* During catabolic reactions, energy is released as chemical bonds are broken. Some of the liberated energy is captured to make ATP, the energy-rich molecule used by body cells to energize all their activities; the balance is lost in the form of thermal energy, or heat. Maintenance of body temperature is critically related to the heat-liberating aspects of metabolism.

Various foodstuffs make different contributions to the process of metabolism. For example, carbohydrates, particularly glucose, are generally broken down or oxidized to make ATP, whereas fats are utilized to form cell membranes and myelin sheaths and to insulate the body with a fatty cushion. (Fats are used secondarily for producing ATP when there are inadequate carbohydrates in the diet.) Proteins and amino acids tend to be carefully conserved by body cells—and understandably so, because most structural elements of the body are proteinaceous in nature.

Thyroid hormone (collectively thyroxine or T_4, and T_3), produced by the thyroid gland, is the single most important hormone influencing the rate of cellular metabolism and body heat production.

Under conditions of excess thyroxine production (hyperthyroidism), an individual's basal metabolic rate (BMR) increases; heat production and oxygen consumption increase; and the individual tends to lose weight, and become heat-intolerant and irritable. Conversely, hypothyroid individuals become obese; are cold-intolerant because of their low BMR; and become mentally and physically sluggish. ■

Many factors other than thyroid hormone levels contribute to metabolic rate (for example, body size and weight, age, and activity level), but the focus of the following experiments is to investigate how differences in thyroid hormone concentration impact on metabolism.

Three groups of laboratory rats will be used. The *control group* animals are assumed to be euthyroid and to have normal metabolic rates for their relative body weights. *Experimental group A* animals have received water containing the chemical *6-n-propylthiouracil (PTU),* which counteracts or antagonizes the effects of thyroxine in the body. *Experimental group B* animals have been fed rat chow containing dried thyroid tissue, which contains thyroid hormone. The rates of oxygen consumption (an indirect means of determining metabolic rate) in the animals of the three groups will be measured and compared to investigate the effects of hyperthyroid, hypothyroid, and euthyroid conditions.

Oxygen consumption will be measured with a simple respirometer-manometer apparatus. Each animal will be placed in a closed chamber containing soda lime. As carbon dioxide is evolved and expired, it will be absorbed by the soda lime; therefore, the pressure changes observed will indicate the volume of oxygen consumed by the animal during the testing interval. Students will work in groups of 3 to 4 to assemble the apparatus, make preliminary weight measurements on the animals, and record the data.

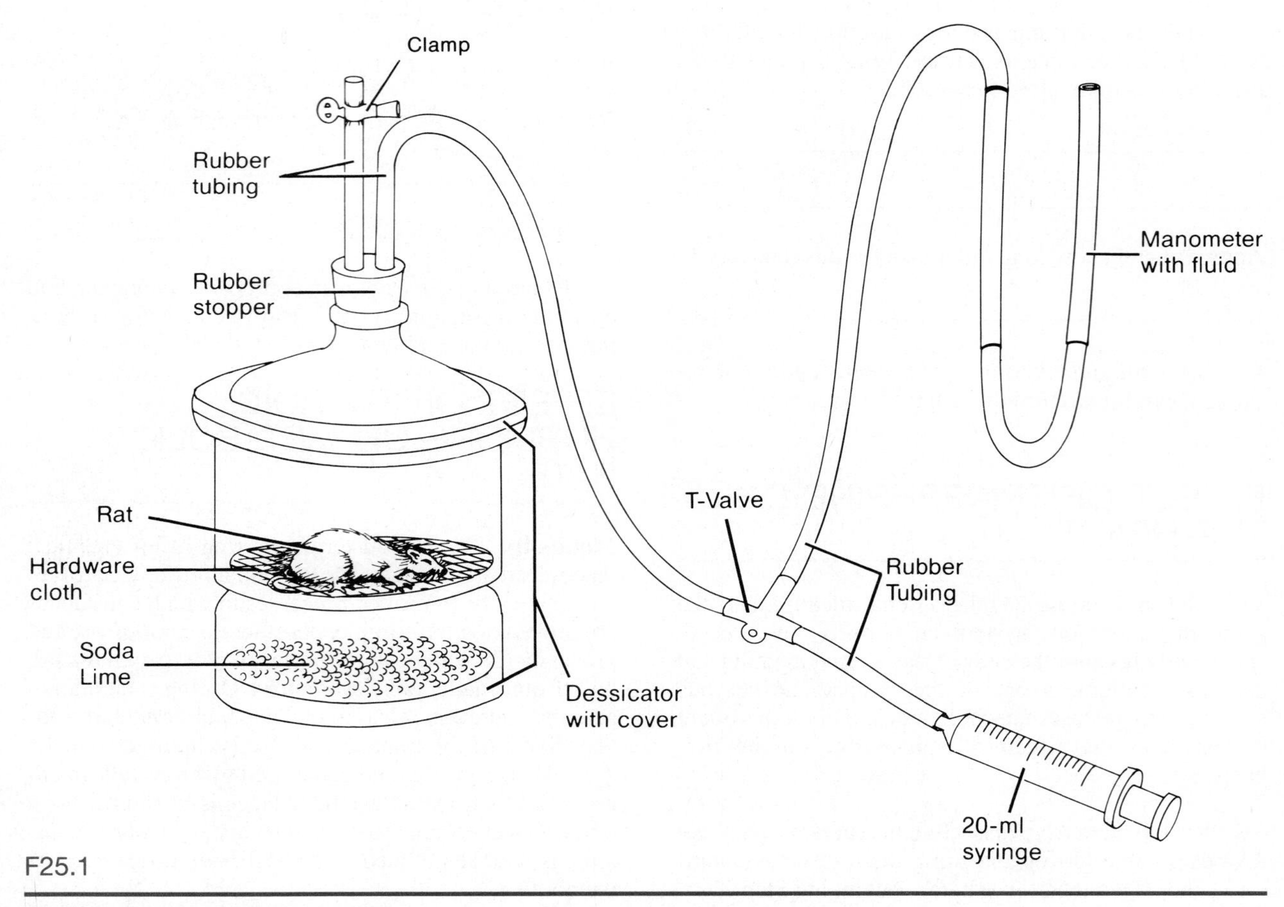

F25.1

Respirometer-manometer apparatus.

Preparation of the Respirometer-Manometer Apparatus

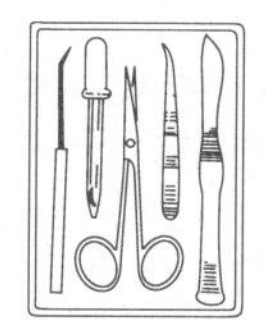

1. Obtain a desiccator, a two-hole rubber stopper, a 20-ml glass syringe, a hardware cloth square, a T-valve, scissors, rubber tubing, two short pieces of glass tubing, and petrolatum; and bring them to your laboratory bench. The apparatus will be assembled as illustrated in Figure 25.1.

2. Shake soda lime into the bottom of the desiccator to thoroughly cover the glass bottom; then place the hardware cloth on the ledge of the desiccator, above the soda lime. The hardware cloth should be well above the soda lime so that the animal will not be able to touch it. Soda lime is quite caustic and can cause chemical burns.

3. Lubricate the ends of the two pieces of glass tubing with petrolatum, and twist them into the holes in the rubber stopper until their distal ends can be seen protruding from the opposite side of the stopper. *Do not plug the tubing with petrolatum.* Place the stopper into the desiccator cover, and temporarily set the cover on the desiccator.

4. Cut off a short (3-in.) piece of rubber tubing, and attach it to the top of one piece of glass tubing extending from the stopper. Cut and attach a 12- to 14-in. piece of rubber tubing to the other glass tubing. Insert the T-valve stem into the distal end of the longer-length tubing.

5. Cut another short piece of rubber tubing; attach one end to the T-valve and the other end to the nib of the 20-ml syringe. Remove the plunger of the syringe, and grease its sides generously with petrolatum. Insert the plunger back into the syringe barrel, and work it up and down to evenly disperse the petrolatum on the inner walls of the syringe. Then pull the plunger out to the 20-ml marking.

6. Cut a piece of rubber tubing long enough to reach from the third arm of the T-valve to one arm of the manometer. (The manometer should be partially filled with water so that a U-shaped water column is seen.) Attach the tubing to the T-valve and the manometer arm.

7. Remove the desiccator cover, and generously grease its bottom edge with petrolatum. Place the cover back on the desiccator, and firmly move it from side to side to spread the lubricant evenly.

8. Test the system for leaks as follows: Firmly clamp the *short* length of rubber tubing extending from the stopper. Now gently push in on the plunger of the syringe. If the system is properly sealed, the fluid in the manometer will move away from the rubber tubing attached to its arm. If there is an air leak, the manometer fluid level will not change; or it will change and then return to its original level. If either of these events occurs, check all glass-to-glass or glass-to-rubber tubing connections. Smear additional petrolatum on suspect areas and test again. The apparatus must be airtight before experimentation can begin.

9. After ensuring that there are no leaks in the system, unclamp the short rubber tubing; and remove the desiccator cover.

Preparation of the Animal

1. Don the heavy animal-handling gloves, and obtain one of the animals as directed by your instructor. Handling it gently, weigh the animal, on the animal balance, to the nearest 0.1 g.

2. Carefully place the animal on the hardware cloth in the desiccator. The objective is to measure oxygen usage at basal levels, so you do not want to prod the rat into high levels of activity, which would produce errors in your measurements.

3. Record the animal's group (control, or experimental group A or B) and its weight in kilograms (that is, weight in grams/1000) on the data sheet on p. 239.

Equilibration of the Chamber

1. Place the lid on the desiccator, and move it slightly from side to side to seal it firmly.

2. Leave the short tubing unclamped for 7 to 10 minutes to allow temperature equilibration in the chamber. (Because the animal's body heat will warm the air in the container, the air will expand initially. This must be allowed to occur before any measurements of oxygen consumption are taken. Otherwise, it would appear that the animal is evolving oxygen rather than consuming it.)

Determination of Oxygen Consumption of the Animal

1. Once again clamp the short rubber tubing extending from the desiccator lid stopper.

2. Check the manometer to make sure that the fluid level is the same in both arms. (If not, manipulate the syringe plunger to make the fluid levels even.) Record the time and the position of the bottom of the plunger (use ml marking) in the syringe.

3. Observe the manometer fluid levels at 1-minute intervals. Each time, carefully push the syringe plunger farther into the barrel to even the fluid levels in the manometer. Determine the amount of oxygen used per minute interval by computing the difference in air volumes within the syringe. For example, if after the first minute you push the plunger from the 20- to the 17-ml marking, the oxygen consumption is 3 ml/min. Then, if the plunger is pushed from 17 ml to 15 ml at the second minute reading, the oxygen usage during minute 2 would be 2 ml.

4. Continue taking readings (and recording, on the data sheet, the oxygen consumption per minute interval) for 10 consecutive minutes or until the syringe plunger has been pushed nearly to the 0-ml mark. Then unclamp the short rubber tubing, remove the desiccator cover, and allow the apparatus to stand open for 2 to 3 minutes to flush out the stale air.

5. Repeat the recording procedures for another 10-minute interval. Make sure that you equilibrate the temperature within the chamber before beginning this second recording series.

6. After you have recorded the animal's oxygen consumption for two 10-minute intervals, unclamp the short rubber tubing, remove the desiccator lid, and carefully return the rat to its cage.

Computation of Metabolic Rate

Metabolic rate calculations are generally reported in terms of $Kcal/m^2/hr$, and require that corrections be made to present the data in terms of standardized pressure and temperature conditions. These more complex calculations will not be used here, because the object is simply to arrive at some generalized conclusions concerning the effect of thyroid hormone on metabolic rate.

1. Obtain the average figure for milliliters of oxygen consumed per 10-minute interval by adding up the minute-interval consumption figures for each 10-minute testing series and dividing the total by 2.

$$\frac{\text{Total ml } O_2 \text{ test series 1} + \text{total ml } O_2 \text{ test series 2}}{2}$$

2. Determine oxygen consumption per hour using the following formula, and record the figure on the data sheet:

$$\frac{\text{Average ml } O_2 \text{ consumed}}{10 \text{ min}} \times \frac{60 \text{ min}}{\text{hr}} = \text{ml } O_2/\text{hr}$$

3. To determine the metabolic rate in milliliters of oxygen consumed per kilogram of body weight per hour so that the results of all experiments can be compared, divide the figure just obtained in procedure 2 by the animal's weight in kilograms.

$$\text{Metabolic rate} = \frac{\text{ml } O_2/\text{hr}}{\text{wt in kg}} = \_\_\_\_\_\_ \text{ ml } O_2/\text{kg}/\text{hr}$$

Record the metabolic rate on the data sheet and also in the appropriate space on the chart on the chalkboard.

4. Once all groups have recorded their final metabolic rate figures on the chalkboard, average the results of each animal grouping to obtain the mean for each experimental group. Also record this information on your data sheet.

Student Data Sheet

Animal used from group: ______________________

14-day prior treatment of animal: ______________________

Body weight in grams: ______________________/1000 = body weight in kg: __________

O_2 consumption/min: Test 1

Syringe reading __________ ml

min 1 __________

min 2 __________

min 3 __________

min 4 __________

min 5 __________

min 6 __________

min 7 __________

min 8 __________

min 9 __________

min 10 __________

__________ Total O_2/10 min

O_2 consumption/min: Test 2

Syringe reading __________ ml

min 1 __________

min 2 __________

min 3 __________

min 4 __________

min 5 __________

min 6 __________

min 7 __________

min 8 __________

min 9 __________

min 10 __________

__________ Total O_2/10 min

Average ml O_2 consumed/10 min: ______________________

Milliliters O_2 consumed/hr: ______________________

Metabolic rate: ______________________ ml O_2/kg/hr

Averaged class results:

Metabolic rate of control animals: __________ ml O_2/kg/hr

Metabolic rate of experimental group A animals (PTU-treated): ______________________ ml O_2/kg/hr

Metabolic rate of experimental group B animals (desiccated thyroid–treated): __________ ml O_2/kg/hr

26 EXERCISE

Blood

OBJECTIVES

1. To name the two major components of blood and to slate their average percentages in whole blood.
2. To cite the composition and functional importance of plasma.
3. To define *formed elements,* list the cell types composing them, cite their relative percentages, and describe their major functions.
4. To identify erythrocytes, basophils, eosinophils, monocytes, lymphocytes, and neutrophils when provided with a microscopic preparation or an appropriate diagram.
5. To define *anemia, polycythemia, leukopenia,* and *leukocytosis.*
6. To conduct the following blood test determinations in the laboratory, and to state their norms and the importance of each:

hematocrit	*differential white blood cell count*
hemoglobin determination	*ABO and Rh blood typing*
clotting time	*RBC count (norm only)*
sedimentation rate	*WBC count (norm only)*

7. To discuss the reason for transfusion reactions resulting from the administration of mismatched blood.

MATERIALS

Compound microscope
Immersion oil
Models/Charts of blood cells

General supply area:

Plasma (obtained from an animal hospital by centrifuging animal (e.g., cattle) blood
Wide range pH paper

Stained smears of human blood or heparinized blood (e.g., dog blood) obtained from an animal hospital (if desired by instructor)
Clean microscope slides
Sterile lancets
Glass stirring rods
Alcohol swabs (wipes)
Absorbent cotton balls
Wright's stain in dropper bottle
Distilled water in dropper bottle
Disposable gloves
Test tubes
Test tube racks
Pipette cleaning solutions—(1) freshly prepared 10% household bleach solution; (2) distilled water; (3) 70% ethyl alcohol; (4) acetone
Battery jar or large beaker containing freshly made 10% bleach solution
Spray bottles containing 10% bleach solution
Disposable autoclave bag

Because many blood tests are to be conducted in this exercise, it seems advisable to set up a number of appropriately labeled supply areas for the various tests. These are designated below.

Hematocrit supply area:

Heparinized capillary tubes
Microhematocrit centrifuge and reading gauge (if the reading gauge is not available, a millimeter ruler may be used)
Seal-ease (Clay Adams Co.) or modeling clay

Hemoglobin-determination supply area:

Tallquist hemoglobin scales and test paper or a hemoglobinometer and hemolysis applicator

Sedimentation-rate supply area:

Landau Sed-rate pipettes with tubing and rack
Wide-mouthed bottle of 5% sodium citrate
Mechanical suction device
Millimeter ruler

Coagulation-time supply area:

Capillary tubes (nonheparinized)
Triangular file

Blood-typing supply area:

Blood-typing sera (anti-A, anti-B, and anti-Rh [D])
Rh-typing box
Wax marker
Toothpicks
Clean microscope slides

Cholesterol-measurement supply area:

Mechanical pipettor
0.1-ml pipettes (3)
10–15-ml graduated cylinder
Test tubes (3)
Cholesterol standard solution (200 mg/100 ml)
Cholesterol reagent (Harleco)
Water bath set at 37°C
Spectrophotometer
Cuvettes for the spectrophotometer
Wax marker

In this exercise you will study the plasma and formed elements of blood, and conduct various hematological tests. These tests are extremely useful diagnostic tools for the physician, because blood composition (number and types of blood cells, and chemical composition) reflects the status of many body functions and malfunctions.

ALERT: The decision to use animal blood for testing or to have students test their own blood will be made by the instructor in accordance with the educational purpose of the student group. For example, for students in the nursing or laboratory technician curricula, learning how to safely handle human blood or human wastes is essential. If blood samples are provided and they are *human* blood samples, gloves should be worn while conducting the blood tests. If human blood is being tested, yours or that obtained from a clinical agency, precautions provided in the text for disposal of human waste **must be observed.** All soiled glassware is to be immersed in household bleach solution immediately after use, and disposable items (lancets, cotton balls, alcohol swabs, etc.) are to be placed in a disposable autoclave bag so that they can be sterilized before disposal.

COMPOSITION OF BLOOD

The blood circulating to and from the body cells, within the vessels of the vascular system, is a rather viscous substance that varies from a bright scarlet to a dull brick red, depending on the amount of oxygen it is carrying. The circulatory system of the average adult contains about 5.5 liters of blood.

Blood is classified as a type of connective tissue, because it is composed of a nonliving fluid matrix (**plasma**) in which living cells (**formed elements**) are suspended. The fibers typical of a connective tissue matrix become visible in blood only when clotting occurs. They then appear as fibrin threads, which form the structural basis for clot formation.

Over 100 different substances are dissolved or suspended in the plasma (Figure 26.1), which is over 90% water. These include nutrients, gases, hormones, various wastes and metabolites, many types of functional proteins, and mineral salts. The composition of plasma varies continuously as cells remove substances from, or add them to, the blood.

Three types of formed elements are present in blood. The most numerous are the **erythrocytes,** or *red blood cells* (*RBCs*), which are literally sacs of hemoglobin molecules that transport the bulk of the oxygen carried in the blood (and a small percentage of the carbon dioxide). **Leukocytes,** or *white blood cells* (*WBCs*), are part of the body's nonspecific defenses and the immune system; and **platelets** function in hemostasis (blood clot formation). Formed elements normally constitute 45% of whole blood, plasma the remaining 55%.

Physical Characteristics of Plasma

Go to the general supply area, and pour a few milliliters of plasma into a test tube. Also obtain some wide-range pH paper, and then return to your laboratory bench to make the following simple observations.

pH OF PLASMA Test the pH of the plasma with wide-range pH paper. Record the pH observed.

___

COLOR AND CLARITY OF PLASMA Hold the test tube up to a source of natural light, and note its color and degree of transparency. Is it clear, translucent, or opaque?

Color ___

Degree of transparency ___

CONSISTENCY Dip your finger and thumb into the plasma, and then press them firmly together for a few seconds. Gently pull them apart. How would you describe the consistency of plasma—slippery, watery, sticky, or granular? Record your observations.

___

Formed Elements of Blood

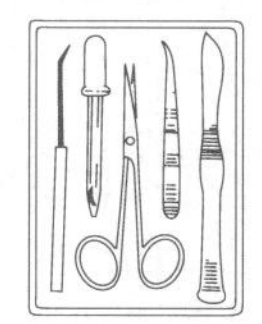

In this section, you will conduct your observations of blood cells on a preprepared (purchased) blood slide, or on a slide prepared from your own blood or blood provided by your instructor. Those using the purchased blood slide are to obtain a slide and begin their observations at step 6. Those testing blood provided by a clinical agency are to obtain a tube of the supplied blood, disposable gloves, and the supplies listed in step 1, except for the lancets and alcohol swabs. After donning the gloves, those students will jump to step 3b to begin their observations. If you are examining your own blood, you will perform all of the steps described below except 3b.

1. Obtain two glass slides, a glass rod, dropper bottles of Wright's stain and distilled water, two or three lancets, cotton balls, and alcohol swabs. Bring this equipment to the laboratory bench. Clean the slides thoroughly and dry them.

2. Open the alcohol swab packet, and scrub your third or fourth finger with the swab. (Because the pricked finger may be a little sore later, it is better to prepare a finger on the hand used less often.) Circumduct your arm (swing it in a cone-shaped path) for 10 to 15 seconds. This will dry the alcohol and cause your fingers to become engorged with blood. Then, open the lancet packet, and grasp the lancet by its blunt end. Quickly

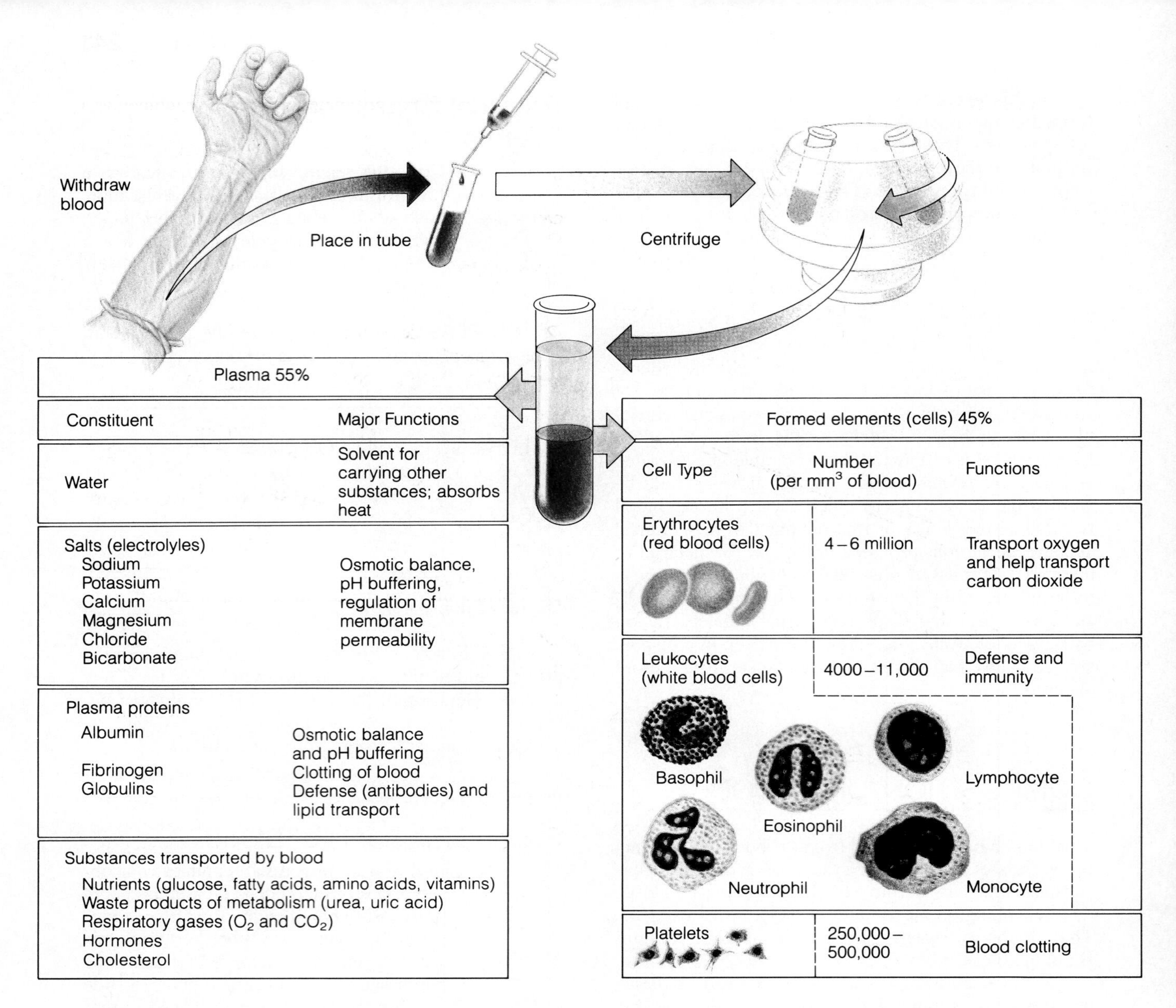

F26.1

The composition of blood.

jab the pointed end into the prepared finger to produce a free flow of blood. It is *not* a good idea to squeeze or "milk" the finger, as this forces out tissue fluid as well as blood. If the blood is not flowing freely, another puncture should be made.

Under no circumstances is a lancet to be used for more than one puncture. Dispose of the lancets in proper containers *immediately* after use.

3. (a) With a cotton ball, wipe away the first drop of blood; then allow another large drop of blood to form. Touch the blood to one of the cleaned slides, approximately ½ inch from the end. Then quickly (to prevent clotting) use the second slide to form a blood smear, as shown in Figure 26.2. When properly prepared, the blood is uniformly thin. If it appears streaked, the blood probably began to clot, or coagulate, before the smear was made, and another slide should be prepared. Continue at step 4.

(b) Dip a glass rod in the blood provided, and transfer a generous drop of blood to the end of a cleaned microscope slide. Then, as described in step 3a, use the second slide to make your blood smear.

4. Dry the slide by waving it in the air. When it is completely dry, it will look dull. Place it on a paper towel; and, counting the number of drops of stain used, flood it with Wright's stain. Allow the stain to remain on the slide for 3 or 4 minutes, and then flood the slide with an equal number of drops of distilled water. Allow the water and Wright's stain mixture to remain on the slide for 4 or 5 minutes, or until a metallic green film or scum is apparent on the fluid surface. Gently blow on the slide every minute or so to keep the water and stain mixed during this interval.

5. Rinse the slide with a stream of distilled water. Then flood it with distilled water, and allow it to lie flat until the slide becomes translucent and takes on a pink cast. Then stand the slide on its long edge on the paper towel, and allow it to dry completely. Once the slide is dry, you can begin your observations.

6. Obtain a microscope, and scan the slide under low power to find the most evenly stained area. Use the oil immersion lens to identify the formed elements. Read the following descriptions of cell types, and find each type in Figure 26.1. (Most are also shown in the color plates inside the front cover.) Then observe the slide carefully to identify each cell type.

ERYTHROCYTES, OR RED BLOOD CELLS Erythrocytes, which range in diameter from 5 to 10 μm (averaging 7.5 μm), are cells varying in color from brick red to pale pink, depending on the effectiveness of the stain. They have a distinctive biconcave disc shape and appear paler in the center than at the edge.

As you observe the slide, note that the red blood cells are by far the most numerous blood cells seen in the field. Their numbers average 4.5 million and 5.0 million cells per cubic millimeter of blood (for women and men, respectively).

Red blood cells differ from the other blood cells because they are anucleate when mature and circulating in the blood. As a result, they are unable to reproduce and have a limited life span of 100 to 120 days, after which they begin to fragment and are destroyed by the spleen and other reticuloendothelial tissues of the body.

In various anemias, the red blood cells may appear pale (an indication of decreased hemoglobin content) or may be nucleated (an indication that the bone marrow is turning out cells prematurely). ■

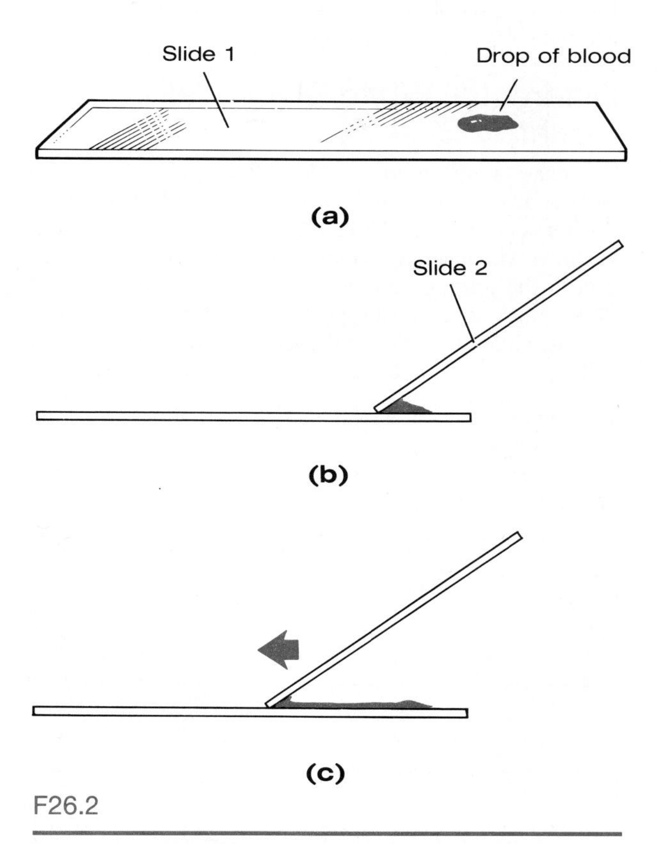

F26.2

Procedure for making a blood smear. (a) Place drop of blood on slide 1 approximately ½ inch from one end; (b) hold slide 2 at a 30° to 45° angle to slide 1 (it should touch the drop of blood); allow blood to spread along entire bottom edge of angled slide; (c) smoothly advance slide 2 to end of slide 1 (blood should run out before reaching the end of slide 1).

LEUKOCYTES Leukocytes, or white blood cells, are nucleated cells that are formed in the bone marrow from the same stem cell (hemocytoblast) as red blood cells. They are much less numerous than the red blood cells, averaging from 4,000 to 11,000 cells per cubic millimeter. Basically, white blood cells are protective, pathogen-destroying cells that are transported to all parts of the body in the blood or lymph. Important to their protective function is their ability to move in and out of blood vessels, a process called **diapedesis,** and to wander through body tissues by **ameboid motion** to reach sites of inflammation or tissue destruction. They are classified into two major groups, depending on whether or not they contain conspicuous granules in their cytoplasm.

Granulocytes comprise the first group. The granules in their cytoplasm stain differentially with Wright's stain; and they have peculiarly lobed nuclei, which often consist of expanded nuclear regions connected by thin strands of nucleoplasm. Additional information about the three types of granulocytes follows:

Neutrophil: Most abundant of the white blood cells (40% to 70% of the leukocyte population); nucleus consists of 3 to 7 lobes, and the pale lilac-appearing cytoplasm contains fine cytoplasmic granules, which are generally indistinguishable and take up both the acidic (red) and basic (blue) dyes (*neutrophil* = neutral loving); functions as an active phagocyte. The numbers of neutrophils increase exponentially during acute infections.

Eosinophil: Represents 1% to 4% of the leukocyte population; nucleus is generally figure 8 or bilobed in shape; contains large cytoplasmic granules (elaborate lysosomes) that stain red-orange with the acid dyes in Wright's stain. Precise function is unknown, but seen to increase during parasitic worm infections and allergies and may selectively phagocytize antigen-antibody complexes.

Basophil: Least abundant leukocyte type, representing less than 1% of the population; large U- or S-shaped nucleus, with two or more indentations. Cytoplasm contains coarse sparse granules (stained deep purple by the basic dyes in Wright's stain). The granules are believed to contain histamine, a vasodilator which is discharged on exposure to antigens and helps mediate the inflammatory response.

Cells of the second group, **agranulocytes,** or **agranular leukocytes,** contain no observable cytoplasmic granules. Although found in the bloodstream, they are much more abundant in lymphoid tissues and lymph. Their nuclei tend to be closer to the norm—that is, spherical, oval, or kidney-shaped. Specific characteristics of the two types of agranulocytes are listed below.

Lymphocyte: Smallest of the leukocytes, approximately the size of a red blood cell. The dark-staining nucleus is generally spherical or slightly indented and accounts for most of the cell mass. Sparse cytoplasm appears as a thin blue rim around the nucleus. Concerned with immunologic responses in the body. One population, the B lymphocytes, oversees the production of antibodies that are released to the blood; the second population, T lymphocytes, plays a regulatory role; destroys grafts, tumors, and virus-infected cells; and activates B lymphocytes. Represents 20% to 45% of the WBC population.

Monocyte: Largest of the leukocytes, approximately twice the size of the red blood cell. Represents 4% to 8% of the leukocyte population. Dark blue nucleus is generally kidney-shaped; abundant cytoplasm stains gray-blue. Functions as an active phagocyte (the "long-term cleanup team"), increasing dramatically in number during chronic infections such as tuberculosis.

Students are often asked to list the leukocytes in order from most abundant to least abundant. The following phrase may help you with this task: **N**ever **l**et **m**onkeys **e**at **b**ananas (neutrophils, lymphocytes, monocytes, eosinophils, basophils).

PLATELETS Platelets are fragments of large multinucleate cells (**megakaryocytes**) formed in the bone marrow. They appear as darkly staining, irregularly shaped bodies interspersed among the blood cells. The normal platelet count in blood is 250,000 to 500,000 per cubic millimeter. Platelets are instrumental in the clotting process that occurs in plasma when blood vessels are ruptured.

After you have identified these cell types on your slide, observe three-dimensional models of blood cells, if these are available. Do not dispose of your slide, because it will be used later for the differential white blood cell count.

HEMATOLOGICAL TESTS

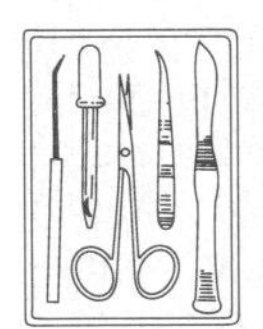

When one enters a hospital as a patient, several hematological studies are routinely done to determine general level of health, as well as the presence of pathological conditions. You will be conducting the most common of these studies in this exercise, *with the exception of total red and white blood cell counts.* The hemacytometer is ordinarily used to make such counts in the college laboratory. This apparatus is now outdated since the advent of the computerized counting apparatus.

Materials such as cotton balls, lancets, and alcohol swabs are used in nearly all of the following diagnostic tests. These supplies are at the general supply area and should be properly disposed of (glassware to the "bleach bucket" and disposable items to the autoclave bag) immediately after use.

Other necessary supplies and equipment are at specific supply areas marked according to the test with which they are used. Because nearly all of the tests require a finger stab to obtain blood, if you are testing your own blood it might be wise to quickly read through the various tests to determine in which instances more than one preparation can be done from the same finger stab. For example, the hematocrit capillary tubes and sedimentation rate samples might be prepared at the same time. A little preplanning will save you the discomfort of multiple finger punctures.

An alternative to using blood obtained by the finger stab technique is using heparinized blood samples supplied by your instructor. The purpose of using heparinized tubes is to prevent the blood from clotting. Thus blood collected and stored in such tubes will be suitable for all tests of whole blood *except* coagulation time testing.

Differential White Blood Cell Count

To make a *differential white blood cell count,* 100 WBCs are counted and classified according to type. Such a count is routine in a physical examination and in sickness, because any abnormality or significant elevation in the normal percentages of the specific types of WBCs may indicate the possible source of pathology. Use the slide prepared for the identification of the blood cells (pp. 242–243) for the count.

1. Begin at the edge of the smear, and move the slide in a systematic manner on the microscope stage—either up and down or from side to side, as indicated in Figure 26.3.

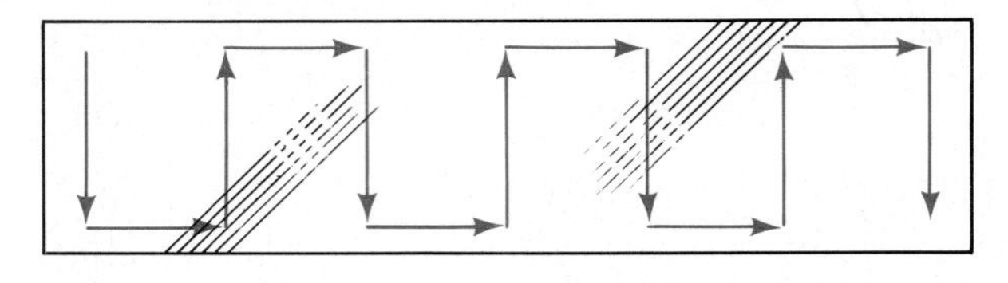

or

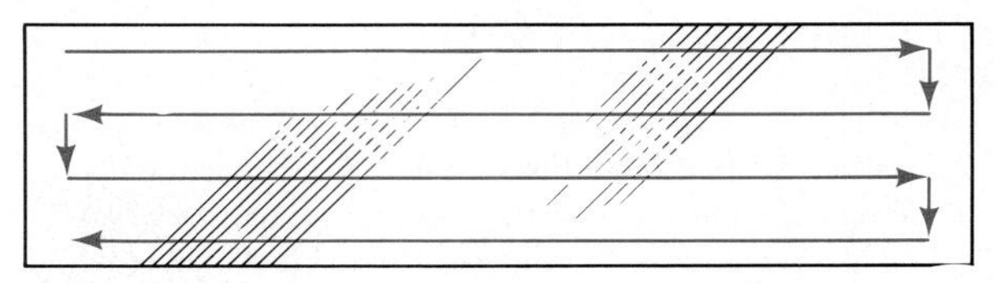

F26.3

Alternative methods of moving the slide for a differential WBC count.

2. Record each type of white blood cell you observe by making a tick mark on the chart below (for example, 𝍸 || = 7 cells), until you have observed and recorded a total of 100 WBCs. Using the equation below, compute the number of each WBC type counted as a percentage, and record the % on the data sheet at the end of this exercise.

$$\text{Percent (\%)} = \frac{\text{\# observed}}{\text{Total \# counted (100)}} \times 100$$

Cell Type	Number Observed
Neutrophils	
Eosinophils	
Basophils	
Lymphocytes	
Monocytes	

How do your observed WBC percentages compare with the percentage figures given for each type on pp. 243 and 244?

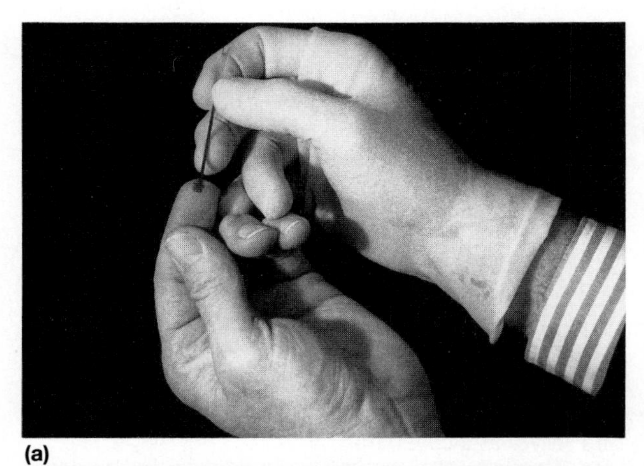

(a)

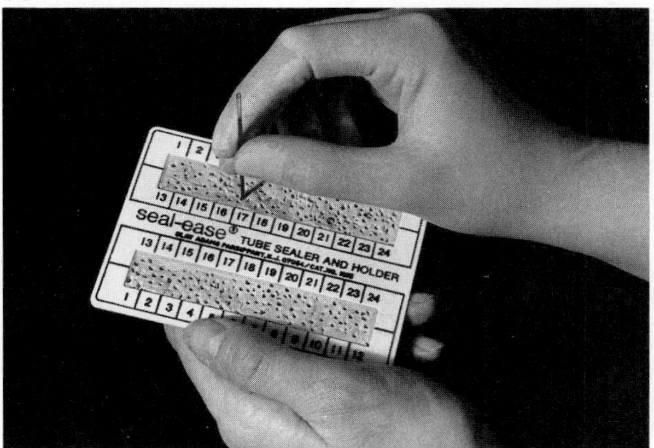

(b)

Hematocrit

The *hematocrit,* or *packed cell volume* (*PCV*), is routinely determined when anemia is suspected. Centrifuging whole blood causes the formed elements to spin to the *bottom* of the tube, leaving plasma as the top layer (see Figure 26.1). Because the blood cell population is primarily RBCs, the PCV is generally considered equivalent to the RBC volume; and this is the only value reported. However, the relative percentage of WBCs can be differentiated, and both WBC and plasma volume will be reported here. Normal PCV values for the male and female, respectively, are 47.0 ± 7 and 42.0 ± 5.

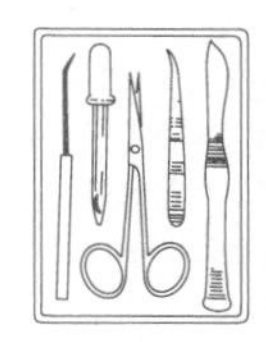

The PCV is determined by the micromethod; so only a drop of blood is needed. If possible, all members of the class should prepare their capillary tubes at the same time so the centrifuge can be properly balanced and run only once.

1. Obtain two heparinized capillary tubes, Seal-ease or modeling clay, a lancet, alcohol swabs, and some cotton balls.

2. Cleanse the finger, and allow the blood to flow freely. Wipe away the first few drops; and, holding the red-line-marked end of the capillary tube to the blood drop, allow the tube to fill at least three-fourths full by capillary action (Figure 26.4a). If the blood is not flowing freely, the end of the capillary tube will not be completely submerged in the blood during filling, air will enter, and you will have to make another sample.

3. Plug the blood-containing end by pressing it into the Seal-ease or clay (Figure 26.4b). Prepare a second tube in the same manner.

F26.4

Steps in the hematocrit determination. (a) Load a heparinized capillary tube with blood; (b) plug the end of the tube with clay; (c) place the tube in a microhematocrit centrifuge. (Centrifuge must be balanced.)

(c)

4. Place the prepared tubes opposite each other in the radial grooves of the microhematocrit centrifuge, with the sealed ends abutting the rubber gasket at the centrifuge periphery (Figure 26.4c). This loading procedure balances the centrifuge and prevents blood from spraying everywhere by centrifugal force. Make a note of the numbers of the grooves in which you have placed your tubes. When all the tubes have been loaded, make sure the centrifuge is properly balanced; and secure the centrifuge cover. Turn the centrifuge on, and set the timer for 4 to 5 minutes.

5. Determine the percentage of RBCs, WBCs, and plasma by using the microhematocrit reader. The RBCs are the bottom layer, the plasma is the top layer, and the WBCs are the buff-colored layer between the two. If the reader is not available, use a millimeter ruler to measure the length of the filled capillary tube occupied by each element, and compute the percentage by using the following formula:

$$\frac{\text{Column height composed of the RBCs* (mm)}}{\text{Original column height of whole blood (mm)}} \times 100$$

Record your calculations on the data sheet.

Usually, WBCs constitute 1% of the total blood volume. How do your blood values compare to this figure and to the normal percentages for RBCs and plasma? (See pp. 243 and 244.)

____

As a rule, the PCV is considered a more accurate test for determining the RBC composition of the blood than the total RBC count. A hematocrit within the normal range is generally indicative of a normal RBC number, whereas an abnormally high or low PCV is cause for concern.

An abnormal increase in the number of RBCs (*polycythemia*) may result from bone marrow cancer or from living at high altitudes, where less oxygen is available. A decrease in the number of RBCs results in anemia. (The term *anemia* simply indicates a decreased oxygen-carrying capacity of the blood that may result from a decrease in RBC number or size, or from a decreased hemoglobin content in the RBCs.) A decrease in RBCs may result suddenly from hemorrhage, or more gradually from conditions that increase RBC destruction or decrease RBC production. ■

Hemoglobin Concentration Determination

As noted earlier, a person can be anemic even with a normal RBC count. Since hemoglobin is the RBC protein responsible for oxygen transport, perhaps the most accurate way of measuring the oxygen-carrying capacity of the blood is to determine its hemoglobin content. Normal blood contains 12 to 16 g of hemoglobin per 100 ml of blood. Hemoglobin content in men is slightly higher (13 to 18 g) than in women (12 to 16 g).

Several techniques have been developed to estimate the hemoglobin content of blood, ranging from the old, rather inaccurate Tallquist method to expensive colorimeters, which are precisely calibrated and yield highly accurate results. Directions for both the Tallquist method and a hemoglobinometer are provided here.

TALLQUIST METHOD

1. Obtain a Tallquist hemoglobin scale, lancets, alcohol swabs, and cotton balls.

2. Use instructor-provided blood or prepare the finger as previously described. (For best results, make sure the alcohol evaporates before puncturing your finger.) Place one good-sized drop of blood on the special absorbent paper provided with the color chart. The blood stain should be larger than the holes on the color chart.

3. As soon as the blood has dried and loses its glossy appearance, match its color, under natural light, with the color standards by moving the specimen under the comparison chart so that the blood stain appears at all the various apertures. (The blood should not be allowed to dry to a brown color, as this will result in an inaccurate reading.) Because the colors on the chart represent 1% variations in hemoglobin content, it may be necessary to estimate the percentage if the color of your blood sample is intermediate between two color standards.

4. On the data sheet, record your results as the percentage of hemoglobin concentration and as grams per 100 ml of blood.

HEMOGLOBINOMETER DETERMINATION

1. Obtain a hemoglobinometer, hemolysis applicator stick, alcohol swab, and lens paper and bring them to your bench. Test the hemoglobinometer light source to make sure it is working; if not, request new batteries before proceeding and test it again.

2. Remove the blood chamber from the slot in the side of the hemoglobinometer and disassemble the blood chamber by separating the glass plates from the metal clip. Notice as you do this that the larger glass plate has an H-shaped depression cut into it that acts as a moat to hold the blood, whereas the smaller glass piece is flat and serves as a coverslip.

3. Clean the glass plates with an alcohol swab and then wipe dry with lens paper. Hold the plates by their sides to prevent smearing during the wiping process.

4. Reassemble the blood chamber (remember: larger glass piece on the bottom with the moat up), but leave the moat plate about halfway out to provide adequate exposed surface to charge it with blood.

5. Obtain a drop of blood (from the provided sample or from your fingertip as before), and place it on the depressed area of the moat plate that is closest to you (Figure 26.5a).

6. Using the wood hemolysis applicator, stir or agitate

* Substitute height of WBC band, and/or plasma to compute those percentages with this equation.

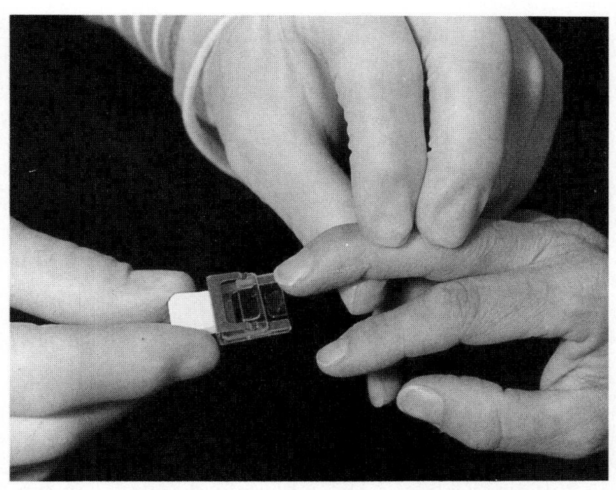

(a) A drop of blood is added to the moat plate of the blood chamber. The blood must flow freely.

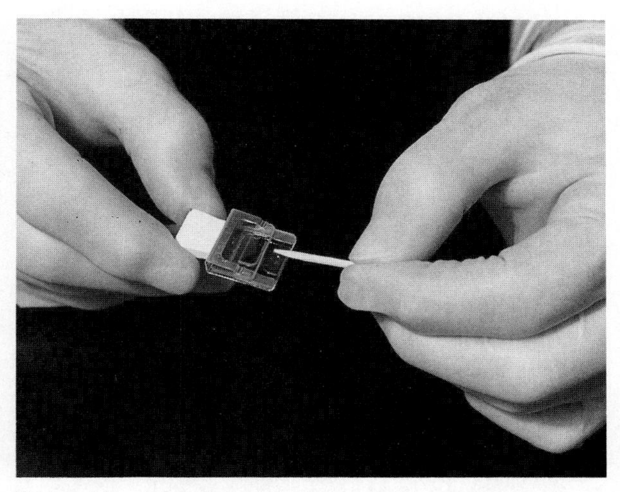

(b) The blood sample is hemolyzed with a wooden hemolysis applicator. Thirty-five to forty-five seconds are required for complete hemolysis.

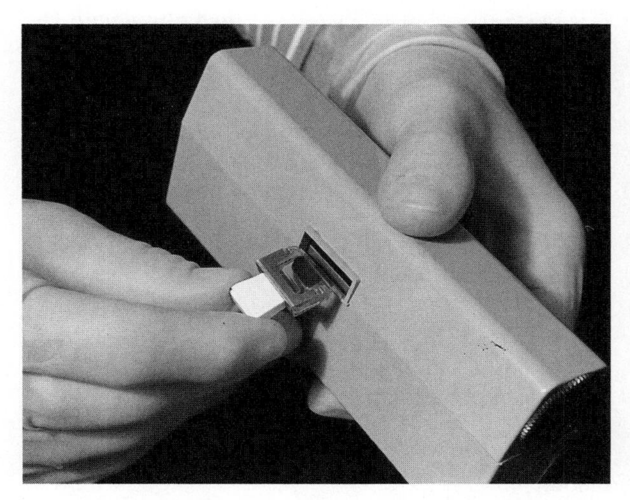

(c) The charged blood chamber is inserted into the slot on the side of the hemoglobinometer.

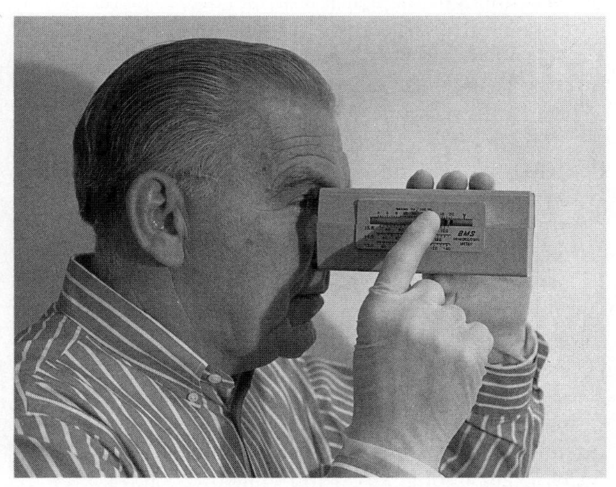

(d) The colors of the green split screen are found by moving the slide with right index finger. When the two colors match in density, the grams/100 ml and %Hb are read on the scale.

F26.5

Hemoglobin determination using a hemoglobinometer.

the blood to rupture (lyse) the RBCs (Figure 26.5b). This usually takes 35 to 45 seconds; hemolysis is complete when the blood appears transparent rather than cloudy.

7. Push the blood-containing glass plate all the way into the metal clip and then firmly insert the charged blood chamber back into the slot on the side of the instrument (Figure 26.5c).

8. Hold the hemoglobinometer in your left hand with your left thumb resting on the light switch located on the underside of the instrument. Look into the eyepiece and notice that there is a green area divided into two halves (a split field).

9. With the index finger of your right hand, slowly move the slide on the right side of the hemoglobinometer back and forth until the two halves of the green field match (Figure 26.5d).

10. Note and record on the data sheet on p. 251 the grams Hb (hemoglobin)/100 ml blood indicated on the uppermost scale by the index mark on the slide. Also record % Hb, indicated by one of the lower scales.

Generally speaking, the relationship between the PCV and grams of hemoglobin per 100 ml of blood is 3:1. How do your values compare?

____

11. Disassemble the blood chambers once again, and carefully place its removable parts (glass plates and clip) into a bleach-containing beaker.

Sedimentation Rate

The speed at which red blood cells settle to the bottom of a vertical tube when allowed to stand is called the *sedimentation rate.* The normal rate for adults is 0 to 6 mm/hr (averaging 3 mm/hr) and for children 0 to 8 mm/hr (averaging 4 mm/hr). Sedimentation of RBCs apparently proceeds in three stages: rouleaux formation, rapid settling, and final packing. Rouleaux formation (alignment of RBCs like a stack of pennies) does not occur with abnormally shaped red blood cells (as in sickle-cell anemia); therefore, the sedimentation rate is decreased. The size and number of RBCs affect the packing phase. In anemia the sedimentation rate increases; in polycythemia the rate decreases. The sedimentation rate is greater than normal during menses and pregnancy, and very high sedimentation rates indicate infectious conditions or tissue destruction occurring somewhere in the body. Although this test is nonspecific, it alerts the diagnostician to the need for further tests to pinpoint the site of pathology. The Landau micromethod, which uses just one drop of blood, is used here.

1. Obtain lancets, cotton balls, alcohol swabs, the Landau Sed-rate pipette and tubing, the Landau rack, a wide-mouthed bottle of 5% sodium citrate, a mechanical suction device, and a millimeter ruler.

2. Use the mechanical suction device to draw up the sodium citrate to the first (most distal) marking encircling the pipette.

3. If using instructor-provided blood, draw the blood into the pipette until the mixture reaches the second encircling line. Keep the pipette tip immersed in the blood to avoid air bubbles. If using your own blood, prepare the finger for puncture, and produce a free flow of blood. Wipe off the first drop, and then draw the blood into the pipette as instructed just above.

4. Thoroughly mix the blood with the citrate (an anticoagulant) by drawing the mixture into the bulb and then forcing it back down into the lumen. Repeat this mixing procedure six times, and then adjust the top level of the mixture as close as possible to the zero marking. *If any air bubbles are introduced during the mixing process, discard the sample and begin again.*

5. Seal the tip of the pipette by holding it tightly against the tip of your index finger, and then carefully remove the suction device from the upper end of the pipette.

6. Stand the pipette in an exactly vertical position on the Sed-rack, with its lower end resting on the base of the rack. Record the time, and allow it to stand for exactly 1 hour.

7. After 1 hour, measure the number of millimeters of visible clear plasma (which indicates the amount of settling of RBCs); record this figure on the data sheet.

8. Clean the pipettes by using the mechanical suction device to draw up each of the cleaning solutions (available at the general supply area) in succession (bleach, distilled water, alcohol, and acetone). Discharge the contents each time before drawing up the next solution. Place the cleaned pipettes in the bleach-containing bucket at the general supply area after first drawing bleach up into the pipette.

Coagulation Time

Blood clotting, or *coagulation,* is a protective device that minimizes blood loss when blood vessels are ruptured. This process requires the interaction of many substances normally present in the plasma (clotting factors, or procoagulants), as well as some released by platelets and injured tissues.

Basically, hemostasis proceeds as follows: The injured tissues and platelets release **thromboplastin** and **PF_3** respectively, which trigger the clotting mechanism, or cascade. Thromboplastin and PF_3 interact with other blood-protein clotting factors and calcium ions to convert **prothrombin** (present in the plasma) to **thrombin.** Thrombin then acts enzymatically to polymerize the soluble **fibrinogen** proteins (present in plasma) into insoluble **fibrin,** which forms a meshwork of strands that traps the RBCs and forms the basis of the clot (Figure 26.6). Normally, blood removed from the body clots within 3 to 6 minutes.

1. Obtain a *nonheparinized* capillary tube, a lancet, cotton balls, a triangular file, and alcohol swabs.

2. Clean and prick the finger to produce a free flow of blood.

3. Place one end of the capillary tube in the blood drop, and hold the opposite end at a lower level to collect the sample.

4. Lay the capillary tube on a paper towel and record the time.

____

5. At 30-second intervals, using the triangular file, make a small nick close to one end of the tube; and then carefully break the tube. Slowly separate the ends to see if a gel-like thread of fibrin spans the gap. When this occurs, record the time.

____

Record, on the data sheet, the time required for coagulation to occur.

6. Dispose of the capillary tube and used supplies in the autoclave bag at the general supply area.

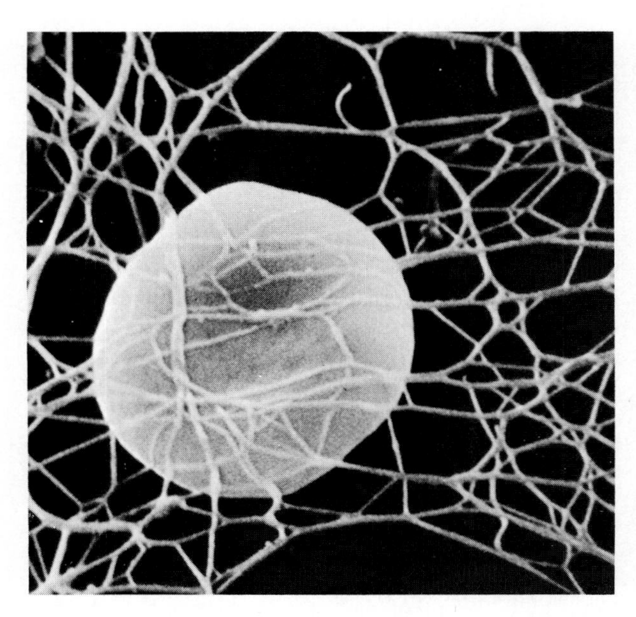

F26.6

Photomicrograph of a RBC trapped in a fibrin mesh.

Blood Typing

Blood typing is a system of blood classification based on the presence of specific glycoproteins on the outer surface of the RBC plasma membrane. Such proteins are called **antigens,** or **agglutinogens,** and are genetically determined. In many cases, these antigens are accompanied by other proteins found in the plasma, **antibodies** or **agglutinins,** that react with RBCs bearing different antigens, causing them to become clumped, agglutinated, and eventually hemolyzed. It is because of this phenomenon that a person's blood must be carefully typed before a whole-blood or packed-cell transfusion.

There are several blood typing systems based on the various possible antigens, but the factors typically typed for are the antigens of the ABO and Rh blood groups which are most commonly involved in transfusion reactions. Other blood factors, such as Kell, Lewis, M, and N, are not routinely typed for, unless the individual is expected to require multiple transfusions. The basis of the ABO typing is shown in the chart below:

Individuals whose red blood cells carry the Rh antigen are considered to be Rh positive (approximately 85% of the U.S. population); those lacking the antigen are Rh negative. Unlike the case of ABO blood groups, the blood of the Rh-positive (Rh^+) and Rh-negative (Rh^-) individuals does not carry *preformed* anti-Rh antibodies. This is understandable in the case of the Rh-positive individual. However, Rh-negative persons who receive transfusions of Rh-positive blood become *sensitized* by the Rh antigens of the donor RBCs; and then their systems begin to produce anti-Rh antibodies. On subsequent exposures to Rh-positive blood, typical transfusion reactions occur, resulting in clumping and hemolysis of the donor blood cells.

1. Obtain two clean microscope slides; a wax marking pencil; anti-A, anti-B, and anti-Rh typing sera; toothpicks; lancets; alcohol swabs; and the Rh-typing box.

2. With the wax marking pencil, divide slide 1 into two equal halves. Label the lower left-hand corner "anti-A" and the lower right-hand corner "anti-B." Label the bottom of slide 2 "anti-Rh."

3. Place one drop of anti-A serum on the *left* side of slide 1. Place one drop of anti-B serum on the *right* side of slide 1. Place one drop of anti-Rh serum in the center of slide 2.

4. Cleanse your finger with an alcohol swab, pierce the finger with a lancet, and wipe away at the first drop of blood. Obtain 3 drops of freely flowing blood, placing one drop on each side of slide 1 and a drop on slide 2.

5. Quickly mix each blood-antiserum sample with a *fresh* toothpick. Then dispose of the toothpicks, lancet, and used alcohol swab in the autoclave bag.

6. Place slide 2 on the Rh-typing box, and gently rock it back and forth. (A slightly higher temperature is required for precise Rh typing than for ABO typing.)

7. After 2 minutes, observe all three blood samples for evidence of clumping. The agglutination that occurs in the positive test for the Rh factor is very fine and difficult to perceive; thus if there is any question, observe the slide under the microscope. Record your observations in the chart on p. 250.

ABO Blood Type	Agglutinogens Present on RBC Membranes	Agglutinins Present in Plasma	% of U.S. Population		
			White	Black	Asian
A	A	Anti-B	41	27	28
B	B	Anti-A	9	20	27
AB	A and B	None	3	4	5
O	Neither	Anti-A and Anti-B	47	49	40

	Observed (+)	Not Observed (−)
Presence of clumping with anti-A		
Presence of clumping with anti-B		
Presence of clumping with anti-Rh		

8. Interpret your results in light of the following information:

Slide 1: If clumping occurs on both sides, your ABO blood group is AB. If clumping occurs only in the blood sample mixed with anti-A serum, you are ABO type A. If clumping occurs only in the blood sample mixed with anti-B serum, you are type B. If clumping was not observed with either serum, you are type O.

Slide 2: If clumping was observed, you are Rh positive. If not, you are Rh negative.

9. Record your blood type on the data sheet.

10. Place the used slides in the bleach-containing bucket at the general supply area.

Measurement of Cholesterol Concentration in Plasma

Atherosclerosis is the disease process in which the body's blood vessels (particularly the coronary and cerebral arteries, and the aorta and its major branch points) become increasingly occluded by plaques. Because the plaques narrow the arteries, they can contribute to hypertensive heart disease. They also serve as focal points for the formation of blood clots (thrombi) which may break away and block smaller vessels farther downstream in the circulatory pathway, causing heart attacks or strokes (myocardial or cerebral infarcts respectively).

Since medical clinicians discovered that cholesterol is a major component of the smooth muscle plaques formed during atherosclerosis, it has had a bad press. Today, virtually no physical examination of an adult is considered complete until data on his or her cholesterol levels are assessed along with their other life-style risk factors. A normal value for plasma cholesterol in adults ranges from 130 to 200 mg/100 ml plasma; you will be making such a determination on the animal plasma provided by the instructor.

Although the total plasma cholesterol concentration is valuable information, it may be misleading, particularly if the person's high-density lipoprotein (HDL) level is high and their low-density lipoprotein (LDL) levels are relatively low. (Cholesterol, being water-insoluble, is transported in the blood complexed to lipoproteins. In general, cholesterol bound into HDLs is destined for degradation by the liver and then elimination from the body, whereas that forming part of the LDLs is traveling to the body's tissue cells. When LDL levels are excessive, cholesterol is deposited in the blood vessel walls; hence, LDLs are considered to carry the "bad" cholesterol.)

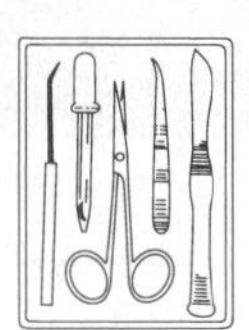

1. Go to the appropriate supply area, put three test tubes in a test tube rack, and add 5 ml of cholesterol reagent into each of the three test tubes. Mark the test tubes 1–3 with a wax marker.

2. Using the mechanical pipettor and a fresh 0.1-ml pipette each time, add 0.1 ml of the following to each of the numbered test tubes:

- Test tube 1: Add 0.1 ml of cholesterol standard (200 mg/100 ml).
- Test tube 2: Add 0.1 ml of plasma (available at the general supply area).
- Test tube 3: Add 0.1 ml of distilled water.

3. Mix each tube by shaking it gently from side to side and then place the tubes (in the rack) into the water bath set at 37°C. (Note: If there is any question about your ability to identify your test tubes, put your initials on them with the wax marker *before* placing them into the water bath.)

Record the time: ______________________

4. Obtain three spectrophotometer cuvettes (containers for the solutions to be "read" by the spectrophotometer). Clean them thoroughly with cuvette brushes and laboratory detergent, rinsing several times with distilled water. Invert them on a paper towel to drain off the distilled water.

5. After the test tubes have been in the water bath for 10 minutes, remove them and transfer their contents to similarly numbered cuvettes. Standardize the spectrophotometer at 625 nm using sample 3 (the distilled water) as the *blank*.

6. Determine and record the absorbance of samples 1 and 2.

Absorbance of 1: ______ Absorbance of 2: ______

7. Calculate the cholesterol concentration (C) in sample 2 (the "unknown" plasma sample) using the following equation:

$$C_{\text{plasma}} = \frac{\text{absorbance}_{\text{plasma}}}{\text{absorbance}_{\text{standard}}} \times C_{\text{standard}}$$

Record the computed cholesterol concentration here and on the data sheet.

Cholesterol concentration: ______ mg/100 ml plasma

8. Clean the pipettes using the mechanical pipettor to draw up each of the pipette cleaning solutions (bleach, distilled water, ethyl alcohol, and acetone) in turn, and then return them to the general supply area.

9. Before leaving the laboratory, use a paper towel saturated with bleach solution to wash down your laboratory bench.

Hematological Test Data Sheet

Differential WBC count:

______% granulocytes ______% agranulocytes

______% neutrophils ______% lymphocytes

______% eosinophils ______% monocytes

______% basophils

Hematocrit (PCV):

______ RBC % of blood volume

______ WBC % of blood volume } not generally reported

______ Plasma % of blood volume }

Hemoglobin content:

Tallquist method:

______ g/100 ml of blood; ______ %

Hemoglobinometer (type: ______)

______ g/100 ml blood; ______ %

Ratio (PCV/grams Hb per 100 ml of blood) ______

Sedimentation rate ______ mm/hr

Coagulation time ______

Blood typing:

ABO group ______ Rh factor ______

Cholesterol concentration: ______ mg/100 ml plasma

27 EXERCISE

Anatomy of the Heart

OBJECTIVES

1. To describe the location of the heart.
2. To name and locate the major anatomical areas and structures of the heart when provided with an appropriate model or diagram, or with a dissected sheep heart, and to explain the function of each.
3. To trace the pathway of blood through the heart.
4. To explain why the heart is called a double pump, and to compare the pulmonary and systemic circuits.
5. To explain the operation of the atrioventricular and semilunar valves.
6. To name and describe the functional blood supply of the heart.
7. To describe the histologic structure of cardiac muscle, and to note the importance of its intercalated discs and spiraling arrangement of its fibers in the heart.

MATERIALS

Torso model or laboratory chart showing heart anatomy
Red and blue pencils
Preserved sheep heart, pericardial sacs intact (if possible)
Dissecting pan and instruments
Protective skin cream or disposable gloves
Pointed glass rods for probes
Compound microscope
Prepared microscopic slide of cardiac muscle (longitudinal section)
Three-dimensional model of cardiac muscle
Heart model (three-dimensional)
X-ray of the human thorax for observation of the position of the heart *in situ;* X-ray viewing box

The major function of the cardiovascular system is transportation. Using blood as the transport vehicle, the system carries oxygen, digested foods, cell wastes, electrolytes, and many other substances vital to the body's homeostasis to and from the body cells. The system's propulsive force is the contracting heart, which can be compared to a muscular pump equipped with one-way valves. As the heart contracts, it forces blood into a closed system of large and small plumbing tubes (blood vessels) within which the blood is confined and circulated. This exercise deals with the structure of the heart, or circulatory pump. The anatomy of the blood vessels is considered separately in Exercise 29.

GROSS ANATOMY OF THE HUMAN HEART

The **heart** is a cone-shaped organ, approximately the size of a fist, and is located within the mediastinum, or medial cavity, of the thorax. It is flanked laterally by the lungs, posteriorly by the vertebral column, and anteriorly by the sternum. Its more pointed **apex** extends slightly to the left and rests on the diaphragm, approximately at the level of the fifth intercostal space. Its broader **base,** from which the great vessels emerge, lies beneath the second rib and points toward the right shoulder. *In situ,* the right ventricle of the heart forms most of its anterior surface. If an X-ray of a human thorax is available, verify the relationships described above.

Figure 27.1 shows two views of the heart—an external anterior view and a frontal section. As its anatomical regions are described in the text, consult the figure. When you have pinpointed all the structures, observe the human heart model; and reidentify the same structures without reference to the figure.

The heart is enclosed within a double-walled fibroserous sac called the pericardium. The thin **visceral pericardium,** or **epicardium,** which is closely applied to the heart muscle, reflects downward at the base of the heart to form its companion serous membrane, the loosely applied **parietal pericardium,** which is attached at the heart apex to the diaphragm. Serous fluid produced by the membranes allows the heart to beat in a relatively frictionless environment. The serous parietal pericardium, in turn, lines the loosely fitting superficial **fibrous pericardium** composed of dense connective tissue.

Inflammation of the pericardium, *pericarditis,* causes painful adhesions between the serous pericardial layers. These adhesions interfere with heart movements. ■

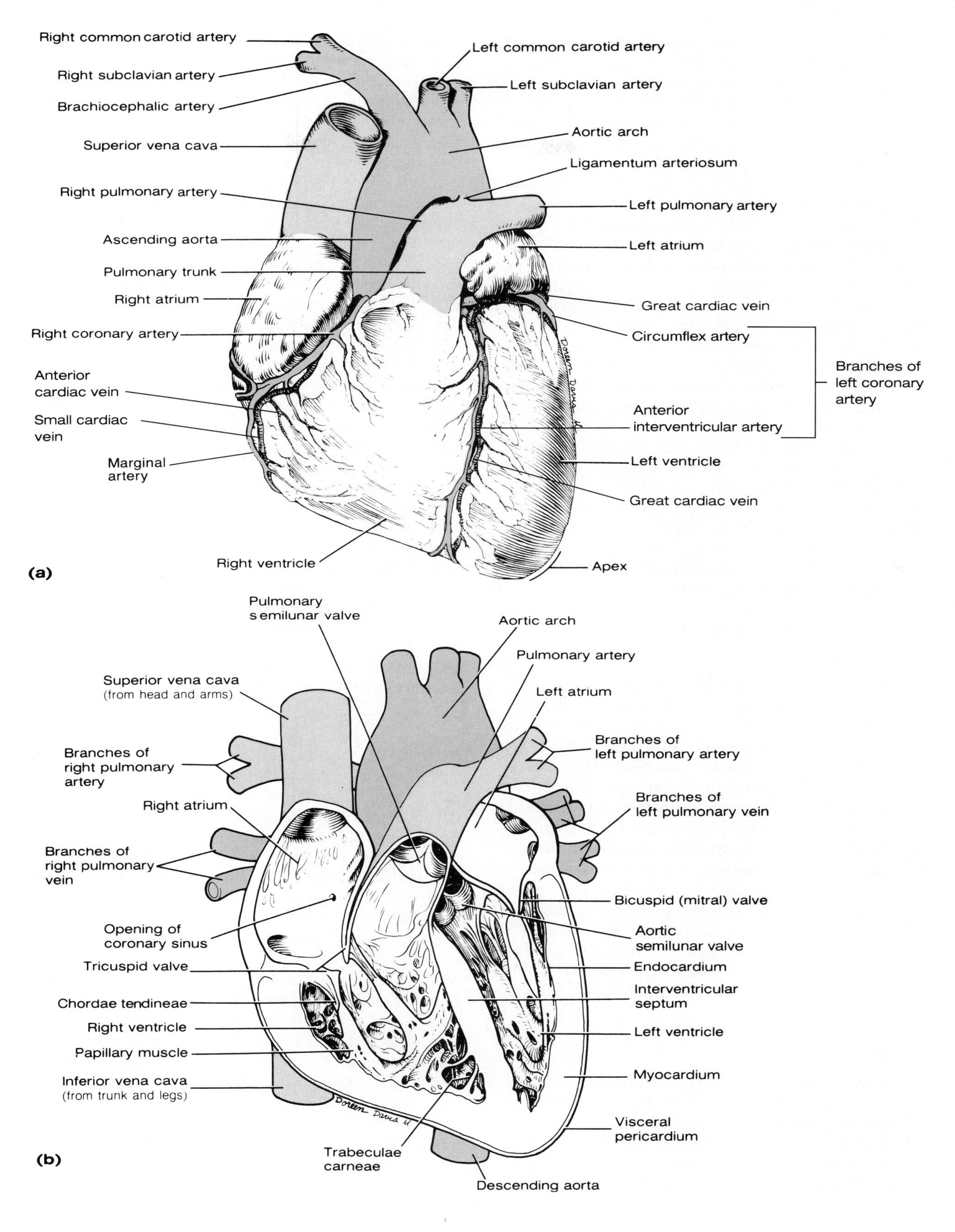

F27.1

Anatomy of the human heart. (a) External anterior view; (b) frontal section.

The walls of the heart are composed mainly of cardiac muscle—the **myocardium**—which is reinforced internally by a dense, fibrous connective tissue network. This network—*the fibrous skeleton of the heart*—is more elaborate and thicker in certain areas—for example, around the valves and at the bases of the great vessels leaving the heart (see Figure 27.2).

The heart is divided into four chambers: two superior **atria** and two inferior **ventricles,** each lined with a thin serous lining called the **endocardium.** The septum that divides the heart longitudinally is referred to as the **interatrial** or **interventricular septum,** depending on which chambers it partitions. Functionally, the atria are receiving chambers and are relatively ineffective as pumps. Blood flows into the atria, under relatively low pressure, from the veins of the body. The right atrium receives relatively oxygen-poor blood from the body via the **superior** and **inferior venae cavae.** Four **pulmonary veins** deliver oxygen-rich blood from the lungs to the left atrium.

The inferior thick-walled ventricles, which form the bulk of the heart, are the discharging chambers that force blood out of the heart and into the large arteries that emerge from its base. The right ventricle pumps blood into the **pulmonary trunk,** which routes the blood to the lungs to be oxygenated. The left ventricle discharges blood into the **aorta,** from which all systemic arteries of the body diverge to supply the body tissues.

Pulmonary and Systemic Circulations

The heart functions as a double pump. The right side serves as the **pulmonary circulation pump,** shunting the carbon dioxide–rich blood entering its chambers to the lungs to unload carbon dioxide and pick up oxygen, and then back to the left side of the heart. The function of this circuit is strictly to provide for gas exchange. The second circuit, which pumps oxygen-rich blood from the left heart through the body tissues and back to the right heart, is called the **systemic circulation.** It supplies the functional blood supply to all body tissues.

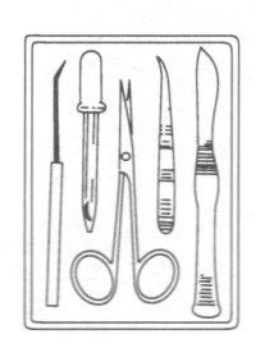

Trace the pathway of blood through the heart by adding arrows to the frontal section diagram (see Figure 27.1b). Use red arrows for the oxygen-rich blood and blue arrows for the less oxygen-rich blood.

Heart Valves

Four valves enforce a one-way blood flow through the heart chambers. The **atrioventricular valves,** located between the atrial and ventricular chambers on each side, prevent backflow into the atria when the ventricles begin to contract. The left atrioventricular valve, also called the **mitral** or **bicuspid valve,** consists of two cusps, or flaps, of endocardium. The right atrioventricular valve, the **tricuspid valve,** has three cusps (see Figure 27.2). Tiny white collagenic cords called the **chordae tendineae** (literally, heart strings) anchor the cusps to the ventricular walls. The chordae tendineae originate from small bundles of cardiac muscle, **papillary muscles,** that project from the myocardial wall.

When blood is flowing passively into the atria and then into the ventricles during **diastole** (the period of ventricular relaxation), the atrioventricular valve flaps hang limply into the ventricular chambers and then are carried passively toward the atria by the accumulating blood. When the ventricles begin to contract (**systole**) and blood in their chambers is compressed, the intraventricular blood pressure begins to rise, causing the valve flaps to be reflected superiorly and close the valve. The chordae tendineae, pulled taut by the contracting papillary muscles, anchor the flaps in a closed position, preventing backflow into the atria during ventricular contraction. If unanchored, the flaps would blow upward into the atria, rather like an umbrella being turned inside out by a strong wind.

The second set of valves, the **semilunar valves,** each composed of three pocketlike cusps, guards the bases of the two large arteries leaving the ventricular chambers. These are referred to as the **pulmonary** and **aortic semilunar valves.** The valve cusps are forced open and flatten against the walls of the artery as the ventricles discharge their blood into the large arteries during systole. However, when the ventricles relax, blood begins to flow backward toward the heart and the cusps fill with blood, closing the valves and preventing arterial blood from reentering the heart.

Cardiac Circulation

Even though the heart chambers are almost continually bathed with blood, this contained blood does not nourish the myocardium. The functional blood supply of the heart is provided by the **right** and **left coronary arteries,** which issue from the base of the aorta, just above the aortic semilunar valve, and encircle the heart in the **atrioventricular groove** at the junction of the atria and ventricles. They then ramify over the heart's surface—the right coronary artery supplying the posterior surface of the ventricles and the lateral aspect of the right side of the heart, largely through its **posterior interventricular** and **marginal artery** branches. The left coronary artery supplies the anterior ventricular walls and the laterodorsal part of the left side of the heart via its two major branches, the **anterior interventricular artery** and the **circumflex artery.** The coronary arteries and their branches are compressed during systole, and fill when the heart is relaxed. The myocardium is drained by the various cardiac veins (see Figure 27.1b), most of which empty into the **coronary sinus,** which in turn empties into the right atrium.

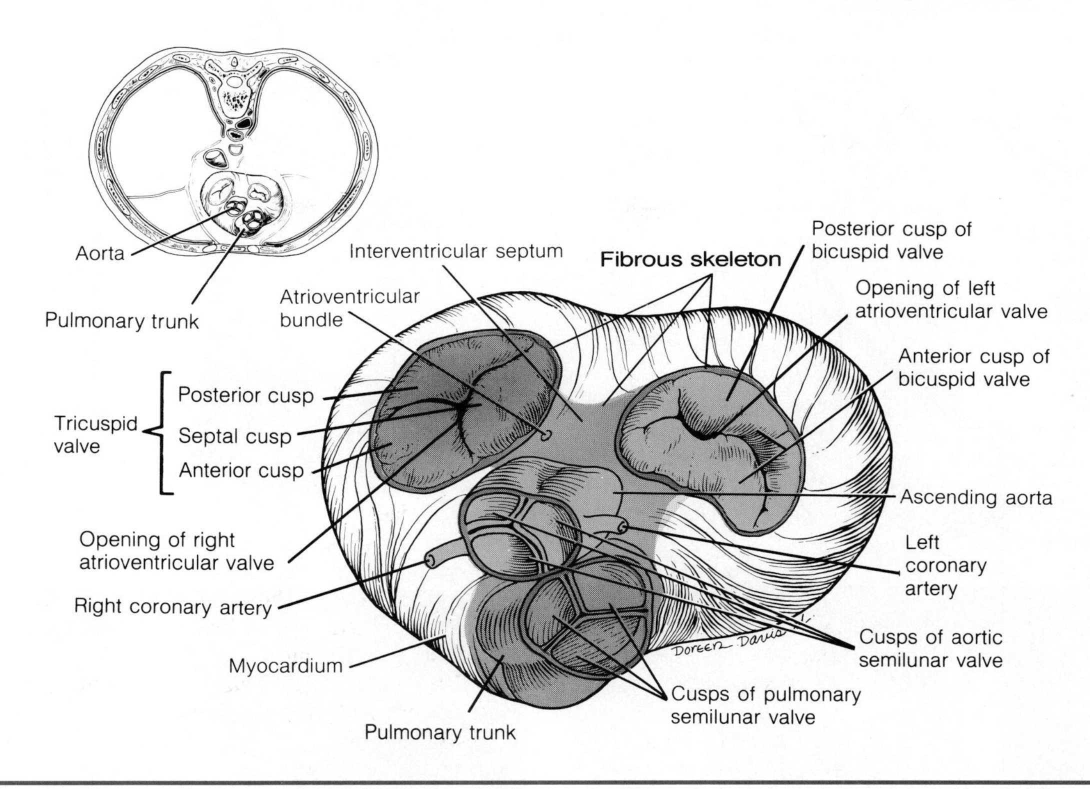

F27.2

Heart valves (superior view).

DISSECTION OF THE SHEEP HEART

Dissection of the sheep heart is valuable because the sheep heart is similar in size and structure to the human heart. Also, a dissection experience allows you to view structures in a way not possible with models and diagrams. Refer to Figure 27.3 as you proceed with the dissection.

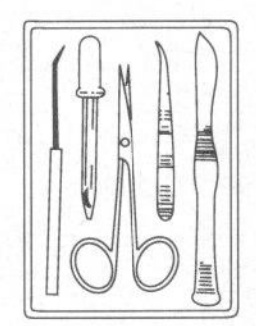

1. Obtain a preserved sheep heart, a dissection tray, and dissecting instruments. Rinse the sheep heart in cold water to remove excess preservatives and to flush out any trapped blood clots. Now you are ready to make your observations.

2. Observe the texture of the pericardium. Also, note its point of attachment to the heart. Where is it attached?

____

3. If the serous pericardial sac is still intact, slit open the parietal pericardium; and cut it from its attachments. Observe the visceral pericardium (epicardium). Using a sharp scalpel, carefully pull a little of this serous membrane away from the myocardium. How do its position, thickness, and apposition to the heart differ from those of the parietal pericardium?

____

____

____

____

4. Examine the external surface of the heart. Notice the accumulation of adipose tissue, which in many cases marks the separation of the chambers and the location of the coronary arteries that nourish the myocardium. With a scalpel, carefully scrape away some of the fat to expose the coronary blood vessels.

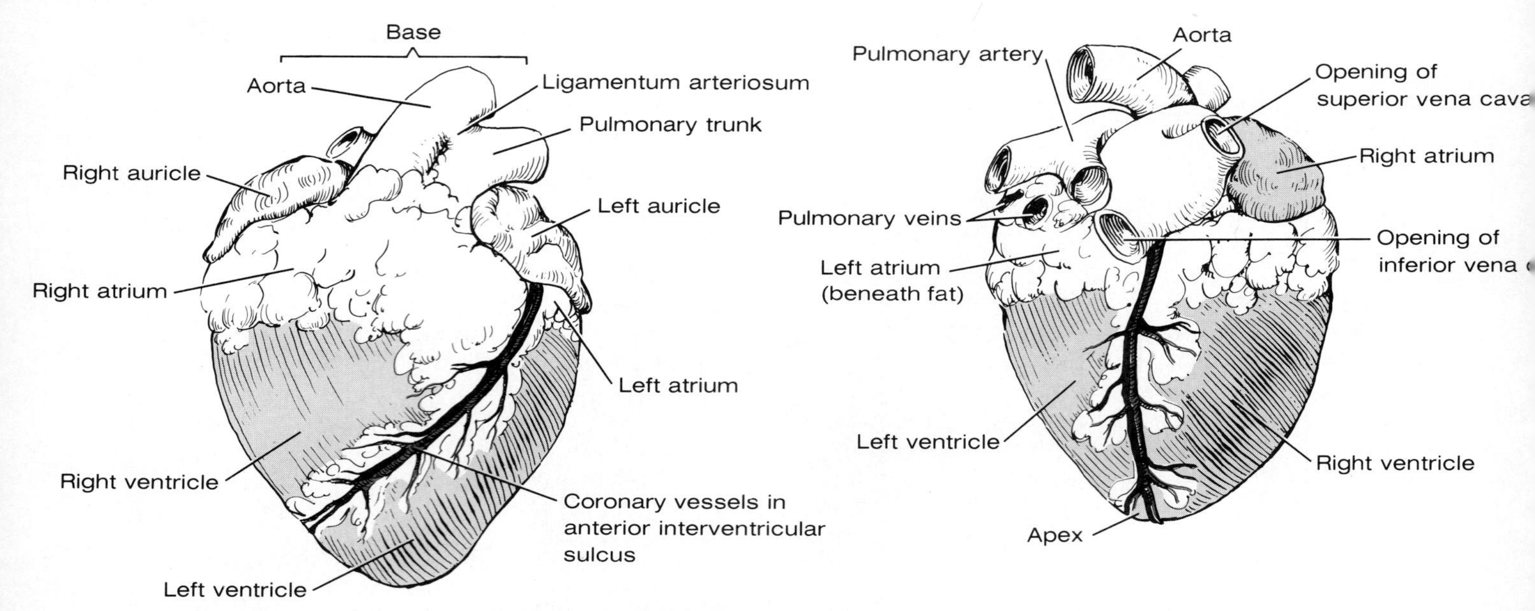

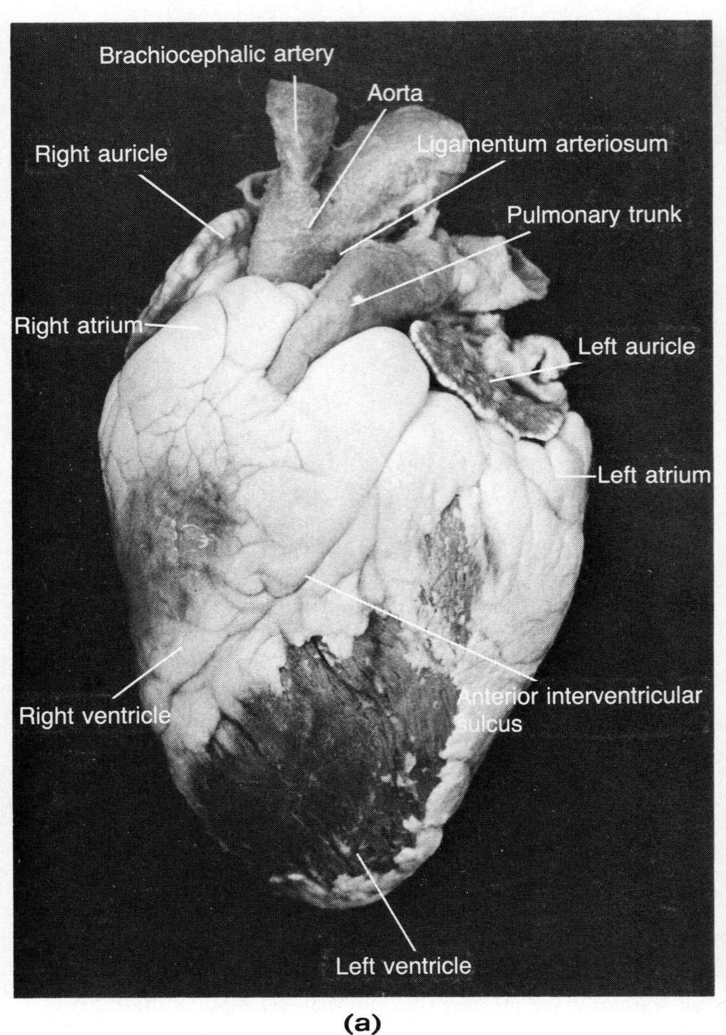

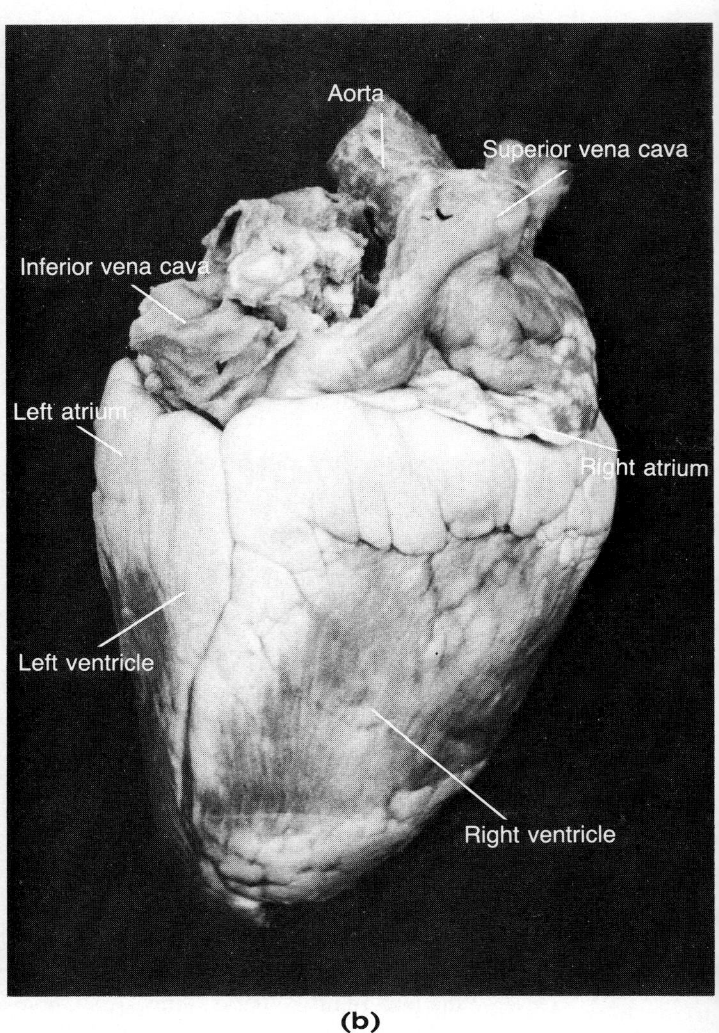

(a) **(b)**

F27.3

Anatomy of the sheep heart. (a) Anterior view. Diagrammatic view at top; photograph at bottom.(b) Posterior view. Diagrammatic view at top; photograph at bottom.

5. Identify the base and apex of the heart; and then identify the two wrinkled **auricles,** earlike flaps of tissue projecting from the atrial chambers. The balance of the heart muscle is ventricular tissue. To identify the left ventricle, compress the ventricular chambers on each side of the longitudinal fissures carrying the coronary blood vessels. The side that feels thicker and more solid is the left ventricle. The right ventricle feels much thinner and somewhat flabby on compression. This difference reflects the greater demand placed on the left ventricle, which must pump blood through the much longer systemic circulation, a pathway with much higher resistance than the pulmonary circulation served by the right ventricle. Hold the heart in its anatomical position (Figure 27.3a), with the anterior surface uppermost. In this position the left ventricle composes the entire apex and the left side of the heart.

6. Identify the pulmonary trunk and the aorta extending from the base of the heart. The pulmonary trunk is more anterior, and you may see its division into the right and left pulmonary arteries if it has not been cut too closely to the heart. The aorta, which is thicker walled and branches almost immediately, is located just beneath the pulmonary trunk. Carefully clear away some of the fat between the pulmonary trunk and the aorta to expose the **ligamentum arteriosum,** a cordlike remnant of the **ductus arteriosus.** (In the fetus, the ductus arteriosus allows blood to pass directly from the pulmonary trunk to the aorta, thus bypassing the nonfunctional fetal lungs.)

7. Cut through the wall of the aorta until you see the aortic semilunar valve. Identify the two openings, just above the valve, to the coronary arteries. Place a probe into one of these holes to see if you can follow the course of a coronary artery across the heart.

8. Turn the heart to view its posterior surface. The heart will appear as shown in Figure 27.3b. Notice that the right and left ventricles appear equal-sized in this view. Identify the four thin-walled pulmonary veins entering the left atrium. (It may or may not be possible to locate the pulmonary veins from this vantage point, depending on how they were cut as the heart was removed.) Identify the superior and inferior venae cavae entering the right atrium. Compare the approximate diameter of the superior vena cava with the diameter of the aorta.

Which is larger? ______________________________

Which has thicker walls? ______________________________

Why do you suppose these differences exist? __________

____

____

____

9. Insert a probe into the superior vena cava, and use scissors to cut through its wall so that you can view the interior of the right atrium. Do *not* extend your cut entirely through the right atrium or into the ventricle. Observe the right atrioventricular valve.

How many flaps does it have? ______________________

Pour some water into the right atrium, and allow it to flow into the ventricle. Slowly and gently squeeze the right ventricle to watch the closing action of this valve. (If you squeeze too vigorously, you'll get a face full of water!) Drain the water from the heart before continuing.

10. Return to the pulmonary trunk, and cut through its anterior wall until you can see the pulmonary semilunar valve. Pour some water into the base of the pulmonary trunk to observe the closing action of this valve. How does its action differ from that of the atrioventricular valve?

____

____

____

Text continues on p. 258

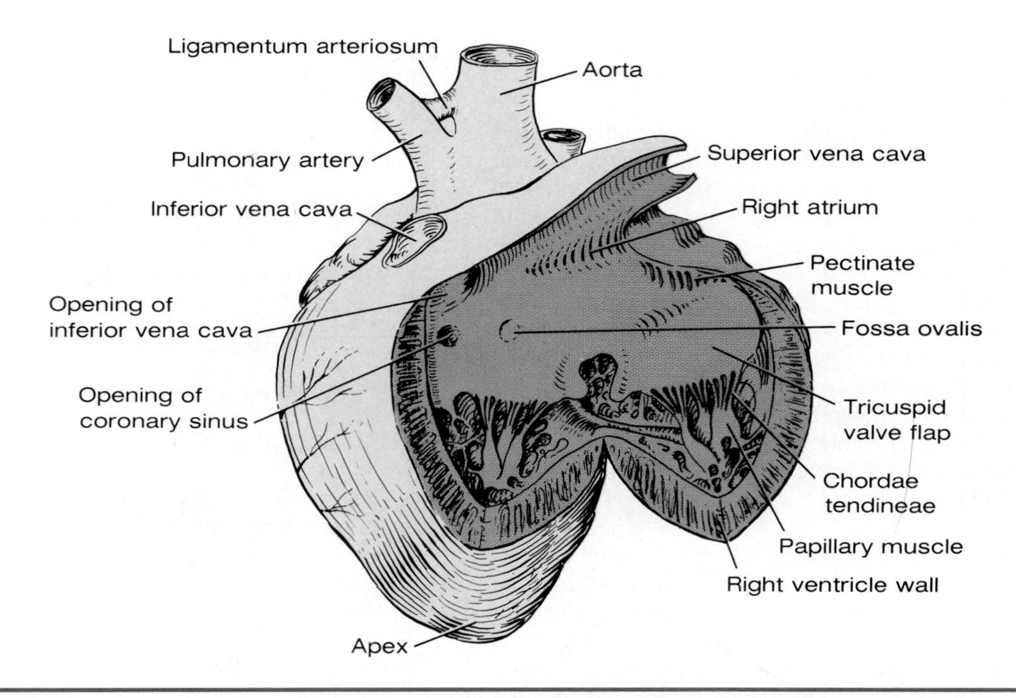

F27.4

Right side of the sheep heart opened and reflected to reveal internal structures (diagrammatic view).

After observing valve action, drain the heart once again. Return to the superior vena cava, and continue the cut made in its wall through the right atrium and right atrioventricular valve into the right ventricle. Parallel the anterior border of the interventricular septum until you "round the corner" to the dorsal aspect of the heart (Figure 27.4).

11. Reflect the cut edges of the superior vena cava, right atrium, and right ventricle to obtain the view seen in Figure 27.4. Observe the comblike ridges of muscle throughout most of the right atrium. This is called **pectinate muscle** (pectin = comb). Identify, on the ventral atrial wall, the large opening of the inferior vena cava; and, with a probe, follow it to its external opening. Notice that the atrial walls in the vicinity of the venae cavae are smooth and do not have the roughened appearance (pectinate musculature) of the other regions of the atrial walls. Just below the inferior vena caval opening, identify the opening of the **coronary sinus,** which returns venous blood of the coronary circulation to the right atrium. Nearby, locate an oval depression, the **fossa ovalis,** in the interatrial septum. This depression marks the site of an opening in the fetal heart, the **foramen ovale,** that allows blood to pass from the right to the left atrium, thus bypassing the fetal lungs.

12. Identify the papillary muscles in the right ventricle, and follow their attached chordae tendineae to the flaps of the tricuspid valve. Also notice the pitted and ridged appearance (**trabeculae carneae**) of the inner ventricular muscle.

13. Make a longitudinal incision through the aorta, and continue it into the left ventricle. Notice how much thicker the myocardium of the left ventricle is compared to that of the right ventricle. Compare the *shape* of the left ventricular cavity to the shape of the right ventricular cavity.

____

____

____

Are the papillary muscles and chordae tendineae observed in the right ventricle also present in the left ventricle?

____

Count the number of cusps in the left atrioventricular valve. How does this compare with the number seen in the right atrioventricular valve?

____

____

How do the sheep valves compare with their human counterparts?

____

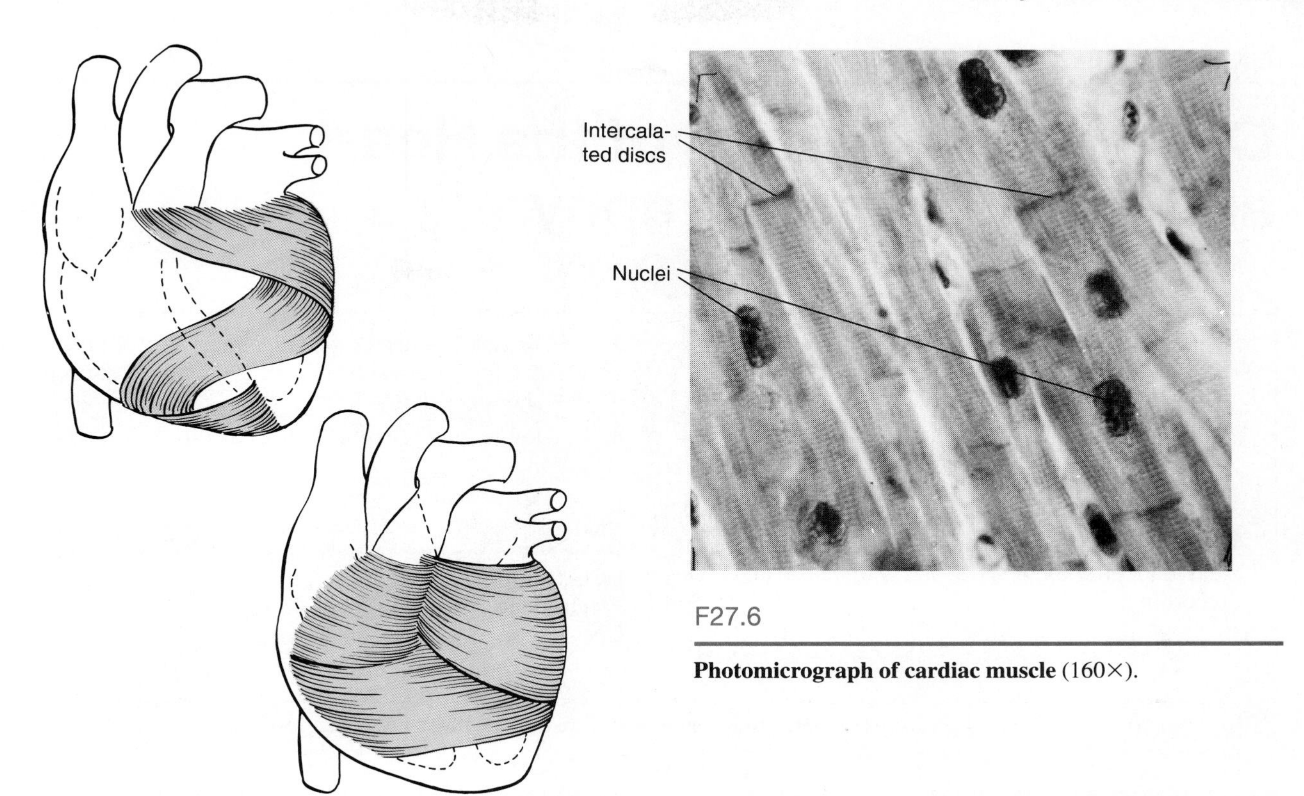

F27.5

Longitudinal view of the heart chambers showing the spiral arrangement of the cardiac muscle fibers.

F27.6

Photomicrograph of cardiac muscle (160×).

14. Continue your incision from the left ventricle superiorly into the left atrium. Reflect the cut edges of the atrial wall, and attempt to locate the entry points of the pulmonary veins into the left atrium. With a probe, follow them to the heart exterior. Note how thin-walled these vessels are.

15. Properly dispose of the organic debris, and clean the dissecting tray and instruments.

MICROSCOPIC ANATOMY OF CARDIAC MUSCLE

Cardiac muscle is found in only one place—the heart. Because the heart acts as a vascular pump, propelling blood to all tissues of the body, cardiac muscle is very important to life. Cardiac muscle is involuntary, thus ensuring a constant blood supply.

The cardiac cells, only sparingly invested in connective tissue, are arranged in spiral or figure 8–shaped bundles (Figure 27.5). When the heart contracts, its internal chambers become smaller (or are temporarily obliterated), forcing the blood into the large arteries leaving the heart.

1. Observe the three-dimensional model of cardiac muscle, noting its branching cells and the areas where the cells interdigitate, the intercalated discs. These two structural features provide a continuity to cardiac muscle not seen in other muscle tissues and allow close coordination of heart activity.

2. Obtain and observe, under high power, a longitudinal section of cardiac muscle; and draw a small section of the tissue in the space below. Label the nucleus, striations, intercalated discs, and sarcolemma. Compare your observations to the photomicrograph shown as Figure 27.6.

28 EXERCISE

Conduction System of the Heart and Electrocardiography

OBJECTIVES

1. To enumerate and localize the elements of the intrinsic conduction, or nodal, system of the heart; and to describe the initiation and conduction of impulses through this system and the myocardium.
2. To interpret the ECG in terms of depolarization and repolarization events occurring in the myocardium; and to identify the P, QRS, and T waves on an ECG recording.
3. To calculate the heart rate, QRS interval, P-R interval, and Q-T interval from an ECG obtained during the laboratory period.
4. To define *tachycardia, bradycardia,* and *fibrillation.*

MATERIALS

ECG recording apparatus
Electrode paste; alcohol swabs
Cot

THE INTRINSIC CONDUCTION SYSTEM

Heart contraction results from a series of electrical potential changes (depolarization waves) that travel through the heart preliminary to each beat. The ability of cardiac muscle to beat is intrinsic—it does not depend on impulses from the nervous system to initiate its contraction and will continue to contract rhythmically even if all nerve connections are severed. However, two types of controlling systems exert their effects on heart activity. One of these, the *extrinsic system,* involves nerves of the autonomic nervous system, which increase or decrease the heart beat rate depending on which division is activated. The second system is the **intrinsic conduction system,** or **nodal system,** of the heart, consisting of specialized noncontractile myocardial tissue. The intrinsic system ensures that heart muscle depolarizes in an orderly and sequential manner (from atria to ventricles) and that the heart beats as a coordinated unit.

The components of the intrinsic conduction system include the **SA (sinoatrial) node,** located in the right atrium, just inferior to the entrance of the superior vena cava; the **AV (atrioventricular) node** in the lower atrial septum, at the junction of the atria and ventricles; the **AV (atrioventricular) bundle (of His)** and right and left **bundle branches,** located in the interventricular septum; and the **Purkinje fibers,** which ramify within the muscle bundles of the ventricular walls. The Purkinje fiber network is much denser and more elaborate in the left ventricle because of the larger size of this chamber (Figure 28.1).

The SA node, which has the highest rate of discharge, provides the stimulus for contraction. Because it sets the rate of depolarization for the heart as a whole, the SA node is often referred to as the *pacemaker.* From the SA node, the impulse spreads throughout the atria and to the AV node. This electrical wave is immediately followed by atrial contraction. At the AV node, the impulse is momentarily delayed (approximately 0.1 sec), allowing the atria to complete their contraction. It then passes through the AV bundle, the right and left bundle branches, and the Purkinje fibers, finally resulting in ventricular contraction. Note that the atria and ventricles are separated from one another by a region of electrically inert connective tissue; so the depolarization wave can be transmitted to the ventricles only via the tract between the AV node and AV bundle. Thus, any damage to the AV node-bundle pathway partially or totally insulates the ventricles from the influence of the SA node. Although autorhythmic cells are found throughout the heart, their rates of spontaneous depolarization differ. The nodal system increases the rate of heart depolarization and synchronizes heart activity.

Various impairments of the intrinsic conduction system result in out-of-phase contractions of the atria and ventricles, or in fibrillation (a shuddering of the heart muscle), which decreases or obliterates the effectiveness of the heart as a pump. ■

ELECTROCARDIOGRAPHY

The conduction of impulses through the heart generates electrical currents that eventually spread throughout the body. These impulses can be detected on the body's surface and recorded with an instrument called an electrocardiograph. The graphic recording of the electrical changes (depolarization and repolarization) occurring during the cardiac cycle is called an **electrocardiogram**

(**ECG**) (Figure 28.2). The typical ECG consists of a series of three recognizable waves called *deflection waves.* The first wave, the **P wave,** is a small wave that indicates the depolarization of the atria immediately before atrial contraction. The large **QRS complex,** resulting from ventricular depolarization, has a complicated shape (primarily because of the variability in size of the two ventricles and the time differences required for these chambers to depolarize). It precedes ventricular contraction. The **T wave** results from currents propagated during ventricular repolarization. The repolarization of the atria, which occurs during the QRS interval, is generally obscured by the large QRS complex.

Abnormalities of the deflection waves and changes in the time intervals of the ECG may be useful in detecting myocardial infarcts or problems with the conduction system of the heart. The P-R (P-Q) interval represents the time between the beginning of atrial depolarization and ventricular depolarization; thus it includes the period during which the depolarization wave passes to the AV node, atrial systole, and the passage of the excitation wave to the balance of the conducting system. Generally, the P-R interval is about 0.16 to 0.18 sec. A longer interval may suggest a partial AV heart block caused by damage to the AV node itself or weak SA nodal impulses. In total heart block, no impulses are transmitted through the AV node, and the atria and ventricles contract independently of each other.

If the QRS interval (normally 0.06 to 0.10 sec) is prolonged, it may indicate a right or left bundle branch

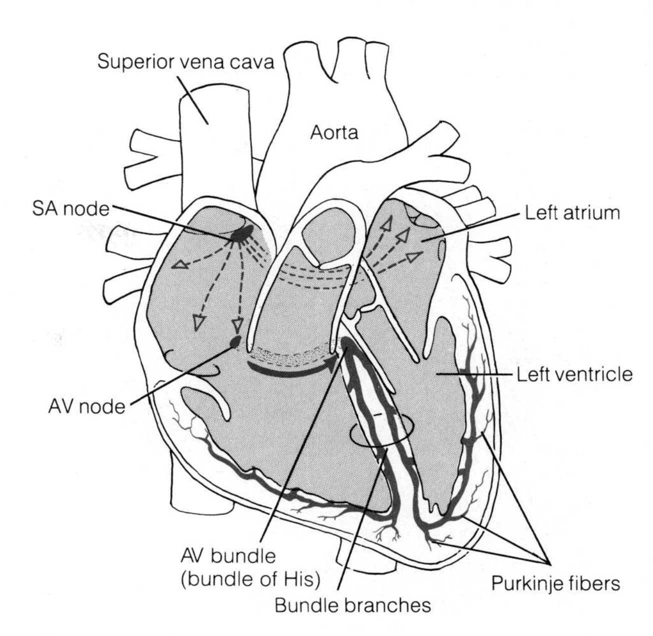

F28.1

The intrinsic conduction system of the heart. Dashed-line arrows indicate transmission of the impulse from the SA node through the atria. Solid arrow indicates transmission of the impulse from the AV node to the AV bundle (of His).

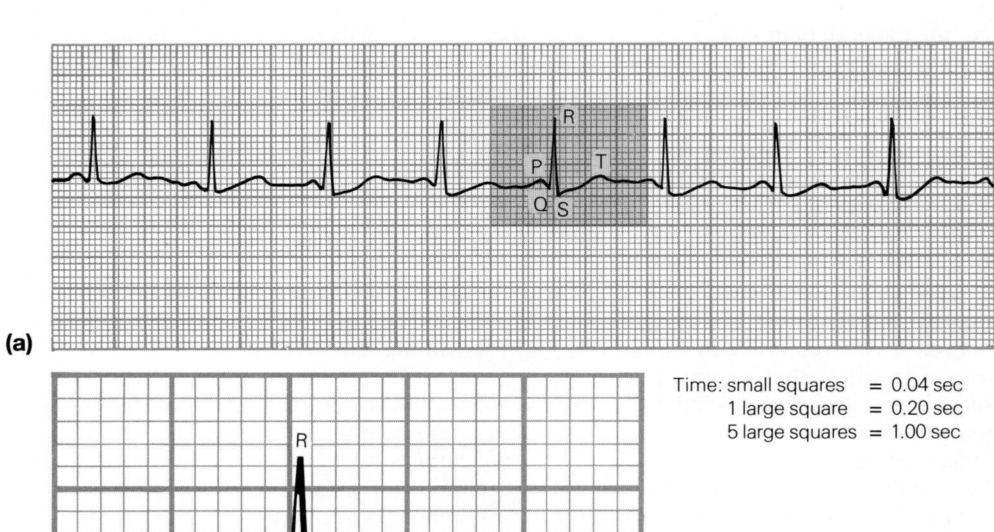

F28.2

The normal electrocardiogram. (a) Regular sinus rhythm; (b) waves, segments, and intervals of a normal ECG.

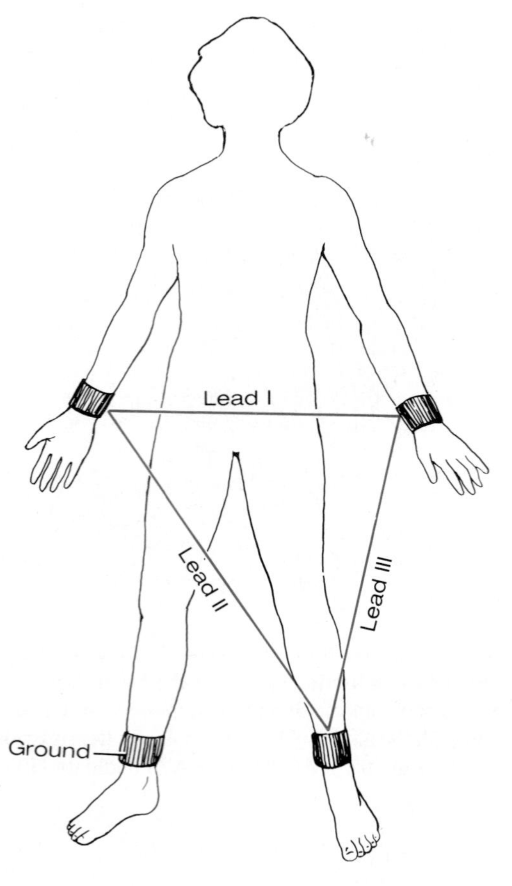

F28.3

ECG recording positions for the standard limb leads.

block in which one ventricle is contracting later than the other. The Q-T interval is the period from the beginning of ventricular depolarization through their repolarization and includes the time of ventricular contraction (i.e., the S-T segment). With a heart rate of 70/min, this interval is normally 0.31 to 0.41 sec. As the rate increases, this interval becomes shorter; conversely, when the heart rate drops, the interval is longer.

A heart rate over 100/min is referred to as **tachycardia;** a rate below 60/min is **bradycardia.** Although neither condition is pathological, prolonged tachycardia may progress to *fibrillation,* a condition of rapid uncoordinated heart contractions. Bradycardia in athletes is a *positive* finding; that is, it indicates an increased efficiency of cardiac functioning. Because stroke volume (the amount of blood ejected by a ventricle with each contraction) increases with physical conditioning, the heart can contract more slowly and still meet circulatory demands.

Twelve standard leads are used to record an ECG for diagnostic purposes. Three of these are bipolar leads that measure the voltage difference between the arms, or an arm and a leg, and nine are unipolar leads. Together the 12 leads provide a fairly comprehensive picture of the electrical activity of the heart. For this investigation, four electrodes are used (Figure 28.3), and results are obtained from the three standard leads (also shown in Figure 28.3). With each of the leads, the electrical changes between two of the electrodes are determined; thus, as shown in the figure, the potential difference between the left and right arms (LA-RA) is determined with lead I, between the right arm and left leg (RA-LL) with lead II, and between the left arm and left leg (LA-LL) with lead III. Many types of polygraphs or ECG recorders are available. Your instructor will provide specific directions on how to set up and use the available apparatus.

Preparation of Subject

1. Place electrode paste on four electrode plates, and position the electrodes at the sites described below after first scrubbing each designated area with an alcohol swab. Attach an electrode to the anterior surface of each forearm, about 2 to 3 in. above the wrist; and secure them with rubber straps. In the same manner, attach an electrode to each leg, approximately 2 to 3 in. above the medial malleolus (inner aspect of the ankle).

2. Attach the appropriate tips of the patient cable to the electrodes. The cable leads are marked RA (right arm), LA (left arm), LL (left leg), and RL (right leg, the ground).

Recording the ECG

The ECG will be recorded first under baseline (resting) conditions and then under conditions of fairly strenuous activity. Finally, recordings will be made while the subject holds his or her breath. The activity and breath-holding recordings will be compared to the baseline recordings, and you will be asked to determine the underlying reasons for the observed differences in the recordings. (It is hoped that this exercise will lead to an increased understanding of the interrelatedness of body systems.)

BASELINE RECORDINGS

1. Position the subject comfortably in a supine position on a cot (if available), or sitting relaxed on a laboratory chair.

2. Turn on the power switch and adjust the sensitivity knob to 1. Set the paper speed to 25 mm/sec and the lead selector to the position corresponding to recording from lead I (measures the voltage difference between the two arms).

3. Set the control knob at the RUN position and record the subject's at-rest ECG from lead I for 2 to 3 minutes or until the recording stabilizes. (You will need a tracing long enough to provide each student in the group with a representative segment.) The subject should try to relax and not move unnecessarily, because the skeletal muscle action potentials will also be picked up and recorded.

4. Stop the recording and mark it "lead I."

5. Repeat the recording procedure for leads II (voltage difference between the right arm and left leg) and III (voltage difference between the left arm and left leg).

6. Each student should take a representative segment of one of the lead recordings and label the record with the name of the subject and the lead used. Identify and label the P, QRS, and T waves. The calculations you perform for your recording should be based on the following information: Because the paper speed was 25 mm/sec, each millimeter of paper corresponds to a time interval of 0.04 sec. Thus, if an interval requires 4 mm of paper, its duration is 4 mm × 0.04 sec/mm = 0.16 sec.

Compute the heart rate. Measure the distance (mm) from the beginning of one QRS complex to the beginning of the next QRS complex, and plug this value into the equation below to find the time for one heartbeat.

_____ mm × 0.04 sec/mm = _____ sec/beat

Now find the beats per minute, or heart rate, by using the figure just computed in the following equation:

$$\text{Beats/min} = \frac{1}{\_\_\_\_\_\_\_\_\_\ \text{(sec/beat)}} \times 60\ \text{sec/min}$$

Beats/min = __________

Is the obtained value within normal limits? __________

Measure the QRS interval and compute its duration.

____

Measure the Q-T interval and compute its duration.

____

Measure the P-R interval and compute its duration.

____

Are the computed values within normal limits?

____

7. At the bottom of this page, attach segments of the ECG recordings from leads I through III. Make sure you note the paper speed, lead, and subject's name on each tracing. To the recording on which you based your previous computations, add your calculations for the duration of the QRS, P-R intervals, and Q-T segment above the respective area of tracing. Also record the heart rate on that tracing.

"RUNNING IN PLACE" RECORDING

1. Make sure the electrodes are securely attached to prevent electrode movement while recording the ECG.

2. Set the paper speed to 25 mm/sec, and prepare to make the recording using lead I.

3. Record the ECG while the subject is running in place for 3 min. Then have the subject sit down, but continue to record the ECG for an additional 4 min. *Mark the recording* at the end of 3 min of running and at 1 min after cessation of activity.

4. Stop the recording; and compute the beats/min during the third minute of running, at 1 min after exercise, and at 4 min after exercise. Record below:

__________________ beats/min while running in place

__________________ beats/min at 1 min after exercise

__________________ beats/min at 4 min after exercise

5. Compare this recording with the previous recording from lead I. Which intervals are shorter in the "running" recording?

____

Does the subject's heart rate return to baseline levels by 4 min after exercise?

____

"BREATH-HOLDING" RECORDING

1. Position the subject comfortably in the sitting position.

2. Using lead I and a paper speed of 25 mm/sec, *begin* the recording. After approximately 10 sec have passed, notify the subject to begin breath holding; and mark the record to indicate the onset of the 1-min breath-holding interval.

3. Stop the recording after 1 min and remind the subject to breathe. Compute the beats/min for the 1-min experimental (breath-holding) period.

Beats/min during breath holding __________________

4. Compare this recording with the lead I recording obtained under baseline conditions.

What differences are seen? __________________

____

Attempt to *explain* the physiologic reason for the differences you have seen. (Hint: a good place to start might be to check "hypoventilation" or the role of the *respiratory system* in acid-base balance of the blood.)

____

____

____

____

29 EXERCISE

Anatomy of Blood Vessels

OBJECTIVES

1. To describe the tunics of arterial and venous walls, and to state the function of each layer.
2. To correlate differences observed in artery, vein, and capillary structures with the functions these vessels perform.
3. To recognize a cross-sectional view of an artery and vein when provided with a microscopic view or appropriate diagram.
4. To list and/or identify the major arteries arising from the aorta, and to indicate the body region supplied by each.
5. To list and/or identify the major veins draining into the superior and inferior venae cavae, and to indicate the body regions drained.
6. To point out and/or discuss the unique features of special circulations in the body:
   - The pattern of blood flow in the hepatic portal system and its importance
   - The vascular supply of the brain and the importance of the circle of Willis
   - Components of the pulmonary circulation
   - Structures unique to the fetal circulation and the importance of each

MATERIALS

Anatomical charts of human arteries and veins (or a three-dimensional model of the human circulatory system)

Anatomical charts of the following specialized circulations: pulmonary circulation, hepatic portal circulation, arterial supply and circle of Willis of the brain (or a brain model showing this circulation), fetal circulation

Compound microscope

Prepared microscope slides showing cross sections of an artery and vein

Blood circulates within the blood vessels, which constitute a closed transport system. As the heart contracts, blood is propelled into the large arteries leaving the heart. It moves into successively smaller arteries and then into the arterioles, which feed the capillary beds in the tissues. Capillary beds are drained by the venules, which in turn empty into veins that ultimately converge on the great veins entering the heart. Thus arteries, carrying blood away from the heart, and veins, which drain the tissues and return blood to the heart, function simply as conducting vessels, or conduits.

Only the tiny capillaries that connect the arterioles and venules and ramify throughout the tissues directly serve the needs of the body's cells. It is through the capillary walls that exchanges between tissue cells and blood occur. Respiratory gases, nutrients, and wastes move along diffusion gradients; thus oxygen and nutrients diffuse from the blood to the tissue cells, and carbon dioxide and metabolic wastes move from the cells to the blood.

In this exercise you will examine the microscopic structure of blood vessels, and the major arteries and veins of the systemic circulation and other special circulations.

MICROSCOPIC STRUCTURE OF THE BLOOD VESSELS

Except for the microscopic capillaries, the walls of blood vessels are constructed of three coats, or tunics (Figure 29.1). The **tunica intima,** or **interna,** which lines the lumen of a vessel, is a single thin layer of endothelium that is continuous with the endocardium of the heart. Its cells fit closely together, forming an extremely smooth blood vessel lining that helps to decrease resistance to blood flow.

The **tunica media** is the more bulky middle coat and is composed primarily of smooth muscle and elastic tissue. The smooth muscle, under the control of the sympathetic nervous system, plays an active role in reducing or increasing the diameter of the vessel, which in turn increases or decreases the peripheral resistance and blood pressure.

The **tunica externa,** or **adventitia,** the outermost tunic, is composed of areolar or fibrous connective tissue. Its function is basically supportive and protective.

In general, the walls of arteries are much thicker than those of veins. The tunica media in particular tends to be considerably heavier, and contains substantially more smooth muscle and elastic tissue. This anatomic

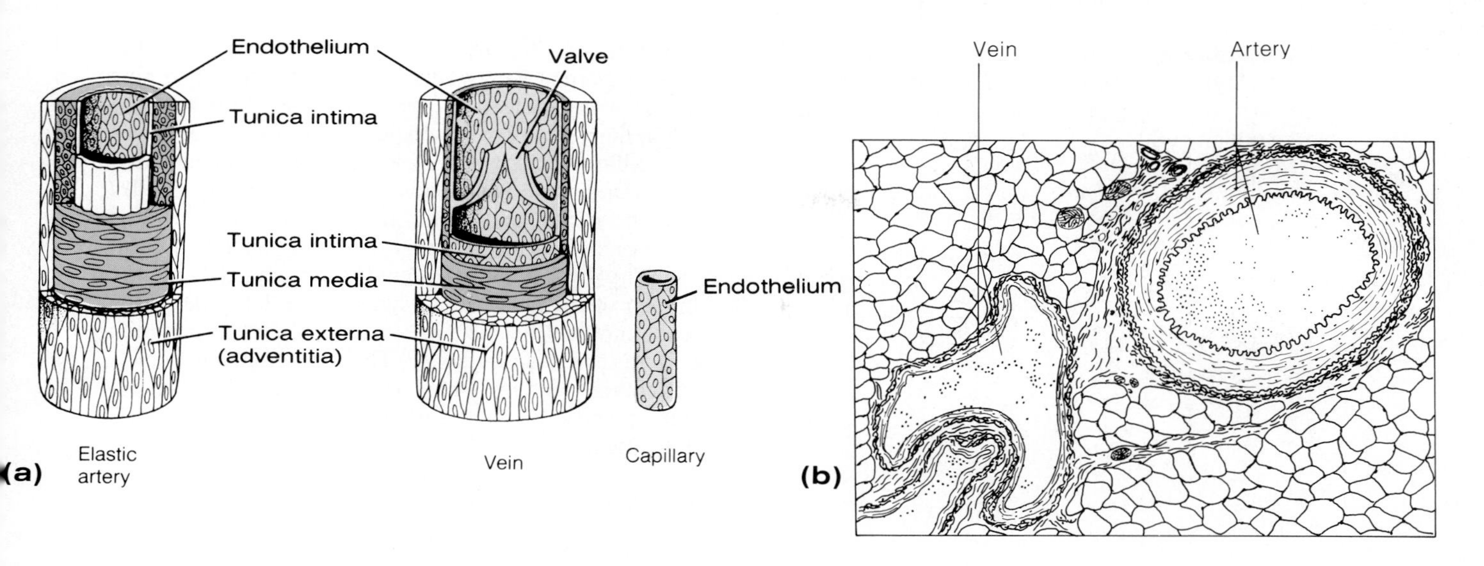

F29.1

Structure of arteries, veins, and capillaries. (a) Diagrammatic views; (b) line drawing of the photomicrograph of an artery and vein, cross-sectional view, shown in plate 26 of the histology atlas.

difference reflects a functional difference in the two types of vessels. Arteries, which are closer to the pumping action of the heart, must be able to expand as an increased volume of blood is propelled into them and then to recoil passively as the blood flows off into the circulation during diastole. Their walls must be sufficiently strong and resilient to withstand such pressure fluctuations. Because these larger arteries have such large amounts of elastic tissue in their media, they are often referred to as *elastic arteries.* Smaller arteries, further along in the circulatory pathway, are exposed to less extreme pressure fluctuations. They have somewhat less elastic tissue but still have substantial amounts of smooth muscle in their media. For this reason, they are called *muscular arteries.*

In contrast, the veins, which are far removed from the heart in the circulatory pathway, are not subjected to such pressure fluctuations and are essentially low-pressure vessels. Thus, veins may be thinner-walled without jeopardy. However, the low-pressure condition itself requires structural modifications to ensure that venous return equals cardiac output; thus, the lumens of veins tend to be substantially larger than those of corresponding arteries.

Because blood returning to the heart often flows against gravity, there are other aids to venous return. Valves in the larger veins function to prevent the backflow of blood in much the same manner as the semilunar valves of the heart. Skeletal muscle activity also promotes venous return. As the skeletal muscles surrounding the veins contract and relax, the blood is "milked" through the veins, toward the heart. (Anyone who has had to stand relatively still for an extended time will be happy to show you their swollen ankles caused by blood pooling in their feet during the period of muscle inactivity!) Finally, pressure changes that occur in the thorax during breathing also facilitate the return of blood to the heart.

- To demonstrate the efficiency of the venous valves in preventing the backflow of blood, perform the following simple experiment. Allow one hand to hang by your side until the blood vessels on the dorsal aspect become distended. Place two fingertips against one of the distended veins. Pressing firmly, move the superior finger proximally along the vein; and then release this finger. The vein will remain flattened and collapsed, despite gravity. Then remove the distal fingertip, and observe the rapid filling of the vein.

The transparent walls of the tiny capillaries are only one cell layer thick, consisting of just the endothelium underlain by a basal lamina, i.e., the tunica intima. Because of this exceptional thinness, exchanges are easily made between the blood and the tissue cells.

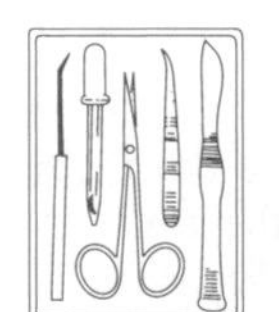

1. Obtain a cross-sectional-view preparation of blood vessels and a microscope.

2. Using Figure 29.1b as a guide, scan the section to locate a thick-walled artery. Very often, but not always, its lumen will appear scalloped, due to the constriction of its walls by the elastic tissue of the media.

3. Identify a vein. Its lumen may appear elongated or irregularly shaped and collapsed, and its walls will be considerably thinner than those of the artery. Notice the difference in the relative amount of elastic fibers in the media of the two vessels. Also note the thinness of the intima layer, which is composed of flat squamous-type cells.

4. Make drawings below of your observations of the two vessel types, and label the tunics. Try to indicate the proper size relationships relative to the wall thickness and the tunic widths.

Artery Vein

MAJOR SYSTEMIC ARTERIES OF THE BODY

The **aorta** is the largest artery in the body. Extending upward as the ascending aorta from the left ventricle, it arches posteriorly and to the left (aortic arch), and courses downward as the descending aorta through the thoracic cavity. It penetrates the diaphragm to enter the abdominal cavity, just anterior to the vertebral column.

Figure 29.2 depicts the course of the aorta and its major branches. As you locate the arteries on the figure, be aware of ways in which you can make your memorization task easier—in many cases the name of the artery reflects the body region traversed (axillary, subclavian, brachial, popliteal), the organ served (renal, hepatic), or the bone followed (tibial, femoral, radial, ulnar). Once you have identified these arteries on the figure, attempt to locate and name them (without a reference) on a large anatomical chart or on a three-dimensional model of the circulatory system vessels.

All arteries described here are shown in the figure; but some that are not described are also named. Ask your instructor which arteries you are required to identify.

Ascending Aorta

The only branches of the ascending aorta are the **right** and **left coronary arteries,** which supply the myocardium. The coronary arteries are described in Exercise 27 in conjunction with heart anatomy.

Aortic Arch

The **brachiocephalic** ("arm-head") **artery** is the first branch of the aortic arch. It persists briefly before dividing into the **right common carotid artery** and the **right subclavian artery.** The common carotid divides to form the **internal carotid artery,** which serves the brain, and the **external carotid artery,** which supplies the extracranial tissues of the neck and head. The subclavian artery gives off three branches to the head and neck—the most important being the **vertebral artery,** which runs up the posterior neck to supply a portion of the brain. In the axillary region, the subclavian artery first becomes the **axillary artery** and then the **brachial artery** as it enters the arm. At the elbow, the brachial artery divides into the **radial** and **ulnar arteries,** which follow the same-named bones to supply the forearm and hand.

The **left common carotid artery,** the second branch of the aortic arch, supplies the left side of the head and neck in the same manner the right common carotid serves the right side. The third branch is the **left subclavian artery,** which supplies the left upper extremity and subdivides as described for the right subclavian artery.

Descending Aorta

Not shown in Figure 29.2 are small branches of the thoracic portion of the descending aorta such as the 9 or 10 pairs of *intercostal arteries* that supply the muscles of the thoracic wall and the *phrenic arteries,* which supply the diaphragm. Other, more major branches of the descending aorta supply the abdominal region. The **celiac trunk** is an unpaired artery that subdivides into three branches: the **left gastric artery,** supplying the stomach; the **splenic artery,** supplying the spleen; and the **common hepatic artery,** which provides the functional blood supply of the liver. The largest branch of the descending aorta, the **superior mesenteric artery,** supplies most of the small intestine and the first half of the large intestine. The small, paired **suprarenal arteries** emerge at approximately the same level as the superior mesenteric artery and run laterally to supply the adrenal glands. (These are not shown in Figure 29.2.) The paired **renal arteries** supply the kidneys; and the **gonadal arteries,** arising from the ventral surface of the aorta, slightly below the renal arteries, run inferiorly to serve the gonads. They are called **ovarian arteries** in the female and **testicular** (or **internal spermatic**) **arteries** in the male. Because, in the male, these vessels must travel through the inguinal canal to supply the testes, they are considerably longer in the male than in the female. The small, unpaired artery supplying the second half of the large intestine is the **inferior mesenteric artery.**

In the pelvic region, the descending aorta divides into the two large **common iliac arteries.** Each of these vessels extends about 2 in. into the pelvis before dividing into the **internal iliac artery,** which supplies the pelvic organs (bladder, rectum, and some reproductive structures), and the **external iliac artery,** which continues into the thigh, where its name changes to the **femoral artery.** A branch of the femoral artery, the **deep femoral artery,** supplies the posterior thigh region. In the knee region, the femoral artery briefly becomes the **popliteal artery;** its subdivisions—the **anterior** and **posterior tibial arteries**—supply the leg, ankle, and foot. The posterior tibial gives off one main branch, the **peroneal artery** (not shown), which supplies the lateral calf (peroneal muscles). The anterior tibial artery terminates at the **dorsalis pedis artery,** which supplies the dorsum of the foot and continues on as the **arcuate ar-**

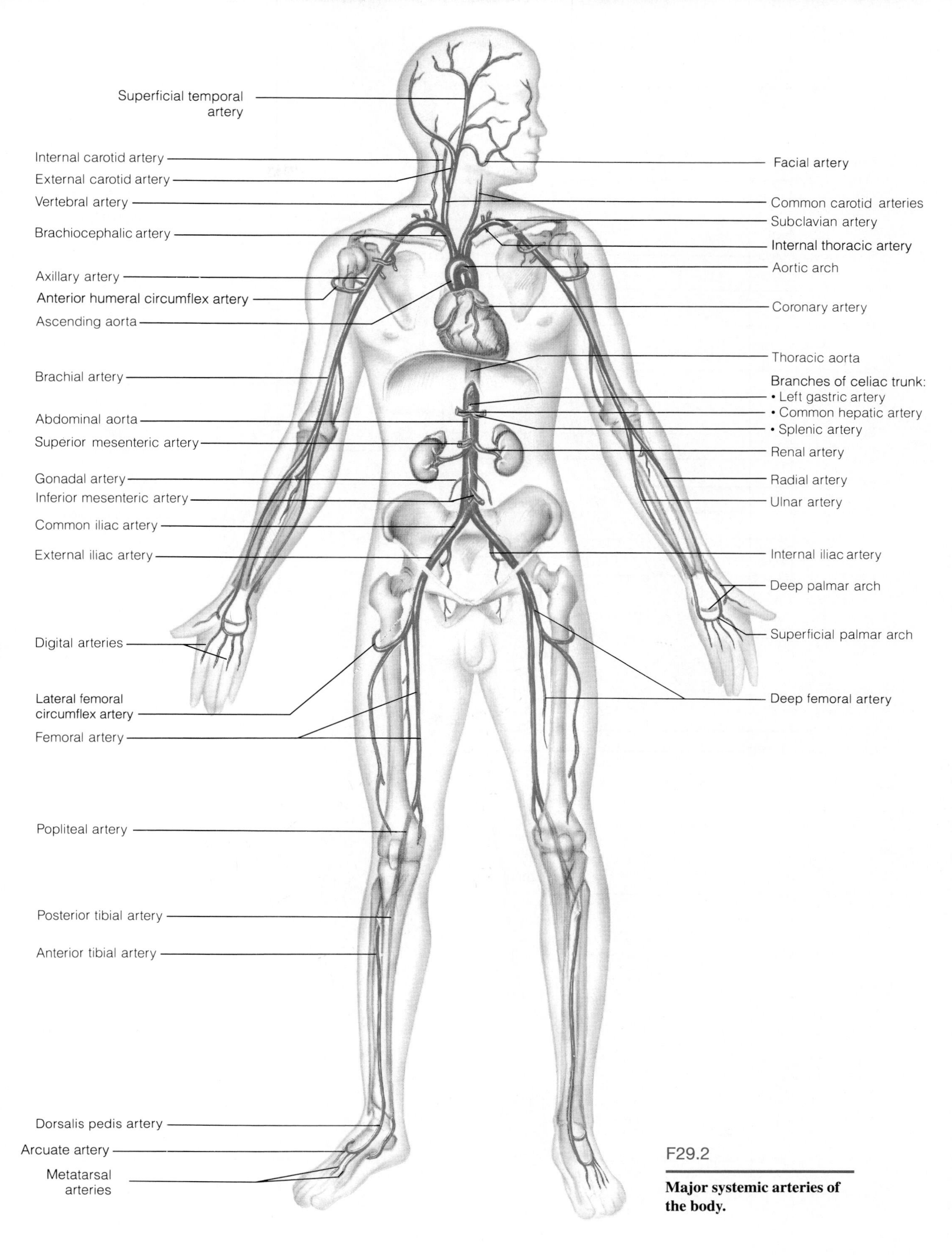

F29.2

Major systemic arteries of the body.

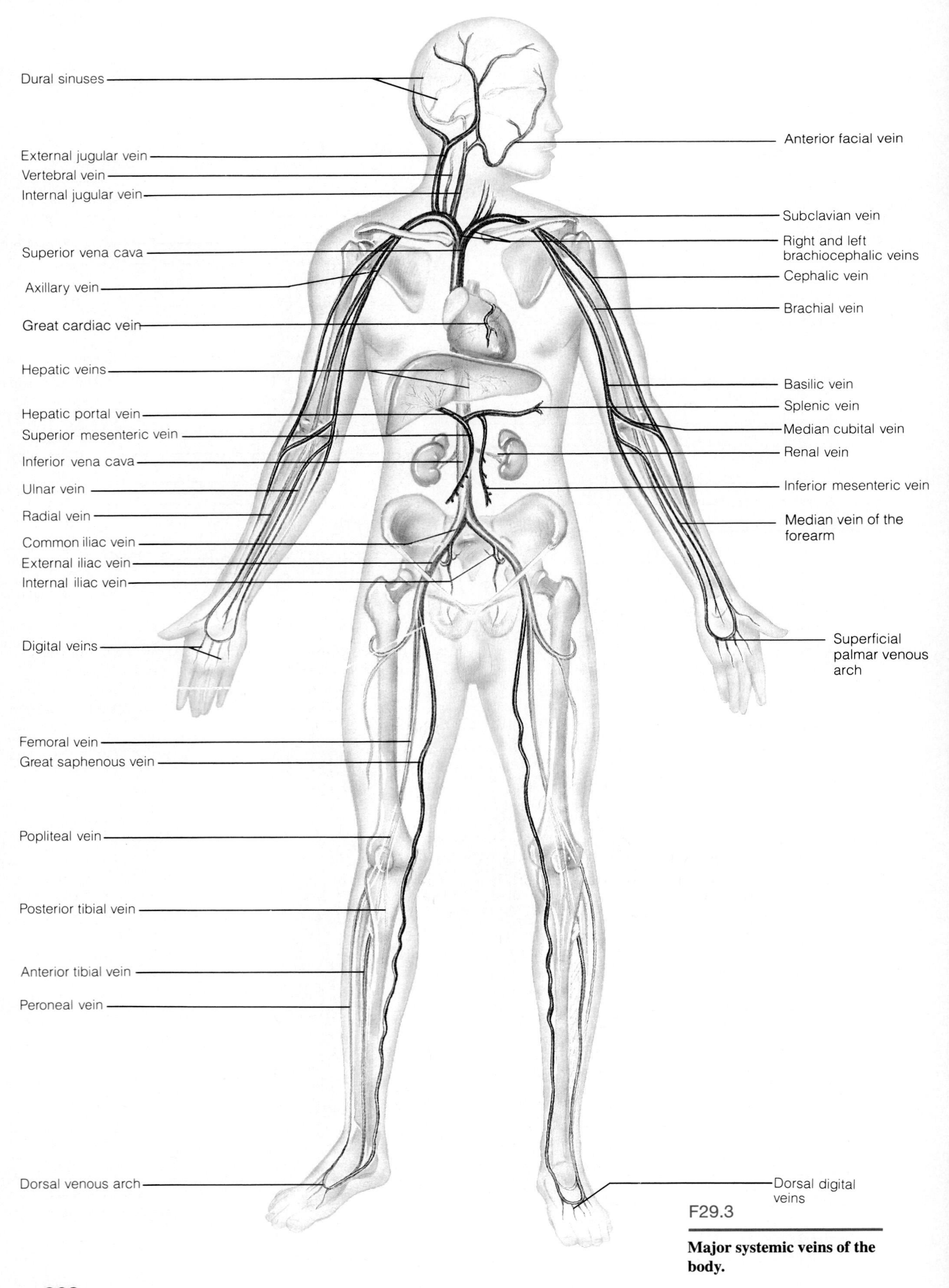

F29.3

Major systemic veins of the body.

tery. The dorsalis pedis is often palpated in patients with circulation problems of the leg to determine the circulation efficiency to the limb as a whole.

- Palpate your own dorsalis pedis artery.

MAJOR SYSTEMIC VEINS OF THE BODY

Whereas arteries are generally located in deep, well-protected body areas, many veins follow a more superficial course, and are often easily visualized and palpated on the body surface (Figure 29.3). Most deep veins parallel the course of the major arteries; thus in many cases the names of the veins and arteries are identical, except for the designation of the vessels as veins. Whereas the major systemic arteries branch off the aorta, the veins tend to converge on the venae cavae, which enter the right atrium of the heart. Veins draining the head and upper extremities empty into the **superior vena cava,** and those draining the lower body empty into the **inferior vena cava.**

Veins Draining into the Superior Vena Cava

Veins draining into the superior vena cava are named from the superior vena cava distally, **but remember that the flow of blood is in the opposite direction.**

The **right** and **left brachiocephalic veins** drain the head, neck, and upper extremities, and unite to form the superior vena cava. (Note that although there is only one brachiocephalic artery, there are two brachiocephalic veins.)

Branches of the brachiocephalic veins include the **internal jugular veins,** large veins that drain the dural sinuses of the brain; the **vertebral veins,** which drain the posterior aspect of the head; and the **subclavian veins,** which receive the venous blood from the upper extremity. The small **external jugular vein** joins the subclavian vein near its origin to return the venous drainage of the extracranial tissues of the head and neck. As the subclavian vein traverses the axilla, it becomes the **axillary vein** and then the **brachial vein** in the arm where it courses along the posterior aspect of the humerus. The brachial vein is formed by the union of the deep **radial** and **ulnar veins** of the forearm. The superficially located venous drainage of the arm includes the **cephalic vein,** which courses along the lateral aspect of the arm and empties into the axillary vein; the **basilic vein,** found on the medial aspect of the arm and entering the brachial vein; and the **median cubital vein,** which runs between the cephalic and basilic veins in the anterior aspect of the elbow (this vein is often the site of choice for removing blood for testing purposes).

The **azygos vein,** part of the *azygos system* that drains the intercostal muscles of the thorax and provides an accessory venous system to drain the abdominal wall, enters the dorsal aspect of the superior vena cava immediately before it enters the right atrium. The azygos system, shown in Figure 29.4, also includes the **hemiazygos** and **accessory hemiazygos veins,** which together drain the left aspect of the thorax and empty into the azygos vein. The azygos vein drains the right aspect of the thorax.

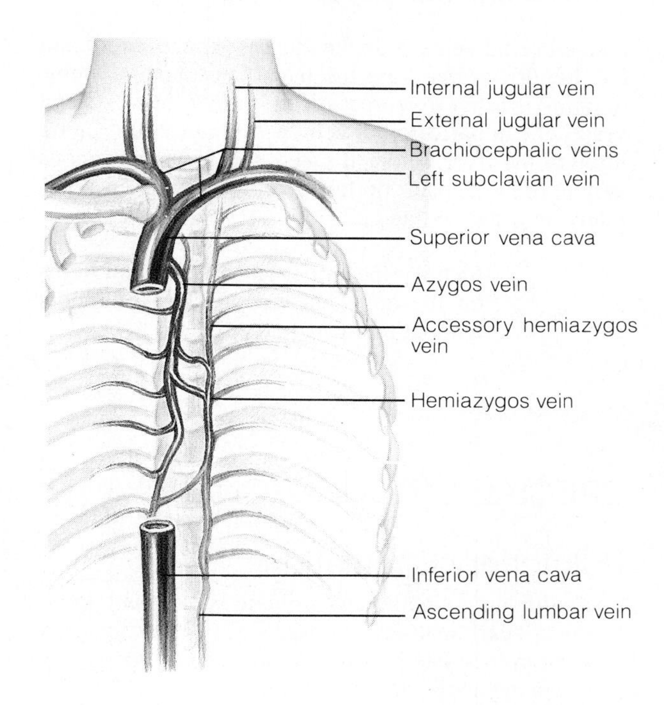

F29.4

The azygos system.

Veins Draining into the Inferior Vena Cava

The inferior vena cava, which is much longer than the superior vena cava, returns blood to the heart from all body regions below the diaphragm. It begins in the lower abdominal region with the union of the paired **common iliac veins,** which drain venous blood from the legs and pelvis. Each common iliac vein in turn is formed by the union of the **internal iliac vein,** draining the pelvis, and the **external iliac vein,** which receives venous blood from the lower limb. Veins of the leg include the **anterior** and **posterior tibial veins,** which supply the calf and foot. The posterior tibial vein becomes the **popliteal vein** in the knee region and the **femoral vein** in the thigh. The femoral vein empties into the external iliac vein in the inguinal region. The **great saphenous vein,** a superficial vein, is the longest vein of the body. Beginning in the foot with the **dorsal venous arch,** it extends up the medial side of the leg, knee, and thigh to empty into the femoral vein. Moving superiorly into the abdominal cavity, the inferior vena cava receives blood from the **right gonadal vein** (testicular, or spermatic, vein in the male, ovarian vein in the female), which drains the right gonad. (The left testicular or ovarian vein drains into the left renal vein.) The **right** and **left renal veins** drain the kidneys, and the **right** and **left hepatic veins** drain the liver. The unpaired veins draining the digestive tract organs empty into a special vessel, the **hepatic portal vein,** which carries this blood through the liver before it enters the systemic venous system. (The hepatic portal system is discussed separately on pages 271 and 272.)

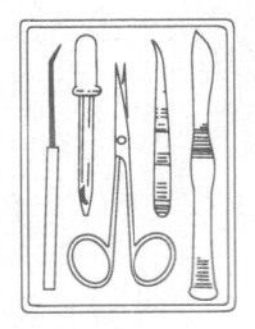

Identify the important arteries and veins on the large anatomical chart or model without referring to the figure.

SPECIAL CIRCULATIONS

Pulmonary Circulation

The pulmonary circulation (discussed previously in relation to heart anatomy on page 254) differs in many ways from the systemic circulation, because it does not serve the metabolic needs of the body tissue with which it is associated, in this case, lung tissue. (The functional blood supply of the lungs is provided by the **bronchial arteries,** which diverge from the thoracic portion of the descending aorta.) It functions instead to bring the blood into close contact with the alveoli of the lungs to permit gaseous exchanges that rid the blood of excess carbon dioxide and that replenish its supply of vital oxygen. The arteries of the pulmonary circulation are structurally much like veins; thus they create a low-pressure bed in the lungs. (If the arterial blood pressure in the systemic circulation is 120/80, the pressure in the pulmonary artery is likely to be approximately 25/10.)

Pulmonary circulation begins with the large **pulmonary trunk,** which leaves the right ventricle and divides into the **right** and **left pulmonary arteries** about 5 cm (2 in.) above its origin (Figure 29.5). The right and left pulmonary arteries plunge into the lungs, where they subdivide into the **lobar arteries** (three on the right and two on the left), which accompany the main bronchi into the lobes of the lungs. Within the lungs, the lobar arteries branch extensively to form the arterioles which finally terminate in the capillary networks surrounding the alveolar sacs of the lungs. Diffusion of the respiratory gases occurs across the walls of the alveoli and **pulmonary capillaries.** The pulmonary capillary beds are drained by venules, which converge to form sequentially larger veins and finally the four **pulmonary veins** (two leaving each lung), which return the blood to the left atrium of the heart.

- Using the terms provided in Figure 29.5 ***label*** all structures provided with leader lines.

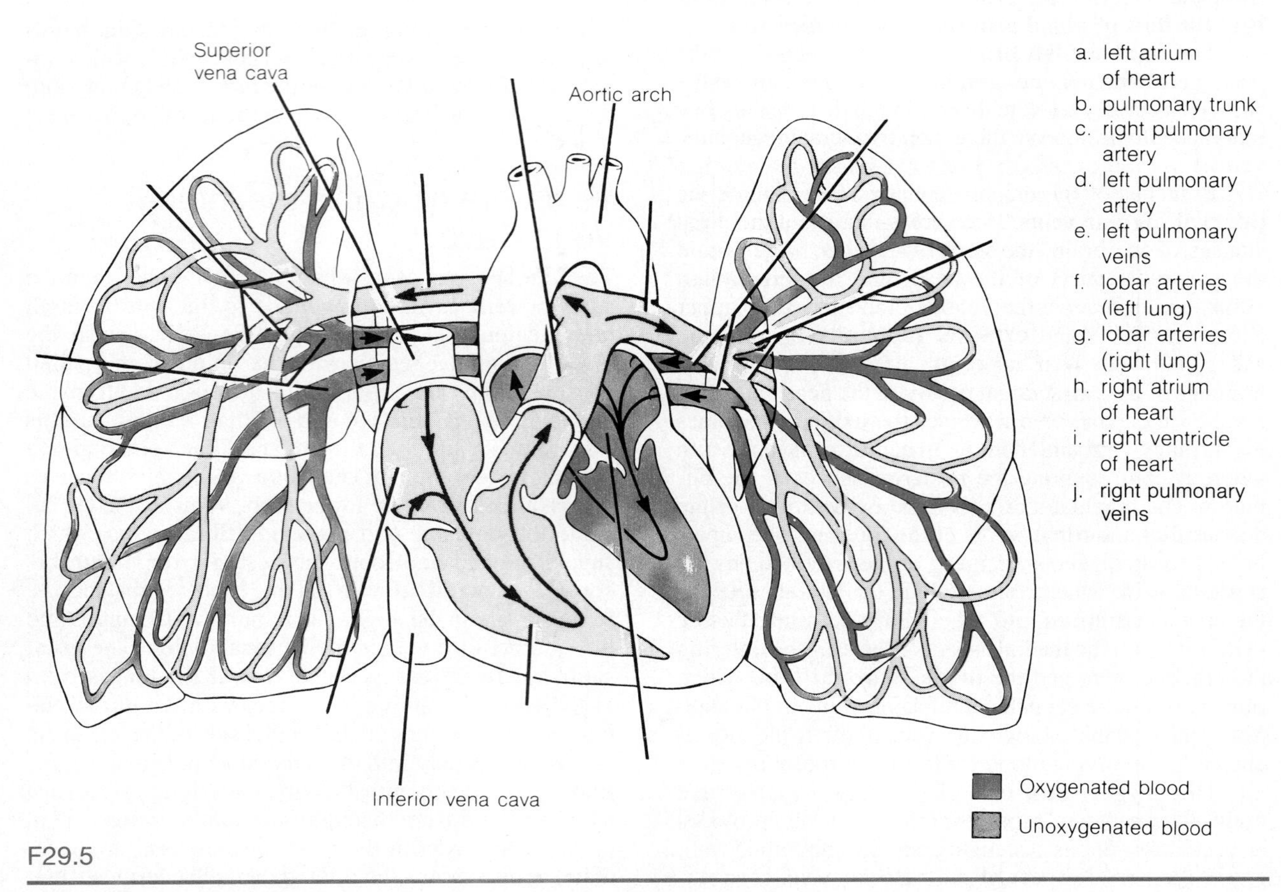

F29.5

The pulmonary circulation.

Arterial Supply of the Brain and the Circle of Willis

A continuous blood supply to the brain is crucial, because deprivation for even a few minutes causes irreparable damage to the delicate brain tissue. The brain is supplied by two pairs of arteries arising from the region of the aortic arch—the *internal carotid arteries* and the *vertebral arteries.*

- Figure 29.6 is a diagram of the brain's arterial supply. The internal carotid and vertebral arteries are labeled. As you read the description of the blood supply below, ***complete the labeling*** of this diagram.

The **internal carotid arteries,** branches of the common carotid arteries, follow a deep course through the neck and along the pharynx, entering the skull through the carotid canals of the temporal bone. Within the cranium, each divides into the **anterior** and **middle cerebral** arteries, which supply the bulk of the cerebrum. The internal carotid arteries also contribute to the formation of the **circle of Willis,** an arterial anastomosis surrounding the pituitary gland and the optic chiasma at the base of the brain, by forming a **posterior communicating artery** on each side. The circle is completed by the **anterior communicating artery,** a short shunt connecting the right and left anterior cerebral arteries.

The paired **vertebral arteries** diverge from the subclavian arteries and pass superiorly through the foramina of the transverse processes of the cervical vertebrae, entering the skull through the foramen magnum. Within the skull, the vertebral arteries unite to form a single **basilar artery,** which continues superiorly along the ventral aspect of the brain stem, giving off branches to the pons, cerebellum, and inner ear. At the base of the cerebrum, the artery divides to form the **posterior cerebral arteries,** which supply portions of the temporal and occipital lobes of the cerebrum, and also become part of the circle of Willis by joining with the posterior communicating arteries.

The uniting of the blood supply of the internal carotid arteries and the vertebral arteries via the circle of Willis is a protective device that theoretically provides an alternate set of pathways for blood to reach the brain tissue in the case of arterial occlusion or impaired blood flow anywhere in the system. In actuality, the communicating arteries are tiny; and in many cases the communicating system is defective.

Hepatic Portal Circulation

Blood vessels of the hepatic portal circulation drain the digestive viscera, spleen, and pancreas, and deliver this

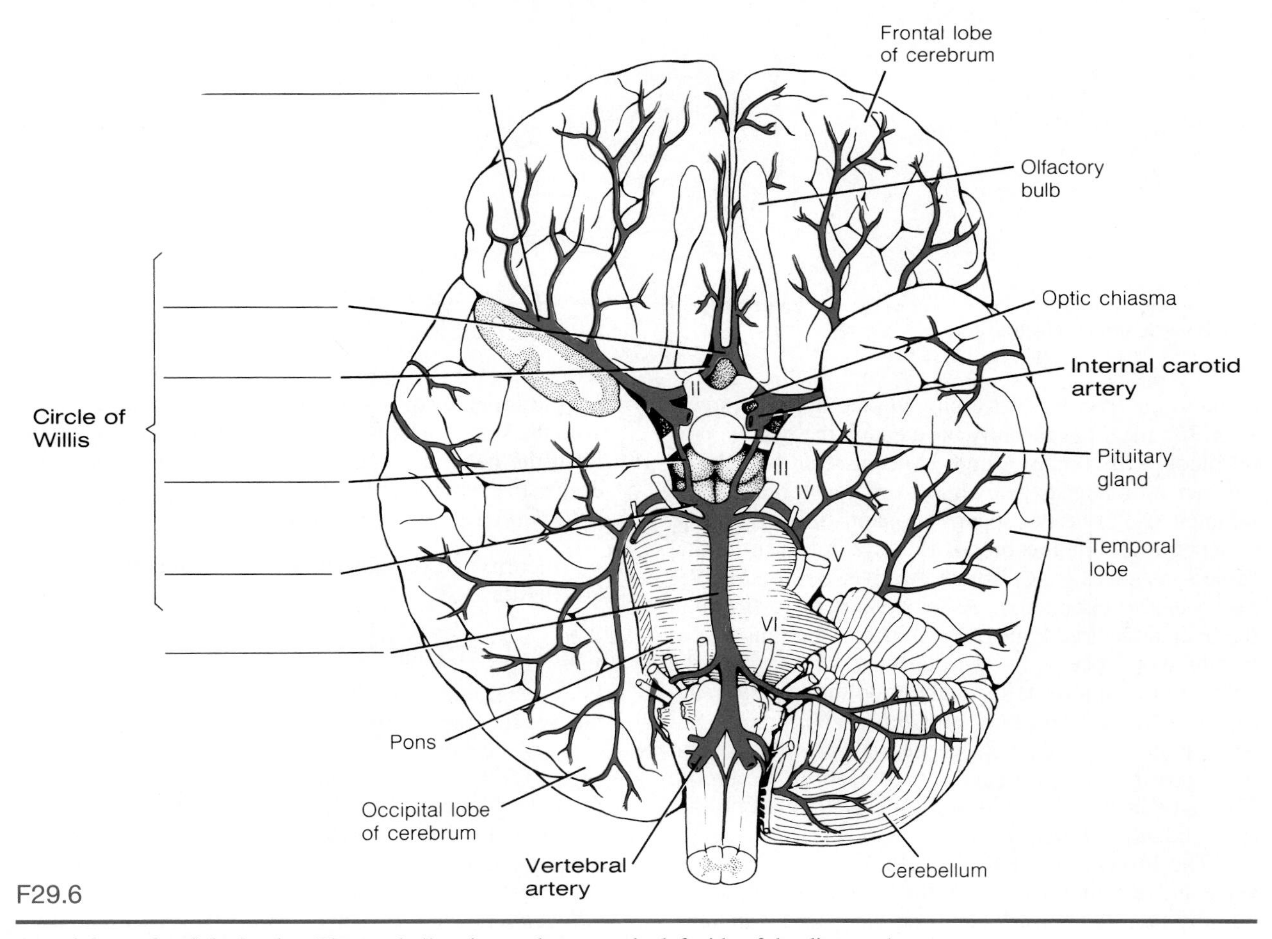

F29.6

Arterial supply of the brain. (The cerebellum is not shown on the left side of the diagram.)

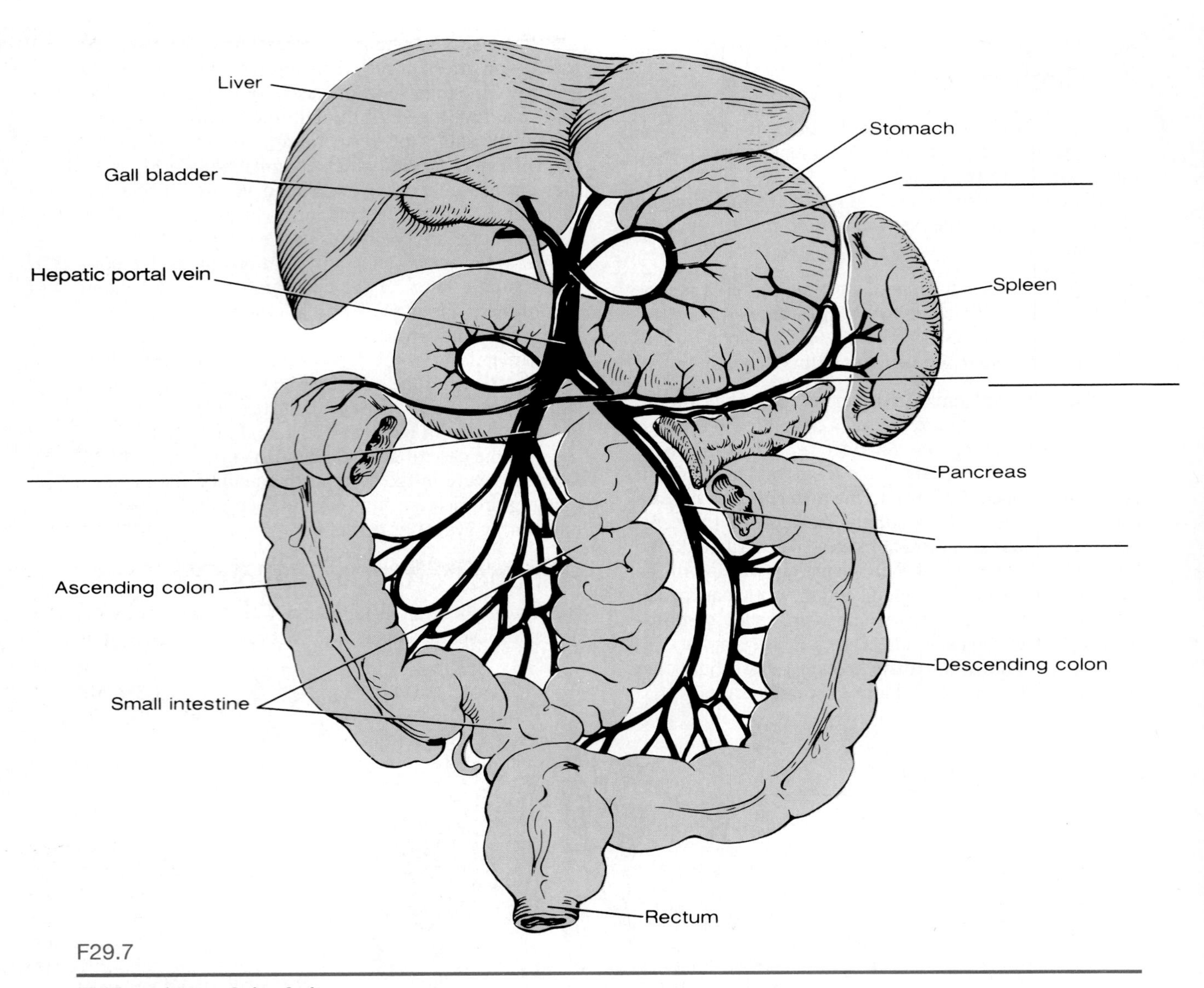

F29.7

The hepatic portal circulation.

blood to the liver for processing via the **hepatic portal vein.** If a meal has recently been eaten, the hepatic portal blood will contain a high concentration of nutrient substances. Because the liver is a key body organ for the maintenance of proper sugar, fatty acid, and amino acid concentrations in the blood, this system ensures that these substances pass through the liver before entering the systemic circulation. As blood percolates through the liver sinusoids, some of the nutrients are removed to be stored or processed in various ways for release to the general circulation. At the same time the hepatic macrophages remove bacteria and other debris from the passing blood, and the hepatocytes detoxify alcohol and other possibly harmful chemicals present in the blood. The liver in turn is drained by the hepatic veins that enter the inferior vena cava.

The **inferior mesenteric vein,** draining the transverse and terminal portions of the large intestine, drains into the **splenic vein,** which drains the spleen, pancreas, and greater curvature of the stomach. The splenic vein and the **superior mesenteric vein,** which receives blood from the small intestine and the ascending colon, join to form the hepatic portal vein. The **gastric vein,** which drains the lesser curvature of the stomach, drains directly into the hepatic portal vein.

- Identify the vessels named above, and ***label*** them in Figure 29.7.

Fetal Circulation

In a developing fetus, the lungs and digestive system are not yet functional; and all nutrient, excretory, and gaseous exchanges must occur through the placenta (see Figure 29.8). Thus nutrients and oxygen move from the mother's blood, across placental barriers, into the fetal blood; and carbon dioxide and other metabolic wastes move from the fetal blood supply to the mother's blood.

Fetal blood travels through the umbilical cord, which contains three blood vessels: two smaller *umbilical arteries* and one large *umbilical vein.* The **umbilical**

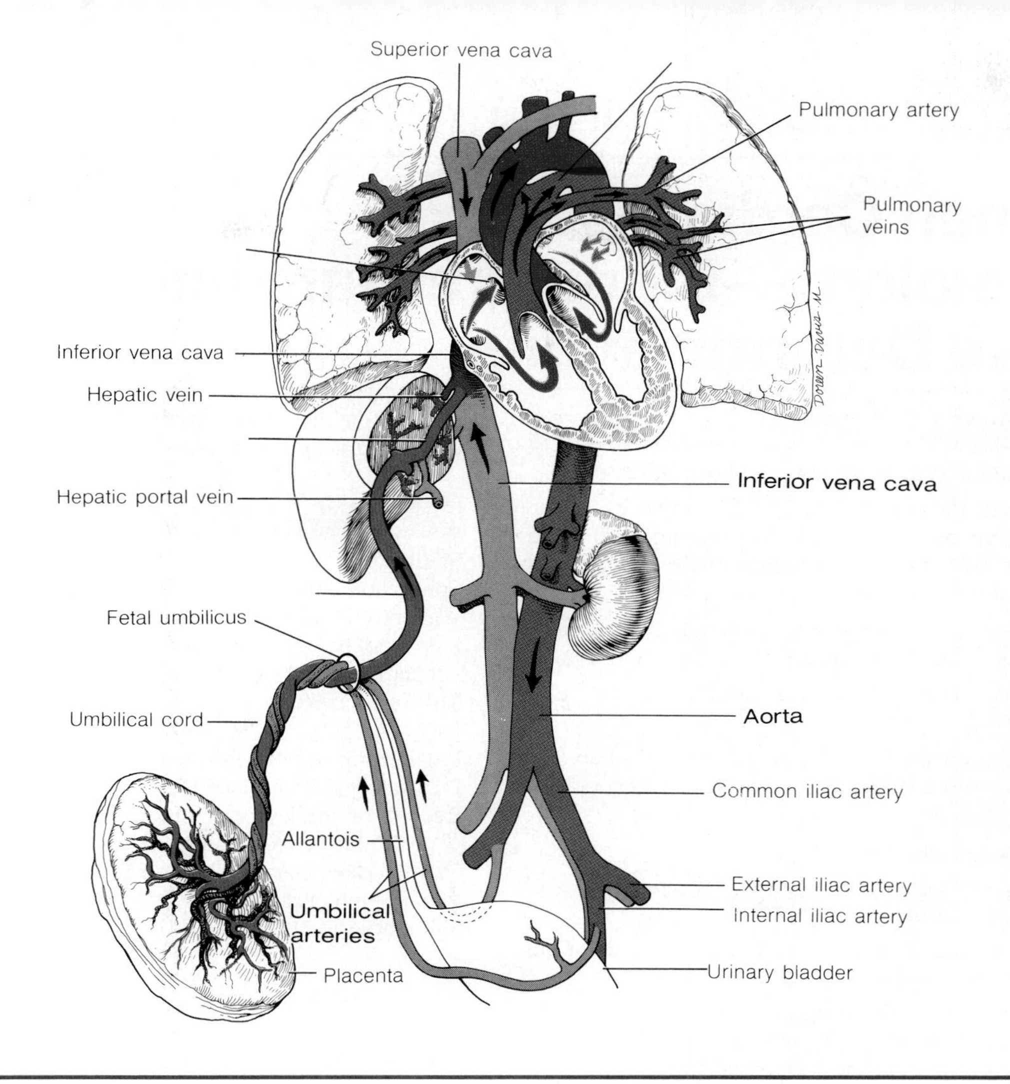

F29.8

The fetal circulation.

vein carries blood rich in nutrients and oxygen to the fetal heart. The **umbilical arteries** carry carbon dioxide and waste-laden blood from the fetus to the placenta. The umbilical arteries meet the umbilical vein at the umbilicus (navel, or belly button) and wrap around the vein within the cord en route to their placental attachments. In the umbilical vein, newly oxygenated blood flows superiorly toward the fetal heart. En route, some of this blood perfuses the liver; but a larger proportion is ducted through the relatively nonfunctional liver to the inferior vena cava via a vessel called the **ductus venosus,** which carries the blood to the right atrium of the heart.

Because fetal lungs are nonfunctional and collapsed, two shunting mechanisms ensure that the lungs are almost entirely bypassed. Much of the blood entering the right atrium is shunted into the left atrium through a flaplike opening in the interatrial septum—the **foramen ovale.** The left ventricle then pumps the blood out the aorta to the systemic circulation. Blood that does enter the right ventricle and is pumped into the pulmonary trunk encounters a second shunt, the **ductus arteriosus,** a short vessel connecting the pulmonary trunk and the aorta. Because the collapsed lungs present an extremely high-resistance pathway, blood more readily enters the systemic circulation through the ductus arteriosus.

The aorta carries blood to the tissues of the body; this blood ultimately finds its way back to the placenta, via the umbilical arteries. The only fetal vessel that carries highly oxygenated blood is the umbilical vein; all other vessels contain varying degrees of oxygenated and deoxygenated blood.

At birth, or shortly thereafter, the foramen ovale closes and becomes the **fossa ovalis;** and the ductus arteriosus collapses and is converted to the fibrous **ligamentum arteriosum.** Lack of blood flow through the umbilical vessels leads to their eventual obliteration, and the circulatory pattern becomes that of the adult. Remnants of the umbilical arteries persist as the **medial umbilical ligaments** on the inner surface of the anterior abdominal wall; of the umbilical vein as the **ligamentum teres,** or **round ligament** of the liver; and of the ductus venosus as a fibrous band called the **ligamentum venosus** on the inferior surface of the liver.

- The pathway of fetal blood flow is indicated with arrows in Figure 29.8. Appropriately ***label*** the specialized fetal circulatory structures provided with leader lines.

30 EXERCISE

Human Cardiovascular Physiology—Blood Pressure and Pulse Determinations

OBJECTIVES

1. To define *systole, diastole,* and *cardiac cycle.*
2. To state the normal length of the cardiac cycle, the relative pressure changes occurring within the atria and ventricles during the cycle, and the timing of valve closure.
3. To use the stethoscope to auscultate heart sounds and to relate heart sounds to cardiac cycle events.
4. To describe the clinical significance of heart sounds.
5. To demonstrate the thoracic locations where the first and second heart sounds are most accurately auscultated.
6. To define *murmur.*
7. To define *pulse, pulse deficit, blood pressure,* and *sounds of Korotkoff.*
8. To accurately determine the apical and radial pulse of a subject.
9. To accurately determine the blood pressure of a subject by using a sphygmomanometer, and to relate systolic and diastolic pressures to events of the cardiac cycle.
10. To note factors affecting and/or determining blood pressure and skin color.

MATERIALS

Stethoscope
Sphygmomanometer
Watch (or clock) with a second hand
Step stool or stair, 18 inches high
Alcohol swabs
Millimeter ruler
Ice
Small basin suitable for the immersion of one hand
Laboratory thermometer (°C)
Record or audiotape: "Interpreting Heart Sounds" (available on free loan from the local chapters of the American Heart Association)
Phonograph or tape deck
Felt marker

Any comprehensive study of human cardiovascular physiology takes much more time than a single laboratory period. However, it is possible to investigate a few phenomena such as pulse determinations, auscultation of heart sounds, and blood pressure measurements, which reflect the heart in action and the function of blood vessels. (The electrocardiogram is studied separately in Exercise 28.) A discussion of the cardiac cycle will provide a basis for understanding and interpreting the various physiologic measurements to be taken.

CARDIAC CYCLE

In the healthy heart, the two atria contract simultaneously. As they begin to relax, simultaneous contraction of the ventricles occurs. According to general usage, the terms **systole** and **diastole** refer to events of ventricular contraction and relaxation, respectively. The **cardiac cycle** is equivalent to one complete heartbeat—during which both atria and ventricles contract, and then relax. It is marked by a succession of changes in blood volume and pressure within the heart. Figure 30.1 is a graphic representation of the events of the cardiac cycle for the left side of the heart. Although pressure changes in the right side are lower than those in the left, the same relationships apply.

We will begin the discussion of the cardiac cycle with the heart in complete relaxation. At this point, pressure in the heart is very low, blood is flowing passively from the pulmonary and systemic circulations into the atria and on through to the ventricles, the semilunar valves are closed, and the AV valves are open. As atrial contraction occurs, atrial pressure increases, forcing residual blood into the ventricles. Shortly thereafter, ventricular systole begins; and the intraventricular pressure increases rapidly, closing the AV valves. When ventricular pressure exceeds that of the large arteries leaving the heart, the semilunar valves are forced open; and the blood in the ventricular chambers is expelled through the valves. During this phase, the aortic pressure reaches approximately 120 mm Hg. During ventricular systole, the atria relax; and their chambers fill with blood, which results in a gradually increasing atrial pressure. At the end of ventricular systole, the ventricles relax; the semilunar valves snap shut, preventing backflow; and momentarily, the ventricles are closed chambers. When the aortic semilunar valve snaps shut, a momentary increase

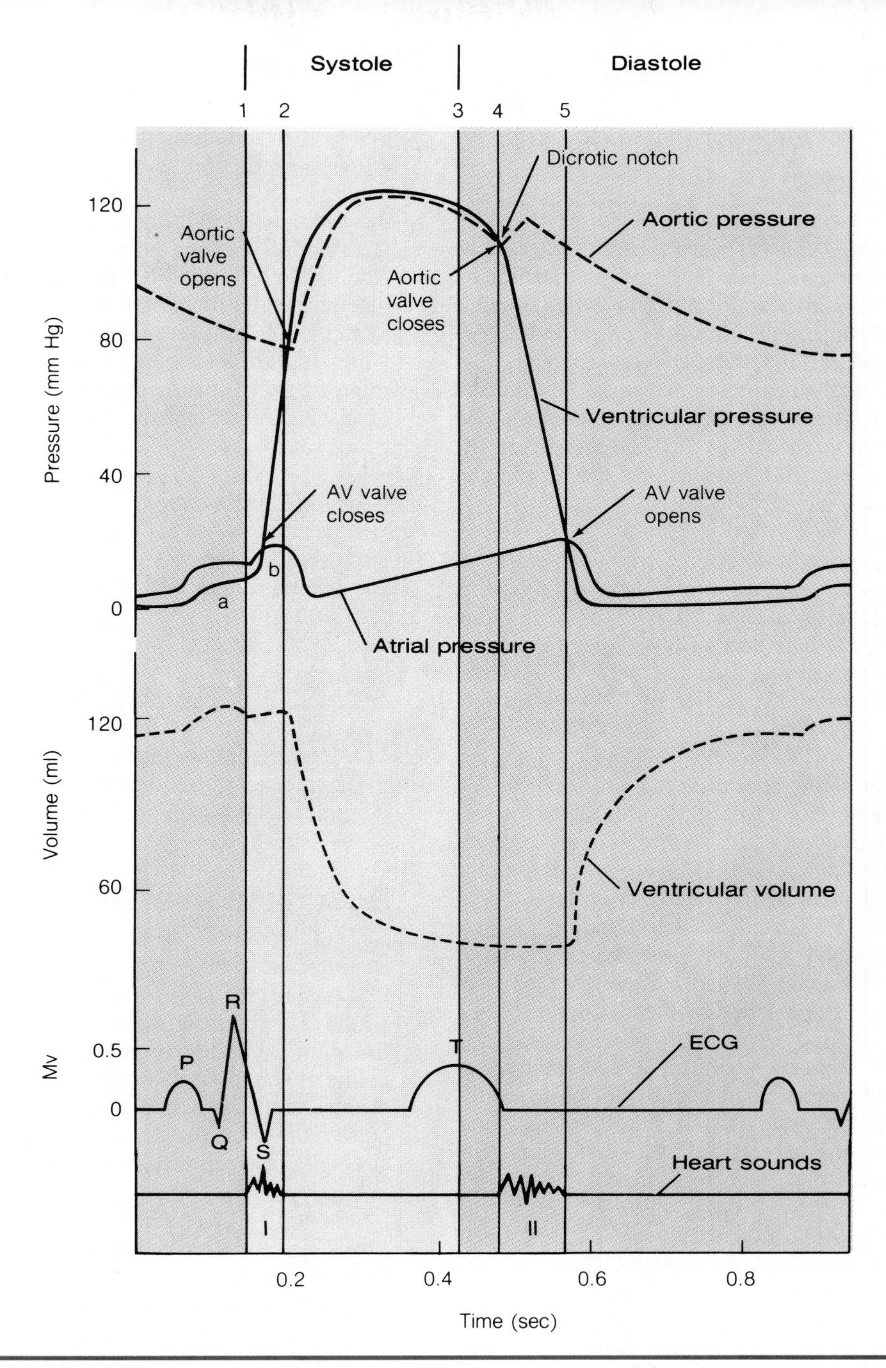

F30.1

Graphic representation of pressure and volume changes in the left ventricle during one cardiac cycle. Depicts valvular events and correlates changes to heart sounds and ECG. (ECG is considered in Exercise 28.)

in the aortic pressure results from the elastic recoil of the aorta after valve closure. This event results in the pressure fluctuation called the **dicrotic notch** (see Figure 30.1). As the ventricles relax, the pressure within them begins to drop. When it is again less than atrial pressure, the AV valves are forced open; and the ventricles again begin to fill with blood. Atrial and aortic pressures decrease; and the ventricles rapidly refill, completing the cycle.

The average heart beats approximately 72 beats per minute, and so the length of the cardiac cycle is about 0.8 second. Of this time period, atrial contraction occupies the first 0.1 second, which is followed by atrial relaxation and ventricular contraction for the next 0.3 second. The remaining 0.4 second is the quiescent, or ventricular relaxation, period. When the heart beats at a more rapid pace than normal, this last period decreases.

Note that two different types of phenomena control the movement of blood through the heart: the alternate contraction and relaxation of the myocardium, and the opening and closing of valves (which are entirely dependent on the pressure changes within the heart chambers).

Study the figure carefully to make sure you understand what has been discussed before continuing with the next portion of the exercise.

AUSCULTATION OF HEART SOUNDS

Two distinct sounds can be heard during each cardiac cycle. These heart sounds are commonly described by the monosyllables "lub" and "dup"; and the sequence is designated lub-dup, pause, lub-dup, pause, and so on. The first heart sound (lub) is associated with the closure of the AV valves at the beginning of systole. The second heart sound (dup) is most commonly associated with the closure of the semilunar valves and corresponds to the end of systole. Figure 30.1 indicates the timing of heart sounds in the cardiac cycle.

Abnormal heart sounds are referred to as *murmurs* and often indicate valvular problems. In valves that do not close tightly, closure is followed by a swishing sound caused by the backflow of blood (regurgitation). Distinct sounds are also associated with the tortuous flow of blood through constricted, or stenosed, valves. ■

Before auscultating your partner's heart sounds, listen to the recording "Interpreting Heart Sounds" so that you may hear both normal and abnormal heart sounds.

In the following procedure, you will auscultate heart sounds with an ordinary stethoscope. A number of more sophisticated heart-sound amplification systems are on the market, and your instructor may prefer to use one such if it is available. If so, directions for the use of this apparatus will be provided by the instructor.

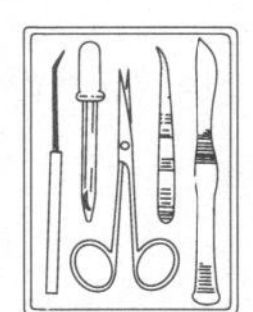

1. Obtain a stethoscope and some alcohol swabs. Heart sounds are best auscultated (listened to) if the subject's outer clothing is removed, so a male subject is preferable.

2. With an alcohol swab, clean the earpieces of the stethoscope. Allow the alcohol to dry. Notice that the earpieces are angled. Hence, for comfort and best auscultation, the earpieces should be angled in a forward direction when placed into the ears.

3. Don the stethoscope; place the diaphragm of the stethoscope on your partner's thorax, just to the sternal side of the left nipple at the fifth intercostal space; and listen carefully for heart sounds. The first sound will be a longer, louder (more booming) sound than the second, which is short and sharp. After listening for a couple of minutes, try to time the pause between the second sound of one heartbeat and the first sound of the subsequent heartbeat.

How long is this interval? ____________________ sec

How does it compare to the interval between the first and second sounds of a single heartbeat?

____

____

4. To differentiate individual valve sounds somewhat more precisely, auscultate the heart sounds over specific thoracic regions. Refer to Figure 30.2 for the positioning of the stethoscope.

Auscultation of AV Valves

As a rule, the mitral valve closes slightly before the tricuspid valve. You can hear the mitral valve more clearly if you place the stethoscope over the apex of the heart, which is at the fifth intercostal space, approximately in line with the middle region of the left clavicle. Listen to the heart sounds at this region; and then move the stethoscope medially to the left margin of the sternum to auscultate the tricuspid valve. Are you able to detect the slight lag between the closure of the mitral and tricuspid valves?

Note that there are normal variations in the site for "best" auscultation of the tricuspid valve. These range from the site depicted in Figure 30.2 (left sternal margin over fifth intercostal space) to over the sternal body in

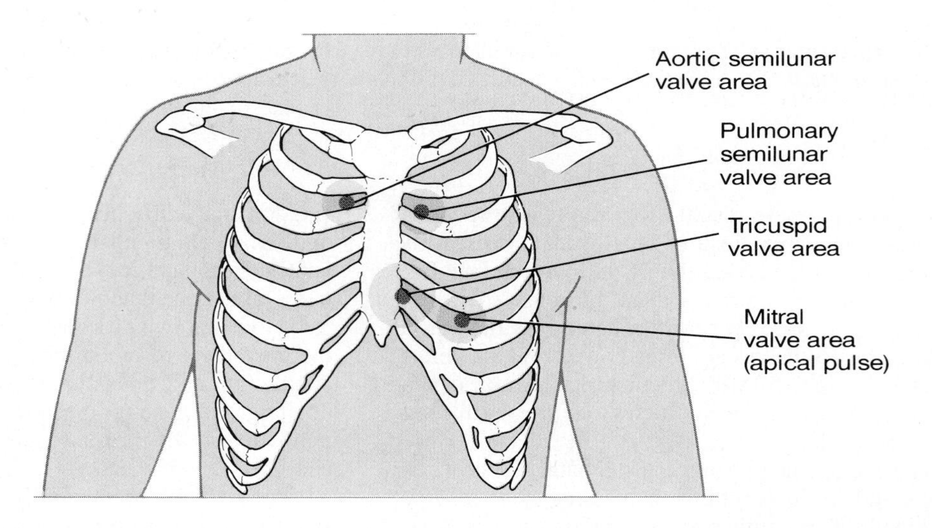

F30.2

Areas of the thorax where valvular sounds can be best detected.

the same plane, to the right sternal margin over the fifth intercostal space. If you are having difficulty hearing closure of the tricuspid valve, try one of these other locations.

Auscultation of Semilunar Valves

Again there is a slight dyssynchrony of valve closure; the aortic semilunar valve normally snaps shut just ahead of the pulmonary semilunar valve. If the subject inspires deeply but gently, the filling of the right ventricle will be delayed slightly (due to the compression of the thoracic blood vessels by the increased intrapulmonary pressure); and the two sounds can be heard more distinctly. Position the stethoscope over the second intercostal space, just to the *right* of the sternum. The aortic valve is best heard at this position. As you listen, have your partner take a deep breath. Then move the stethoscope to the *left* side of the sternum, in the same line; and auscultate the pulmonary valve. Listen carefully; try to hear the "split" between the closure of these two valves in the second heart sound.

Although at first it may seem a bit odd that the pulmonary valve issuing from the *right* heart is heard most clearly to the *left* of the sternum and the aortic valve of the left heart is easiest to auscultate at the right sternal border, this is easily explained by reviewing your heart anatomy. Because the heart is twisted, with the right ventricle forming most of the anterior ventricular surface, the pulmonary trunk actually crosses to the right as it issues from the right ventricle. Similarly, the aorta issues from the left ventricle at the left side of the pulmonary trunk before arching up and over that vessel.

PALPATION OF THE PULSE

The term **pulse** refers to the alternating surges of pressure (expansion and then recoil) in an artery that occur with each contraction and relaxation of the left ventricle. Normally the pulse rate (pressure surges per minute) equals the heart rate (beats per minute), and the pulse averages 70 to 76 beats per minute in the resting state.

Parameters other than pulse rate are also useful clinically. You may also assess the regularity (or rhythmicity) of the pulse, and its amplitude and/or tension—does the blood vessel expand and recoil (sometimes visibly) with the pressure waves? Can you feel it strongly, or is it difficult to detect? Is it regular like the ticking of a clock, or does it seem to skip beats?

Superficial Pulse Points

The pulse may be felt easily on any superficial artery when the artery is compressed over a bone or firm tissue. Palpate the following pulse or pressure points on your partner by placing the fingertips of the first two or three fingers of one hand over the artery. It is generally helpful to compress the artery firmly as you begin your palpation and then immediately ease up on the pressure slightly. In each case, note the regularity of the pulse; and assess the degree of tension or amplitude. Figure 30.3 illustrates the superficial pulse points to be palpated.

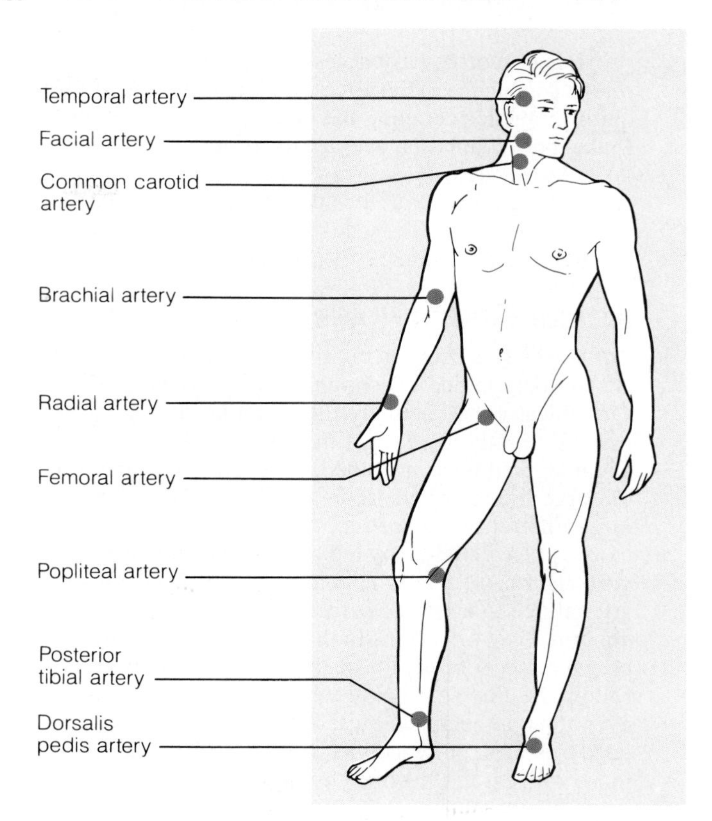

F30.3

Body sites where the pulse is most easily palpated.

Common carotid artery: at the side of the neck
Temporal artery: anterior to the ear, in the temple region
Facial artery: clench the teeth, and palpate the pulse just anterior to the masseter muscle on the mandible (in line with the corner of the mouth)
Brachial artery: in the antecubital fossa, at the point where it bifurcates into the radial and ulnar arteries
Radial artery: at the lateral aspect of the wrist, above the thumb
Femoral artery: in the groin
Popliteal artery: at the back of the knee
Posterior tibial artery: just above the medial malleolus
Dorsalis pedis artery: on the dorsum of the foot

Which pulse point had the greatest amplitude?

____

Which the least? ______________________________

Can you offer any explanation for this? ______________

____

____

Because of its easy accessibility, the pulse is most often taken on the radial artery. With your partner sitting quietly, practice counting the radial pulse for 1 minute. Make three counts and average the results.

count 1 ______________ count 2 ______________

count 3 ______________ average ______________

Apical-Radial Pulse

The correlation between the apical and radial pulse rates can be determined by simultaneously counting them. The apical pulse (actually the counting of heartbeats) may be slightly more rapid than the radial because of a slight lag in time as the blood rushes from the heart into the large arteries where it can be palpated. However, any *large* difference between the values observed, referred to as a **pulse deficit,** may indicate cardiac impairment (a weakened heart that is unable to pump blood into the arterial tree to a normal extent—low cardiac output) or abnormal heart rhythms. In the case of atrial fibrillation or ectopic heartbeats, for instance, the second beat may follow the first so quickly that no second pulse is felt even though the apical pulse can still be heard (auscultated). Apical pulse counts are routinely ordered for those with cardiac decompensation.

With the subject sitting quietly, one student, using a stethoscope, should determine the apical pulse rate while another simultaneously counts the radial pulse rate. The stethoscope should be positioned over the fifth left intercostal space. The person taking the radial pulse should determine the starting point for the count and give the stop-count signal exactly 1 minute later. Record your values below.

apical count ______________ beats/min

radial count ______________ pulses/min

pulse deficit ______________ /min

BLOOD PRESSURE DETERMINATIONS

Blood pressure is defined as the pressure the blood exerts against any unit area of the blood vessel walls, and it is generally measured in the arteries. Because the heart alternately contracts and relaxes, the resulting rhythmic flow of blood into the arteries causes the blood pressure to rise and fall during each beat. Thus you must take two blood pressure readings: the **systolic pressure,** which is the pressure in the arteries at the peak of ventricular ejection, and the **diastolic pressure,** which reflects the pressure during ventricular relaxation. Blood pressures are reported in millimeters of mercury (mm Hg), with the systolic pressure appearing first; 120/80 translates to 120 over 80, or a systolic pressure of 120 mm Hg and a diastolic pressure of 80 mm Hg. Normal blood pressure varies considerably from one person to another.

In this procedure, you will measure arterial and venous pressures by indirect means and under various conditions. You will investigate and demonstrate factors affecting blood pressure, the rapidity of blood pressure changes, and the large differences between arterial and venous pressures.

Indirect Measurement of Arterial Blood Pressure

The *sphygmomanometer* is an instrument used to obtain blood pressure readings by the auscultatory method. It consists of an inflatable cuff with an attached pressure gauge. The cuff is placed around the arm and inflated to a pressure higher than systolic pressure to occlude circulation to the forearm. Cuff pressure is gradually released, and the examiner listens with a stethoscope for characteristic sounds called the **sounds of Korotkoff,** which indicate the resumption of blood flow into the forearm. The pressure at which the first soft tapping sounds can be detected is recorded as the systolic pressure. As the pressure is further reduced, blood flow becomes more turbulent; and the sounds become louder. As the pressure is reduced still further, below the diastolic pressure, the artery is no longer compressed and blood flows freely and without turbulence. At this point, the sounds of Korotkoff can no longer be detected. The pressure at which the sounds disappear is recorded as the diastolic pressure.

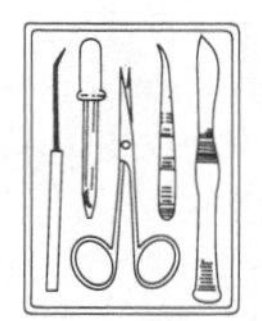

1. Work in pairs to obtain radial artery blood pressure readings. Obtain a stethoscope, alcohol swabs, and a sphygmomanometer. Clean the earpieces of the stethoscope with alcohol swabs, and check the cuff for the presence of trapped air by compressing it against the laboratory table. (A partially inflated cuff will cause erroneous measurements.)

2. The subject should sit in a comfortable position with one arm resting on the laboratory table (approximately at heart level if possible). Wrap the cuff around the subject's elevated arm, just above the elbow, with the inflatable area on the medial arm surface. The cuff may be marked with an arrow; if so, the arrow should be positioned over the brachial artery (Figure 30.4). Secure the cuff by tucking the distal end under the wrapped portion or by bringing the Velcro areas into apposition.

3. Palpate the brachial pulse, and lightly mark its position with a felt pen. Don the stethoscope, and place its diaphragm over the pulse point. The cuff should not be kept inflated for more than 1 minute; so if you have any trouble obtaining a reading within this time, deflate the cuff, wait 1 or 2 minutes, and try again. (A prolonged upsetting of BP homeostasis can lead to fainting.)

4. Inflate the cuff to approximately 160 mm Hg pressure, and slowly release the pressure valve. Watch the pressure gauge as you listen carefully for the first soft thudding sounds of the blood spurting through the par-

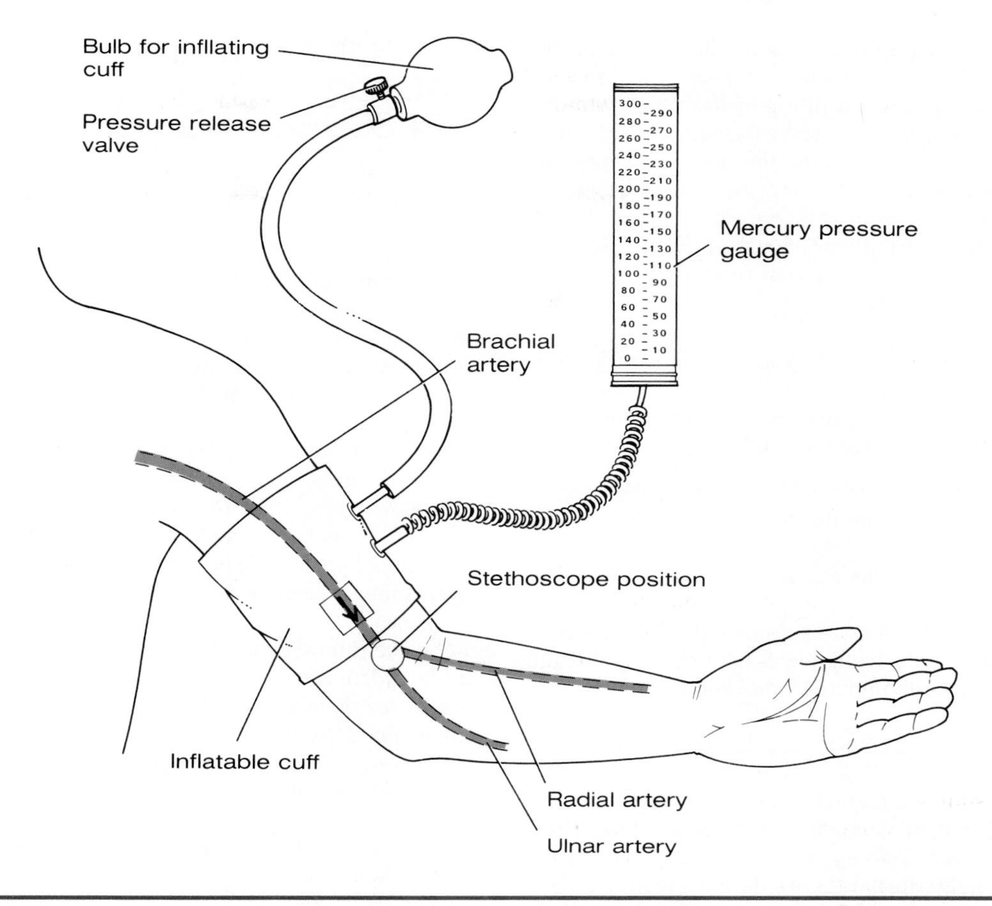

F30.4

Positioning of the sphygmomanometer cuff and stethoscope when auscultating blood pressure in the brachial artery.

tially occluded artery. Note this pressure (systolic pressure), and continue to release the cuff pressure. You will notice first an increase, then a muffling, of the sound. Note, as the diastolic pressure, the pressure at which the sound becomes muffled or disappears. Controversy exists over which of the two points should be recorded as the diastolic pressure; so in some cases you may see readings such as 120/80/78, which indicates the systolic pressure, followed by the *first* and *second diastolic end points.* The first diastolic end point is the pressure at which the sound muffles; the second is the pressure at which the sound disappears. It makes little difference here as to which of the two diastolic pressures is recorded, but be consistent. Make two blood pressure determinations, and record your results below.

First trial:

systolic pressure ______ diastolic pressure ______

Second trial:

systolic pressure ______ diastolic pressure ______

5. Compute the **pulse pressure** for each trial. The pulse pressure is the difference between the systolic and diastolic pressures, and indicates the amount of blood forced from the heart during systole, or the actual "working" pressure.

Pulse pressure:

first trial ______________ second trial ___________

6. Compute the **mean arterial pressure** (**MAP**) for each trial using the following equation:

$$\text{MAP} = \text{diastolic pressure} + \frac{\text{pulse pressure}}{3}$$

first trial ______________ second trial ___________

Estimation of Venous Pressure

It is not possible to measure venous pressure with the sphygmomanometer; and the methods available for measuring it produce estimates at best, because venous pressures are so much lower than arterial pressures. The difference in pressure in the veins versus the arteries becomes obvious when these vessels are cut. If a vein is cut, the blood flows evenly from the cut; a lacerated artery produces rapid spurts of blood.

1. Ask your lab partner to stand with her right side toward the blackboard, arms hanging freely at her sides. Mark, on the board, the approximate level of the right atrium (just slightly higher than the point at which you auscultated the apical pulse).

2. Observe the superficial veins on the dorsum of the subject's right hand as she alternately raises and lowers it. Note the collapsing and filling of the veins with the changing internal pressures. Have the subject repeat this action until you can determine the point at which the veins have just collapsed. Then measure, in millimeters, the distance in the vertical plane from this point to the level of the right atrium (previously marked). Record this value. Distance of right arm from right atrium at point of venous collapse:

__________ mm

3. Compute the venous pressure (P_v), in millimeters of mercury, with the following formula:

$$P_v = \frac{1.056 \text{ (specific gravity of blood)} \times \text{mm (measured)}}{13.6 \text{ (specific gravity of Hg)}}$$

venous pressure computed: ________________ mm Hg

Normal venous pressure varies from approximately 3 to 10 mm Hg; that of the hand ranges between 5 and 7 mm Hg. How does your computed value compare?

____

4. Because of the relative thinness of the venous walls, pressure within them is readily affected by external factors such as muscle activity, deep pressure, and pressure changes occurring in the thorax during breathing. The **Valsalva maneuver,** which increases intrathoracic pressure, can be used to demonstrate the effect of thoracic pressure changes on venous pressure.

To perform this maneuver take a deep breath, and then mimic the motions of exhaling forcibly, but without actually exhaling. In reaction to this, the glottis will close; and intrathoracic pressure will increase. (Most of us have performed this maneuver unknowingly in acts of defecation in which there is "straining at stool.") Measure the height of the hand at the point of venous collapse while the subject is performing the Valsalva maneuver. Compute the venous pressure and record it below. Distance of the right arm from right atrium at point of venous collapse while performing the Valsalva maneuver:

__________ mm

Venous pressure computed: ________________ mm Hg

How does this value compare with the venous pressure measurement computed for the relaxed state?

____

Explain: ______________________________________

____

____

EFFECT OF VARIOUS FACTORS ON BLOOD PRESSURE AND HEART RATE

Arterial blood pressure can be shown to be directly proportional to cardiac output (amount of blood pumped out of the left ventricle per unit time) and peripheral resistance to blood flow. Peripheral resistance is increased by constriction of blood vessels (most importantly, the arterioles), by an increase in blood viscosity or volume, and by a loss of elasticity of the arteries (seen in arteriosclerosis). Thus, any factor that increases either the cardiac output or the peripheral resistance causes an almost immediate reflex rise in blood pressure. A close examination of these relationships reveals that many factors—age, weight, time of day, exercise, body position, emotional state, and various drugs, for example—alter blood pressure. The influence of a few of these factors is investigated here.

The following tests are done most efficiently if one student acts as the subject, two are examiners (one taking the radial pulse and the other auscultating the brachial blood pressure), and a fourth student is the data collector/recorder. The sphygmomanometer cuff should be left on the subject's arm throughout the experiments (in a deflated state, of course) so that, at the proper times, the blood pressure can be taken quickly. In each case, take the measurements at least twice.

Posture

To monitor circulatory adjustments to changes in position, take blood pressure and pulse measurements under the conditions noted in Chart 1 (p. 281). Also record your results on that chart.

Exercise

Blood pressure and pulse changes occurring during and after exercise provide a good yardstick for measuring one's overall cardiovascular fitness. Although there are more sophisticated and more accurate tests that evaluate fitness according to a specific point system, the bench-step test described here and in Chart 2 is a quick way to compare the relative fitness level of a group of people.

You will be working in groups of four, duties assigned as indicated above, except that student 4, in addition to his or her recording duties, will act as the timer and call the cadence.

Any student with a known heart problem should refuse to participate as the subject.

All four students may participate as the subject in turn, if desired, but the bench stepping is to be performed *at least twice* in each group—once with a well-conditioned person acting as the subject, and once with a poorly conditioned subject.

Bench stepping is the following series of movements repeated sequentially:

1. Place one foot on the step.

Chart 1 Posture

	Trial 1 BP	Pulse	Trial 2 BP	Pulse
Sitting quietly				
Reclining (after 2 to 3 min)				
Immediately on standing from the reclining position ("at attention" stance)				
After standing for 3 min				

Chart 2 Exercise

Bench Step for 2 min at 30/min	Baseline		Interval Following Test									
			Immediately		1 min		2 min		3 min		4 min	
	BP	P	BP	P	BP	P	BP	P	BP	P	BP	P
Well-conditioned individual												
Poorly conditioned individual												

2. Step up with the other foot so that both feet are on the platform. Straighten the legs and back.
3. Step down with one foot.
4. Bring the other foot down.

The pace for the stepping will be set by the "timer" (student 4), who will repeat "Up-2-3-4, up-2-3-4" at such a pace that each "up-2-3-4" sequence takes 2 seconds.

1. Student 4 should obtain the 18-inch step while baseline measurements are being obtained on the subject.
2. Once the baseline pulse and blood pressure measurements have been recorded, the subject is to stand quietly for 2 min to allow his or her blood pressure to stabilize before beginning to step.
3. The subject is to perform the bench stepping for 2 minutes according to the cadence called by the timer.
4. After the 2-minute exercise period, the subject is to sit down and blood pressure and pulse are to be measured immediately and thereafter at 1-minute intervals until blood pressure stabilizes.
5. Record the test values on Chart 2, and repeat the testing and recording procedure with the second subject.

When did you notice a greater elevation of blood pressure and pulse?

____

Explain. ______________________________

____

Did you note a sizable difference between the after-exercise values for well-conditioned and poorly conditioned individuals?

__________ Explain: ______________________________

____

____

Did the diastolic pressure also increase? __________

Explain: ______________________________

____

____

Nicotine (Optional)

To investigate the effects of nicotine on blood pressure and pulse, take a baseline blood pressure and pulse of a smoker in a standing position. Ask the subject to light up and smoke in the usual manner. After 2 minutes, take blood pressure and pulse readings. Repeat the measurements at 1-minute intervals for the next 2 minutes, and record your values on Chart 3.

How do the effects of nicotine bring about the changes noted? (Hint: nicotine has vasoconstrictor effects initially.)

____

____

Chart 3 Nicotine

Baseline BP	P	After Smoking for 2 min BP	P	After Smoking for 3 min BP	P	After Smoking for 4 min BP	P
______	______	______	______	______	______	______	______

Chart 4 A Noxious Sensory Stimulus (Cold)

Baseline BP	P	1 min BP	P	2 min BP	P	3 min BP	P
______	______	______	______	______	______	______	______

A Noxious Sensory Stimulus (Cold)

There is little question that blood pressure is affected by emotional disturbances and pain. This lability of blood pressure will be investigated through use of the **cold pressor test,** in which one hand will be immersed in unpleasantly (even painfully) cold water.

Measure the blood pressure and pulse of the subject as he sits quietly. Obtain a basin, fill it with ice cubes, and add water. When the temperature of the ice bath has reached 5°C, immerse the subject's other hand in the ice water. With the hand still immersed, take blood pressure and pulse readings at 1-minute intervals for a period of 3 minutes, and record the values on Chart 4.

How did the blood pressure change during cold exposure?

____

____

Was there any change in pulse? ________________

____

Subtract the respective baseline readings of systolic and diastolic blood pressure from the highest single reading of systolic and diastolic pressure obtained during cold immersion (that is, if the highest experimental reading is 140/88 and the baseline reading is 120/70, then the differences in blood pressure would be indicated as: systolic pressure, 20 mm Hg; and diastolic pressure, 18 mm Hg). These differences are called the **index of response.** According to their index of response, subjects can be classified as follows:

hyporeactors: (stable blood pressure)—exhibit a rise of diastolic and/or systolic pressure ranging from 0 to 22 mm Hg; or a drop in pressures

hyperreactors: (labile blood pressure)—exhibit a rise of 23 mm Hg or more in the diastolic and/or systolic blood pressure

Is the subject tested a hypo- or hyperreactor? ________

SKIN COLOR AS AN INDICATOR OF LOCAL CIRCULATORY DYNAMICS

Careful observation of skin color under carefully controlled conditions reveals (with surprising accuracy) the state of the local circulation, and can enable inferences concerning the larger blood vessels and the circulation as a whole. The experiments on local circulation outlined below illustrate a number of fundamental factors that affect blood flow to the tissues.

Clinical expertise in anyone conducting a physical examination depends upon good observation skills, accurate recording of data, and logical interpretation of the findings. A single example will be given to demonstrate this statement: A massive hemorrhage may be internal and hidden (thus, not obvious), but will still threaten the blood delivery to the brain and other vital organs. One of the earliest compensatory reactions of the body to such a threat is constriction of cutaneous blood vessels, which reduces blood flow to the skin and diverts it into the circulatory mainstream to serve other, more vital tissues. As a result, the skin of the face and particularly of the extremities becomes pale, cold, and finally moist with perspiration. Therefore, a pale, cold, clammy skin should immediately lead the careful diagnostician to suspect that the circulation is dangerously inefficient. Other conditions, such as local arterial obstruction and venous congestion, as well as certain pathologies of the heart and lungs, also alter skin texture, color, and circulation in characteristic ways.

The local blood supply to the skin (indeed, to any tissue) is influenced by (1) local metabolites, (2) oxygen supply, (3) local temperature, (4) autonomic nervous system impulses, (5) local vascular reflexes, (6) certain hormones, and (7) substances released by injured tissues. A number of these factors are examined in the simple experiments that follow. Each experiment should be conducted by students in groups of three or four. One student will act as the subject; the others will conduct the testing, and make and record observations.

Vasodilation and Flushing of the Skin Resulting from Local Metabolites

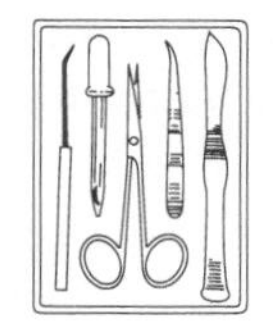

1. Obtain a blood pressure cuff (sphygmomanometer) and stethoscope. A watch with a second hand will also be needed.

2. The subject should bare both arms by rolling up both sleeves as high as possible and then lay his forearms side by side on the bench top.

3. Observe the general color of the subject's forearm skin, and the normal contour and size of his veins. Notice whether skin color is bilaterally similar. Record your observations:

____

____

4. Apply the blood pressure cuff, and inflate it to 250 mm Hg. Keep it inflated for 1 minute. During this period, repeat the observations made above and record the results:

____

____

5. Release the pressure in the cuff (leaving the deflated cuff in position), and again record the forearm skin color and the condition of the forearm veins. Make this observation immediately after deflation and then again 30 seconds later.

immediately after deflation ____________________

____

30 seconds after deflation ____________________

____

The above observations constitute your baseline information. Now conduct the following tests.

6. Instruct the subject to raise his cuffed arm above his head and to clench his fist as tightly as possible. While his hand and forearm muscles are tightly contracted, rapidly inflate the cuff to 240 mm Hg or more. (This maneuver partially empties the hand and forearm of blood, and stops most blood flow to the hand and forearm.) Once the cuff has been inflated, the subject is to relax his fist and return his forearm to the bench top so that it can easily be compared to the other forearm.

7. Leave the cuff inflated for exactly 1 minute. During this interval, compare the skin color in the "ischemic" (blood-deprived) hand to that of the "normal" (non-cuffed-limb) hand; and then quickly release the pressure immediately after the 1-minute count.

What are the subjective effects* of stopping blood flow to the arm and hand for 1 minute?

____

____

What are the objective effects (color of skin and venous condition)?

____

____

How long does it take for the subject's ischemic hand to resume its normal color?

____

Effects of Venous Congestion

1. Again, but with a different subject, observe and record the appearance of the skin and veins on the forelimbs resting on the bench top. This time, pay particular attention to the color of the fingers, the distal phalanx, and the nail beds. Record this information:

____

____

2. Wrap the blood pressure cuff around one of the subject's arms, and inflate it to 40 mm Hg. Maintain this pressure for 5 minutes. Make a record of the subjective and objective findings just before the 5 minutes are up, and then again immediately after release of the pressure at the end of 5 minutes.

subjective (arm cuffed) ____________________

____

objective (arm cuffed) ____________________

____

subjective (pressure released) ________________

____

objective (pressure released) ________________

____

3. With still another subject, conduct the following simple experiment: Hold one arm above the head, and let the other hang by the side for 1 minute. After 1 minute, quickly lay both arms on the bench top; and compare their color.

* Subjective effects are sensations, such as pain, coldness, warmth, tingling, and weakness, experienced by the subject. They are "symptoms" of a change in function.

color of raised arm ______________________

color of dependent arm ______________________

From this and the two preceding observations, analyze the factors that determine tint of color (pink or blue) and intensity of skin color (deep pink or blue as opposed to light pink or blue). Record your conclusions.

____

____

____

Collateral Blood Flow

In many diseases, blood flow to an organ through one or more arteries may be completely and irreversibly obstructed. Fortunately, in most cases a given body area is supplied both by one main artery and by anastomosing channels connecting the main artery with one or more neighboring blood vessels. As a result, an organ may remain viable even though its main arterial supply is occluded, as long as these **collateral vessels** are still operative.

The effectiveness of collateral blood flow in preventing ischemia can be easily demonstrated.

1. Check the subject's hands to be sure they are *warm* to the touch. If they are not, choose another, "warm-handed" subject; or warm the subject's hands in 35°C water for 10 minutes before beginning.

2. Palpate the subject's radial and ulnar arteries approximately 1 inch above the wrist flexure, and mark their locations with a felt marker.

3. Instruct the subject to supinate one forearm and to hold it in a partially flexed (about a 30° angle) position, with the elbow resting on the bench top.

4. Face the subject; and grasp his forearm with both of your hands, the thumb and fingers of one hand compressing the marked radial artery and the thumb and fingers of the other hand compressing the ulnar artery. Maintain total pressure for 5 minutes, noting the progression of the subject's hand to total ischemia.

5. At the end of 5 minutes, release the pressure suddenly; and record the subject's sensations, and the intensity and duration of the flush in the previously occluded hand. (Use the other hand as a comparison baseline.)

____

____

____

6. Allow the subject to relax for 5 minutes; then repeat the maneuver, but this time *compress only the radial artery.* Record your observations:

____

____

How do the observations from the first test differ from those of the second test with respect to color changes during compression, and to the intensity and duration of reactive hyperemia (redness of the skin)?

____

____

____

7. Again follow the subject to relax for 5 minutes, and then repeat the maneuver, *with only the ulnar artery compressed.* Record your observations:

____

____

____

What can you conclude about the relative sizes of, and hand areas served by, the radial and ulnar arteries?

____

____

Effect of Mechanical Stimulation of Blood Vessels of the Skin

With moderate pressure, draw the blunt end of your pen across the skin of a subject's forearm. Wait 3 minutes to observe the effects, and then repeat with firmer pressure.

What changes in skin color do you observe with light to moderate pressure?

____

With heavy pressure? ______________________

The redness, or flare, observed after mechanical stimulation of the skin is the result of a local inflammatory response promoted by chemical mediators released by injured tissues. These mediators promote increased blood flow into the area and leaking of fluid (from the capillaries) into the local tissues. (Note: people differ considerably in skin sensitivity. Those most sensitive will show **dermographism,** a condition in which the direct line of stimulation will swell quite obviously. This excessively swollen area is called a *wheal.*)

Frog Cardiovascular Physiology

OBJECTIVES

1. To compare the relative length of the refractory period of cardiac muscle with that of skeletal muscle and to explain why it is not possible to tetanize cardiac muscle.
2. To define *extrasystole* and to explain at what point in the cardiac cycle (and on an ECG tracing) an extrasystole can be induced.
3. To describe the effect of the following on heart rate: cold; heat; vagal stimulation; digitalis; pilocarpine, atropine sulfate; epinephrine; and potassium, sodium, and calcium ions.
4. To define *vagal escape* and to discuss its value.
5. To define *ectopic pacemaker.*
6. To define *partial* and *total heart block.*

MATERIALS

Freshly pithed frog*
Dissecting pan and instruments
Petri dishes
Medicine dropper
Millimeter ruler
Frog Ringer's solutions (at room temperature, 5°C, and 32°C)
Apparatus A or B:
A: Kymograph, kymograph paper (smoking stand, burner and glazing fluid if using a smoke-writing apparatus), heart lever, signal marker, electronic stimulator, platinum electrodes
B: Physiograph (polygraph), polygraph paper and ink, myograph transducer, transducer coupler, stimulator output extension cable, electrodes
Thread; rubber bands; fine common pins
Frog board
Ring stand; right-angle clamps (two)
Cotton balls
Freshly prepared solutions (using frog Ringer's solution as the solvent) of the following in dropper bottles:
2.5% pilocarpine
5% atropine sulfate
2% digitalis
1% epinephrine
5% potassium chloride (KCl)
1% calcium chloride ($CaCl_2$)
0.7% sodium chloride (NaCl)

* Instructor will double-pith frogs as required for student experimentation.

Investigation of heart and blood vessel function in human subjects is most interesting, but many areas obviously do not lend themselves to human experimentation. It would be tantamount to murder to inject a human being with various drugs to observe their effects on heart activity, and to expose the human heart in order to study the length of its refractory period. However, this type of investigation can be done on frogs or small laboratory animals and provides valuable data, because the physiological mechanisms in these animals are similar if not identical to those in humans.

RECORDING FROG HEART ACTIVITY UNDER VARIOUS CONDITIONS

The heart's effectiveness as a pump is dependent on intrinsic (within the heart) and extrinsic (external to the heart) controls. In this experiment, you will investigate some of these factors.

The nodal system, in which the "pacemaker" imposes its depolarization rate on the rest of the heart, is one intrinsic factor that influences the heart's pumping action. If its impulses fail to reach the ventricles (as in heart block), the ventricles continue to beat, but at the inherent rate of their own autorhythmic tissue, which is much slower than that usually imposed on them. Although heart contraction does not depend on nervous impulses, its rate can be modified by extrinsic impulses reaching it through the autonomic nerves. Additionally, cardiac activity is modified by various chemicals, hormones, ions, and metabolites, the effects of which will be examined in the next experimental series.

The frog heart has two atria and a single, incompletely divided ventricle (Figure 31.1). The pacemaker is located in the sinus venosus, an enlarged region between the venae cavae and the right atrium. Many researchers believe that the SA node of mammals evolved from the sinus venosus.

To record frog heart activity, work in groups of four—two students handling the equipment setup and two preparing the frog for experimentation. Two sets of instructions are provided for apparatus setup—one for the kymograph (Figure 31.2), the other for the physiograph (Figure 31.3). Follow the procedure outlined for the apparatus you will be using.

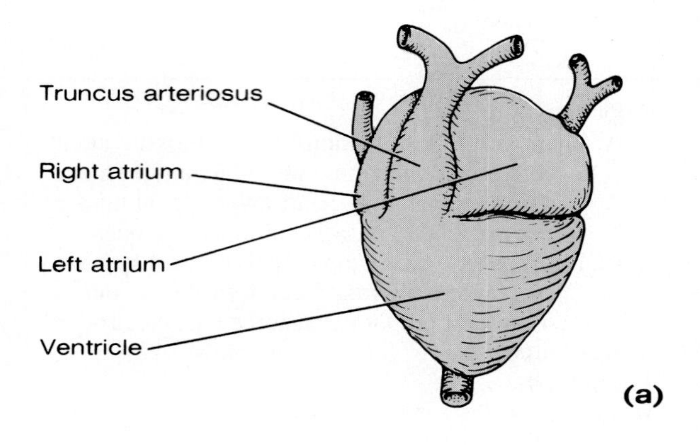

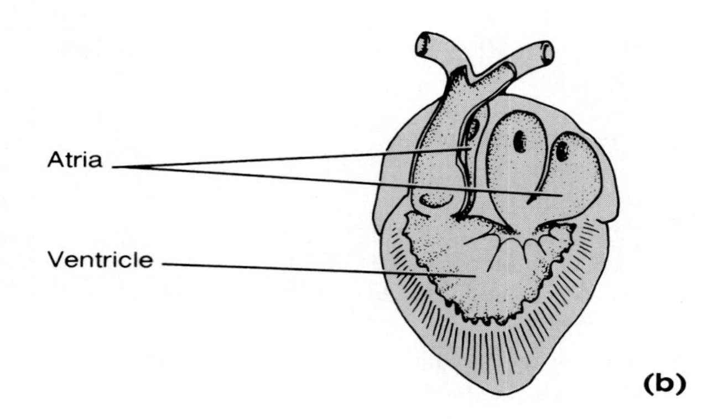

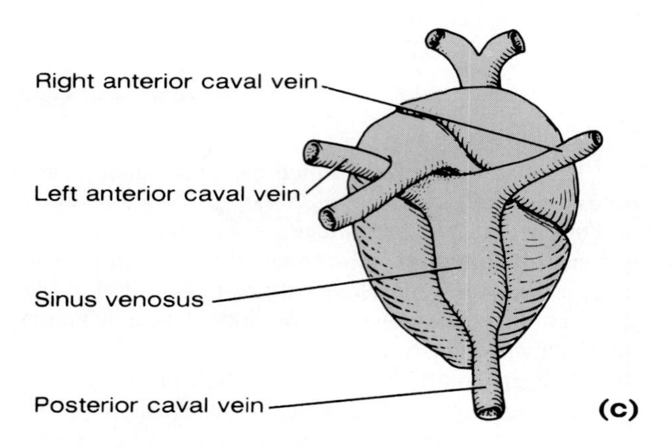

F31.1

Anatomy of the frog heart. (a) Ventral view showing the single truncus arteriosus leaving the undivided ventricle; (b) longitudinal section showing the two atrial and single ventricular chambers; (c) dorsal view showing the sinus venosus (pacemaker).

Kymograph Apparatus Setup

1. Obtain a kymograph, recording paper, ring stand, frog board, two right-angle clamps, signal magnet, stimulator, platinum electrodes, and a heart lever.

2. If you're not using an ink-writing kymograph apparatus, smoke the drum lightly; it is important that the recording be made with minimum pressure against the drum.

3. According to the setup depicted in Figure 31.2, connect the electrodes to the two output terminals on the stimulator; and attach the signal magnet to the two binding posts on the back of the stimulator. With a clamp, connect the signal magnet to the ring stand so that it will record close to the bottom of the drum when recording is done. Attach the heart lever to the stand, above and in line with the signal magnet.

Physiograph Apparatus Setup

1. Obtain a myograph transducer, transducer cable, and stand and bring them to the recording site.
2. Attach the myograph transducer to the transducer stand as shown in Figure 31.3.
3. Then attach the transducer cable to the transducer coupler (input) on the amplifier of the physiograph and to the myograph transducer.
4. Attach the stimulator output extension cable to output on the stimulator panel (red to red, black to black).

PREPARATION OF THE FROG

1. Obtain room-temperature frog Ringer's solution, a medicine dropper, dissecting instruments and pan, fine common pins, and some thread, and bring them to your laboratory bench.

2. Obtain a double pithed frog from your instructor.

3. Make a longitudinal incision through the abdominal and thoracic walls with scissors, and then cut through the sternum to expose the heart.

4. Grasp the pericardial sac with forceps, and cut it open so that the beating heart can be observed.

Is the sequence an atrial-ventricular one? ___________

5. Locate the vagus nerve, which runs down the side of the neck and parallels the trachea and carotid artery. (In good light, it appears to be striated.) Slip an 18-inch length of thread under the vagus nerve so that it can later be lifted away from the surrounding tissues by the thread. Then place a saline-soaked cotton ball over the nerve to keep it moistened until you are ready to stimulate it later in the procedure.

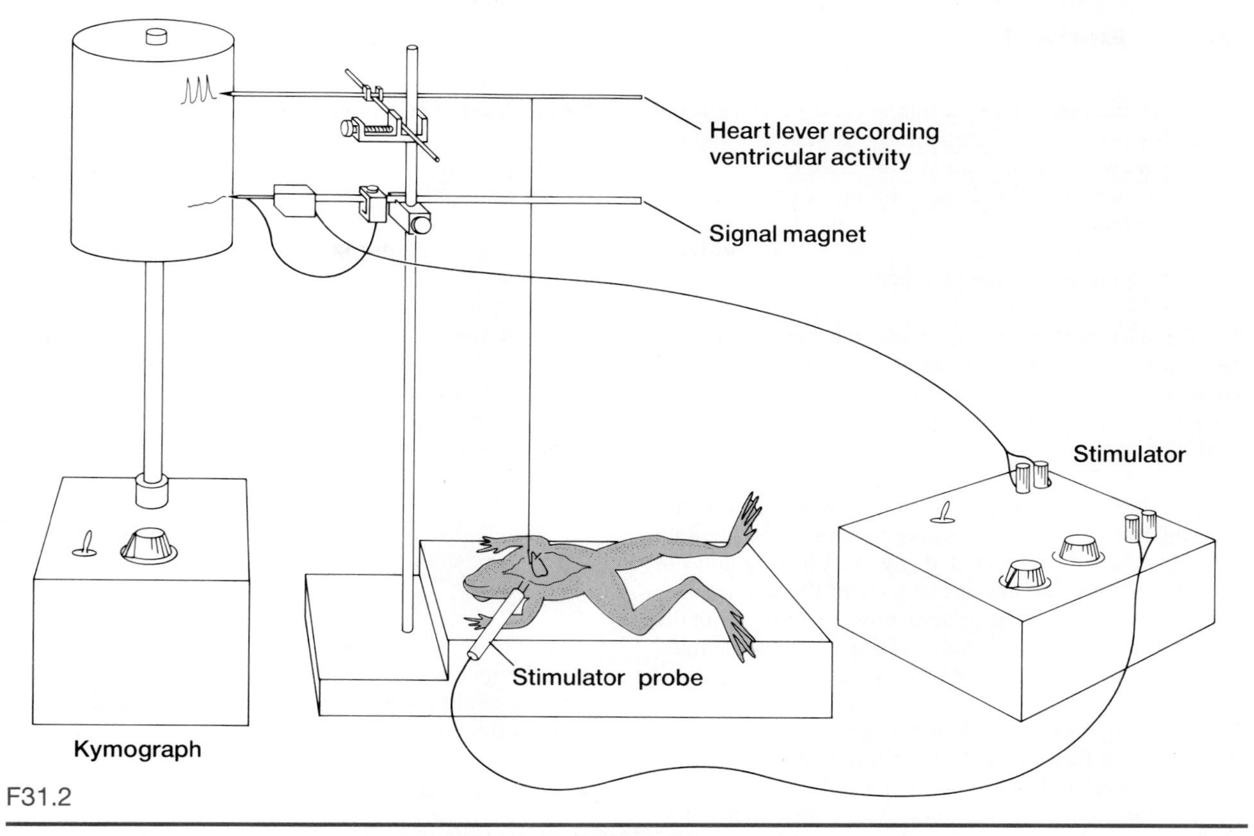

Kymograph apparatus setup for recording the activity of the frog heart.

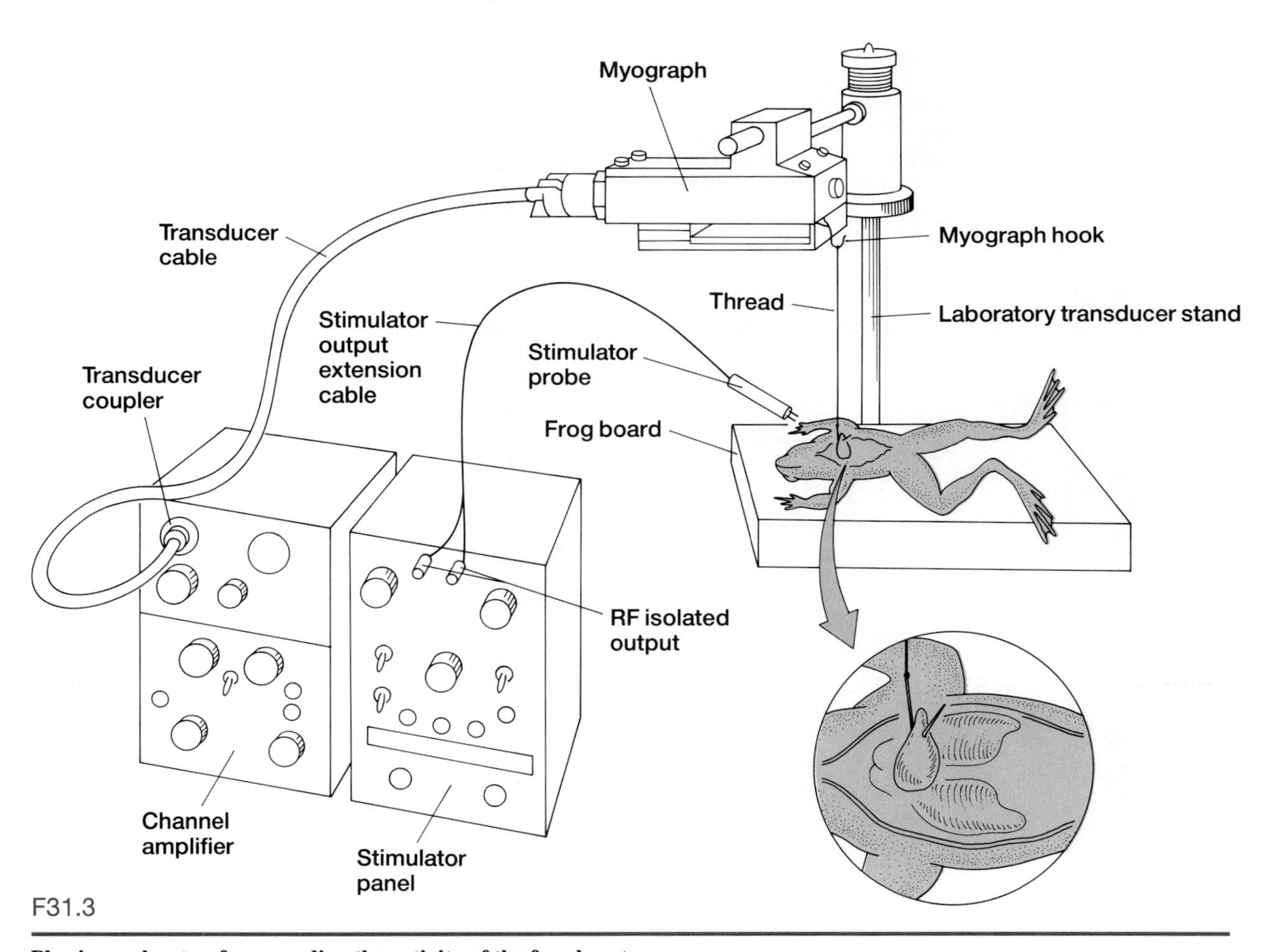

Physiograph setup for recording the activity of the frog heart.

6. Flush the heart with the saline (Ringer's) solution. From this point on, the heart must be kept continually moistened with room-temperature Ringer's solution, unless other solutions are being used for the experimentation.

7. Attach the frog to the frog board.

8. Bend a common pin to a 90-degree angle, and tie an 18- to 20-inch length of thread to its head. Force the pin through the apex of the heart (do not penetrate the ventricular chamber) until the apex is well secured in the angle of the pin.

9. Tie the thread to the end of the muscle lever directly over the heart if using the kymograph, or to the hook on the myograph transducer if using the physiograph apparatus. Do not pull the thread too tightly. It should be taut enough to lift the heart apex upward, away from the thorax, but should *not* stretch the heart. Adjust the muscle lever or myograph transducer as necessary.

10. If using a kymograph, adjust the writing tip of the heart lever so that it is directly in line with the stylus of the signal marker. Leave enough room between the two writing tips to give each adequate working distance. When properly set up, the heart lever should be in a horizontal position.

Recording Base Line Frog Heart Activity

1. Kymograph users should set the drum speed at 7, and briefly turn on the kymograph to ensure that the styluses are recording properly. If using the physiograph, turn the amplifier on and balance the apparatus according to instructions provided by your instructor. Set the paper speed at 0.5 cm/sec. Press the record and paper advance buttons.

2. Set the signal magnet or time marker at 1/sec.

3. Record several normal heartbeats (12 to 15). Be sure you can distinguish atrial and ventricular contractions (see Figure 31.4). Then adjust the paper speed so that the peaks of ventricular contractions are approximately 2 cm apart. (Peaks indicate systole; troughs indicate diastole.) Note the relative force of heart contractions while recording.

4. Count the number of ventricular contractions per minute and record.

______________________ beats/min

5. Compute the A-V interval (the period from the beginning of atrial contraction to the beginning of ventricular contraction).

______________________ sec

How do the two tracings compare in time?

____

____

6. Mark, on the recording, the atrial and ventricular systoles and diastoles. Remember to keep the heart moistened with Ringer's solution.

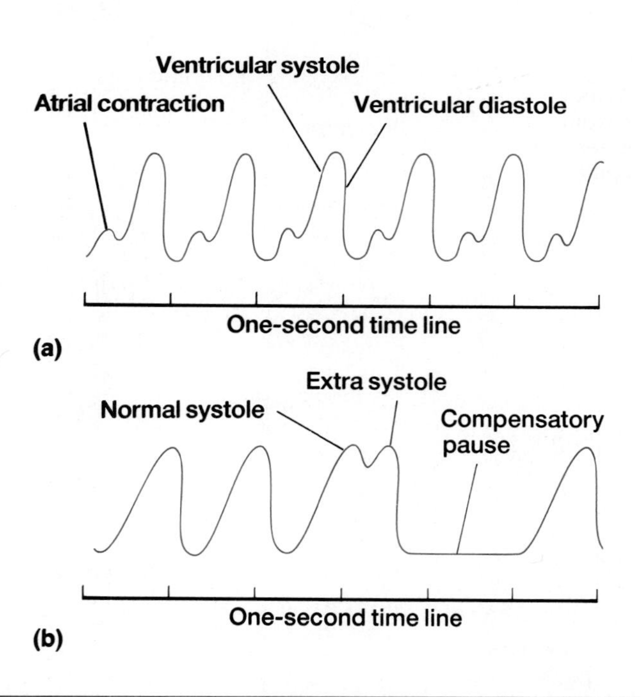

F31.4

Recording of contractile activity of a frog heart. (a) Normal heartbeat; (b) induction of an extrasystole.

REFRACTORY PERIOD OF CARDIAC MUSCLE

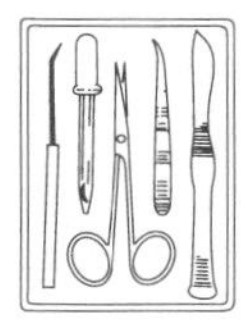

When you investigated the physiology of skeletal muscle in Exercise 14A, you saw that repeated rapid stimuli could cause the muscle to remain in the contracted state. In other words, the muscle was tetanized. This was possible because of the relatively short refractory period of skeletal muscle. In this experiment, you will investigate the refractory period of cardiac muscle and its response to stimulation. During the procedure, one student should keep the stimulating electrodes in constant contact with the frog heart ventricle.

1. Set the stimulator to deliver 20-V shocks of 2-msec duration, and begin recording.

2. Deliver single shocks at the beginning of ventricular contraction, at the peak of ventricular contraction, and then later and later in the cardiac cycle.

3. Observe the recording for **extrasystoles,** which are extra beats that show up riding on the ventricular contraction peak. Also note the **compensatory pause,** which allows the heart to get back on schedule after an extrasystole.

In which portion of the cardiac cycle was it possible to induce an extrasystole?

___

___

___

4. Attempt to tetanize the heart by applying a tetanizing stimulation of 20 to 30 impulses per second. What were the results?

___

___

___

Considering the function of the heart, why is it important that heart muscle cannot be tetanized?

___

___

___

___

PHYSICAL AND CHEMICAL FACTORS MODIFYING HEART RATE

Now that you have observed normal frog heart activity, you will have an opportunity to investigate various modifying factors and their effects on heart activity. In each case, record a few normal heartbeats before introducing the modifying factor. After removing the agent, allow the heart to return to its normal rate before continuing with the testing. On each record, indicate the points of introduction and removal of the modifying agent.

Temperature

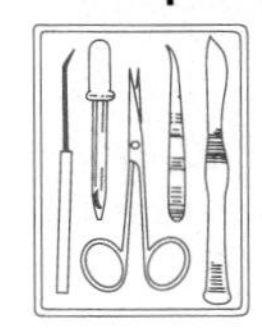

1. Obtain 5°C and 32°C frog Ringer's solutions and medicine droppers.

2. Increase the speed of the drum or the physiograph paper speed so that heartbeats appear as spikes 4 to 5 mm apart.

3. Flood the heart with 5°C saline, and continue to record until the recording indicates a change in cardiac activity.

4. Stop recording, pipette off the cold Ringer's solution, and flood the heart with room-temperature saline.

5. Start recording again to determine the resumption of the normal heart rate. When this has been achieved, flood the heart with 32°C Ringer's solution, and again record until a change is noted.

6. Stop the recording, pipette off the saline, and once again flood the heart with room-temperature Ringer's solution.

What change occurred with the cold (5°C) Ringer's solution?

___

___

What change occurred with the warm (32°C) Ringer's solution?

___

___

7. Count and record the heart rate at the two temperatures.

_______ beats/min at 5°C; _______ beats/min at 32°C

Vagus Nerve Stimulation

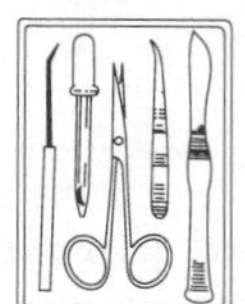

The vagus nerve carries parasympathetic impulses to the heart, which modify heart activity.

1. Remove the cotton placed over the vagus nerve. Using the previously tied thread, lift the nerve away from the tissues; and place the nerve on the platinum electrodes.

2. Stimulate the nerve at a rate of 50/sec for 0.5 msec at a voltage of 1 mV. Continue nerve stimulation until the heart stops momentarily and then begins to beat again (vagal escape). If no effect is observed, increase the stimulus intensity; and try again.

3. Discontinue stimulation after you observe vagal escape, and flush the heart with saline until the normal heart rate resumes. What effect on the heart rate did vagal stimulation have?

____

____

The phenomenon of vagal escape, just observed, demonstrates that many factors are involved in heart regulation and that any deleterious factor (in this case, excessive vagal stimulation) will be overcome, if possible, by other physiological mechanisms such as activation of the sympathetic division of the ANS.

Pilocarpine

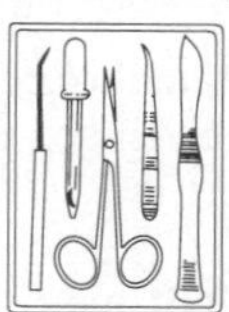

Flood the heart with a 2.5% solution of pilocarpine. Record until a change in the pattern of the ECG is noticed. Pipette off the excess pilocarpine solution, and proceed immediately to the next test, which uses atropine as the testing solution. What happened when the heart was bathed in the pilocarpine solution?

____

Pilocarpine simulates the effect of parasympathetic nerve (hence, vagal) stimulation by enhancing acetylcholine release; such drugs are called parasympathomimetic drugs.

Atropine Sulfate

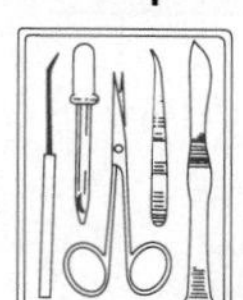

Apply a few drops of atropine sulfate to the frog's heart, and observe the recording. If no changes are observed within 2 minutes, apply a few more drops. When you observe a response, pipette off the excess atropine sulfate; and flood the heart with Ringer's solution. What happens when the atropine sulfate is added?

____

____

Atropine is a drug that blocks the effect of the neurotransmitter acetylcholine,which is liberated by the parasympathetic nerve endings. Do your results accurately reflect this effect of atropine?

____

Are pilocarpine and atropine agonists or antagonists in their effects on heart activity?

____

Epinephrine

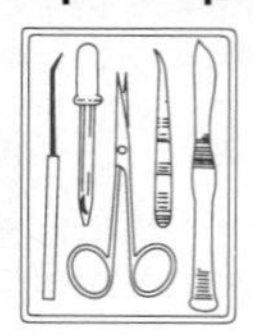

Flood the frog heart with epinephrine solution, and continue to record until a change in heart activity is noted.

What are the results? ______________________________

____

Which division of the autonomic nervous system does its effect imitate?

____

Digitalis

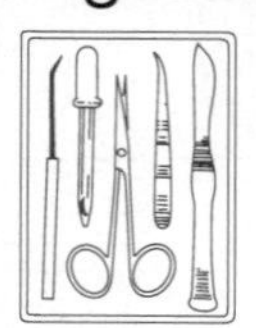

Pipette off the excess epinephrine solution, and rinse the heart with room-temperature Ringer's solution. Continue recording, and when the heart rate returns to baseline values, bathe it in digitalis solution. What is the effect of digitalis on the heart?

____

____

Digitalis is a drug commonly prescribed for heart patients with congestive heart failure. It slows heart rate, providing more time for venous return and decreasing the load on the weakened heart. These effects are thought to be due to its inhibition of the Na^+-K^+ pump and its enhancement of Ca^{2+} entry into the myocardial fibers.

Effect of Various Ions

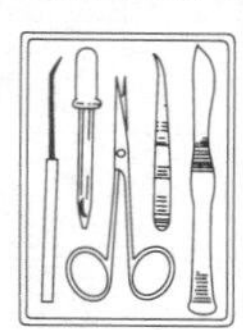

To test the effect of various ions on the heart, apply the indicated solution until you observe a change in heart rate or in strength of contraction. Then pipette off the solution; flush the heart with Ringer's solution; and allow the heart to resume its normal rate before continuing. Do not allow the heart to stop. If the heart rate decreases dramatically, flood the heart with Ringer's solution.

Effect of Ca^{2+} (use 1% $CaCl_2$) ____________________

____

Effect of Na^+ (use 0.7% NaCl) ____________________

____

Effect of K^+ (use 5% KCl) ____________________

____

Potassium ion concentration is normally higher within the cells than in the extracellular fluid. Hyperkalemia tends to decrease the resting potential of the cell membranes, thus reducing the difference in potential between the inside and outside of the cell, and decreasing the force of heart contraction. In some cases, the conduction rate of the heart is so depressed that **ectopic pacemakers** (pacemakers appearing erratically and at abnormal sites in the heart muscle) appear in the ventricles; and fibrillation may occur. Was there any evidence of premature beats in the recording of potassium ion effects?

____

Was arrhythmia produced with any of the ions tested?

____

If so, which? ____________________

Disturbance of the Intrinsic Conduction System (Heart Block)

1. Moisten a 10-inch length of thread, and make a Stannius ligature (loop the thread around the heart at the junction of the atria and the ventricle).

2. Decrease the drum or paper speed to achieve intervals of approximately 2 cm between the ventricular contractions, and record a few normal heartbeats.

3. Tighten the ligature in a stepwise manner while observing the atrial and ventricular contraction curves. As heart block occurs, the atria and ventricle will no longer show a 1:1 contraction ratio. Record a few beats each time you observe a different degree of heart block—a 2:1 atria-to-ventricle ratio, 3:1, 4:1, and so on. As long as you can continue to count a whole number ratio between the two chamber types, the heart is in **partial heart block.** When you can no longer count a whole number ratio, the heart is in **total,** or **complete, heart block.**

4. When total heart block occurs, release the ligature to see if the normal A-V rhythm is reestablished.

What is the result? ____________________

____

5. Attach properly labeled recordings (or copies of the recordings), made during this procedure, to this sheet for future reference.

32 EXERCISE

The Lymphatic System and Immune Response

OBJECTIVES

1. To name the components of the lymphatic system.
2. To relate the function of the lymphatic system to that of the blood vascular system.
3. To describe the formation and composition of lymph, and to describe how it is transported through the lymphatic vessels.
4. To relate the importance of immunologic memory, specificity, and differentiation of self from nonself to immune function.
5. To differentiate between the roles of B cells and T cells in the immune response.
6. To describe the structure and function of lymph nodes; and to indicate the localization of T cells, B cells, and macrophages in a typical lymph node.
7. To describe (or draw) the structure of the immunoglobulin monomer, and to name the five immunoglobulin subclasses.
8. To discuss the immunologic "abnormality" evidenced by many rheumatoid arthritis sufferers.

MATERIALS

Large anatomical chart of the human lymphatic system
Prepared microscope slides of lymph nodes
Compound microscope
Microscope slide
Wax marking pencil
Applicator sticks
Arthritis screening test (Wampole)
pH paper
Plastic (disposable) gloves
Disposable autoclave bag

THE LYMPHATIC SYSTEM

General Description

The lymphatic system consists of a network of successively larger lymphatic vessels (lymphatics), the lymph nodes, and a number of other lymphoid organs, such as the tonsils, thymus, and spleen (Figure 32.1). We will focus on the lymphatic vessels and lymph nodes in this section. The overall function of the lymphatic system is twofold. It returns excess tissue fluid (lymph) to the blood vessels. Because lymph flows only toward the heart, it is a one-way system. In addition, it protects the body by removing foreign material such as bacteria from the lymphatic stream and by serving as a site for lymphocyte "policing" of body fluids and lymphocyte multiplication. (The function of these white blood cells in body immunity is described later in this exercise.)

Distribution and Function of Lymphatic Vessels and Lymph Nodes

As blood circulates through the body, the hydrostatic and osmotic pressures operating at the capillary beds result in an outward flow of fluid at the arterial end of the bed and in its return at the venous end. However, not all of the lost fluid is returned to the bloodstream by this mechanism; and the fluid that lags behind in the tissue spaces must eventually return to the blood if the vascular system is to operate properly. (If it does not, fluid accumulates in the tissues, producing a condition called edema.) It is the microscopic, blind-ended **lymphatic capillaries,** which ramify through all the tissues of the body, that pick up this leaked fluid (primarily water and a small amount of dissolved proteins) and carry it through successively larger vessels—**lymphatic collecting vessels,** to **lymphatic trunks**—until the lymph finally returns to the venous system through one of the two large ducts in the thoracic region. The **right lymphatic duct** drains the lymph from the right upper extremity, head, and thorax; the large **thoracic duct** receives lymph from the rest of the body. In humans, both ducts empty the lymph into the venous circulation at the junction of the internal jugular vein and the subclavian vein, on their respective sides of the body (Figure 32.1).

Like veins of the blood vascular system, the lymphatic collecting vessels have three tunics and are equipped with valves. However, the lymphatics tend to be thinner-walled, to have *more* valves, and to anastomose more than veins. Since the lymphatic system is a pumpless system, lymph transport depends mainly on the milking action of the skeletal muscles and on pressure changes within the thorax during breathing.

As lymph is transported, it filters through oval or bean-shaped **lymph nodes,** which cluster along the lym-

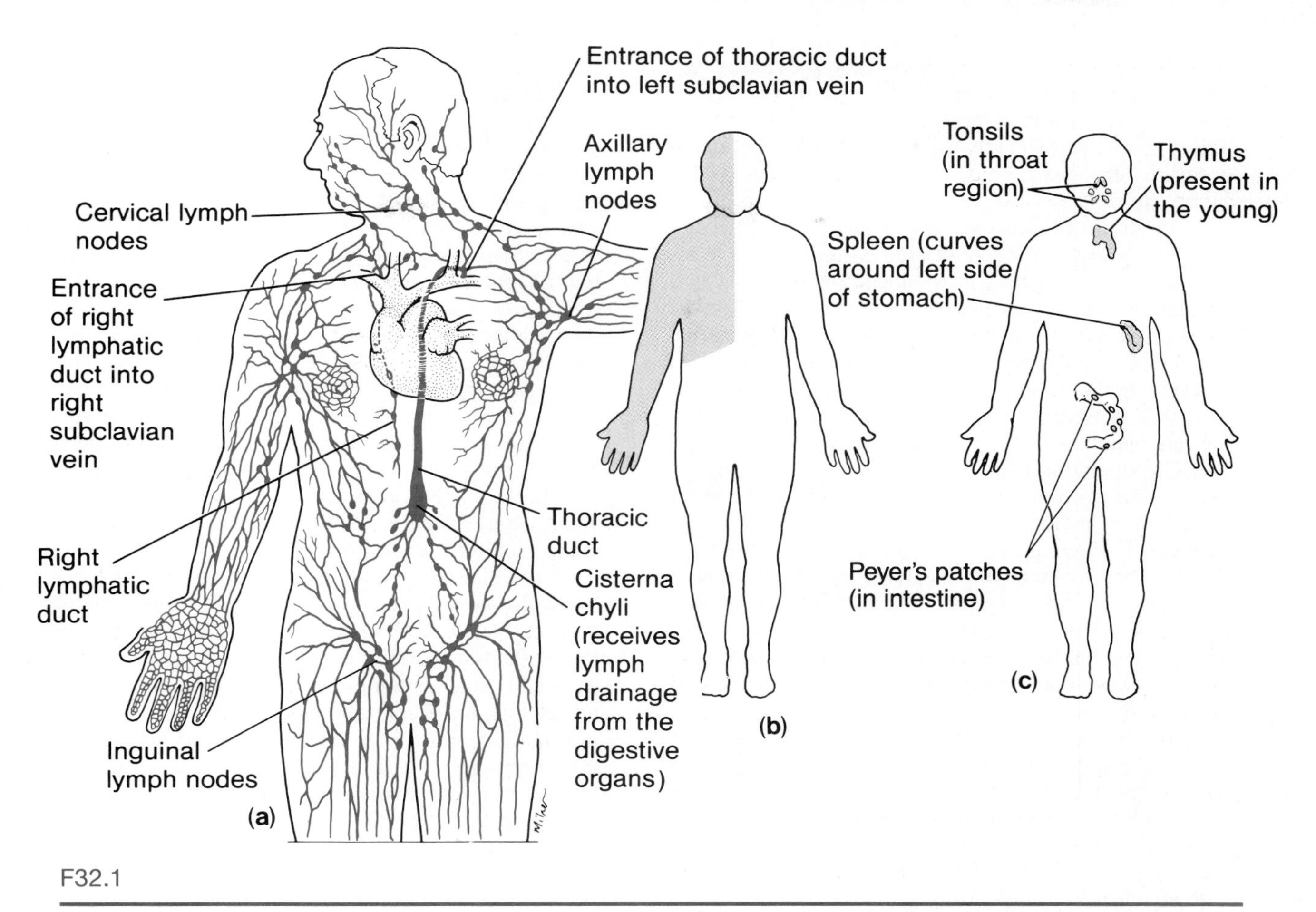

F32.1

Lymphatic system. (a) Distribution of lymphatic vessels and lymph nodes. (b) Shading represents body area drained by the right lymphatic duct; the rest of the body is drained by the thoracic duct. (c) Body location of the tonsils, thymus, spleen, and Peyer's patches.

phatic vessels of the body. There are hundreds of lymph nodes; but because they are usually embedded in connective tissue, they are not ordinarily seen. Particularly large collections of lymph nodes are found in the inguinal, axillary, and cervical regions of the body. Within the lymph nodes are **phagocytic cells** (**macrophages**), which destroy bacteria, cancer cells, and other foreign matter in the lymphatic stream, thus rendering many harmful substances or cells harmless before the lymph enters the bloodstream. Although we are not usually aware of the filtering and protective nature of the lymph nodes, most of us have experienced "swollen glands" during an active infection. This swelling is a manifestation of the trapping function of the nodes.

The tonsils, thymus, and spleen are generally considered to be lymphoid organs because they resemble the lymph nodes histologically, and house similar cell populations (lymphocytes and macrophages).

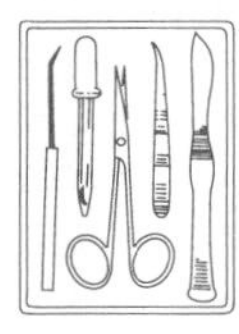

Study the large anatomical chart to observe the general plan of the lymphatic system. Note the distribution of the lymph nodes, the various lymphatics, the lymphatic trunks, and the location of the right lymphatic duct and the thoracic duct. Also identify the **cisterna chyli,** the enlarged terminus of the thoracic duct that receives fat-rich lymph from the digestive viscera.

THE IMMUNE RESPONSE

The **immune system** is a functional system that recognizes something as foreign, and acts to immobilize, neutralize, or metabolize it. This is known as the **immune response.** When operating effectively, the immune response protects us from bacterial and viral infections, cancer, and certain toxins. When it fails or malfunctions, the body is quickly devastated by pathogens or its own assaults.

Major Characteristics of the Immune Response

The most important characteristics of the immune response are its (1) **memory,** (2) **specificity,** and (3) **ability to differentiate self from nonself.**

Most of us are aware that once we have suffered through an episode of chicken pox, for instance, we are unlikely to have a recurrence of that disease because the immune system has a "memory" for previously encountered *antigens* (molecules foreign to the body). Further, this memory is remarkably accurate, having the ability to distinguish between even very closely related antigenic structures. Thus, the immune system is also highly specific.

An almost limitless variety of macromolecules are antigenic—that is, capable of provoking an immune response and reacting with its products. Nearly all foreign proteins, many polysaccharides, and many small molecules (*haptens*), when linked to our own body proteins, exhibit this capability. The cells that recognize antigens and initiate the immune response are lymphocytes. (As described in Exercise 26, lymphocytes are the most numerous members of the leukocyte or white blood cell [WBC] population.) Each immunocompetent lymphocyte is virtually monospecific; that is, it has receptors on its surface allowing it to bind with only one or a few antigens. Following such binding, the lymphocyte proliferates; and its members differentiate—some forming memory cells and others becoming effector or regulatory cells. Upon subsequent meetings with the same antigen, the immune response proceeds considerably faster because the "troops are already mobilized and awaiting further orders," so to speak.

As a rule, our own proteins are tolerated, a fact that reflects the ability of the immune system to distinguish our own tissues (self) from foreign antigens (nonself). Nevertheless, an inability to recognize self can and does occasionally happen and our own tissues are attacked by the immune system. This phenomenon is called *autoimmunity.*

The immune response is genetically controlled by the immune response (Ir) gene complex, which is located on the same chromosome(s) as the MHC genes that direct the production of cell-surface proteins that identify as "self" the cells upon which they are located. Together the Ir and MHC genes form part of the *major histocompatibility complex,* a gene segment crucial in the regulation of immune system capability and activity.

Organs, Cells, and Cell Interactions of the Immune Response

The immune system utilizes as part of its arsenal, the **lymphoid organs,** including the thymus, lymph nodes, spleen, tonsils, and bone marrow, and the cells which travel between these regions and the rest of the body. Of these, the thymus and bone marrow are considered to be the *primary lymphoid organs;* the others are *secondary lymphoid areas.*

The stem cells which give rise to the immune system arise in the bone marrow, and their subsequent differentiation into one of the two populations of immunocompetent lymphocytes occurs in the primary lymphoid organs. The **B cells** differentiate in the bone marrow, and the **T cells** differentiate in the thymus. While in their "programming organs," the lymphocytes become *immunocompetent,* an event indicated by the appearance of specific cell-surface proteins that enable the lymphocytes to respond (by binding) to a particular antigen. Although the mechanism of this specialization is still far from clear, it is known that the mechanism is *not* antigen-driven at this point.

After differentiation, the B and T cells leave the bone marrow and thymus, respectively; enter the bloodstream; and travel to peripheral (secondary) lymphoid organs, where they await the antigen challenge. It is in the secondary lymphoid organs that clonal selection occurs. **Clonal selection** results when an antigen binds to the specific cell-surface receptors of a T or B cell—an event which causes the lymphocyte to proliferate rapidly, forming a clone of like cells, all bearing the same antigen-specific receptors. Then, in the presence of certain regulatory signals, the members of the clone specialize. In the case of B cell clones, some become memory B cells; the others form antibody-producing **plasma cells.** Because the B cells act indirectly through the antibodies that are released into the bloodstream, they are said to provide **humoral immunity.**

T cell clones are more diverse. Although all T cell clones also contain memory cells, some clones contain *killer,* or *cytotoxic, T cells* (effector cells that directly attack virus-infected tissue cells). Others contain regulatory cells such as the *helper cells* (that interact with, and activate, the B cells and killer T cells), and still others contain *suppressor cells* that can act to inhibit antibody production or killer cell activity. Because the killer T cells act directly to destroy virus-infected cells, many bacteria, and cancer cells, and to reject foreign grafts, T cells are said to mediate **cellular immunity.**

Absence or failure of thymic differentiation of T lymphocytes results in a marked depression of both antibody and cell-mediated immune functions. Additionally, the observation that the thymus naturally involutes with age has been correlated with the relatively immune-deficient status of elderly individuals. ■

All lymphoid tissues except the thymus and bone marrow contain both T and B cell–dependent regions. The lymph node is used as the example in the following microscopic study.

MICROSCOPIC ANATOMY OF A LYMPH NODE

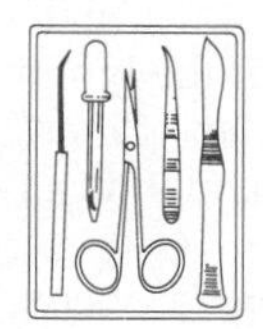

Obtain a prepared slide of a lymph node and a compound microscope. As you examine the slide, notice the following anatomical features, depicted in Figure 32.2. The node is enclosed within a fibrous **capsule,** from which connective tissue septa (**trabeculae**) extend inward to divide the node into several compartments. Very fine strands of reticular connective tissue issue from the trabeculae, forming the stroma of the gland within which cells are found.

In the outer region of the node, the **cortex,** some of the cells are arranged in globular masses, which are referred to as **germinal centers.** The germinal centers contain rapidly dividing B lymphocytes. The rest of the cortical cells are primarily T lymphocytes that circulate

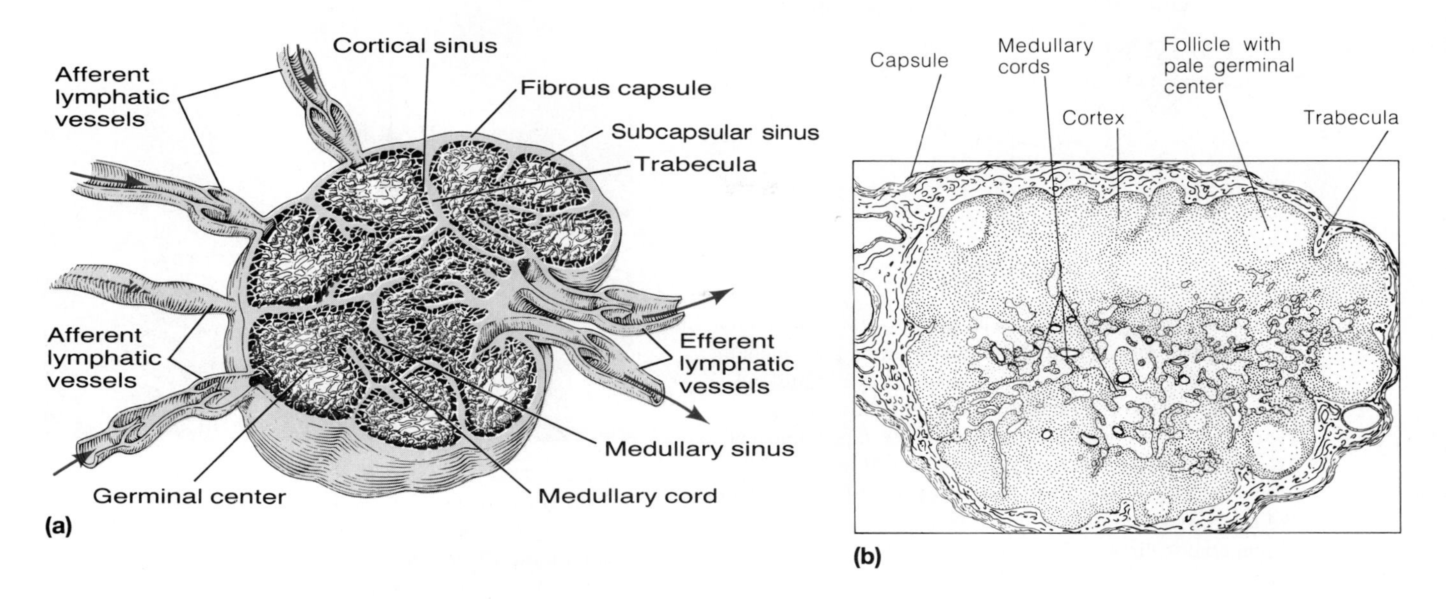

F32.2

Structure of lymph nodes. (a) Cross section of a lymph node, diagrammatic view. Note that the afferent vessels outnumber the efferent vessels, which slows the rate of lymph flow. The arrows indicate the direction of the lymph flow. (b) Line drawing of a photomicrograph (plate 27 of the histology atlas) showing part of a lymph node.

continuously, moving from the blood into the node and then exiting from the node in the lymphatic stream.

In the internal portion of the gland, the **medulla,** the cells are arranged in cordlike fashion. Most of the medullary cells are macrophages. Macrophages are important not only for their phagocytic function, but also because they play an essential role in "presenting" the antigens to the T cells.

Lymph enters the node through a number of afferent vessels, circulates through sinuses within the node, and leaves the node through efferent vessels at the **hilus.** Since each node has fewer efferent than afferent vessels, the lymph flow stagnates somewhat within the node, allowing time for the generation of an immune response and for the macrophages to remove debris from the lymph before it reenters the blood vascular system.

In the space provided here, draw a pie-shaped section of a lymph node, showing the detail of cells in a germinal center, sinusoids, and afferent and efferent vessels. Label all elements.

Before leaving the topic of the lymphoid organs, compare and contrast the structure of the lymph node examined here with the microscopic anatomy of the spleen and the tonsils, illustrated respectively in plates 28 and 29 of the histology atlas.

ANTIBODIES AND TESTS FOR THEIR PRESENCE

Antibodies, produced by sensitized B cells and their progeny plasma cells in response to an antigen, are a heterogeneous group of proteins that comprise the general class of plasma proteins called **gamma globulins.** The antibodies can be further categorized into their immunoglobulin subgroups, which are found not only in plasma, but also (to greater or lesser extents) in all body secretions. Five major classes of immunoglobulins (Igs) have been identified: IgM, IgG, IgD, IgA, and IgE. The immunoglobulin classes share a common basic structure, but differ functionally and in their localization in the body.

All Igs are composed of one or more monomers (structural units). A *monomer* consists of four protein chains bound together by disulfide bridges (Figure 32.3). Two of the chains are quite large and have a high molecular weight; these are the **heavy chains.** The other two chains have a low molecular weight and are called the **light chains.** The two heavy chains have a constant (C) region, in which the amino acid sequence is identical in both chains, and a variable (V) region, which differs in the Igs formed in response to different antigens. The same is true of the two light chains; each has a constant and a variable region.

The intact Ig molecule has a three-dimensional shape that generally looks like a Y. Together, the variable regions of the light and heavy chains in each "arm" construct one antigen-binding site. Thus, each Ig monomer bears two identical sites that bind to a specific (and the same) antigen. Binding of the immunoglobulins to

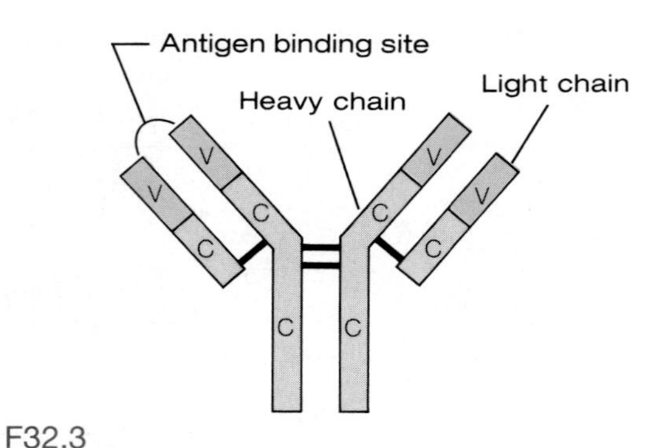

F32.3

Structure of an immunoglobulin monomer. Each monomer is composed of four protein chains (two heavy chains and two light chains) connected by disulfide bonds. Both the heavy and light chains have regions of constant amino acid sequence (C regions) and regions of variable amino acid sequence (V regions). The variable regions differ in each type of antibody and construct the antigen-binding sites. Each immunoglobulin monomer has two such antigen-specific sites.

their complementary antigen(s) effectively immobilizes the antigens until they can be phagocytized or lysed by complement fixation.

Although the role of the immune system is to protect the body, symptoms of certain diseases may reflect excessively high antibody synthesis (as in multiple myeloma, a cancer of the bone marrow and adjacent bony structures) and/or the production of abnormal antibodies (such as **rheumatoid factor** present in the blood of many rheumatoid arthritis sufferers). The simple experiment described below tests for the presence of the rheumatoid factor in plasma.

Test for Rheumatoid Factor

Although the precise cause of rheumatoid arthritis is not yet known, the disease process itself involves an astonishing number of the body's immune elements—T cells, rheumatoid factor (an abnormal antibody), and antirheumatoid antibodies, to name a few. Thus, rheumatoid arthritis is believed to involve an autoimmune reaction; and rheumatoid factor (while not fully diagnostic for the disease) is found in the serum of many rheumatoid arthritis victims. Perhaps the initial disease trigger is a genetic defect that leads to the local formation of the abnormal antibody—which, in turn, prompts the clustering, at certain body joints, of T cells, protective antibodies, and other inflammatory cells.

Although rheumatoid arthritis is a systemic disease that may involve many body organs and tissues, it is characterized by the manner in which the joints are affected. The synovial membranes become inflamed and then thicken (Figure 32.4). This is followed by the formation of **pannus,** an abnormal tissue that clings to the articular cartilages and eventually destroys them. The final stage involves the elaboration of fibrous tissue that connects the bone ends. Because this fibrous tissue eventually ossifies, the bone ends become firmly fused, and often deformed. Not all cases progress to joint immobilization (the severely crippling stage), but all cases do involve some restriction of joint movement and extreme pain.

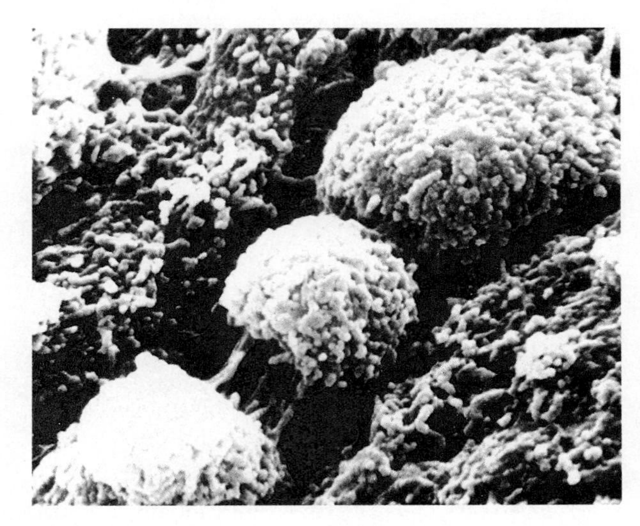

F32.4

Scanning electron micrograph of the synovial membrane from a joint affected by rheumatoid arthritis. Notice the inflammatory cells, exhibiting numerous microvilli, that have accumulated on the membrane surface (3500×).

The test for rheumatoid factor involves mixing a drop of plasma with serum containing antibodies against the abnormal antibody (that is, antirheumatoid-factor antibodies). If rheumatoid factor is present, agglutination will occur, indicating the formation of antigen-antibody complexes.

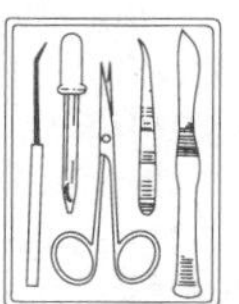

1. Obtain the arthritis screening test materials, a microscope slide, a wax marker, two applicator sticks, and plastic gloves.

2. With the wax marker, label one end of the slide "control" and the other end "test." Don the plastic gloves.

3. Place a drop of test plasma on the end of the slide marked "test" and a drop of the positive control plasma on the end marked "control."

4. Add one drop of reagent 1 to each sample, and mix with an applicator stick. (Use opposite ends of the applicator stick to mix each sample, and then discard the applicator stick in the disposable autoclave bag provided at the supply area.)

5. Add 2 drops of reagent 2 to each sample. Using opposite ends of a fresh applicator stick, mix each sample thoroughly; and then spread the samples over a 1-inch-square area on the slide.

6. Gently rock the slide back and forth for 3 minutes, and then check each sample for the presence of agglutination.

Which sample gave the positive test for the presence of rheumatoid factor?

___

7. Dispose of the second applicator stick and the gloves in the autoclave bag.

EXERCISE 33

Anatomy of the Respiratory System

OBJECTIVES

1. To define *pulmonary ventilation, external respiration,* and *internal respiration.*
2. To label the major respiratory system structures on a diagram (or identify them on a model or dissection) and to describe the function of each.
3. To recognize the histological structure of the trachea (cross section) and lung tissue on prepared slides, and to describe the functions the observed structural modifications serve.

MATERIALS

Human torso model
Respiratory organ system model and/or chart of the respiratory system
Larynx model (if available)
Sheep pluck (preserved, or fresh from the slaughterhouse)
Dissection trays and instruments
Protective skin cream or disposable plastic gloves
Source of compressed air*
2-foot length of laboratory rubber tubing
Alcohol swabs
Histological slides of trachea (cross section) and lung tissue, both normal and pathological (e.g., sections from lung tissues exhibiting bronchitis, emphysema, lung cancer)
Compound and dissecting microscopes

* If a compressed air source is not available, cardboard mouthpieces that fit the cut end of the rubber tubing should be available for student use. Disposable autoclave bags should also be provided for discarding the mouthpiece.

To carry out their vital processes, body cells require an abundant and continuous supply of oxygen. As the cells use oxygen, they release carbon dioxide, a waste product that the body must get rid of. These oxygen-using cellular processes, collectively referred to as *cellular respiration,* are more appropriately described in conjunction with the topic of cellular metabolism. The major role of the **respiratory system,** our focus in this exercise, is to supply the body with oxygen and dispose of carbon dioxide. For it to fulfill this role, at least four distinct processes, collectively referred to as **respiration,** must occur:

Pulmonary ventilation: the tidelike movement of air into and out of the lungs so that the gases in the alveoli are continuously changed and refreshed. Also more simply called *ventilation,* or *breathing.*

External respiration: the gas exchanges to and from the pulmonary circuit blood that occur in the lungs (oxygen loading/carbon dioxide unloading).

Transport of respiratory gases: the transport of respiratory gases between the lungs and tissue cells of the body accomplished by the cardiovascular system, using blood as the transport vehicle.

Internal respiration: exchange of gases to and from the blood capillaries of the systemic circulation (oxygen unloading and carbon dioxide loading).

Although only the first two processes are the exclusive province of the respiratory system, all four must occur for the respiratory system to "do its job." Hence, the respiratory and circulatory system are irreversibly linked. Should either system fail, the cells begin to die from oxygen starvation and the accumulation of carbon dioxide. If uncorrected, this situation soon causes death of the entire organism.

UPPER RESPIRATORY SYSTEM STRUCTURES

The upper respiratory system structures—the nose, pharynx, and larynx—are shown in Figure 33.1 and described below. As you read through the descriptions, identify each structure by referring to this figure.

Air generally passes into the respiratory tract through the **external nares** (nostrils), and enters the **nasal cavity** (divided by the **nasal septum**). It then flows posteriorly over three pairs of lobelike structures, the **inferior, superior,** and **middle nasal conchae,** which increase the air turbulence. As the air passes through the nasal cavity, it is also warmed, moistened, and filtered by the nasal mucosa. The air that flows directly beneath the upper part of the nasal cavity may chemically stimulate the olfactory receptors located in the mucosa of that region. The nasal cavity is surrounded by the **paranasal sinuses** in the frontal, sphenoid, ethmoid, and maxillary bones. These sinuses act as resonance chambers in speech and, like the nasal mu-

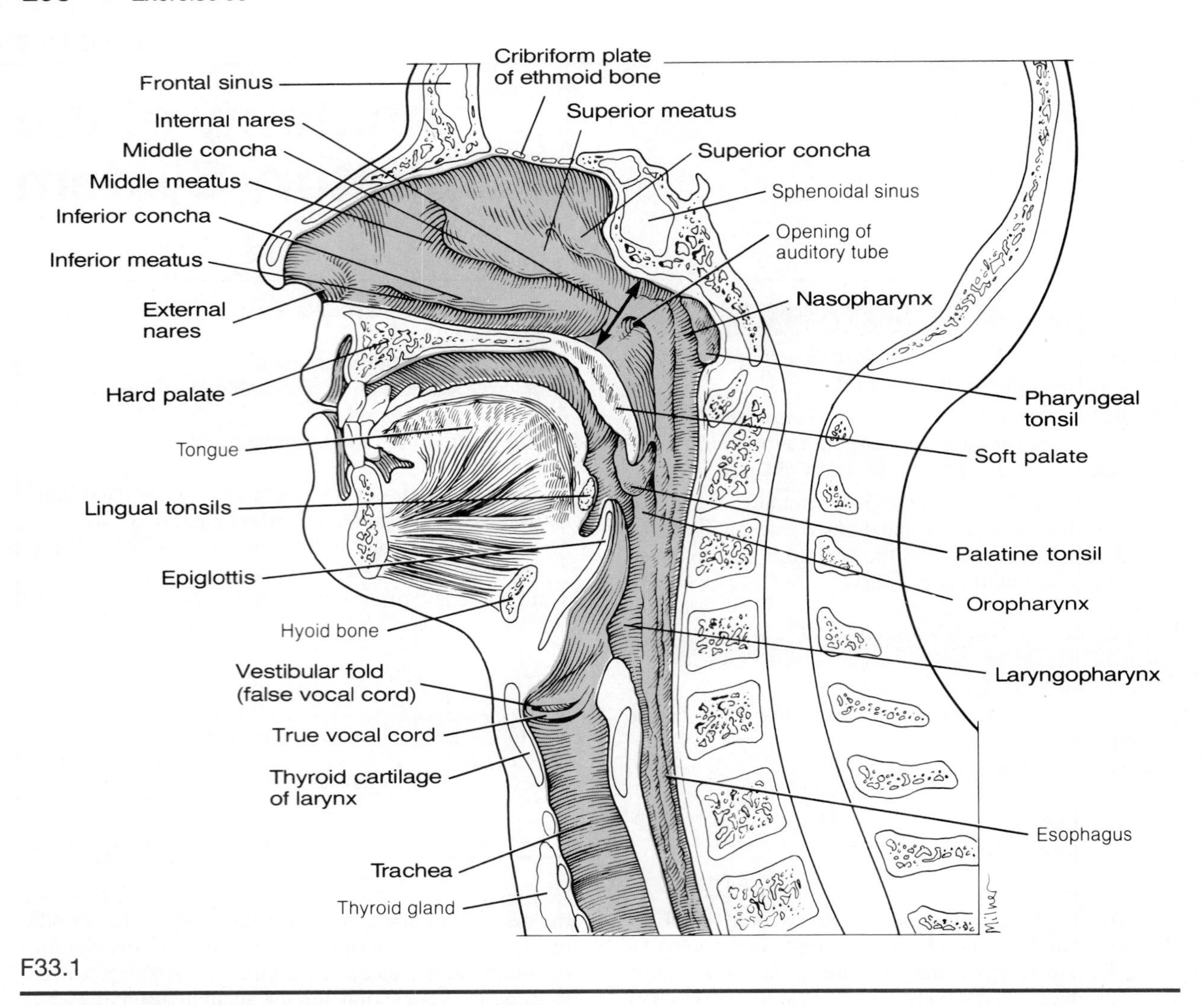

F33.1

Structures of the upper respiratory tract (sagittal section).

cosa, warm and moisten the incoming air. The nasal passages are separated from the oral cavity below by a partition composed of the **hard palate** anteriorly and the **soft palate** posteriorly

The genetic defect called *cleft palate* (failure of the palatine bones and/or the palatine processes of the maxillary bones to fuse medially) causes difficulty in breathing and in oral cavity functions such as sucking and, then later, mastication and speech. ■

Needless to say, air may also enter the body via the mouth, and pass through the oral cavity to move into the pharynx posteriorly, where the oral and nasal cavities are joined temporarily.

Commonly called the *throat,* the funnel-shaped **pharynx** connects the nasal and oral cavities to the larynx and esophagus inferiorly. It has three named parts:

1. The **nasopharynx** lies posterior to the nasal cavity and is continuous with it via the **internal nares.** It lies above the soft palate; hence, it serves *only* as an air passage. High on its posterior wall are the *pharyngeal tonsils,* paired masses of lymphoid tissue that help to protect the respiratory passages from invading pathogens. The *auditory tubes,* which allow middle ear pressure to become equalized to atmospheric pressure, drain into the lateral aspects of the nasopharynx.

Because of the continuity of the middle ear and nasopharyngeal mucosae, nasal infections may invade the middle ear cavity causing *otitis media,* which is difficult to treat. ■

2. The **oropharynx** is continuous posteriorly with the oral cavity. Since it extends from the soft palate to the epiglottis of the larynx inferiorly, it serves as a common conduit for food and air. In its lateral walls are the *palatine tonsils;* the *lingual tonsils* cover the base of the tongue.

3. The **laryngopharynx,** like the oropharynx, accommodates both ingested food and air. It lies directly posterior to the upright epiglottis and extends to the larynx, where the common pathway divides into the respiratory and digestive channels. From the laryngopharynx, air

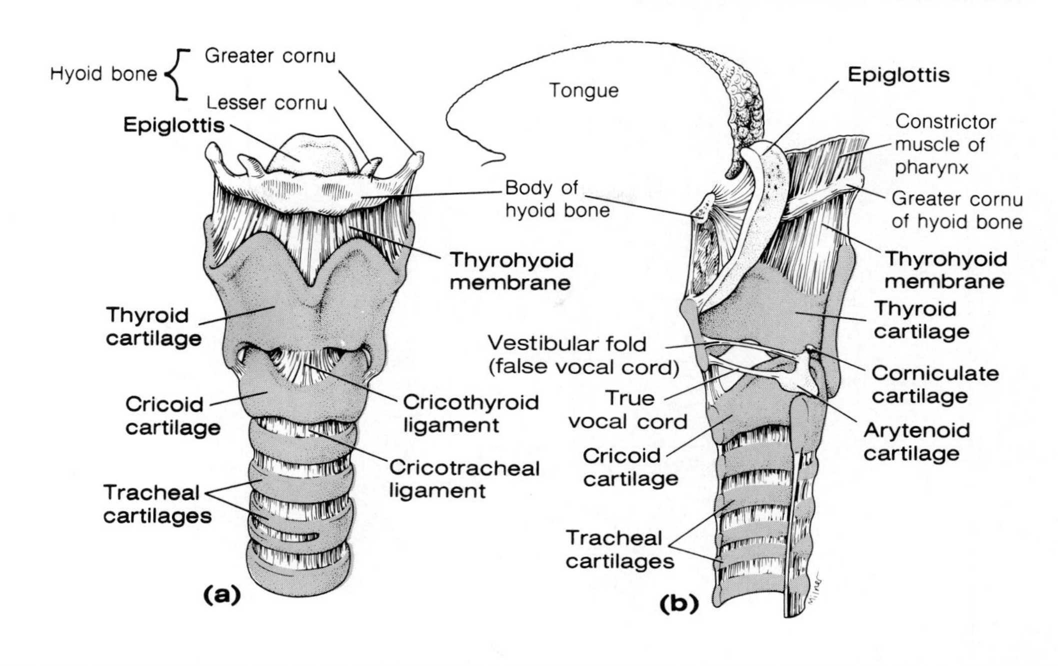

F33.2

Structure of the larynx. (a) Anterior view; (b) sagittal section.

enters the lower respiratory passageways by passing through the larynx (voice box) and into the trachea below.

The **larynx** (Figure 33.2) consists of nine cartilages. The two most prominent are the large shield-shaped **thyroid cartilage,** the anterior medial prominence of which is commonly referred to as the Adam's apple, and the inferiorly located, ring-shaped **cricoid cartilage,** the widest dimension of which faces posteriorly. All the laryngeal cartilages are composed of hyaline cartilage except the flaplike **epiglottis,** a flexible elastic cartilage located superior to the opening of the larynx. The epiglottis, sometimes referred to as the "guardian of the airways," forms a lid over the larynx when we swallow. This closes off the respiratory passageways to incoming food or drink, which is routed into the posterior esophagus, or food chute.

- Palpate your larynx by placing your hand on the anterior neck surface, approximately halfway down its length. Swallow. Can you feel the cartilaginous larynx rising?

If anything other than air enters the larynx, a cough reflex attempts to expel the substance. Note that this reflex operates only when a person is conscious; thus you should never try to feed or pour drink down the throat of an unconscious person.

The mucous membrane of the larynx is thrown into two pairs of folds—the upper **vestibular folds,** also called the **false vocal cords,** and the lower **vocal folds,** or **true vocal cords,** which vibrate with expelled air for speech. The vocal cords are attached posterolaterally to the small triangular **arytenoid cartilages.** The slitlike passageway between the folds is called the **glottis.**

LOWER RESPIRATORY SYSTEM STRUCTURES

Air entering the **trachea,** or *windpipe,* from the larynx travels down its length (about 11.5 cm) to the level of the *sternal angle* (or the disc between the fourth and fifth thoracic vertebrae). There the passageway divides into the right and left **primary bronchi,** which plunge into their respective **lungs** at an indented area called the **hilus** (Figure 33.3). The right primary bronchus is wider, shorter, and more vertical than the left; as a result, foreign objects that enter the respiratory passageways are more likely to become lodged in it.

The trachea is lined with a ciliated mucus-secreting pseudostratified columnar epithelium, as are many of the other respiratory system passageways. The cilia beat in unison and propel mucus (produced by goblet cells) laden with dust particles, bacteria, and other debris away from the lungs and toward the throat, where it can be expectorated or swallowed. The walls of the trachea are reinforced with C-shaped cartilage rings, the incomplete portion being located posteriorly. These C-shaped cartilages serve a double function: The incomplete parts allow the esophagus to expand anteriorly when a large food bolus is swallowed; the solid portions reinforce the trachea walls to maintain its open passageway regardless of the pressure changes that occur during breathing.

The primary bronchi further divide into smaller and smaller branches (the secondary bronchi, tertiary bronchi, and so on), finally becoming the **bronchioles,** which have terminal branches called **respiratory bronchioles** (Figure 33.3). All but the most minute branches contain cartilaginous reinforcements in their walls, usually in the form of small plates of hyaline cartilage rather

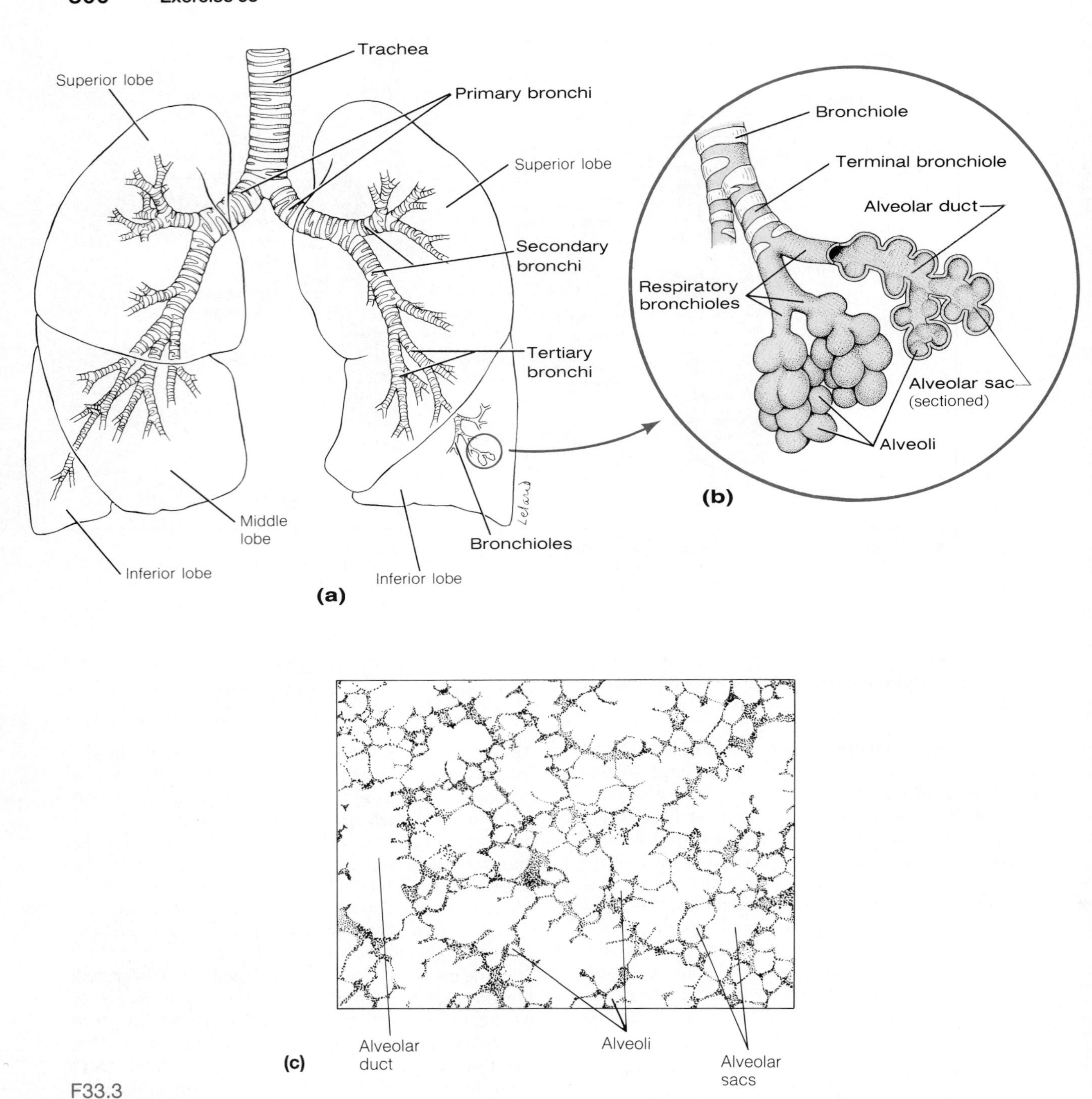

F33.3

Structures of the lower respiratory tract. Inset (b) shows enlarged view of alveoli. Line drawing of a photomicrograph (c) shows respiratory zone structures. (Corresponding plate 30 appears in the histology atlas.)

than cartilaginous rings. As the respiratory tubes get smaller and smaller, the relative amount of smooth muscle in their walls increases as the amount of cartilage declines and finally disappears. The complete layer of smooth muscle present in the bronchioles enables them to provide considerable resistance to air flow under certain conditions (asthma, hayfever, etc.). The respiratory bronchioles in turn subdivide into several **alveolar ducts,** which terminate in **alveolar sacs** that rather resemble clusters of grapes. **Alveoli,** minute balloonlike expansions along the alveolar sacs and occasionally found protruding from the alveolar ducts and respiratory bronchioles, are composed of a single, very thin layer of squamous epithelium overlying a scant basal lamina. The external surfaces of the alveoli are densely spiderwebbed with a network of pulmonary capillaries (Figure 33.4). Together, the alveolar and capillary walls and their fused basal laminas form the **respiratory mem-**

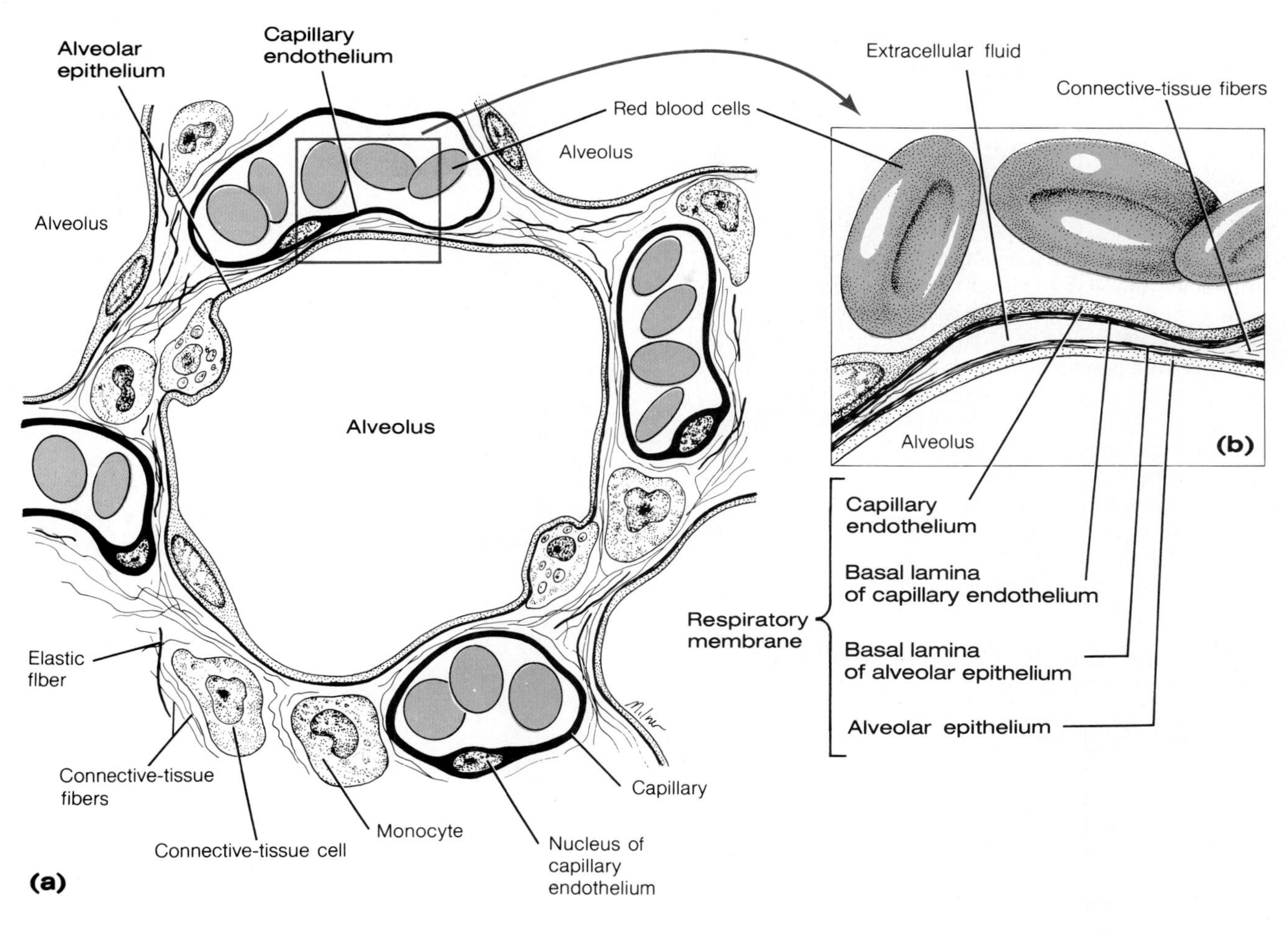

F33.4

Diagrammatic view of the relationship between the alveoli and pulmonary capillaries involved in gas exchange. (a) One alveolus surrounded by capillaries; (b) enlargement of the respiratory membrane.

brane. Because gas exchanges occur by simple diffusion across the respiratory membrane—oxygen passing from the alveolar air to the capillary blood and carbon dioxide leaving the capillary blood to enter the alveolar air—the alveolar sacs, alveolar ducts, and respiratory bronchioles are referred to collectively as **respiratory zone structures.** All other respiratory passageways (from the nasal cavity to the bronchioles), which simply serve as access or exit routes to and from these gas exchange chambers, are called **conducting zone structures.** Because the conducting zone structures have no exchange function, they are also termed *anatomical dead space.*

The continuous branching of the respiratory passageways in the lungs is often referred to as the **respiratory tree.** The comparison becomes much more meaningful if you observe a resin cast of the respiratory passages (Figure 33.5).

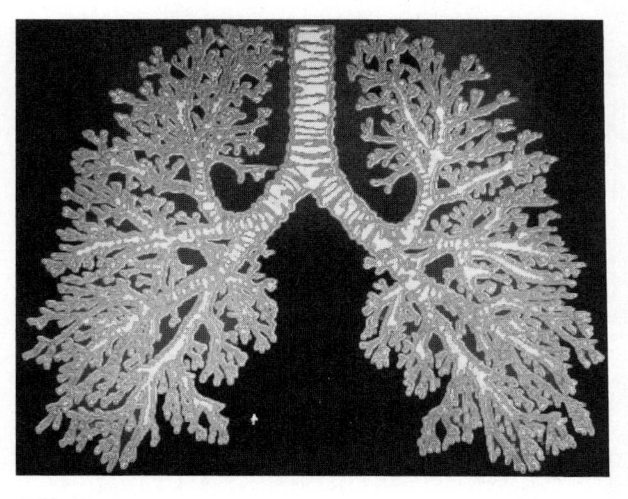

F33.5

A resin cast showing the extensive branching of the human respiratory tree.

The paired lungs are separated from one another by the structures of the mediastinum (heart, trachea, bronchi, major blood vessels, and esophagus). The substance of the lungs, other than the respiratory passageways that make up the bulk of their volume, is primarily elastic connective tissue that allows the lungs to recoil passively during expiration. Each lung is enclosed in a double-layered sac of serous membrane called the **pleura.** The outer layer, the **parietal pleura,** is attached to the thoracic walls and the **diaphragm;** the inner layer, covering the lung tissue, is the **visceral pleura.** The two pleural layers are separated by the *pleural space,* which is more of a potential space than an actual one. The parietal and visceral layers produce lubricating serous fluid that causes them to adhere closely to one another, holding the lungs to the thoracic wall and allowing them to move easily against one another in a frictionless environment during the movements of breathing.

Before proceeding, be sure to locate, on the torso model, the thoracic cavity structures model, or an anatomical chart, all the respiratory structures described.

EXAMINATION OF PREPARED SLIDES OF LUNG AND TRACHEA TISSUE

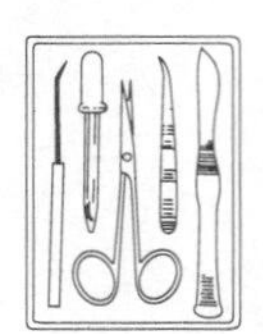

1. Obtain and examine a cross section of trachea tissue. Identify the smooth muscle layer, the hyaline cartilage supporting rings, and the pseudostratified ciliated epithelium (see Figure 33.6a). Also try to identify a few goblet cells in the epithelium. In the space at the top of the next column, draw a section of the trachea wall; and label all tissue layers.

2. Obtain a slide of lung tissue for examination. The alveolus is the main structural and functional unit of the lung, and is the actual site of gas exchange (see Figure 33.3c). Identify the thin squamous epithelium of the alveolar walls, a bronchiole (see Figure 33.6b), and if possible one of the smaller bronchi. Draw your observations of a small section of the alveolar tissue in the space below. Label the alveoli.

3. Examine slides of pathologic lung tissues (if available), and compare them to the normal lung specimens.

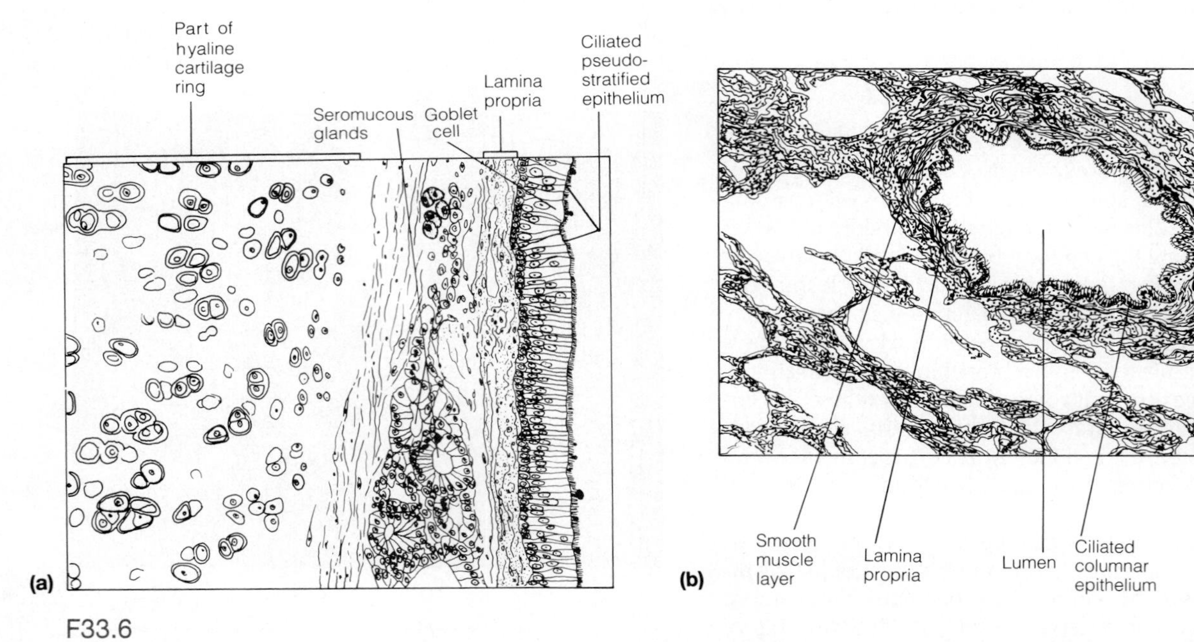

F33.6

Microscopic structure of the tracea and a bronchiole. (a) Cross-section through trachea, see corresponding plate 31 of the histology atlas; (b) cross-sectional view of a bronchiole, see plate 32 of the histology atlas.

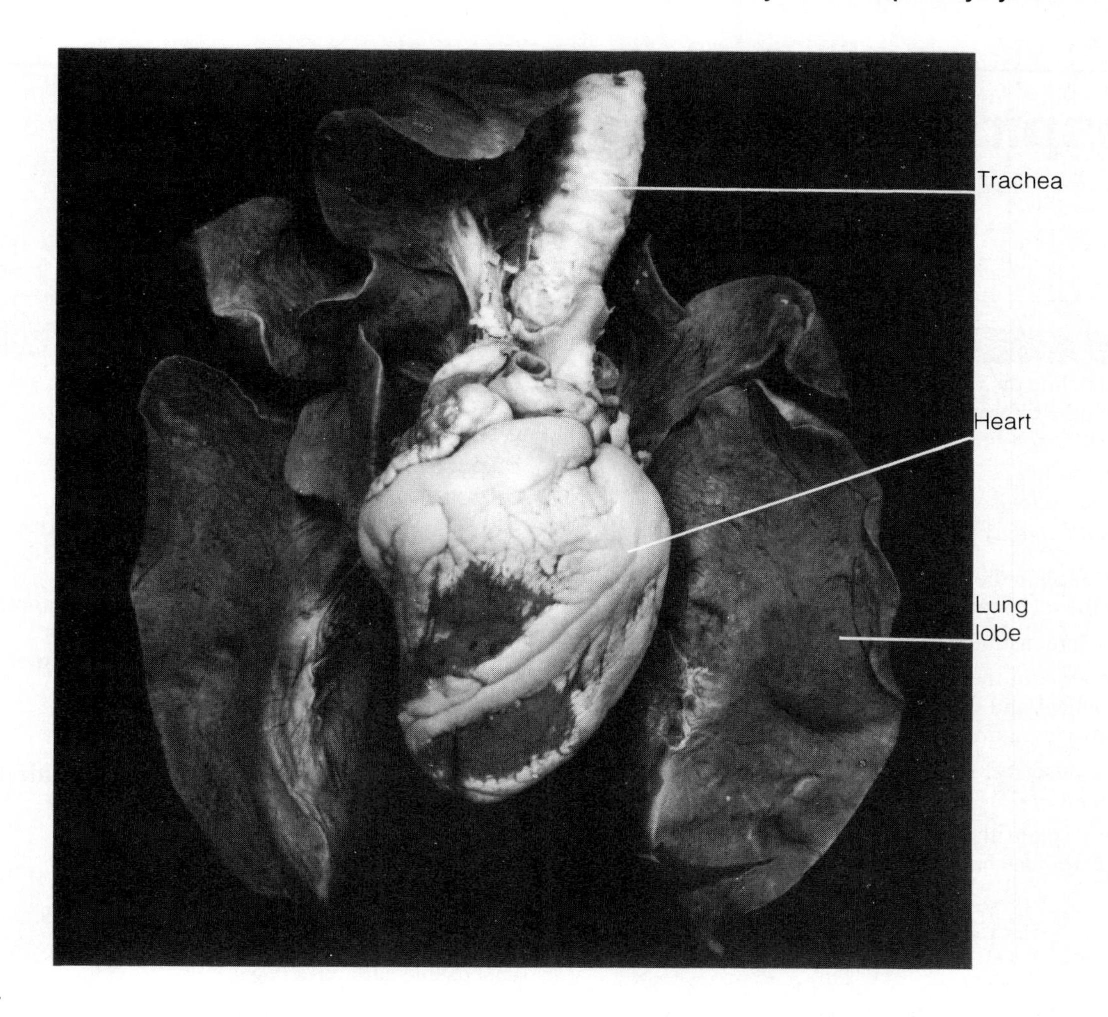

F33.7

Sheep pluck.

SHEEP PLUCK DEMONSTRATION

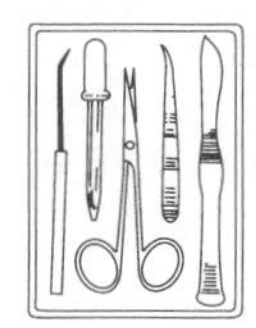

A sheep pluck (Figure 33.7) includes the larynx, the trachea with attached lungs, the heart and pericardium, and portions of the major blood vessels found in the mediastinum (aorta, pulmonary artery and vein, and vena cava). Obtain a sheep pluck and identify the lower respiratory system organs. Once you have completed your observations, insert a hose from an air compressor (vacuum pump) into the trachea; and alternately allow air to flow in and out of the lungs. Notice how the lungs inflate. This observation is educational in a preserved pluck but is a spectacular sight in a fresh one. Another advantage of using a fresh pluck is that the lung pluck changes color (becomes redder) as hemoglobin in trapped RBCs becomes loaded with oxygen. If air compressors are not available, the same effect may be obtained by using a length of laboratory rubber tubing to blow into the trachea. Obtain a cardboard mouthpiece and fit it into the cut end of the laboratory tubing before attempting to inflate the lungs by mouth.

Dispose of the mouthpiece in the autoclave bag immediately after use.

34 EXERCISE

Respiratory System Physiology

OBJECTIVES

1. To define the following (and be prepared to provide volume figures, if applicable):

inspiration	*expiratory reserve volume*
expiration	*expiratory end point*
tidal volume	*inspiratory reserve volume*
vital capacity	*minute respiratory volume*

2. To explain the role of muscles and volume changes in the mechanical process of breathing.
3. To demonstrate proper usage of the spirometer.
4. To explain the relative importance of various mechanical and chemical factors in producing respiratory variations.
5. To describe bronchial and vesicular breathing sounds.
6. To explain the importance of the carbonic acid–bicarbonate buffer system in maintaining blood pH.

MATERIALS

Model lung (bell jar demonstrator)
Tape measure
Spirometer
Disposable mouthpieces
Nose clips
Alcohol swabs
Paper bag
Pneumograph and recording attachments
Recording apparatus (kymograph or physiograph)
Stethoscope
Table (on chalkboard) prepared for recording of class data
pH meter (standardized with buffer of pH 7)
Buffer solution (pH 7)
Concentrated HCl and NaOH (in dropper bottles)
0.01 *M* HCl
250-ml beakers; 50-ml beakers
Graduated cylinder (100 ml)
Glass stirring rod
Plastic wash bottles containing distilled water
Animal plasma
Disposable autoclave bag

MECHANICS OF RESPIRATION

Pulmonary ventilation, or **breathing,** consists of two phases: **inspiration,** during which air is taken into the lungs, and **expiration,** during which air passes out of the lungs. As the inspiratory muscles (external intercostals and diaphragm) contract during inspiration, the size of the thoracic cavity increases. The diaphragm moves from its relaxed dome shape to a flattened position, increasing the superoinferior volume, and the external intercostals lift the rib cage, increasing the anteroposterior and lateral dimensions (Figure 34.1). Since the lungs adhere to the thoracic walls like flypaper because of the cohesive character of the pleurae, the intrapulmonary (within the lungs) volume also increases, lowering the air (gas) pressure inside the lungs. The gases then expand to fill the available space creating a partial vacuum, which causes air to flow into the lungs—constituting the act of inspiration. During expiration, the inspiratory muscles relax, and the natural tendency of the elastic lung tissue to recoil acts to decrease the intrathoracic and intrapulmonary volumes. As the gas molecules within the lungs are forced closer together, the intrapulmonary pressure rises to a point higher than atmospheric pressure. This causes gases to flow from the lungs to equalize the pressure inside and outside the lungs—the act of expiration.

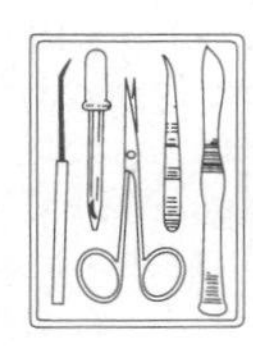

Observe the model lung, which demonstrates the principles involved in the filling and emptying of the lungs. It is a simple apparatus with a bottle "thorax," a rubber membrane "diaphragm," and balloon "lungs."

1. At the demonstration area, work the model lung by moving the rubber diaphragm up and down, noting changes in balloon (lung) size as the volume of the thoracic cavity is alternately increased and decreased.

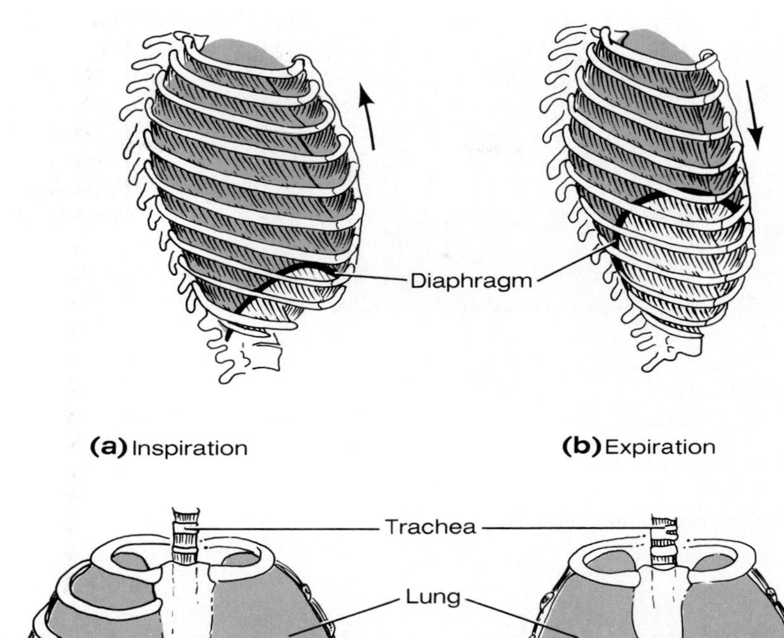

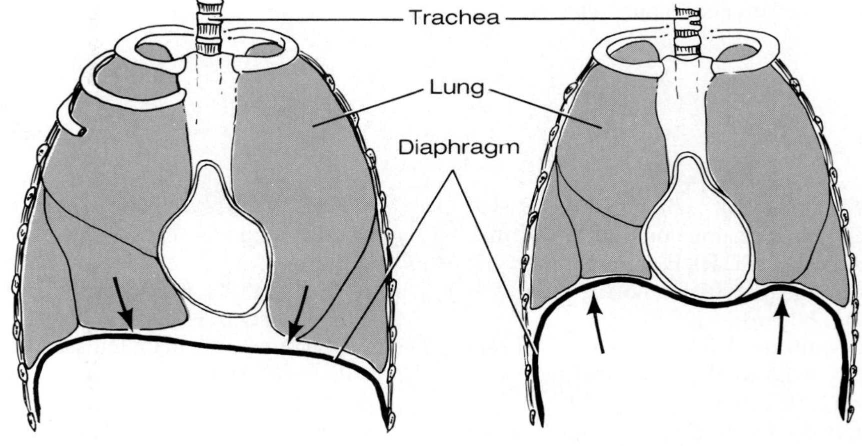

F34.1

Rib cage and diaphragm positions during breathing. (a) At the end of a normal inspiration; chest expanded, diaphragm depressed; (b) at the end of a normal expiration; chest depressed, diaphragm elevated.

2. Before proceeding, check the appropriate columns concerning these observations in the chart in the Lab Review section for Exercise 34.

3. After observing the operation of the model lung, conduct the following test on your lab partner. Use the tape measure to determine chest circumference by placing the tape around the chest as high up as possible under the armpits. Record the measurements, in inches, in the appropriate space for each of the conditions below.

Quiet breathing:

inspiration ____________ expiration ____________

Forced breathing:

inspiration ____________ expiration ____________

Do the results coincide with what you expected on the basis of what you have learned thus far?

____

RESPIRATORY VOLUMES AND CAPACITIES—SPIROMETRY

A person's size, sex, age, and physical condition produce variations in respiratory volumes. Normal quiet breathing moves about 500 ml of air in and out of the lungs with each breath. As you have seen in the previous experiment, a person can usually forcibly inhale or exhale much more air than is normally exchanged in quiet breathing. The terms given to the measurable respiratory volumes are defined just below. These terms and their normal values for an adult male should be memorized.

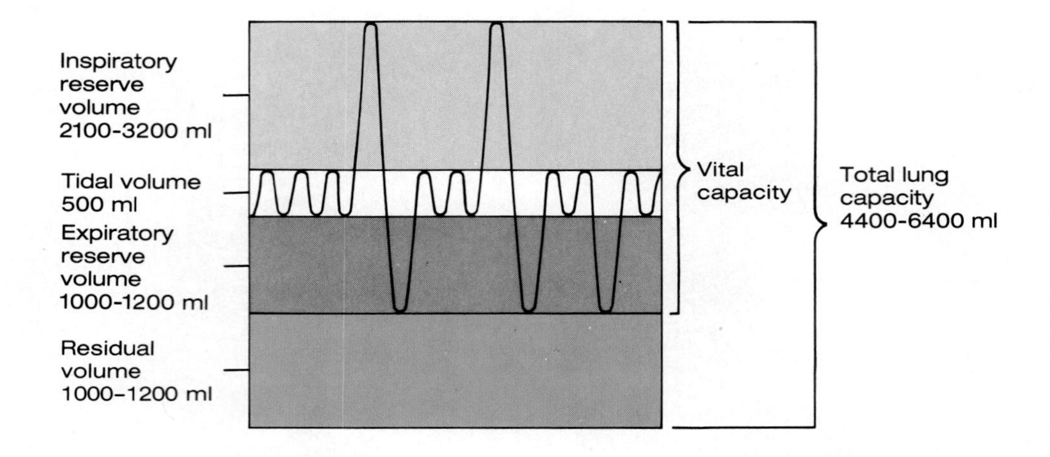

F34.2

Idealized tracing of the various respiratory volumes.

Tidal volume (TV): amount of air inhaled or exhaled with each breath under resting conditions (500 ml).

Inspiratory reserve volume (IRV): amount of air that can be forcefully inhaled after a normal tidal volume exhalation (2800 ml).

Expiratory reserve volume (ERV): amount of air that can be forcefully exhaled after a normal tidal volume exhalation (1200 ml).

Vital capacity (VC): maximum amount of air that can be exhaled after a maximal inspiratory effort; VC = TV + IRV + ERV; (4800 ml).

An idealized tracing of the various respiratory volumes and their relationships to each other are shown in Figure 34.2.

Respiratory volumes will be measured with an apparatus called a **spirometer.** A *wet spirometer,* such as a Collins spirometer, consists of a cylinder within a tank into which air can be added or removed. The outer tank contains water and has a tube running through it to carry air above the water level. The floating bottomless inner cylinder is inverted over the water-containing tank and connected to a volume indicator. The volume of air exhaled can be read off the indicator, which moves to indicate the changes in air volume within the apparatus (Figure 34.3). Alternatively, a *hand-held dry,* or *wheel, spirometer* may be used and will yield comparable results.

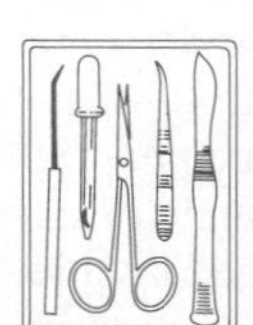

1. Examine a spirometer indicator scale before beginning, to ensure proper reading of the scale. On a Collins wet spirometer, each number from 1 to 6 indicates the number of liters, and each smaller marking between the liter marks is equivalent to 0.1 liter, or 100 ml. On the hand-held dry spirometer, volumes are indicated in ml and each small division represents 100 ml.

2. Work in pairs, with one person recording data while the other conducts the volume determinations. Make sure the indicator is set at zero before beginning each trial. Obtain a disposable cardboard mouthpiece. Insert it in the open end of the flexible tube of the wet spirometer apparatus (over the fixed stem of the hand-held dry spirometer). Before beginning, practice exhaling through the mouthpiece without exhaling through the nose, or use the nose clips (clean them first with an alcohol swab). If you are using the dry spirometer make sure its dial faces upward so that the volumes can be easily read.

3. Conduct the test three times for each required measurement. Record the data here, and then find the average volume figure for that respiratory measurement. After you have completed the trials and computed the averages, enter the average values on the table prepared on the chalkboard for tabulation of class data;* and copy all averaged data onto the chart provided in the laboratory review section.

4. Tidal volume (TV) measurement: The volume of air inhaled and exhaled with each normal respiration is approximately 500 ml. To conduct this test, inhale a normal breath, and then exhale a normal breath of air into the spirometer mouthpiece. (Do not force the expiration!) Record the volume, and repeat the test two more times.

* Note to the instructor: The format of class data tabulation can be similar to that shown here. However, it would be interesting to divide the class into smokers and nonsmokers, and then compare the mean average VC and ERV for each group. Such a comparison might help to determine if smokers are handicapped in any way. It also might be a good opportunity for an informal discussion of the early warning signs of bronchitis and emphysema, which are primarily smokers' diseases.

trial 1 ______________ ml trial 2 ______________ ml

trial 3 ______________ ml average TV __________ ml

5. Inspiratory reserve volume (IRV) measurement: The volume of air that can be forcibly inhaled following a normal inspiration is substantial, ranging from 2100 to 3200 ml.

Breathe normally two or three times. Then make the deepest possible inspiration; and exhale, into the spirometer, only to the expiratory end point. Do not force the expiration. (The *expiratory end point* is the position of the rib cage at the end of a normal expiration.) Record the volume, and repeat the test two more times. Subtract the TV average from each figure to find the IRV value, and then compute the average IRV. (Respiratory volume − TV average = IRV.)

trial 1 ______________ ml trial 2 ______________ ml

trial 3 ______________ ml average IRV __________ ml

The IRV is very difficult to measure as described above because of the problem in stopping the tidal volume breath precisely at the expiratory end point. If your values are quite different from the anticipated values, wait until you have obtained results for your vital capacity, described in instruction 7, and use the following equation to compute your IRV:

IRV =

VC (measured) − ((TV (measured) + ERV (measured))

Record your computed value.

6. Expiratory reserve volume (ERV) measurement: The volume of air that can be forcibly exhaled after a normal expiration normally ranges between 1000 and 1200 ml. The ERV is dramatically reduced when the elasticity of the lungs is decreased by a chronic obstructive pulmonary disease (COPD) such as *emphysema.* Because energy must be used to *deflate* the lungs in such conditions, expiration is physically exhausting for individuals suffering from COPD. Inhale and exhale normally two or three times; then insert the spirometer mouthpiece, and forcibly exhale as much of the additional air as you can. Record your results, and repeat the test twice again.

trial 1 ______________ ml trial 2 ______________ ml

trial 3 ______________ ml average ERV __________ ml

7. Vital capacity (VC) measurement: The total exchangeable air of the lungs (the sum of TV + IRV + ERV) is normally 4500 to 4800 ml. Breathe in and out normally two or three times, and then bend over and exhale all the air possible. Then, as you raise yourself to the upright position, inhale as fully as possible. (It is very important to *strain* to inhale the maximum amount of air that you can.) Quickly insert the mouthpiece, and exhale as forcibly as possible. Record your results, and repeat the test two more times.

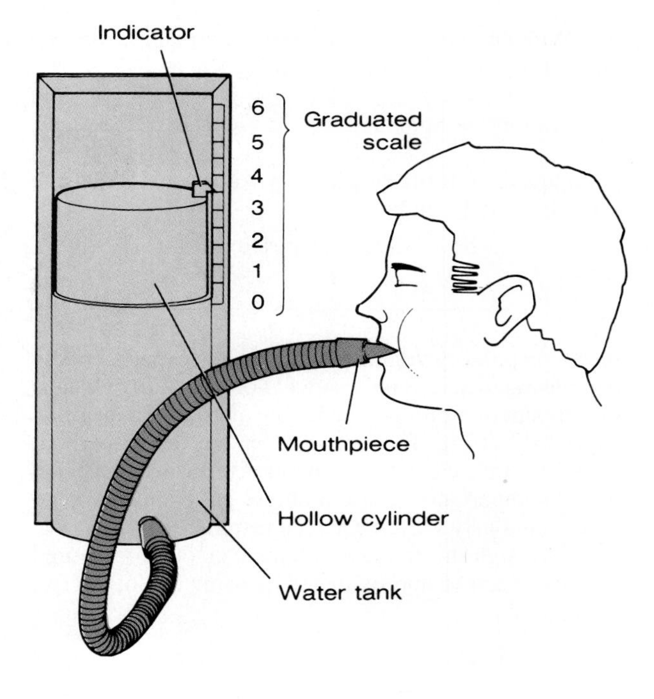

F34.3

Operation of a wet spirometer. Breathing into the mouthpiece causes the inner cylinder to rise because of the increase in volume (and pressure) of air within the cylinder. Water provides an airtight seal.

trial 1 ______________ ml trial 2 ______________ ml

trial 3 ______________ ml average VC __________ ml

Figure out how closely your measured average vital capacity volume compares with the *predicted values* for someone your age, sex, and height. Obtain the predicted figure either from Appendix C male values or female values. Notice that you will have to convert your height in inches to centimeters (cm) to find the corresponding value. This is easily done by multiplying your height in inches by 2.54.

Computed height: ______________________________ cm

Predicted VC value (obtained from the appropriate table):

____ ml

Use the following equation to compute your VC as a percentage of the predicted VC value:

$$\% \text{ of predicted VC} = \frac{\text{average measured VC}}{\text{predicted value}} \times 100$$

% predicted VC value: ______________________________ %

8. Without using the spirometer, count and record the subject's normal respiratory rate.

Respirations per min: ______________________________

Compute the **minute respiratory volume** (**MRV**) using the following formula:

$$MRV = TV \times \text{respirations/min}$$

MRV = ______________________________ ml/min

9. A respiratory volume that cannot be experimentally demonstrated here is the residual volume (RV), which is the amount of air remaining in the lungs after a maximal expiratory effort. The presence of air (usually about 1200 ml) that cannot be voluntarily flushed from the lungs is important because it allows gas exchange to go on continuously—even between respirations.

Although the residual volume cannot be measured directly, it can be approximated by using one of the following factors:

For ages 16–34 factor = 0.250

For ages 35–49 factor = 1.305

For ages 50–69 factor = 1.445

Compute your predicted RV using the following equation:

$$RV = VC \times \text{factor}$$

10. Dispose of the used cardboard mouthpieces in the autoclave bag before continuing.

Many disease states alter the volumes described and tested in this exercise; however, the small variations you are likely to see during this lab are of little significance.

USE OF THE PNEUMOGRAPH TO DETERMINE FACTORS INFLUENCING RATE AND DEPTH OF RESPIRATION*

The neural centers that control respiratory rhythm are located in the medulla and pons. Oscillations between the two centers maintain a rate of 12 to 18 respirations per minute. On occasion, input from the stretch receptors in the lungs (via the vagus nerve to the medulla) modifies the respiratory rate, as in cases of overinflation or extreme deflation of the lungs (Hering-Breuer reflex).

Death occurs when medullary centers are completely suppressed, as from an overdose of sleeping pills or gross overindulgence in alcohol, and respiration ceases completely. ■

Although these nervous system centers initiate the basic rhythm of breathing, there is no question that physical phenomena—such as talking, yawning, coughing, and exercise—and chemical factors—such as changes in oxygen and carbon dioxide concentrations in the blood, and fluctuations in blood pH—can modify the rate and depth of respiration. Changes in blood carbon dioxide levels seem to act directly on the medulla control centers, whereas changes in oxygen concentrations are monitored by chemoreceptor regions in the aortic and carotid bodies, which in turn send input to the medulla. The experimental sequence in this section is designed to test the relative importance of various physical and chemical factors to the process of respiration.

The **pneumograph,** an apparatus that records variations in breathing patterns, is the best means of observing respiratory variations resulting from physical and chemical factors. The chest pneumograph is a coiled rubber hose that is attached around the thorax. As the subject breathes, chest movements produce pressure changes within the pneumograph that are transmitted to a recorder.

The instructor will demonstrate how to set up the pneumograph and will discuss the interpretation of results. Work in pairs so that one person can mark the record to identify the tests for later interpretation. Ideally, to prevent voluntary modification of the record, the student being tested should not observe the recording process.

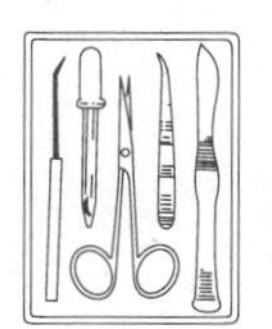

1. Attach the pneumograph tubing firmly, but not restrictively, around the thoracic cage, at the level of the sixth rib, leaving room for chest expansion during testing. If the subject is female, position the tubing above the breasts to prevent slippage during testing. Set the pneumograph speed at 1 or 2, and the time signal at 10-second intervals. Record quiet breathing for 1 minute, with the subject in a sitting position. Record breaths per minute.

____

2. Record the subject's breathing rate as he or she performs activities from the list below. Make sure the record is marked accurately to identify each test conducted. Record your results in the laboratory review section.

talking
yawning
laughing
standing
swallowing water
coughing
lying down
running in place
doing a math problem (concentrating)

3. Without recording, have the subject breathe normally for 2 minutes, then inhale deeply, and hold his breath for as long as he can.

Time the breath-holding interval. ____________ sec

As the subject begins to expire, turn on the recording apparatus and record the recovery period (time to return to normal breathing—usually slightly over 1 minute).

time of recovery period.__________ sec

* Note to instructor: This exercise may also be done without using the recording apparatus by simply having the students count the respiratory rate visually.

During breath holding, was the subject's urge to inspire *or* to expire?

____

Repeat the above experiment, without recording, but exhale completely and forcefully after taking the deep breath. What was observed this time?

____

____

Explain the results. (Hint: the vagus nerve is the sensory nerve of the lungs and plays a role here.)

____

____

4. Without recording, have the subject hyperventilate (breathe deeply and forcefully at the rate of 1 breath/4 sec) for about 30 seconds.* Record after hyperventilation. Is the respiratory rate faster or slower than during normal, quiet breathing?

____

5. Repeat the above test; but after the hyperventilating, have the subject hold his breath as long as he can. Can the breath be held for a longer or shorter period of time after hyperventilation?

____

6. Without recording, have the subject breathe into a paper bag for 3 minutes; then record breathing movements.

Caution: during the bag-breathing exercise, the subject's partner should watch the subject carefully for any untoward reactions.

Is the breathing rate faster or slower than that recorded during normal, quiet breathing?

____

After hyperventilating? _______________

7. Run in place for 2 minutes; then have your partner determine the duration of your breath holding.

_______________ sec

* A sensation of dizziness may develop. (As the carbon dioxide is washed out of the blood by overventilation, the blood pH increases, leading to a decrease in blood pressure and reduced cerebral circulation.) The subject may experience a lack of desire to breathe after forced breathing is stopped. (If the period of breathing cessation—**apnea**—is extended, cyanosis of the lips may occur.)

8. To prove that respiration has a marked effect on circulation, conduct the following test. Have your lab partner note the rate and relative force of your radial pulse before beginning.

rate __________ beats/min relative force __________

Inspire forcibly. Immediately close your mouth and nose to retain the inhaled air, and then make a forceful and prolonged expiration. Your lab partner should observe and record the condition of the blood vessels of your neck and face, and again immediately palpate the radial pulse.

Observations: _______________

____

radial pulse ________ beats/min relative force ________

Explain the changes observed. _______________

____

____

Dispose of the paper bag in the autoclave bag. Keep the pneumograph records to interpret results, and hand them in if requested by the instructor. Observation of the test results should enable you to determine which chemical factor, carbon dioxide or oxygen, has the greater effect on modifying the respiratory rate and depth.

RESPIRATORY SOUNDS

As air flows in and out of the respiratory tree, it produces two characteristic sounds that can be picked up with a stethoscope. The **bronchial sounds** are produced by air rushing through the large respiratory passageways (the trachea and the bronchi). The second sound type, **vesicular breathing sounds,** apparently results from air filling the alveolar sacs, and resembles the sound of a rustling or muffled breeze.

1. Place the diaphragm of the stethoscope on the throat of the test subject, just below the larynx. Listen for bronchial sounds on inspiration and expiration.
2. Move the stethoscope downward toward the bronchi until you can no longer hear sounds.
3. Place the stethoscope over the following chest areas, and listen for vesicular sounds during respiration (heard primarily during inspiration):
   - At various intercostal spaces
   - At the *triangle of auscultation* (a small depressed area of the back where the muscles fail to cover the rib cage. Located just medial to the inferior part of the scapula)
   - Under the clavicle

Diseased respiratory tissue, mucus, or pus can produce abnormal chest sounds such as rales (a rasping sound) and wheezing (a whistling sound). ■

ROLE OF THE RESPIRATORY SYSTEM IN ACID-BASE BALANCE OF BLOOD

As you have already learned, respiratory ventilation is necessary for the continuous oxygenation of the blood and for the removal of carbon dioxide (a waste product of cellular respiration) from the blood. Blood pH must be relatively constant for the cells of the body to function optimally; therefore the carbonic acid–bicarbonate buffer system of the blood is extremely important because it helps stabilize arterial blood pH at 7.4 ± 0.02.

When carbon dioxide diffuses from the tissue cells into the blood, much of it enters the red blood cells, where it combines with water to form carbonic acid:

$$H_2O + CO_2 \xrightarrow[\text{enzyme present in RBC}]{\text{carbonic anhydrase}} H_2CO_3$$

Some carbonic acid formation also occurs in the plasma, but its formation there is very slow because of the lack of carbonic anhydrase enzyme. Shortly after it forms, carbonic acid dissociates to release bicarbonate (HCO_3^-) and hydrogen (H^+) ions. The hydrogen ions that remain in the cells are neutralized, or buffered, when they combine with the hemoglobin molecules. If they were not neutralized, the intracellular pH would become very acidic due to the accumulation of H^+ ions. Once formed, the bicarbonate ions diffuse out of the red blood cells into the plasma, where they become part of the carbonic acid–bicarbonate buffer system. As HCO_3^- follows its concentration gradient into the plasma, an electrical imbalance develops in the RBCs that draws Cl^- into them from the plasma. This exchange phenomenon is called the *chloride shift.*

Acids (more precisely, H^+) released into the blood by the body cells tend to lower the pH of the blood and cause the blood to become too acidic. On the other hand, basic substances that enter the blood tend to cause the blood to become too alkaline and the pH to rise. Both of these tendencies are resisted in large part by the carbonic acid–bicarbonate buffer system. If H^+ concentration in the blood begins to increase, the H^+ ions combine with bicarbonate ions to form carbonic acid (a weak acid that does not tend to dissociate at physiologic or acid pH) and are thus removed:

$$H^+ + HCO_3^- \longrightarrow H_2CO_3$$

Likewise, as blood H^+ concentration drops below what is desirable and blood pH rises, H_2CO_3 dissociates to release the bicarbonate ions (weak bases that are poorly functional under alkaline conditions) and H^+ ions to the blood. The released H^+ lowers the pH again:

$$H_2CO_3 \longrightarrow H^+ + HCO_3^-$$

In the case of excessively slow or shallow breathing (hypoventilation) or fast deep breathing (hyperventilation), the amount of carbonic acid in the blood can be greatly modified—it may increase dramatically during hypoventilation and decrease substantially during hyperventilation. In either situation, the buffering ability of the blood may be inadequate, and respiratory acidosis or alkalosis can result. It is therefore important to maintain the normal rate and depth of breathing for proper control of blood pH.

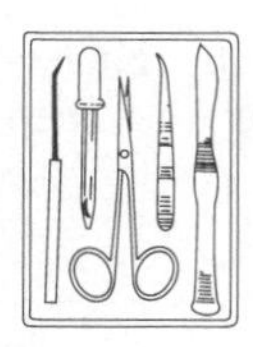

To observe the ability of a buffer system to stabilize the pH of a solution, obtain five 250-ml beakers and a wash bottle containing distilled water. Set up the following experimental samples:

Beaker 1
(150 ml distilled water) pH ____________

Beaker 2
(150 ml distilled water and
1 drop concentrated HCl) pH ____________

Beaker 3
(150 ml distilled water and
1 drop concentrated NaOH) pH ____________

Beaker 4
(150 ml standard buffer
solution, pH 7, and 1 drop
concentrated HCl) pH ____________

Beaker 5
(150 ml standard buffer
solution, pH 7, and 1 drop
concentrated NaOH) pH ____________

Using a pH meter standardized with a buffer solution of pH 7, determine the pH of the contents of each beaker; and record your results in the space provided. After each and every pH recording, the pH meter switch should be turned to "standby," and the electrodes rinsed thoroughly with a stream of distilled water from the wash bottle.

Add an additional 3 drops of concentrated HCl to beaker 4, stir, and record the pH:

____

Add an additional 3 drops of concentrated NaOH to beaker 5, stir, and record the pH:

____

How successful was the buffer solution in resisting pH changes when a strong acid (HCl) or a strong base (NaOH) was added?

____

____

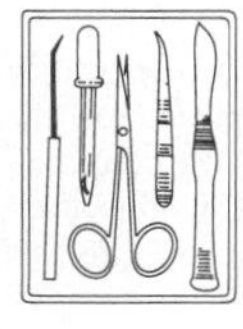

To observe the ability of the carbonic acid–bicarbonate buffer system of blood to resist pH changes, perform the following simple experiment.

Obtain two small beakers (50 ml), animal plasma, graduated cylinder, glass stirring rod, and a dropper bottle of 0.01 *M* HCl. Using the pH meter standardized with the buffer solution of pH 7.0, measure the pH of the animal plasma. Use only enough plasma to allow immersion of the electrodes and measure the volume used carefully.

pH of the animal plasma: ______________________

Add 2 drops of the 0.01 *M* HCl solution to the plasma; stir, and measure the pH again.

pH of plasma plus 2 drops of HC1: ______________

Turn the pH meter switch to standby, rinse the electrodes, and then immerse them in a quantity of distilled water (pH 7) exactly equal to the amount of animal plasma used. Measure the pH of the distilled water:

pH of distilled water: ______________________

Add 2 drops of 0.01 *M* HCl, swirl and measure the pH again.

pH of distilled water plus the two drops of HCl: ______

Is the plasma a good buffer? ______________________

What component of the plasma carbonic acid–bicarbonate buffer system was operating to counteract a change in pH when the HCl was added?

____

35 EXERCISE

Anatomy of the Digestive System

OBJECTIVES

1. To state the overall function of the digestive system.
2. To identify, on an appropriate diagram and/or torso model, the organs comprising the alimentary canal; and to name their subdivisions, if any.
3. To describe the general histological structure of the wall of the alimentary canal.
4. To describe the general functions of the digestive system organs or structures, relative to digestive system activity.
5. To list and explain the anatomical specializations of the stomach and small intestine that contribute to their functional roles.
6. To name and/or identify the accessory digestive organs, and to state their function(s) relative to digestive system activities.
7. To name human deciduous and permanent teeth, and to describe the anatomy of a generalized tooth.
8. To recognize, by microscopic inspection, or by viewing an appropriate diagram or photomicrograph, the histological structure of the following organs:

   small intestine, pancreas, stomach, salivary glands, tooth

MATERIALS

Dissectible torso model
Anatomical chart of the human digestive system
Model of a villus and liver (if available)
Jaw model or human skull
Prepared microscope slides of mixed salivary glands and pancreas; cross sections of the stomach, duodenum, and ileum; longitudinal sections of a tooth and of the gastroesophageal junction
Compound microscope

The **digestive system** provides the body with the nutrients, water, and electrolytes essential for metabolic processes and health. The organs of this system are responsible for food ingestion, digestion, and absorption, and for the elimination of the undigested remains as feces.

The digestive system consists of a hollow tube extending from the mouth to the anus, into which various accessory organs or glands empty their secretions. Food material within this tube, the alimentary canal, is technically outside the body because it has contact only with the cells lining the tract. For the ingested food to become available to the body cells, it must first be broken down physically (chewing and churning) and chemically (enzymatic hydrolysis) into its smaller, diffusible molecules—a process called **digestion.** The digested end products can then pass through the epithelial cells lining the tract and into the blood for distribution to the body cells—a process termed **absorption.** In one sense, the digestive tract can be viewed as a disassembly line, in which food is carried from one stage of its digestive processing to the next by muscular activity and its nutrients are made available to the cells of the body en route.

The organs of the digestive system are traditionally separated into two major groups: the **alimentary canal,** or **gastrointestinal tract,** and the **accessory digestive organs.** The alimentary canal is approximately 30 feet long in a cadaver but considerably less in a living person. It consists of the mouth, pharynx, esophagus, stomach, small and large intestines, rectum, and anus. The accessory structures include the salivary glands, gallbladder, liver, and pancreas, which secrete their products into the alimentary canal. These individual organs are described shortly.

GENERAL HISTOLOGICAL PLAN OF THE ALIMENTARY CANAL

Because the alimentary canal has a shared basic structure (particularly from the esophagus to the anus), it makes sense to review that structure as we begin studying this group of organs. Once done, all that need be emphasized as the individual organs are described is their specializations for unique functions in the digestive process.

Essentially the alimentary canal walls have four basic **tunics** (layers). From the lumen outward, these are the *mucosa,* the *submucosa,* the *muscularis externa,* and the *serosa* or *adventitia* (Figure 35.1). Each of these tunics has a predominant tissue type and a specific function in the digestive process.

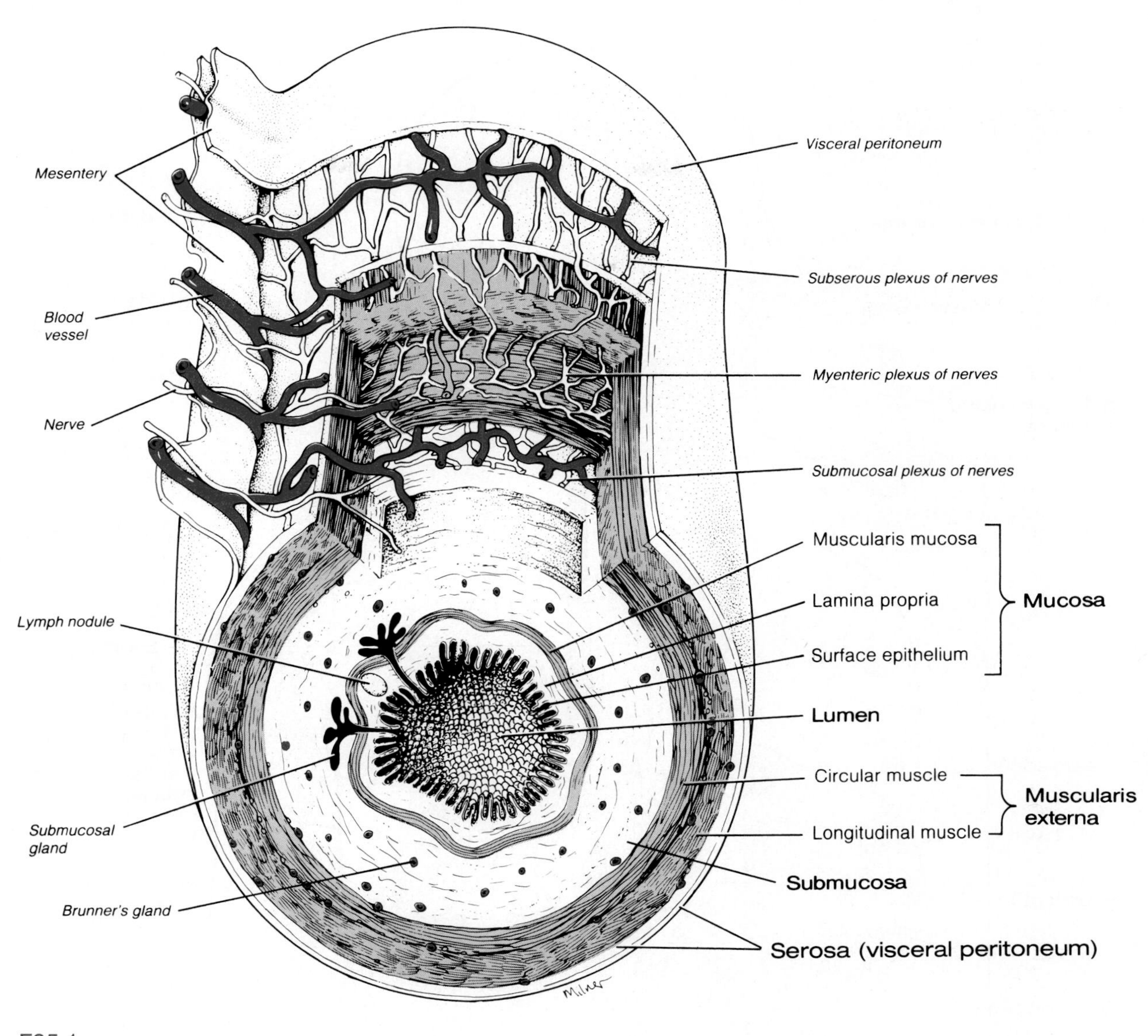

F35.1

Basic structure of the alimentary canal wall.

Mucosa (mucous membrane): The mucosa is the wet epithelial membrane abutting the alimentary canal lumen. It consists of a surface *epithelium* (in most cases, a simple columnar), a *lamina propria* (areolar connective tissue on which the epithelial layer rests), and a *muscularis mucosae* (a scant layer of smooth muscle fibers that provide for local movements of the mucosa). The major functions of the mucosa are secretion (of enzymes, mucus, hormones, etc.), absorption of digested foodstuffs, and protection (against bacterial invasion). A particular mucosal region may be involved in one or all three functions.

Submucosa: Superficial to the mucosa, the submucosa is moderately dense connective tissue containing blood and lymphatic vessels, scattered lymph nodules, and nerve fibers. Its intrinsic nerve supply is called the *submucosal plexus.* Its major functions are nutrition and protection.

Muscularis externa: The muscularis externa, also simply called the *muscularis,* typically is a bilayer of smooth muscle, with the deeper layer running circularly and the more superficial layer running longitudinally. Another intrinsic nerve plexus, the *myenteric plexus,* is associated with this tunic and is a major regulator of GI motility effected by the smooth muscle muscularis.

Serosa: The outermost serosa is equal to the *visceral peritoneum.* It consists of mesothelium associated with a thin layer of areolar connective tissue. The *subserous plexus* is associated with this tunic. In areas *outside* the abdominopelvic cavity, the serosa is replaced by an **adventitia,** a layer of coarse fibrous connective tissue that binds the organ to sur-

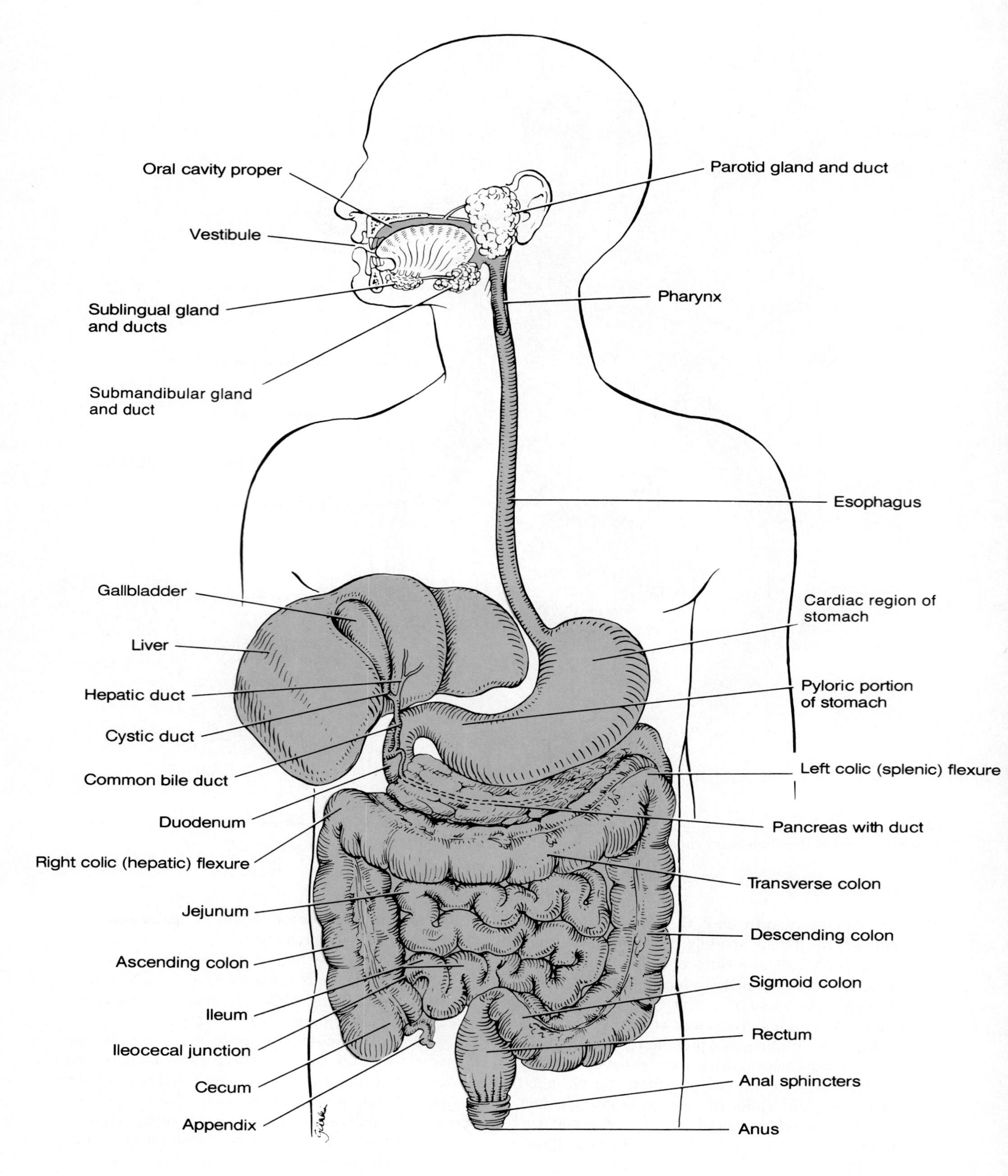

F35.2

The human digestive system: alimentary canal and accessory organs. (Liver and gallbladder are reflected superiorly and to the right.)

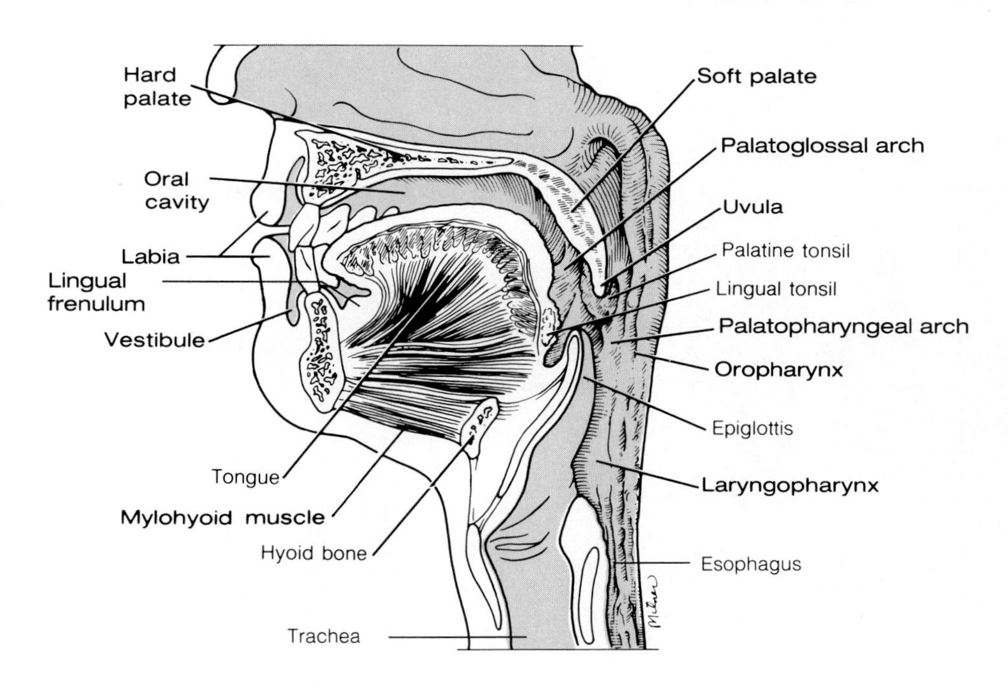

F35.3

Sagittal view of the head showing oral, nasal, and pharyngeal cavities.

rounding tissues. (Such is the case with the esophagus.) The major function of the serosa is to reduce friction as the mobile GI tract organs work and slide across one another and the cavity walls. The adventitia anchors and protects the surrounded GI tract organ.

ORGANS OF THE ALIMENTARY CANAL

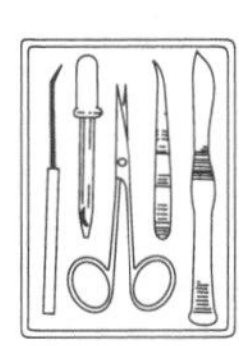

The sequential pathway and fate of food as it passes through the alimentary canal organs are described in the next sections. Identify each structure in Figure 35.2 and on the torso model as you work.

Oral Cavity, or Mouth

Food enters the digestive tract through the **oral cavity,** or **mouth.** Within this mucous membrane–lined cavity are the gums, teeth, tongue, and openings of the ducts of the salivary glands. The **lips (labia)** protect the opening of the chamber anteriorly; the **cheeks** form its lateral walls; and the **palate,** its roof. The anterior portion of the palate is referred to as the **hard palate,** because bone (the palatine processes of the maxillae and the palatine bones) underlies it. The posterior **soft palate** is a fibromuscular structure that is unsupported by bone. The **uvula,** a fingerlike projection of the soft palate, extends downward from its posterior margin. The soft palate rises to close off the oral cavity from the nasal and pharyngeal passages during swallowing. The floor of the oral cavity is occupied primarily by the muscular **tongue,** which is largely supported by the **mylohyoid muscle** and is attached to the hyoid bone, mandible, styloid processes, and pharynx. A membrane called the **lingual frenulum** secures the inferior midline of the tongue to the floor of the mouth. The space between the lips and cheeks and the teeth is the **vestibule;** the area containing the teeth, which is posterior to the alveolar arches, is the **oral cavity** proper. (Figures 35.3 and 35.4 depict the structures of the oral cavity.)

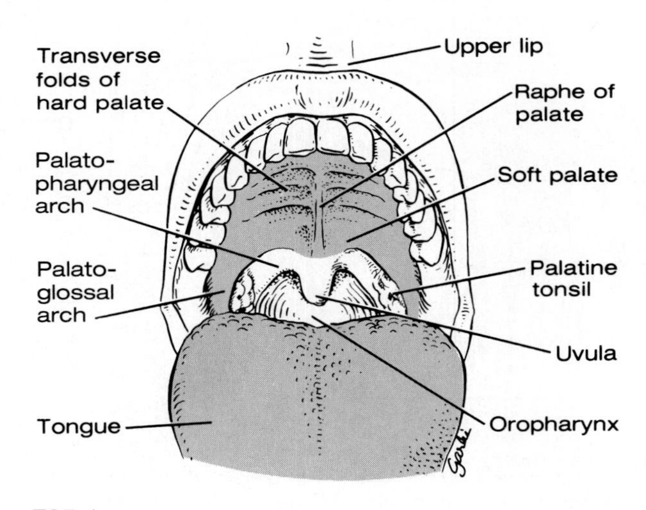

F35.4

Anterior view of the oral cavity.

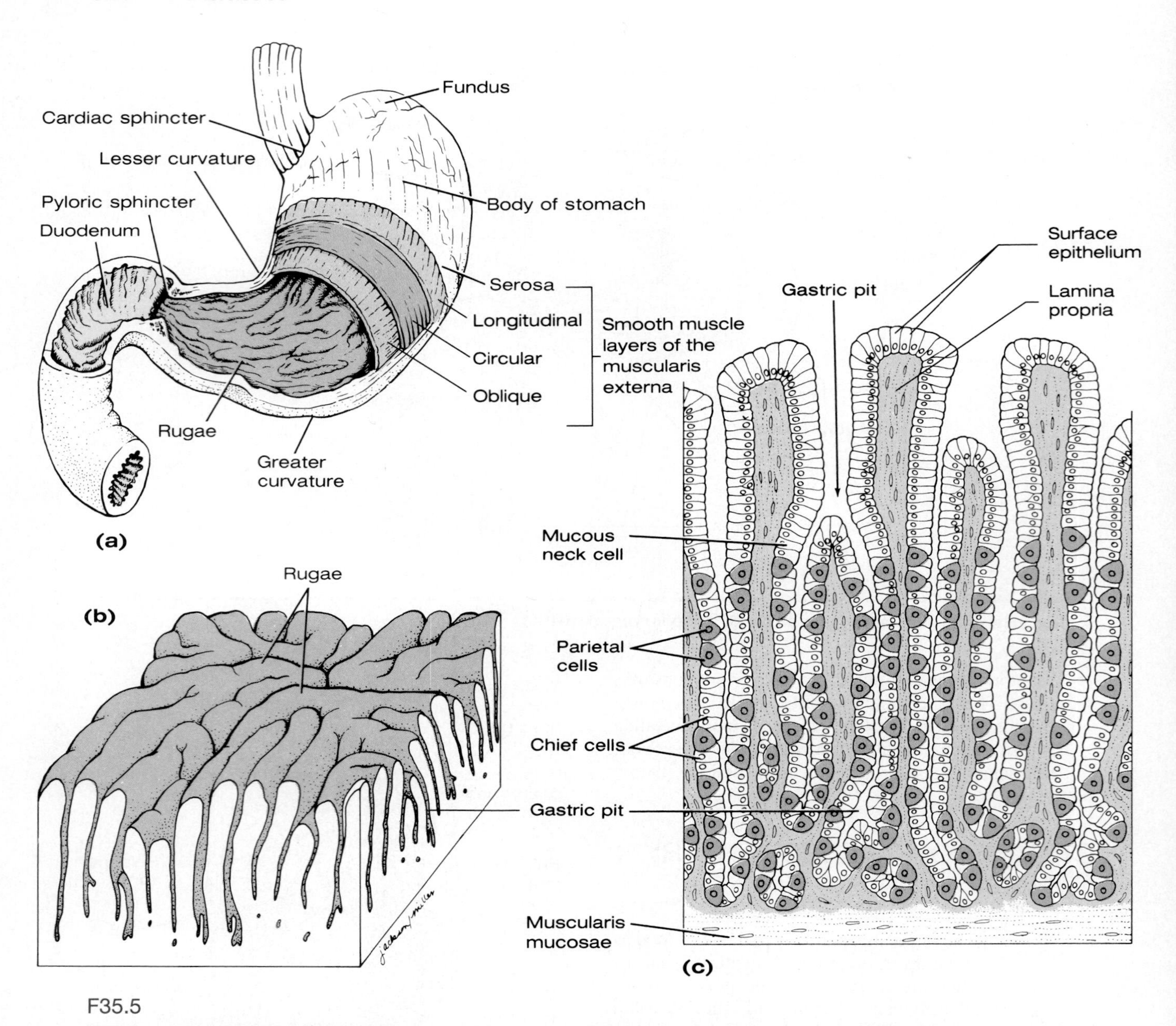

F35.5

Anatomy of the stomach. (a) Gross internal and external anatomy; (b) section of the stomach wall showing rugae and gastric pits; (c) enlarged view of the stomach mucosa (surface epithelium, lamina propria, and muscularis mucosa) showing the gastric pits (longitudinal section).

On each side of the mouth, at its posterior end, are masses of lymphatic tissue, the **palatine tonsils.** Each lies in a concave area bounded anteriorly and posteriorly by membranes, the **palatoglossal arch** (anterior membrane) and the **palatopharyngeal arch** (posterior membrane). The tonsils, in common with other lymphoid tissues, are part of the body's defense system. Another mass of lymphoid tissue, the **lingual tonsil,** covers the base of the tongue, posterior to the oral cavity proper.

Very often in young children, the palatine tonsils become inflamed and enlarge, partially blocking the entrance to the pharynx posteriorly, and making swallowing difficult and painful. This condition is called *tonsillitis.* ■

Three pairs of salivary glands duct their secretion, saliva, into the oral cavity. One component of saliva, salivary amylase, begins the digestion of starchy foods within the oral cavity. (The salivary glands are discussed in more detail on page 322.)

As food enters the mouth, it is mixed with saliva and masticated (chewed). The cheeks and lips help hold the food between the teeth during mastication, and the highly mobile tongue manipulates the food for chewing and initiates swallowing. Thus the mechanical and chemical breakdown of food begins before the food has left the oral cavity. As noted in Exercise 23, the surface of the tongue is covered with papillae, many of which contain taste buds, the receptors for taste sensation. Thus, in addition to its manipulative function, the tongue provides for the enjoyment and appreciation of the food ingested.

Pharynx

When the tongue initiates swallowing, the food passes posteriorly into the *pharynx,* a common passageway for food, fluid, and air. The pharynx is subdivided anatomically into the **nasopharynx** (behind the nasal cavity); the **oropharynx** (behind the oral cavity, extending from the soft palate to the epiglottis); and the **laryngopharynx** (extending from the epiglottis to the base of the larynx), which is continuous with the esophagus.

The walls of the pharynx consist largely of two layers of skeletal muscles, which initiate wavelike contractions that propel the food inferiorly into the esophagus. Its mucosa, like that of the oral cavity, contains a friction-resistant stratified squamous epithelium.

Esophagus

The **esophagus,** or gullet, extends from the pharynx, through the diaphragm, to the cardiac sphincter in the superior aspect of the stomach. It is approximately 25 cm (10 inches) long in humans and is essentially a food passageway that conducts food to the stomach with a wavelike peristaltic motion. The esophagus has no digestive or absorptive function. The walls at the superior end of the esophagus contain skeletal muscle, which is replaced by smooth muscle in the area nearing the stomach. Since the esophagus is located in the thoracic rather than the abdominal cavity, its outermost layer is an *adventitia,* not a serosa.

Stomach

The **stomach** (Figure 35.5) is on the left side of the abdominal cavity, and is hidden by the liver and diaphragm. The regions of the saclike stomach are the **cardiac region** (the area surrounding the **cardiac sphincter,** through which food enters the stomach from the esophagus), the **fundus** (the expanded portion of the stomach, superolateral to the cardiac region), the **body** (midportion of the stomach, inferior to the fundus), and the **pylorus** (the terminal part of the stomach, which is continuous with the small intestine through the **pyloric sphincter**).

The concave medial surface of the stomach is called the **lesser curvature,** and its convex lateral surface is called the **greater curvature.** Extending from these curvatures are two mesenteries, called *omenta.* The **lesser omentum** extends from the liver to attach to the lesser curvature of the stomach. The **greater omentum,** a saclike mesentery, extends from the greater curvature of the stomach, reflects downward over the abdominal contents to cover them in an apronlike fashion, and then blends with the **mesocolon** attaching the transverse colon to the posterior body wall. Figure 35.6 illustrates the omenta, as well as the other peritoneal attachments of the abdominal organs.

The stomach is a temporary storage region for food, as well as a site for the mechanical and chemical breakdown of food. The stomach contains a third, obliquely oriented layer of smooth muscle in its muscularis externa that allows it to churn, mix, and pummel the food, physically reducing it to smaller fragments. Gastric glands of its mucosa secrete hydrochloric acid (HCl) and hydrolytic enzymes (primarily pepsinogen, a protein-digesting enzyme), which begin the enzymatic, or chemical, breakdown of protein foods. The mucosal glands also secrete a viscous mucus that prevents the stomach itself from being eroded by HCl and the proteolytic enzymes. Most of its digestive activity occurs in the pyloric region of the stomach. After the food has been processed in the stomach, it resembles a creamy mass (chyme), which enters the small intestine through the pyloric sphincter.

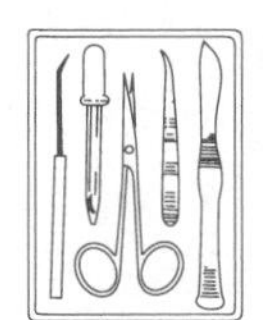

Obtain a microscope and the following slides, in preparation for the histological study you will be conducting now and later in the lab: salivary glands (submandibular or sublingual); pancreas; cross sections of the duodenum, ileum, and stomach; and longitudinal sections of a tooth and the gastroesophageal junction.

1. **Stomach:** The stomach slide will be viewed first. Refer to Figure 35.7a as you scan the tissue under low power to locate the muscularis externa; then move to high power to more closely examine this layer. Try to pick out the three smooth muscle layers. How does the extra (oblique) layer of smooth muscle found in the

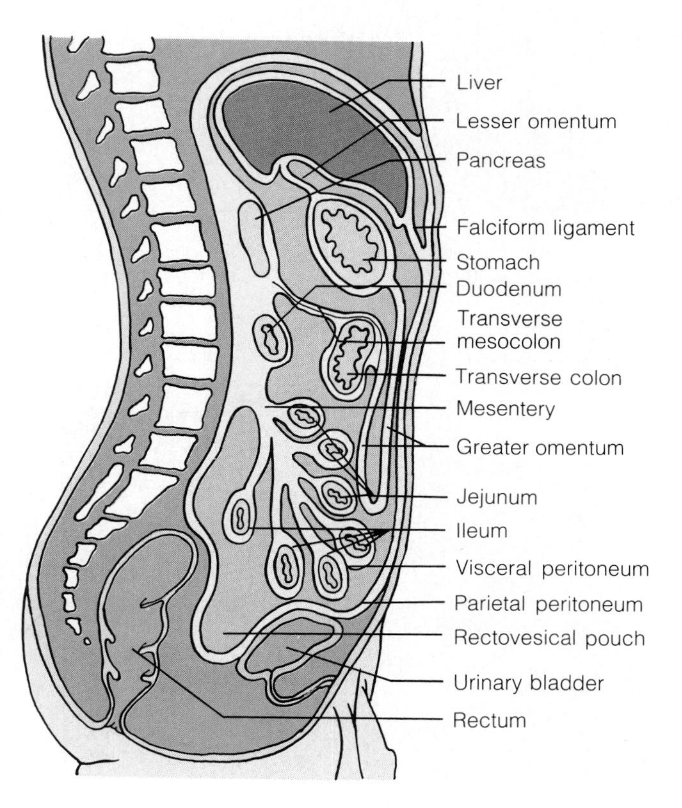

F35.6

Peritoneal attachments of the abdominal organs; sagittal view of a male torso.

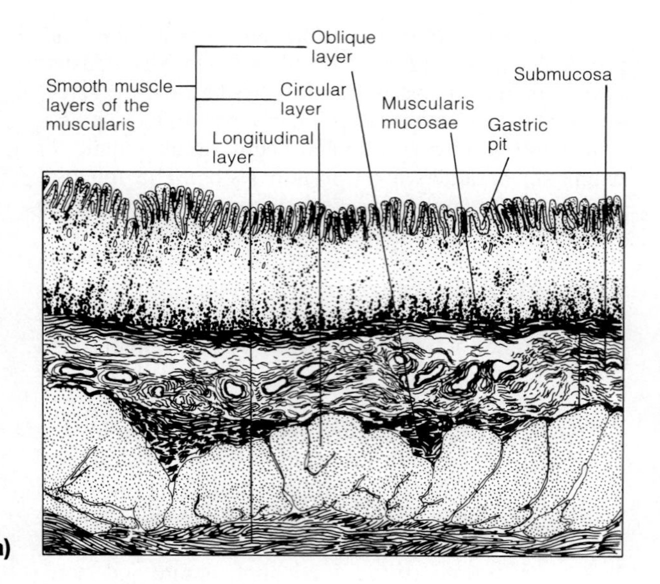

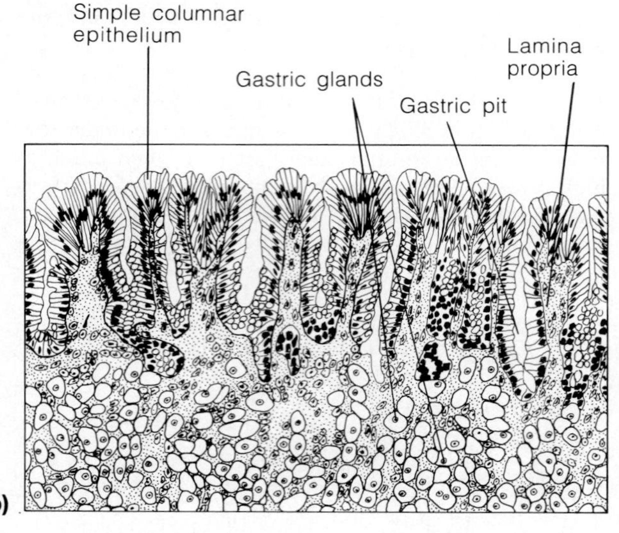

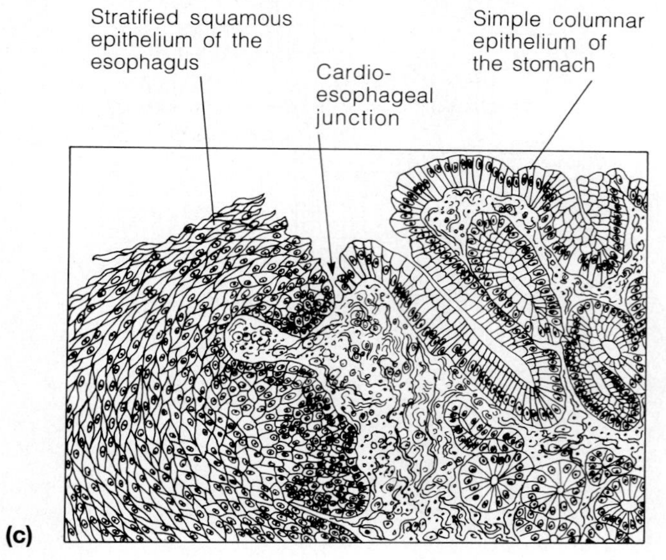

F35.7

Histology of selected regions of the stomach and gastroesophageal junction. (a) Low-power view of stomach wall; (b) high-power view of gastric pits and glands; (c) gastroesophageal junction, longitudinal section. (See corresponding plates 33, 34, and 35 in the histology atlas.)

stomach correlate with the stomach's churning movements?

____

____

____

Identify the gastric glands and the gastric pits (see Figure 35.5 and 35.7b). If the section is taken from the stomach fundus and is appropriately stained, you can identify, in the gastric glands, the blue-staining **chief,** or **zymogenic, cells,** which produce pepsinogen, and the red-staining **parietal cells,** which secrete HCl. Draw a small section of the stomach wall and label it appropriately.

2. **Gastroesophageal junction:** Examine the slide under low power; and scan it to locate the junction between the end of the esophagus and the beginning of the stomach, the gastroesophageal junction. Compare your observations to Figure 35.7c. What is the functional importance of the epithelial differences seen in the two organs?

____

____

____

Small Intestine

The **small intestine** is a convoluted tube, some 6 to 7 meters (m) long, extending from the pyloric sphincter to the ileocecal valve. The small intestine is suspended from the posterior abdominal wall by a double layer of peritoneum, the fan-shaped **mesentery** (see Figure 35.6); and it lies, framed laterally and superiorly by the large intestine, in the abdominal cavity. The small intestine has three subdivisions: the **duodenum,** extending from the pyloric sphincter for about 25 cm (10 inches) and curving around the head of the pancreas; the **jejunum,** continuous with the duodenum and extending for 2.5 to 3 m (about 8 feet), and the **ileum,** the terminal portion of the small intestine. The ileum is 4 m (about 12 feet) long and joins the large intestine at the **ileocecal valve.**

Brush border enzymes, hydrolytic enzymes bound to the microvilli of the columnar epithelial cells and, more importantly, enzymes produced by the pancreas and ducted into the duodenum via the **pancreatic duct** complete the enzymatic digestion process in the small intestine. Bile (formed in the liver) also enters the duodenum in the same area via the **common bile duct** (see Figure 35.2). At the duodenum, the ducts join to form the bulblike **hepatopancreatic ampulla** and empty their products into the duodenal lumen through the **duodenal papilla,** an orifice controlled by a muscular valve called the **hepatopancreatic sphincter (of Oddi).**

Nearly all nutrient absorption occurs in the small intestine, where three structural modifications that increase the mucosal absorptive area appear—the **microvilli, villi,** and **plicae circulares.** Microvilli are minute projections of the surface plasma membrane of the columnar epithelial lining cells of the mucosa. The villi are the fingerlike projections of the mucosa tunic that give it a velvety appearance and texture (Figure 35.8). The plicae circulares are deep folds of the mucosa and submucosa layers that extend partially or totally around the intestine. These structural modifications, which increase the surface area, decrease in frequency and elaboration toward the end of the small intestine. Any residue remaining undigested and unabsorbed at the terminus of the small intestine enters the large intestine through the ileocecal valve. In contrast, the amount of lymphoid tissue in the submucosa of the small intestine (**Peyer's patches**) increases along the length of the small intes-

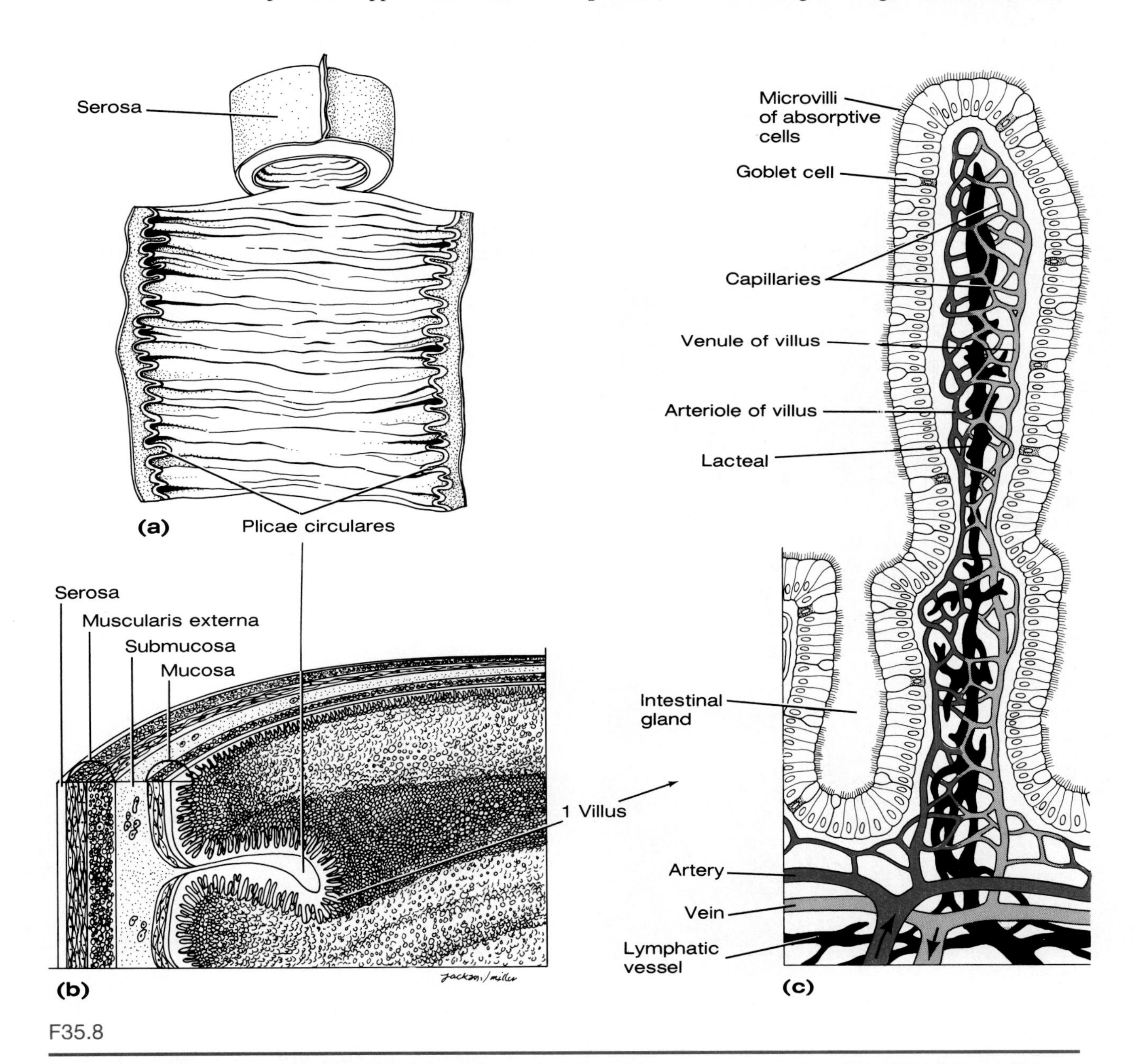

F35.8

Structural modifications of the small intestine. (a) Plicae circulares (circular folds) seen on the inner surface of the small intestine; (b) enlargement of one plica circulare to show villi; (c) detailed anatomy of a villus.

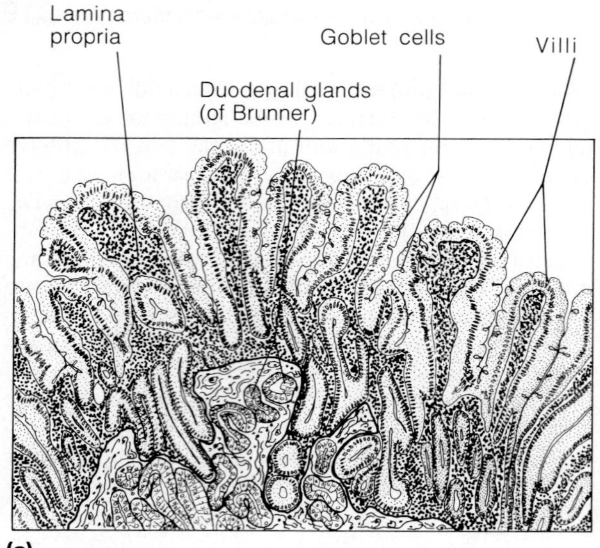

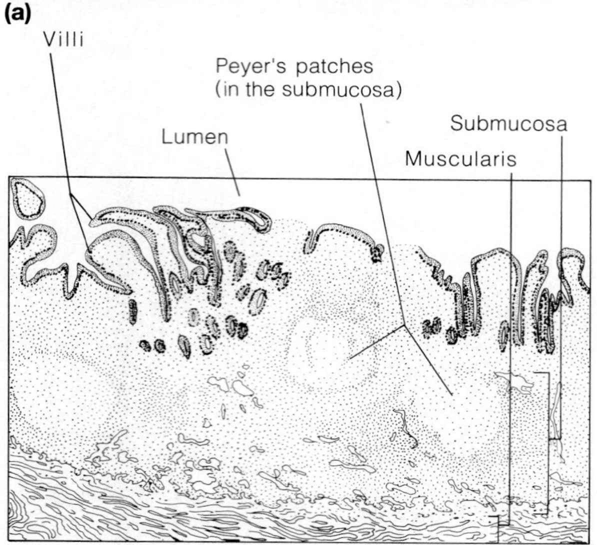

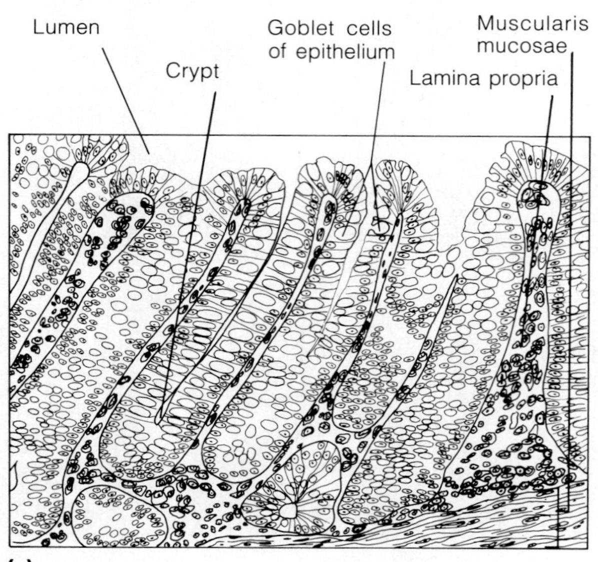

F35.9

Histology of selected regions of the small and large intestines. (a) Duodenum of the small intestine (XS); (b) ileum of the small intestine (XS); (c) large intestine (XS). (See corresponding plates 36, 37, and 38 in the histology atlas.)

tine and is very apparent in the ileum. This reflects the fact that the undigested food residue contains large numbers of bacteria that must be prevented from entering the bloodstream.

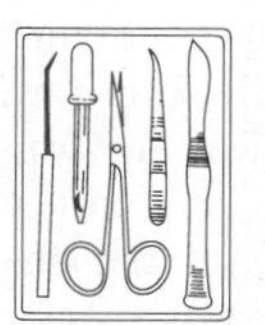

1. **Duodenum:** Secure the slide of the duodenum (cross section) to the microscope stage. Observe the tissue under low power to identify the four basic tunics of the intestinal wall—that is, the three layers of the **mucosal** lining, the **submucosa** (areolar connective tissue layer deep to the mucosa), the **muscularis externa** (composed of circular and longitudinal smooth muscle layers), and the **serosa** (the outermost layer, also called the *visceral peritoneum*). (See Figure 35.9a). Identify the scattered mucous-producing **duodenal glands** (Brunner's glands) in the submucosa.

What type of epithelium do you see here? __________

____

Examine the large leaflike *villi,* which increase the surface area for absorption. Note also the *intestinal crypts* (crypts of Lieberkühn), invaginated areas of the mucosa between the villi where the cells are involved in the production of intestinal juice, a watery mucus-containing mixture that serves as a carrier fluid for absorption of nutrients from the chyme. Sketch and label a small section of the duodenal wall, showing all layers and villi.

2. **Ileum:** The structure of the ileum is similar to that of the duodenum, except that the villi tend to be less elaborate (most of the absorption has occurred by the time the ileum is reached). Obtain and secure a slide of the ileum for viewing. Observe the villi, and identify the four layers of the wall and the large generally spherical Peyer's patches (Figure 35.9b). What tissue comprises Peyer's patches?

____

3. If a *villus model* is available, identify its following cells or regions before continuing: absorptive epithelium, goblet cells, lamina propria, slips of the muscularis mucosae, capillary bed, and lacteal. If possible, also identify the intestinal crypts which lie between the villi.

Large Intestine

The **large intestine** (see Figure 35.2) is slightly more than 1.5 m (about 5 feet) long and extends from the ileocecal valve to the anus. It encircles the small intestine on three sides and consists of the following subdivisions: the **cecum, appendix, colon, rectum,** and **anal canal.**

The blind, tubelike appendix, which hangs from the cecum, is a trouble spot in the large intestine. Because it is generally twisted, it provides an ideal location for bacteria to accumulate and multiply. Inflammation of the appendix, or *appendicitis,* is the result. ■

The colon is divided into several distinct regions (Figure 35.10). The **ascending colon** travels up the right side of the abdominal cavity and makes a right-angle turn at the **right colic,** or **hepatic flexure,** crossing the abdominal cavity as the **transverse colon.** It then turns at the **left colic,** or **splenic flexure,** and continues, as the **descending colon,** down the left side of the abdominal cavity, where it takes an S-shaped course as the **sigmoid colon.** The sigmoid colon, rectum, and anal canal lie in the pelvis, anterior to the sacrum, and thus are not considered abdominal cavity structures. Except for the transverse and sigmoid colons, which are secured to the dorsal body wall by mesocolons (see Figure 35.6), the colon is retroperitoneal.

The anal canal terminates in the **anus,** the opening to the exterior of the body. The anus, which has an external sphincter of skeletal muscle (the voluntary sphincter) and an internal sphincter of smooth muscle (the involuntary sphincter), is normally closed except during defecation, when the undigested remains of food and bacteria are eliminated from the body as feces.

In the large intestine, the longitudinal muscle layer of the muscularis externa is reduced to three longitudinal muscle bands called the **teniae coli.** Because these bands are shorter than the rest of the wall of the large intestine, they cause the wall to pucker into small pocketlike sacs called **haustra.**

The major function of the large intestine is to consolidate and propel the unusable fecal matter toward the anus and eliminate it from the body. While it does that "chore," it (1) provides a site for the manufacture, by

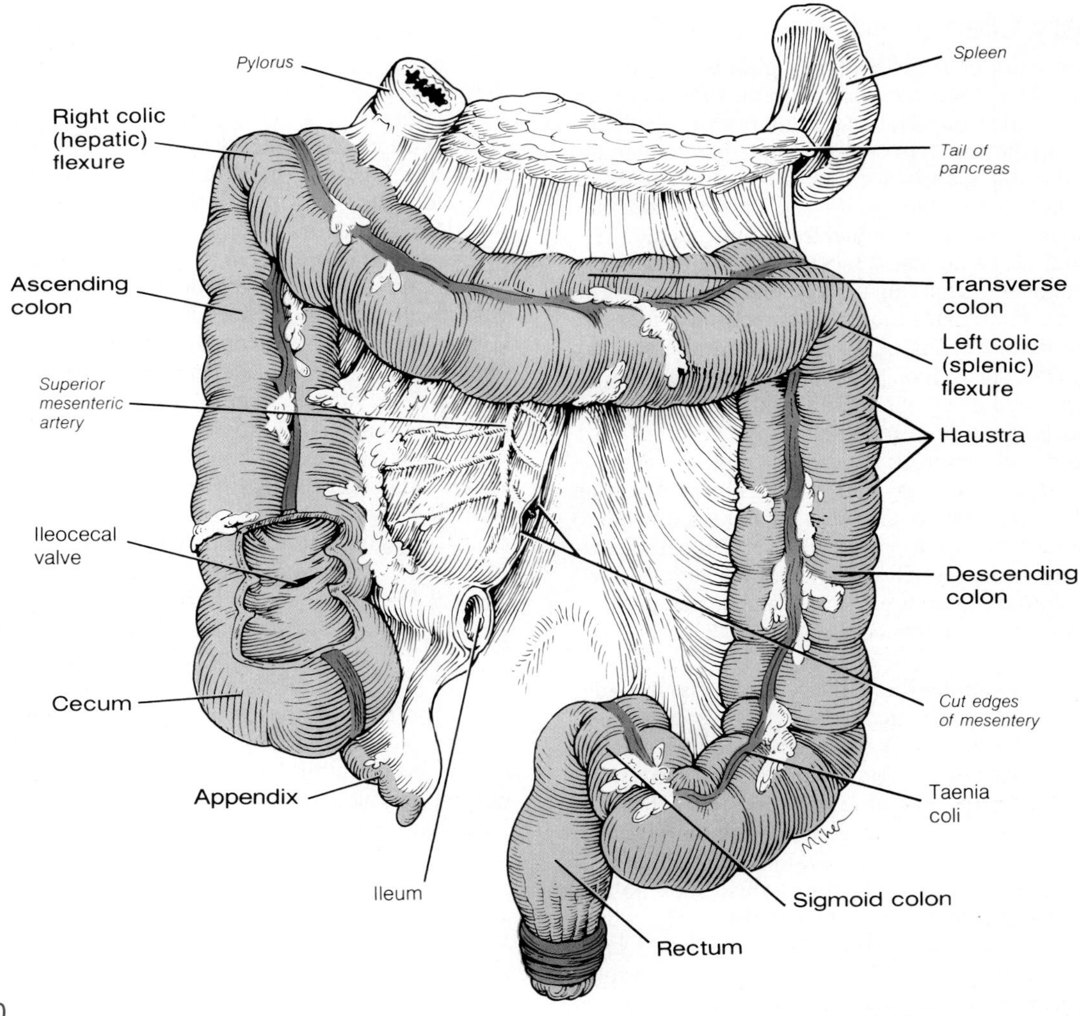

F35.10

The large intestine. (Section of the cecum removed to show the ileocecal valve.)

intestinal bacteria, of some vitamins (B vitamins and vitamin K), which it then absorbs into the bloodstream; and (2) reclaims most of the remaining water (and some of the electrolytes) from undigested food, thus conserving body water.

Watery stools, or diarrhea, result from any condition that rushes undigested food residue through the large intestine before it has had sufficient time to absorb the water (as in irritation of the colon by bacteria). Conversely, when food residue remains in the large intestine for extended periods (as with an atonic colon or failure of the defecation reflex), excessive water is absorbed and the stool becomes hard and difficult to pass (constipation). ■

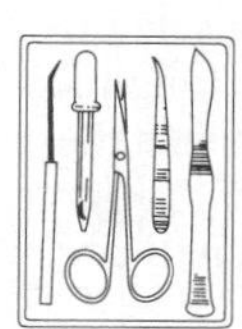

Examine Figure 35.7c to compare the histology of the large intestine to that of the small intestine just studied.

ACCESSORY DIGESTIVE ORGANS

Salivary Glands

The three pairs of major **salivary glands** (see Figure 35.2) that empty their secretions into the oral cavity are the large **parotid glands,** located anterior to the ear and ducting into the mouth over the second upper molar; the **submandibular glands,** located inside the mandibular arch in the floor of the mouth and ducting under the tongue to the base of the lingual frenulum; and the small **sublingual glands,** located most anteriorly in the floor of the mouth and emptying under the tongue via several small ducts. Food in the mouth and mechanical pressure (chewing rubber bands or wax) stimulate the salivary glands to secrete saliva. Saliva consists primarily of mucin (a viscous glycoprotein), which moistens the food and helps bind it together into a mass called a **bolus,** and a clear serous fluid containing the enzyme salivary amylase. Salivary amylase begins the digestion of starch (a large polysaccharide), breaking it down into disaccharides, or double sugars. The secretion of the parotid glands is mainly serous, whereas the submandibular and sublingual glands are mixed glands that produce mucin and serous components.

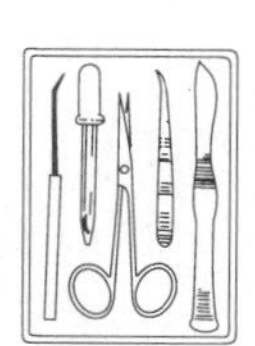

Examine the salivary gland tissue under low power and then high power to become familiar with the appearance of a glandular tissue. Note the clustered arrangement of the cells around their ducts. The cells are basically triangular, with their pointed ends facing the duct orifice. If possible, differentiate between the mucous-producing cells, which look hollow or have a clear cytoplasm, and the serous cells, which produce the clear, enzyme-containing fluid and have granules in their cytoplasm. The serous cells often form "caps" (*demilunes*) around the more central mucous cells. Figure 35.11 may be helpful in this task.

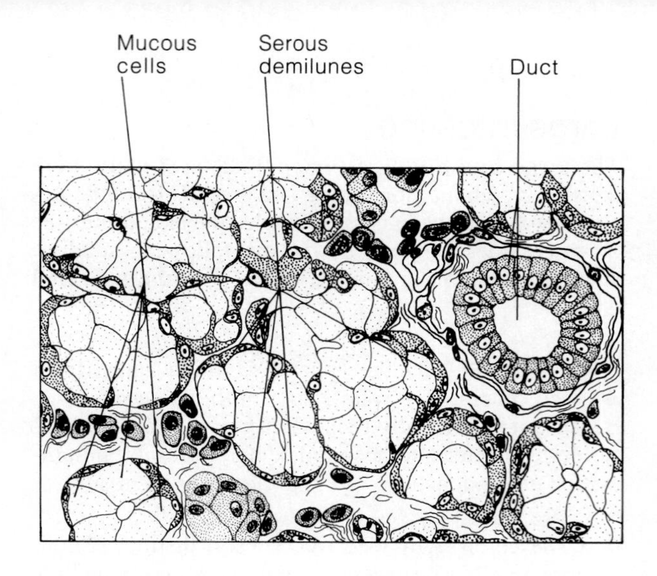

F35.11

A mixed salivary gland. Corresponding micrograph is plate 39 of the histology atlas.

Pancreas

The **pancreas** is a soft, triangular gland that extends horizontally across the posterior abdominal wall from the spleen to the duodenum (see Figure 35.2). Like the duodenum, it is a retroperitoneal organ (see Figure 35.6). As noted in Exercise 24, the pancreas has both an endocrine function (it produces the hormones insulin and glucagon) and an exocrine (enzyme-producing) function. It produces a whole spectrum of hydrolytic enzymes, which it secretes in an alkaline fluid into the duodenum through the pancreatic duct. Pancreatic juice is very alkaline. Its high concentration of bicarbonate ion (HCO_3^-) neutralizes the acidic chyme entering the duodenum from the stomach, enabling the pancreatic and intestinal enzymes to operate at their optimal pH. (Optimal pH for digestive activity to occur in the stomach is very acidic and results from the presence of HCl; that for the small intestine is slightly alkaline.)

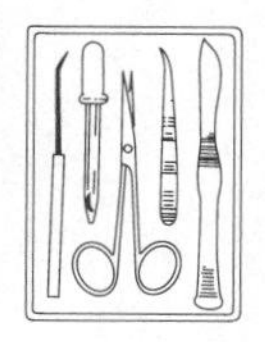

Observe the pancreas tissue under low power and then high power to distinguish between the lighter-staining, endocrine-producing islets of Langerhans and the deeper-staining acinar cells, which produce the hydrolytic enzymes and form the major portion of the pancreatic tissue (see Figure 35.12). Note the arrangement of the exocrine cells around their central ducts. Draw a small portion of the enzyme-producing pancreatic tissue and appropriately label your drawing (i.e., acinar cells and ducts).

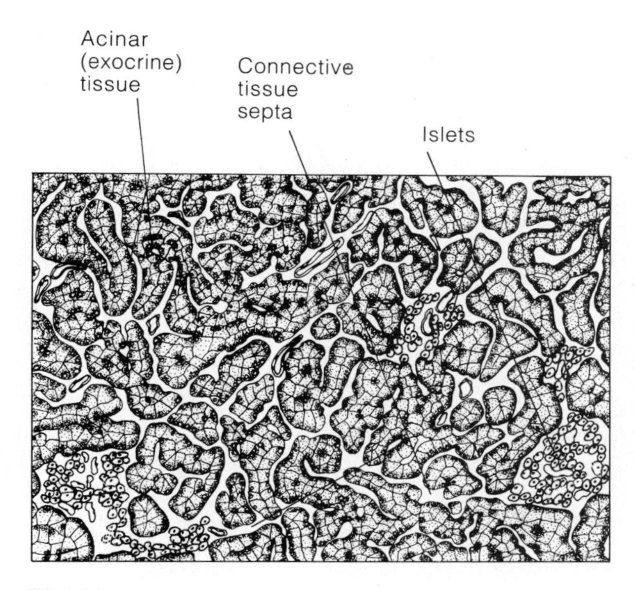

F35.12

Histology of the pancreas. The pancreatic islet cells produce insulin and glucagon (hormones). The acinar cells synthesize digestive enzymes for "export" to the duodenum. See also plate 40 in the histology atlas.

Liver and Gallbladder

The **liver** (see Figure 35.2), the largest gland in the body, is located inferior to the diaphragm, more to the right side of the body. As noted earlier, it hides the stomach from view in a superficial observation of abdominal contents. The human liver has four lobes, and is suspended from the diaphragm and anterior abdominal wall by the **falciform ligament** (see Figure 35.6).

The liver is one of the body's most important organs, and it performs many metabolic roles. However, its digestive function is to produce *bile,* which leaves the liver through the **hepatic duct** and then enters the duodenum through the **common bile duct.** Bile has no enzymatic action, but emulsifies fats (spreads thin or breaks up large particles into smaller ones), thus creating a larger surface area for more efficient lipase activity. Without bile, little fat digestion or absorption occurs.

When digestive activity is not occurring in the digestive tract, bile backs up the cystic duct and enters the **gallbladder,** a small, green sac on the inferior surface of the liver. It is stored there until needed for the digestive process. While in the gallbladder, bile is concentrated by the removal of water and some ions. When fat-rich food enters the duodenum, a hormonal stimulus causes the gallbladder to contract, releasing the stored bile and making it available to the duodenum.

If the hepatic or common bile duct is blocked (for example, by wedged gallstones), bile is prevented from entering the small intestine, begins to accumulate, and eventually backs up into the liver. This exerts pressure on the liver cells, and bile begins to enter the bloodstream. As the bile circulates through the body, the tissues become yellow, or *jaundiced.*

Blockage of the ducts is just one cause of jaundice; more often it results from actual liver problems such as *hepatitis* (an inflammation of the liver) or *cirrhosis,* a condition in which the liver is severely damaged, becoming hard and fibrous. Cirrhosis is almost guaranteed in those who drink excessive alcohol for many years. ■

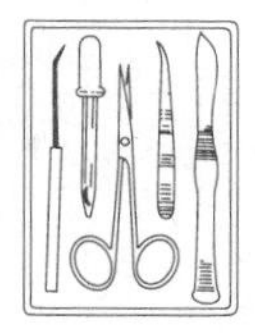

If a three-dimensional model of the liver is available, examine it and identify the four liver lobes, falciform ligament, hepatic duct, and adjoining gallbladder. Also identify the region called the *porta hepatis* (literally the "gateway to the liver"), a deep fissure that contains the hepatic portal vein, hepatic artery, and hepatic duct.

Teeth

By the age of 21, two sets of teeth have developed (Figure 35.13). The initial set—called **deciduous,** or **milk, teeth**—normally appear between the ages of 6 months and $2\frac{1}{2}$ years. The first of these to erupt are the lower central incisors, an event which is usually applauded by the child's parents. The child begins to shed the deciduous teeth around the age of 6. The second set of teeth, the **permanent teeth,** gradually replace them. As the deeper, permanent teeth progressively enlarge and develop, the roots of the deciduous teeth are resorbed, leading to their final shedding; and during the sixth to twelfth years, the child has mixed dentition—both permanent and deciduous teeth. Generally by the age of 12, all of the deciduous teeth have been shed, or exfoliated.

The teeth are classified as **incisors, canines** (eye teeth), **premolars** (bicuspids), and **molars;** and dentition is described by means of a **dental formula,** which designates the numbers, types, and position of the teeth in one side of the jaw. (Because tooth arrangement is bilaterally symmetrical, it is only necessary to designate one side of the jaw.) The dental formula for the deciduous teeth, proceeding posteriorly from the medial aspect of the jaw, is as follows:

$$\frac{\text{upper jaw: 2 incisors, 1 canine, 0 premolars, 2 molars}}{\text{lower jaw: 2 incisors, 1 canine, 0 premolars, 2 molars}} \times 2$$

This formula is generally abbreviated to read as follows:

$$\frac{2, 1, 0, 2}{2, 1, 0, 2} \times 2 = 20 \text{ (number of deciduous teeth)}$$

The 32 permanent teeth are then described by the following dental formula:

$$\frac{2, 1, 2, 3}{2, 1, 2, 3} \times 2 = 32 \text{ (number of permanent teeth)}$$

Although 32 is designated as the normal number of permanent teeth, not everyone develops a full complement. In many people, the No. 3 molars, commonly called the *wisdom teeth,* never erupt.

Teeth names reflect differences in relative structure and function. The incisors are chisel shaped and exert a shearing action used in biting. The canines are cone shaped or fanglike, the latter description being much more applicable to the canines of animals whose teeth

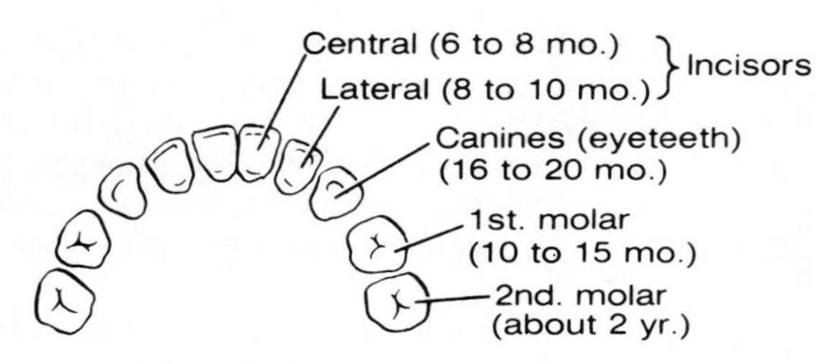

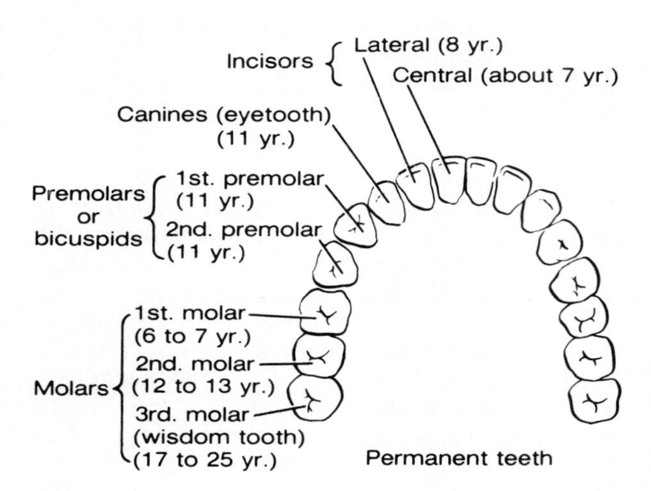

F35.13

Human deciduous and permanent teeth. (Approximate time of teeth eruption shown in parentheses.)

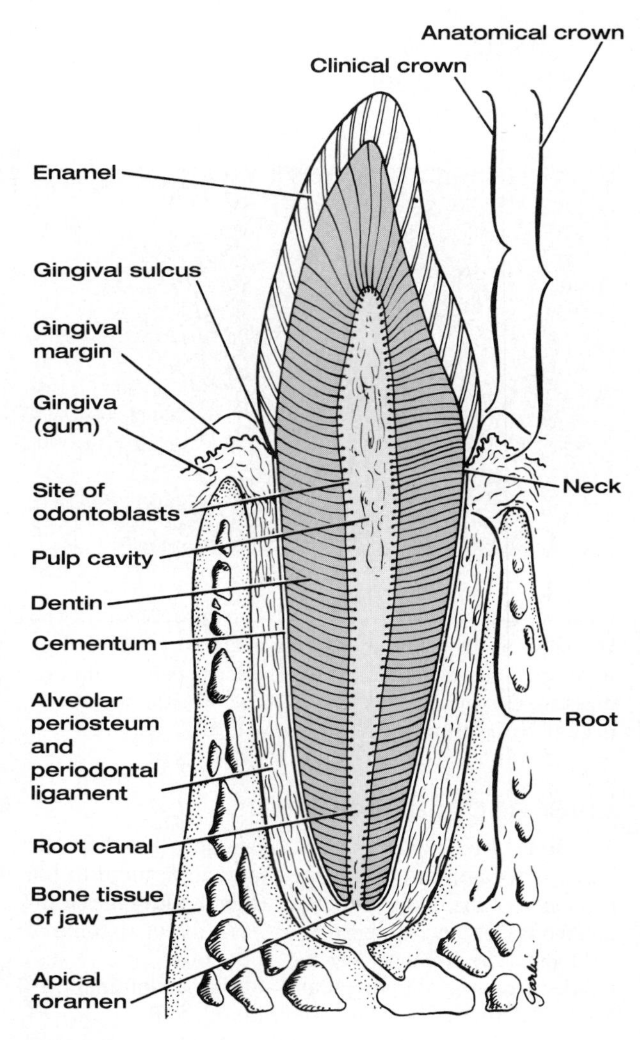

F35.14

Longitudinal section of human canine tooth.

are used for the tearing of food. Incisors, canines, and premolars typically have single roots, though the first upper premolars may have two. The lower molars have two roots but the upper molars usually have three. The premolars have two cusps (grinding surfaces); the molars have relatively flat, broad superior surfaces specialized for the fine grinding of food.

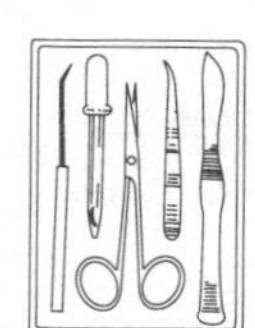

Identify the four types of teeth (incisors, canines, premolars, and molars) on the jaw model or human skull.

A tooth consists of two major regions, the **crown** and the **root.** A longitudinal section made through a tooth shows the following basic anatomical plan (Figure 35.14): The crown is the superior portion of the tooth; the portion of the crown visible above the **gum,** or **gingiva,** is referred to as the **clinical crown.** The entire area covered by **enamel** is called the **anatomical crown.** The crevice between the end of the anatomical crown and the upper margin of the gingiva is referred to as the **gingival sulcus,** and its apical border as the **gingival margin.** Enamel is the hardest substance in the body and is fairly brittle. It consists of 95% to 97% inorganic calcium salts (chiefly $CaPO_4$) and thus is heavily mineralized. That portion of the tooth embedded in the alveolar portion of the jaw is the root; and the root and crown are connected by a slight constriction called the **neck.** The outermost surface of the root is covered by **cementum,** which is similar to bone in composition and less brittle than enamel. The cementum attaches the tooth to the **periodontal ligament,** which holds the tooth in the alveolar socket and exerts a cushioning effect. **Dentin,** bonelike material which comprises the bulk of the tooth, lies deep to the enamel and cementum. The **pulp cavity** occupies the central portion of the tooth. **Pulp,** connective tissue liberally supplied with blood vessels, nerves, and lymphatics, occupies this cavity and provides for tooth sensation and supplies nutrients to the tooth tissues. Specialized cells, **odontoblasts,** line the pulp cavity and produce the dentin. As the pulp cavity extends into the root, it becomes the **root canal.** An opening at the root apex, the **apical foramen,** provides a route of entry into the tooth for the blood vessels, nerves, and other structures from the tissues beneath.

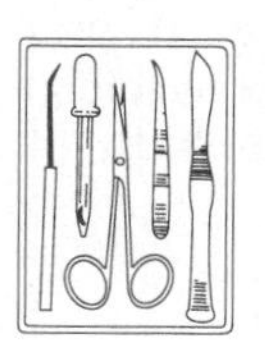

Observe a slide of a longitudinal section of a tooth, and compare your observations with the structures detailed in Figure 35.14. Identify as many of these structures as possible.

EXERCISE 36

Chemical Breakdown of Foodstuffs: Enzymatic Action

OBJECTIVES

1. To summarize the digestive system enzymes involved in the digestion of proteins, fats, and carbohydrates; and to state their site of origin and the environmental conditions promoting their optimal functioning.
2. To name the end products of digestion of proteins, fats, and carbohydrates.
3. To perform the appropriate chemical tests to determine if digestion of a particular foodstuff has occurred.
4. To cite the function(s) of bile in the digestive process.
5. To discuss the role of temperature and pH in the regulation of enzyme activity.
6. To define *enzyme, catalyst, control, substrate,* and *hydrolase.*

MATERIALS

General supply area:

- 10-ml and 50-ml graduated cylinders
- Test tubes and test tube rack
- Glass stirring rods
- 100-ml beakers
- Wide-range pH paper
- Wax markers
- Distilled water
- Water bath set at 37°C
- Chart on chalkboard for recording class results

Supply area 1:

- 0.1% alpha-amylase solution*
- 1% boiled starch solution, freshly prepared†
- 1% maltose solution
- 1 N HCl and 1 N NaOH in dropper bottles
- IKI solution (Lugol's iodine) and Benedict's solutions in dropper bottles
- 500-ml beaker
- Hot plate
- Ice bath

Supply area 2:

- 2% pepsin solution
- Alternative substrates:
  - *Procedure A:* albumin solution. Either a freshly prepared solution consisting of one part egg white to three parts water or a 1% solution of commercially made albumin powder
  - *Procedure B:* hard-boiled egg white
- 0.5 to 1.0% HCl and 10 N NaOH in dropper bottles
- 20% NaOH and 10% $CuSO_4$ in separate stoppered flasks
- Medicine dropper
- Single-edge razor blades

Supply area 3:

- 1.5% pancreatin
- Phenol red pH indicator
- 0.2% NaOH
- 0.1 N HCl
- 2% bile solution or "artificial bile" (10% solution of Tween-20)**
- Vegetable oil in dropper bottle

* The alpha-amylase should be a low-maltose preparation for best results.

† Prepare by adding 1 g starch to 100 ml distilled water; boil and cool; add a pinch of salt (NaCl). Prepare fresh daily.

** This emulsifier, Tween-20 (polyoxyethylene-20-sorbitan monolaurate) available from Fisher-Biotech, allows better visualization of color change than the bile solution.

Because nutrients can only be absorbed when broken down to their monomer forms, food digestion is a prerequisite to food absorption. You have already studied mechanisms of passive and active absorption in Exercise 5. Before proceeding, review that material on pages 32–37.

Enzymes are large protein molecules produced by body cells. They are biological catalysts, which increase the rate of a chemical reaction without themselves becoming part of the product. The digestive enzymes are hydrolytic enzymes, or *hydrolases,* which break down organic food molecules by adding water to the molecular bonds, thus cleaving the bonds between the subunits, or monomers.

The various hydrolytic enzymes are highly specific in their action. Each enzyme hydrolyzes only one or a small group of substrate molecules, and very specific environmental conditions are necessary for it to function optimally. Since digestive enzymes actually function outside the body cells, in the digestive tract, their hydrolytic activity can also be studied in a test tube. Such an *in vitro* study provides a convenient laboratory environment for investigating the effect of such variations on enzymatic activity.

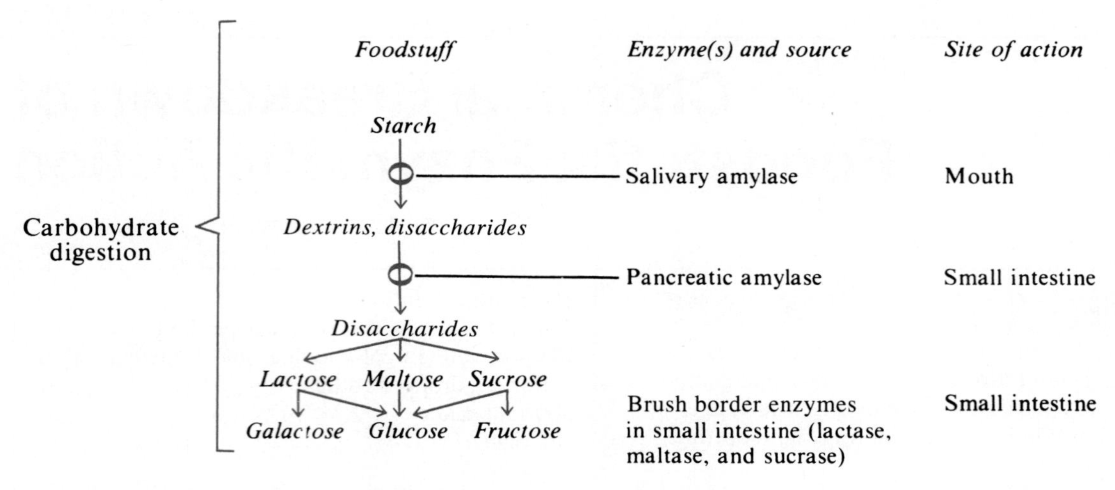

Absorption: The monosaccharides (glucose, galactose, and fructose) are absorbed into the capillary blood in the villi and transported to the liver via the hepatic portal vein

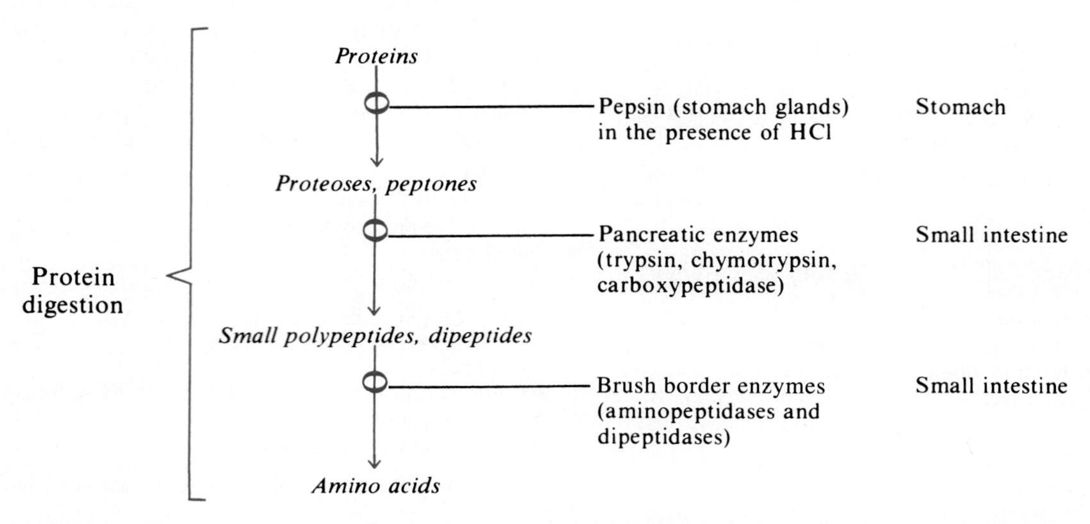

Absorption: The amino acids are absorbed into the capillary blood in the villi and transported to the liver via the hepatic portal vein

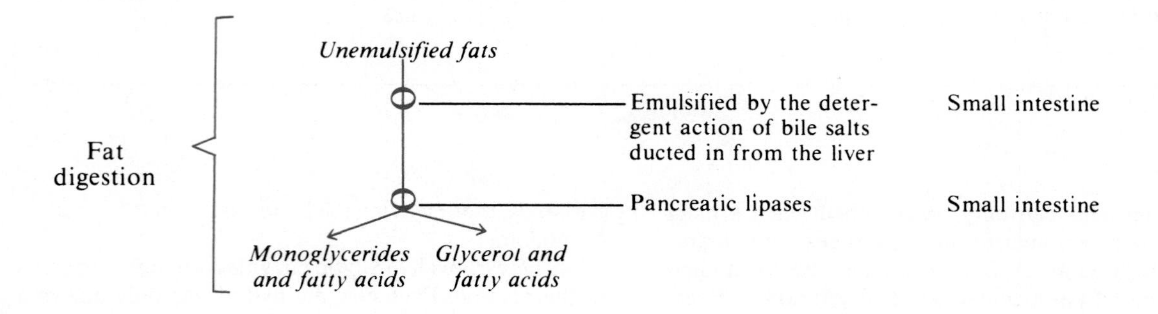

Absorption: Absorbed primarily into the lacteals of the villi and transported to the systemic circulation via the lymph in the thoracic duct. (Glycerol and short-chain fatty acids are absorbed into the capillary blood in the villi and transported to the liver via the hepatic portal vein.)

F36.1

Flow chart of digestion and absorption of foodstuffs.

Figure 36.1 is a flow chart of the progressive digestion of proteins, fats, and carbohydrates, and indicates specific enzymes involved and their site of formation. Acquaint yourself with this flow sheet before beginning this experiment; and refer to it, as necessary, during the laboratory session.

Work in groups of three or four, with each group taking responsibility for setting up and conducting one of the following experiments. Each group should then communicate its results to the rest of the class by recording them in a chart on the chalkboard. Additionally, all members of the class should observe the controls, as well as the positive and negative examples of all experimental results. All members of the class should be able to explain the tests used and the results observed and anticipated for each experiment.

STARCH DIGESTION BY SALIVARY AMYLASE

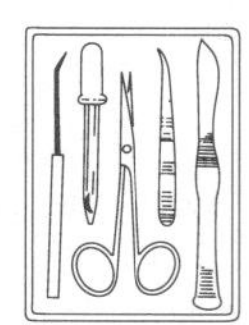

From the general supply area, obtain a test tube rack, 10 test tubes, 10-ml graduated cylinder, wide-range pH paper, a glass stirring rod, and wax marking pencils. From supply area 1, obtain a hot plate and a 500-ml beaker, dropper bottles of 1 N NaOH and HCl, 10 ml of starch solution, 10 ml of amylase solution, and dropper bottles of IKI and Benedict's solutions. Two students should prepare the controls (steps 1 to 4) while the other two prepare the experimental samples (steps 5 and 6).

Since in this experiment you will investigate the hydrolysis of starch to maltose by **salivary amylase** (the enzyme produced by the salivary glands and secreted into the mouth), it is important to be able to identify the presence of these substances to determine to what extent the enzymatic activity has occurred. Thus controls must be prepared to provide a known standard against which comparisons can be made. Starch decreases and sugar increases as digestion occurs according to the following equation:

$$\text{Starch} + \text{water} \xrightarrow{\text{amylase}} X \text{ maltose}$$

1. With the wax marker, mark a test tube No. 1. Place 1 ml of starch solution in the test tube, and add 2–3 drops of IKI solution. The presence of a blue-black color when IKI is added indicates the presence of starch and is referred to as a **positive starch test.** As the starch is progressively hydrolyzed, the color of the solution will change from blue-black to blue-red, and then to faint red, and finally disappears when all the starch has been digested.

2. To obtain a *negative starch test,* place 1 ml of distilled water in test tube 2; and add 2–3 drops of IKI. The unchanged color of the solution indicates the absence of starch. What color did you obtain with water and IKI?

3. To obtain a **positive sugar test,** place 1 ml of maltose solution in test tube 3; and add 5 drops of Benedict's solution. Mix well. Place the test tube in a water bath (a beaker of water on a hot plate), and boil until a color change is noted (but no longer than 5 min). The presence of a bright yellow to deep red precipitate (cuprous oxide) indicates a positive test for maltose, sucrose, or any other reducing sugar. (A change to green is also considered a positive test for sugar but indicates the presence of a smaller amount.)

4. To obtain a *negative sugar test,* place 1 ml of distilled water in test tube 4, add 5 drops of Benedict's solution, and boil for 5 minutes. Note that the solution's color remains unchanged from the blue of Benedict's solution, indicating the absence of sugar.

5. Mark six test tubes with the numbers 5 through 10, and prepare them for incubation as described in Chart 1. You can assume that all of the experimental samples at the time of preparation will give a *negative* test for sugar. Since the enzyme solution in sample 8 must be boiled before incubation begins, the preparation of that sample should be started first. Note and record the time that incubation begins in each case.

6. As the tubes become colorless (negative starch test), discontinue their incubation. Perform *Benedict's test* on each sample. (Add 1 ml of the incubated solution to a clean test tube, and then add 5 drops of Benedict's solution to the tube. Boil until a color change is noted [or for a maximum of 5 minutes].) After 1 hour of incubation, discontinue the incubation of any remaining tubes; and conduct Benedict's test on their contents. Record the results on Chart 1 and on the chalkboard.

PEPSIN DIGESTION OF PROTEIN

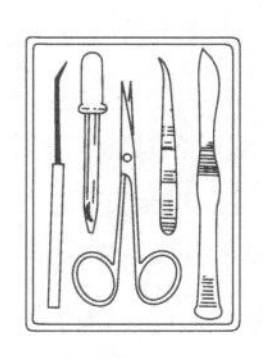

Obtain a test tube rack, seven test tubes, wax marking pencils, and wide-range pH paper from the general supply area, and one of each of the materials/supplies listed under *Supply area 2.* If albumin solution is provided, you will be conducting Procedure A. If hard-boiled egg white is provided, follow the instructions for Procedure B.

Pepsin, produced by the chief cells of the stomach glands, hydrolyzes proteins to smaller fragments (peptides and polypeptides), i.e.:

$$\text{Protein} + \text{water} \xrightarrow{\text{pepsin}} X \text{ peptides}$$

Procedure A:
One student should prepare the control sample according to the directions in step 1 while the other members of the group set up the experimental samples (steps 2 and 3).

Chart 1 Salivary Amylase Digestion of Starch

Tube no.	Additives	Incubation condition	Time of initiation	Time for negative IKI test*	Benedict's test	
					+	–
5	4 drops IKI, 1 ml amylase solution	37°C				
6	4 drops IKI; 1 ml starch solution; 1 ml amylase solution	37°C				
7	4 drops IKI; 1 ml starch solution; 1 ml amylase solution	0°C (ice bath)				
8	5 ml amylase solution; boil 4 min before adding 4 drops IKI and 1 ml starch, and incubating	37°C				
9	4 drops IKI; 1 ml starch solution; 1 ml amylase solution; add 1 N HCl until a pH of 3 is achieved (stir after each addition); incubate	37°C				
10	4 drops IKI; 1 ml starch solution; 1 ml amylase solution; add 1 N NaOH until a pH of 9 is achieved (stir after each addition); incubate	37°C				

* If the sample is still positive for the presence of starch at the end of the incubation period, put a (+) in this column.

1. To prepare a control sample that indicates the presence of protein, perform a *biuret test.* Add 3 ml of 1% albumin solution or diluted egg white to a test tube marked A, and then add 10 drops of 20% NaOH and 5 drops of 10% $CuSO_4$. Mix well with a glass stirring rod. If the mixture turns *violet,* protein is present. In the presence of polypeptides, the biuret test yields a *pink* color. A negative biuret test (no color change) indicates complete protein digestion.

2. To prepare the experimental samples, mark five test tubes with numbers 1 to 5. Preparation and incubation conditions for the experimental samples are listed in Chart 2. Determine the pH of each sample <u>before</u> adding the albumin solution and record the resultant pH on Chart 2. Except for test tube 4, which is to be incubated at room temperature, place all tubes in the water bath set at 37°C and continue incubation for 2 hours. Shake the tubes occasionally.

3. Perform the biuret test on the contents of each test tube after the 2-hour incubation period. Record the results on Chart 2 and on the chalkboard. A – indicates a negative test (complete protein digestion); + indicates a test that is slightly positive (incomplete protein digestion; polypeptides present); ++ indicates a strongly positive test (virtually no protein digestion has occurred).

Procedure B:

In this procedure, no control sample will be prepared. Instead, extent of protein digestion will be estimated by visual inspection only.

1. Mark five test tubes with the numbers 1 to 5.

2. Using a razor blade, cut five slices of egg white. The slices should be fingernail size and paper thin. <u>The thinness of the egg white is critical to the success of this experiment.</u>

Chart 2 Pepsin Digestion of Protein

			Procedure A			Procedure B			
			Biuret test			Transparency			
Tube no.	Additives	pH of sample	–	+	+ +	–	+	+ +	+ + +
1	2 ml pepsin solution; 2 ml 0.5%–1.0% HCl; 2 ml albumin solution or 1 slice egg white; incubate at 37°C								
2	2 ml pepsin solution; 2 ml distilled water; 2 ml albumin solution or 1 slice egg white; incubate at 37°C								
3	2 ml pepsin solution; 2 drops 10 N NaOH; 2 ml albumin solution or 1 slice egg white; incubate at 37°C								
4	2 ml pepsin solution; 2 ml 0.5%–1.0% HCl; 2 ml albumin solution or 1 slice egg white; incubate at room temperature								
5	2 ml distilled water; 2 ml 0.5%–1.0% HCl; 2 ml albumin solution or 1 slice egg white; incubate at 37°C								

3. The preparation and incubation conditions for the experimental samples are identical to those indicated for Procedure A, except that slices of egg white instead of albumin solution are to be used as the substrate. Determine the pH of each sample *before* adding the slice of egg white, and record it in Chart 2.

Except for Sample 4, which is to be incubated at room temperature, place all the samples in a water bath set at 37°C and continue incubation for 2 hours. Shake the tubes occasionally, and determine if there is any evidence of digestive activity. The egg white will become increasingly transparent and will decrease in mass as digestion proceeds.

4. At the end of 2 hours of incubation, observe each sample, and record its relative transparency on Chart 2 according to the following scale:

+ + + no egg white observed
+ + egg white present, but decreased in mass and very transparent
+ egg white slightly transparent
– appearance of egg white exhibits no difference from untreated egg white

PANCREATIC LIPASE DIGESTION OF FATS AND THE ACTION OF BILE

The treatment that fats and oils go through during digestion in the small intestine is a bit more complicated than that of carbohydrates or proteins—it requires pretreatment with bile to physically emulsify the fats first. Hence, two sets of reactions are required:

First:

$$\text{Fats/oils} \xrightarrow[\text{(emulsification)}]{\text{bile}} \text{minute fat/oil droplets}$$

Then:

$$\text{Fat/oil droplets} \xrightarrow{\text{lipase}} \text{monoglycerides and (2) fatty acids}$$

The term **pancreatin** describes the enzymatic product of the pancreas, which includes protein, carbohydrate, nucleic acid, and fat-digesting enzymes. It is used here to investigate the properties of **pancreatic lipase,** which hydrolyzes fats and oils to their component monoglycerides and two fatty acids (and occasionally to glycerol and three fatty acids).

Chart 3 Pancreatic Lipase Digestion of Fats

Tube no.	Additives	Incubation		Change in color	Change in odor
		Began	Ended		
2	3 ml pancreatin, 3 ml distilled water, 3 drops phenol red; add 0.2% NaOH until pink; 5 drops of oil; incubate at 37°C				
3	3 ml pancreatin, 3 ml bile solution, 3 drops phenol red; add 0.2% NaOH until pink; 5 drops of oil; incubate at 37°C				
4	3 ml pancreatin, 3 ml bile solution, 3 drops phenol red; add 0.2% NaOH until pink; 5 drops of oil; incubate at room temperature				
5	3 ml distilled water, 3 ml bile solution, 3 drops phenol red; add 0.2% NaOH until pink; 5 drops of oil; incubate at 37°C				

The fact that some of the end products of fat digestion (fatty acids) are organic acids that decrease the pH provides an easy way to recognize that digestion is ongoing or completed. You will be using a pH indicator called *phenol red* to follow these changes; it changes from *red* to *yellow* as the test tube contents become acid.

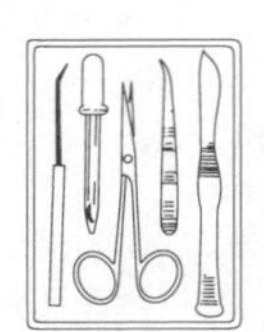

1. From the general supply area, obtain a test tube rack, seven test tubes, a glass stirring rod, and a 10-ml graduated cylinder. Also obtain one sample each of the substances and supplies listed under *Supply area 3.* One student should prepare the control (step 2), another should set up the demonstration of the action of bile on fats (step 3), while the other two group members prepare the experimental samples (step 4).

2. To prepare the control, add 3 ml of 1.5% pancreatin solution, 3 ml of bile (real or artificial) solution, and 3 drops of phenol red indicator to test tube 1. Add 0.2% NaOH dropwise, stirring with a glass rod after each addition until the content of the tube turns pink. Put in 5 drops of vegetable oil, and then add 0.1 N HCl, drop by drop (stirring after each addition) until the tube contents turn yellow. This color change indicates that the test tube contains an acidic product and will identify the tubes in which fat hydrolysis has occurred. Mark the tube "control" with a wax marker and put it in the test tube holder for future reference.

3. Although bile, a secretory product of the liver, is not an enzyme, it is important to fat digestion because of its emulsifying action (the physical breakdown of larger particles into smaller ones) on fats. Emulsified fats provide a larger surface area for enzymatic activity. To demonstrate the action of bile on fats, prepare two test tubes; and label them A and B. To tube A, add 6 ml H_2O and 5 drops of vegetable oil. To tube B, add 3 ml of H_2O, 3 ml of bile solution, and 5 drops of vegetable oil. Shake each tube vigorously, and allow the tubes to stand in a test tube rack at room temperature for 10 to 15 minutes. Observe both tubes. If emulsification has *not* occurred, the oil will be floating on the surface of the water. If emulsification *has* occurred, the fat droplets will be suspended throughout the water, forming an emulsion.

In which tube has emulsification occurred? __________

4. Prepare the experimental samples as indicated in Chart 3. (Remember to stir when adding the sodium hydroxide [NaOH] solution to turn the phenol red indicator pink.) Then shake each test tube well, before beginning its incubation at room temperature or in a 37°C water bath. Incubate until a color change (pink to yellow) becomes apparent. Shake the tubes occasionally during the incubation period. Note the time incubation begins, the time it ends, and any changes in color and odor of the samples. Record results on the chalkboard and in Chart 3. After 1 hour, discontinue incubation of any remaining samples, whether or not a color change has occurred. (If there was no color change, write N.C.)

Anatomy of the Urinary System

OBJECTIVES

1. To describe the overall function of the urinary system.
2. To identify on an appropriate diagram or torso model the urinary organs, and to describe the general function of each.
3. To define *micturition* and to explain the controls of the internal and external urethral sphincters.
4. To compare the course and length of the urethra in males and females.
5. To identify the following regions of the dissected kidney (longitudinal section): hilus, cortex, medulla, medullary pyramids, major and minor calyces, pelvis, and renal columns.
6. To trace the blood supply of the kidney from the renal artery to the renal vein.
7. To define the nephron as the physiological unit of the kidney and to describe its anatomy.
8. To define *glomerular filtration, tubular reabsorption,* and *tubular secretion,* and to note the nephron areas involved in these processes.
9. To recognize microscopic or diagrammatic views of the histological structure of the kidney and bladder.

MATERIALS

Human dissectible torso model and/or anatomical chart of the human urinary system
Three-dimensional model of the cut kidney and of the nephron (if available)
Dissection tray and instruments
Pig or sheep kidney, doubly or triply injected
Prepared histological slides of a longitudinal section of kidney and of a cross section of the bladder
Compound microscope
Disposable gloves or protective skin cream

Metabolism of nutrients by the body produces wastes (carbon dioxide, nitrogenous wastes, ammonia, and so on) that must be eliminated from the body if normal function is to continue. Although excretory processes involve several organ systems (the lungs excrete carbon dioxide, and the skin glands excrete salts and water), it is the **urinary system** that is primarily concerned with the removal of nitrogenous wastes from the body. In addition to this purely excretory function, the kidney maintains the electrolyte, acid-base, and fluid balances of the blood and is thus a major, if not *the* major, homeostatic organ of the body.

To perform its functions, the kidney acts first as a blood "filter" and then as a blood "processor." It allows toxins, metabolic wastes, and excess ions to leave the body in the urine while simultaneously retaining needed substances and returning them to the blood. Malfunction of the urinary system, particularly of the kidneys, leads to a failure in homeostasis, which, unless corrected, results in death.

GROSS ANATOMY OF THE HUMAN URINARY SYSTEM

The urinary system (Figure 37.1) consists of the paired kidneys and ureters, and the single urinary bladder and urethra. The kidneys perform the functions described above and manufacture urine in the process. The remaining organs of the system provide temporary storage reservoirs or transportation channels for urine.

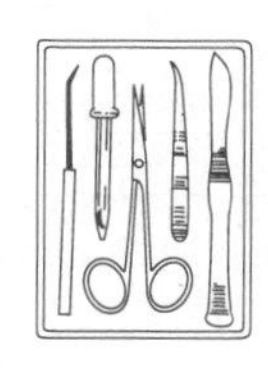

Examine the human torso model, a large anatomical chart, or a three-dimensional model of the urinary system to locate and study the anatomy and relationships of the urinary organs. As you work, refer to Figure 37.1.

1. Locate the paired **kidneys** on the dorsal body wall in the superior lumbar region. Note that they are not positioned at exactly the same level; because it is "crowded" by the liver, the right kidney is slightly lower than the left kidney. In the living person, fat deposits hold the kidneys in place in a retroperitoneal position.

When the fatty material surrounding the kidneys is reduced or deficient in amount (in cases of rapid weight loss or in very thin individuals), the kidneys are less securely anchored to the body wall and may drop to a lower or more inferior position in the abdominal cavity. This phenomenon is called **ptosis.** ■

2. Observe the **renal arteries** as they diverge from the descending aorta and plunge into the indented medial region (**hilus**) of each kidney. Note also the **renal veins,**

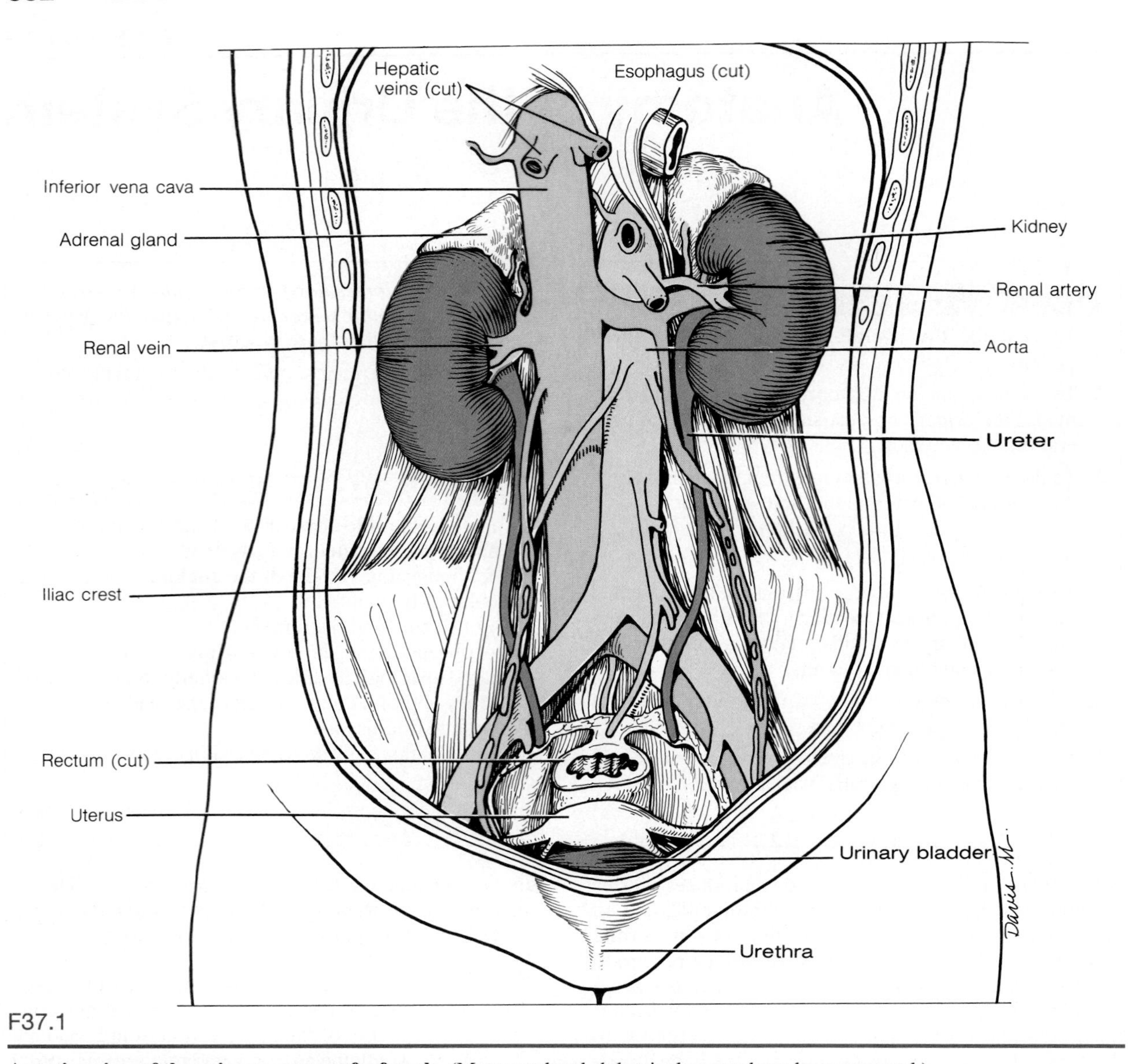

Anterior view of the urinary organs of a female. (Most unrelated abdominal organs have been removed.)

which drain the kidneys (circulatory drainage), and the two **ureters,** which drain the urine from the kidneys and conduct it by peristalsis to the bladder for temporary storage.

3. Locate the **urinary bladder,** and observe the point of entry of the two ureters into this organ. Also locate the single **urethra,** which drains the bladder. The triangular region of the bladder, which is delineated by these three openings (two ureteral and one urethral orifice), is referred to as the **trigone** (Figure 37.2). Although the formation of urine by the kidney is a continuous process, urine is usually removed from the body when voiding is convenient. In the meantime, the bladder provides temporary storage for the urine.

Voiding, or **micturition,** is the process in which urine empties from the bladder. Two sphincter muscles or valves, the **internal urethral sphincter** (more superiorly located) and the **external urethral sphincter** (more inferiorly located), control the emptying of urine from the bladder. Ordinarily the bladder continues to collect urine until about 200 ml have accumulated, at which time the stretching of the bladder wall activates stretch receptors. Impulses transmitted to the central nervous system subsequently produce reflex contractions of the bladder wall through parasympathetic nervous system pathways (i.e., via the pelvic splanchnic nerves). As the contractions increase in force and frequency, the stored urine is forced past the internal sphincter, which is a smooth-muscle involuntary sphincter, and into the superior part of the urethra. It is then that a person feels the urge to void. The inferior external sphincter consists of skeletal muscle and is voluntarily controlled. If it is not convenient to void, the opening of this sphincter can be inhibited. Conversely, if the time is convenient, the sphincter may be relaxed, and the stored urine flushed from the body. If voiding is inhibited, the reflex contractions of the bladder cease temporarily; and urine continues to accumulate in the bladder. After 200 to 300 ml more have been collected, the *micturition reflex* will again be initiated.

Lack of voluntary control over the external sphincter is referred to as *incontinence.* Incontinence is a normal phenomenon in children 2 years old or younger, as they have not yet gained control over the voluntary sphincter. In adults and older children, incontinence is generally a result of spinal cord injury, emotional problems, bladder irritability, or some other pathologic condition of the urinary tract. ■

4. Follow the course of the urethra to the body exterior. In the male, it is approximately 20 cm (8 inches) long, travels the length of the **penis,** and opens at its tip. Its three named regions—the *prostatic, membranous,* and *spongy (penile) urethrae*—are described in more detail in Exercise 39 and illustrated in Figure 39.1 (p. 345). The urethra of males has a dual function: It is a urine conduit to the body exterior, and it provides a passageway for the ejaculation of the semen. Thus in the male, the urethra is part of both the urinary and reproductive systems. In females, the urethra is very short, approximately 4 cm (1½ inches) long. There are no common urinary-reproductive pathways in the female, and the female urethra serves only to transport urine to the body exterior. Its external opening, the **external urethral orifice,** lies anterior to the vaginal opening.

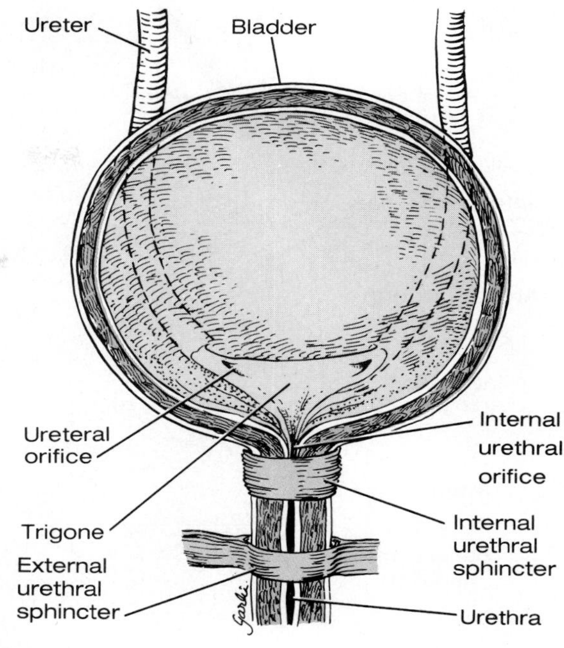

Detailed structure of the urinary bladder and urethral sphincters.

GROSS INTERNAL ANATOMY OF THE PIG OR SHEEP KIDNEY

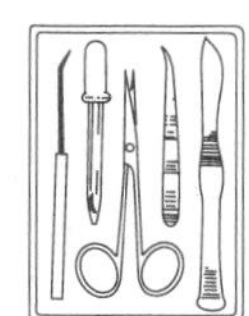

1. Obtain a preserved sheep or pig kidney, a dissecting pan, and instruments. Observe the kidney to identify the **renal capsule,** a smooth transparent membrane that adheres tightly to the kidney tissue.

2. Find the ureter, renal vein, and renal artery at the hilus (indented) region. The renal vein has the thinnest wall and will be collapsed. The ureter is the largest of these structures and has the thickest wall.

3. Make a cut through the longitudinal axis (frontal section) of the kidney, and locate the anatomical areas described below and depicted in Figure 37.3.

Kidney cortex: the outer kidney region, which is lighter in color. (If the kidney is double-injected with latex, you will see a predominance of red and blue latex specks in this region, indicating its rich vascular supply.)

Medullary region: deep to the cortex; a darker, reddish-brown color. The medulla is segregated into triangular regions that have a striped appearance—the **medullary (renal) pyramids.** The base of each pyramid faces toward the cortex; its **apex,** or **papilla,** points to the innermost kidney region.

Renal columns: areas of tissue, more like the cortex in appearance, that segregate and dip downward between the pyramids.

Renal pelvis: medial to the hilus; a relatively flat, basinlike cavity that is continuous with the **ureter,** which exits from the hilus region. Fingerlike extensions of the pelvis should be visible. The larger, or primary, extensions are called the **major calyces;** subdivisions of the major calyces are the **minor calyces.** Note that the minor calyces terminate in cuplike areas that enclose the apexes of the medullary pyramids and collect urine draining from the pyramidal tips into the pelvis.

4. If the preserved kidney is doubly or triply injected, follow the renal blood supply from the renal artery to the **glomeruli.** The glomeruli appear as little red and blue specks in the cortex region. (See Figures 37.3 and 37.4, and the discussion below.)

Because the kidneys continuously cleanse the blood and adjust its composition, it is not surprising that they have a rich vascular supply. Approximately a fourth of the total blood flow of the body is delivered to the kidneys each minute by the large **renal arteries.** As a renal artery approaches the kidney, it breaks up into five branches called **segmental arteries** which enter the hilus. Each segmental artery, in turn, divides into several **lobar arteries.** The lobar arteries branch to form **inter-**

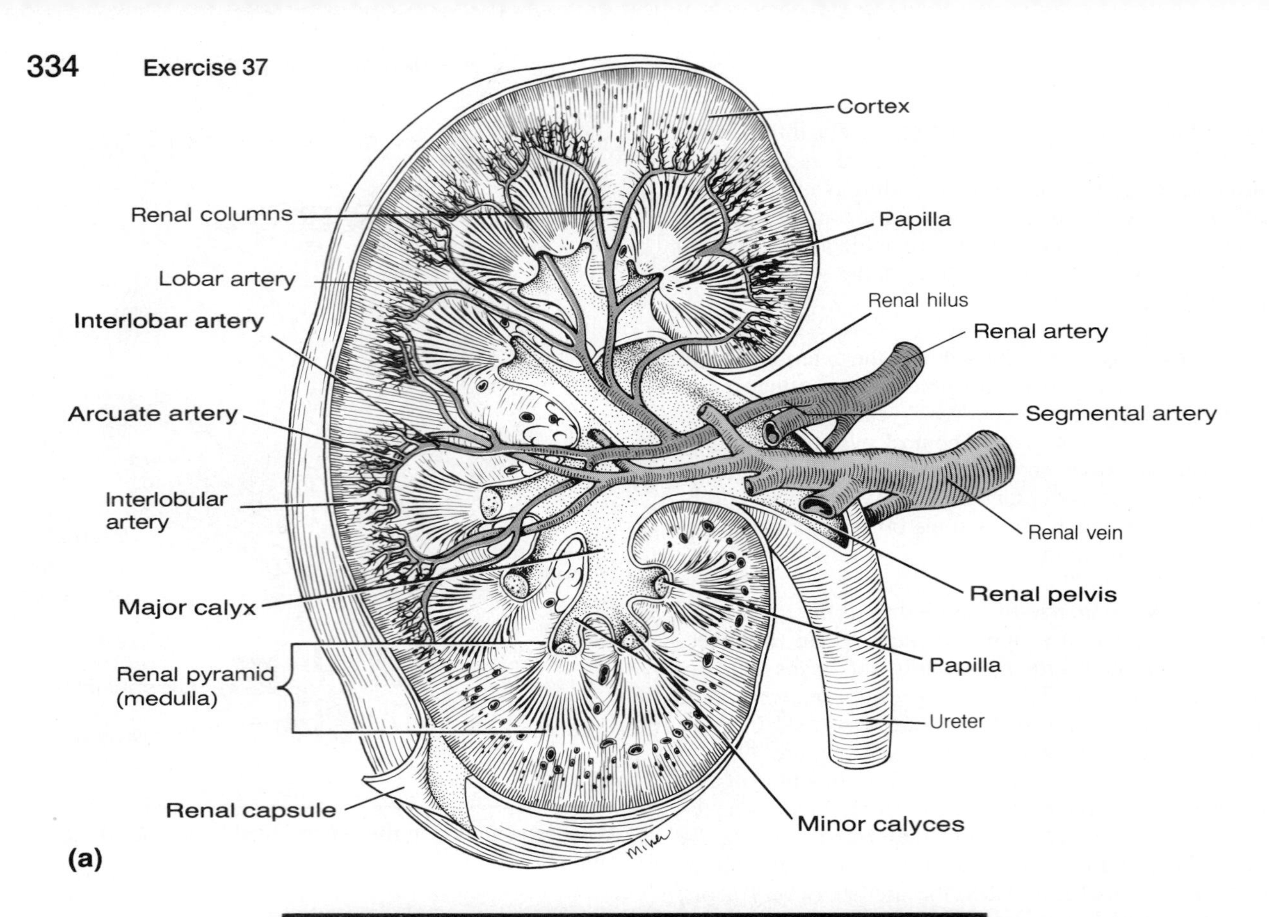

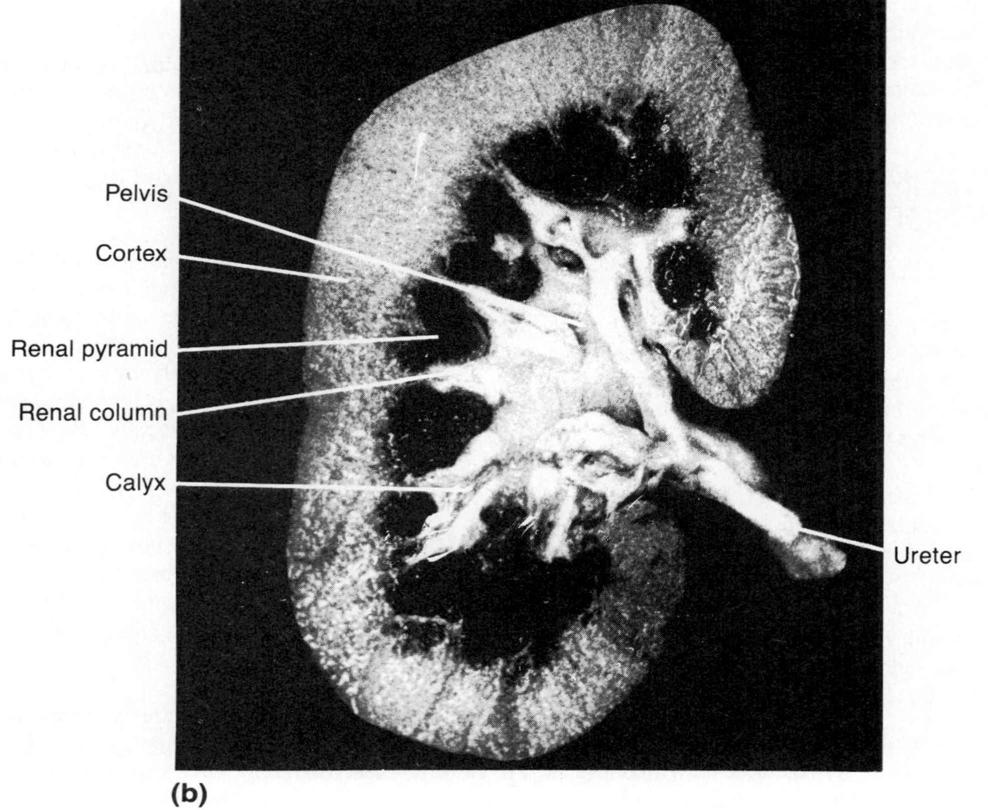

F37.3

Frontal section of a kidney. (a) Diagrammatic view, showing the arterial supply of the kidney tissue; (b) photograph of a pig kidney.

lobar arteries, which ascend toward the cortex in the renal columns. At the top of the medullary region, these arteries give off arching branches, the **arcuate arteries,** which curve over the bases of the medullary pyramids. Small **interlobular arteries** branch off the arcuate arteries and ascend into the cortex, giving off the individual **afferent arterioles,** which provide the capillary networks (**glomeruli** and **peritubular capillary beds**) that supply the nephrons, or functional units, of the kidney. Blood draining from the nephron capillary networks in the cortex enters the **interlobular veins,** and then drains through the **arcuate veins** and the **interlobar veins** to finally enter the **renal vein** in the pelvis region. (There are no lobar or segmental veins.)

MICROSCOPIC ANATOMY OF THE KIDNEY AND BLADDER

Obtain prepared slides of kidney and bladder tissue, and a compound microscope.

Kidney

Each kidney contains over one million **nephrons,** which are the anatomical units responsible for forming urine. Figure 37.4 depicts the detailed structure and the relative positioning of the nephrons in the kidney.

Each nephron consists of two major structures: a **glomerulus** (a capillary knot) and a **renal tubule.** Dur-

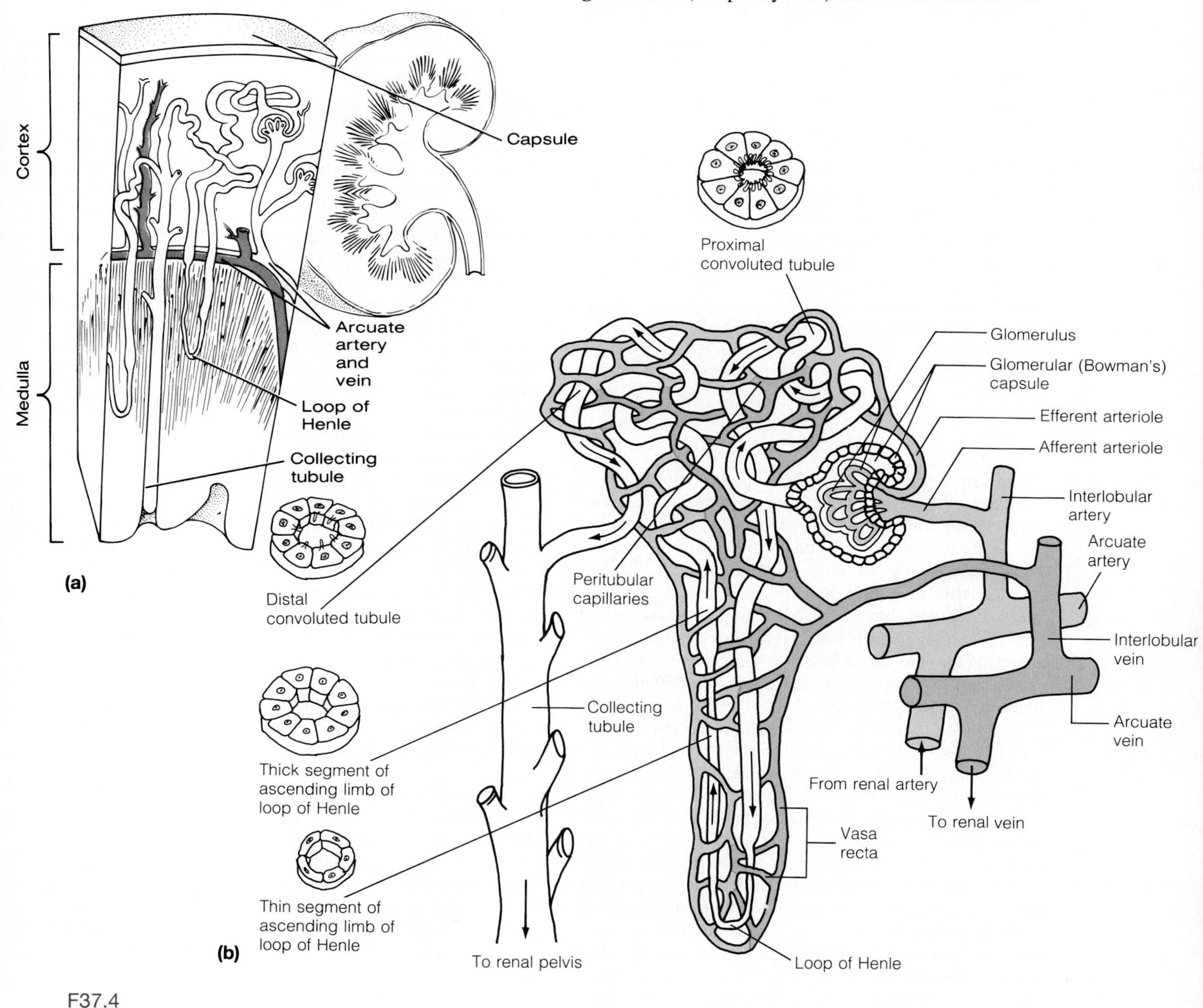

F37.4

Structure of a nephron. (a) Wedge-shaped section of kidney tissue indicating the position of the nephrons in the kidney; (b) detailed nephron anatomy and associated blood supply.

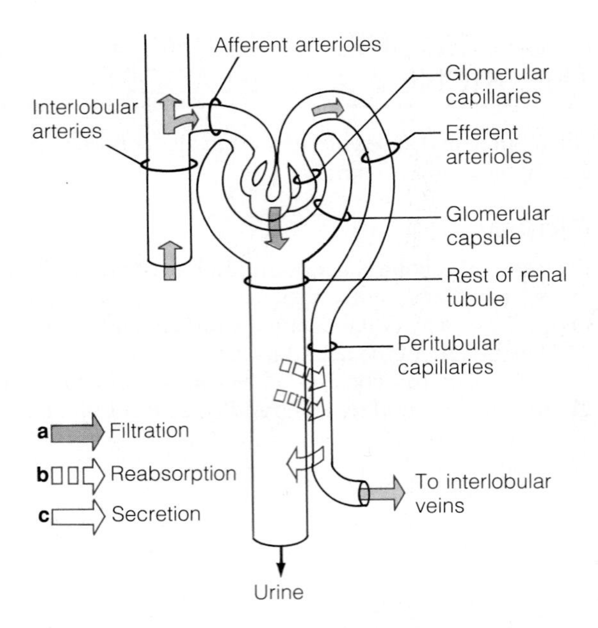

F37.5

The kidney depicted as a single, large nephron. A kidney actually has millions of nephrons acting in parallel. The three major mechanisms by which the kidneys adjust the composition of plasma are **(a)** glomerular filtration, **(b)** tubular reabsorption, and **(c)** tubular secretion. Red arrows show the path of blood flow through the renal microcirculation.

ing embryological development, each renal tubule begins as a blind-ended tubule that gradually encloses an adjacent capillary cluster, or glomerulus. The enlarged end of the tubule encasing the glomerulus is the **glomerular (Bowman's) capsule;** and its inner, or visceral, wall consists of highly specialized cells called **podocytes.** Podocytes have long branching processes (*foot processes*) that interdigitate with those of other podocytes and cling to the endothelial wall of the glomerular capillaries, thus forming a very porous epithelial membrane surrounding the glomerulus. The glomerulus-capsule complex is sometimes called the **renal corpuscle.**

The rest of the tubule is approximately 3 cm (1.25 inches) long. As it emerges from the glomerular capsule, it becomes highly coiled and convoluted; drops down into a long hairpin loop; and then again twists and coils before entering a collecting tubule. In order from the glomerular capsule the anatomical areas of the renal tubule are: the **proximal convoluted tubule,** the **loop of Henle** (descending and ascending limbs), and the **distal convoluted tubule.** The wall of the renal tubule is composed almost entirely of cuboidal epithelial cells, with the exception of the descending limb (and sometimes part of the ascending limb) of the loop of Henle, which is simple squamous epithelium. The lumen surfaces of the cuboidal cells in the proximal convoluted tubule have dense microvilli (a cellular modification that greatly increases the surface area exposed to the lumen contents, or filtrate). Microvilli also occur on cells of the ascending limb of the loop of Henle and of the distal convoluted tubule, but in reduced numbers revealing their less significant role in reclaiming filtrate contents.

Most nephrons, called **cortical nephrons,** are located entirely within the cortex. However, parts of the loops of Henle of the **juxtamedullary nephrons** (located close to the cortex-medulla junction), penetrate well into the medulla. The **collecting tubules,** each of which receives urine from many nephrons, run downward through the medullary pyramids, giving them their striped appearance. As the collecting tubules approach the renal pelvis, they fuse to form larger *papillary ducts,* which empty the final urinary product into the calyces and pelvis of the kidney.

The function of the nephron depends on several unique features of renal circulation. The capillary vascular supply consists of two distinct capillary beds, the **glomerulus** and the **peritubular capillary bed.** Vessels leading to and from the glomerulus, the first capillary bed, are both arterioles: The **afferent arteriole** feeds the bed, and the **efferent arteriole** drains it. The glomerular capillary bed has no parallel elsewhere in the body. It is a high-pressure bed along its entire length. Its high pressure is a result of two major factors: (1) The bed is *fed and drained* by arterioles (arterioles are high-resistance vessels, as opposed to venules, which are low-resistance vessels), and (2) the afferent feeder arteriole is larger in diameter than the efferent arteriole that drains the bed. The high hydrostatic pressure created by these two anatomical features forces out fluid and blood components smaller than proteins from the glomerulus into the glomerular capsule; that is, it forms the filtrate which is processed by the nephron tubule.

The peritubular capillary bed arises from the efferent arteriole draining the glomerulus. This set of capillaries cling intimately to the renal tubule and empty into the interlobular veins that leave the cortex. The peritubular capillaries are *low-pressure* very porous capillaries adapted for absorption rather than filtration and readily take up the solutes and water reabsorbed from the filtrate by the tubule cells. The juxtamedullary nephrons have additional looping vessels, called the **vasa recta** ("straight vessels"), that parallel their long loops of Henle in the medulla (see Figure 37.4). Hence, the two capillary beds of the nephron have very different, but complementary, roles: The glomerulus produces the filtrate and the peritubular capillaries reclaim most of that filtrate.

Urine formation is a result of three processes: *filtration, reabsorption,* and *secretion* (Figure 37.5). Filtration is the role of the glomerulus and is largely a passive process in which a portion of the blood passes from the glomerular bed into the glomerular capsule. This filtrate then enters the proximal convoluted tubule, where tubular reabsorption and secretion begin. During tubular reabsorption, many of the filtrate components move through the tubule cells and return to the blood in the peritubular capillaries. Some of this reabsorption is passive, such as that of water which passes by osmosis; but the reabsorption of most substances depends on active transport processes and is highly selective. Which substances are reabsorbed at a particular time depends on

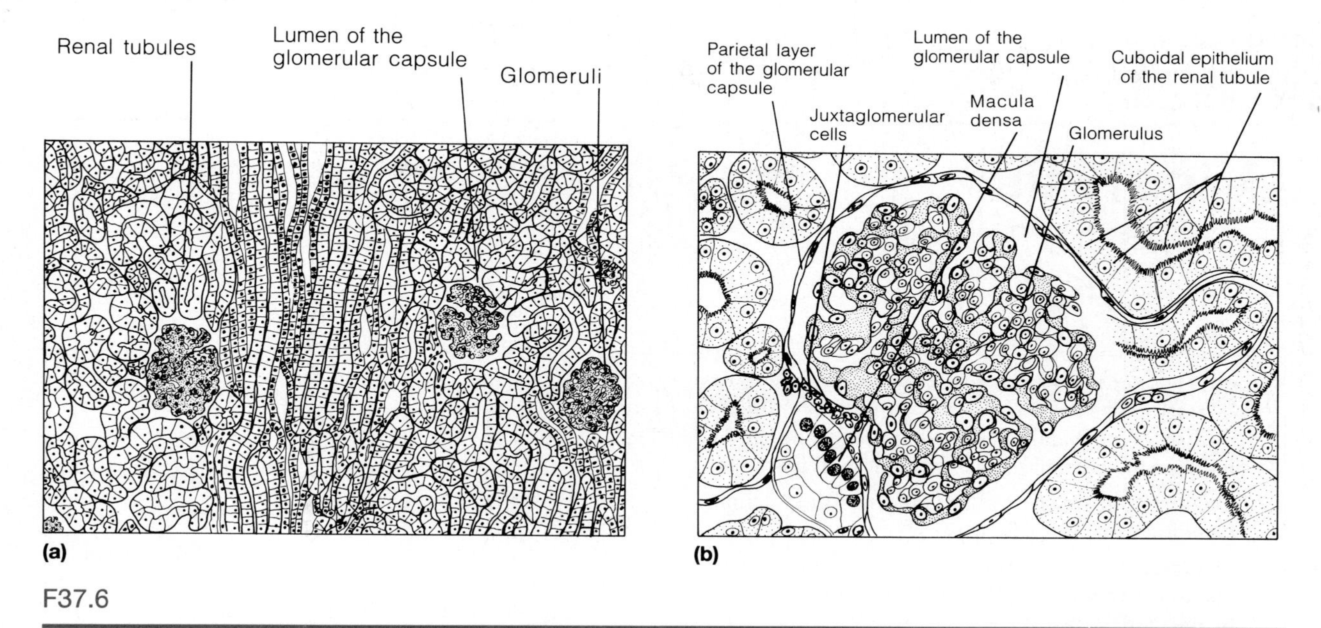

F37.6

Microscopic structure of kidney tissue. (a) Low-power view of the renal cortex; (b) detailed structure of the glomerulus. (See corresponding plates 43 and 44 in the histology atlas.)

the composition of the blood and needs of the body at that time. Substances that are almost entirely reabsorbed from the filtrate include water, glucose, and amino acids. Various ions are selectively reabsorbed or allowed to go out in the urine, according to what is required to maintain appropriate blood pH and electrolyte composition. Waste products (urea, creatinine, uric acid, and drug metabolites) are reabsorbed to a much lesser degree or not at all. Most (75% to 80%) of tubular reabsorption occurs in the proximal convoluted tubule; the balance occurs in other areas, especially in the distal convoluted and collecting tubules.

Tubular secretion is essentially the reverse process. Substances such as hydrogen and potassium ions, creatinine, and ammonia move either from the blood of the peritubular capillaries through the tubular cells or from the tubular cells into the filtrate to to be disposed of in the urine. This process is particularly important for the disposal of substances not already in the filtrate (such as drug metabolites), and as an adjunct method for controlling pH.

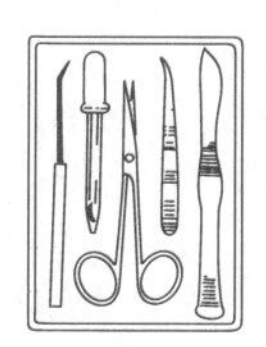

Observe a model of the nephron before continuing with the microscope study of the kidney.

Hold the longitudinal section of the kidney up to the light to identify the cortical and medullary areas. Then secure the slide on the microscope stage and scan the slide under low power. Move the slide so that you can see the cortical area. Identify a glomerulus, which appears as a ball of tightly packed material containing many small nuclei (Figure 37.6). It is usually delineated by a vacant-appearing region (corresponding to the space between the visceral and parietal layers of the glomerular capsule) that surrounds it. Notice that the renal tubules are cut at various angles. Also try to differentiate between the thin-walled loop of Henle portion of the tubules and the cuboidal epithelium of the proximal convoluted tubule, which has microvilli.

Bladder

Scan the bladder tissue. Identify its three layers—mucosa, muscular layer, and fibrous adventitia. Observe the mucosa, with its highly specialized transitional epithelium. The plump, transitional epithelial cells have the ability to slide over one another, thus decreasing the thickness of the mucosa layer as the bladder fills and stretches to accommodate the increased urine volume. Depending on the degree of stretching of the bladder, the mucosa may be from three to eight cell layers thick. Scan the heavy muscular wall (detrusor muscle), which consists of three irregularly arranged muscular layers. The innermost and outermost muscle layers are arranged longitudinally; the middle layer is arranged circularly. Attempt to differentiate the three muscle layers.

Draw a small section of the bladder wall, and label all regions or tissue areas.

Compare your sketch of the bladder wall to the structure of the ureter wall shown in plate 45 of the histology atlas. How are the two organs similar histologically?

____

____

____

What is/are the most obvious difference(s)?

____

38 EXERCISE

Urinalysis

OBJECTIVES

1. To list the physical characteristics of urine, and to indicate the normal pH and specific gravity ranges.
2. To list substances that are normal urinary constituents.
3. To conduct various urinalysis tests and procedures, and to use them to determine the substances present in a urine specimen.
4. To define the following urinary conditions:

calculi	*hematuria*
glycosuria	*hemoglobinuria*
albuminuria	*pyuria*
ketonuria	*casts*

5. To explain the implications and possible causes of the conditions listed in objective 4.

MATERIALS

Urine samples (collected in clean bottles by the students before the laboratory begins) or "normal" urine specimens provided by the instructor
Pathological urine specimens provided by the instructor and numbered
Disposable plastic gloves
Urinometer
Hot plate
1000-ml beakers
Wide-range pH paper
Dip sticks: individual (Clinistix, Ketostix, Albustix, Bilistix, and Hemastix) or combination (Chemstrip or Multistix)
Ictotest reagent
Test reagents for sulfates (10% barium chloride solution; dilute HCl [hydrochloric acid])
Test reagent for glucose: Clinitest tablets
Test reagents for phosphates (dilute nitric acid, dilute ammonium molybdate)
Test reagent for urea (concentrated nitric acid in a dropper bottle)
Test reagent for Cl (3.0% silver nitrate solution [$AgNO_3$])
Test tubes, test tube rack, and test tube holders
10-cc graduated cylinders
Glass stirring rods
Medicine droppers
Wax marking pencils

Microscope slides
Coverslips
Compound microscope
Centrifuge and centrifuge tubes
Sedi-stain
Flasks and buckets containing 10% bleach solution
Disposable autoclave bag

Blood composition depends on three major factors: dietary intake, cellular metabolism, and urinary output. In 24 hours, the kidneys' two million nephrons filter approximately 150 to 180 liters of blood plasma through their glomeruli into the tubules, where it is selectively processed by tubular reabsorption and secretion. In the same period, urinary output, which contains by-products of metabolism and excess ions, is 1.0 to 1.8 liters. In healthy individuals, the kidneys can maintain blood constancy despite wide variations in diet and metabolic activity. With certain pathological conditions, urine composition often changes dramatically.

CHARACTERISTICS OF URINE

Freshly voided urine is generally clear and is pale yellow to amber in color. This normal yellow color is due to *urochrome,* a pigment metabolite arising from the body's destruction of hemoglobin (via bilirubin or bile pigments). As a rule, color variations from pale yellow to deeper amber indicate the relative concentration of solutes to water in the urine. The greater the solute concentration, the deeper the color. Abnormal urinary color may be due to certain foods, such as beets, various drugs, bile, or blood.

The odor of freshly voided urine is characteristic and slightly aromatic, but bacterial action gives it an ammonialike odor when left standing. Some drugs, vegetables (such as asparagus), and various disease processes (such as diabetes mellitus) alter the characteristic odor of urine. For example, the urine of a person with uncontrolled diabetes (and elevated levels of ketones) smells fruity or acetonelike.

The pH of urine ranges from 4.5 to 8.0; but its average value, 6.0, is slightly acidic. Diet may markedly influence the pH of the urine. For example, a diet high in protein (meat, eggs, and cheese) and whole wheat products increases the acidity of the urine. Such foods are called *acid ash* foods. On the other hand, a vegetarian diet (*alkaline ash* diet) increases the alkalinity of the urine. A bacterial infection of the urinary tract may also result in urine with a high pH.

Specific gravity is the relative weight of a specific volume of liquid compared with an equal volume of distilled water. The specific gravity of distilled water is 1.000, because 1 ml weighs 1 g. Because urine contains dissolved solutes, it weighs more than water; and its customary specific gravity ranges from 1.001 to 1.030. Urine with a specific gravity of 1.001 contains few solutes and is very dilute. Dilute urine commonly results when a person drinks excessive amounts of water, uses diuretics, or suffers from diabetes insipidus or chronic renal failure. Conditions that produce urine with a high specific gravity include limited fluid intake, fever, and a kidney inflammation called *pyelonephritis.* If the urine becomes excessively concentrated, some of the substances normally held in solution begin to precipitate or crystallize, forming **kidney stones,** or **renal calculi.**

Normal constituents of urine (in order of decreasing concentration) include water, urea,* sodium,† potassium, phosphate and sulfate ions, creatinine,* and uric acid.* Much smaller but highly variable amounts of calcium, magnesium, and bicarbonate ions are also found in the urine. Abnormally high concentrations of any of these urinary constituents may indicate a pathological condition.

ABNORMAL URINARY CONSTITUENTS

Abnormal urinary constituents are substances not normally present in the urine when the body is operating properly.

Glucose

The presence of glucose in the urine, a condition called **glycosuria,** is indicative of abnormally high blood sugar levels. Normally blood sugar levels are maintained between 80 and 120 mg/100 ml of blood. At this level all glucose in the filtrate is reabsorbed by the tubular cells and returned to the blood. Glycosuria may result from carbohydrate intake so excessive that normal physiologic and hormonal mechanisms cannot clear it from the blood quickly enough. In such cases of glycosuria, the active transport reabsorption mechanisms of the tubules for glucose are exceeded, but only temporarily.

Pathologic glycosuria occurs in conditions such as uncontrolled diabetes mellitus, in which the body cells are unable to absorb glucose from the blood because the pancreatic islet cells produce inadequate amounts of the hormone insulin or there is some abnormality of the insulin receptors. Under such circumstances, the body cells increase their metabolism of fats, and the excess and unusable glucose spills out in the urine.

Albumin

Albuminuria, or the presence of albumin in the urine, is an abnormal finding. Albumin is the single most abundant blood protein and is very important in maintaining the osmotic pressure of the blood. Albumin, like other blood proteins, normally is too large to pass through the glomerular filtration membrane; thus albuminuria is generally indicative of an abnormally increased permeability of the glomerular membrane. Certain nonpathological conditions, such as excessive exertion, pregnancy, or overabundant protein intake, can temporarily increase the membrane permeability, leading to **physiologic albuminuria.** Pathologic conditions resulting in the appearance of albumin in the urine include events that damage the glomerular membrane, such as kidney trauma due to blows, the ingestion of heavy metals, bacterial toxins, glomerulonephritis, and hypertension.

Ketone Bodies

Ketone bodies (acetoacetic acid, beta-hydroxybutyric acid, and acetone) normally appear in the urine in very small amounts. **Ketonuria** or acetonuria, the presence of these intermediate products of fat metabolism in excessive amounts, usually indicates that abnormal metabolic processes are occurring. The result may be acidosis and its complications. Ketonuria is an expected finding during starvation, when inadequate food intake forces the body to use its fat stores. Ketonuria coupled with a finding of glycosuria is generally diagnostic for diabetes mellitus.

Red Blood Cells

Hematuria, the appearance of red blood cells, or erythrocytes, in urine, almost always indicates pathology of the urinary tract, since erythrocytes are too large to pass through the glomerular pores. Possible causes include irritation of the urinary tract organs by calculi (kidney stones), which produces frank bleeding, infections, or physical trauma to the urinary organs. In healthy menstruating females, it may reflect accidental contamination with the menstrual flow.

Hemoglobin

Hemoglobinuria, the presence of hemoglobin in the urine, is a result of fragmentation, or hemolysis, of red blood cells and liberation of the hemoglobin into the plasma with its subsequent appearance in the kidney filtrate. Hemoglobinuria may indicate various pathologic conditions, including hemolytic anemias, transfusion reactions, burns, or renal disease.

Bile Pigments

Bilirubinuria, the appearance of bilirubin (bile pigments) in the urine, is an abnormal finding and most often reflects liver pathology such as hepatitis or cirrho-

* Urea, uric acid, and creatinine are the most important nitrogenous wastes found in urine. Urea is an end product of protein breakdown; uric acid is a metabolite of purine breakdown; and creatinine is associated with muscle metabolism of creatine phosphate.

† Sodium ions appear in relatively high concentration in the urine because of reduced urine volume, not because large amounts are being secreted. Sodium is the major positive ion in the plasma; under normal circumstances, most of it is actively reabsorbed.

sis. Bilirubinuria is indicated by a yellow foam that results when the urine sample is shaken.

White Blood Cells

Pyuria is the presence of white blood cells or other pus constituents in the urine. It indicates an inflammatory process in the urinary tract.

Casts

Any complete discussion of the varieties and implications of casts is beyond the scope of this exercise; however, because they always represent some pathology of the kidney or urinary tract, they should be mentioned. **Casts** are hardened cell fragments, usually cylindrical, which are flushed out of the urinary tract. *White blood cell casts* are a common finding with pyelonephritis, *red blood cell casts* are commonly seen with glomerulonephritis, and *fatty casts* indicate severe renal damage.

ANALYSIS OF URINE SAMPLES

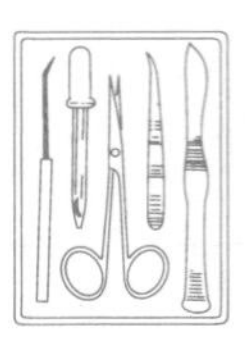

In this part of the exercise, you will use various types of prepared dip sticks and perform chemical tests to determine the characteristics of normal urine as well as to identify abnormal urinary components. You will investigate your own (student-collected) specimen or a "normal" urine sample provided by your instructor *and* an unknown specimen of urine. Make the following determinations on both samples and record your results by circling the appropriate item or description or by adding data to complete Table 38.1. If you have more than one unknown sample, accurately identify each sample used by number.

Obtain and wear plastic disposable gloves throughout this laboratory session. When you have completed the laboratory procedures: (1) dispose of the gloves and used pH paper strips in the autoclave bag; (2) put used glassware in the bleach-containing laboratory bucket; (3) wash the lab bench down with 10% bleach solution.

Determination of the Physical Characteristics of Urine

1. Determine the color, transparency, and odor of your sample, and one of the numbered pathologic samples; and circle the appropriate descriptions in Table 38.1.

2. Obtain a roll of wide-range pH paper to determine the pH of each sample. Use a fresh piece of paper for each test; and before comparing the color obtained with the chart on the dispenser, dip the strip two or three times into the urine to be tested. Record your results in Table 38.1. (If you will be using one of the combination dip sticks—Chemstrip or Multistix—this pH determination can be done later.)

3. To determine specific gravity, obtain a urinometer cylinder and float. Mix the urine well, and fill the urinometer cylinder about two-thirds full with urine. Examine the urinometer float to determine how to read its markings. (In most cases, the scale has numbered lines separated by a series of unnumbered lines. The numbered lines give the reading for the first two decimal places. You must determine the third decimal place by reading the stem of the float at the lower edge of the meniscus—the curved surface representing the urine-air junction.)

4. Carefully lower the urinometer float into the urine. Make sure it is floating freely before attempting to take the reading. Record, in the table, the specific gravity of both samples. *Do not dispose of this urine if the samples that you have are less than 200 ml in volume,* because you will need to make several more determinations.

Determination of Inorganic Constituents in Urine

SULFATES Add 5 ml of urine to a test tube, and then add a few drops of dilute hydrochloric acid and 2 ml of 10% barium chloride solution. The appearance of a white precipitate (barium sulfate) indicates the presence of sulfates in the sample. Clean the test tubes thoroughly after use. Record your results. Are sulfates a normal constituent of urine?

PHOSPHATES Add 5 ml of urine to a test tube, and then add 3 or 4 drops of dilute nitric acid and 3 ml of ammonium molybdate. Mix well with a glass stirring rod, and then heat gently in a hot-water bath. The formation of a yellow precipitate indicates the presence of phosphates in the sample.

CHLORIDES Place 5 ml of urine in a test tube, and add several drops of silver nitrate ($AgNO_3$). The appearance of a white precipitate (silver chloride) is a positive test for chlorides. Record your results.

Determination of Organic Constituents in Urine

Individual dip sticks or combination dip sticks (Chemstrip or Multistix) may be used for many of the tests in this section. If the combination dip sticks are used, be prepared to take the readings on several factors (pH, protein [albumin], glucose, ketones, blood/hemoglobin, and leukocytes [pus]) at the same time. Generally speaking, results for all of these tests may be read *during* the second minute after immersion, but readings taken after 2 minutes have passed should be considered invalid. Pay careful attention to the directions for method and time of immersion and disposal of excess urine from the strip, regardless of the dip stick used.

UREA Put 2 drops of urine on a clean microscope slide, and carefully add 1 drop of concentrated nitric acid to the urine. Slowly warm the mixture on a hot plate until it begins to dry at the edges, but do not allow it to

TABLE 38.1 Urinalysis Results*

Observation or test	Normal values	Student urine specimen	Unknown specimen (#)
Physical characteristics			
Color	Pale yellow	Yellow: pale medium dark other ______	Yellow: pale medium dark other ______
Transparency	Transparent	Clear slightly cloudy cloudy	Clear slightly cloudy cloudy
Odor	Characteristic	Describe ______	Describe ______
pH	4.5–8.0	______	______
Specific gravity	1.001–1.030	______	______
Inorganic components			
Sulfates	Present	Present Absent	Present Absent
Phosphates	Present	Present Absent	Present Absent
Chlorides	Present	Present Absent	Present Absent
Organic components			
Urea	Present	Present Absent	Present Absent
Glucose Dip stick ______	Negative	Record results: ______	Record results: ______
Clinitest	Negative	Record results: ______	Record results: ______
Albumin Dip stick: ______	Negative	Record results: ______	Record results: ______
Ketone bodies Dip stick: ______	Negative	Record results: ______	Record results: ______
RBCs/hemoglobin Dip stick: ______	Negative	Record results: ______	Record results: ______
WBCs (leukocytes) Dip stick: ______	Negative	Record results: ______	Record results: ______
Bilirubin Dip stick: ______	Negative	Record results: ______	Record results: ______
Ictotest	Negative (no color change)	Negative Positive (purple)	Negative Positive (purple)

*In recording urinalysis data, circle the appropriate description if provided; otherwise record the results you observed. Identify dipsticks used.

boil or to evaporate to dryness. When the slide has cooled, examine the edges of the preparation under low power to identify the rhombic or hexagonal crystals of urea nitrate, which form when urea and nitric acid react chemically. Keep the light low for best contrast. Record your results.

GLUCOSE Use a combination dip stick or obtain a vial of Clinistix, and conduct the dip stick test according to the instructions on the vial. Record the results in Table 38.1.

Because the Clinitest reagent is often used in clinical agencies for pediatric patients (children), it is worthwhile to conduct this test for glucose as well. Obtain the Clinitest tablets and the associated color chart. Using a medicine dropper, put 5 drops of urine into a test tube; then rinse the medicine dropper and add 10 drops of water to the tube. Add a Clinitest tablet. Wait 15 seconds and then compare the color obtained to the color chart. Record the results in Table 38.1.

ALBUMIN Use a combination dip stick or obtain the Albustix dip sticks, and conduct the determinations as indicated on the vial. Record your results.

KETONES Use the Ketostix or combination dip stick and carry out the test as indicated on the vial. Record your results.

BLOOD/HEMOGLOBIN Test your urine samples for the presence of blood/hemoglobin by using a Hemastix dip stick or a combination dip stick according to the directions on the vial. (Usually a short drying period is required before making the reading, so read the directions carefully.) Record your results.

BILIRUBIN Using a Bilistix dip stick, determine if there is any bilirubin in your urine samples. Record your results.

Also conduct the Ictotest for the presence of bilirubin. Using a medicine dropper, place 1 drop of urine in the center of one of the special test mats provided with the Ictotest reagent tablets. Place one of the reagent tablets over the drop of urine; and then add 2 drops of water directly to the tablet. If the mixture turns purple when you add water, bilirubin is present. Record your results.

MICROSCOPIC ANALYSIS OF URINE SEDIMENT (OPTIONAL)

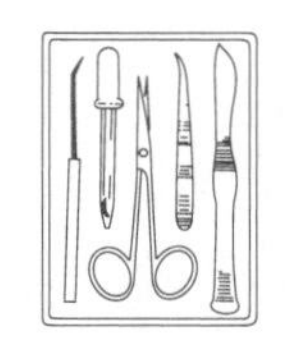

If a centrifuge is available, a microscopic analysis of the urine sediment can be done. Your instructor will provide directions for use of the centrifuge. After centrifugation, pour off the supernatant (the fluid that appears above the sediments at the bottom of the tube); and place a drop of the sediment on a clean glass slide. Add a drop of Sedi-stain, and cover the specimen with a coverslip.

Using the lowest light source possible, examine the slide under low power to determine if any of the sediments illustrated in Figures 38.1 and 38.2 (and described here) can be seen.

Unorganized sediments (Figure 38.1): chemical substances that form crystals or precipitate from solution; for example, calcium oxalates and carbonates, uric acid, ammonium ureates, and cholesterol. Also, if an individual has been taking antibiotics or certain drugs, such as sulfa drugs, these may often be detected in crystalline form in the urine. Normal urine contains very small amounts of crystals, but conditions such as urinary retention or urinary tract infection may result in the appearance of much larger amounts (and in their possible consolidation into calculi). High-power examination may be necessary to view the various crystals, which tend to be much more minute than the organized (cellular) sediments.

Organized sediments (Figure 38.2): include epithelial cells (rarely of any pathologic significance), pus cells (white blood cells), red blood cells, and casts. Urine is normally negative for organized sediments, and the presence of sediments of the last three categories mentioned (other than trace amounts) always indicates kidney pathology (infectious or noninfectious in nature).

Draw, below, examples of your observations; and attempt to identify them by comparing them with Figures 38.1 and 38.2.

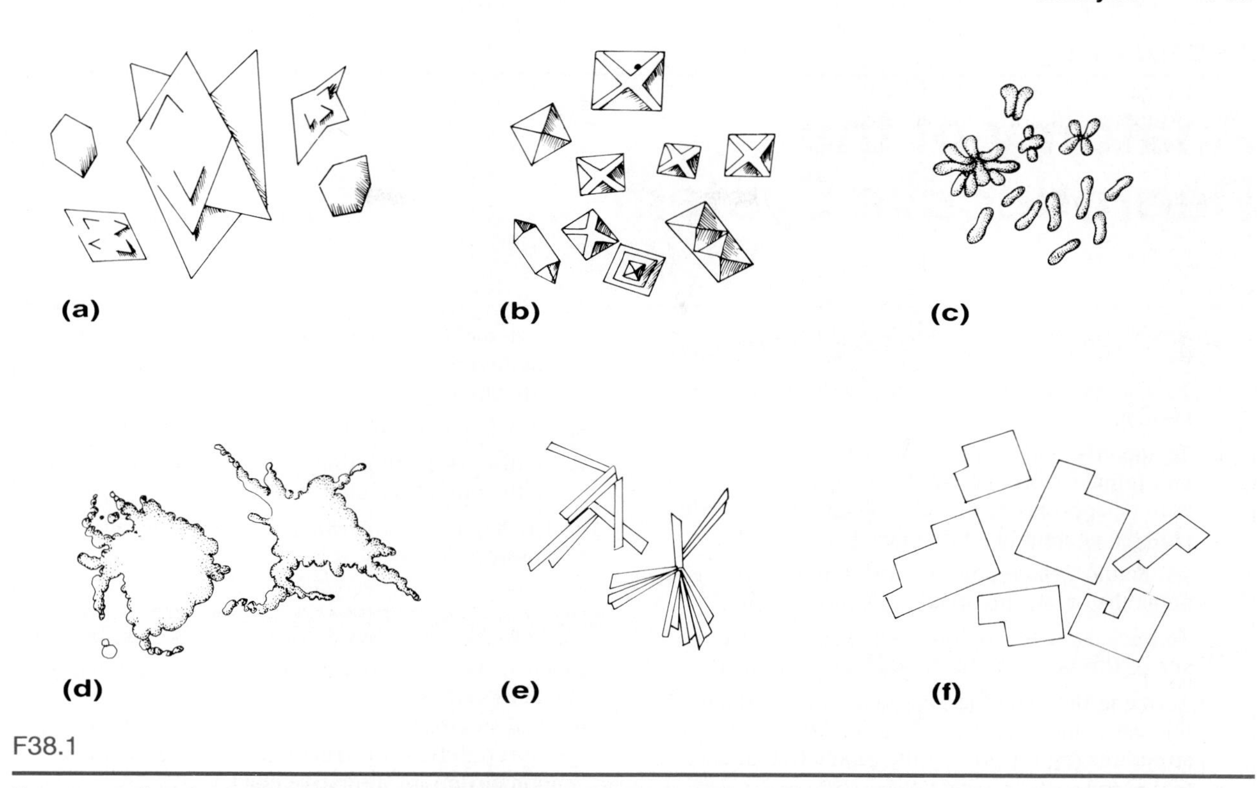

F38.1

Some examples of unorganized sediments. (a) Uric acid crystals; (b) calcium oxalate crystals; (c) calcium carbonate crystals; (d) ammonium ureate crystals; (e) calcium phosphate crystals; (f) cholesterol crystals.

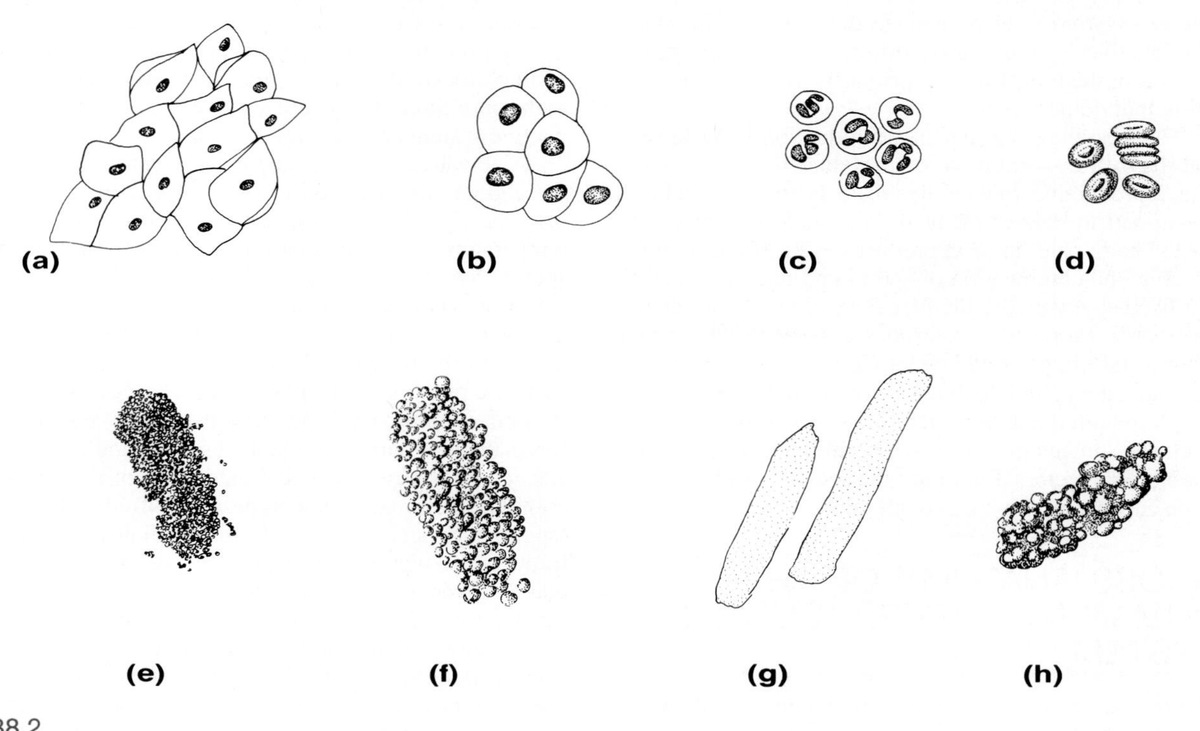

F38.2

Some examples of organized sediments. (a) Squamous epithelial cells; (b) transitional epithelial cells; (c) white blood cells (pus); (d) red blood cells; (e) granular casts; (f) red blood cell casts; (g) hyaline casts; (h) fatty casts.

39

EXERCISE

Anatomy of the Reproductive System

OBJECTIVES

1. To discuss the general function of the reproductive system.
2. To identify and name the structures of the male and female reproductive systems when provided with an appropriate model or diagram, and to discuss the general function of each.
3. To define *semen,* discuss its composition, and name the organs involved in its production.
4. To trace the pathway followed by a sperm from its site of formation to the external environment.
5. To name the exocrine and endocrine products of the testes and ovaries, indicating the cell types or structures responsible for the production of each.
6. To identify homologous structures of the male and female systems.
7. To describe the microscopic structure of the penis, epididymis, uterine tube, and uterus, and to relate structure to function.
8. To define *ejaculation, erection,* and *gonad.*
9. To discuss the function of the fimbriae and ciliated epithelium of the uterine tubes.
10. To define *endometrium, myometrium,* and *ovulation.*

MATERIALS

Models or large laboratory charts of the male and female reproductive tracts
Prepared microscope slides of cross sections of the penis, epididymis, uterine tube, and uterus showing endometrium (proliferative phase)
Compound microscope

Most simply stated, the biological function of the **reproductive system** is to perpetuate the species. Thus the reproductive system is unique because the other organ systems of the body function primarily to sustain the existing individual.

The essential organs of reproduction—the testes and the ovaries—are those that produce the germ cells. The reproductive role of the male is to manufacture sperm and to deliver them to the female reproductive tract. The female, in turn, produces eggs. If the time is suitable, the combination of sperm and egg produces a fertilized egg, which is the first cell of a new individual. Once fertilization has occurred, the female uterus provides a nurturing, protective environment in which the embryo, later called the fetus, develops until birth.

Although the drive to reproduce is strong in all animals, in humans this drive is also intricately related to nonbiologic factors. Emotions and social considerations often enhance or thwart its expression.

GROSS ANATOMY OF THE HUMAN MALE REPRODUCTIVE SYSTEM

The primary reproductive organs of the male are the **testes,** the male *gonads,* which have both an exocrine (sperm production) and an endocrine (testosterone production) function. All other reproductive structures are *accessory structures:* conduits or sources of secretions, which aid in the safe delivery of the sperm to the body exterior or to the female reproductive tract.

As the following organs and structures are described, locate them on Figure 39.1; and then identify them on a three-dimensional model of the male reproductive system or on a large laboratory chart.

The paired oval testes lie in the **scrotal sac,** outside the abdominopelvic cavity. The temperature there, approximately 34°C (93°F), is slightly lower than body temperature, a requirement for the production of viable sperm.

The accessory structures forming the *duct system* are the epididymis, the ductus deferens, the ejaculatory duct, and the urethra. The **epididymis** is an elongated structure running up the posterolateral aspect of the testis and capping its superior aspect. The epididymis forms the first portion of the duct system and provides a site for immature sperm that enter it from the testis to complete their maturation process. The **ductus deferens** (sperm duct) arches upward from the epididymis, passes through the inguinal canal into the pelvic cavity, and courses over the superior aspect of the bladder. In life, the ductus deferens (also called the *vas deferens*) is enclosed, along with blood vessels and nerves, in a connective tissue sheath called the **spermatic cord.** The terminus of the ductus deferens enlarges to form the region called the **ampulla,** which empties into the **ejaculatory duct.** Contraction of the ejaculatory duct propels the sperm through the prostate gland to the **prostatic urethra,** which in turn empties into the **membranous ure-**

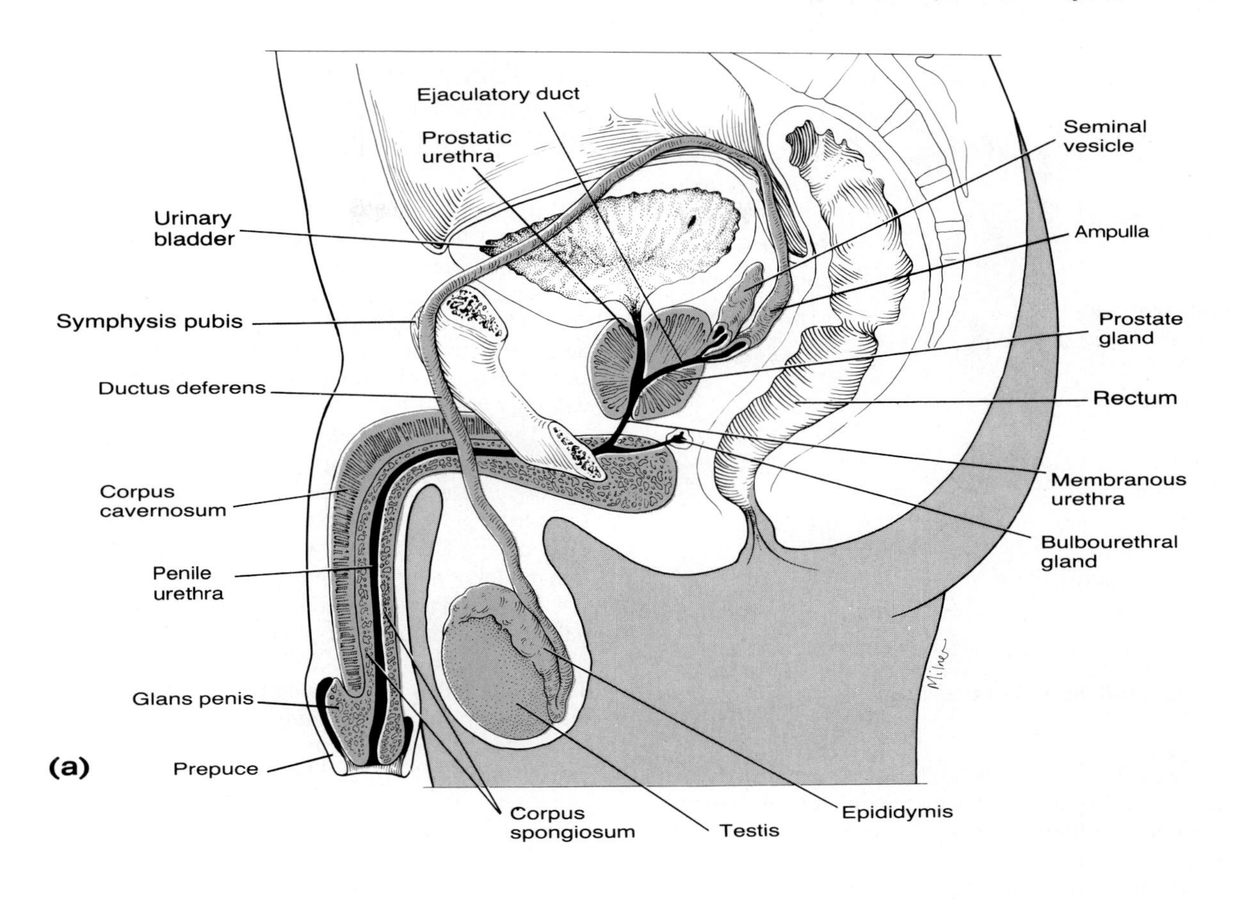

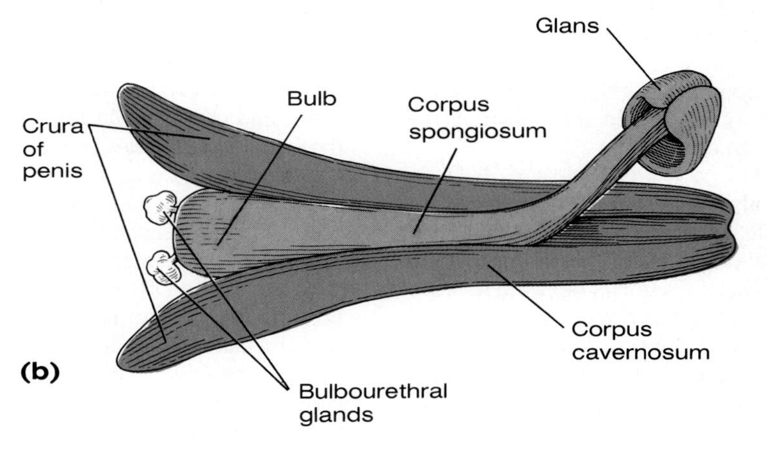

F39.1

Reproductive system of the human male. (a) Median sagittal section; (b) inferior intact view of the penis.

thra and then into the **penile urethra,** which runs through the length of the penis to the body exterior.

The spermatic cord is easily palpated through the skin of the scrotum. When a *vasectomy* is performed, a small incision is made in each side of the scrotum; and each ductus deferens is cut through or cauterized. Although sperm are still produced, they can no longer reach the body exterior; and thus a man is sterile after this procedure (and 12 to 15 ejaculations to clear the conducting tubules).

The *accessory glands* include the prostate gland, the paired seminal vesicles, and the bulbourethral glands. These glands produce **seminal fluid,** the liquid medium in which sperm leave the body. The **seminal vesicles,** which produce about 60% of seminal fluid, lie at the posterior wall of the urinary bladder close to the terminus of the ductus deferens. They produce a viscous secretion containing fructose (a simple sugar) and other substances that nourish the sperm passing through the tract or promote the fertilizing capability of sperm in

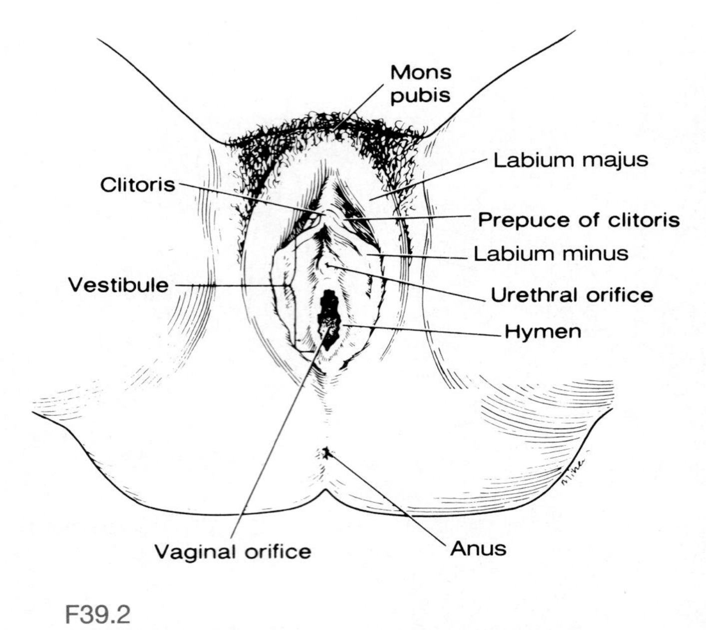

F39.2

External genitalia of the human female.

some way. The duct of each seminal vesicle merges with a ductus deferens to form the ejaculatory duct (mentioned above); thus sperm and seminal fluid enter the urethra together.

The **prostate gland** encircles the urethra just inferior to the bladder. It secretes a milky alkaline fluid into the urethra that plays a role in activating the sperm.

Hypertrophy of the prostate gland, a troublesome condition commonly seen in old age, constricts the urethra, making urination difficult. ■

The **bulbourethral glands** (**Cowper's glands**) are tiny, pea-shaped glands inferior to the prostate. They produce a thick, clear alkaline mucus that drains into the membranous urethra. This secretion is believed to wash the residual urine out of the urethra when ejaculation of **semen** (sperm plus seminal fluid) occurs. The relative alkalinity of all of these glandular secretions also buffers the sperm against the acidity of the female reproductive tract.

The **penis,** part of the external genitalia of the male along with the scrotal sac, is the copulatory organ of the male and is designed to deliver sperm into the female reproductive tract. It consists of a shaft that terminates in an enlarged tip, the **glans** (see Figure 39.1b). The skin covering the penis is loosely applied; and it reflects downward to form a circular fold of skin, the **prepuce,** or **foreskin,** around the proximal end of the glans. (The foreskin is removed in the surgical procedure called *circumcision.*) Internally, the penis consists primarily of three elongated cylinders of erectile tissue, which become engorged with blood during sexual excitement. This causes the penis to become rigid and enlarged so that it may more adequately serve as a penetrating device. This event is called **erection.** The paired dorsal cylinders are the **corpora cavernosa.** The single ventral **corpus spongiosum** surrounds the penile urethra.

GROSS ANATOMY OF THE HUMAN FEMALE REPRODUCTIVE SYSTEM

The **ovaries** (female gonads) are the primary reproductive organs of the female. Like the testes of the male, the ovaries produce both an exocrine product (the eggs, or ova) and endocrine products (estrogens and progesterone). The other accessory structures of the female reproductive system transport, house, nurture, or otherwise serve the needs of the reproductive cells and/or the developing fetus.

The reproductive structures of the female are generally considered in terms of internal organs and external organs, or external genitalia. As you read the descriptions of these structures, locate them in Figures 39.2 and 39.3, and then on the female reproductive system model or a large laboratory chart.

The **external genitalia** (**vulva**) consist of the mons pubis, labia majora and minora, clitoris, urethral and vaginal orifices, hymen, and greater vestibular glands. The **mons pubis** is a rounded fatty eminence overlying the pubic symphysis. Running inferiorly and posteriorly from the mons pubis are two elongated, pigmented, hair-covered skin folds, the **labia majora,** which enclose two smaller, hair-free folds, the **labia minora.** (Terms indicating only one of the two folds in each case are *labium majus* and *minus,* respectively.) The labia majora are homologous to the scrotum of the male. The labia minora, in turn, enclose a region called the **vestibule,** which contains many structures: the clitoris, most anteriorly, followed by the urethral orifice and the vaginal orifice. The diamond-shaped region between the anterior end of the labial folds, the anus posteriorly, and the ischial tuberosities laterally is called the **perineum.**

The **clitoris** is a small, protruding structure, homologous to the male penis, and like its counterpart is composed of highly sensitive, erectile tissue. It is hooded by skin folds of the anterior labia minora, referred to as the **prepuce of the clitoris.** The urethral orifice, which lies posterior to the clitoris, is the outlet for the urinary system and has no reproductive function in the female. The vaginal opening is partially closed by a thin fold of mucous membrane called the **hymen** and is flanked by the mucus-secreting **greater vestibular** (**Bartholin's**) **glands,** which lubricate the distal end of the vagina during coitus. (These glands are not depicted in the illustrations.)

The internal female organs include the vagina; uterus; uterine, or fallopian, tubes; and ovaries—and the ligaments and supporting structures that suspend these organs in the pelvic cavity. The **vagina** extends for about 10 cm (4 inches) from the vestibule superiorly to the uterus. It serves as the copulatory organ and the birth canal, and permits passage of the menstrual flow. The pear-shaped **uterus,** situated between the bladder and the rectum, is a highly muscular organ with its narrow end, the **cervix,** directed inferiorly. The major portion of the uterus is referred to as the **body;** its superior rounded region above the entrance of the uterine tubes is called the **fundus.** The fertilized egg is implanted in the uterus, housing the embryo or fetus during its development.

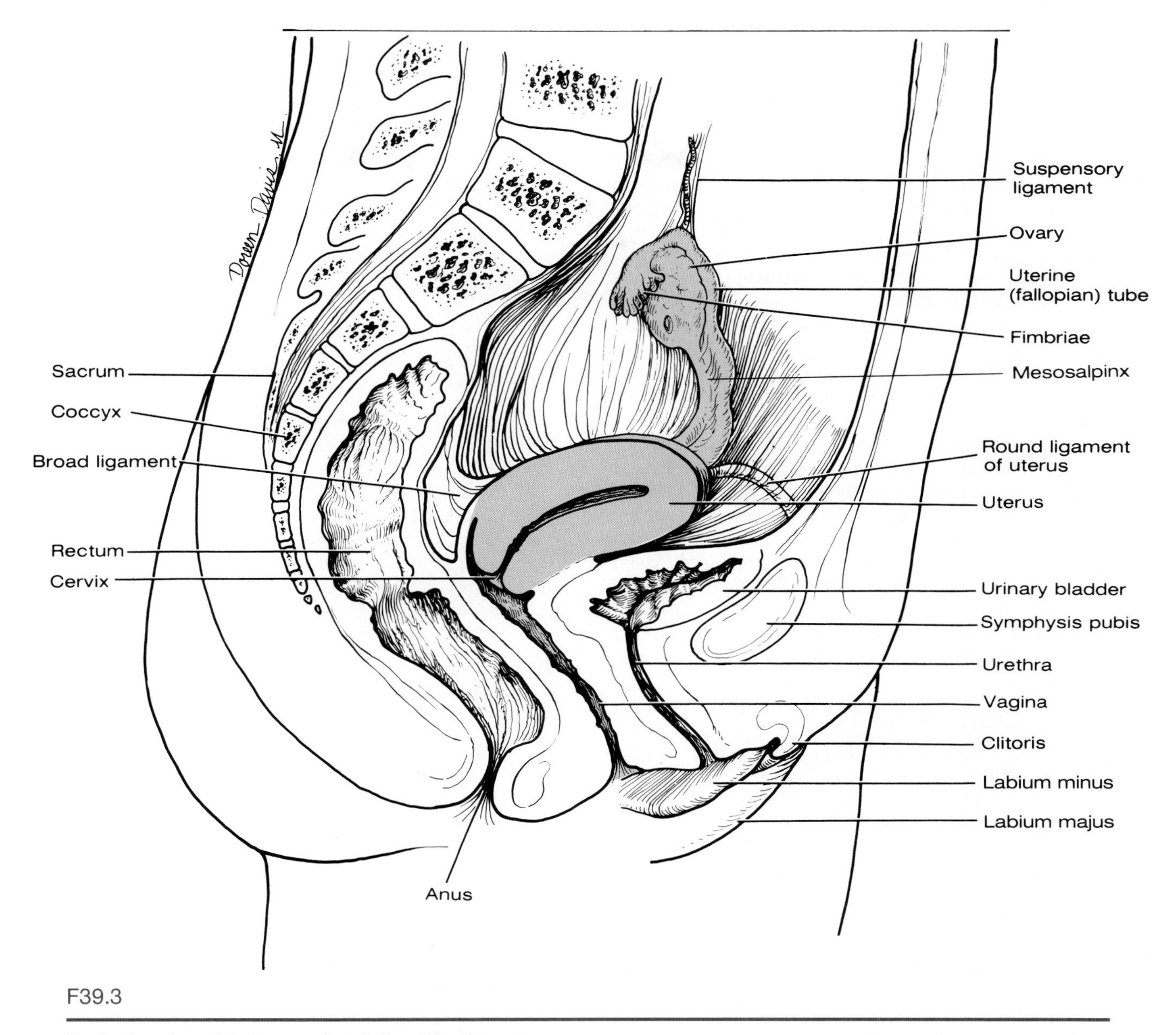

F39.3

Sagittal section of the human female reproductive system.

In some cases, the fertilized egg may be implanted in a uterine tube or even on the abdominal viscera, resulting in an *ectopic pregnancy.* Such implantations are usually unsuccessful and may even endanger the mother's life, because the uterine tubes cannot accommodate the increasing size of the fetus. ■

The apical portion of the **endometrium,** the thick mucosal lining of the uterus, sloughs off periodically (usually about every 28 days) in response to cyclic changes in the levels of ovarian hormones in the woman's blood. This sloughing-off process, which is accompanied by bleeding, is referred to as the **menstrual flow,** or **menses.**

The **uterine,** or **fallopian, tubes** enter the superior region of the uterus and extend laterally for about 10 cm (4 inches) toward the **ovaries** in the peritoneal cavity. The distal ends of these tubes are funnel shaped and have fingerlike projections called **fimbriae.** Unlike the male duct system, there is no actual contact between the female gonad and the initial part of the female duct system, the uterine tube.

Because of this open passageway between the female reproductive organs and the peritoneal cavity, reproductive system infections, such as gonorrhea, can spread to cause widespread inflammations of the pelvic viscera, a condition called *pelvic inflammatory disease,* or PID. ■

The flattened, almond-shaped ovaries lie adjacent to the uterine tubes but are not connected to them. As an egg is expelled from the ovary, an event called **ovulation,** it enters the pelvic cavity. The waving fimbriae of the uterine tubes create fluid currents that (if successful) draw the egg into the lumen of the uterine tube, where it begins its passage to the uterus, propelled by the cilia of the tubule walls. The usual and most desirable site of fertilization is the uterine tube, because the journey to the uterus takes 3–4 days and an egg is viable for up to 24 hours after it is expelled from the ovary. Thus, sperm must swim upward through the vagina and uterus, and into the uterine tubes to reach the egg. This must be an arduous journey, because they must swim against the downward current created by ciliary action—rather like swimming against the tide!

The internal female organs are all retroperitoneal, except the ovaries. They are supported and suspended somewhat freely by ligamentous folds of peritoneum. The fold that encloses the uterine tubes and uterus, and secures them to the lateral body walls, is referred to as the **broad ligament.** The portion of the broad ligament specifically anchoring the uterus is called the **mesometrium,** and that anchoring the uterine tubes, the **mesosalpinx.** The **round ligaments,** fibrous cords that run from the uterus to the labia majora, also aid in attachment of the uterus to the body wall. The ovaries are supported medially by the **ovarian ligament** extending from the uterus to the ovary, laterally by the **suspensory ligaments,** and posteriorly by a fold of the broad ligament, the **mesovarium.**

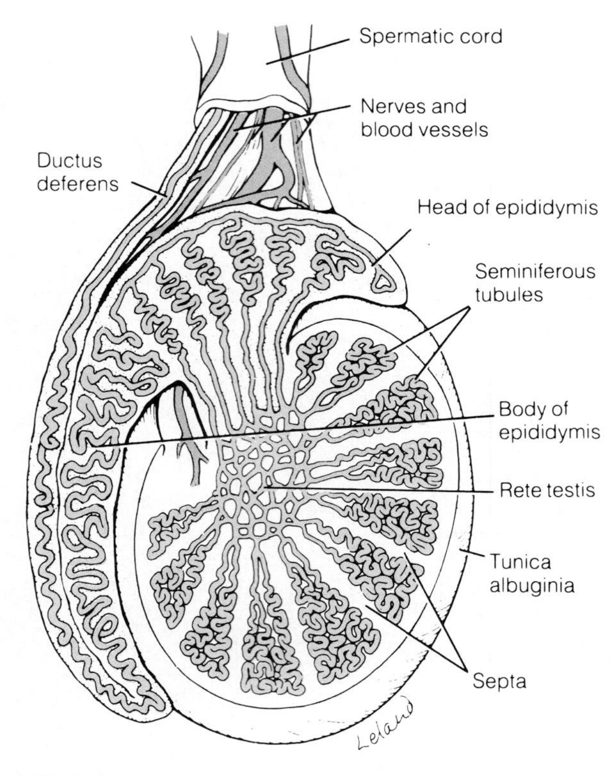

F39.4

Structure of the human testis; sagittal section, showing internal structure. Epididymis and ductus deferens also shown.

MICROSCOPIC ANATOMY OF SELECTED REPRODUCTIVE ORGANS*

Testis

Each testis is covered by a dense connective tissue capsule called the **tunica albuginea** (literally, white tunic). Extensions of this sheath enter the testis, dividing it into a number of lobes—each of which houses one to four highly coiled **seminiferous tubules,** the sperm-forming factories (Figure 39.4). The seminiferous tubules of each lobe converge to empty the sperm into another tubular region, the **rete testis,** at the mediastinum of the testis. Sperm traveling through the rete testis then enter the epididymis, located on the exterior aspect of the testis, as previously described. Lying between the seminiferous tubules and softly padded with connective tissue are the **interstitial cells,** which produce testosterone, the hormonal product of the testis. You will conduct a microscopic study of testis tissue in Exercise 40.

Epididymis

Obtain a cross section of the epididymis. Notice the abundant tubule cross sections resulting from the fact that the coiling epididymis tubule has been cut through many times in the specimen (Figure 39.5a). Look for sperm in the lumen of the tubule. Examine the composition of the tubule wall carefully. Identify the stereocilia of the pseudostratified columnar epithelial lining. (These nonmotile microvilli absorb excess fluid and pass nutrients to the sperm in the lumen.) Now identify the smooth muscle layer. What do you think the function of the smooth muscle is?

____

____

Penis

Obtain a cross section of the penis. Scan the tissue under low power to identify the urethra and the cavernous bodies. Compare your observations to Figure 39.5b. Carefully observe the lumen of the urethra. What type of epithelium do you see?

____

Explain the function of this type of epithelium.

____

____

* A microscopic study of the ovary is described in Exercise 40, pp. 354 and 355. If that exercise is not to be conducted in its entirety, the instructor might want to include the ovary study in the scope of this exercise.

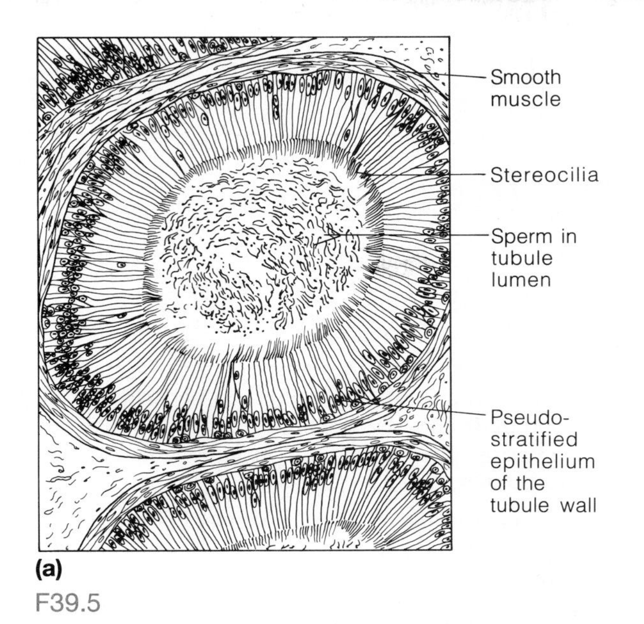

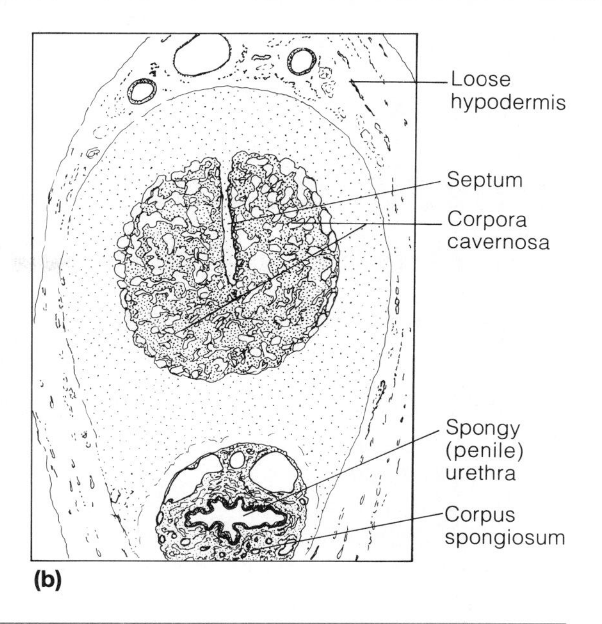

F39.5

Microscopic anatomy of the epididymis and penis. (a) Epididymis (see corresponding plate 46 in the histology atlas); (b) cross-sectional view of penis (see corresponding plate 47).

Uterine Tube

Obtain a prepared slide of a cross-sectional view of a uterine (fallopian) tube for examination. Note the highly folded mucosa (the folds nearly fill the tubule lumen) as illustrated in Figure 39.6 and then switch to high power to examine the ciliated secretory epithelium. Draw your observations of the tubule mucosa below.

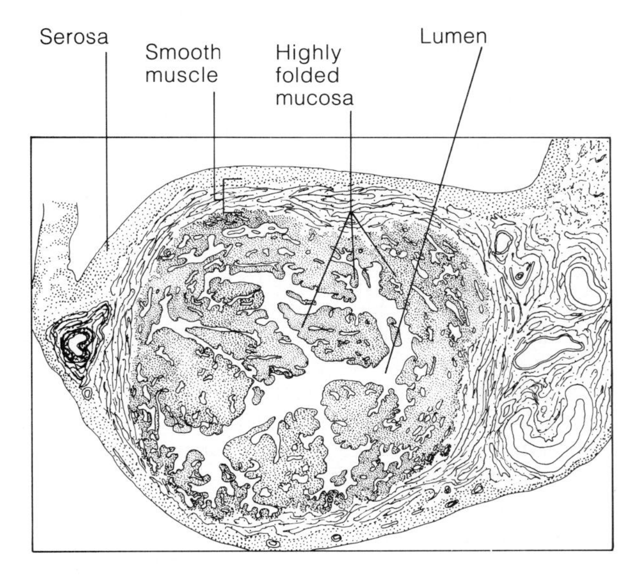

F39.6

Transverse section of the uterine tube. Notice its highly folded mucosa (see corresponding plate 48 in the histology atlas).

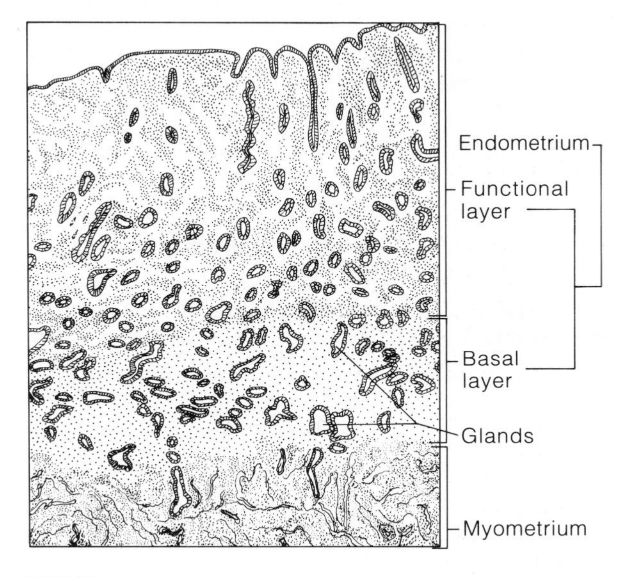

F39.7

Structure of the endometrium showing its stratum functionalis and stratum basalis. (See also the plates of the endometrium inside the back cover.)

Wall of the Uterus

Obtain a cross-sectional view of the uterine wall. Identify, sketch, and label the three layers of the uterine wall—the endometrium, myometrium, and serosa. Also identify the two strata of the endometrium, the functional layer (stratum functionalis), which sloughs off during menses, and the basal layer (stratum basalis), which re-forms a new functional layer each month. Figure 39.7 may be of some help in this study. As you study the slide, notice that the bundles of smooth muscle are oriented in several different directions. What is the function of the myometrium (smooth muscle layer) during the birth process?

____

40 EXERCISE

Physiology of Reproduction: Gametogenesis and the Female Cycles

OBJECTIVES

1. To define *meiosis, gametogenesis, oogenesis, spermatogenesis, synapsis, haploid, diploid,* and *menses,* and to state the similarities and differences between spermatogenesis and oogenesis.
2. To relate the stages of spermatogenesis to the cross-sectional structure of the seminiferous tubule.
3. To discuss the microscopic structure of the ovary and be prepared to identify primary, secondary, and vesicular follicles and the corpus luteum, and to state the hormonal products of the last two structures.
4. To relate the stages of oogenesis to follicular development in the ovary.
5. To cite similarities and differences between mitosis and meiosis.
6. To describe the anatomical structure of the sperm and relate it to function.
7. To discuss the phases and control of the menstrual cycle.
8. To describe and explain the effect of pituitary hormones on the ovary.
9. To describe the effect of FSH and LH on the ovary and to discuss the feedback relationship between anterior pituitary gonadotropins and ovarian hormones.
10. To describe the effects of FSH and LH (ICSH) on testicular function.

MATERIALS

Prepared microscope slides of testis, ovary, human sperm, and uterine endometrium (showing menses, proliferative, and secretory stages)
Three-dimensional models illustrating meiosis, spermatogenesis, and oogenesis
Compound microscope
Demonstration: Microscopes set up to demonstrate the following stages of oogenesis in *Ascaris megalocephala:*

1. Primary oocyte with fertilization membrane, sperm nucleus, and aligned tetrads apparent
2. Formation of the first polar body
3. Secondary oocyte with dyads aligned
4. Formation of the ovum and second polar body
5. Fusion of the male and female pronuclei to form the fertilized egg

MEIOSIS

Every human being, so far, has developed from the union of gametes. The gametes, produced only in the testis or ovary, are unique cells, because they have only half the normal chromosome number (designated as **n,** or the **haploid complement**) seen in all other body cells. In humans, gametes have 23 chromosomes instead of the 46 of other tissue cells. Theoretically, every gamete has a full set of genetic instructions, a conclusion borne out by the observation that some animals can develop from an egg that is artificially stimulated, as by a pinprick rather than by sperm entry. (This is less true of the sperm, because it has an incomplete sex chromosome.) The reduction of the chromosome number by half is important because it maintains the characteristic chromosomal number of the species generation after generation. Otherwise there would be a doubling of chromosome number with each succeeding generation, and the cells would become so chock-full of genetic material there would be little room for anything else.

Egg and sperm chromosomes that carry genes for the same traits are called **homologous chromosomes.** When the sperm and egg fuse to form the **zygote,** or fertilized egg, it is said to contain 23 pairs of homologous chromosomes, or the **diploid** (**2n**) chromosome number of 46. The zygote, once formed, then divides to produce the cells needed to construct the multicellular human body. All cells of the developing human body have a chromosome content identical in quality and quantity to that of the fertilized egg; this is assured by the nuclear division process called **mitosis.** (Mitosis was considered in depth in Exercise 4. You may want to review it at this time.)

To produce the reduced (haploid) chromosomal number, **meiosis,** a specialized type of nuclear division, occurs in the ovaries and testes. Before meiosis begins, the chromosomes are replicated in the *mother cell,* or stem cell, just as they are for mitosis. As a result, the mother cell briefly has twice the normal diploid genetic complement. The stem cell then undergoes two consecutive nuclear divisions, termed *meiosis I* and *II,* or the

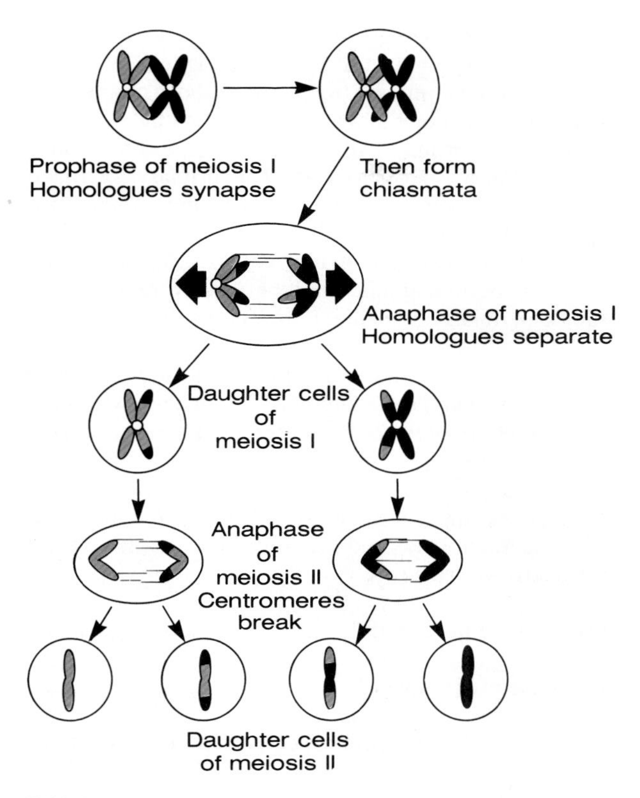

F40.1

Events of meiosis. One pair of homologous chromosomes; male homologue is dark; female homologue is red.

first and second maturation divisions, without replicating the chromosomes before the second division. The result is that four haploid daughter cells are produced, rather than two diploid daughter cells produced by mitotic division.

The entire process of meiosis is quite complex, and is discussed here only to the extent necessary to reveal important points of difference between this type of nuclear division and mitosis. Essentially, each meiotic division involves the same phases and events observed in mitosis (prophase, metaphase, anaphase, and telophase); but in the first maturation division (meiosis I), an event not seen in mitosis occurs as the mother cell goes into prophase. The homologous chromosomes, each now a duplicated structure, begin to pair so that they are closely aligned along their entire length. This pairing is called **synapsis.** As a result, 23 **tetrads** (groupings of four chromatids) form, become attached to the spindle fibers, and begin to align themselves on the spindle equator. While in synapsis, two of the four strands (one from each homologue) in each tetrad wrap and coil around each other, forming many points of **crossover,** or **chiasmata.** (Perhaps this could be called the conjugal bed of the cell!) When anaphase of meiosis I begins, the homologues separate from one another, breaking and exchanging parts at points of crossover, and move apart toward opposite poles of the cell. The centromeres, holding the "sister" chromatids or *dyads* together, do *not* break at this point (Figure 40.1).

During the second maturation division, events parallel those in mitosis, except that the daughter cells do not replicate their chromosomes before this division and each daughter cell has only half of the homologous chromosomes, rather than a complete set. The crossover events and the way in which the homologues align on the spindle equator during the first maturation division introduce an immense variability in the resulting gametes, which explains why we are all unique.

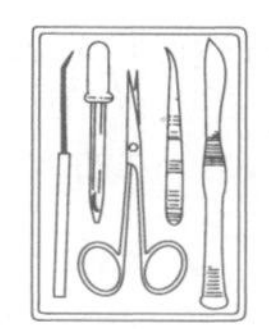

Obtain a model depicting the events of meiosis, and follow the sequence of events during the first and second maturation divisions. Identify prophase, metaphase, anaphase, and telophase in each; the tetrads and chiasmata during the first maturation division; and the dyads in the second maturation division. Note ways in which the daughter cells resulting from meiosis I differ from the mother cell and how the gametes differ from both cell populations. (Use the key on the model, your textbook, or an appropriate reference as necessary to aid you in these observations.)

If your instructor wishes you to observe meiosis in *Ascaris* to provide cellular material for comparison, continue with the microscopic study described next. Otherwise, skip to the study of spermatogenesis on page 352.

DEMONSTRATION OF OOGENESIS IN ASCARIS (OPTIONAL)

Generally speaking, oogenesis (egg production) in mammals is difficult to demonstrate in a laboratory situation. However, the process of oogenesis and mechanics of meiosis may be studied rather easily in the transparent eggs of *Ascaris megalocephala,* an invertebrate roundworm parasite found in the intestine of mammals. Such a study is significant because the process in *Ascaris* is much the same* as in humans; and because its diploid chromosome number is 4, the chromosomes are easily counted.

Go to the demonstration area where the slides are set up and make the following observations:

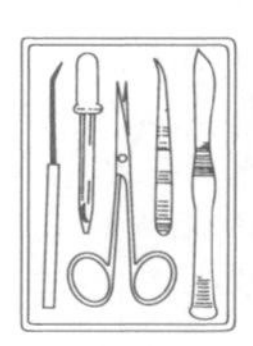

1. Scan the first demonstration slide to identify a primary oocyte, the cell type that begins the meiotic process. It will have what appears to be a relatively thick cell membrane; this is the fertilization membrane that the oocyte produces after sperm penetration. Find and study a primary oocyte that is undergoing the first maturation division. Look for a barrel-shaped spindle with two tetrads (two groups of four beadlike chromosomes) on it. Most often the spindle is located at the periphery of the cell. (The sperm nucleus may or may not be seen, depending on how the cell was cut.)

* In *Ascaris,* meiosis does not begin until the sperm has penetrated the primary oocyte, whereas in humans, meiosis I occurs before sperm penetration.

2. Observe slide 2. Locate a cell in which half of each tetrad (a dyad) is being extruded from the cell surface into the first polar body.

3. On slide 3, attempt to locate a secondary oocyte (a daughter cell produced during meiosis I) undergoing the second maturation division. In this view, two dyads (two groups of two beadlike chromosomes) will be seen on the spindle.

4. Locate, on slide 4, a cell in which the second polar body is being formed. In this case, both it and the ovum will now contain two chromosomes, the haploid number for *Ascaris.*

5. On the fifth slide, identify a fertilized egg, or a cell in which the sperm and ovum nuclei (actually *pronuclei*) are fusing to form a single nucleus containing four chromosomes.

SPERMATOGENESIS

Human sperm production begins at puberty and continues without interruption throughout life. The average male ejaculation contains between a quarter and a half-billion sperm. In light of the fact that only one sperm fertilizes an ovum, it would seem that nature has tried to assure that the perpetuation of the species will not be endangered by a lack of sperm.

As explained in Exercise 39, **spermatogenesis** occurs in the seminiferous tubules of the testes. The primitive stem cells, or **spermatogonia,** found at the tubule periphery, undergo extensive mitotic activity to build up and retain the stem cell line. Before puberty, all divisions are mitotic divisions that produce more spermatogonia. At puberty, however, under the influence of FSH secreted by the anterior pituitary gland, each mitotic division of a spermatogonium results in the production of one spermatogonium and another cell, now called a **primary spermatocyte,** which is destined to undergo meiosis. As meiosis occurs, the dividing cells approach the lumen of the tubule. Thus the progression of meiotic events can be followed from the tubule periphery to the lumen. It is important to recognize that the **spermatids,** which are actually the product of meiosis (they are haploid cells), are not functional gametes; they are nonmotile cells and have too much excess baggage to function well in a reproductive capacity. Another process, called **spermiogenesis,** which follows meiosis, strips away the extraneous cytoplasm from the spermatid and converts it to a motile, streamlined sperm.

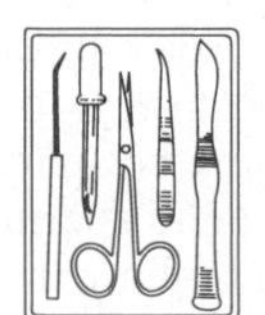

1. Obtain a slide of the testis and a microscope. Examine the slide under low power to identify the cross-sectional views of the cut seminiferous tubules (Figure 40.2). Then rotate the h.p. lens into position, and observe the wall of one of the cut tubules. As you work, refer to Figure 40.2 and to the photomicrograph of seminiferous tubules inside the back cover to make the following identifications.

2. Scrutinize the cells at the periphery of the tubule. The cells in this area are the spermatogonia, which undergo frequent mitoses to increase their number and maintain their population. About half of the spermatogonia's "offspring," called *type B cells,* differentiate to form primary spermatocytes. The primary spermatocytes begin meiosis, which leads to the formation of spermatids, which have half the usual genetic composition. The remaining daughter cells from mitotic divisions of spermatogonia, the type *A cells,* remain at the tubule periphery to maintain the germ cell line.

3. Observe the cells in the middle of the tubule wall. There you should see a large number of cells (spermatocytes) that are obviously undergoing a nuclear division process. Look for the chromosomes, visible only during nuclear division, that have the appearance of coiled springs. Attempt to differentiate between the larger primary spermatocytes and the somewhat smaller secondary spermatocytes.

Can you see the tetrads? ______________________

Evidence of crossover? ______________________

Where would you expect to see the tetrads, closer to the

spermatogonia or to the lumen? ______________________

In the primary or secondary spermatocytes? ____________

4. Examine the cells at the tubule lumen. Identify the small, round-nucleated spermatids, many of which may appear lopsided and look as though they are starting to lose their cytoplasm. See if you can find a spermatid embedded in an elongated cell type, a **sustentacular,** or **Sertoli, cell,** which extends inward from the periphery of the tubule. Sustentacular cells nourish the spermatids as they begin their transformation into sperm. Also in the adluminal area, locate sperm, which can be identified by their tails. The sperm develop directly from the spermatids by the loss of extraneous cytoplasm and the development of a propulsive tail.

5. Identify the testosterone-producing **interstitial cells** lying external to and between the seminiferous tubules.

In the next stage of sperm development, spermiogenesis, all the superficial cytoplasm is sloughed off, and the remaining cell organelles are compacted into the three regions of the mature sperm. At the risk of oversimplifying, these anatomical regions are the *head,* the *midpiece,* and the *tail,* which correspond roughly to the *activating and genetic region,* the *metabolic region,* and the *locomotor region* respectively. The mature sperm is a streamlined cell with an organ of locomotion and a high rate of metabolism, which enable it to move long distances in jig time to get to the egg. It is a prime example of the correlation of form and function.

The sperm head contains the DNA, or genetic material, of the chromosomes. Essentially, it is the nucleus of the spermatid. Anterior to the nucleus is the **acro-**

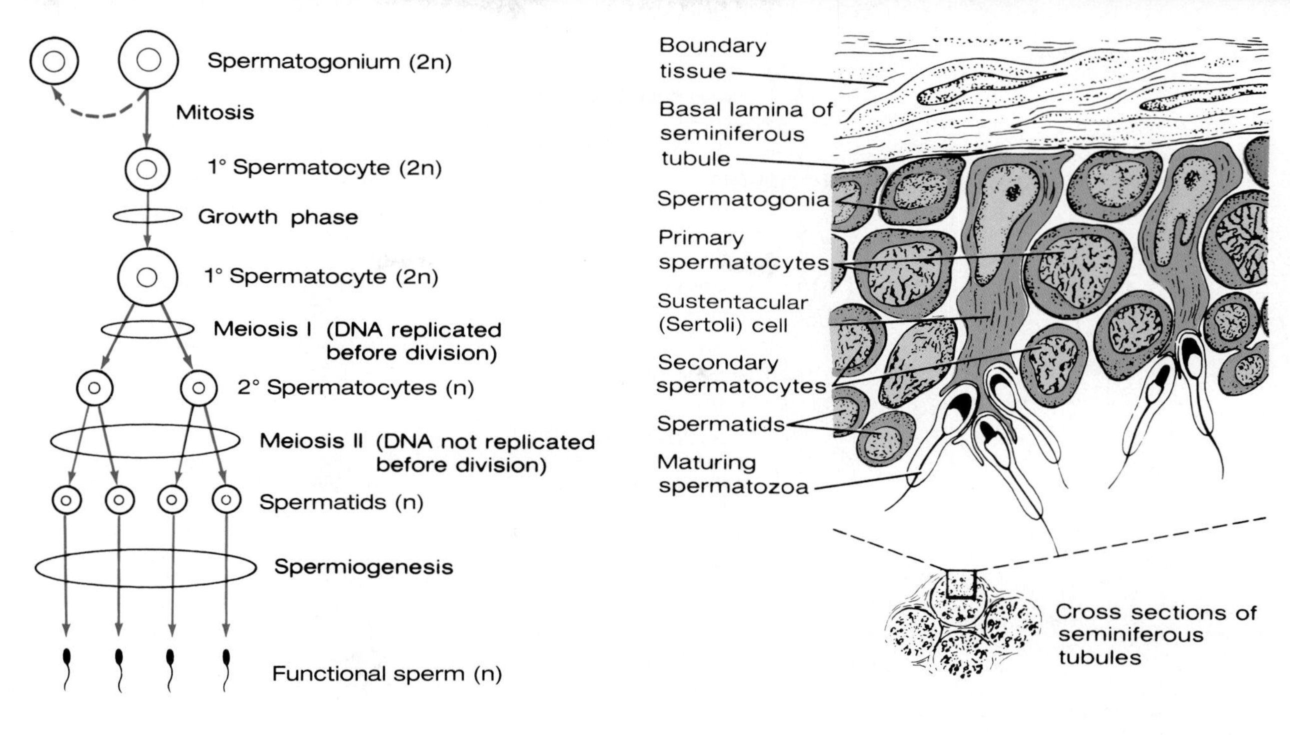

F40.2

Spermatogenesis. Left, flow chart of meiotic and spermiogenesis events; right, diagrammatic view of a portion of the wall of a seminiferous tubule. Redrawn with permission from C. R. Leeson and T. S. Leeson, *Histology,* 4th ed. (Philadelphia: W. B. Saunders, 1981).

some, which contains enzymes involved in sperm penetration of the egg.

In the midpiece of the sperm is a centriole from which arise the filaments that structure the sperm tail. Wrapped tightly around the centriole are mitochondria, which apparently provide the ATP needed for the contractile activity of the tail filaments.

The tail is composed of filaments that arise from the centriole and are constructed of contractile proteins, much like those in muscle. The filaments, when powered by ATP, propel the sperm.

6. Obtain a prepared slide of human sperm, and view it under high power or oil immersion. See the photo of sperm inside the back cover. Identify the head, acrosome, and tail regions.

7. Examine the model of spermatogenesis to identify the spermatogonia, the primary and secondary spermatocytes, the spermatids, and the functional sperm.

OOGENESIS AND THE OVARIAN CYCLE

The gonadotropic hormones produced by the anterior pituitary influence the development of ova in the ovaries and their cyclic production of female sex hormones. Within an ovary, each immature ovum develops within a saclike structure called a *follicle,* where it is encased by one or more layers of smaller cells called *follicle cells* (when one layer is present) or *granulosa cells* (when there is more than one layer).

The process of **oogenesis,** or female gamete formation, which occurs in the ovary, is very similar to spermatogenesis occurring in the testis; but there are some important differences. The process, schematically outlined in Figure 40.3, begins with the primitive stem cells called **oogonia,** located in the vascular connective tissue of the ovarian cortices of the developing female fetus. During fetal development, the oogonia undergo mitosis thousands of times until their number reaches approximately 700,000. They then become encapsulated by a single layer of squamouslike follicle cells and form the **primordial follicles** of the ovary. By the time the female child is born, most of her oogonia have increased in size and have become **primary oocytes,** which are in the prophase stage of meiosis I. Thus, at birth, the total potential for the production of germ cells in the female is already determined because the primitive stem-cell line no longer exists or will exist for only a brief period after birth.

From birth until puberty, the primary oocytes are quiescent. Then, under the influence of FSH, one or sometimes more of the follicles begin to undergo maturation approximately every 28 days.

As a follicle grows, it begins to produce estrogen; and the primary oocyte completes its first maturation division, producing two haploid daughter cells that are very disproportionate in size. One of these is the **secondary oocyte,** which contains nearly all of the cytoplasm from the primary oocyte; the other is the tiny **first polar body.** The first polar body then completes the second maturation division, producing two more polar bodies. These eventually disintegrate for lack of sustaining cytoplasm.

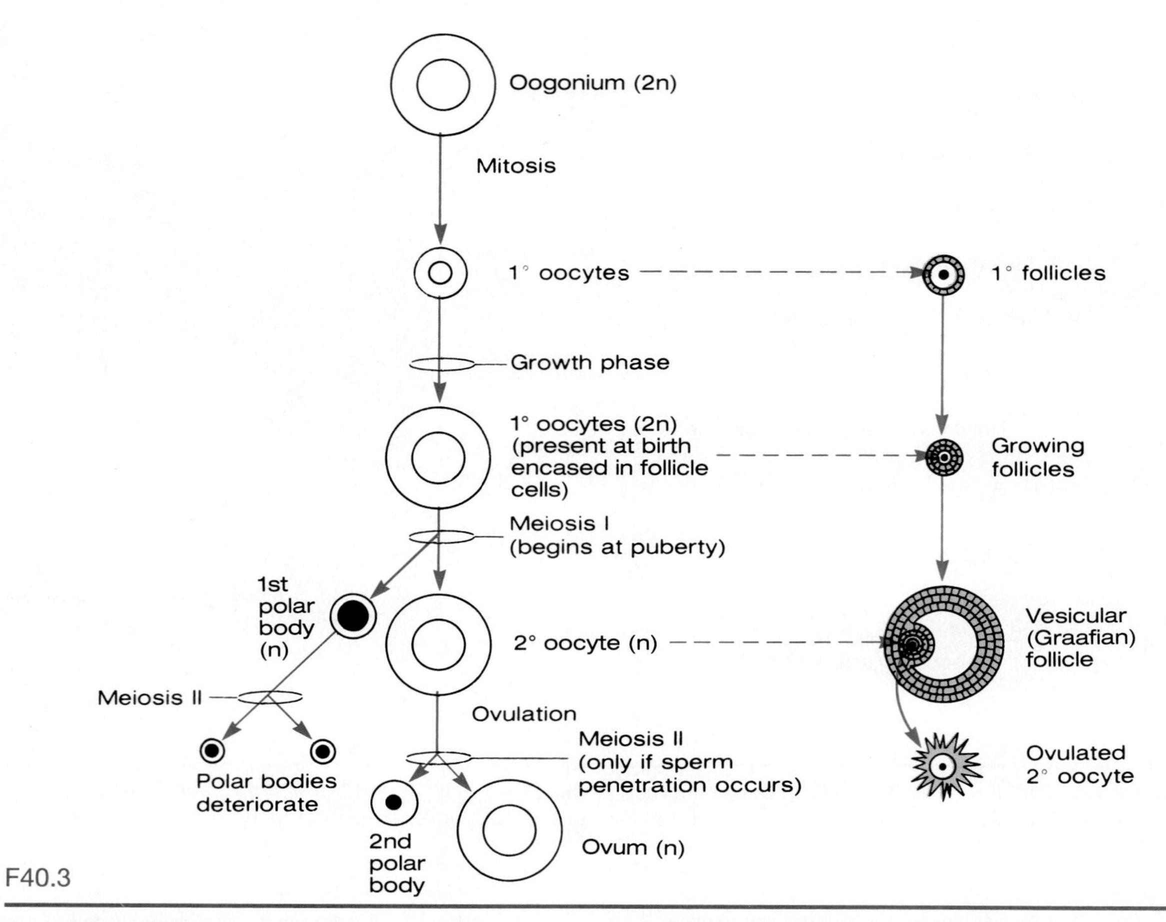

Oogenesis. Left, flow chart of meiotic events; right, correlation with follicular development and ovulation in the ovary.

As the follicle containing the secondary oocyte continues to enlarge, blood levels of estrogen rise. Initially, estrogen exerts a negative feedback influence on the release of gonadotropins by the anterior pituitary. However, approximately in the middle of the 28-day cycle, as the follicle reaches the mature **vesicular,** or **Graafian, follicle** stage, rising estrogen levels become highly stimulatory and a sudden burstlike release of LH (and, to a lesser extent, FSH) by the anterior pituitary triggers ovulation. The secondary oocyte is extruded and begins its journey down the uterine tube to the uterus. If penetrated en route by a sperm, the secondary oocyte will undergo meiosis II, producing one large **ovum** and a tiny second polar body. When the second maturation division is complete, the chromosomes of the egg and sperm combine to form the diploid nucleus of the fertilized egg. If sperm penetration does not occur, the secondary oocyte simply disintegrates without ever producing the female gamete in human females.

Thus, in the female, meiosis produces only one functional gamete, in contrast to the four produced in the male. Other major differences are the relative size and structure of the functional gametes. The sperm are tiny and are equipped with tails for locomotion. They have few organelles and virtually no nutrient-containing cytoplasm; hence the nutrients contained in semen are essential to their survival. In contrast, the egg is a relatively large, nonmotile cell that is well stocked with cytoplasmic reserves that nourish the developing embryo until implantation can be accomplished. Essentially all the zygote's organelles are "delivered" by the egg.

Once the secondary oocyte has been extruded from the ovary, LH transforms the ruptured follicle into the **corpus luteum,** which begins producing progesterone. Like estrogen, progesterone inhibits FSH release by the anterior pituitary. As FSH declines, its stimulatory effect on follicular production of estrogen ends, and estrogen blood levels begin to decline. Because the increased estrogen levels triggered LH release by the anterior pituitary, lower estrogen levels result in declining levels of LH in the blood. Because corpus luteum secretory function is maintained by high blood levels of LH, as LH blood levels begin to drop toward the end of the 28-day cycle, progesterone production ends and the corpus luteum begins to degenerate and is replaced by scar tissue (**corpus albicans**). The graphs on p. 356 depict the hormonal relationships described here.

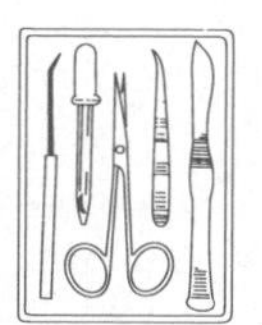

Because many different stages of ovarian development exist within the ovary at any one time, a single microscopic preparation will contain follicles at many different stages of development. Obtain a cross section of ovary tissue, and identify the following structures. Refer to Figure 40.4 as you work.

Germinal epithelium: outermost layer of the ovary.

Primary follicle: follicle with one or a few layers of cuboidal follicle cells surrounding the larger central developing ovum.

Secondary (growing) follicles: follicles consisting of several layers of follicle (granulosa) cells surround-

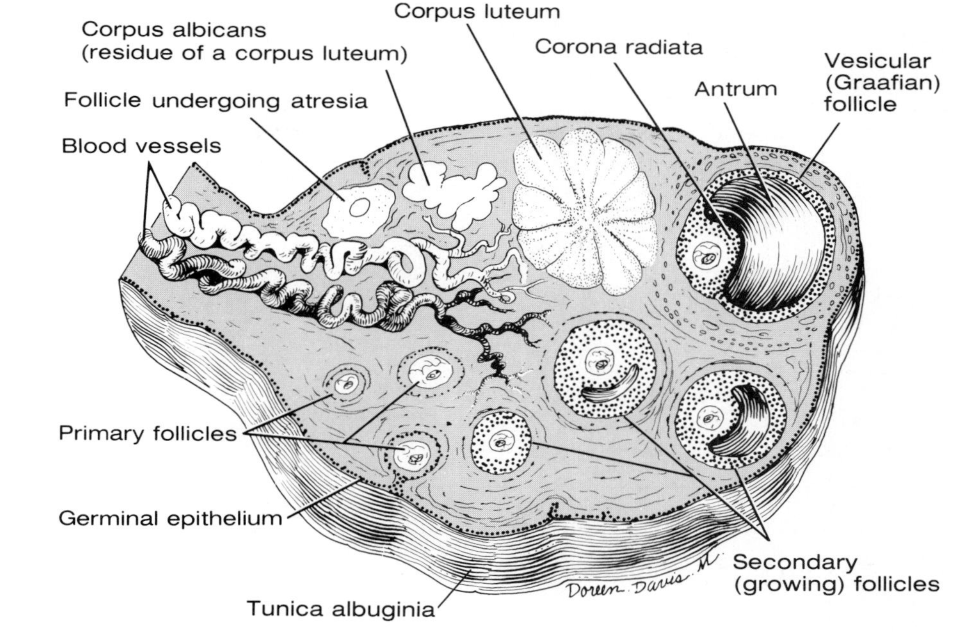

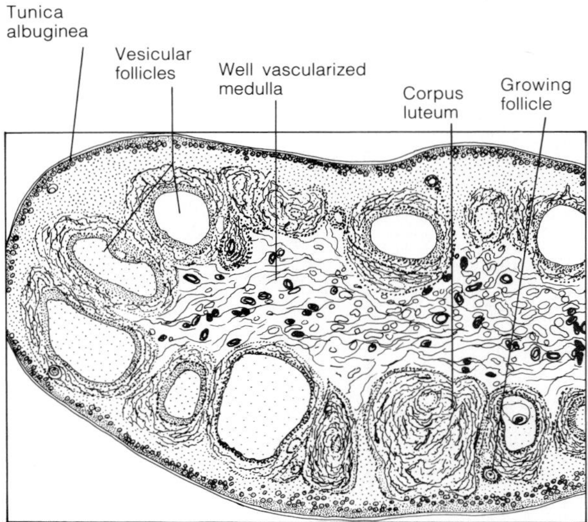

F40.4

Anatomy of the ovary. (a) Diagrammatic view of a human ovary; (b) line drawing of a photomicrograph of the ovary (See corresponding plate 51 inside front cover).

ing the central developing ovum and beginning to show evidence of fluid accumulation and **antrum** (central cavity) formation.

Vesicular (Graafian) follicle: at this stage of development, the follicle has a large antrum containing fluid produced by the granulosa cells. The developing ovum is pushed to one side of the follicle and is surrounded by a capsule of several layers of granulosa cells called the **corona radiata** (radiating crown). (When the immature ovum is released, it enters the uterine tubes with its corona radiata intact.) The connective tissue stroma (background tissue) adjacent to the mature follicle forms a capsule that encloses the follicle and is called the **theca.** (See also plate 24 in the histology atlas.)

Corpus luteum: a solid glandular structure or a structure containing a scalloped lumen that develops from the ovulated follicle. (See plate 25 of the histology atlas.)

Examine the model of oogenesis, and compare it with the spermatogenesis model. Note differences in the size and structure of the functional gametes.

THE MENSTRUAL CYCLE

The **menstrual cycle,** sometimes referred to as the **uterine cycle,** is hormonally controlled by the ovarian production of estrogen and progesterone. It is normally divided into three stages: menstrual, proliferative, and secretory. These stages, shown in the bottom portion of Figure 40.5, are described as follows:

Menstrual stage (menses): approximately days 1 through 5; sloughing off of the thick functionalis portion of the endometrial lining of the uterus, accompanied by bleeding.

Proliferative stage: approximately days 6 through 14. Under the influence of estrogens produced by the growing follicle of the ovary, the endometrium is repaired, glands and blood vessels proliferate, and the endometrium thickens. Ovulation occurs at the end of this stage.

Secretory stage: approximately days 15 through 28. Under the influence of progesterone produced by the corpus luteum, the vascular supply to the endometrium increases further. The glands increase in size and begin to secrete nutrient substances, which sustain a developing embryo, if present, until implantation can occur. If fertilization has occurred, the embryo will produce a hormone much like LH,

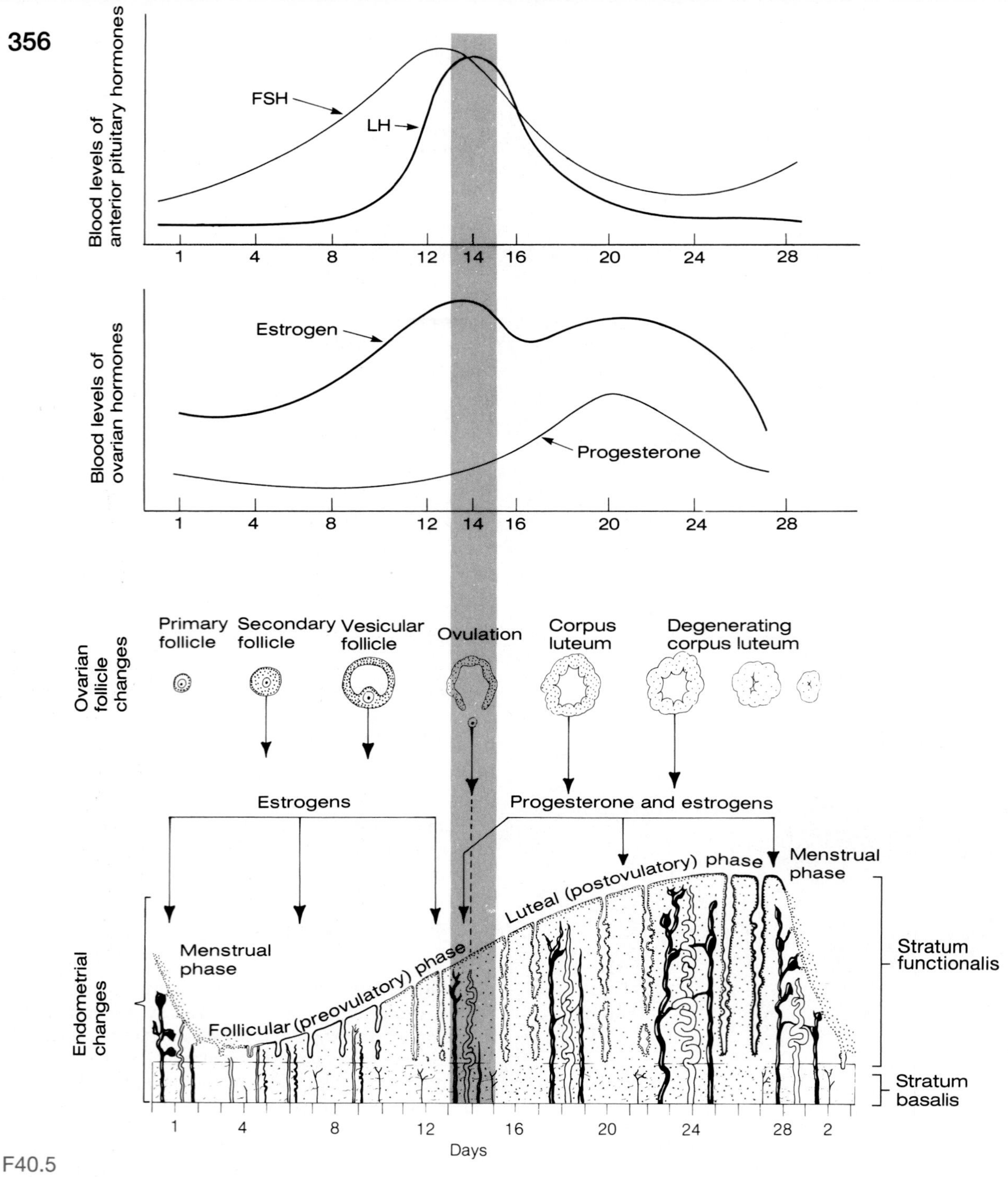

F40.5

Hormonal interactions of the female cycles. Relative levels of anterior pituitary hormones correlated with follicular and hormonal changes of the ovary. The menstrual cycle (endometrial changes) is depicted at the bottom of the figure.

which will maintain the function of the corpus luteum. Otherwise, as the corpus luteum begins to deteriorate, lack of ovarian hormones in the blood causes blood vessels supplying the endometrium to kink and become spastic, setting the stage for menses to begin by the 28th day.

Although the foregoing explanation assumes a classic 28-day cycle, the length of the menstrual cycle is highly variable, sometimes as short as 21 days or as long as 38. Only one interval is relatively constant in all females: The time from ovulation to the onset of menstruation is almost always 14 days.

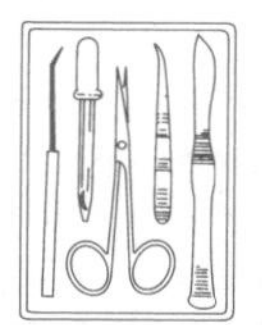

Obtain slides showing the menstrual, secretory, and proliferative phases of the uterine endometrium. Carefully observe each, comparing their relative thicknesses and vascularity. As you work, refer to the corresponding photomicrographs of these phases provided inside the back cover.

EXERCISE 41

Principles of Heredity

OBJECTIVES

1. To define *allele, dominance, genotype, heterozygous, homozygous, incomplete dominance, phenotype,* and *recessiveness.*
2. To gain practice working out simple genetics problems, using a Punnet square.
3. To become familiar with the basic laws of probability.
4. To observe selected human phenotypes and determine their genotype basis.

MATERIALS

Pennies (for coin tossing)
PTC (phenylthiocarbamide) taste strips
Sodium benzoate taste strips
Chart drawn on chalkboard for tabulation of class results of human phenotype/genotype determinations
Blood-typing supplies: anti-A and anti-B sera, slides, toothpicks, wax pencils, sterile lancets, alcohol swabs
Beaker containing 10% bleach solution
Disposable autoclave bag

The field of genetics is currently bristling with excitement. Complex gene-splicing techniques have allowed researchers to precisely isolate genes coding for specific proteins, and then to insert those genes into bacterial or tumor cells so that large amounts of particular proteins can be harvested. At present, growth hormone, insulin, and interferon produced by these genetic engineering techniques are available for clinical use.

Comprehension of the complex science of genetics in relation to such studies requires arduous training. However, a basic understanding and appreciation of how genes regulate our various traits (dimples and hair color, for example) can be gained by anyone. The thrust of this exercise is to provide a "genetics sampler," or relatively simple introduction to the principles of heredity.

INTRODUCTION TO THE LANGUAGE OF GENETICS

In humans, all cells, except eggs and sperm, contain 46 chromosomes—that is, the diploid number. This number is established when fertilization occurs as a result of the fusion of the ovum and sperm, and the 23 chromosomes (or haploid complement) each is carrying combine. The diploid chromosomal number is maintained throughout life in nearly all cells of the body by the precise process of mitosis. As explained in Exercise 40, the diploid chromosomal number actually represents two complete (or nearly complete) sets of genetic instructions—one from the egg and the other from the sperm—or 23 pairs of **homologous chromosomes.**

Genes coding for the same traits on each pair of homologous chromosomes are called **alleles.** The alleles may be identical or different in their influence. For example, the members of the gene pair, or alleles, coding for hairline shape on your forehead may specify either straight across or widow's peak. When both alleles in a homologous chromosome pair have the same expression, the individual is said to be **homozygous** for that trait. When the alleles differ in their expression, the individual is **heterozygous** for the given trait; and typically only one of the alleles, called the **dominant gene,** will exert its effects. The allele with less potency, the **recessive gene,** while still present, will be masked. Whereas dominant genes, or alleles, exert their effects in both homozygous and heterozygous conditions, as a rule recessive alleles *must* be present in double dose (homozygosity) to exert their influence. An individual's actual genetic makeup—that is, whether one is homozygous or heterozygous for the various alleles—is called **genotype.** The manner in which genotype is expressed (for example, the presence of a widow's peak or not, blue vs. brown eyes) is referred to as **phenotype.**

The complete story of heredity is much more complex than just outlined, and in actuality the expression of many traits (for example, eye color) is determined by the interaction of many allele pairs. However, our emphasis here will be to investigate only the less complex aspects of genetics.

DOMINANT-RECESSIVE INHERITANCE

One of the best ways to master the terminology and to learn the principles of heredity is to work out the solutions to some genetic crosses, much in the manner Mendel did in his classic experiments with pea plants. To work out the various simple monohybrid (one pair of alleles) crosses in this exercise, you will be given the genotype of the parents. You will then determine the possible genotypes of their offspring by using the *Punnet square,* and you will record both genotype and phenotype percentages. To illustrate this procedure, an example of one of Mendel's pea plant crosses is outlined next.

Alleles: T (determines *tallness:* dominant)
t (determines *dwarfness:* recessive)
Genotypes of parents: TT (♂) × tt (♀)
Phenotypes of parents: Tall × dwarf

To use the Punnet, or checkerboard, square, write the alleles (actually gametes) of one parent across the top and the gametes of the other parent down the left side. Then combine the gametes across and down to achieve all possible combinations, as shown below:

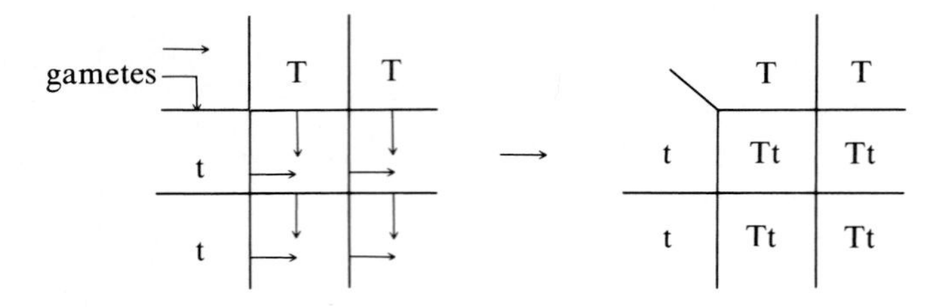

Results: Genotypes 100% Tt (all heterozygous)
Phenotypes 100% tall (because T, which determines tallness, is dominant and all contain the T allele).

1. Using the technique outlined above, determine the genotypes and phenotypes of the offspring of the following crosses:

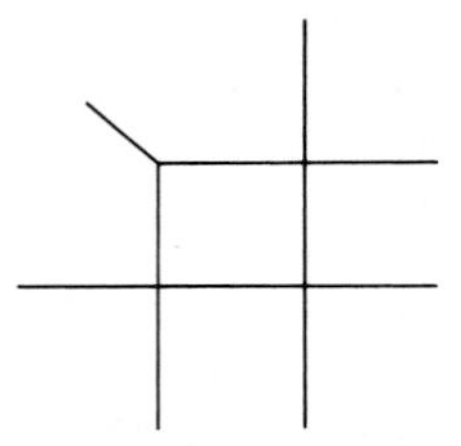

a. Genotypes of parents: Tt (♂) × tt (♀)

% of each genotype: ____________

% of each phenotype: ____________ % tall

____________ % dwarf

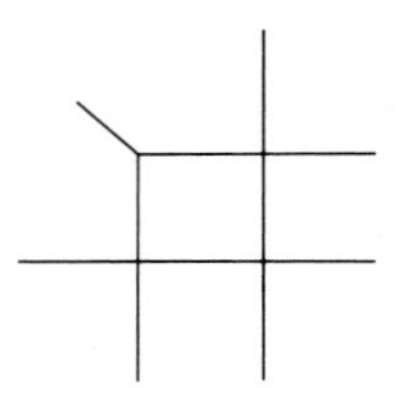

b. Genotypes of parents: Tt (♂) × Tt (♀)

% of each genotype: ____________

% of each phenotype: ____________ % tall

____________ % dwarf

2. In guinea pigs, rough coat (R) is dominant over smooth coat (r). What will be the genotypes and phenotypes of the following monohybrid crosses?

a. Genotypes of the parents: RR × rR

% of each genotype: ____________

% of each phenotype: ____________ % rough

____________ % smooth

b. Genotypes of the parents: Rr × rr

% of each genotype: ____________

% of each phenotype: ____________ % rough

____________ % smooth

c. Genotypes of the parents: RR × rr

% of each genotype: ____________

% of each phenotype: ____________ % rough

____________ % smooth

INCOMPLETE DOMINANCE

In actuality, the concepts of dominance and recessiveness are somewhat arbitrary and artificial in some instances, because so-called dominant genes may be expressed differently in homozygous and heterozygous individuals. This gives rise to a condition called *incomplete dominance* or *intermediate inheritance.* In such cases, both alleles express themselves in the offspring. The crosses are worked out in the same manner as indicated previously; but the heterozygous offspring exhibit a phenotype intermediate between that of the homozygous individuals. Some examples follow.

1. The inheritance of flower color in snapdragons illustrates the principle of incomplete dominance. The genotype RR is expressed as a red flower. Rr yields pink flowers, and rr produces white flowers. Work out the following crosses to determine the phenotypes seen, and both genotype and phenotype percentages.

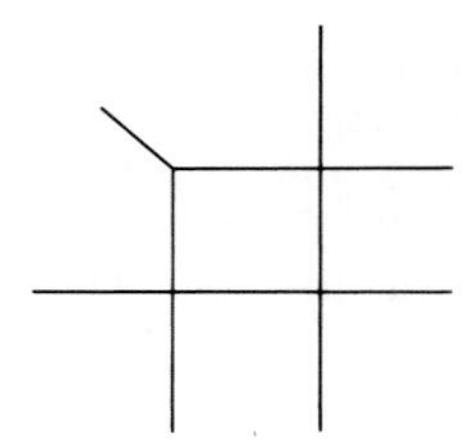

a. Genotypes of parents: RR × rr

Genotypes and %: ______________________

Phenotypes and %: ______________________

____

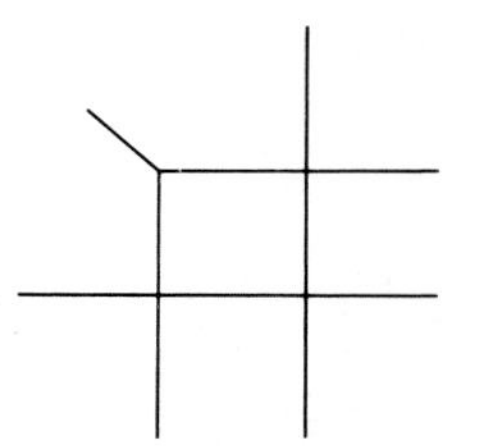

b. Genotypes of parents: Rr × rr

Genotypes and %: ______________________

Phenotypes and %: ______________________

____

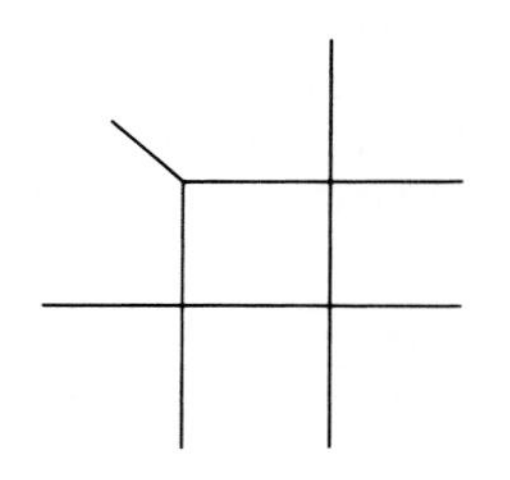

c. Genotypes of parents: Rr × Rr

Genotypes and %: ______________________

Phenotypes and %: ______________________

____

2. In humans, the inheritance of sickle-cell anemia/trait is determined by a single pair of alleles that exhibit incomplete dominance. Individuals homozygous for the sickling gene(s) are said to have *sickle-cell anemia.* In double dose (ss) the sickling gene causes the production of a very abnormal hemoglobin, which crystallizes and becomes sharp and spiky under conditions of oxygen deficit. This, in turn, leads to the clumping and hemolysis of red blood cells in the circulation, which causes a great deal of pain and can be fatal. Heterozygous individuals (Ss) are said to have the *sickle-cell trait,* which is much less severe; however, they are carriers of the abnormal gene and may pass it on to their offspring. Individuals with the genotype SS form normal hemoglobin. Work out the following crosses.

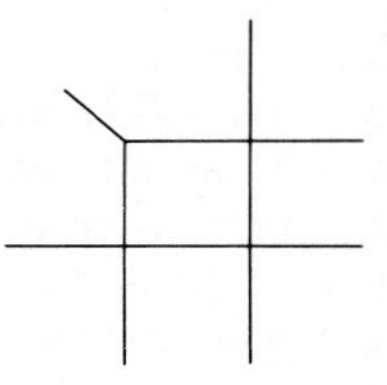

a. Parental genotypes: SS × ss

Genotypes and %: ______________________

Phenotypes and %: ______________________

____

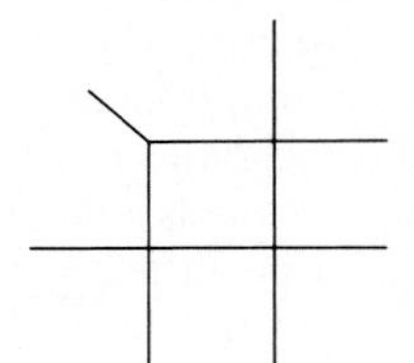

b. Parental genotypes: Ss × Ss

Genotypes and %: ______________________

Phenotypes and %: ______________________

____

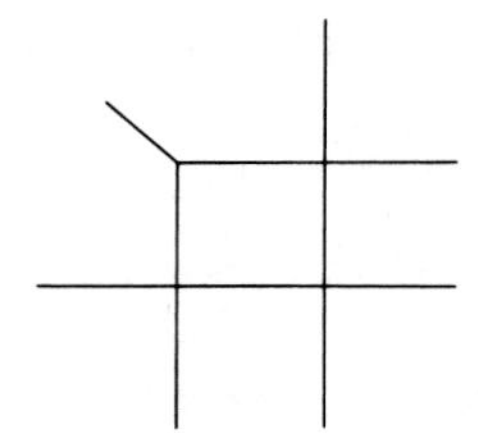

c. Parental genotypes: ss × Ss

Genotypes and %: ____________________

Phenotypes and %: ____________________

____

SEX-LINKED INHERITANCE

Of the 23 pairs of homologous chromosomes, 22 pairs are referred to as **autosomes.** Autosomes contain genes that determine most body (somatic) characteristics. The 23rd pair, called the **sex chromosomes,** determine the sex of an individual—that is, whether an individual will be male or female. Normal females possess two sex chromosomes that look alike, the X chromosomes. Males possess two dissimilar sex chromosomes, referred to as X and Y. Possession of the Y chromosome determines maleness. A photomicrograph of a male's chromosome complement (male karyotype) is shown in Figure 41.1.

The Y sex chromosome is only about a third as large as the X sex chromosome and lacks many of the genes (directing characteristics other than sex) that are found on the X. Genes present *only* on the X sex chromosome are called *sex-linked* (or X-linked) genes. Some examples of X-linked genes include those that determine normal color vision (or, conversely, color blindness), and normal clotting ability (as opposed to hemophilia, or bleeder's disease). The alleles that determine color blindness and hemophilia are recessive alleles. In females, *both* X chromosomes must carry the recessive alleles for a woman to express either of these conditions; and thus they tend to be infrequently seen. However, should a male receive even one sex-linked recessive allele for these conditions, he will exhibit the recessive phenotype because his Y chromosome lacks any genes that might dominate, or mask, the recessive allele.

The critical understanding of X-linked inheritance is the *absence* of male-to-male (that is, father-to-son) transmission of X-linked genes. The X of the father *will* pass to each of his daughters but to none of his sons; males always inherit sex-linked conditions from their mothers (through the X chromosome).

1. A heterozygous woman carrying the recessive gene for color blindness marries a man who is color-blind. Assume the dominant gene is X^C (allele for normal color vision) and the recessive gene is X^c (determines color blindness). The mother's genotype is $X^C X^c$, and the father's is $X^c Y$. Do a Punnet square to determine the answers to the following questions.

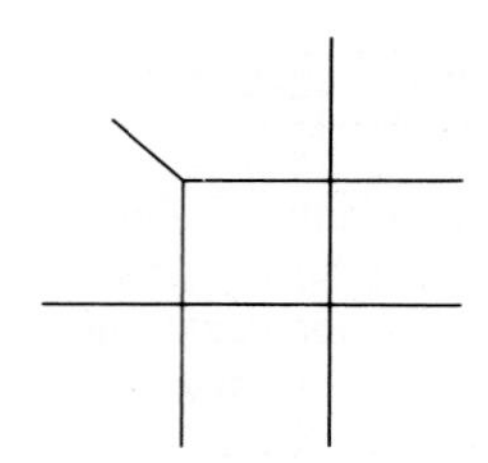

According to the laws of probability, what percentage of their children will be color-blind?

____________ %

What is the proportion of color-blind individuals by sex? ____________ male; ____________ female

What percentage will be carriers? ____________ %

What is the sex of the carriers? ____________

2. A heterozygous woman carrying the recessive gene for hemophilia marries a man who is not a hemophiliac. Assume the dominant gene is X^H and the recessive gene is X^h. The woman's genotype is $X^H X^h$, and her husband's genotype is $X^H Y$. Of each sex of their offspring, what is the potential percentage that will be hemophiliacs?

____________ % males; ____________ % females

What percentage can be expected neither to exhibit nor to carry the allele for hemophilia?

____________ %

What is the anticipated sex and percentage of individuals that will be carriers for hemophilia?

____________ %; ____________ sex

PROBABILITY

Because the segregation (or parceling out) of chromosomes to daughter cells (gametes) during meiosis and the combination of egg and sperm are random or chance events, the possibility that certain genomes will arise and be expressed is based on the laws of probability. The randomness of gene recombination from each parent determines individual uniqueness and explains why siblings, however similar, never have totally corresponding traits (unless, of course, they are identical twins). The Punnet square method that you have been using to work out the genetics problems actually provides information on the *probability* that certain genotypes will appear, considering all possible events. Probability (P) is defined as:

$$P = \frac{\text{number of specific events/cases}}{\text{total number of events/cases}}$$

If an event is certain to happen, its probability is 1. If

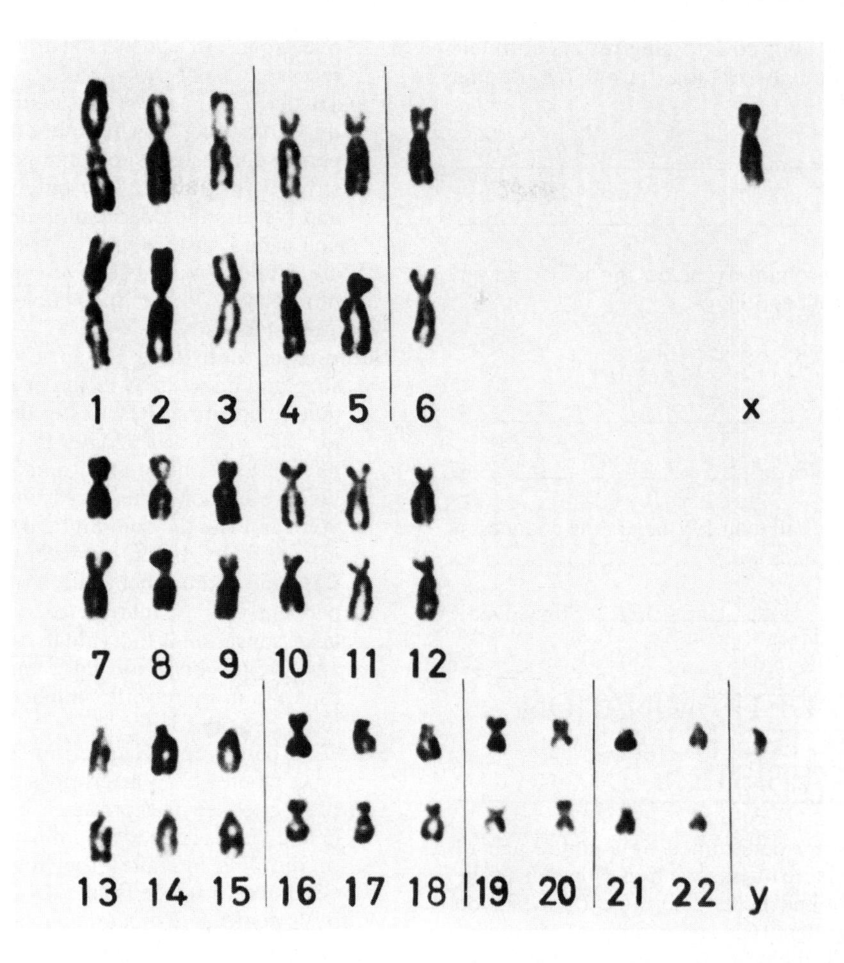

F41.1

Karyotype of human male chromosomes. Each pair of homologous chromosomes is numbered (1–22), except for the sex chromosomes, which are identified by their letters, X and Y.

it happens one out of every two times, its probability is ½; if one out of 4 times, its probability is ¼, and so on.

When figuring the probability of separate events occurring together (or consecutively), the probabilities of each event must be multiplied together to get the final probability figure. For example, the probability of a penny coming up "heads" in each toss is ½ (because it has two sides—heads and tails). Therefore, the probability of a tossed penny coming up heads four times in a row is: $\frac{1}{2} \times \frac{1}{2} \times \frac{1}{2} \times \frac{1}{2} = \frac{1}{16}$.

1. Obtain two pennies, and perform the following simple experiment to explore the laws of probability.

a. Toss one penny into the air 10 times, and record the number of heads/tails observed.

______________ heads ______________ tails

Probability: ______/10ths heads; ______/10ths tails

b. Now simultaneously toss two pennies into the air for 24 tosses; and record, below, the results of each toss. In each case, report the probability in the lowest fractional terms.

HH ______________ Probability ________

HT ______________ Probability ________

TT ______________ Probability ________

Does the first toss have any influence on the second?

____

Does the third toss have any influence on the fourth?

____

c. Do a Punnet square, using HT for one coin and HT for the alleles of the other.

Probability of HH: ________

Probability of HT: ________

Probability of TT: ________

How closely do your coin-tossing results correlate with the percentages obtained from the Punnet square results?

___

___

2. Determine the probability of having a boy or girl offspring for each conception.

Parental genotypes: XY × XX

Probability of a male: ___ %

Probability of a female: ___ %

3. Dad wants a baseball team! What are the chances of his having nine sons in a row?

___ (Sorry, Dad!)

GENETIC DETERMINATION OF SELECTED HUMAN CHARACTERISTICS

Many human traits are determined by a single pair of alleles easily identifiable by observation. For each of the characteristics described here, determine (as best you can) both your own phenotype and genotype, and record this information in Table 41.1. Because it is impossible to know whether you are homozygous or heterozygous for a trait when you exhibit its dominant expression, you are to record your genotype as A− (or B−, and so on, depending on the letter used to indicate the alleles) in such cases. If you exhibit the recessive trait, you are homozygous for the recessive allele and should record it accordingly as aa (bb, cc, and so on). When you have completed your observations, also record your data in the chart for tabulation of class results, which has been set up on the chalkboard.

Tongue rolling: Extend your tongue and attempt to roll it into a U-shape longitudinally. People with this ability have the dominant allele for this trait. Use T for the dominant allele and t for the recessive allele (see Figure 41.2).

Attached earlobes: Have your lab partner examine your earlobes. If no portion of the lobe hangs free inferior to its point of attachment to the head, you are homozygous recessive (ee) for attached earlobes. If part of the lobe hangs free below the point of attachment, you possess at least one dominant gene (E) (see Figure 41.2).

Interlocking fingers: Clasp your hands together by interlocking your fingers. Now observe your clasped hands. Which thumb is uppermost? If the left thumb is uppermost, you possess a dominant allele (I) for this trait. If you clasped your right thumb over your left, you are illustrating the homozygous recessive (ii) phenotype.

PTC taste: Obtain a PTC taste strip. PTC, or phenylthiocarbamide, is a harmless chemical that some people can taste and others find tasteless. Chew the strip. If it tastes slightly bitter, you are a "taster" and possess the dominant gene (P) for this trait. If you cannot taste anything, you are a nontaster and are homozygous recessive (pp) for the trait. Approximately 70% of the people in the United States are tasters.

Sodium benzoate taste: Obtain a sodium benzoate taste strip and chew it. A different pair of alleles (from that determining PTC taste) determines the ability to taste sodium benzoate. If you can taste it, you have at least one of the dominant alleles (S). If not, you are homozygous recessive (ss) for the trait. If you can taste the sodium benzoate, record whether it tastes salty, bitter, or sweet to you. Even though PTC and sodium benzoate taste are inherited independently, they interact to determine a person's taste sensations. Individuals who find PTC bitter and sodium benzoate salty tend to be devotees of sauerkraut, buttermilk, spinach, and other slightly bitter or salty foods.

Sex: The genotype XX determines the female phenotype, whereas XY determines the male phenotype.

Dimpled cheeks: The presence of dimples in one or both cheeks is due to a dominant gene (D). Absence of dimples indicates the homozygous recessive condition (dd) (see Figure 41.2).

Widow's peak: A distinct downward V-shaped hairline at the middle of the forehead is referred to as a widow's peak. It is determined by a dominant allele (W), whereas the straight or continuous forehead hairline is determined by the homozygous recessive condition (ww) (see Figure 41.2).

Bent little finger: Examine the little finger on each of your hands. If its terminal phalanx angles toward the ring finger, you are dominant for this trait. If one or both terminal digits are essentially straight, you are homozygous recessive for the trait. Use L for the dominant allele and l for the recessive allele.

Double-jointed thumb: A dominant gene determines a condition of loose ligaments that allows one to throw the thumb out of joint. The homozygous recessive condition determines tight joints. Use J for the dominant allele and j for the recessive allele.

Middigital hair: Critically examine the dorsum of the middle segment (phalanx) of fingers 3 and 4. If *no* hair is obvious, you are recessive (hh) for this condition. If hair is present, you have the dominant gene (H) for this trait (which, however, is determined by multigene inheritance) (see Figure 41.2).

Freckles: The appearance of freckles is the result of a dominant gene. Use F as the dominant allele and f as the recessive allele (see Figure 41.2).

Blaze: A lock of hair different in color from the rest of scalp hair is called a blaze; it is determined by a dominant gene. Use B for the dominant gene and b for the recessive gene.

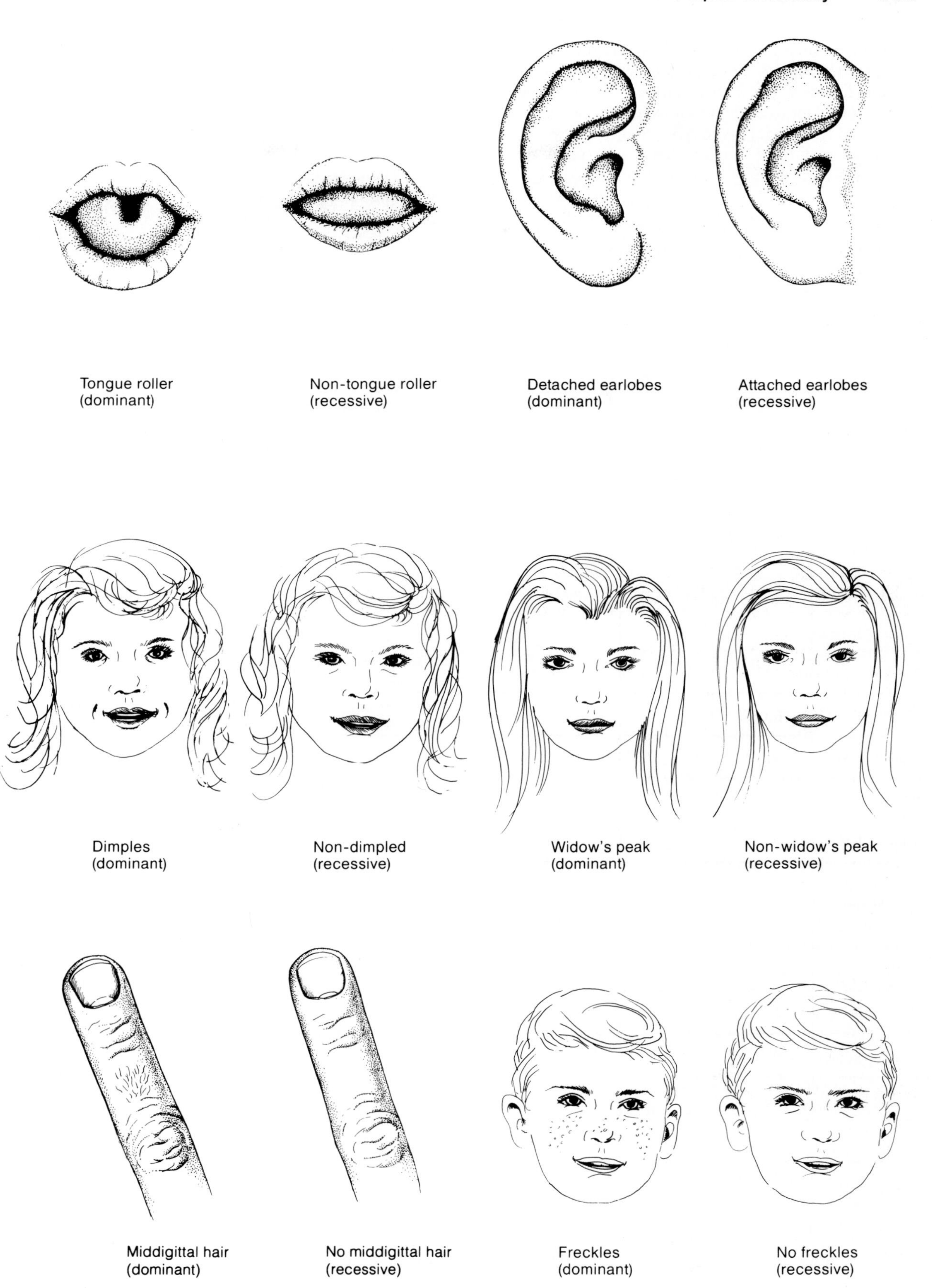

F41.2

Selected examples of human phenotypes.

Blood type: Inheritance of ABO blood type is based on the existence of three alleles designated as I^A, I^B, and i. Both I^A and I^B are dominant over i, but neither is dominant over the other. Thus the possession of I^A and I^B will yield type AB blood, whereas the possession of the I^A and i alleles will yield type A blood, and so on, as explained in Exercise 26.

There are four ABO blood groups, or phenotypes, A, B, AB, and O; and their correlation to genotype is indicated as follows:

ABO blood group	Genotype
A	I^AI^A or I^Ai
B	I^BI^B or I^Bi
AB	I^AI^B
O	ii

Assuming you have previously typed your blood, record your phenotype and genotype on the chart. If you have not, type your blood following the instructions on page 249; and then enter your results in the table.

Dispose of your blood-soiled supplies by placing the glassware in the bleach-containing beaker and all other items in the autoclave bag.

Once class data have been tabulated, scrutinize the results. Is there a single trait that is expressed in an identical manner by all members of the class?

____

Because all human beings have 23 pairs of homologues and each pair segregates independently at meiosis, the number of possible combinations at segregation is over 8 million! On the basis of this information, what would you guess are the chances of *any* two individuals in the class having identical phenotypes for all 14 traits investigated?

TABLE 41.1 Record of Human Genotypes/Phenotypes

Characteristic	Phenotype	Genotype
Tongue rolling (T,t)		
Attached earlobes (E,e)		
Interlocking fingers (I,i)		
PTC taste (P,p)		
Sodium benzoate taste (S,s)		
Sex (X,Y)		
Dimples (D,d)		
Widow's peak (W,w)		
Bent little finger (L,l)		
Double-jointed thumb (J,j)		
Middigital hair (H,h)		
Freckles (F,f)		
Blaze (B,b)		
ABO blood type (I^A,I^B,i)		

Surface Anatomy Roundup

OBJECTIVES

1. Define surface anatomy, and explain why it is an important field of study. Define *palpation.*
2. Describe and palpate the major surface features of the cranium, face, and neck.
3. Describe the easily palpated bony and muscular landmarks of the back. Locate the vertebral spines on the living body.
4. List the bony surface landmarks of the thoracic cage, and explain how they relate to the major soft organs of the thorax. Explain how to find any rib (second to eleventh).
5. Name and palpate the important surface features on the anterior abdominal wall, and explain how to palpate a full bladder.
6. Define and explain the following: *linea alba, umbilical hernia,* examination for an inguinal hernia, *linea semilunaris,* and *McBurney's point.*
7. Locate and palpate the main surface features of the upper limb.
8. Explain the significance of the cubital fossa, pulse points in the distal forearm, and the anatomical snuff box.
9. Describe and palpate the surface landmarks of the lower limb.
10. Explain exactly where to administer an injection in the gluteal region and in the other major sites of intramuscular injection.

MATERIALS

Articulated skeletons
Charts or models of the skeletal muscles of the body
Washable markers
Hand mirror
Stethoscope
Alcohol swabs

Surface anatomy is an extremely valuable branch of anatomical and medical science. True to its name, surface anatomy does indeed study the *external surface* of the body, but more importantly, it also studies *internal* organs as they relate to external surface landmarks and as they are seen and felt through the skin. Feeling internal structures through the skin with the fingers is called **palpation** (literally, "touching").

Surface anatomy is living anatomy, better studied in live people than in cadavers. It can provide a great deal of information about the living skeleton (almost all bones can be palpated) and about the muscles and vessels that lie near the body surface. Furthermore, a skilled examiner can learn much about the heart, lungs, and other deep organs by performing a surface assessment. Thus, surface anatomy serves as the basis of the standard physical examination. For those of you planning a career in the health sciences or physical education, a study of surface anatomy will show you where to take pulses, where to insert tubes and needles, where to locate broken bones and inflamed muscles, and where to listen for the sounds of the lungs, heart, and intestines.

We will take a regional approach to surface anatomy, exploring the head first and proceeding to the trunk and the limbs. You will be observing and palpating your own body as you work through the exercise, because your body is the best learning tool of all. To aid your exploration of living anatomy, skeletons and muscle models are provided around the lab so that you can review the bones and muscles you will encounter. Where you are asked to mark with a washable marker, areas of skin that are personally unreachable, it probably would be best to choose a male student as a subject.

THE HEAD

The head (Figures 42.1 and 42.2) is divided into the cranium and the face.

Cranium

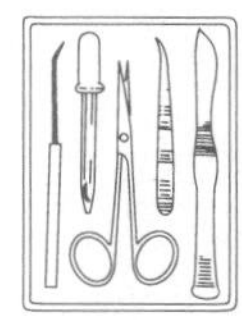

1. Run your fingers over the superior surface of your head. Notice that the underlying cranial bones lie very near the surface. Proceed to your forehead and palpate the *superciliary arches* ("brow ridges") directly superior to your orbits (Figure 42.1).

2. Move your hand to the posterior surface of your skull, where you can feel the knoblike *external occipital protuberance.* Run your finger directly laterally from this protuberance to feel the ridgelike *superior nuchal line* on the occipital bone. This line, which marks the superior extent of the muscles of the posterior neck,

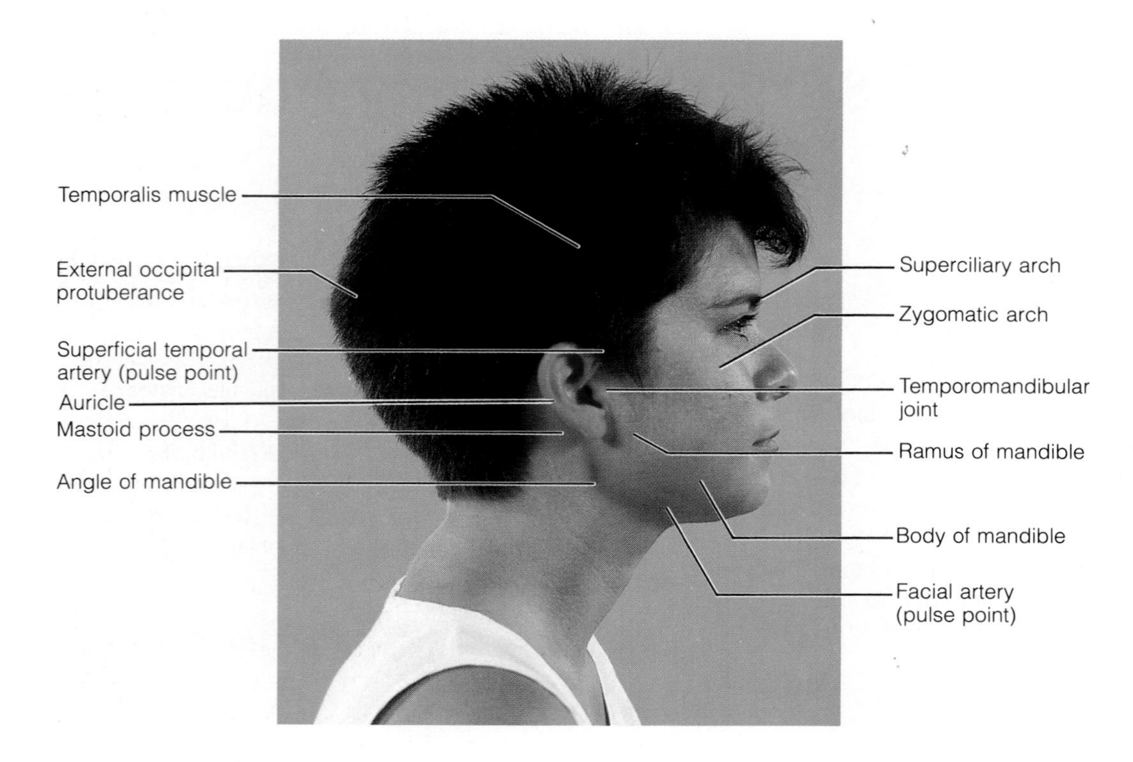

F42.1

Surface anatomy of the lateral aspect of the head.

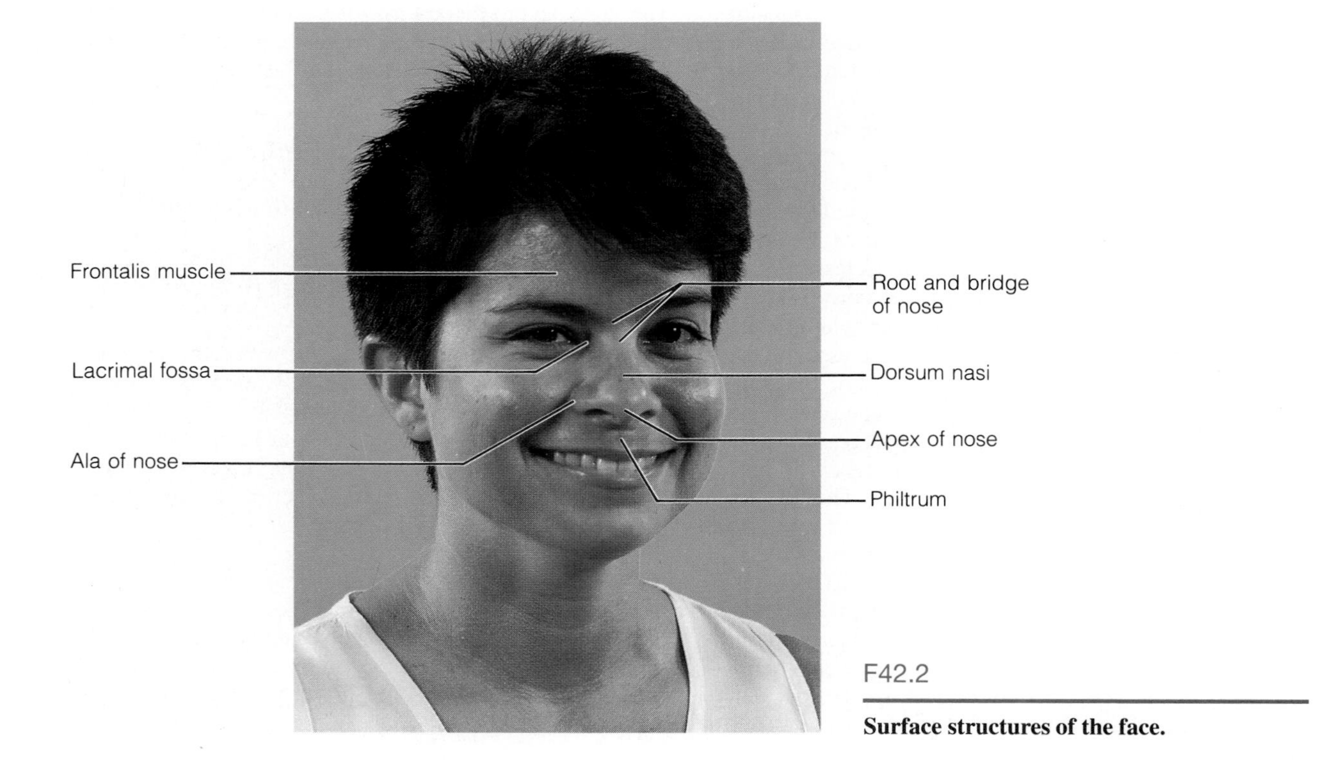

F42.2

Surface structures of the face.

serves as the boundary between the head and the neck. Now feel the prominent *mastoid process* on each side of the cranium just posterior to your ear.

3. Place a hand on your temple, and clench your teeth together in a biting action. You should be able to feel the *temporalis muscle* bulge as it contracts. Next, raise your eyebrows, and feel your forehead wrinkle, an action produced by the *frontalis muscle.* The frontalis inserts superiorly onto a broad aponeurosis called the *galea aponeurotica* (Table 13.1, p. 108), which covers the superior surface of the cranium. This aponeurosis binds tightly to the overlying subcutaneous tissue and skin to form the true **scalp.** Push on your scalp, and confirm that it slides freely over the underlying cranial bones. Because the scalp is only loosely bound to the skull, people can easily be "scalped" (in industrial accidents, for example). The scalp is richly vascularized by a large number of arteries running through its subcutaneous tissue. Most arteries of the body constrict and close after they are cut or torn, but those in the scalp are unable to do so because they are held open by the dense connective tissue surrounding them.

What do these facts suggest about the amount of bleeding that accompanies scalp wounds?

____

____

____

Face

The surface of the face is divided into many different regions, including the *orbital, nasal, oral* (mouth), and *auricular* (ear) areas (Figure 42.2).

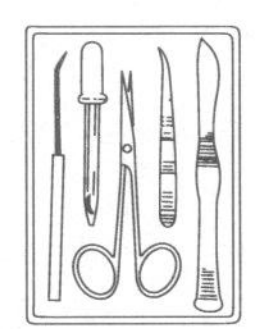

1. Trace a finger around the entire bony margin of an orbit. The *lacrimal fossa,* which contains the tear-gathering lacrimal sac, may be felt on the medial side of the eye socket.

2. Touch the most superior part of your nose, its **root,** between the eyebrows (Figure 42.2). Just inferior to this, between your eyes, is the **bridge** of the nose formed by the nasal bones. Continue your finger's progress inferiorly along the nose's anterior margin, the **dorsum nasi,** to its tip, the **apex.** Place one finger in a *nostril* and another finger on the flared wing, the **ala,** that defines the nostril's lateral border. Then feel the **philtrum,** the shallow vertical groove on the upper lip below the nose.

3. Grasp your *auricle,* the shell-like part of the external ear that surrounds the opening of the *external auditory canal* (Figure 42.1). Now trace the ear's outer rim, or *helix,* to the *lobule* (earlobe) inferiorly. The lobule is easily pierced, and since it is not highly sensitive to pain, it provides a convenient place to hang an earring or obtain a drop of blood for clinical blood analysis. Next, place a finger on your temple just anterior to the auricle (Figure 42.1). There, you will be able to feel the pulsations of the *superficial temporal artery,* which ascends to supply the scalp.

4. Run your hand anteriorly from your ear toward the orbit, and feel the *zygomatic arch* just deep to the skin. This bony arch is easily broken by blows to the face. Next, place your fingers on the skin of your face, and feel it bunch and stretch as you contort your face into smiles, frowns, and grimaces. You are now monitoring the action of several of the subcutaneous *muscles of facial expression* (Table 13.1).

5. On your lower jaw, palpate the parts of the bony *mandible:* its anterior *body* and its posterior ascending *ramus.* Press on the skin over the mandibular ramus, and feel the *masseter muscle* bulge when you clench your teeth. Palpate the anterior border of the masseter, and trace it to the mandible's inferior margin. At this point, you will be able to detect the pulse of your *facial artery* (Figure 42.1). Finally, to feel the *temporomandibular joint,* place a finger directly anterior to the external auditory canal of your ear, and open and close your mouth several times. The bony structure you feel moving is the *head of the mandible.*

THE NECK

Bony Landmarks

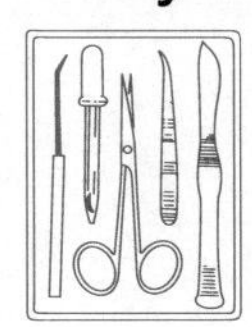

1. Run your fingers inferiorly along the back of your neck, in the posterior midline, to feel the *spinous processes* of the cervical vertebrae. The spine of C_7, the *vertebra prominens,* is especially prominent.

2. Now, beginning at your chin, run a finger inferiorly along the anterior midline of your neck (Figure 42.3). The first hard structure you encounter will be the U-shaped *hyoid bone,* which lies in the angle between the floor of the mouth and the vertical part of the neck (Figure 42.4). Directly inferior to this, you will feel the *laryngeal prominence* (Adam's apple) of the thyroid cartilage. Just inferior to the laryngeal prominence, your finger will sink into a soft depression (formed by the *cricothyroid ligament*) before proceeding onto the rounded surface of the *cricoid cartilage.* Now swallow several times, and feel the whole larynx move up and down.

3. Continue inferiorly to the trachea. Attempt to palpate the *isthmus of the thyroid gland,* which feels like a spongy cushion over the second to fourth tracheal rings. Then, try to palpate the two soft lateral *lobes* of your thyroid gland along the sides of the trachea.

4. Move your finger all the way inferiorly to the root of the neck, and rest it in the *jugular notch,* the depression in the superior part of the sternum between the two clavicles. By pushing deeply at this point, you can feel the cartilage rings of the trachea.

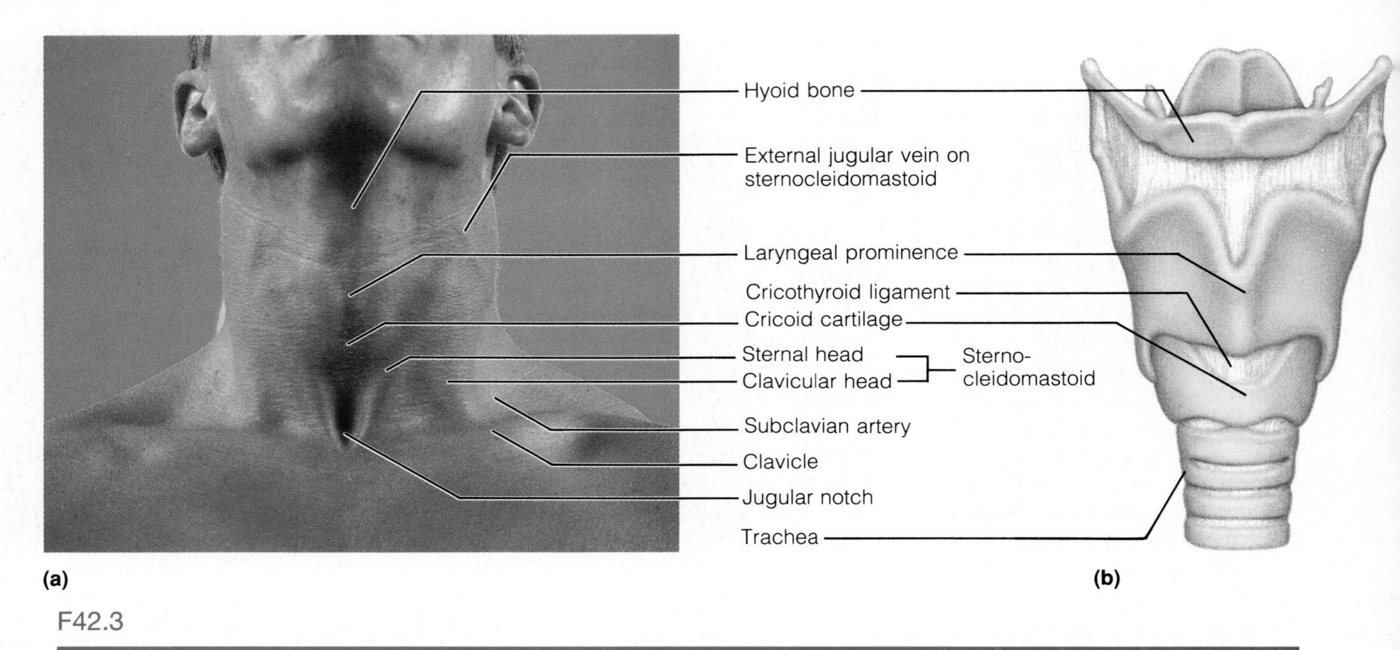

F42.3

Anterior surface of the neck. (a) Photograph; (b) diagram of the underlying skeleton of the larynx.

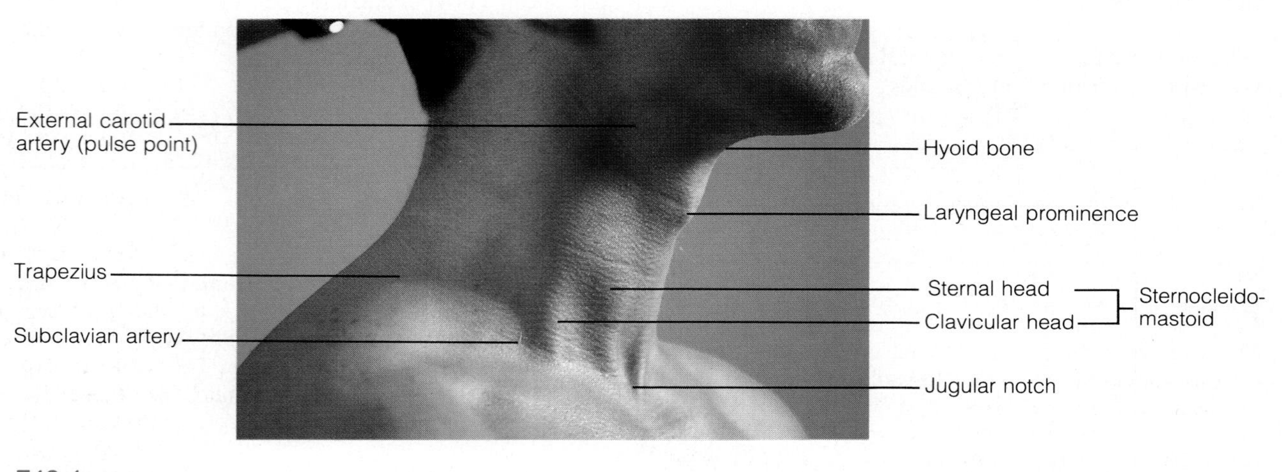

F42.4

Lateral surface of the neck.

Muscles

The *sternocleidomastoid* is the most prominent muscle in the neck and the neck's most important surface landmark. You can best see and feel it when you turn your head to the side.

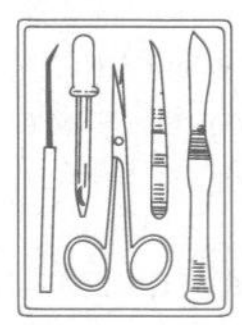

1. Obtain a hand mirror, hold it in front of your face, and turn your head sharply from right to left several times. You will be able to see both heads of this muscle, the *sternal head* medially and the *clavicular head* laterally (Figures 42.3 and 42.4).

Several important structures lie beside or beneath the sternocleidomastoid:

- The *cervical lymph nodes* lie both superficial and deep to this muscle. (Swollen cervical nodes provide evidence of infections or cancer of the head and neck.)
- The *common carotid artery* and *internal jugular vein* lie just deep to the sternocleidomastoid, a relatively superficial location that exposes these vessels to danger in slashing wounds to the neck.
- Just lateral to the inferior part of the sternocleidomastoid is the large *subclavian artery* on its way to supply the upper limb. By pushing on the subclavian artery at this point, one can stop the bleeding from a wound anywhere in the associated limb.
- Just anterior to the sternocleidomastoid, superior to the level of your larynx, you can feel a carotid

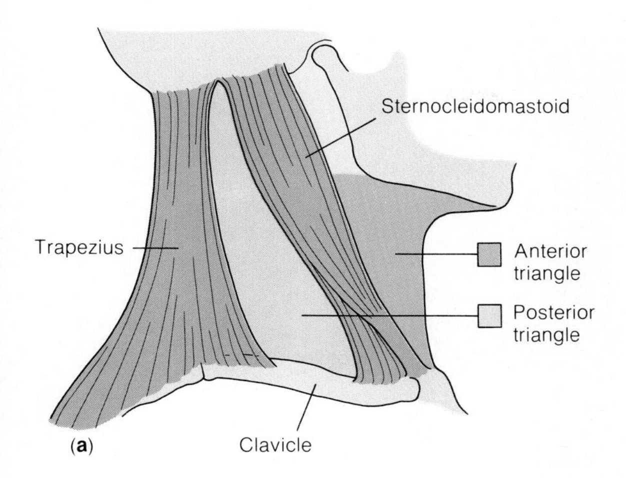

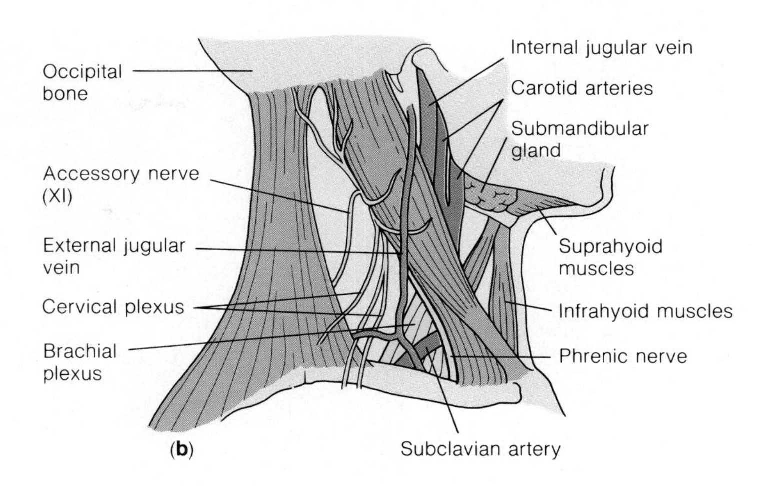

F42.5

Anterior and posterior triangles of the neck. (a) Boundaries of the triangles. (b) Some contents of the triangles.

pulse—the pulsations of the *external carotid artery* (Figure 42.4).

- The *external jugular vein* descends vertically, just superficial to the sternocleidomastoid and deep to the skin (Figure 42.3). To make this vein "appear" on your neck, stand before the mirror, and gently compress the skin superior to your clavicle with your fingers.

2. Another large muscle in the neck, on the posterior aspect, is the *trapezius* (Figure 42.4). You can feel this muscle contract just deep to the skin as you shrug your shoulders.

Triangles of the Neck

The sternocleidomastoid muscles divide each side of the neck into the posterior and anterior triangles (Figure 42.5a).

1. The **posterior triangle** is defined by the sternocleidomastoid anteriorly, the trapezius posteriorly, and the clavicle inferiorly. Palpate the muscular borders of the posterior triangle.

The **anterior triangle** is defined by the inferior margin of the mandible superiorly, the midline of the neck anteriorly, and the sternocleidomastoid posteriorly.

2. The contents of these two triangles are shown in Figure 42.5. The posterior triangle contains many important nerves and blood vessels, including the *accessory nerve* (cranial nerve XI), most of the *cervical plexus,* and the *phrenic nerve.* In the inferior part of the triangle are the external jugular vein, the trunks of the *brachial plexus,* and the subclavian artery. These structures are relatively superficial and are easily cut or injured by wounds to the neck. In the neck's anterior triangle, important structures include the *submandibular gland,* the *suprahyoid* and *infrahyoid muscles,* and parts of the carotid arteries and jugular veins that lie superior to the sternocleidomastoid.

- Palpate your carotid pulse.

A wound to the posterior triangle of the neck can lead to long-term loss of sensation in the skin of the neck and shoulder, as well as a partial paralysis of the sternocleidomastoid and trapezius muscles. Can you explain these effects?

____

____

____ ■

THE TRUNK

The trunk of the body consists of the thorax, abdomen, pelvis, and perineum. The *back* includes parts of all of these regions, but for convenience it is treated separately.

The Back

BONES

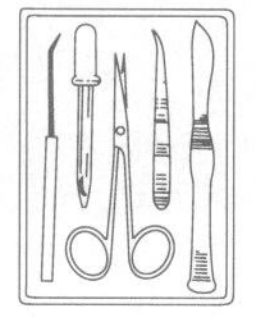

1. The vertical groove in the center of the back is called the **posterior median furrow** (Figure 42.6a). The *spinous processes* of the vertebrae are visible in this furrow when the spinal column is flexed. Palpate a few of these processes on your partner's back (C_7 and T_1 are the most prominent and the easiest to find). Also palpate the posterior parts of some ribs, as well as the prominent *spine of the scapula* and the scapula's long *medial border.* The scapula lies superficial to ribs 2 to 7, its *inferior angle* is at the level

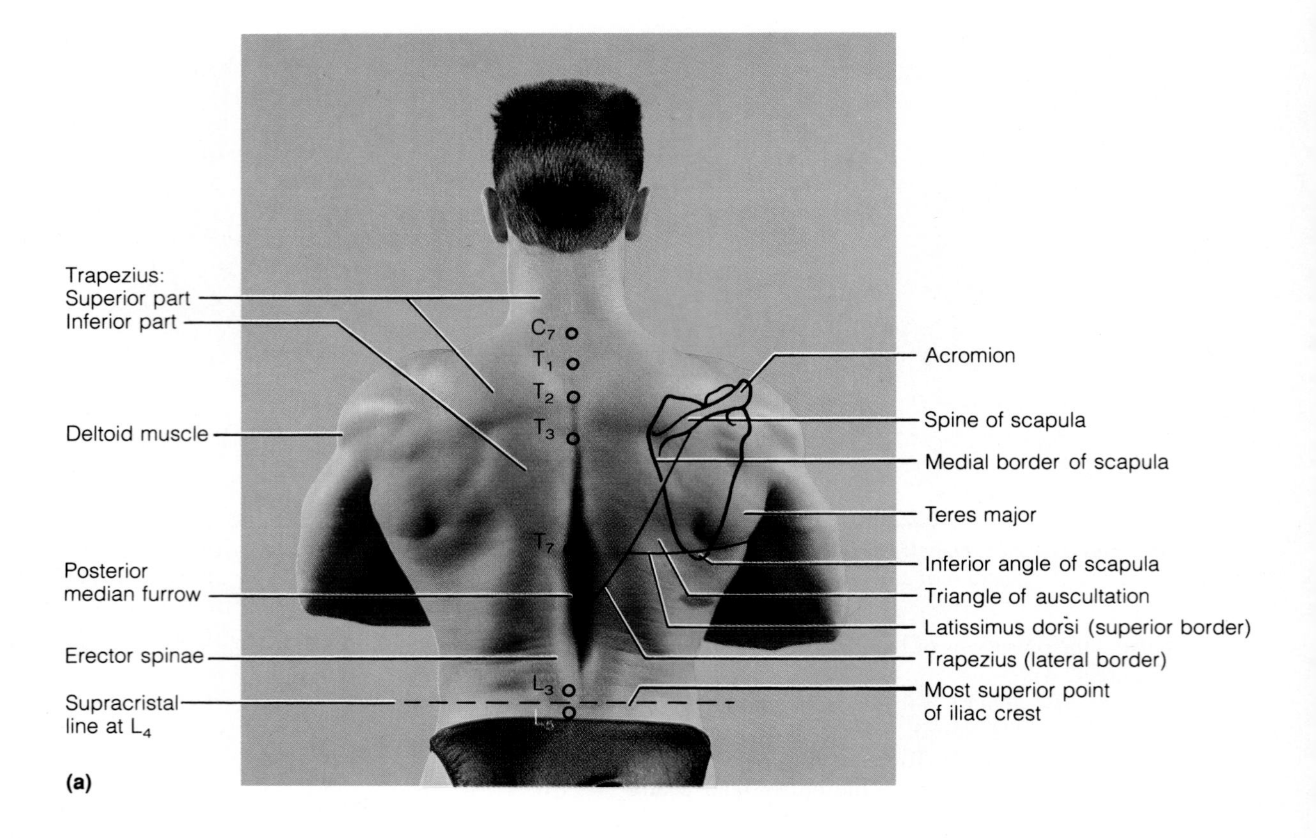

F42.6

Surface anatomy of the back. Two different poses, (a) and (b).

of the spinous process of vertebra T_7, and the medial end of the scapular spine lies opposite the spinous process of T_3.

2. Now feel the *iliac crests* (superior margins of the iliac bones) in your own lower back. You can find these crests effortlessly by resting your hands on your hips. Locate the most superior point of each crest, a point that lies roughly halfway between the posterior median furrow and the lateral side of the body (Figure 42.6a). A horizontal line through these two superior points, the **supracristal line,** intersects L_4, providing a simple way to locate that vertebra. The ability to locate L_4 is essential for performing a *lumbar puncture,* a procedure in which the clinician inserts a needle into the vertebral canal of the spinal column directly superior or inferior to L_4 and withdraws cerobrospinal fluid.

3. The *sacrum* is easy to palpate just superior to the cleft in the buttocks, and you can feel the *coccyx* in the extreme inferior part of that cleft, just posterior to the anus.

MUSCLES The largest superficial muscles of the back are the *trapezius* superiorly and *latissimus dorsi* inferiorly (Figure 42.6b). Furthermore, the deeper *erector spinae* muscles are very evident in the lower back, flanking the vertebral column like thick vertical cords.

Feel your partner's erector spinae muscles contract and bulge as he extends his spine from a slightly bent-over position.

The superficial muscles of the back fail to cover a small area of the rib cage called the **triangle of auscultation** (Figure 42.6a). This triangle lies just medial to the inferior part of the scapula. Its three boundaries are formed by the trapezius medially, the latissimus dorsi inferiorly, and the scapula laterally. The physician places a stethoscope over the skin of this triangle to listen for lung sounds (*auscultation* = listening). To hear the lungs clearly, the doctor first asks the patient to fold the arms together in front of the chest and then flex the trunk.

What do you think is the precise reason for having the patient take this action?

____

____

____

Have your partner assume the position just described. After cleaning the earpieces with an alcohol swab, use the stethoscope to auscultate his lung sounds. Compare the clarity of the lung sounds heard over the triangle of auscultation to that over other areas of the back.

The Thorax

BONES

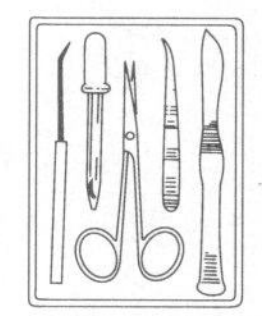

1. Start exploring the anterior surface of your partner's bony *thoracic cage* (Figures 42.7 and 42.8) by defining the extent of the *sternum.* Use a finger to trace the sternum's triangular *manubrium,* its flat *body,* and the tongue-shaped *xiphoid process.* Now palpate the ridgelike *sternal angle,* where the manubrium meets the body of the sternum. Locating the

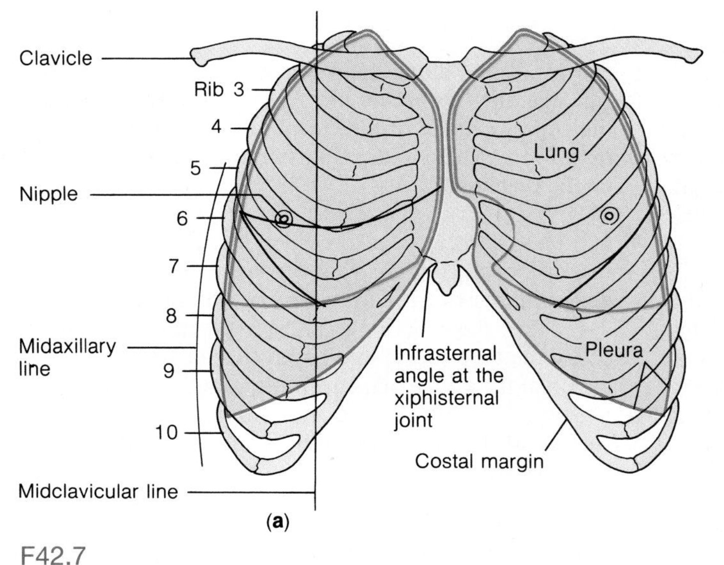

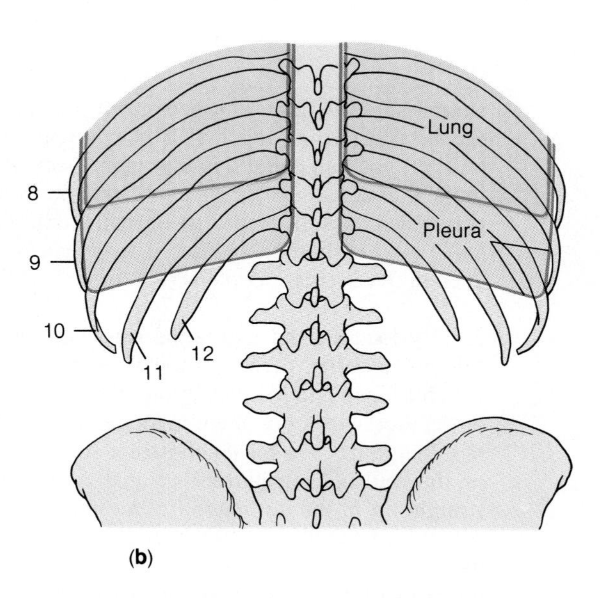

F42.7

The bony rib cage as it relates to the underlying lungs and pleural cavities. Both the lungs (red) and the pleural cavities (blue) are outlined. (a) Anterior view; (b) posterior view.

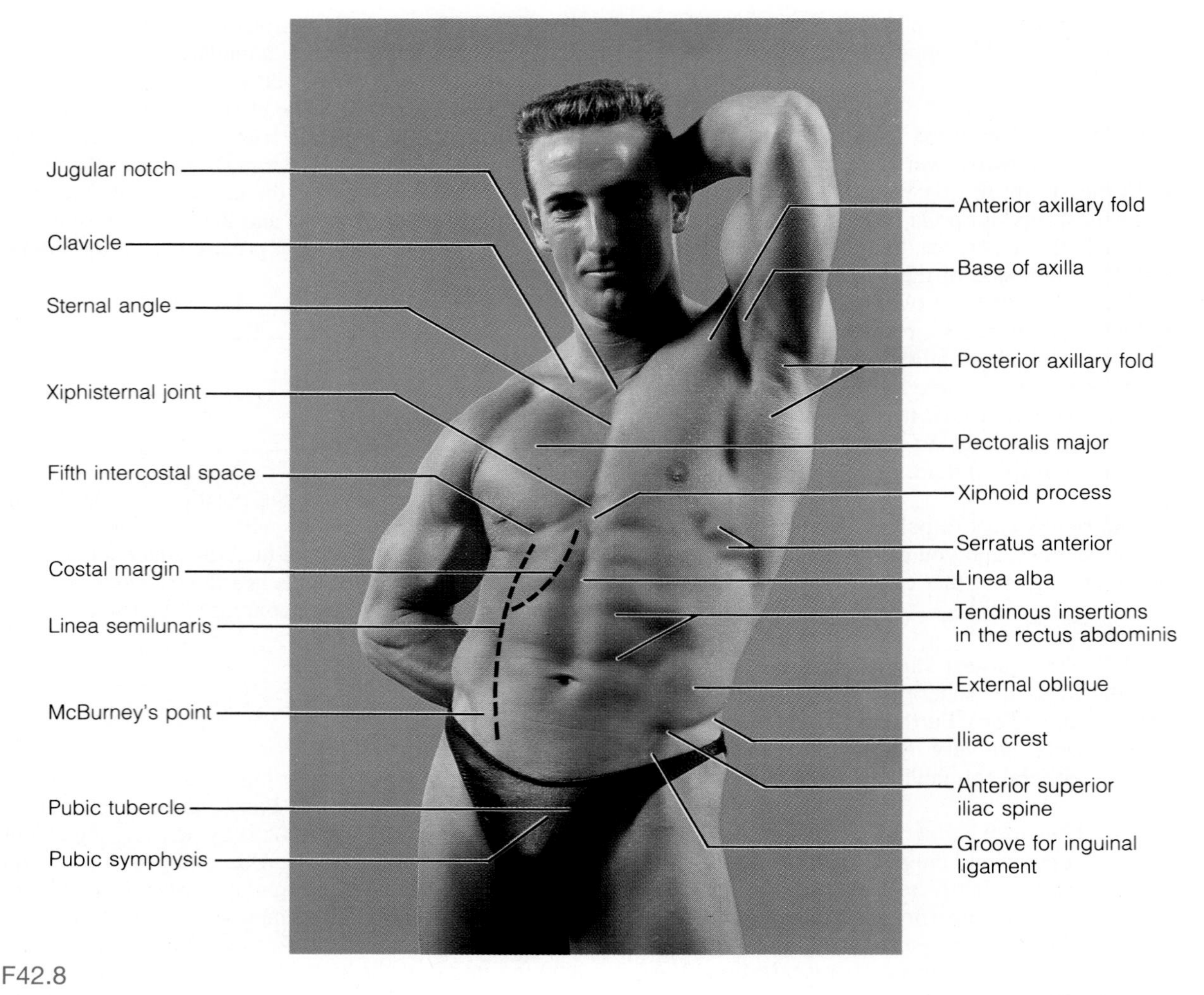

F42.8

The anterior thorax and abdomen.

sternal angle is important because it directs you to the second ribs (which attach to it). Once you find the second rib, you can count down to identify every other rib in the thorax (except the first and sometimes the twelfth rib, which lie too deeply to be palpated). The sternal angle is a highly reliable landmark—it is easy to locate, even in overweight people.

2. By locating the individual ribs, you can mentally "draw" a series of horizontal lines of "latitude" by which to map and locate the underlying visceral organs of the thoracic cavity. Such mapping also requires lines of "longitude," so let us construct some vertical lines on the wall of your partner's trunk. As he lifts an arm straight up in the air, extend a line inferiorly from the center of the axilla onto his lateral thoracic wall. This is the **midaxillary line** (Figure 42.7a). Now estimate the midpoint of his *clavicle,* and run a vertical line inferiorly from that point toward the groin. This is the **midclavicular line,** and it will pass about 1 cm medial to the nipple.

3. Next, feel along the V-shaped inferior edge of the rib cage, the *costal margin.* At the **infrasternal angle,** the superior angle of the costal margin, lies the *xiphisternal joint.* Deep to the xiphisternal joint, the heart lies on the diaphragm.

4. The thoracic cage provides many valuable landmarks for locating the vital organs of the thoracic and abdominal cavities. On the anterior thoracic wall, ribs 2–6 define the superior-to-inferior extent of the female breast, and the fourth intercostal space indicates the location of the *nipple* in men, children, and small-breasted women. The right costal margin runs across the anterior surface of the liver and gallbladder. Surgeons must be aware of the inferior margin of the *pleural cavities* because if they accidentally cut into one of these cavities, a lung collapses. The inferior pleural margin lies adjacent to vertebra T_{12} near the posterior midline (Figure 42.7b) and runs horizontally across the back to reach rib 10 at the midaxillary line. From there, the pleural margin ascends to rib 8 in the midclavicular line (Figure

42.7a) and to the level of the xiphisternal joint near the anterior midline. The *lungs* do not fill the inferior region of the pleural cavity. Instead, their inferior borders run at a level that is two ribs superior to the pleural margin, until they meet that margin near the xiphisternal joint.

5. The relationship of the *heart* to the thoracic cage is considered in Exercise 27 (p. 252). We will review that information here. In essence, the superior right corner of the heart lies at the junction of the third rib and the sternum; the superior left corner lies at the second rib, near the sternum; the inferior left corner lies in the fifth intercostal space in the midclavicular line; and the inferior right corner lies at the sternal border of the sixth rib. You may wish to outline the heart on your chest or that of your lab partner by connecting the four corner points with a washable marker.

MUSCLES The main superficial muscles of the anterior thoracic wall are the *pectoralis major* and the anterior slips of the *serratus anterior.*

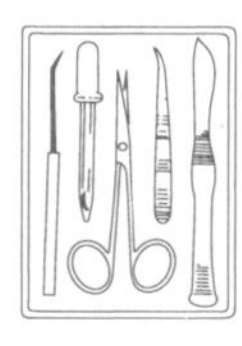

Using Figure 42.8 as a guide, try to palpate these two muscles on your chest. They both contract during push-ups, and you can confirm this by pushing yourself up from your desk with one arm while palpating the muscles with your opposite hand.

The Abdomen

BONY LANDMARKS

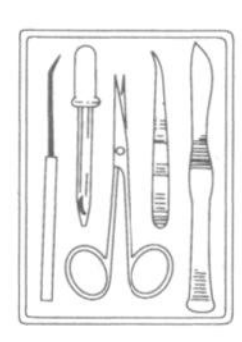

The anterior abdominal wall (Figure 42.8) extends inferiorly from the costal margin to an inferior boundary that is defined by several landmarks. Palpate these landmarks as they are described below.

1. **Iliac crest.** Recall that the iliac crests are the superior margins of the iliac bones, and you can locate them by resting your hands on your hips.

2. **Anterior superior iliac spine.** Representing the most anterior point of the iliac crest, this spine is a prominent landmark. It can be palpated in everyone, even those who are overweight. Run your fingers anteriorly along the iliac crest to its end.

3. **Inguinal ligament.** The inguinal ligament, indicated by a groove on the skin of the groin, runs medially from the anterior superior iliac spine to the pubic tubercle of the pubic bone.

4. **Pubic crest.** You will have to press deeply to feel this crest on the pubic bone near the median *pubic symphysis.* The *pubic tubercle,* the most lateral point of the pubic crest, is easier to palpate, but you will still have to push deeply.

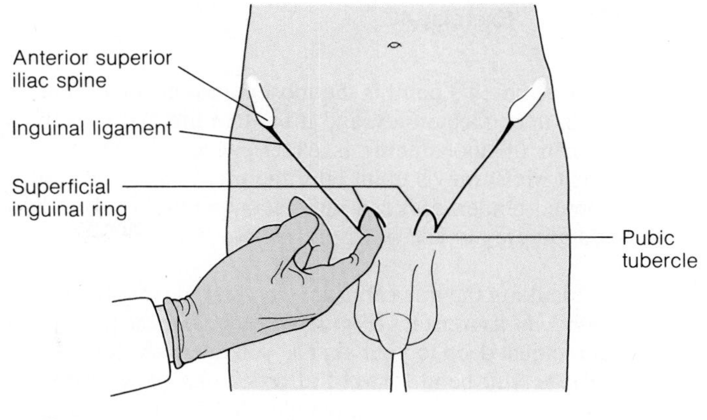

F42.9

Clinical examination for an inguinal hernia in a male. The examiner palpates the patient's pubic tubercle, pushes superiorly to invaginate the scrotal skin into the superficial inguinal ring and asks the patient to cough. If an inguinal hernia exists, it will push inferiorly and touch the examiner's fingertip.

Inguinal hernias occur immediately superior to the inguinal ligament and may exit from a medial opening called the *superficial inguinal ring.* To locate this ring, one would palpate the pubic tubercle (Figure 42.9). The procedure used by the physician to test whether a male has an inguinal hernia is depicted in Figure 42.9. ■

MUSCLES AND OTHER SURFACE FEATURES

The central landmark of the anterior abdominal wall is the *umbilicus* (navel). Running superiorly and inferiorly from the umbilicus is the *linea alba* ("white line"), represented in the skin of lean people by a vertical groove (Figure 42.8). The linea alba is a tendinous line that extends from the xiphoid process to the pubic symphysis, just medial to the rectus abdominis muscles (Table 13.3, p. 113). The linea alba is a favored site for surgical entry into the abdominal cavity because the surgeon can make a long cut through this line with no muscle damage and a minimum of bleeding.

Several kinds of hernias involve the umbilicus and the linea alba. In an **acquired umbilical hernia,** the linea alba weakens until intestinal coils push through it just superior to the navel. The herniated coils form a bulge just deep to the skin.

Another type of umbilical hernia is a **congenital umbilical hernia,** present in some infants: The umbilical hernia is seen as a cherry-sized bulge deep to the skin of the navel that enlarges whenever the baby cries. Congenital umbilical hernias are usually harmless, and most correct themselves automatically before the child's second birthday. ■

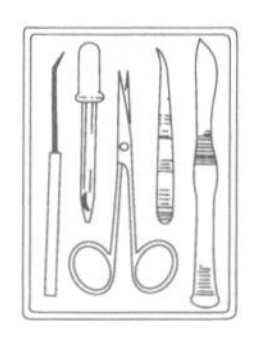

1. **McBurney's point** is the spot on the anterior abdominal skin that is directly superficial to the base of the appendix (Figure 42.8). It is located one-third of the way along a line between the right anterior superior iliac spine and the umbilicus. Try to find it on your body.

McBurney's point is the most common site of incision in appendectomies, and it is often the place where the pain of appendicitis is experienced most acutely. Pain at McBurney's point after the pressure is removed (rebound tenderness) can indicate appendicitis. This is not a *precise* method of diagnosis, however.

2. Flanking the linea alba are the vertical straplike *rectus abdominis* muscles (Figure 42.8). Feel these muscles contract just deep to your skin as you do a bent-knee sit-up (or as you bend forward after leaning back in your chair). In the skin of lean people, the lateral margin of each rectus muscle makes a groove known as the **linea semilunaris** ("half-moon line"). On your right side, estimate where your linea semilunaris crosses the costal margin of the rib cage. The *gallbladder* lies just deep to this spot, so this is the standard point of incision for gallbladder surgery. In muscular people, three horizontal grooves can be seen in the skin covering the rectus abdominis. These grooves represent the *tendinous insertions* (or *intersections* or *inscriptions*), fibrous bands that subdivide the rectus muscle. Because of these subdivisions, each rectus abdominis muscle presents four distinct bulges. Try to identify these insertions on yourself or your partner.

3. The only other major muscles that can be seen or felt through the anterior abdominal wall are the lateral *external obliques.* Feel these muscles contract as you cough, strain, or raise your intra-abdominal pressure in some other way.

4. Recall that the anterior abdominal wall can be divided into four quadrants (Figure 42.10). A clinician listening to a patient's **bowel sounds** places the stethoscope over each of the four abdominal quadrants, one after another. Normal bowel sounds, which result as peristalsis moves air and fluid through the intestine, are high-pitched gurgles that occur every 5 to 15 seconds.

- Use the stethoscope to listen to your own or your partner's bowel sounds.

Abnormal bowel sounds can indicate intestinal disorders. Absence of bowel sounds indicates a halt in intestinal activity, which follows a long-term obstruction of the intestine, surgical handling of the intestine, peritonitis, or other conditions. Loud tinkling or splashing sounds, by contrast, indicate an increase in intestinal activity. Such loud sounds may accompany gastroenteritis (inflammation and upset of the GI tract) or a partly obstructed intestine. ■

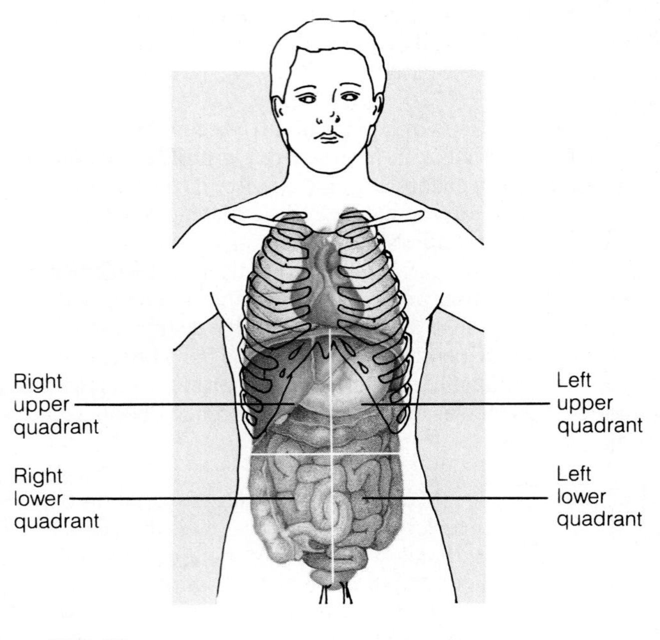

F42.10

The four abdominopelvic quadrants.

The Pelvis and Perineum

The bony surface features of the *pelvis* are best studied with the abdomen or the lower limb, so they are considered later. Most *internal* pelvic organs are not palpable through the skin of the body surface. A full *bladder,* however, becomes firm and can be felt through the abdominal wall just superior to the pubic symphysis. A bladder that can be palpated more than a few centimeters above this symphysis is retaining urine and dangerously full, and it should be drained by catheterization.

UPPER LIMB

Axilla

The **base of the axilla** is the groove in which the underarm hair grows (Figure 42.8). Deep to this base lie the *axillary lymph nodes* (which swell and can be palpated in breast cancer), the large *axillary vessels* serving the upper limb, and much of the brachial plexus. The base of the axilla forms a "valley" between two thick, rounded ridges, the **axillary folds.** Just anterior to the base, clutch your **anterior axillary fold,** formed by the pectoralis major muscle. Then grasp your **posterior axillary fold.** This fold is formed by the latissimus dorsi and teres major muscles of the back as they course toward their insertions on the humerus.

Shoulder

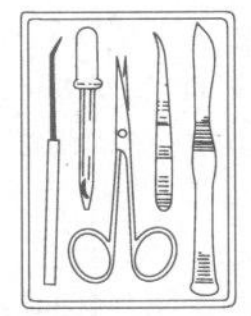

1. Relocate the prominent spine of the scapula posteriorly (Figure 42.11). Follow the spine to its lateral end, the flattened *acromion* on the shoulder's summit. Then, palpate the *clavicle* anteriorly, tracing this bone from the sternum to the shoulder. Notice the clavicle's curved shape.

2. Now locate the junction between the clavicle and the acromion on the superolateral surface of your shoulder, at the *acromioclavicular joint.* To find this joint, thrust your arm anteriorly repeatedly until you can palpate the precise point of pivoting action.

3. Next, place your fingers on the *greater tubercle* of the humerus. This is the most lateral bony landmark on

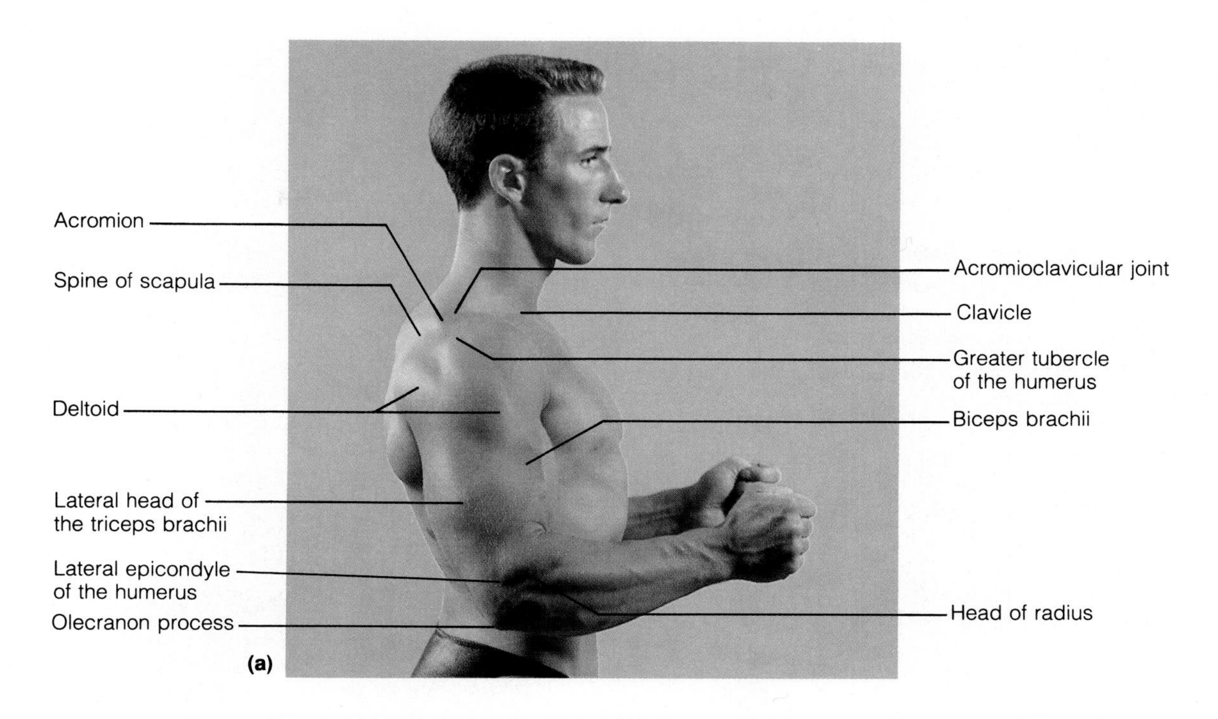

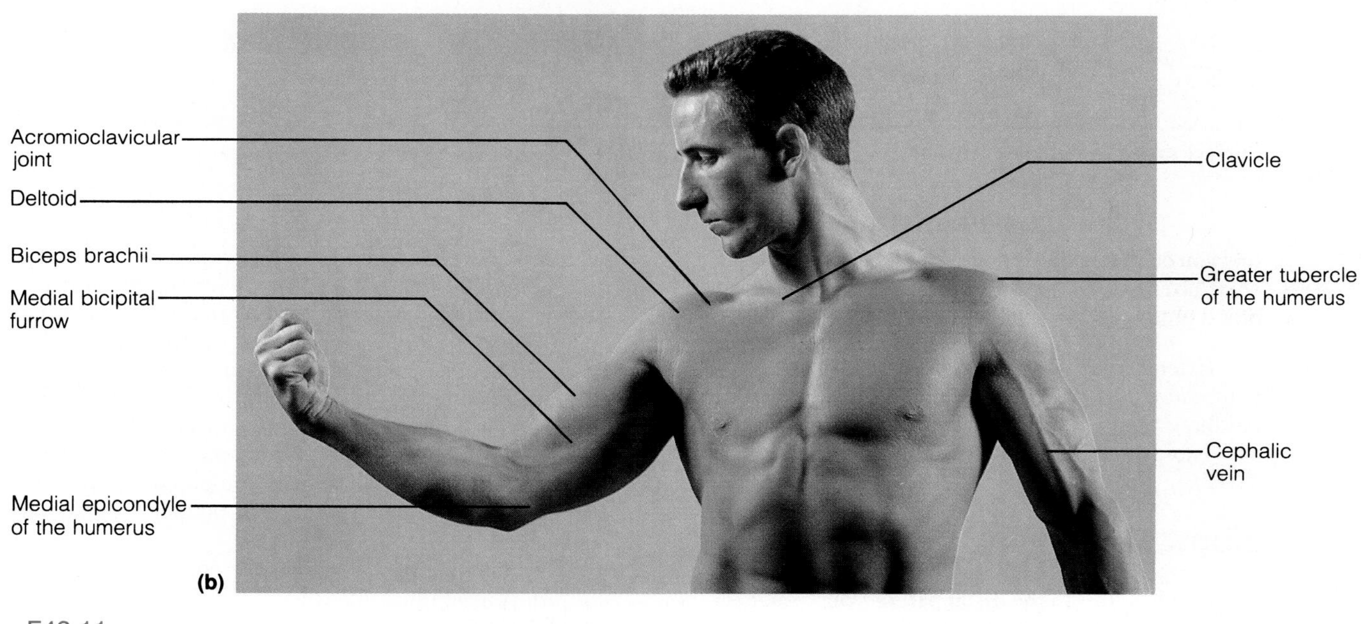

F42.11

The shoulder and arm. (a) Lateral view; (b) anterior and medial view.

the superior surface of the shoulder. It is covered by the thick *deltoid muscle,* which forms the rounded superior part of the shoulder. Intramuscular injections are often given into the deltoid, about 5 cm (2 inches) inferior to the greater tubercle (refer to Figure 42.19, p. 380).

Arm

Remember, according to anatomists, the arm runs only from the shoulder to the elbow, and not beyond.

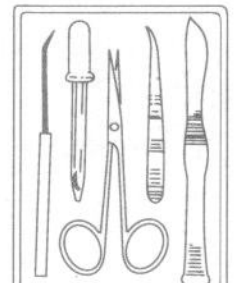

1. In the arm, palpate the *humerus* along its entire length, especially along its medial and lateral sides.

2. Feel the *biceps brachii* muscle contract on your anterior arm when you flex your forearm against resistance. The medial boundary of the biceps is represented by the **medial bicipital furrow** (Figure 42.11b). This groove contains the large *brachial artery,* and by pressing on it with your finger-

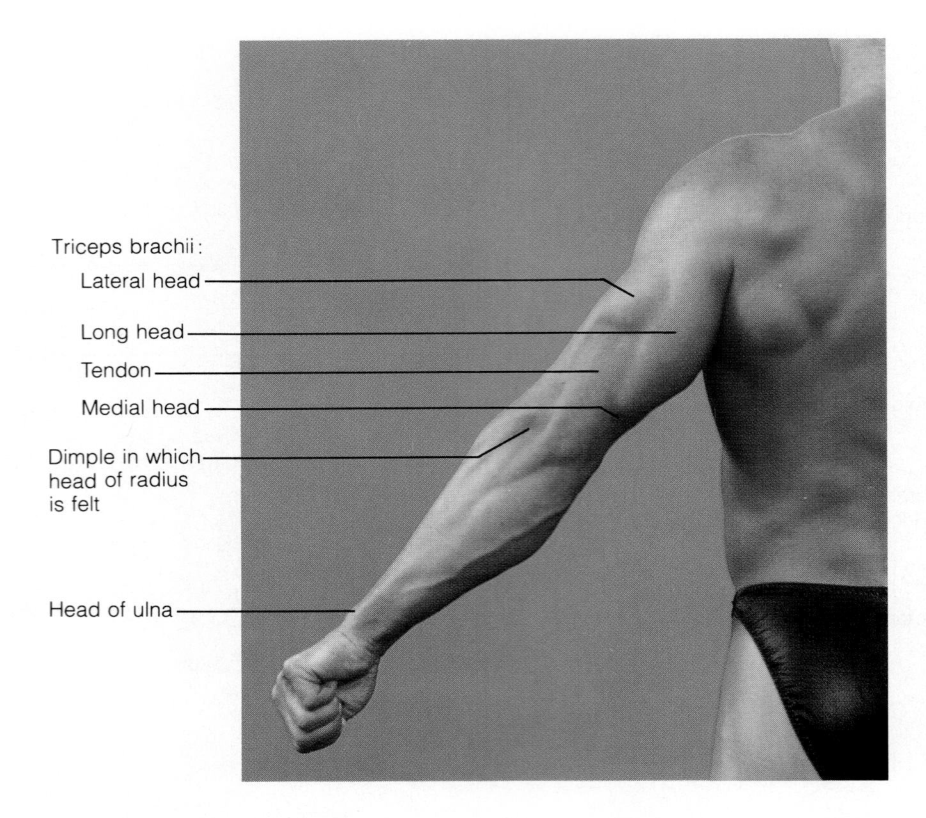

F42.12

Surface anatomy of the upper limb, posterior view.

tips you can feel your *brachial pulse*. Recall that the brachial artery is the artery routinely used in measuring blood pressure with a sphygmomanometer.

3. Extend your forearm against resistance, and feel your *triceps brachii* muscle bulge in the posterior arm. All three heads of the triceps (lateral, long, and medial) are visible through the skin of a muscular person (Figure 42.12).

Elbow Region

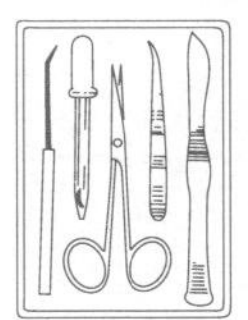

1. In the distal part of your arm, near the elbow, palpate the two projections of the humerus, the *lateral* and *medial epicondyles* (Figure 42.11). Midway between the epicondyles, on the posterior side, feel the *olecranon process* of the ulna, which forms the point of the elbow.

2. Confirm that the two epicondyles and the olecranon all lie in the same horizontal line when the elbow is extended. If these three bony processes do not line up, the elbow is dislocated.

3. Now feel along the posterior surface of the medial epicondyle: You are palpating your ulnar nerve.

4. On the anterior surface of the elbow is a triangular depression called the **cubital fossa** or **antecubital fossa** (Figure 42.13). The triangle's superior *base* is formed by a horizontal line between the humeral epicondyles; its two inferior sides are defined by the *brachioradialis* and *pronator teres* muscles (Figure 42.13b). Try to define these boundaries on your own limb: To find the brachioradialis muscle, flex your forearm against resistance, and watch this muscle bulge through the skin of your lateral forearm. To feel your pronator teres contract, palpate the cubital fossa as you pronate your forearm against resistance. (Have your partner provide the resistance.)

The cubital fossa contains the superficial *median cubital* vein (Figure 42.13a). Clinicians often draw blood from this superficial vein and insert intravenous (IV) catheters into it to administer medications, transfused blood, and nutrient fluids. The large *brachial artery* lies just deep to the median cubital vein (Figure 42.13b), so a needle must be inserted into the vein from a shallow angle (almost parallel to the skin) to avoid puncturing the artery. Other structures that lie deep in the fossa are also shown in Figure 42.13b.

5. The median cubital vein interconnects the larger cephalic and basilic veins of the upper limb. Recall from Exercise 29 (p. 269) that the *cephalic vein* ascends along the lateral side of the forearm and arm, whereas the *basilic vein* ascends through the limb's medial side. These veins are visible through the skin of lean people (Figure 42.13a). Examine your arm to see if your cephalic and basilic veins are visible.

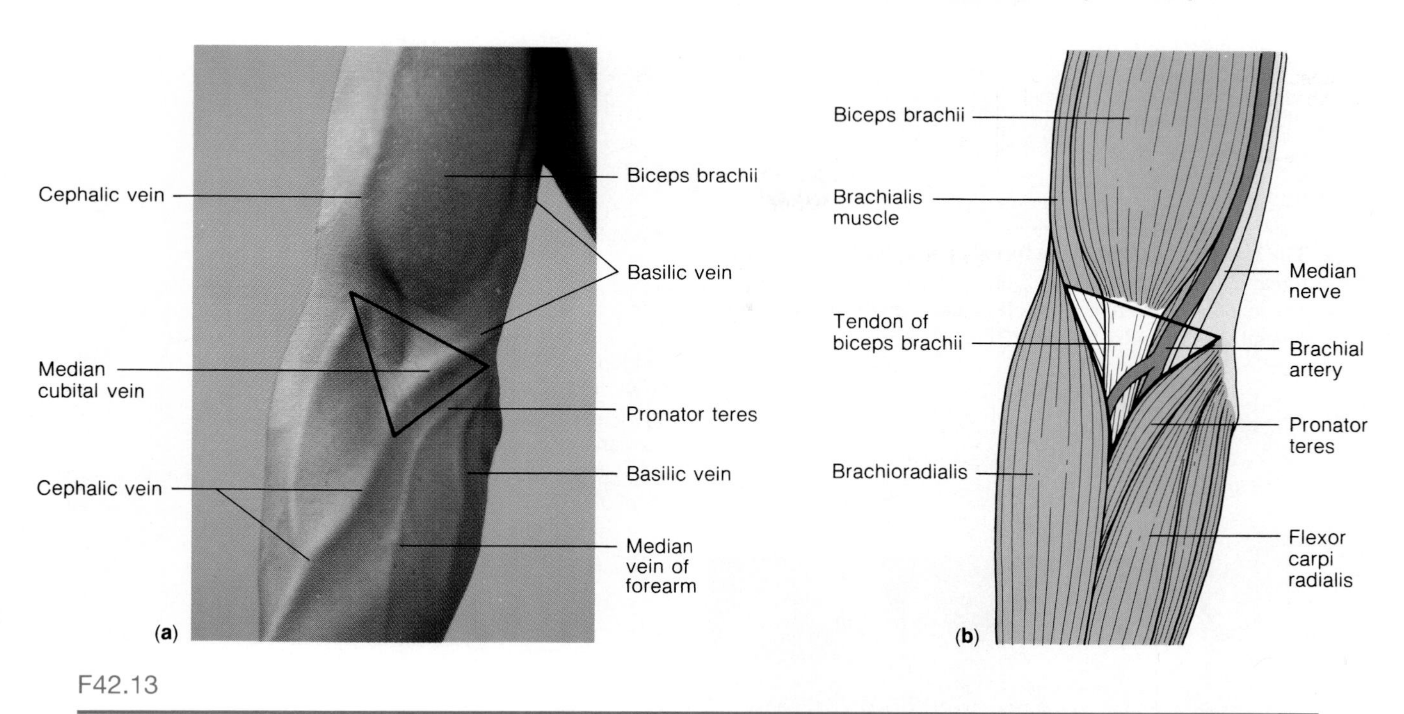

F42.13

The cubital (antecubital) fossa on the anterior surface of the right elbow (outlined by the triangle). (a) Photograph; (b) diagram of deeper structures in the fossa.

Forearm and Hand

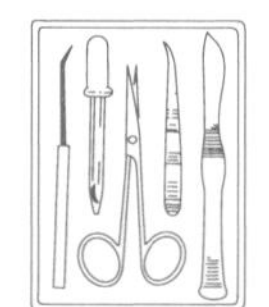

The two parallel bones of the forearm are the medial *ulna* and the lateral *radius.*

1. Feel the ulna along its entire length as a sharp ridge on the posterior forearm (confirm that this ridge runs inferiorly from the olecranon process). As for the radius, you can feel its distal half, but most of its proximal half is covered by muscle. You can, however, feel the rotating *head* of the radius. To do this, extend your forearm, and note that a dimple forms on the posterior lateral surface of the elbow region (Figure 42.12). Press three fingers into this dimple, and rotate your free hand as if you were turning a doorknob. You will feel the head of the radius rotate as you perform this action.

2. Both the radius and ulna have a knoblike *styloid process* at their distal end. Figure 42.14 shows a way to locate these processes. Do not confuse the ulna's styloid process with the conspicuous *head of the ulna,* from which the styloid process stems. Confirm that the styloid process of the radius lies about 1 cm (0.4 inch) distal to that of the ulna.

Colles' fracture of the wrist is an impacted fracture in which the distal end of the radius is pushed proximally into the shaft of the radius. This usually occurs when someone falls on outstretched hands, and it most often happens to elderly women with osteoporosis. Colles' fracture bends the wrist into curves that resemble those on a fork. ■

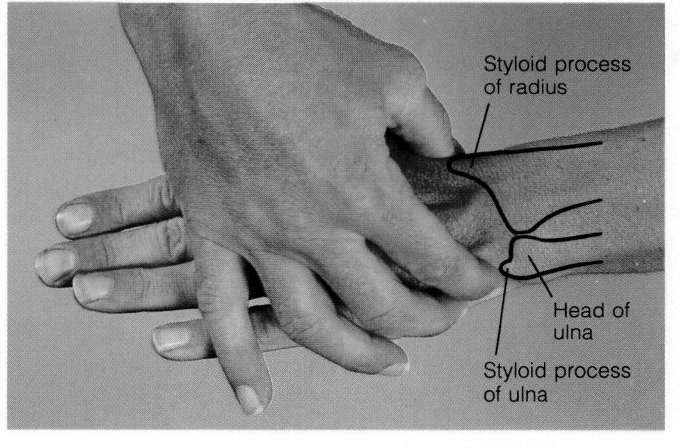

F42.14

A way to locate the styloid processes of the ulna and radius. The right hand is palpating the left hand in this illustration. Note that the head of the ulna is not the same as its styloid process. The styloid process of the radius lies about 1 cm distal to the styloid process of the ulna.

Can you deduce how physicians use palpation to diagnose a Colles' fracture?

____

____

3. Next, feel the major groups of muscles within your forearm. Flex your hand and fingers against resistance, and feel the anterior *flexor muscles* contract. Then ex-

F42.15

The anterior surface of the forearm and fist. (a) The entire forearm. (b) Enlarged view of the distal forearm and hand. The tendons of the flexor muscles guide the clinician to several sites for pulse taking.

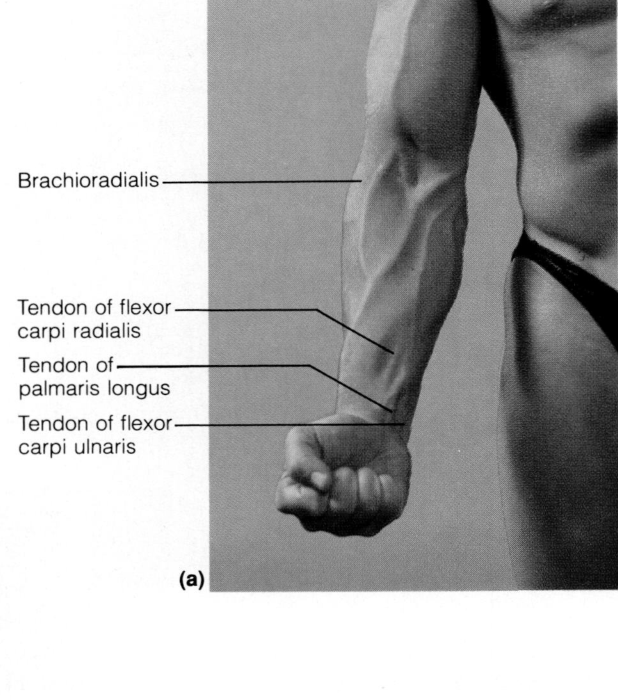

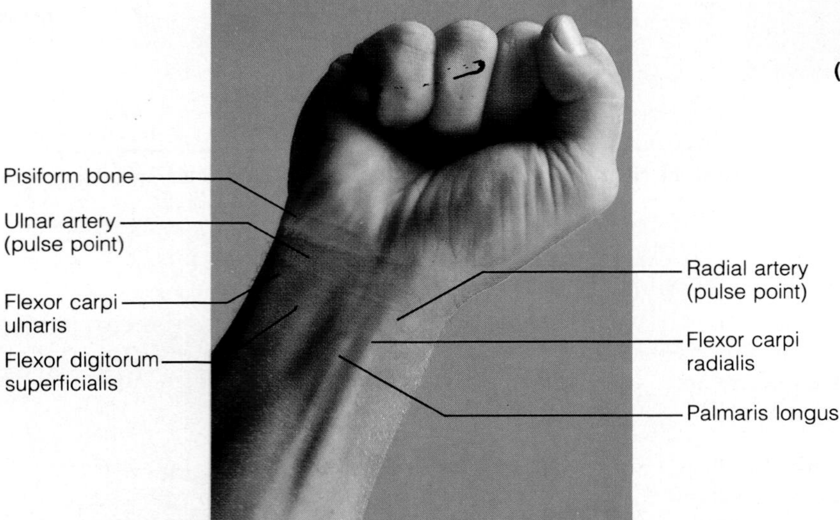

tend your hand at the wrist, and feel the tightening of the posterior *extensor muscles.*

4. Near the wrist, the anterior surface of the forearm reveals many significant features (Figure 42.15). Flex your fist against resistance; the tendons of the main wrist flexors will bulge the skin of the distal forearm. The tendons of the *flexor carpi radialis* and *palmaris longus* muscles are most obvious. (The palmaris longus, however, is absent from at least one arm in 30% of all people, so your forearm may exhibit just one prominent tendon instead of two.) The *radial artery* lies just lateral to (on the thumb side of) the flexor carpi radialis tendon, where the pulse is easily detected (Figure 42.15b). Feel your radial pulse here. The *median nerve* (which innervates the thumb) lies deep to the palmaris longus tendon. Finally, the *ulnar artery* lies on the medial side of the forearm, just lateral to the tendon of the *flexor carpi ulnaris.* Using Figure 42.15b as a guide, locate and feel your ulnar arterial pulse.

5. Extend your thumb and point it posteriorly to form a triangular depression in the base of the thumb on the back of your hand. This is the **anatomical snuff box** (Figure 42.16). Its two elevated borders are defined by the tendons of the thumb extensor muscles, *extensor pollicis brevis* and *extensor pollicis longus.* The radial artery runs within the snuff box, so this is another site for taking a radial pulse. The main bone on the floor of the snuff box is the scaphoid bone of the wrist, but the styloid process of the radius is also present here. (If displaced by a bone fracture, the radial styloid process will be felt outside of the snuff box rather than within it.) The "snuff box" took its name from the fact that people once put snuff (tobacco for sniffing) in this hollow before lifting it up to the nose.

6. On the dorsum of your hand, observe the superficial veins just deep to the skin. This is the *dorsal venous network,* which drains superiorly into the cephalic vein. This venous network provides a site for drawing blood

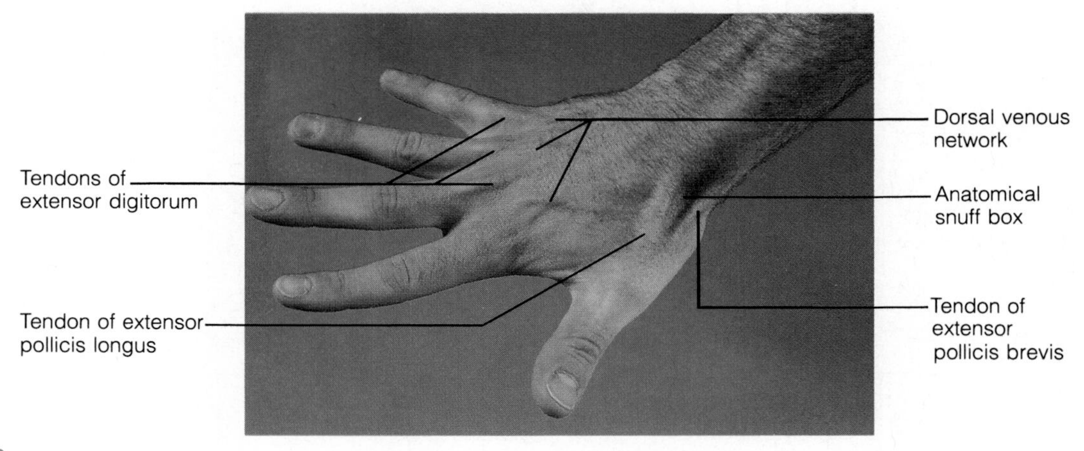

F42.16

The dorsum of the hand. Note especially the anatomical snuff box and the dorsal venous network.

and inserting intravenous catheters and is preferred over the median cubital vein for these purposes. Next, extend your hand and fingers, and observe the tendons of the *extensor digitorum.*

7. The anterior surface of the hand also contains some features of interest (Figure 42.17). These features include the *epidermal ridges* ("fingerprints") and many *flexion creases* in the skin. Grasp your *thenar eminence* (the bulge on the palm that contains the thumb muscles) and your *hypothenar eminence* (the bulge on the medial palm that contains muscles that move the little finger).

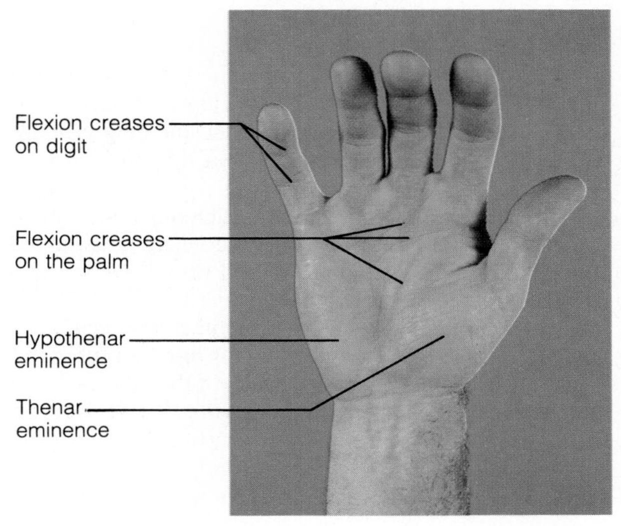

F42.17

The palmar surface of the hand.

LOWER LIMB

Gluteal Region

Dominating the gluteal region are the two *prominences* ("cheeks") of the buttocks. These are formed by subcutaneous fat and by the thick *gluteus maximus* muscles. The midline groove between the two prominences is called the **natal cleft** (natal=rump) or *gluteal cleft.* The inferior margin of each prominence is the horizontal **gluteal fold,** which roughly corresponds to the inferior margin of the gluteus maximus.

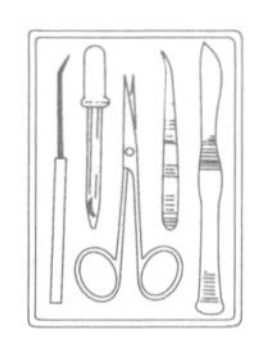

1. Try to palpate your *ischial tuberosity* just above the medial side of each gluteal fold (it will be easier to feel if you sit down or flex your thigh first). The ischial tuberosities are the robust inferior parts of the ischial bones, and they support the body's weight during sitting.

2. Next, palpate the *greater trochanter* of the femur on the lateral side of your hip (Figures 42.18 and 42.20). This trochanter lies just anterior to a hollow and about 10 cm (one hand's breadth) inferior to the iliac crest. To confirm that you have found the greater trochanter, alternately flex and extend your thigh. Because this trochanter is the most superior point on the lateral femur, it moves with the femur as you perform this movement.

3. To palpate the sharp *posterior superior iliac spine* (Figure 42.18), locate your iliac crests again, and trace each to its most posterior point. You may have difficulty feeling this spine, but it is indicated by a distinct dimple in the skin that is easy to find. This dimple lies two to three fingers' breadths lateral to the midline of the back. The dimple also indicates the position of the *sacroiliac joint,* where the hip bone attaches to the sacrum of the spinal column. (You can check *your* "dimples" out in the privacy of your home.)

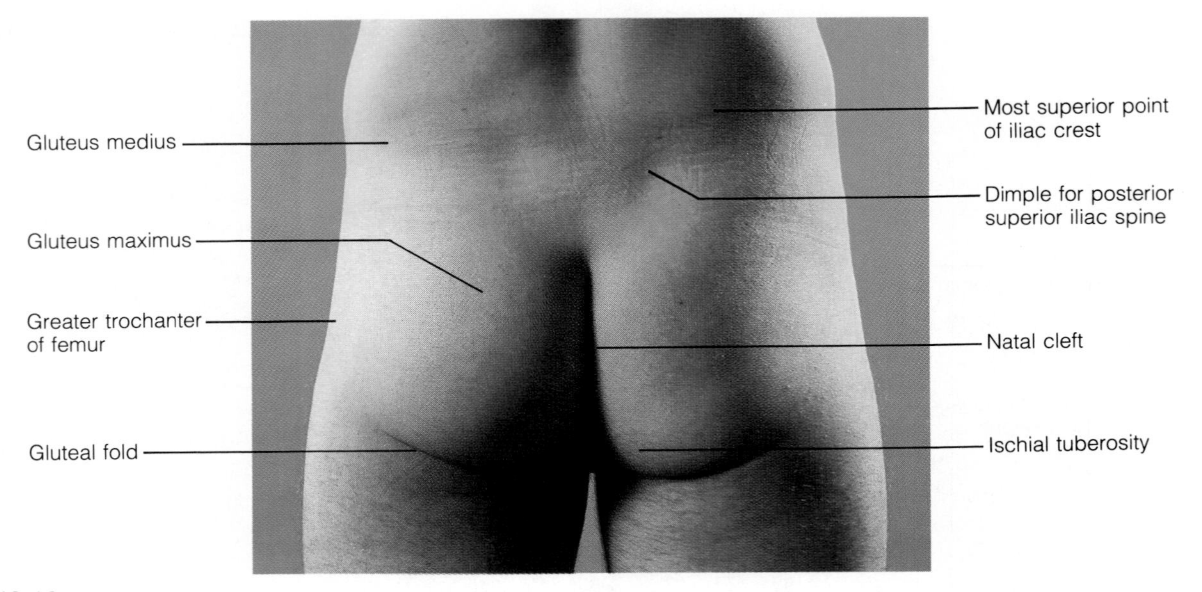

F42.18

The gluteal region. This region extends from the iliac crests superiorly to the gluteal folds inferiorly. Hence, it includes more than just the prominences (cheeks) of the buttocks.

The gluteal region is a major site for administering intramuscular injections. When giving such injections, extreme care must be taken to avoid piercing a major nerve that lies just deep to the gluteus maximus muscle. Can you guess what nerve this is?

It is the thick *sciatic nerve,* which innervates much of the lower limb. Furthermore, the needle must avoid the gluteal nerves and gluteal blood vessels, which also lie deep to the gluteus maximus.

To avoid harming these structures, the injections are most often applied to the gluteus *medius* (not maximus) muscle superior to the cheeks of the buttocks, in a safe area called the **ventral gluteal site** (Figure 42.19b). To locate this site, mentally draw a line laterally from the posterior superior iliac spine (dimple) to the greater trochanter and then proceed to give the injection 5 cm (2 inches) superior to the midpoint of that line. Another safe way to locate the ventral gluteal site is to approach the lateral side of the patient's left hip with your extended right hand (or the right hip with your left hand), then place your thumb on the anterior superior iliac spine and your index finger as far posteriorly on the iliac crest as it can reach. The heel of your hand comes to lie on the greater trochanter, and the needle is inserted in the angle of the V formed between your thumb and index finger about 4 cm (1.5 inches) inferior to the iliac crest.

Gluteal injections are not given to small children because their "safe area" is too small to locate with certainty and because the gluteal muscles are thin at this age. Instead, infants and toddlers receive intramuscular shots in the prominent vastus lateralis muscle of the thigh.

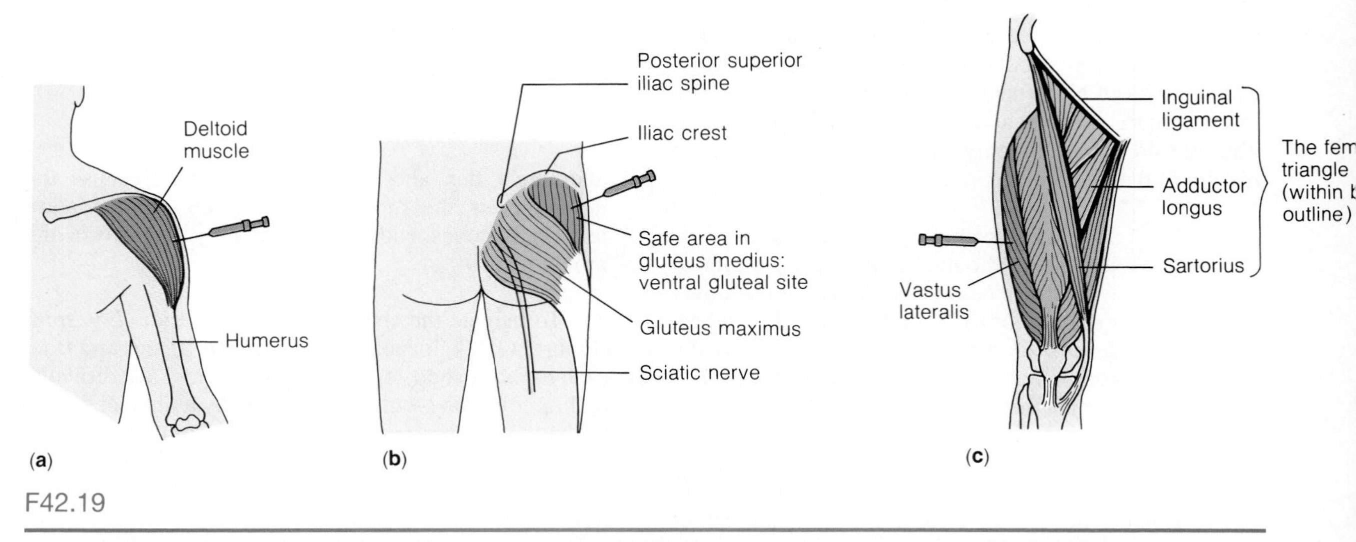

F42.19

Three major sites of intramuscular injection administration. (a) Deltoid muscle of the arm. (b) Ventral gluteal site (gluteus medius). (c) Vastus lateralis in the lateral thigh. Part (c) also shows the femoral triangle.

Thigh

The thigh is pictured in Figures 42.20, 42.21, and 42.22. Much of the femur is clothed by thick muscles, so the thigh has few palpable bony landmarks.

1. Distally, feel the *medial* and *lateral condyles of the femur* and the *patella* anterior to the condyles (see Figure 42.21c and a.)

2. Next, palpate your three groups of thigh muscles (Figure 42.20): the *quadriceps femoris muscles* anteriorly, the *adductor muscles* medially, and the *hamstrings* posteriorly. The *vastus lateralis,* the lateral muscle of the quadriceps group, is a site for intramuscular injections. Such injections are administered about halfway down the length of this muscle (see Figure 42.19c).

3. The anterosuperior surface of the thigh exhibits a three-sided depression called the **femoral triangle** (Figure 42.21a). As shown in Figure 42.19c, the superior border of this triangle is formed by the inguinal ligament, and its two inferior borders are defined by the *sartorius* and *adductor longus* muscles. The large *femoral artery* and *vein* descend vertically through the center of the femoral triangle. To feel the pulse of your femoral artery, press inward just inferior to your midinguinal point (halfway between the anterior superior iliac spine and the pubic tubercle). Be sure to push hard, because the artery lies somewhat deep. By pressing very hard on this point, one can stop the bleeding from a hemorrhage in the lower limb. The femoral triangle also contains most of the *inguinal lymph nodes* (which are easily palpated if swollen).

Leg and Foot

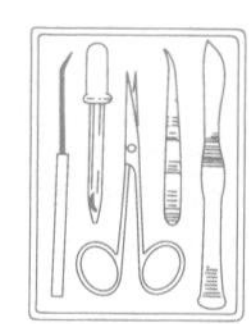

1. Locate your patella again, then follow the thick *patellar ligament* inferiorly from the patella to its insertion on the superior tibia (Figure 42.21c). Here you can feel a rough projection, the *tibial tuberosity.* Continue running your fingers inferiorly along the tibia's sharp *anterior border* and its flat *medial*

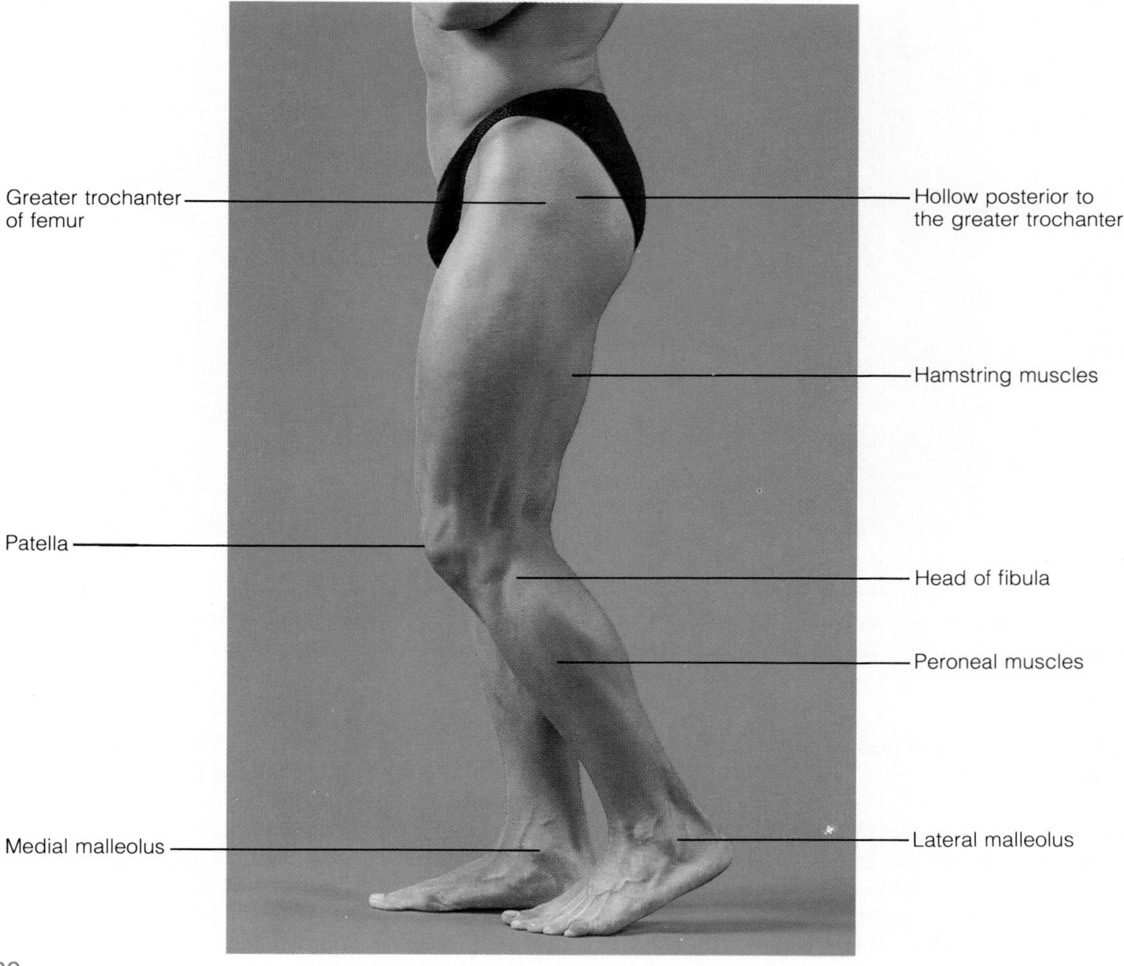

F42.20

Lateral surface of the lower limb.

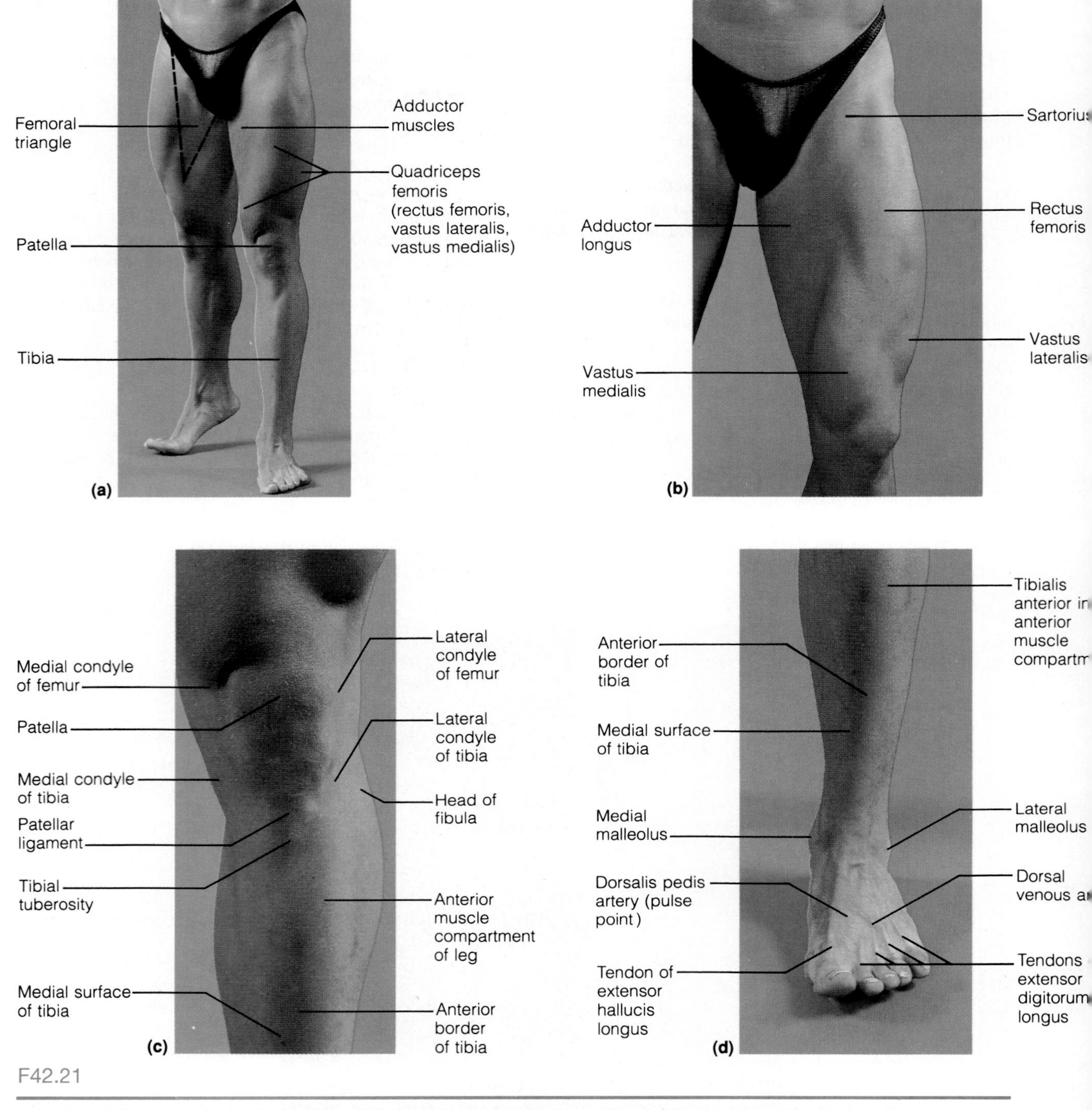

Anterior surface of the lower limb. (a) Both limbs, with the right limb revealing its medial aspect. The femoral triangle is outlined on the right limb. (b) Enlarged view of the left thigh. (c) The left knee region. (d) The dorsum of the left foot.

surface—bony landmarks that lie very near the surface throughout their length.

2. Now, return to the superior part of your leg, and palpate the expanded *lateral* and *medial condyles of the tibia* just inferior to the knee. (You can distinguish the tibial condyles from the *femoral* condyles because you can feel the tibial condyles move with the tibia during knee flexion.) Feel the bulbous *head of the fibula* in the superolateral region of the leg (Figures 42.20 and 42.21c). Try to feel the *common peroneal nerve* (nerve to the anterior leg and foot) where it wraps around the fibula's *neck* just inferior to its head. This nerve is often bumped against the bone here and damaged.

3. In the most distal part of the leg, feel the *lateral malleolus* of the fibula as the lateral prominence of the ankle (Figure 42.21d). Notice that this lies slightly inferior to the *medial malleolus* of the tibia, which forms the ankle's medial prominence. Place your finger just posterior to the medial malleolus to feel the pulse of your *posterior tibial artery.*

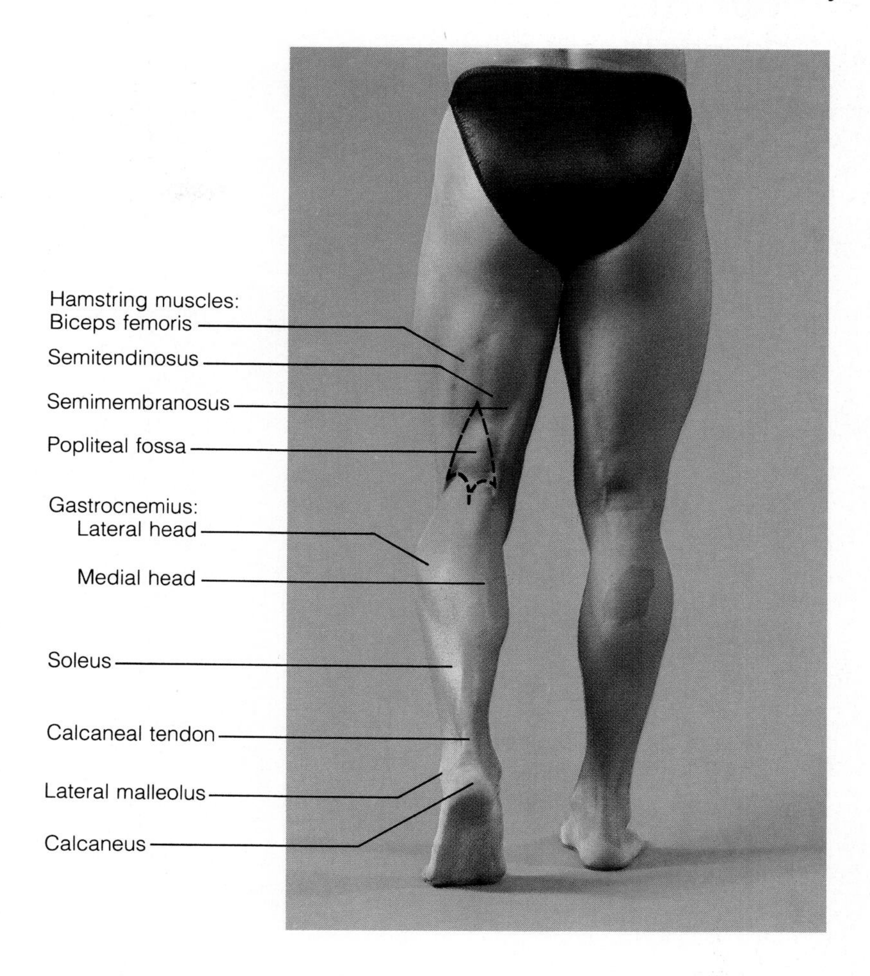

F42.22

Posterior surface of the lower limb. Notice the diamond-shaped popliteal fossa posterior to the knee.

4. On the posterior aspect of the knee is a diamond-shaped hollow called the **popliteal fossa** (Figure 42.22). Palpate the large muscles that define the four borders of this fossa: The *biceps femoris* forming the superolateral border, the *semitendinosus* and *semimembranosus* defining the superomedial border, and the two heads of the *gastrocnemius* forming the inferior border. The *popliteal artery* and *vein* (main vessels to the leg) lie deep within this fossa. To feel a popliteal pulse, flex your leg at the knee and push your fingers firmly into the popliteal fossa. If a physician is unable to feel a patient's popliteal pulse, the femoral artery may be narrowed by atherosclerosis.

5. Next, palpate the main muscle groups of your leg, starting with the calf muscles posteriorly (Figure 42.22). Standing on tiptoes will help you feel the *lateral* and *medial heads of the gastrocnemius* and, inferior to these, the broad *soleus* muscle. Also feel the tension in your *calcaneal* (*Achilles*) *tendon* and at the point of insertion of this tendon onto the calcaneus bone of the foot.

6. Return to the anterior surface of the leg, and palpate the *anterior muscle compartment* (Figure 42.21c and d) while alternately dorsiflexing then plantar flexing your foot. You will feel the tibialis anterior and extensor digitorum muscles contracting then relaxing. Then, palpate the *peroneal muscles* that cover most of the fibula laterally (Figure 42.20). The tendons of these muscles pass posterior to the lateral malleolus and can be felt at a point posterior and slightly superior to that malleolus.

7. Observe the dorsum (superior surface) of your foot. You may see the superficial *dorsal venous arch* overlying the proximal part of the metatarsal bones (Figure 42.21d). This arch gives rise to both saphenous veins (the main superficial veins of the lower limb). Visible in lean people, the *great saphenous vein* ascends along the medial side of the entire limb (see Figure 29.3, p. 268), and the *small saphenous vein* ascends through the center of the calf.

As you extend your toes, observe the tendons of the *extensor digitorum longus* and *extensor hallucis longus* muscles on the dorsum of the foot. Finally, place a finger on the extreme proximal part of the space between the first and second metatarsal bones. Here you should be able to feel the pulse of the *dorsalis pedis artery.*

STUDENT NAME ______________________________

LAB TIME/DATE ______________________________

EXERCISE 1 The Language of Anatomy

Body Orientation, Direction, Planes, and Sections

Several incomplete statements are listed below. Correctly complete each statement by choosing the appropriate anatomical term from the key. Record the key letters and/or terms on the correspondingly numbered blanks below.

Key:
a. anterior
b. distal
c. frontal
d. inferior
e. lateral
f. medial
g. posterior
h. proximal
i. sagittal
j. superior
k. transverse

In the anatomical position, the face and palms are on the __1__ body surface; the buttocks and shoulder blades are on the __2__ body surface; and the top of the head is the most __3__ part of the body. The ears are __4__ and __4__ to the shoulders and __5__ to the nose. The heart is __6__ to the vertebral column (spine) and __7__ to the lungs. The elbow is __8__ to the fingers but __9__ to the shoulder. The abdominopelvic cavity is __10__ to the thoracic cavity and __11__ to the spinal cavity. In humans, the dorsal surface can also be called the __12__ surface; however, in quadruped animals, the dorsal surface is the __13__ surface. If an incision cuts the heart into right and left parts, the section is a __14__ section; but if the heart is cut so that the superior and inferior portions result, the section is a __15__ section. You are told to cut a dissection animal along two planes so that the kidneys are observable in both sections. The two sections that meet this requirement are the __16__ and __17__ sections.

1. ______________________
2. ______________________
3. ______________________
4. ______________________
5. ______________________
6. ______________________
7. ______________________
8. ______________________
9. ______________________
10. ______________________
11. ______________________
12. ______________________
13. ______________________
14. ______________________
15. ______________________
16. ______________________
17. ______________________

Surface Anatomy

Match each of the following descriptions with a key equivalent, and record the key letter or term in front of the description.

Key:
a. buccal
b. calcaneal
c. deltoid
d. digital
e. patellar
f. scapular

______________________ cheek

______________________ pertaining to the fingers

______________________ shoulder blade region

______________________ anterior aspect of knee

______________________ heel of foot

______________________ curve of shoulder

2. Indicate the following body areas on the accompanying diagram by placing the correct key letter at the end of each line.

Key:

a. abdominal
b. antecubital
c. axillary
d. brachial
e. cervical
f. femoral
g. gluteal
h. inguinal
i. lumbar
j. occipital
k. oral
l. popliteal
m. pubic
n. sural
o. thoracic
p. umbilical

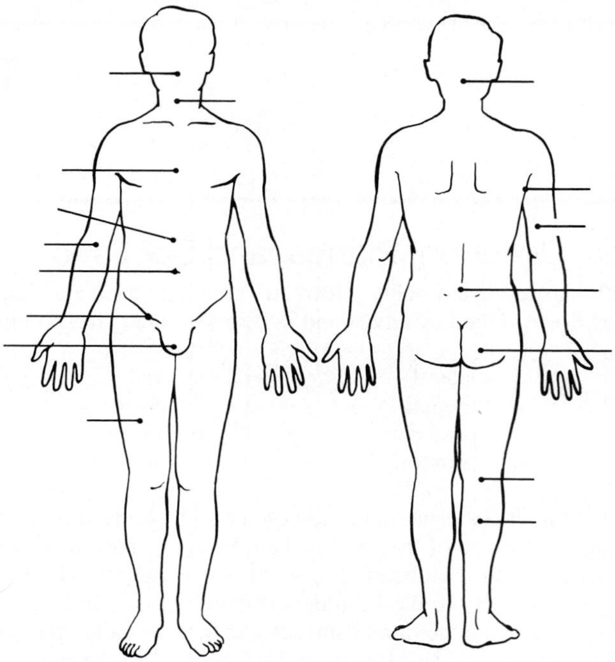

Body Cavities

1. Which body cavity would have to be opened for the following types of surgery? (Insert letter of key choice in same numbered blank.)

Key: a. abdominopelvic c. dorsal e. thoracic
b. cranial d. spinal f. ventral

1. coronary bypass surgery
2. removal of the uterus or womb
3. removal of a brain tumor
4. appendectomy
5. stomach ulcer operation

1. ________
2. ________
3. ________
4. ________
5. ________

The abdominopelvic and thoracic cavities are subdivisions of the __6__ body cavity, while the cranial and spinal cavities are subdivisions of the __7__ body cavity. The __8__ body cavity is totally surrounded by bone, and thus affords its contained structures very good protection.

6. ________
7. ________
8. ________

2. Name the serous membranes covering the lungs (#9), the heart (#10), and the organs of the abdominopelvic cavity (#11), and insert your responses in the blanks on the right.

9. ________
10. ________
11. ________

3. What muscle subdivides the ventral body cavity? (#12)

12. ________

4. What are the bony landmarks of the abdominopelvic cavity? ________________________________

____

5. Which body cavity affords the least protection to its internal structures? ________________

6. What is the function of the serous membranes of the body? ________________________________

____

STUDENT NAME ____________________

LAB TIME/DATE ____________________

EXERCISE 2

Organ Systems Overview

1. Use the key below to indicate the body systems that perform the following functions for the body.

Key:

a.	cardiovascular	e.	integumentary	i.	reproductive
b.	digestive	f.	lymphatic	j.	respiratory
c.	endocrine	g.	muscular	k.	skeletal
d.	immune	h.	nervous	l.	urinary

____________ rids the body of nitrogen-containing wastes

____________ is affected by removal of the thyroid gland

____________ provides support and levers on which the muscular system acts

____________ includes the heart

____________ causes the onset of the menstrual cycle

____________ protects underlying organs from drying out and from mechanical damage

____________ protects the body; destroys bacteria and tumor cells

____________ breaks down ingested food into its building blocks

____________ removes carbon dioxide from the blood

____________ delivers oxygen and nutrients to the tissues

____________ moves the limbs; facilitates facial expression

____________ conserves body water or eliminates excesses

____________ and ____________ facilitate conception and childbearing

____________ controls the body by means of chemical molecules called hormones

____________ is damaged when you cut your finger or get a severe sunburn

2. Using the above key, choose the *organ system* to which each of the following sets of organs or body structures belong:

____________ thymus, spleen, lymphatic vessels

____________ bones, cartilages, tendons

____________ pancreas, pituitary, adrenals

____________ trachea, bronchi, alveoli

____________ kidneys, bladder, ureters

____________ testis, vas deferens, urethra

____________ esophagus, large intestine, rectum

____________ arteries, veins, heart

3. Using the key below, place the following organs in their proper body cavity.

Key: a. abdominopelvic b. cranial c. spinal d. thoracic

______ 1. stomach

______ 2. small intestine

______ 3. large intestine

______ 4. spleen

______ 5. liver

______ 6. spinal cord

______ 7. bladder

______ 8. heart

______ 9. lungs

______ 10. brain

______ 11. rectum

4. Using the organs listed in item 3 above, record, by number, which would be found in the abdominal regions listed below:

______ hypogastric region

______ right lumbar region

______ umbilical region

______ epigastric region

______ left iliac region

______ left hypochondriac region

5. The five levels of organization of a living body are cell, ______,

______, ______, and organism.

6. Define *organ.* ______

______

______

7. During the course of this laboratory exercise, a rat was dissected. What is the *value* of observing the anatomy of a rat (or any other small mammal) when *human anatomy* is the actual topic of study?

______

______

______

STUDENT NAME ____________________

LAB TIME/DATE ____________________

EXERCISE 3

The Microscope

Care and Structure of the Compound Microscope

1. The following statements are true or false. If true, write *T* on the answer blank. If false, correct the statement by writing on the blank the proper word or phrase to replace that underlined.

____________________ The microscope lens may be cleaned with any soft tissue.

____________________ The coarse adjustment knob may be used in focusing with all three objectives.

____________________ The microscope should be stored with the oil immersion lens in position over the stage.

____________________ When beginning to focus, the low-power lens should be used.

____________________ In low power, always focus toward the specimen.

____________________ A coverslip should always be used with the high-power and oil lenses.

____________________ The greater the amount of light delivered to the objective lens, the less the resolution.

2. Match the microscope structures given in column B with the statements in column A that identify or decribe them:

Column A

_______ platform on which the slide rests for viewing

_______ lens located at the superior end of the body tube

_______ secure(s) the slide to the stage

_______ delivers a concentrated beam of light to the specimen

_______ used for precise focusing once initial focusing has been done

_______ carries the objective lenses; rotates so that the different objective lenses can be brought into position over the specimen

_______ used to increase the amount of light passing through the specimen

Column B

a. coarse adjustment knob

b. condenser

c. fine adjustment knob

d. iris diaphragm

e. mechanical stage or spring clips

f. movable nosepiece

g. objective lenses

h. ocular

i. stage

3. Explain the proper technique for transporting the microscope.

____

____

Viewing Objects Through the Microscope

1. Complete, or respond to, the following statements:

________________ The distance from the bottom of the objective lens in use to the specimen is called the ________.

________________ The resolution of the human eye is ________ μm.

________________ The area of the specimen seen when looking through the microscope is the ________.

________________ If a microscope has a 10× ocular and the total magnification at a particular time is 950×, the objective lens in use at that time is ________×.

________________ Why should the light be dimmed when looking at living (nearly transparent) cells?

________________ If, after focusing in low power, only the fine adjustment need be used to focus the specimen at the higher powers, the microscope is said to be ________.

________________ Imagine that you are observing an object in the low-power field. When you switch to high power, the object is not visible. The most likely explanation for this is that ________.

________________ If, when using a 10× ocular and a 15× objective, the field size is 1.5 mm, the approximate field size with a 30× objective is ________ mm.

________________ If the size of the high-power field is 1.2 mm, an object that occupies approximately a third of that field has an estimated diameter of ________ mm.

________________ Assume there is an object on the left side of the field that you want to bring to the center (that is, toward the apparent right). In what direction would you move your slide?

________________ If the object is in the top of the field and you want to move it downward to the center, you would move the slide ________.

2. Indicate whether the following factors increase or decrease as one moves to higher magnifications with the microscope:

depth of field ________________ working distance ________________

amount of light needed ________________

3. You have been asked to prepare a slide with letter *k* on it (as below). In the circle below, draw the *k* as seen in the low-power field.

k

STUDENT NAME ___________________________

LAB TIME/DATE ___________________________

EXERCISE 4

The Cell— Anatomy and Division

Anatomy of the Composite Cell

Identify the following cell parts:

___________________ external boundary of cell; regulates flow of materials into and out of the cell; site of cell signalling

___________________ contains digestive enzymes of many varieties; "suicide sac" of the cell

___________________ scattered throughout the cell; major site of ATP synthesis

___________________ slender extensions of the plasma membrane that increase its surface area

___________________ stored glycogen granules, crystals, pigments, and so on

___________________ membranous system consisting of flattened sacs and vesicles; packages proteins for export

___________________ control center of the cell; necessary for cell division and cell life

___________________ two rod-shaped bodies near the nucleus; direct formation of the mitotic spindle

___________________ dense, darkly staining nuclear body; packaging site for ribosomes

___________________ membranous system involved with synthesis of steroidal hormones

___________________ membranous system; involved in intracellular transport of proteins and synthesis of membrane lipids

___________________ attached to membrane systems or scattered in the cytoplasm; synthesize proteins

___________________ threadlike structures in the nucleus; contain genetic material (DNA)

Observing Differences and Similarities in Cell Structure

List *one* important structural characteristic (a) of each of the following cell types observed in the laboratory, and then give the function (b) that structure complements or ensures:

squamous epithelium a. ___________________________

b. ___________________________

sperm a. ___________________________

b. ___________________________

smooth muscle a. ___________________________

b. ___________________________

red blood cell a. ___________________________

b. ___________________________

Cell Division: Mitosis and Cytokinesis

1. Using the key, categorize each of the events described below according to the phase in which it occurs.

   Key: a. prophase b. anaphase c. telophase d. metaphase e. none of these

   ____________ Chromatin coils and condenses, forming chromosomes.

   ____________ The chromosomes (chromatids) are V-shaped.

   ____________ The nuclear membrane re-forms.

   ____________ Chromosomes stop moving toward the poles.

   ____________ Chromosomes line up in the center of the cell.

   ____________ The nuclear membrane fragments.

   ____________ The mitotic spindle forms.

   ____________ DNA synthesis occurs.

   ____________ Centrioles replicate.

   ____________ Chromosomes first appear to be duplex structures.

   ____________ Chromosomal centromeres attach to the spindle fibers.

   ____________ Cleavage furrow forms.

   ____________ The nuclear membrane(s) is absent.

2. Identify the phases of mitosis in the following diagrams.

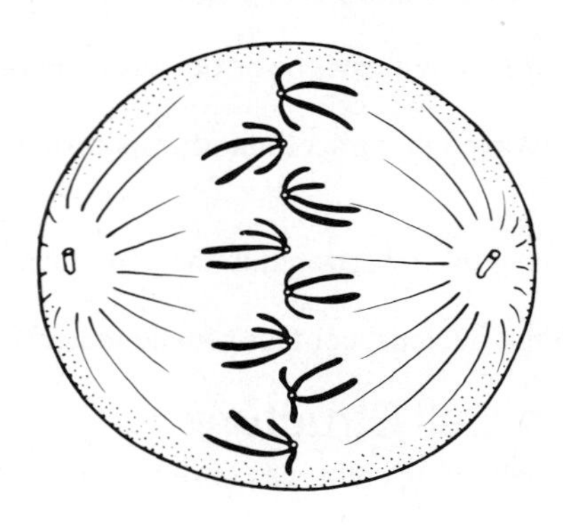

____

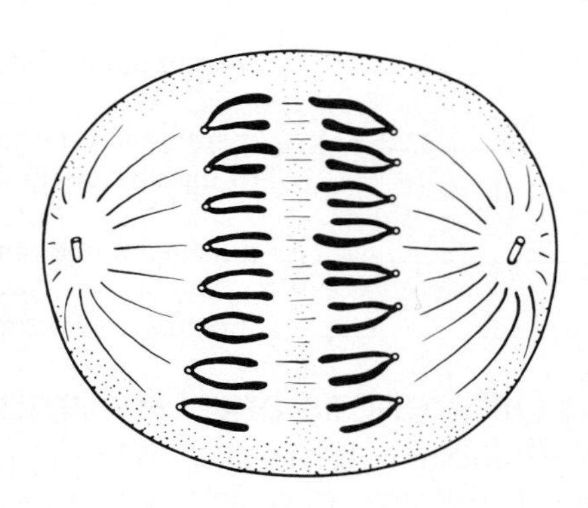

____

3. Complete the following statements:

   Nuclear division is referred to as __1__. Division of the cytoplasm is called __2__. If a cell undergoes mitosis but not cytokinesis, the product is __3__. __4__ is the period of cell life when the cell is not involved in division. Two cell populations in the body that do not undergo cell division are __5__ and __6__. The implication of an inability of a cell population to divide is that when some of its members die, they are replaced by __7__.

   1. ____________
   2. ____________
   3. ____________
   4. ____________
   5. ____________
   6. ____________
   7. ____________

STUDENT NAME ______________________

LAB TIME/DATE ______________________

EXERCISE 5

The Cell—Transport Mechanisms and Cell Permeability

Choose all answers that apply to items 1 and 2, and place their letters on the response blanks to the right.

1. Brownian motion: ______________________

   a. reflects the kinetic energy of the smaller solvent molecules
   b. reflects the kinetic energy of the larger solute molecules
   c. is ordered and predictable
   d. is random and erratic

2. Kinetic energy ______________________

   a. is higher in larger molecules
   b. is lower in larger molecules
   c. increases with increasing temperature
   d. decreases with increasing temperature
   e. is reflected in the speed of molecular movement

3. The following refer to the laboratory experiment using dialysis sacs 1 through 4 to study diffusion through nonliving membranes:

   Sac 1: 40% glucose suspended in distilled water

   Did glucose pass out of the sac? ______________________

   Test used to determine presence of glucose: ______________________

   Did the sac weight change? ______________________

   If so, explain the reason for its weight change: ______________________

   ____

   ____

   Sac 2: 40% glucose suspended in 40% glucose

   Was there net movement of glucose in either direction? ______________________

   Explanation: ______________________

   Did the sac weight change? ______________________ Explanation: ______________________

   ____

   Sac 3: 10% NaCl in distilled water

   Was there net movement of NaCl out of the sac? ______________________

   Test used to determime the presence of NaCl: ______________________

   ____

   Direction of net osmosis: ______________________

Sac 4: Boiled starch in distilled water

Was there net movement of starch out of the sac? ______________________

Test used to determine presence of starch: ______________________

Direction of net osmosis: ______________________

4. What single characteristic of the semipermeable membranes used in the laboratory determines the substances that can pass through them? ______________________

   In addition to this characteristic, what other factors influence the passage of substances through living membranes?

   ____

   ____

5. A semipermeable sac containing 4% NaCl, 9% glucose, and 10% albumin is suspended in a solution with the following composition: 10% NaCl, 10% glucose, and 40% albumin. Assume that the sac is permeable to all substances except albumin. State whether each of the following will (a) move into the sac, (b) move out of the sac, or (c) not move.

   glucose ______________________ albumin ______________________

   water ______________________ NaCl ______________________

6. The diagrams below represent three microscope fields containing red blood cells. Arrows show the direction of net osmosis. Which field contains a hypertonic solution? ________ The cells in this field are said to be ______________________. Which field contains an isotonic bathing solution? ________ Which field contains a hypotonic solution? ________ What is happening to the cells in this field? ______________________

   ____

(a) (b) (c)

7. What determines whether a transport process is active or passive? ______________________

   ____

8. Characterize passive and active transport as fully as possible by choosing all the phrases that apply and inserting their letters on the answer blanks.

   Passive transport: ________ Active transport: ________

   a. accounts for the movement of fats and respiratory gases through the plasma membrane
   b. explains solute pumping, phagocytosis, and pinocytosis
   c. includes osmosis, diffusion, dialysis, and filtration
   d. may occur against concentration and electrical gradients
   e. uses hydrostatic pressure or molecular energy as the driving force
   f. moves ions, amino acids, and some sugars across the plasma membrane

STUDENT NAME ______________________________

LAB TIME/DATE ______________________________

EXERCISE 6 — Classification of Tissues

Tissue Structure and Function—General Review

1. Define *tissue:* ______________________________

____

2. Use the key choices to identify the *major* tissue types described below.

Key: a. connective tissue b. epithelium c. muscle d. nervous tissue

______________ lines body cavities and covers the body's external surface

______________ pumps blood, flushes urine out of the body, allows one to swing a bat

______________ transmits electrochemical impulses

______________ anchors, packages, and supports body organs

______________ cells may absorb, secrete, and protect

______________ most involved in regulating and controlling body functions

______________ major function is to contract

______________ synthesizes hormones

______________ the most durable tissue type

______________ abundant nonliving extracellular matrix

______________ most widespread tissue in the body

______________ forms nerves and the brain

Epithelial Tissue

1. Describe the general characteristics of epithelial tissue. ______________________________

____

____

____

2. On what bases are epithelial tissues classified? ______________________________

3. What are the major functions of epithelium in the body? (Give examples.) ______________________________

____

____

____

4. Respond to the following with the key choices.

Key:
a. pseudostratified ciliated columnar
b. simple columnar
c. simple cuboidal
d. simple squamous
e. stratified squamous
f. transitional

_______________ lining of the esophagus

_______________ lining of the stomach and small intestine

_______________ lung tissue, alveolar sacs

_______________ collecting tubules of the kidney

_______________ epidermis of the skin

_______________ lining of bladder; peculiar cells that have the ability to slide over each other

_______________ forms the thin serous membranes; a single layer of flattened cells

Connective Tissue

1. Using the key, choose the best response to identify the connective tissues described below.

Key:
a. adipose connective tissue
b. areolar connective tissue
c. dense connective tissue
d. elastic cartilage
e. fibrocartilage
f. hematopoietic tissue
g. hyaline cartilage
h. osseous tissue

_______________ attaches bones to bones and muscles to bones

_______________ acts as a storage depot for fat

_______________ the dermis of the skin

_______________ makes up the intervertebral discs

_______________ forms your hip bone

_______________ composes basement membranes; a soft packaging tissue with a jelly-like matrix

_______________ forms the larynx, the costal cartilages of the ribs, and the embryonic skeleton

_______________ provides a flexible framework for the external ear

_______________ firm, structurally amorphous matrix heavily invaded with fibers; appears glassy and smooth

_______________ matrix hard owning to calcium salts; provides levers for muscles to act on

_______________ insulates against heat loss

2. What are the general characteristics of connective tissues? _______________

____

____

3. What functions are performed by connective tissue? ______________________

4. How are the functions of connective tissue reflected in its structure? ______________________

Muscle Tissue

The three types of muscle tissue exhibit similarities as well as differences. Check the appropriate space in the chart below to indicate which muscle types exhibit each characteristic.

Characteristic	Skeletal	Cardiac	Smooth
Voluntarily controlled			
Involuntarily controlled			
Striated			
Has a single nucleus in each cell			
Has several nuclei per cell			
Found attached to bones			
Allows you to direct your eyeballs			
Found in the walls of the stomach, uterus, and arteries			
Contains spindle-shaped cells			
Contains branching cylindrical cells			
Contains long, nonbranching cylindrical cells			
Has intercalated discs			
Concerned with locomotion of the body as a whole			
Changes the internal volume of an organ as it contracts			
Tissue of the heart			

Nervous Tissue

1. What two physiologic characteristics are highly developed in nervous tissue? ______________________

2. In the space below sketch a neuron, recalling in your diagram the most important aspects of its structure. Below the diagram, describe how its particular structure relates to its function in the body.

____

____

For Review

Label the following tissue types here and on the next pages, and identify all major structures.

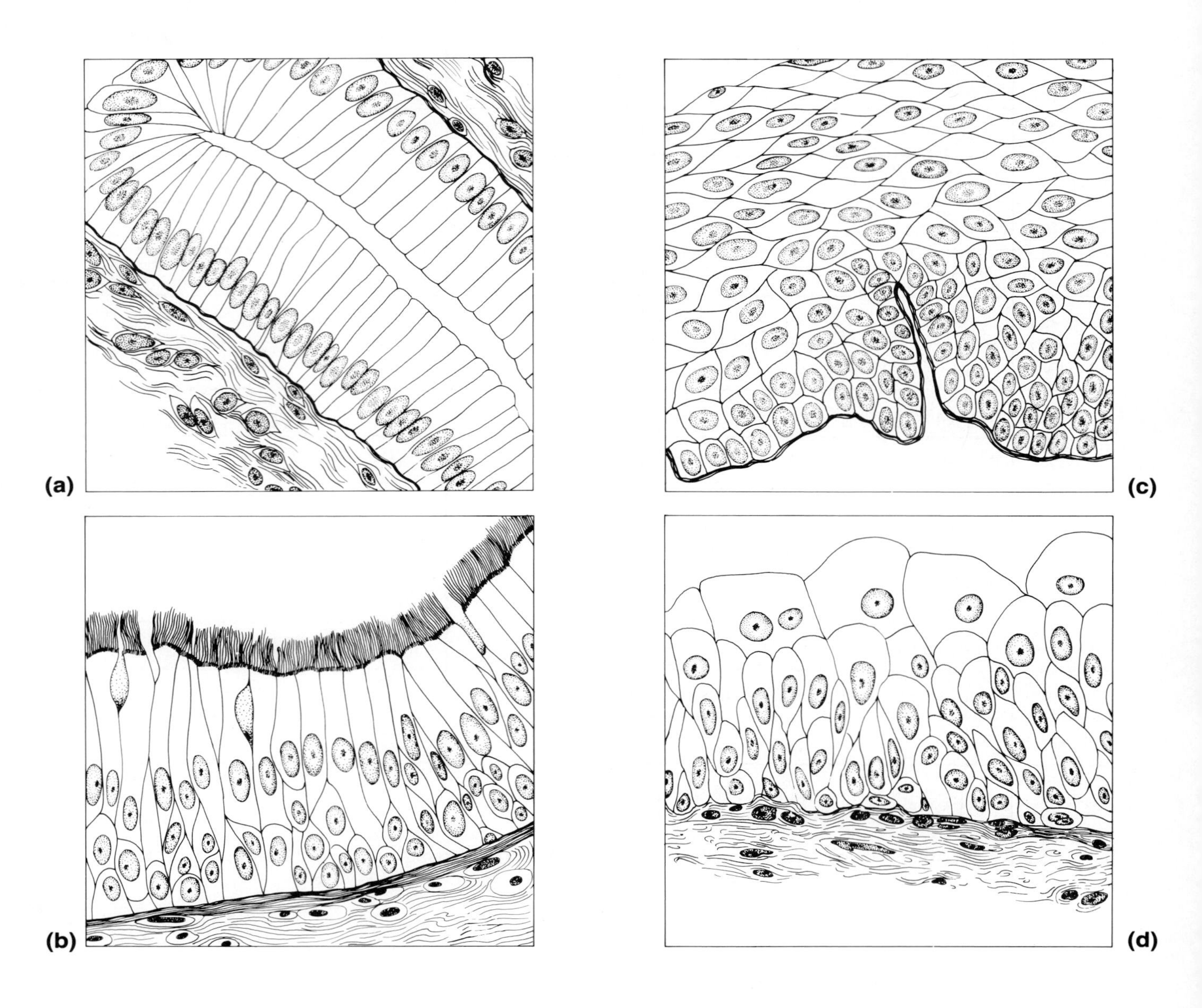

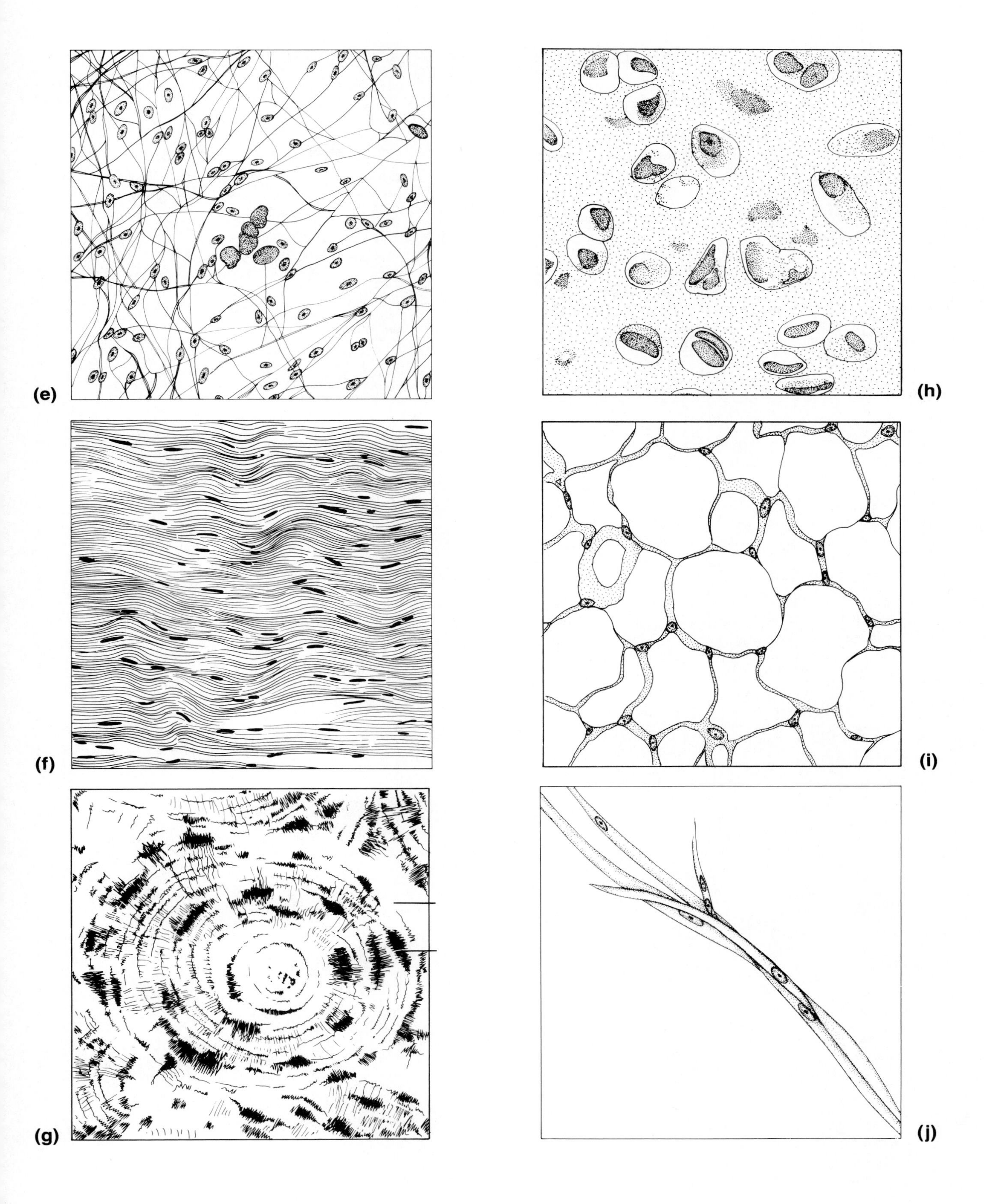
(e)
(f)
(g)
(h)
(i)
(j)

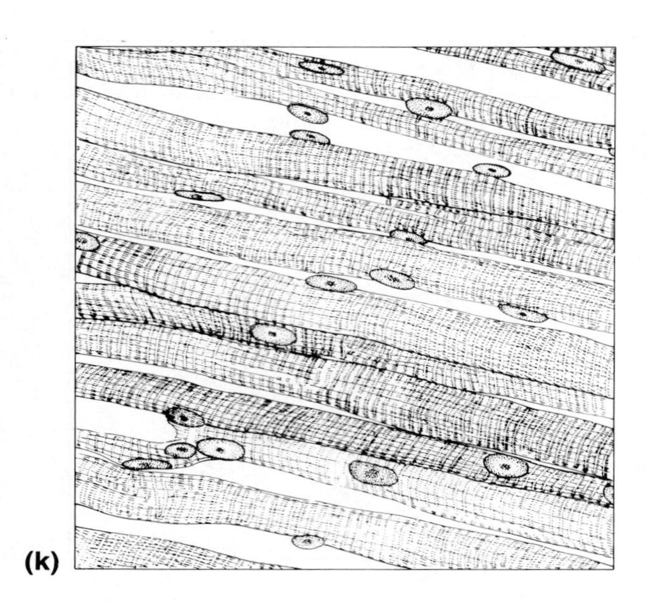

(k)

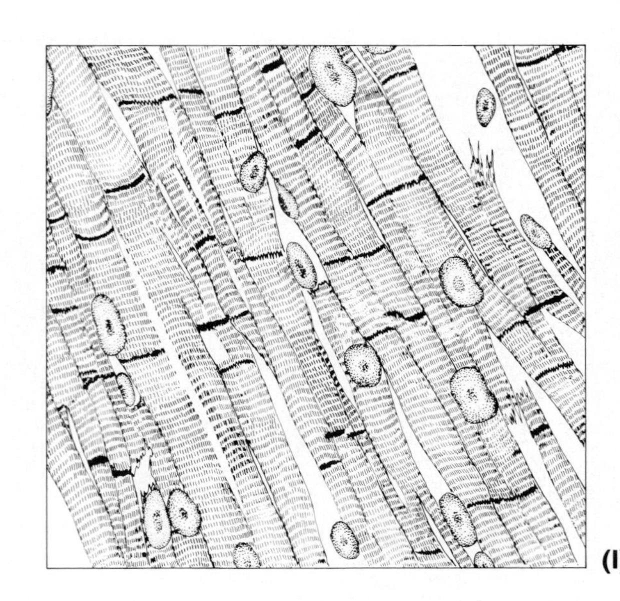

(l)

STUDENT NAME ____________________

LAB TIME/DATE ____________________

EXERCISE 7 The Skin and Other Body Membranes

Basic Structure of the Skin

1. Complete the following statements by writing the appropriate word or phrase on the correspondingly numbered blanks:

The two basic tissues of which the skin is composed are dense irregular connective tissue, which makes up the dermis, and __1__, which forms the epidermis. The tough waterproofing protein found in the epidermal cells is called __2__. The pigments melanin and __3__ contribute to skin color. A localized concentration of melanin is referred to as a __4__.

1. ____________________
2. ____________________
3. ____________________
4. ____________________

2. Four (4) protective functions of the skin are ____________________

____

3. Label the skin structures and areas indicated in the accompanying diagram.

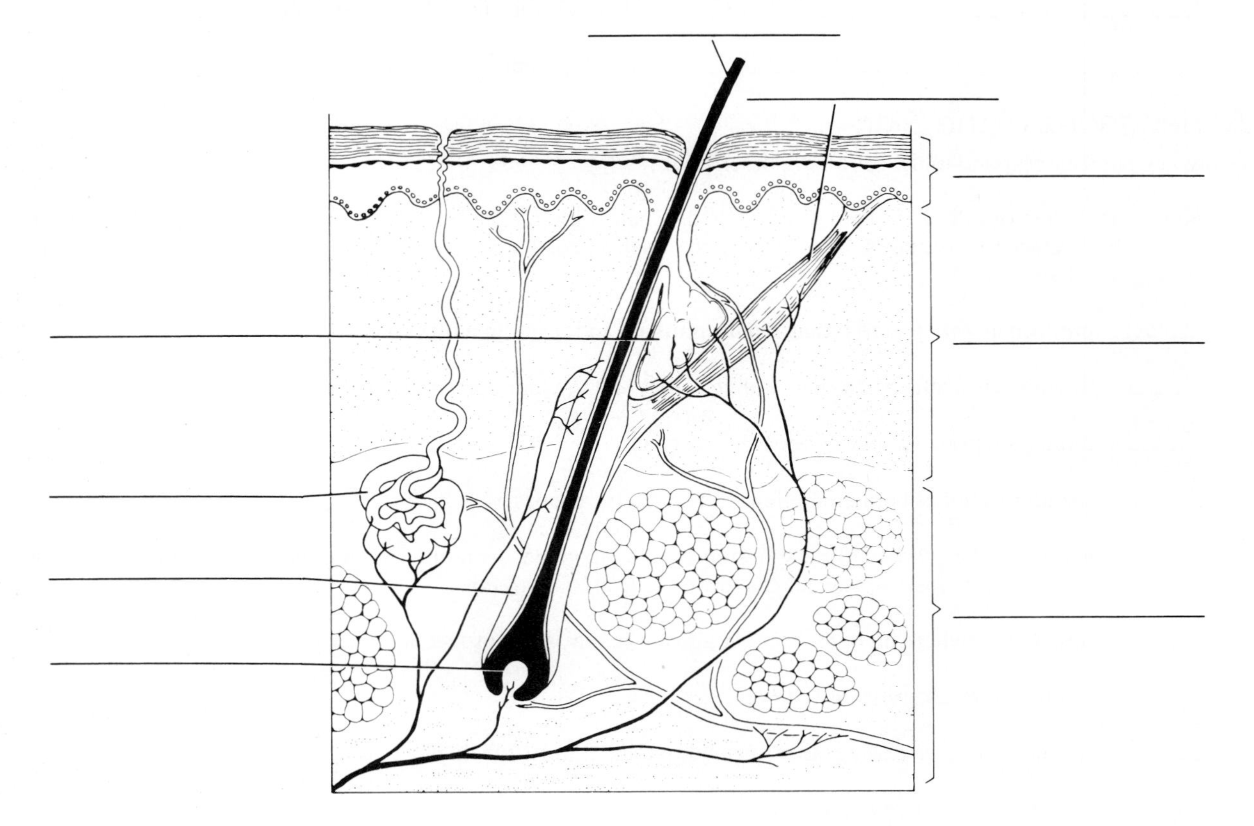

4. A nurse tells the doctor that a patient is cyanotic. What is cyanosis and what does its presence imply?

____

5. Using the key choices, choose all responses that apply to the following descriptions.

Key:
a. stratum corneum
b. stratum germinativum
c. stratum granulosum
d. stratum lucidum
e. papillary layer
f. reticular layer
g. epidermis as a whole
h. dermis as a whole

______________________ translucent cells containing keratin fibrils

______________________ dead cells

______________________ dermis layer responsible for fingerprints

______________________ vascular region

______________________ major skin area where derivatives (nails and hair) arise

______________________ epidermal region exhibiting rapid cell division

______________________ scalelike dead cells, full of keratin, that constantly slough off

______________________ actually composed of two strata, the basale and spinosum

______________________ site of elastic and collagenic fibers

______________________ site of melanin formation

Appendages of the Skin

Using key choices, respond to the following descriptions.

Key:
a. arrector pili
b. cutaneous receptors
c. hair
d. hair follicle
e. sebaceous glands
f. sweat gland—apocrine
g. sweat gland—eccrine

_______ produce an oily material that can accumulate and result in a blackhead

_______ tiny muscles, attached to hair follicles, that pull the hair upright during fright or cold

_______ more numerous variety of perspiration gland

_______ sheath formed of both epithelial and connective tissues

_______ less numerous variety of perspiration-producing gland; secretion (often milky in appearance) contains proteins and other substances that promote bacterial growth

_______ found everywhere on body except palms of hands, soles of feet, and lips

_______ primarily dead/keratinized cells

_______ specialized nerve endings that respond to temperature, touch, etc.

_______ become more active at puberty

_______ part of the heat-liberating apparatus of the body

Plotting the Distribution of Sweat Glands

1. With what substance in the bond paper does the iodine painted on the skin react? ______________________

2. Which skin area—the forearm or palm of hand—has more sweat glands? ______________________

   Which other body areas would, if tested, prove to have a high density of sweat glands? ______________________

   ____

3. What organ system controls the activity of the eccrine sweat glands? ______________________

   ____

Classification of Body Membranes

1. Complete the following chart:

Membrane	Tissue type (epithelial/connective)	Common locations	Functions
cutaneous			
mucous			
serous			
synovial			

2. Respond to the following statements by choosing an answer from the key.

Key: a. cutaneous b. mucous c. serous d. synovial

____________ membrane type associated with skeletal system structures

____________ always formed of simple squamous epithelium

____________, ____________ membrane types *not* found in the ventral body cavity

____________ the only membrane type in which goblet cells are found

____________ the only *external* membrane

____________ "wet" membranes

____________ adapted for absorption and secretion

____________ has parietal and visceral layers

STUDENT NAME ______________________

LAB TIME/DATE ______________________

EXERCISE 8

Bone Classification, Structure, and Relationships: An Overview

Bone Markings

1. Match the terms in column B with the appropriate description in column A:

	Column A		Column B
______________	sharp, slender process*	a.	condyle
______________	small rounded projection*	b.	crest
______________	narrow ridge of bone*	c.	epicondyle
______________	large rounded projection*	d.	fissure
______________	structure supported on neck†	e.	foramen
______________	armlike projection†	f.	fossa
______________	rounded, convex projection†	g.	head
______________	narrow depression or opening‡	h.	meatus
______________	canallike structure‡	i.	ramus
______________	opening through a bone‡	j.	sinus
______________	shallow depression†	k.	spine
______________	air-filled cavity	l.	trochanter
______________	large, irregularly shaped projection*	m.	tubercle
______________	raised area of a condyle*	n.	tuberosity

Classification of Bones

1. The four major anatomic classifications of bones are long, short, flat, and irregular. Which category has the least amount of spongy bone relative to its total volume? ______________________

* A site of muscle attachment.

† Takes part in joint formation.

‡ A passageway for nerves or blood vessels.

2. Classify each of the bones in the next chart into one of the four major categories by checking the appropriate column. Use appropriate references as necessary.

	Long	Short	Flat	Irregular
humerus				
phalanx				
parietal				
calcaneus				
rib				
vertebra				
ulna				

Gross Anatomy of the Typical Long Bone

1. Using the terms to the right, characterize the following statements:

_______________ contains spongy bone in adults — a. diaphysis

_______________ made of compact bone — b. endosteum

_______________ site of blood cell formation — c. epiphyseal plate

_______________ major submembranous site of osteoclasts — d. epiphysis

_______________ scientific term for bone shaft — e. periosteum

_______________ contains fat in adult bones — f. red marrow cavity

_______________ growth plate — g. yellow marrow cavity

2. What differences between compact and spongy bone can be seen with the naked eye? _______________

____

3. What is the function of the periosteum? _______________

____

Microscopic Structure of Compact Bone

1. Trace the route taken by nutrients through a bone, starting with the periosteum and ending with an osteocyte in a lacuna. Periosteum ____

____ osteocyte

2. Several descriptions of bone structure are given in column B. Identify the structure involved by choosing the appropriate term from column A and placing the corresponding letter in the correct blank.

Column A		Column B
a. central canal	________	layers of bony matrix around a central canal
b. concentric lamellae	________	site of osteocytes
c. lacunae	________	longitudinal canal carrying blood vessels, lymphatics, and nerves
d. canaliculi	________	nonliving, structural part of bone
e. matrix	________	minute canals connecting lacunae

3. On the photomicrograph of bone below, identify all structures named in column A in question 2 above.

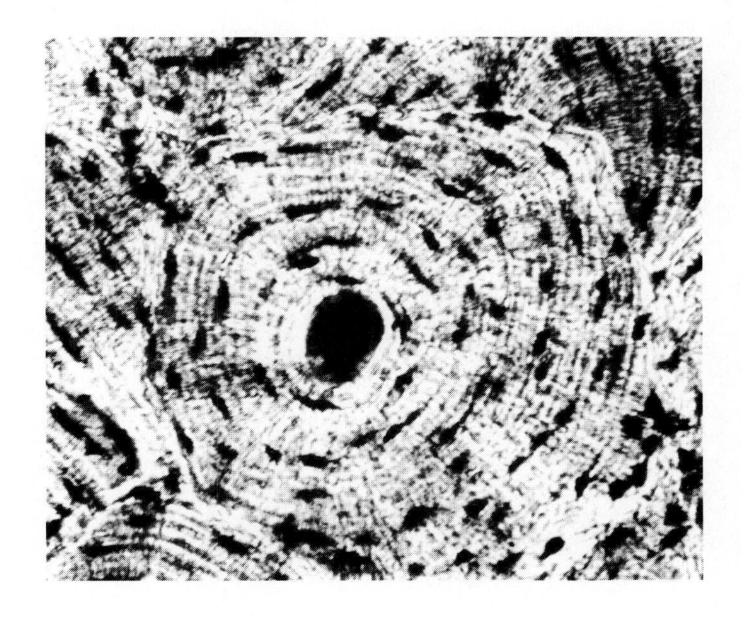

Chemical Composition of Bone

1. What is the function of the organic matrix in bone? ________________________________

2. Name the important organic bone components. ________________________________

3. Calcium salts form the bulk of the inorganic material in bone. What is the function of the calcium salts?

____

4. Which is responsible for bone structure? (circle the appropriate response)

inorganic portion　　　organic portion　　　both contribute

STUDENT NAME ____________________________

LAB TIME/DATE ____________________________

EXERCISE 9

The Axial Skeleton

The Skull

1. The skull is one of the major components of the axial skeleton. Name the other two:

   ____________________ and ____________________

2. With one exception, the skull bones are joined by sutures. Name the exception. ____________________

   ____

3. Match the bone names in column B with the descriptions in column A. (Descriptions continue on p. RS26.)

	Column A	Column B
________	forehead bone	a. ethmoid
________	cheekbone	b. frontal
________	lower jaw	c. hyoid
________	bridge of nose	d. lacrimals
________	posterior part of hard palate	e. mandible
________	much of the lateral and superior cranium	f. maxillae
________	most posterior part of cranium	g. nasals
________	single, irregular, bat-shaped bone forming part of the cranial floor	h. occipital
________	tiny bones bearing tear ducts	i. palatines
________	anterior part of hard palate	j. parietals
________	superior and medial nasal conchae formed from its projections	k. sphenoid
________	site of mastoid process	l. temporals
________	site of sella turcica	m. vomer
________	site of cribriform plate	n. zygomatic
________	site of mental foramen	
________	site of styloid processes	

________, ________, ________,

________ four bones containing paranasal sinuses

______________ condyles here articulate with the atlas

______________ foramen magnum contained here

______________ small U-shaped bone in neck, where many tongue muscles attach

______________ middle ear found here

______________ nasal septum

______________ bears an upward protrusion, the "cock's comb," or crista galli

4. Give two possible functions of the sinuses. ______________________________

____

5. What is the orbit? ______________________________

What bones contribute to the formation of the orbit? ______________________________

____

6. Why can the sphenoid bone be called the keystone of the cranial floor? ______________________________

____

7. Identify all bones and bone markings provided with leader lines in the two diagrams that follow:

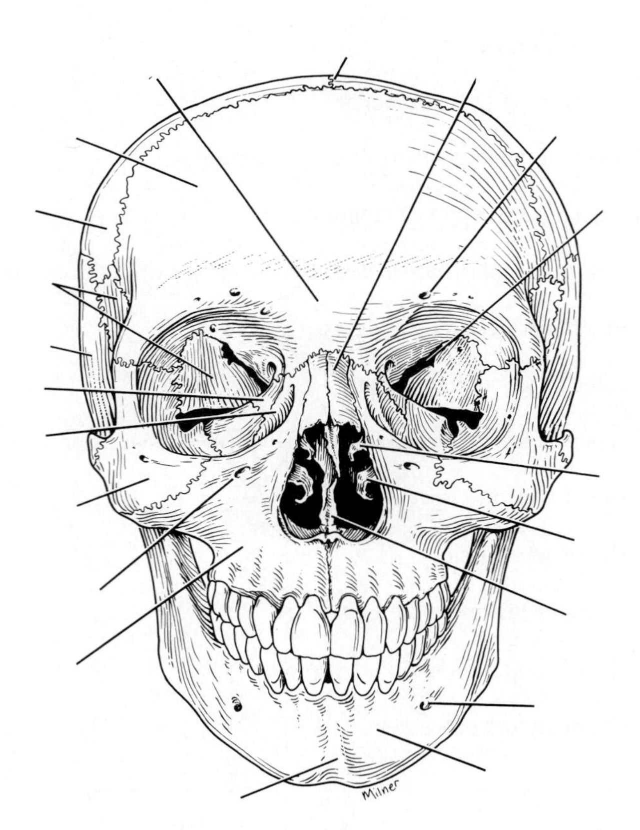

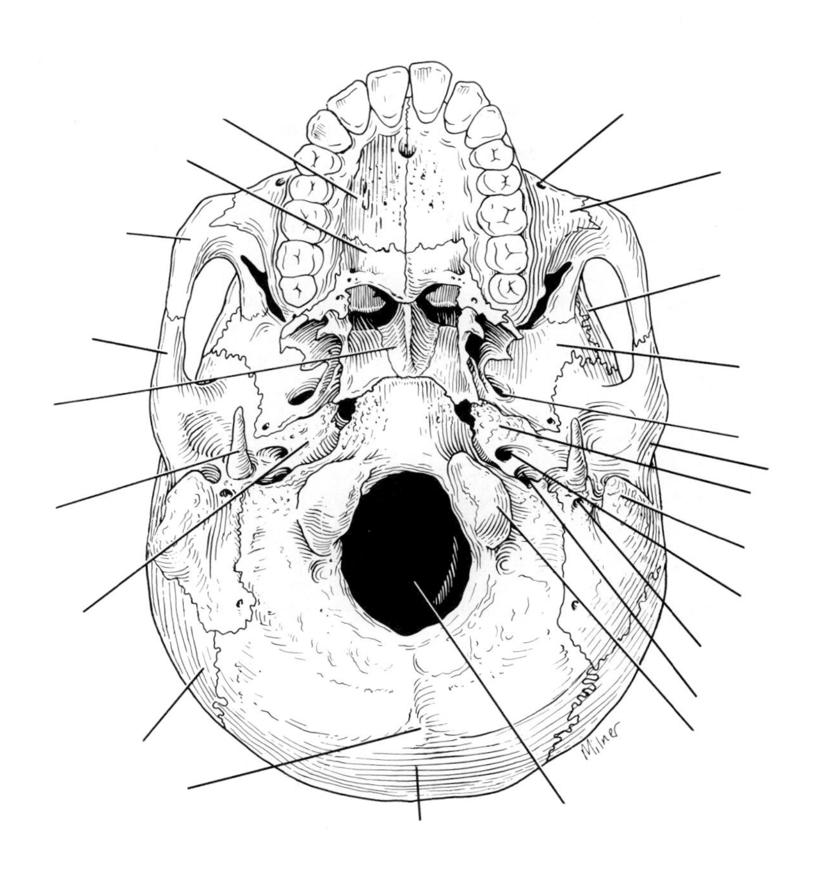

The Vertebral Column

1. Using the key, correctly identify the vertebral parts/areas described below. (More than one choice may apply in some cases.)

Key: a. body; b. intervertebral foramina; c. spinous process; d. superior articular process; e. transverse process; f. vertebral arch

_______ structures enclosing the nerve cord

_______ weight-bearing portion of the vertebra

_______ provide levers against which muscles pull

_______ provide an articulation point for the ribs

_______ openings providing for exit of spinal nerves

2. Of what kind of tissue are the intervertebral discs composed? _______________

3. What is a herniated disc? _______________

What problems might it cause? _______________

____

4. Name two factors/structures that allow for flexibility of the vertebral column:

_______________ and _______________

5. The distinguishing characteristics of the vertebrae composing the vertebral column are noted below. Correctly identify each described structure/region by choosing a response from the key.

Key:
a. atlas
b. axis
c. cervical vertebra—typical
d. coccyx
e. lumbar vertebra
f. sacrum
g. thoracic vertebra

__________ vertebral type containing foramina in the transverse processes, through which the vertebral arteries ascend to reach the brain

__________ dens here provides a pivot for rotation of the first cervical vertebra (C_1)

__________ transverse processes faceted for articulation with ribs; spinous process pointing sharply downward

__________ composite bone; articulates with the hip bone laterally

__________ massive vertebrae; weight-sustaining

__________ "tail bone"; vestigeal fused vertebrae

__________ supports the head; allows a rocking motion in conjunction with the occipital condyles

__________ seven components; unfused

__________ twelve components; unfused

6. The abnormal spinal curvatures are described below. Identify each:

__________ exaggeration of the lumbar curvature

__________ abnormal lateral displacement of the vertebral column

__________ exaggerated thoracic curvature

7. Which two spinal curvatures are present at birth? __________ and __________

Under what conditions do the secondary curvatures develop? __________

____

____

The Bony Thorax

1. The major components of the thorax (excluding the vertebral column) are the *ribs* __________

and the __________.

2. Differentiate between a true rib and a false rib. __________

____

____

Is a floating rib a true or a false rib? __________

3. What is the general shape of the thoracic cage? __________

STUDENT NAME ______________________

LAB TIME/DATE ______________________

EXERCISE 10 The Appendicular Skeleton

Bones of the Shoulder Girdle and Upper Extremity

1. Match the bone names or markings in column B with the descriptions in column A.

Column A	Column B
______ raised area on lateral surface of humerus to which deltoid muscle attaches	a. acromion
______ arm bone	b. capitulum
______, ______ bones of the shoulder girdle	c. carpals
______, ______ forearm bones	d. clavicle
______ scapular region to which the clavicle connects	e. coracoid process
______ shoulder girdle bone that is unattached to the axial skeleton	f. coronoid fossa
______ shoulder girdle bone that articulates anteriorly with the sternum	g. deltoid tuberosity
______ depression in the scapula that articulates with the humerus	h. glenoid cavity
______ process above the glenoid cavity that permits muscle attachment	i. humerus
______ commonly called the collarbone	j. metacarpals
______ distal condyle of the humerus that articulates with the ulna	k. olecranon fossa
______ medial bone of forearm in anatomic position	l. olecranon process
______ rounded knob on the humerus; adjoins the radius	m. phalanges
______ anterior depression, superior to the trochlea, which receives part of the ulna when the forearm is flexed	n. radial tuberosity
______ forearm bone involved in formation of the elbow joint	o. radius
______ wrist bones	p. scapula
______ fingers bones	q. sternum
______ heads of these bones form the knuckles	r. styloid process
______, ______ bones that articulate with the clavicle	s. trochlea
	t. ulna

2. Why does the clavicle often fracture when a person falls on his or her shoulder? ______

______

3. Why is there generally no problem in the arm clearing the widest dimension of the thoracic cage?

______

______

4. What is the total number of phalanges in the hand? ______

5. What is the total number of carpals in the wrist? ______

In the proximal row, the carpals are (medial to lateral) ______

In the distal row, they are (medial to lateral) ______

Bones of the Pelvic Girdle and Lower Extremity

1. Compare the pectoral and pelvic girdles by choosing appropriate descriptive terms from the key.

Key:
a. flexibility most important
b. massive
c. lightweight
d. insecure axial and limb attachments
e. secure axial and limb attachments
f. weightbearing most important

Pectoral: ______, ______, ______

Pelvic: ______, ______, ______

2. What organs are protected, at least in part, by the pelvic girdle? ______

______

3. Distinguish between the true pelvis and the false pelvis. ______

______

______

4. Name five differences between the male and female pelves. ______

______

______

______

______

______

5. Deduce why the pelvic bones of a four-legged animal such as the cat or pig are much less massive than those of the human. ______

______

6. A person instinctively curls over his abdominal area in times of danger. Why? ______

______

7. For what anatomic reason do many women appear to be slightly knock-kneed? ______________________________

____

____

____

8. What does *fallen arches* mean? ____

____

9. Match the bone names and markings in column B with the descriptions in column A (cont. on p. RS32).

Column A	Column B
______________, ______________, and ______________ fuse to form the coxal bone	a. acetabulum
______________ inferoposterior part of the coxal bone	b. calcaneus
______________ point where the coxal bones join anteriorly	c. femur
______________ superiormost margin of the coxal bone	d. fibula
______________ deep socket in the coxal bone that receives the head of the thigh bone	e. gluteal tuberosity
______________ joint between the axial skeleton and the pelvic girdle	f. greater sciatic notch
______________ longest, strongest bone in body	g. greater and lesser trochanters
______________ thin lateral leg bone	h. iliac crest
______________ heavy medial leg bone	i. ilium
______________, ______________ bones forming the knee joint	j. ischial tuberosity
______________ point where the patellar ligament attaches	k. ischium
______________ kneecap	l. lateral malleolus
______________ shin bone	m. lesser sciatic notch
______________ medial ankle projection	n. linea aspera
______________ lateral ankle projection	o. medial malleolus
______________ largest tarsal bone	p. obturator foramen
______________ ankle bones	q. metatarsals
______________ bones forming the instep of the foot	r. patella
______________ opening in hip bone formed by the pubic and ischial rami	s. pubic symphysis
	t. pubis
	u. sacroiliac joint
	v. talus

________________ and ________________ sites of muscle attachment on the proximal end of the femur

________________ tarsal bone that "sits" on the calcaneus

w. tarsals

x. tibia

y. tibial tuberosity

Summary of Skeleton

1. Identify all indicated bones (or groups of bones) in the diagram of the articulated skeleton.

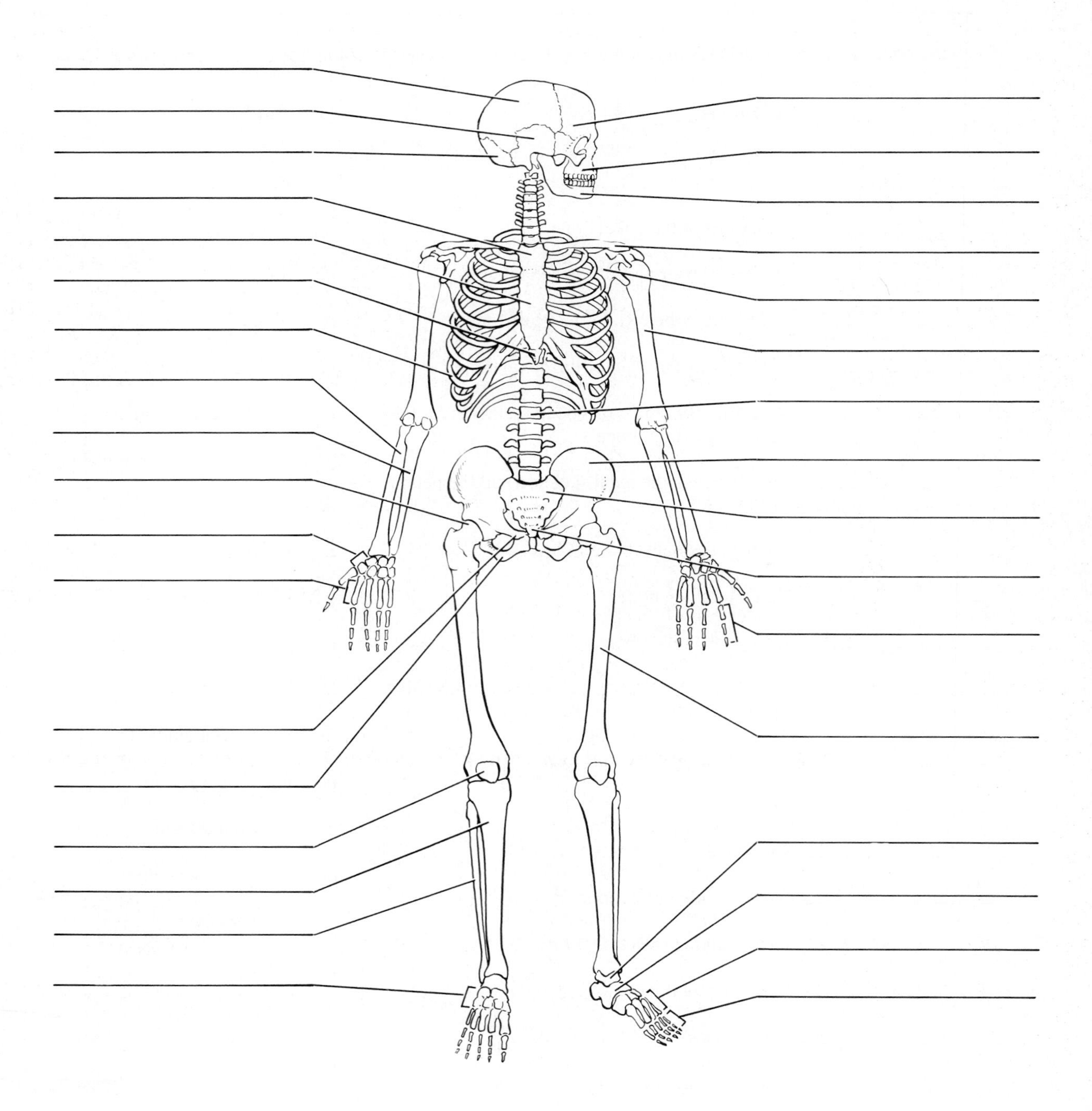

STUDENT NAME ____________________

LAB TIME/DATE ____________________

EXERCISE 11 Articulations and Body Movements

Types of Joints

1. Use key responses to identify the joint types described below.

Key: a. cartilaginous b. fibrous c. synovial

____________ typically allows a slight degree of movement

____________ includes joints between the vertebral bodies and the pubic symphysis

____________ essentially immovable joints

____________ sutures are the most remembered examples

____________ characterized by cartilage connecting the bony portions

____________ all characterized by a fibrous articular capsule lined with a synovial membrane surrounding a joint cavity

____________ all are freely movable or diarthrotic

____________ bone regions are united by fibrous connective tissue

____________ include the hip, knee, and elbow joints

2. Match the joint subcategories in column B with their descriptions in column A, and place an asterisk (*) beside all choices that are examples of synovial joints.

	Column A	Column B
____________	joint between skull bones	a. ball and socket
____________	joint between the axis and atlas	b. condyloid
____________	hip joint	c. gliding
____________	intervertebral joints (between articular processes)	d. hinge
____________	joint between forearm bones and wrist	e. pivot
____________	elbow	f. saddle
____________	interphalangeal joints	g. suture
____________	intercarpal joints	h. symphysis
____________	joint between tarsus and tibia/fibula	i. synchondrosis
____________	joint between skull and vertebral column	j. syndesmosis
____________	joint between jaw and skull	k. symphysis
____________	joints between proximal phalanges and metacarpal bones	
____________	epiphyseal plate of a child's long bone	
____________	a multiaxial joint	
____________, ____________	biaxial joints	
____________, ____________	uniaxial joints	

3. What characteristics do all joints have in common? ____________________

____

4. Describe the structure and function of the following structures or tissues in relation to a synovial joint and label the structures indicated by leader lines in the diagram.

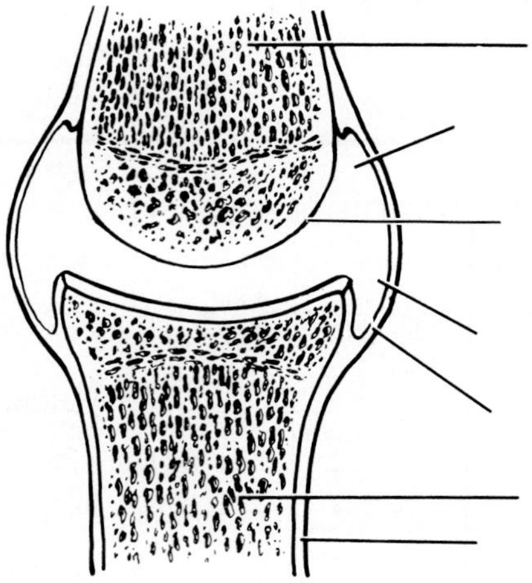

ligament ____________________

____

tendon ____________________

____

hyaline cartilage ____________________

____

synovial membrane ____________________

____

____

bursa ____________________

5. Which joint, the hip or the knee is more stable? ____________________

Name two important factors that contribute to the stability of the hip joint

____________________ and ____________________

Name two important factors that contribute to the stability of the knee.

____________________ and ____________________

6. What structural joint changes are common to the elderly? ____________________

____

____

Body Movements

Complete the following statements:

The movable attachment of a muscle is called its __1__, and its stationary attachment is called the __2__. Winding up for a pitch (as in baseball) can properly be called __3__. To keep your seat when riding a horse, the tendency is to __4__ your thighs. In running, the action at the hip joint is __5__ in reference to the leg moving forward and __6__ in reference to the leg in the posterior position. In kicking a football, the action at the knee is __7__. In climbing stairs, the hip and knee of the forward leg are both __8__. You have just touched your chin to your chest. This is __9__ of the neck. Using a screwdriver with your arm straight requires __10__ of the arm. Consider all the movements of which the arm is capable. One often used for strengthening all the upper arm and shoulder muscles is __11__. Movement of the head that signifies "no" is __12__. Standing on your toes, as in ballet, requires __13__ of the foot. Action that moves the distal end of the radius across the ulna is __14__. Raising the arms laterally away from the body is called __15__ of the arms. Walking on one's heels is __16__.

1. ____________________
2. ____________________
3. ____________________
4. ____________________
5. ____________________
6. ____________________
7. ____________________
8. ____________________
9. ____________________
10. ____________________
11. ____________________
12. ____________________
13. ____________________
14. ____________________
15. ____________________
16. ____________________

STUDENT NAME ______________________________

LAB TIME/DATE ______________________________

EXERCISE 12

Microscopic Anatomy, Organization, and Classification of Skeletal Muscle

Skeletal Muscle Cells and Their Packaging into Muscles

1. Use the items on the right to correctly identify the structures described on the left:

______________	connective tissue ensheathing a bundle of muscle cells	a.	endomysium
______________	bundle of muscle cells	b.	epimysium
______________	contractile unit of muscle	c.	fascicle
______________	a muscle cell	d.	myofiber
______________	thin reticular connective tissue investing each muscle cell	e.	myofilament
______________	plasma membrane of the muscle cell	f.	myofibril
______________	a long filamentous organelle with a banded appearance found within muscle cells	g.	perimysium
		h.	sarcolemma
______________	actin- or myosin-containing structure	i.	sarcomere
______________	cordlike extension of connective tissue beyond the muscle, serving to attach it to a bone	j.	sarcoplasm
		k.	tendon

2. The diagram illustrates a small portion of a muscle myofibril. Using letters from the key, correctly identify each structure indicated by a leader line. Also add a bracket to delineate the extent of one sarcomere.

Key:
a. actin filament
b. A band
c. I band
d. myosin filament
e. Z line

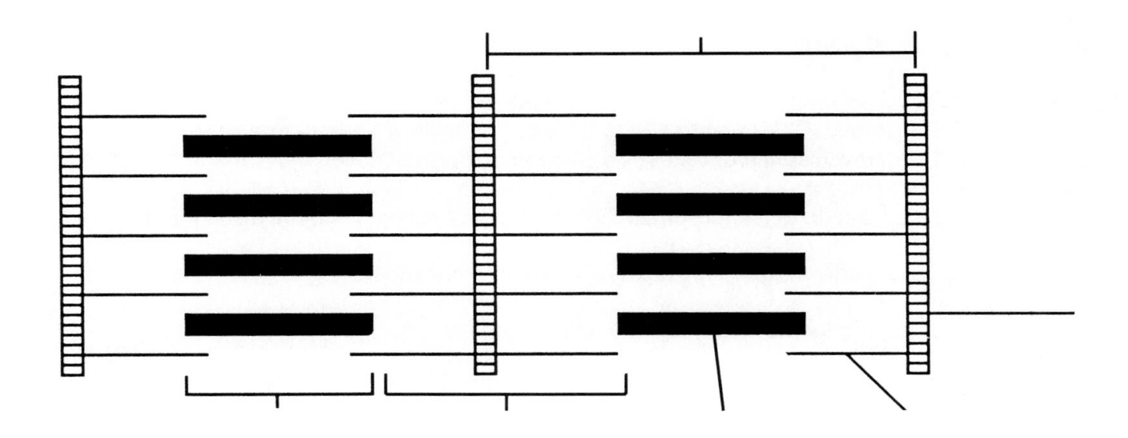

The Neuromuscular Junction

Complete the following statements:

The junction between a motor neuron's axon and the muscle cell membrane is called a neuromuscular junction or a __1__ junction. A motor neuron and all of the skeletal muscle cells it stimulates is called a __2__. The axonal terminals of each motor axon have numerous projections called __3__. The actual gap between the axonal terminal and the muscle cell is called a __4__. Within the axonal terminals are many small vesicles containing a neurotransmitter substance called __5__. When the __6__ reaches the ends of the axon, the neurotransmitter is released and diffuses to the muscle cell membrane to combine with receptors there. The combining of the neurotransmitter with the muscle membrane receptors causes the membrane to become permeable to sodium, which results in the influx of sodium ions and __7__ of the membrane. Then contraction of the muscle cell occurs. Before a muscle cell can be stimulated to contract again, __8__ must occur.

1. ______________________
2. ______________________
3. ______________________
4. ______________________
5. ______________________
6. ______________________
7. ______________________
8. ______________________

Classification of Skeletal Muscles

1. Several criteria were given relative to the naming of muscles. Match the criteria (column B) to the muscle cell names (column A). Note that more than one criterion may apply in some cases:

Column A	Column B
________ gluteus maximus	a. action of the muscle
________ adductor magnus	b. shape of the muscle
________ biceps femoris	c. location of the origin and/or insertion of the muscle
________ abdominis transversus	d. number of origins
________ extensor carpi ulnaris	e. location of muscle relative to a bone or body region
________ trapezius	f. direction in which the muscle fibers run relative to some imaginary line
________ rectus femoris	g. relative size of the muscle
________ external oblique	

2. When muscles are discussed relative to the manner in which they interact with other muscles, the terms shown in the key are often used. Match the key terms with the appropriate definitions.

Key: a. antagonist   b. fixator   c. prime mover   d. synergist

________________ agonist

________________ postural muscles, for the most part

________________ reverses and/or opposes the action of a prime mover

________________ stabilizes a joint so that the prime mover may act at more distal joints

________________ performs the same movement as the prime mover

________________ immobilizes the origin of a prime mover

STUDENT NAME ______________________

LAB TIME/DATE ______________________

EXERCISE 13

Identification of Human Muscles

Muscles of the Head and Neck

Identify the major muscles described below:

______________________ used in smiling

______________________ used to suck in your cheeks

______________________ used in blinking and squinting

______________________ used in pouting (pulls the corners of the mouth downward)

______________________ raises your eyebrows for a questioning look

______________________ used to form the vertical frown crease on the forehead

______________________ your "kisser"

______________________ prime mover to raise the mandible

Muscles of the Trunk

Identify the major muscles described below:

______________________ a major spine flexor

______________________ prime mover for pulling the arm posteriorly

______________________ prime mover for shoulder flexion

__________, __________ assume major responsibility for forming the abdominal, girdle (three pairs of muscles)

____

______________________ pulls the shoulder backward and downward

______________________ prime mover of shoulder abduction

__________, __________ important in shoulder adduction; antagonists of the shoulder abductor (two muscles)

______________________ moves the scapula forward and downward

______________________ small, inspiratory muscles between the ribs; elevate the ribs

______________________ extends the head

______________________ pull the scapulae medially

Arm Muscles

Identify the muscles described below:

____________________________________ places the palm upward (two muscles)

____________________________________ flexes the forearm and supinates the hand

_________________, _________________ forearm flexors; no role in supination (two muscles)

____________________________________ elbow extensor

____________________________________ power wrist flexor and hand abductor

____________________________________ flexes wrist and distal phalanges

_________________, _________________ pronate the hand (two muscles)

____________________________________ flexes the thumb

_________________, _________________ extend and abduct the wrist (three muscles)

____

____________________________________ extend the wrist and digits

Muscles of the Lower Extremity

Identify the muscles described below:

____________________________________ moves the thigh laterally to take the "at ease" stance (2)

____________________________________ used to extend the hip when climbing stairs

____________________________________ prime movers of ankle plantar flexion

____________________________________ major foot inverters (3)

____

____________________________________ allow you to draw your legs to the midline of your body, as when standing at attention

____________________________________ "tailor's muscle"

_________________, _________________, extend thigh and flex knee (three muscles)

____

____________________________________ extends knee and flexes thigh

Muscle Descriptions: General Review

Identify the muscles described below by completing the statements:

____________________, ____________________, and ____________________ are commonly used for intramuscular injections (three muscles).

The insertion tendon of the ____________________ group contains a large sesamoid bone, the patella.

The triceps surae insert in common into the ____________________ tendon.

The bulk of the tissue of a muscle tends to lie ____________________ to the part of the body it causes to move.

The extrinsic muscles of the hand originate on the ____.

Most flexor muscles are located on the ______________________ aspect of the body; most extensors are located ______________________. An exception to this generalization is the extensor-flexor musculature of the ______________________.

Muscle Recognition: General Review

1. Identify the numbered muscles in the diagram of the human anterior superficial musculature by matching the number with one of the following muscle names:

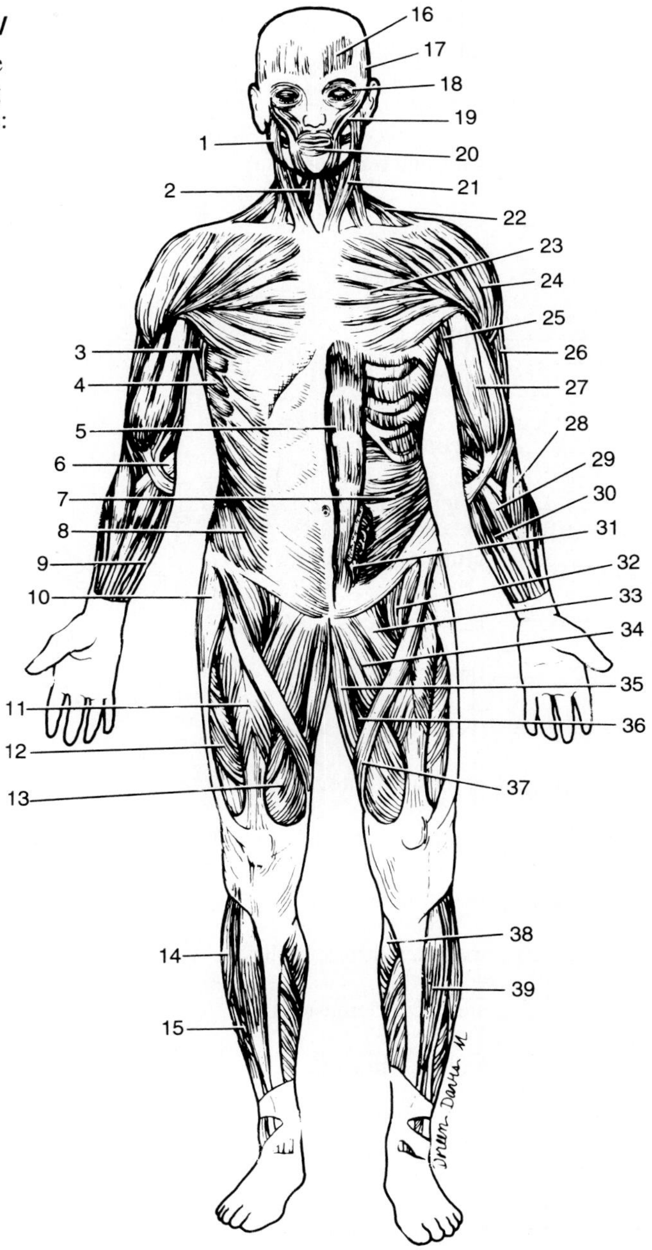

________ orbicularis oris

________ pectoralis major

________ external oblique

________ sternocleidomastoid

________ biceps brachii

________ deltoid

________ vastus lateralis

________ brachioradialis

________ frontalis

________ rectus femoris

________ pronator teres

________ rectus abdominis

________ sartorius

________ gracilis

________ flexor carpi ulnaris

________ adductor longus

________ palmaris longus

________ flexor carpi radialis

________ latissimus dorsi

________ orbicularis oculi

________ gastrocnemius

________ masseter

________ trapezius

________ tibialis anterior

________ extensor digitorum longus

________ tensor fasciae latae

________ pectineus

________ sternohyoid

________ serratus anterior

________ adductor magnus

________ vastus medialis

________ transversus abdominis

________ peroneus longus

________ iliopsoas

________ temporalis

________ zygomaticus

________ coracobrachialis

________ triceps brachii

________ internal oblique

2. Identify each of the numbered muscles in this diagram of the human posterior superficial musculature by matching its number to one of the following muscle names:

_______ gluteus maximus

_______ semimembranosus

_______ gastrocnemius

_______ latissimus dorsi

_______ deltoid

_______ iliotibial tract (tendon)

_______ teres major

_______ semitendinosus

_______ trapezius

_______ biceps femoris

_______ triceps brachii

_______ external oblique

_______ gluteus medius

_______ gracilis

_______ flexor carpi ulnaris

_______ extensor carpi ulnaris

_______ extensor digitorum communis

_______ extensor carpi radialis longus

_______ occipitalis

_______ extensor carpi radialis brevis

_______ sternocleidomastoid

_______ adductor magnus

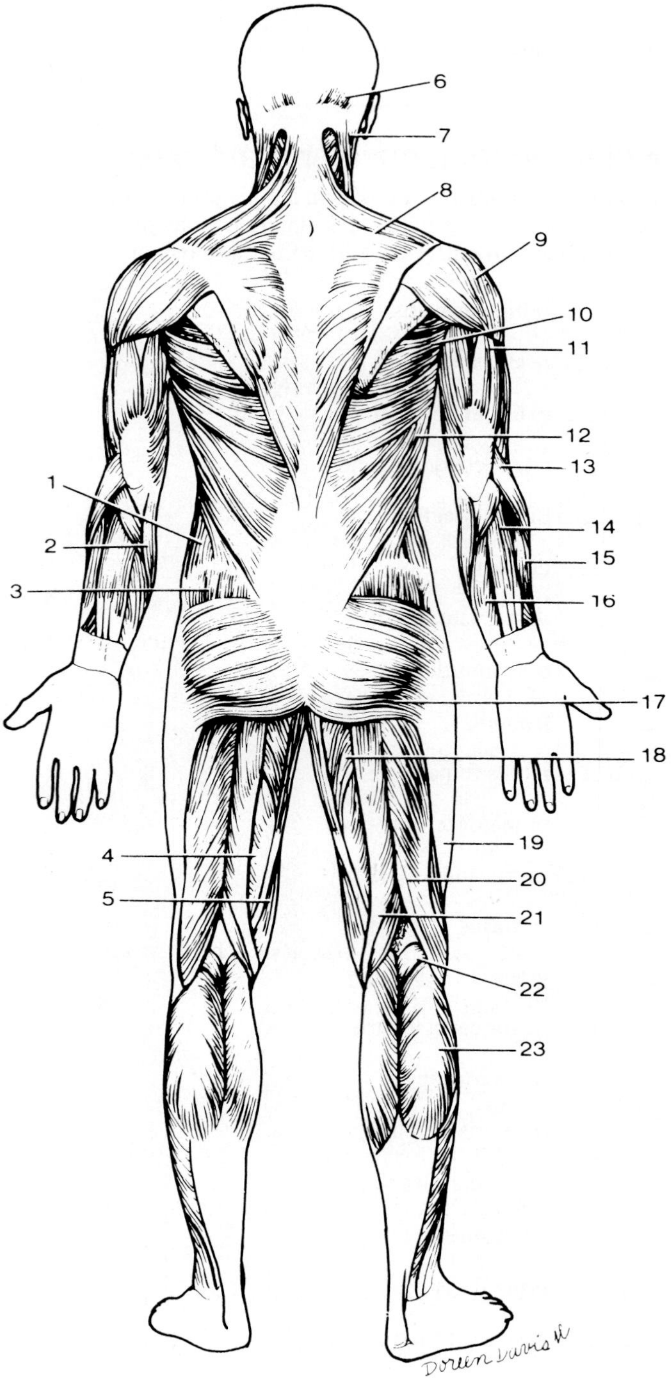

STUDENT NAME ______________________________

LAB TIME/DATE ______________________________

EXERCISE 14A

Muscle Physiology (Frog Experimentation)

Muscle Activity

1. The following group of incomplete statements refers to a muscle cell in the resting or polarized state just before stimulation. Complete each statement by choosing the correct response from the key items below.

Key:
a. Na^+ diffuses out of the cell
b. K^+ diffuses out of the cell
c. Na^+ diffuses into the cell
d. K^+ diffuses into the cell
e. inside the cell
f. outside the cell
g. relative ionic concentrations on the two sides of the membrane
h. electrical conditions
i. activation of the sodium-potassium pump, which moves K^+ into the cell and Na^+ out of the cell
j. activation of the sodium-potassium pump, which moves Na^+ into the cell and K^+ out of the cell

There is a greater concentration of Na^+ ________; there is a greater concentration of K^+ ________. When the stimulus is delivered, the permeability of the membrane at that point is changed; and ________ initiating the depolarization of the membrane. Almost as soon as the depolarization wave has begun, a repolarization wave follows it across the membrane. This occurs as ________. Repolarization restores the ________ of the resting cell membrane. The ________ is (are) reestablished by ________.

2. Number the following statements in the proper sequence to describe the contraction mechanism in a skeletal muscle cell. Number 1 has already been designated.

________ Acetylcholine is released into the neuromuscular junction by the axonal terminal.

________ The action potential, carried deep into the cell, causes the sarcoplasmic reticulum to release calcium ions.

________ The muscle cell relaxes and lengthens.

________ Acetylcholine diffuses across the neuromuscular junction and binds to receptors on the sarcolemma.

________ The calcium ion concentrations at the myofilaments increase; the myofilaments slide past one another, and the cell shortens.

________ Depolarization occurs, and the action potential is generated.

________ The concentration of the calcium ions at the myofilaments decreases as they are actively reabsorbed into the sarcoplasmic reticulum.

3. Muscle contraction is commonly explained by the sliding filament hypothesis. What are the essential points of this hypothesis? ______________________________

____

____

4. Relative to your observations of muscle fiber contraction (pp. 131–132):

   a. What percentage of contraction was observed with the solution containing ATP, K^+, and Mg^{2+}? ______%

      With *just* ATP? ______% With *just* Mg^{2+} and K^+? ______%

   b. *Explain* your observations fully. ______________________________

      ____

      ____

   c. What zones or bands disappear when the muscle cell contracts? ______________________________

   d. *Draw* a relaxed and a contracted sarcomere below.

Relaxed

Contracted

Induction of Contraction in the Frog Gastrocnemius Muscle

1. Why is it important to destroy the brain and spinal cord of a frog before conducting physiologic experiments on muscle contraction? ______________________________

   ____

2. What sources of stimuli, other than electrical shocks, cause a muscle to contract? ______________________________

   ____

3. What is the most common stimulus for muscle contraction in the body? ______________________________

4. Name the three phases of the muscle twitch, and state what happens during each phase:

   ______________, ______________________________

   ____

   ______________, ______________________________

   ______________, ______________________________

5. Use the terms given on the right to identify the conditions described on the left:

______________	sustained contraction without any evidence of relaxation	a.	maximal stimulus
______________	stimulus that results in no perceptible contraction	b.	multiple motor unit summation
______________	stimulus at which the muscle first contracts perceptibly	c.	subthreshold or subliminal stimulus
______________	increasingly stronger contractions in the absence of increased stimulus intensity	d.	tetanus
______________	increasingly stronger contractions owing to stimulation at a rapid rate	e.	threshold stimulus
______________	increasingly stronger contractions owing to increased stimulus strength	f.	treppe
______________		g.	wave summation
______________	weakest stimulus at which all muscle cells in the muscle are contracting		

6. With brackets and labels, identify the portions of the tracing below that best correspond to three of the phenomena listed in the preceding key. Assume that only the timing of the stimulus has changed.

7. Complete the following statements by writing the appropriate words on the correspondingly numbered blanks at the right.

The "all or none" law applies to skeletal muscle functions at the __1__ level. When a weak but smooth muscle contraction is desired, a few motor units are stimulated at a __2__ rate. Treppe is referred to as the "warming up" process. It is believed that muscles contract more strongly after the first few contractions because the __3__ become more efficient. If blue litmus paper is pressed to the cut surface of a fatigued muscle, the paper color changes to red, indicating low pH. This situation is caused by the accumulation of __4__ in the muscle. Within limits, as the load on a muscle is increased, the muscle contracts __5__ strongly. The refractory period is the time when the muscle cell will not respond to a stimulus because __6__ is occurring.

1. ______________
2. ______________
3. ______________
4. ______________
5. ______________
6. ______________

8. During the experiment on muscle fatigue, how did the muscle contraction pattern change as the muscle began to fatigue? ______________

____

How long was stimulation continued before fatigue was apparent? ______________

If the sciatic nerve that stimulates the living frog's gastrocnemius muscle had been left attached to the muscle and the stimulus had been applied to the nerve rather than the muscle, would fatigue have become apparent

sooner or later? ______________________________

Explain your answer. ______________________________

____

9. Explain how the weak but sustained (smooth) muscle contractions of precision movements are produced.

____

10. What do you think happens to a muscle in the body when its nerve supply is destroyed or badly damaged?

____

____

11. Explain the relationship between the load on a muscle and its strength of contraction. ______________________________

____

____

12. The skeletal muscles are maintained in a slightly stretched condition for optimal contraction. How is this accomplished? ______________________________

Why does overstretching a muscle drastically reduce its ability to contract? (Include an explanation of the events at the level of the myofilaments.) ______________________________

____

____

13. If the length but not the tension of a muscle is changed, the contraction is called an isotonic contraction. In an isometric contraction the tension is increased but the muscle does not shorten. Which type of contraction did you observe most often during the laboratory experiments? ______________________________

What is the role of isometric contractions in normal body functioning? ______________________________

____

____

STUDENT NAME ______________________________

LAB TIME/DATE ______________________________

EXERCISE 14B

Muscle Physiology (Computerized Simulation)

Electrical Stimulation

1. Complete the following statements.

A motor unit consists of a __1__ and all the __2__ it innervates. If a single motor unit is stimulated, it will respond in a(n) __3__ fashion, whereas whole muscle contraction is a(n) __4__ response. In order for muscles to work in a practical sense, __5__ is the method used to produce a slow steady increase in muscle force.

When we see the slightest evidence of force production on a tracing, the stimulus applied must have reached __6__.

The weakest stimulus that will elicit the strongest contraction that a muscle is capable of is called the __7__, and that level of contraction is called the __8__.

When the __9__ of stimulation is so high that the muscle tracing shows fused twitch peaks, __10__ has been reached.

1. ______________________________
2. ______________________________
3. ______________________________
4. ______________________________
5. ______________________________
6. ______________________________
7. ______________________________
8. ______________________________
9. ______________________________
10. ______________________________

2. Name and describe what is happening in each phase of the typical muscle twitch.

(1) ______________________, ____

____

(2) ______________________, ____

(3) ______________________, ____

____

3. What are the two ways in which mode of stimulation can affect the force a muscle produces?

______________________ and ______________________

Explain. ____

____

____

____

Isometric Contraction

1. Identify the following conditions by choosing one of the key terms listed on the right.

Key:

a. total force

b. passive force

c. resting force

d. active force

_______ is generated by muscle tissue when it is being stretched

_______ requires the input of energy

_______ is directly measured by recording instrumentation *before* electrical stimulation

_______ is measured by recording instrumentation *during* contraction

2. Using the following statements, correctly label the area on the given Force-Length curves.

   a. an increase in resting length produced an *increase* in the active force generated

   b. an increase in resting length produced a *decrease* in active force generated

   c. an increase in resting length produced an *increase* in resting force

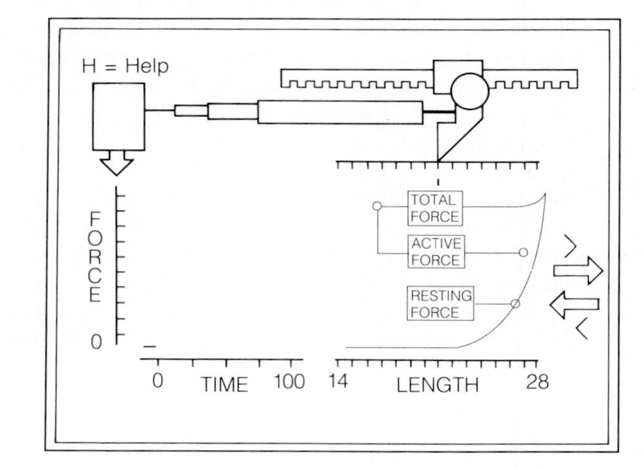

3. Explain what happens to muscle force production at extremes of length (too short or too long).

Muscle too short: ____

____

Muscle too long: ____

____

Isotonic Contraction

1. Assuming a fixed starting length, describe the effect afterload has on the initial velocity of shortening, and explain why.

____

____

____

____

2. A muscle has just been stimulated under conditions that will allow both isometric and isotonic contractions. Describe what is happening in terms of length and force.

   Isometric: ____

   ____

   Isotonic: ____

   ____

Terms

1. Select the condition from column B that most correctly identifies the term in column A.

Column A		Column B
_______	muscle twitch	a. response is all or none
_______	wave summation	b. affects the force a muscle can generate
_______	motor unit summation	c. a single contraction of intact muscle
_______	resting length	d. recruitment
_______	afterload	e. increasing force produced by increasing frequency
_______	initial velocity of shortening	f. muscle length changing due to relaxation
_______	isotonic shortening	g. caused by application of maximal stimulus
_______	isotonic lengthening	h. weight
_______	motor unit	i. exhibits graded response
_______	whole muscle	j. high values with low afterloads
_______	tetany	k. changing muscle length due to active forces

Attach any tracings required by your instructor to the reverse side of this sheet.

STUDENT NAME ______________________________

LAB TIME/DATE ______________________________

EXERCISE 15

Neuron Anatomy and the Nerve Impulse

1. The cellular unit of the nervous system is the neuron. What is the major function of this cell type?

____

2. Name four types of neuroglia and list at least four functions of these cells. (You will need to consult your textbook for this.)

Types	Functions
______________________	______________________________
______________________	______________________________
______________________	______________________________
	______________________________
______________________	______________________________
	______________________________

3. Match each statement with a response chosen from the key.

Key:
a. afferent neuron
b. association neuron
c. central nervous system
d. efferent neuron
e. ganglion
f. neuroglia
g. neurotransmitters
h. nerve
i. nuclei
j. peripheral nervous system
k. synapse
l. tract

________ the brain and spinal cord collectively

________ specialized supporting cells in the CNS

________ junction or point of close contact between neurons

________ a bundle of nerve processes inside the central nervous system

________ neuron serving as part of the conduction pathway between sensory and motor neurons

________ spinal and cranial nerves and ganglia

________ collection of nerve cell bodies found outside the CNS

________ neuron that conducts impulses away from the CNS to muscles and glands

________ neuron that conducts impulses toward the CNS from the body periphery

________ chemicals released by neurons that stimulate other neurons, muscles, or glands

Neuron Anatomy

1. Match the following anatomic terms (column B) with the appropriate description or function (column A).

	Column A	Column B
________	contains enlarged region of the cell body from which the axon originates	a. axon
________	secretes neurotransmitters	b. axonal terminal
________	receptive region of a neuron	c. axon hillock
________	increases the speed of impulse transmission and insulates the nerve fibers	d. dendrite
________	is site of the nucleus	e. myelin sheath
________	may be involved in the transport of substances within the neuron	f. neuronal cell body
________	essentially rough endoplasmic reticulum, important metabolically	g. neurofibril
________	impulse generator and transmitter	h. Nissl bodies

2. Draw a "typical" neuron in the space below. Include and label the following structures on your diagram: cell body, nucleus, Nissl bodies, dendrites, axon, axon collaterals, myelin sheath, and nodes of Ranvier.

3. How is one-way conduction at synapses assured? ____

____

4. What anatomical characteristic determines whether a particular neuron is classified as unipolar, bipolar, or multipolar? ____

Make a simple line drawing of each type here.

Unipolar neuron Bipolar neuron Multipolar neuron

5. Describe how Schwann cells form the myelin sheath and the neurilemma encasing the neuron processes. (You may want to diagram the process.) ____________________

____

____

____

The Nerve Impulse

1. Match each of the terms in column B to the appropriate definition in column A.

	Column A	Column B
______	period of repolarization of the neuron membrane during which it cannot respond to a second stimulus	a. action potential
______	state of reversal of the resting potential owing to an influx of sodium ions	b. depolarization
______	period during which potassium ions diffuse out of the neuron owing to a change in membrane permeability	c. refractory period
______	self-propagated transmission of the depolarization wave along the neuron membrane	d. repolarization
______	process during which ATP is used to move sodium out of the cell and potassium into the cell; restores the resting conditions of the membrane potential and intracellular ionic concentrations	e. sodium-potassium pump

2. Respond appropriately to each statement below either by completing the statement or by answering the question raised. Insert your responses in the corresponding numbered blanks on the right.

1. The cellular unit of the nervous system is the neuron. What is the major function of this cell type?

2 and 3. What characteristics are highly developed to allow the neuron to perform this function?

4. Would a substance that decreases membrane permeability to sodium increase or decrease the probability of generating a nerve impulse?

1. ____________________

2. ____________________

3. ____________________

4. ____________________

3. Why don't the terms *depolarization* and *action potential* mean the same thing? (*Hint:* Under which conditions will a local depolarization *not* lead to the action potential?) ____________________

____

____

4. A nerve generally contains many thickly myelinated fibers that typically exhibit nodes of Ranvier. An action potential is generated along these fibers by "saltatory conduction." Use an appropriate reference to explain how saltatory conduction differs from conduction along unmyelinated fibers.

____

____

____

Physiology of Nerve Fibers

Respond appropriately to each question posed below. Insert your responses in the corresponding numbered blanks to the right.

1–3. Name three types of stimuli that resulted in the generation of action potentials in the sciatic nerve of the frog during the laboratory experiments.

4. Which of the stimuli resulted in the most effective nerve stimulation?

5. Which of the stimuli employed in that experiment might represent types of stimuli to which nerves in the human body are subjected?

6. What is the usual mode of stimulus transfer in neuron-to-neuron interactions?

7. Since the action potentials themselves were not visualized with an oscilloscope when these experiments were conducted, how did you recognize that impulses were being transmitted?

1. ______________________________

2. ______________________________

3. ______________________________

4. ______________________________

5. ______________________________

6. ______________________________

7. ______________________________

Visualization of the Action Potential with an Oscilloscope

1. What is a stimulus artifact? ______________________________

____

2. Explain why the amplitude of the action potential recorded from the frog sciatic nerve increased when the voltage of the stimulus was increased above the threshold value. ______________________________

____

3. What was the effect of cold temperature (flooding the nerve with iced Ringer's solution) on the functioning of the sciatic nerve tested? ______________________________

____

4. When the nerve was reversed in position, was the impulse conducted in the opposite direction? ______________

How can this result be reconciled with the concept of one-way conduction in neurons? ______________

____

____

____

____

____

STUDENT NAME ________________________________

LAB TIME/DATE ________________________________

EXERCISE 16

Gross Anatomy of the Brain and Cranial Nerves

The Human Brain

1. Match the letters on the diagram of the human brain (right lateral view) to the appropriate terms listed at the left:

_______ frontal lobe

_______ parietal lobe

_______ temporal lobe

_______ precentral gyrus

_______ parieto-occipital sulcus

_______ postcentral gyrus

_______ lateral sulcus

_______ central sulcus

_______ cerebellum

_______ medulla _______ occipital lobe _______ pons

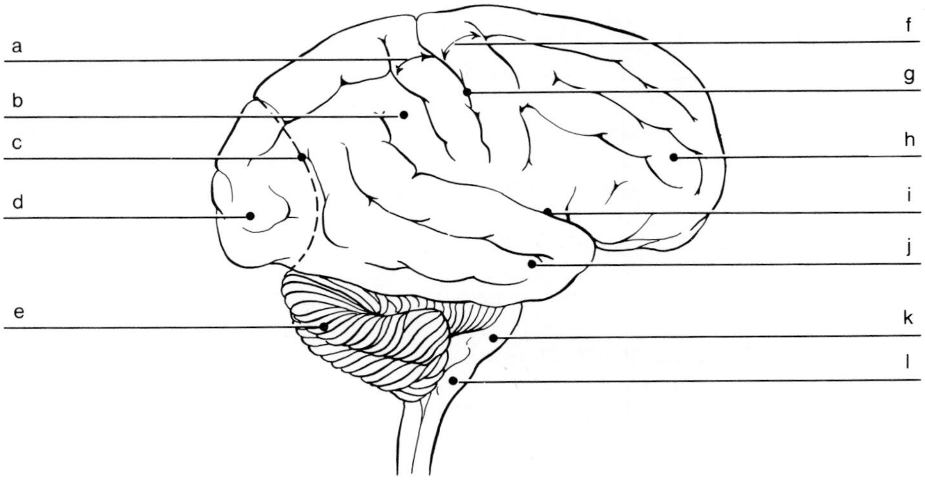

2. In which of the cerebral lobes would the following functional areas be found?

auditory area ________________________ olfactory area ________________________

primary motor area ________________________ visual area ________________________

primary sensory area ________________________ Broca's area ________________________

3. Which of the following structures are *not* part of the brain stem? (Circle the appropriate response or responses.)

cerebral hemispheres pons midbrain cerebellum medulla diencephalon

4. Complete the following statements by writing the proper word or phrase on the corresponding blanks at the right.

A __1__ is an elevated ridge of cerebral tissue. The convolutions seen in the cerebrum are important because they increase the __2__. Gray matter is composed of __3__. White matter is composed of __4__. A fiber tract that provides for communication between different parts of the same cerebral hemisphere is called a(n) __5__, whereas one that carries impulses to and from the cerebrum from and to lower CNS areas is called a(n) __6__ tract. The lentiform nucleus along with the amygdaloid and caudate nuclei are collectively called the __7__.

1. ________________________
2. ________________________
3. ________________________
4. ________________________
5. ________________________
6. ________________________
7. ________________________

5. Identify the structures on the following sagittal view of the human brain by matching the lettered areas to the proper terms at the left:

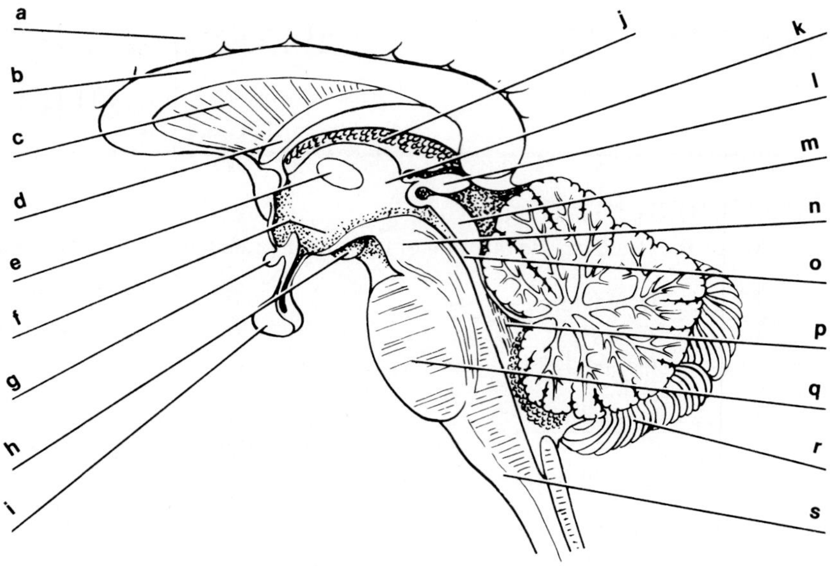

______ cerebellum

______ cerebral aqueduct

______ cerebral hemisphere

______ cerebral peduncle

______ choroid plexus

______ corpora quadrigemina

______ corpus callosum

______ fornix

______ fourth ventricle

______ hypothalamus

______ mammillary bodies

______ massa intermedia

______ medulla oblongata

______ optic chiasma

______ pineal body

______ pituitary gland

______ pons

______ septum pellucidum

______ thalamus

6. Using the letters from the diagram in item 5, match the appropriate structures with the descriptions given below:

______ site of regulation of body temperature and water balance; most important autonomic center

______ consciousness depends on the function of this part of the brain

______ located in the midbrain; contains reflex centers for vision and audition

______ responsible for regulation of posture and coordination of complex muscular movements

______ important synapse site for afferent fibers traveling to the sensory cortex

______ contains autonomic centers regulating blood pressure, heart rate, and respiratory rhythm, as well as coughing, sneezing, and swallowing centers

______ large commissure connecting the cerebral hemispheres

______ fiber tract involved with olfaction

______ connects the third and fourth ventricles

______ encloses the third ventricle

7. What is the function of the basal nuclei? ______________________________

____

8. What is the corpus striatum, and how is it related to the fibers of the internal capsule? ______________________________

____

____

9. A brain hemorrhage within the region of the right internal capsule results in paralysis of the left side of the body.

   Explain why the left side (rather than the right side) is affected. ______________________________

   ____

10. Explain why trauma to the base of the brain is often much more dangerous than trauma to the frontal lobes. (*Hint:* Think about the relative functioning of the cerebral hemispheres and the brain stem structures. Which contain centers more vital to life?)

    ____

    ____

Meninges of the Brain

Identify the meningeal (or associated) structures described below:

______________________________ outermost meninx covering the brain; composed of tough fibrous connective tissue

______________________________ innermost meninx covering the brain; delicate and highly vascular

______________________________ structures instrumental in returning cerebrospinal fluid to the venous blood in the dural sinuses

______________________________ structure that forms the cerebrospinal fluid

______________________________ middle meninx; like a cobweb in structure

______________________________ its outer layer forms the periosteum of the skull

______________________________ a dural fold that attaches the cerebrum to the crista galli of the skull

______________________________ a dural fold separating the cerebrum from the cerebellum

Cerebrospinal Fluid

Fill in the following flow sheet by delineating the circulation of cerebrospinal fluid from its formation site (assume that this is one of the lateral ventricles) to the site of its reabsorption into the venous blood:

Lateral ventricle --------------> ______________________ --------------> Third ventricle --->

-------------------------------> ______________________ --------------> ____________--->

____________ --------------> ____________ \
              \--------------> ____________ --->

______________________ surrounding the brain and cord --------------> Arachnoid villi --------->
(and central canal of the cord)

-------------------------------> ______________________ containing venous blood

Cranial Nerves

Provide the name and number of the cranial nerves involved in each of the following activities, sensations, or disorders:

____________________ shrugging the shoulders

____________________ smelling a flower

____________________ raising the eyelids; focusing the lens of the eye for accommodation; and pupillary constriction

____________________ slows the heart; increases the mobility of the digestive tract

____________________ involved in Bell's palsy (facial paralysis)

____________________ chewing food

____________________ listening to music; seasickness

____________________ secretion of saliva; tasting well-seasoned food

____________________ involved in "rolling" the eyes (three nerves—provide numbers only)

____________________ feeling a toothache

____________________ reading *Playgirl* or *Playboy* magazine

____________________ purely sensory in function (three nerves—provide numbers only)

Dissection of the Sheep Brain

1. In your own words, describe the relative hardness of the sheep brain tissue as observed when cutting into it.

   ____

   Because formalin hardens all tissue, what conclusions might you draw about the relative hardness and texture of living brain tissue? ____________________

2. How does the relative size of the cerebral hemispheres compare in sheep and human brains? __________

   ____

   What is the significance? ____________________

3. What is the significance of the fact that the olfactory bulbs are much larger in the sheep brain than in the human brain? ____________________

   ____

STUDENT NAME ________________

LAB TIME/DATE ________________

EXERCISE 17 Electroencephalography

Brain Wave Patterns and the Electroencephalogram

1. Define *EEG*. ________________

2. What are the four major types of brain wave patterns? ________________

   Match each statement below to a type of brain wave pattern:

   ________ below 3 cps; slow, large waves; normally seen during deep sleep

   ________ rhythm generally apparent when an individual is in a relaxed, nonattentive state with the eyes closed

   ________ correlated to the alert state; usually about 15 to 30 cps

   ________ large, irregular, low-frequency waves; uncommon in adults but common in children

3. What is meant by the term *alpha block*? ________________

   ____

   ____

4. List at least four types of brain lesions that may be determined by EEG studies. ________________

   ____

5. What is the common result of hypoactivity or hyperactivity of the brain neurons? ________________

Observing Brain Wave Patterns

1. How was alpha block demonstrated in the laboratory experiment? ________________

   ____

   What was the effect of mental concentration on the brain wave pattern? ________________

   ____

   ____

3. What effect on the brain wave pattern did hyperventilation have? ________________

   Why? ________________

   ____

STUDENT NAME ______________________________

LAB TIME/DATE ______________________________

EXERCISE 18

Spinal Cord, Spinal Nerves, and the Autonomic Nervous System

Anatomy of the Spinal Cord

1. Match the descriptions given below to the proper anatomic term:

   a. cauda equina   b. conus medullaris   c. filium terminale   d. foramen magnum

   _____ most superior boundary of the spinal cord

   _____ meningeal extension beyond the spinal cord terminus

   _____ spinal cord terminus

   _____ collection of spinal nerves traveling in the vertebral canal below the terminus of the spinal cord

2. Choose the proper answer from the following key to respond to the descriptions relating to spinal cord anatomy.

   Key: a. afferent   b. efferent   c. both afferent and efferent   d. association

   _____ neuron type found in dorsal horn   _____ fiber type in ventral root

   _____ neuron type found in ventral horn   _____ fiber type in dorsal root

   _____ neuron type in dorsal root ganglion   _____ fiber type in spinal nerve

3. Where in the vertebral column is a lumbar puncture generally done? ______________________________

   Why is this the site of choice? ______________________________

   ____

4. The spinal cord is enlarged in two regions, the ______________ and the ______________

   regions. What is the significance of these enlargements? ______________________________

   ____

5. How does the position of the gray and white matter differ in the spinal cord and the cerebral hemispheres?

   ____

   ____

   ____

   ____

Structure of a Nerve

1. What is a nerve? ______________________________

____

2. State the location of each of the following connective tissue coverings:

endoneurium ______________________________

perineurium ______________________________

epineurium ______________________________

3. Correctly identify all indicated parts of the nerve section.

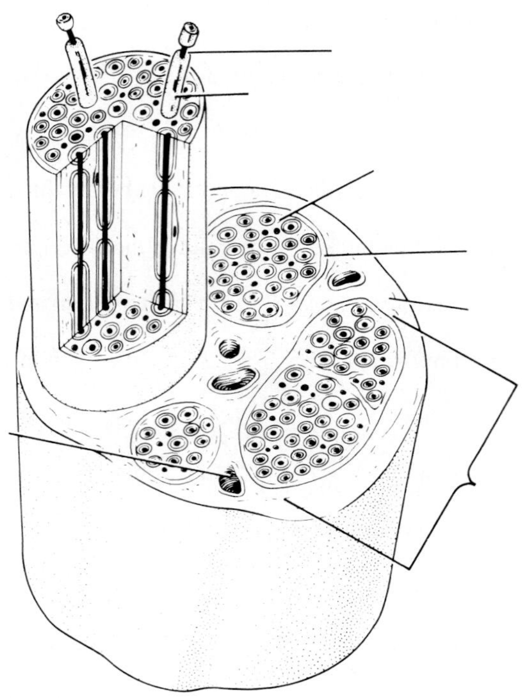

Spinal Nerves and Nerve Plexuses

1. In the human, there are 31 pairs of spinal nerves named according to the region of the vertebral column from which they issue. The spinal nerves are named below; note, by number, the vertebral level at which they emerge:

cervical nerves ______________ sacral nerves ______________

lumbar nerves ______________ thoracic nerves ______________

2. The ventral rami of spinal nerves C_1 through T_1 and T_{12} through S_4 take part in forming ______________, which serve the ______________ of the body. The ventral rami of T_1 through T_{12} run between the ribs to serve the ______________. The posterior rami of the spinal nerves serve ______________.

3. What would happen if the following structures were damaged or transected? (Use key choices for responses.)

Key: a. loss of motor function b. loss of sensory function c. loss of both motor and sensory function

________ dorsal root of a spinal nerve

________ ventral root of a spinal nerve

________ anterior ramus of a spinal nerve

4. Name the major nerves that serve the following body areas:

________________________________ head, neck, shoulders (name plexus only)

________________________________ diaphragm

________________________________ posterior thigh

________________________________ leg and foot (name two)

________________________________ most anterior forearm muscles

________________________________ arm muscles (name two)

________________________________ abdominal wall (name plexus only)

________________________________ anterior thigh

________________________________ medial side of the hand

The Autonomic Nervous System

1. For the most part, sympathetic and parasympathetic fibers serve the same organs and structures. How can they exert antagonistic effects? (After all, nerve impulses are nerve impulses—aren't they?)

____

2. Name three structures that receive sympathetic but not parasympathetic innervation.

____

____

3. The pelvic nerve contains (circle one):

(a) preganglionic sympathetic fibers

(b) postganglionic sympathetic fibers

(c) preganglionic parasympathetic fibers

(d) postganglionic parasympathetic fibers

4. The following chart states a number of conditions. Use a check mark to show which division of the autonomic nervous system is involved in each.

Sympathetic Division	Condition	Parasympathetic Division
	Secretes norepinephrine; adrenergic fibers	
	Secretes acetylcholine; cholinergic fibers	
	Long preganglionic axon; short postganglionic axon	
	Short preganglionic axon; long postganglionic axon	
	Arises from cranial and sacral nerves	
	Arises from spinal nerves T_1 through L_3	
	Normally in control	
	"Fight or flight" system	
	Has more specific control (Look it up!)	

5. You are alone in your home late in the evening, and you hear an unfamiliar sound in your backyard. List four physiological events promoted by the sympathetic nervous system that would aid you in coping with this rather frightening situation:

____

____

____

____

____

6. Often after surgery, people are temporarily unable to urinate, and bowel sounds are absent. What division of the ANS is affected by the anesthesia? _______________

STUDENT NAME ______________________

LAB TIME/DATE ______________________

EXERCISE 19

General Sensation

Structure of Sensory Receptors

1. Differentiate between interoceptors and exteroceptors in terms of anatomical location and stimulus source:

   Interoceptor: ______________________

   Exteroceptor: ______________________

2. A number of activities and sensations are listed in the chart below. For each, check whether the receptors would be exteroceptors or interoceptors; and then name the specific receptor types. (Because visceral receptors were not described in detail in this exercise, you need only indicate that the receptor is a visceral receptor if it falls into that category.)

Activity or Sensation	Exteroceptor	Interoceptor	Specific Receptor Type
Backing into a sun-heated iron railing			
Someone steps on your foot			
Reading a book			
Leaning on your elbows			
Doing sit-ups			
The "too full" sensation			
Seasickness			

Receptor Physiology

1. Explain how the sensory receptors act as transducers: ______________________

   ____

2. Define *stimulus.* ______________________

   ____

3. What was demonstrated by the two-point discrimination test? ______________________

   ____

   How did the accuracy of the subject's tactile localization correlate with the results of the two-point discrimination test? ______________________

   ____

4. Define *punctate distribution:* ____________________

____

5. Several questions regarding general sensation are posed below. Answer each by placing your response in the appropriately numbered blanks to the right.

1.	Which cutaneous receptors are the most numerous?	1.	____________
2–3.	Which two body areas tested were most sensitive to touch?	2.	____________
		3.	____________
4–5.	Which two body areas tested were least sensitive to touch?	4.	____________
		5.	____________
6.	Which appears to be more numerous—receptors that respond to cold or heat?	6.	____________
7–9.	Where would referred pain appear if the following organs were receiving painful stimuli—gallbladder (#7), kidneys (#8), and appendix (#9)? (Use your textbook if necessary.)	7.	____________
		8.	____________
		9.	____________
10.	Where was referred pain felt when the elbow was immersed in ice water during the laboratory experiment?	10.	____________
11.	What region of the cerebrum interprets the kind and intensity of stimuli that cause cutaneous sensations?	11.	____________

6. Define *adaptation:* ____________________

____

7. Why is it advantageous to have pain receptors that are sensitive to all vigorous stimuli, whether heat, cold, or pressure? ____________________

Why is the nonadaptability of pain receptors important? ____________________

____

8. Imagine yourself without any cutaneous sense organs. Why might this be very dangerous? ____________________

____

____

9. Define *referred pain:* ____________________

____

What is the probable explanation for referred pain? (Consult your textbook or an appropriate reference if necessary.) ____________________

____

____

STUDENT NAME ____________________

LAB TIME/DATE ____________________

EXERCISE 20

Human Reflex Physiology

The Reflex Arc

1. Define *reflex.* ____________________

____

2. Name five essential components of a reflex arc: ____________, ____________

____________, ____________, and ____________

3. In general, what is the importance of reflex testing in a routine physical examination? ____________

____

____

Somatic and Autonomic Reflexes

1. Use the key terms to complete the statements given below.

a. abdominal reflex	d. corneal reflex	g. patellar reflex
b. Achilles jerk	e. crossed extensor reflex	h. plantar reflex
c. ciliospinal reflex	f. gag reflex	i. pupillary light reflex

Reflexes classified as somatic reflexes include ______, ______, ______, ______, ______, ______,

and ______. Of these, the simple stretch reflexes are ______ and ______, and the superficial cord reflexes

are ______ and ______. Reflexes classified as autonomic reflexes include ______ and ______.

2. In what way do cord-mediated reflexes differ from those involving higher brain centers? ____________

____

Name two cord-mediated reflexes: ____________ and ____________

____

Name two somatic reflexes in which the higher brain centers participate: ____________

and ____________________

3. Can the stretch reflex be elicited in a pithed animal? ____________

Explain your answer. ____________________

____

4. Trace the reflex arc, naming efferent and afferent nerves, receptors, effectors, and integration centers, for the following reflexes:

patellar reflex ____________________

____

Achilles reflex ____

____

5. Three factors that influence the rapidity and effectiveness of reflex arcs were investigated in conjunction with patellar reflex testing—mental distraction, effect of simultaneous muscle activity in another body area, and fatigue.

   Which of these factors increases the excitatory level of the spinal cord? ______________________

   Which factor decreases the excitatory level of the muscles? ______________________

   When the subject was concentrating on an arithmetic problem, did the change noted in the patellar reflex indicate that brain activity is necessary for the patellar reflex or only that it may modify it? ______________________

   ____

6. Name the division of the autonomic nervous system responsible for each of the following reflexes:

   ciliospinal reflex ______________________ salivary reflex ______________________

   pupillary light reflex ______________________

7. The pupillary light reflex, the crossed extensor reflex, and the corneal reflex illustrate the purposeful nature of reflex activity. Describe the protective aspect of each:

   pupillary light reflex ____

   corneal reflex ____

   crossed extensor reflex ____

   ____

   ____

   ____

   ____

Reaction Time of Unlearned Responses

1. Name at least three factors that may modify reaction time to a stimulus. ______________________

   ____

2. In general, how did the response time for the unlearned activity performed in the laboratory compare to that for the simple patellar reflex? ______________________

3. Did the response time without verbal stimuli decrease with practice? __________ Explain the reason for this.

   ____

4. Explain, in detail, why response time increased when the subject had to react to a word stimulus.

   ____

   ____

STUDENT NAME ________________________

LAB TIME/DATE ________________________

EXERCISE 21

Special Senses: Vision

Anatomy of the Eye

1. Three accessory eye structures contribute to the formation of tears and/or aid in lubrication of the eyeball. Name each and then name its major secretory product. Indicate which has antibacterial properties by circling the correct secretory product.

Accessory Structures	Product

2. Match the key responses with the descriptive statements that follow. (Statements continue on p. RS68.)

Key:
a. aqueous humor
b. choroid
c. ciliary body
d. ciliary processes of the ciliary body
e. cornea
f. fovea centralis
g. iris
h. lens
i. optic disc
j. retina
k. sclera
l. scleral venous sinus
m. suspensory ligament
n. vitreous humor

________________ attaches the lens to the ciliary body

________________ fluid filling the anterior segment of the eye

________________ the "white" of the eye

________________ part of the retina that lacks photoreceptors

________________ modification of the choroid; controls the shape of the crystalline lens

________________ contains the ciliary muscle

________________ drains the aqueous humor from the eye

________________ tunic containing the rods and cones

________________ substance occupying the posterior segment of the eyeball

________________ forms the bulk of the heavily pigmented vascular tunic

________________, ________________ smooth muscle structures

________________ area of critical focusing and discriminatory vision

________________ form (by filtration) the aqueous humor

________________, ________________, ________________,

________________ light-bending media of the eye

________________ anterior continuation of the sclera—your "window on the world"

________________ composed of tough, white, opaque fibrous connective tissue

3. You would expect the pupil to be dilated in which of the following circumstances? (Circle the correct response(s).)

   a. in brightly lighted surroundings
   b. in dimly lit surroundings
   c. during focusing for near vision
   d. in observing distant objects

4. The intrinsic eye muscles are under the control of which of the following? (Circle the correct response.)

   autonomic nervous system   somatic nervous system

Dissection of the Cow (Sheep) Eye

1. What modification of the choroid that is not present in humans is found in the cow eye? ________________

   ____

   What is its function? ________________________________

   ____

   ____

2. What is the anatomical appearance of the retina? ________________________________

   ____

   At what point is it attached to the posterior aspect of the eyeball? ________________

Microscopic Anatomy of the Retina

1. The two major layers of the retina are the epithelial and nervous layers. In the nervous layer, the neuron populations are arranged as follows from the epithelial layer to the vitreous humor. (Circle all proper responses.)

   bipolar cells, ganglion cells, photoreceptors
   photoreceptors, ganglion cells, bipolar cells
   ganglion cells, bipolar cells, photoreceptors
   photoreceptors, bipolar cells, ganglion cells

2. The axons of the ________________ cells form the optic nerve, which exits from the eyeball.

3. Complete the following statements by writing either *rods* or *cones,* on each blank:

   The dim light receptors are the ________________. Only ________________ are found in the fovea centralis, whereas mostly ________________ are found in the periphery of the retina. ________________ are the photoreceptors that operate best in bright light and allow for color vision.

Visual Tests and Experiments

1. Match the terms in column B with the descriptions in column A:

	Column A	Column B
____________	light bending	a. accommodation
____________	ability to focus for close (under 20 ft) vision	b. astigmatism
____________	normal vision	c. convergence
____________	inability to focus well on close objects (farsightedness)	d. emmetropia
____________	nearsightedness	e. hyperopia
____________	blurred vision due to unequal curvatures of the lens or cornea	f. myopia
____________	medial movement of the eyes during focusing on close objects	g. refraction

2. Complete the following statements:

In farsightedness, the light is focused __1__ the retina. The lens required to treat myopia is a __2__ lens. The "near point" increases with age because the __3__ of the lens decreases as we get older. A convex lens, like that of the eye, produces an image that is upside down and reversed from left to right. Such an image is called a __4__ image.

1. ____________
2. ____________
3. ____________
4. ____________

3. Use terms from the key to complete the statements concerning near and distance vision.

Key: (a) contracted (b) decreased (c) increased (d) relaxed (e) taut

During distance vision: The ciliary muscle is ____, the suspensory ligament is ____, the convexity of the lens is ____, and light refraction is ____. During close vision: The ciliary muscle is ____, the suspensory ligament is ____, lens convexity is ____, and light refraction is ____.

4. Explain why each eye is tested separately when using the Snellen eye chart. ____________

____

Explain 20/40 vision. ____________

____

Explain 20/10 vision. ____________

____

5. To which wavelengths of light do the three cone types of the retina respond maximally?

____________, ____________, and ____________

6. How can you explain the fact that we see a great range of colors even though only three cone types exist?

7. Explain the difference between binocular and panoramic vision.

What is the advantage of binocular vision?

8. In the experiment on the convergence reflex, what happened to the position of the eyeballs as the object was moved closer to the subject's eyes?

What extrinsic eye muscles control the movement of the eyes during this reflex?

What is the value of this reflex?

9. In the experiment on the photopupillary reflex, what happened to the pupil of the eye exposed to light?

What happened to the pupil of the nonilluminated eye?

Explanation?

10. Why is the ophthalmoscopic examination an important diagnostic tool?

11. Many college students struggling through mountainous reading assignments are told that they need glasses for "eyestrain." Why is it more of a strain on the extrinsic and intrinsic eye muscles to look at close objects than at far objects?

12. Describe and explain:

positive afterimage:

negative afterimage:

STUDENT NAME ______________________________

LAB TIME/DATE ______________________________

EXERCISE 22

Special Senses: Hearing and Equilibrium

Anatomy of the Ear

1. Select the terms from column B that apply to the column-A descriptions. Some terms are used more than once.

Column A	Column B
______________, ______________, ______________ structures comprising the outer or external ear	a. anvil (incus)
______________, ______________, ______________ structures composing the inner ear	b. cochlea
______________, ______________, ______________ collectively called the ossicles	c. endolymph
______________, ______________ ear structures not involved with audition	d. eustachian tube
______________ involved in equalizing the pressure in the middle ear with atmospheric pressure	e. external auditory canal
______________ vibrates at the same frequency as sound waves hitting it; transmits the vibrations to the ossicles	f. hammer (malleus)
______________, ______________ contain receptors for the sense of balance	g. oval window
______________ transmits the vibratory motion of the stirrup to the fluid in the scala vestibuli of the inner ear	h. perilymph
______________ acts as a pressure relief valve for the increased fluid pressure in the scala tympani; bulges into the tympanic cavity	i. pinna
______________ passage between the throat and the tympanic cavity	j. round window
______________ fluid contained within the membranous labyrinth	k. semicircular canals
______________ fluid contained within the osseous labyrinth and bathing the membranous labyrinth	l. stirrup (stapes)
	m. tympanic membrane
	n. vestibule

2. Sound waves hitting the eardrum initiate its vibratory motion. Trace the pathway through which vibrations and fluid currents are transmitted to finally stimulate the hair cells in the organ of Corti. (Name the appropriate ear structures in their correct sequence.) Eardrum → ______________________________

____

____

3. Identify all indicated structures and ear regions in the following diagram.

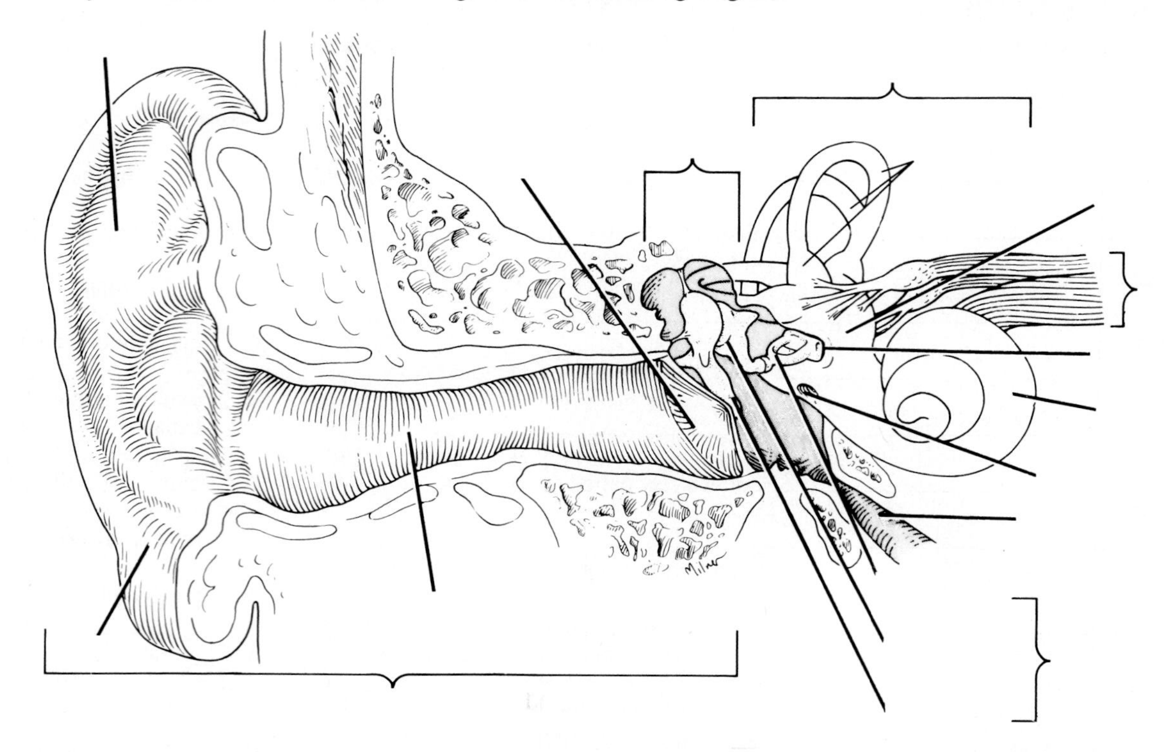

4. Match the membranous labyrinth structures listed in column B with the descriptive statements in column A:

Column A	Column B
________, ________ sacs found within the vestibule	a. ampulla
________ contains the organ of Corti	b. basilar membrane
________, ________ sites of the maculae	c. cochlear duct
________ positioned in all spatial planes	d. cochlear nerve
________ hair cells of organ of Corti rest on this membrane	e. cupula
________ gelatinous membrane overlying the hair cells of the organ of Corti	f. otoliths
________ contains the crista ampullaris	g. saccule
________, ________, ________, ________, function in static equilibrium	h. semicircular ducts
________, ________, ________, ________, function in dynamic equilibrium	i. tectorial membrane
________ carries auditory information to the brain	j. utricle
________ gelatinous cap overlying hair cells of the crista ampullaris	k. vestibular nerve
________ grains of calcium carbonate in the maculae	

5. Describe how sounds of different frequency (pitch) are differentiated in the cochlea. ______________________

____

____

____

Laboratory Tests

1. Was the auditory acuity measurement made during the experiment on page 217 the same or different for both ears? ______________________ What factors might account for a difference in the acuity of the two ears?

   ____

   ____

2. During the sound localization experiment on page 218, in which position(s) was the sound least easily located?

   ____

   How can this phenomenon be explained? ____

   ____

3. In the frequency range of hearing experiment on page 218, which tuning fork was the most difficult to hear?

   ____________ cps

   What conclusion can you draw? ____

   ____

4. When the tuning fork handle was pressed to your forehead during the Weber test, where did the sound seem to originate? ____

   Where did it seem to originate when one ear was plugged with cotton? ______________________

   How do sound waves reach the cochlea when conduction deafness is present? ______________________

   ____

5. Indicate whether the following conditions relate to conduction deafness (C) or sensorineural (central) deafness (S):

   ________ can result from the fusion of the ossicles

   ________ can result from a lesion on the cochlear nerve

   ________ sound heard in one ear but not in the other during bone and air conduction

   ________ can result from otitis media

   ________ can result from impacted cerumen or a perforated eardrum

   ________ can result from a blood clot in the auditory cortex

6. The Rinne test evaluates an individual's ability to hear sounds conducted by air or bone. Which is more indicative of normal hearing? ____

7. Define *nystagmus.* ____________________

____

Define *vertigo.* ____________________

8. The Barany test investigated the effect that rotatory acceleration had on the semicircular canals. Explain *why* the subject still had the sensation of rotation immediately after being stopped. ____________________

____

9. What is the usual reason for conducting the Romberg test? ____________________

____

Was the degree of sway greater with the eyes open or closed? ____________________

Why? ____________________

10. Normal balance, or equilibrium, depends on input from a number of sensory receptors. Name them.

____

____

STUDENT NAME ______________________

LAB TIME/DATE ______________________

EXERCISE 23

Special Senses: Taste and Olfaction

Localization and Anatomy of Taste Buds

1. Name three sites where receptors for taste are found, and circle the predominant site:

   ______________________, ______________________, and ______________________

2. Describe the cellular makeup and arrangement of a taste bud. (Use a diagram, if helpful.) ______________________

Localization and Anatomy of the Olfactory Receptors

1. Describe the cellular composition and the location of the olfactory epithelium. ______________________

2. How and why does sniffing improve your sense of smell? ______________________

Laboratory Experiments

1. Taste and smell receptors are both classified as ______________________, because they both respond to ______________________.

2. Why is it impossible to taste substances with a dry tongue? ______________________

3. State the most important sites of the taste-specific receptors, as determined during the plotting exercise in the laboratory:

   salt ______________________ sour ______________________

   bitter ______________________ sweet ______________________

4. The basic taste sensations are elicited by specific chemical substances or groups. Name them:

   salt ______________________ sour ______________________

   bitter ______________________ sweet ______________________

5. Name three factors that influence our appreciation of foods. Substantiate each choice with an example from the laboratory experience.

_______________ Substantiation _______________

____

____

_______________ Substantiation _______________

____

____

_______________ Substantiation _______________

____

____

Which of the factors chosen is most important? _______________

Substantiate your choice with an example from everyday life. _______________

____

Expand on your explanation and choices by explaining why a cold, greasy hamburger is unappetizing to most people. _______________

____

6. Babies tend to favor bland foods, whereas adults tend to like highly seasoned foods. What is the basis for this phenomenon? _______________

____

____

7. How palatable is food when you have a cold? _______________

Explain. _______________

____

8. What is the mechanism of olfactory adaptation? _______________

____

____

____

In your opinion, is olfactory adaptation desirable? _______________ Explain your answer.

____

____

STUDENT NAME ______________________________

LAB TIME/DATE ______________________________

EXERCISE 24 Anatomy and Basic Function of the Endocrine Glands

Gross Anatomy and Basic Function of the Endocrine Glands

1. Both the endocrine and nervous systems are major regulating systems of the body; however, the nervous system has been compared to an airmail delivery system and the endocrine system to the pony express. Briefly explain this comparison.

____

____

____

2. Define *hormone.* ______________________________

____

3. Chemically, hormones belong chiefly to two molecular groups, the ______________________________

and the ______________________________.

4. What do all hormones have in common? ______________________________

____

5. Define *target organ:* ______________________________

____

6. Why don't all tissues respond to all hormones? ______________________________

____

7. Identify the endocrine organ described by the following statements:

______________ located in the throat; bilobed gland connected by an isthmus

______________ found close to the kidney

______________ a mixed gland, located close to the stomach and small intestine

______________ paired glands suspended in the scrotum

______________ ride "horseback" on the thyroid gland

______________ found in the pelvic cavity of the female, concerned with ova and female hormone production

______________ found in the upper thorax overlying the heart; large during youth

______________ found in the roof of the third ventricle

8. For each statement describing hormonal effects, identify the hormone(s) involved by choosing a number from key A, and note the hormone's site of production with a letter from key B. More than one hormone may be involved in some cases.

Key A

1. ACTH
2. ADH
3. aldosterone
4. cortisone
5. epinephrine
6. estrogen
7. FSH
8. glucagon
9. GH
10. insulin
11. LH
12. melatonin
13. MSH
14. oxytocin
15. progesterone
16. prolactin
17. PTH
18. serotonin
19. testosterone
20. thymosin
21. thyrocalcitonin/calcitonin
22. T_4/T_3
23. TSH

Key B

a. adrenal cortex
b. adrenal medulla
c. anterior pituitary
d. hypothalamus
e. ovaries
f. pancreas
g. parathyroid glands
h. pineal gland
i. posterior pituitary
j. testes
k. thymus gland
l. thyroid gland

________, ________ basal metabolism hormone

________, ________ programming of T lymphocytes

________, ________ and ________, ________ regulation of blood calcium levels

________, ________ and ________, ________ released in response to stressors

________, ________ and ________, ________ development of secondary sexual characteristics

________, ________, ________, ________, ________, ________; and ________, ________ regulate the function of another endocrine gland

________, ________ mimics the sympathetic nervous system

________, ________ and ________, ________ regulate blood glucose levels; produced by the same "mixed" gland

________, ________ and ________, ________ directly responsible for regulation of the menstrual cycle

________, ________ and ________, ________ regulation of the ovarian cycle

________, ________ and ________, ________ maintenance of salt and water balance in the ECF

________, ________ and ________, ________ directly involved in milk production and ejection

________, ________ questionable function; may stimulate the melanocytes of the skin

9. Although the pituitary gland is often referred to as the master gland of the body, recent studies show that the hypothalamus exerts some control over the pituitary gland. How does the hypothalamus control both anterior and posterior pituitary functioning?

____

____

____

____

____

10. Indicate whether the release of the hormones listed below is stimulated by A, another hormone; B, the nervous system (neurotransmitters, or releasing factors); or C, humoral factors (the concentration of specific nonhormonal substances in the blood or extracellular fluid):

_______ T_4/T_3 _______ parathyroid hormone _______ ADH

_______ insulin _______ testosterone _______ TSH, FSH

_______ estrogen _______ epinephrine _______ aldosterone

11. Name the hormone that would be produced in *inadequate* amounts under the following conditions:

_______________________________ sexual immaturity

_______________________________ tetany

_______________________________ excessive diuresis without high blood glucose levels

_______________________________ polyurea, polyphagia, and polydipsia

_______________________________ abnormally small stature, normal proportions

_______________________________ miscarriage

_______________________________ lethargy, hair loss, low BMR, obesity

12. Name the hormone that is produced in *excessive* amounts in the following conditions:

_______________________________ lantern jaw and large hands and feet in the adult

_______________________________ bulging eyeballs, nervousness, increased pulse rate

_______________________________ demineralization of bones, spontaneous fractures

Microscopic Anatomy of Selected Endocrine Glands (optional)

1. Choose a response from the key below to name the hormone(s) produced by the cell types listed:

Key:
a. insulin
b. GH, prolactin
c. T_4/T_3
d. calcitonin
e. TSH, ACTH, FSH, LH
f. mineralocorticoids
g. glucagon
h. PTH
i. glucocorticoids

_______ parafollicular cells of the thyroid

_______ follicular epithelial cells of the thyroid

_______ beta cells of the islets of Langerhans

_______ alpha cells of the islets of Langerhans

_______ basophil cells of the anterior pituitary

_______ zona fasciculata cells

_______ zona glomerulosa cells

_______ chief cells

_______ acidophil cells of the anterior pituitary

2. Five diagrams of the microscopic structures of the endocrine glands are presented here. Identify each and name all indicated structures.

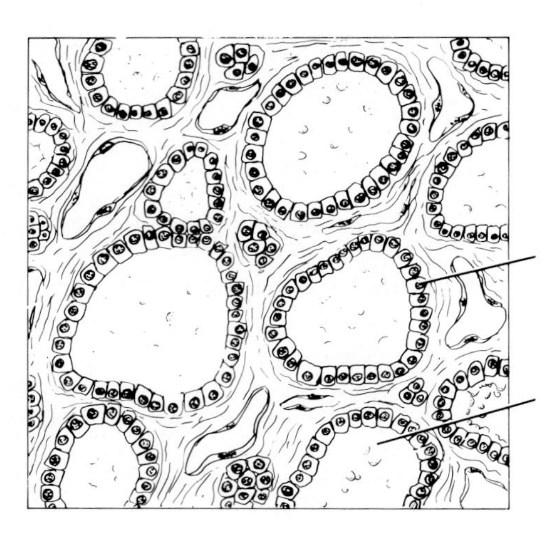

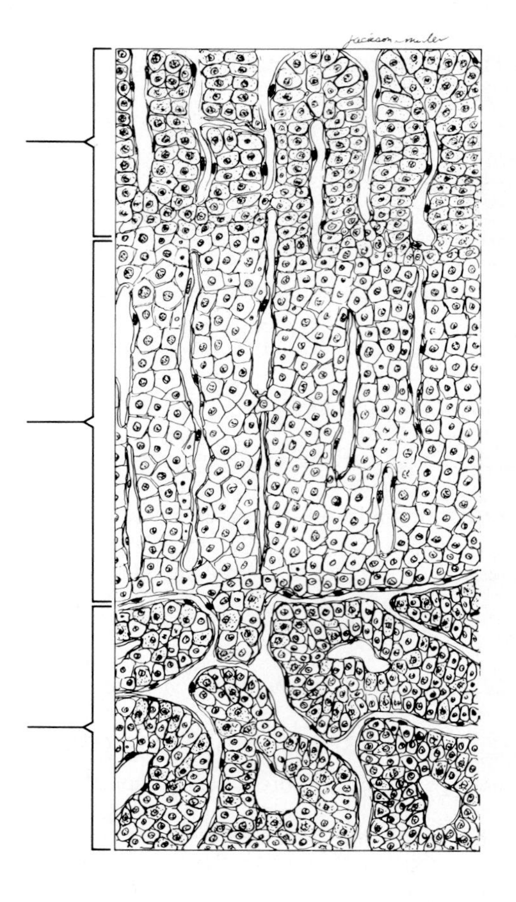

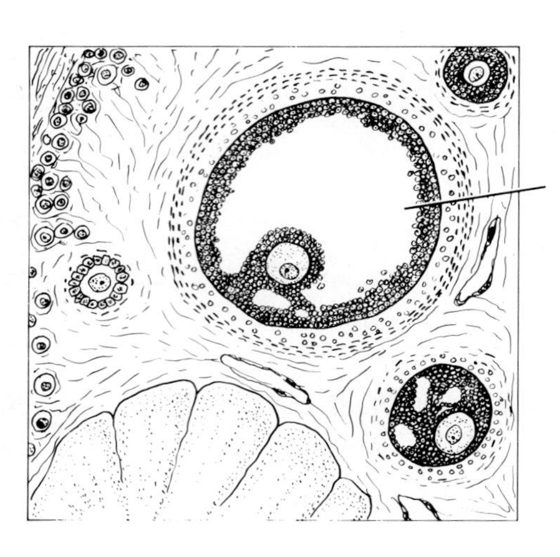

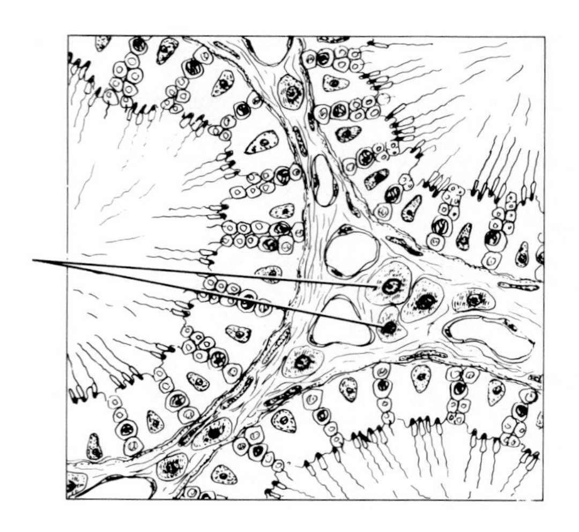

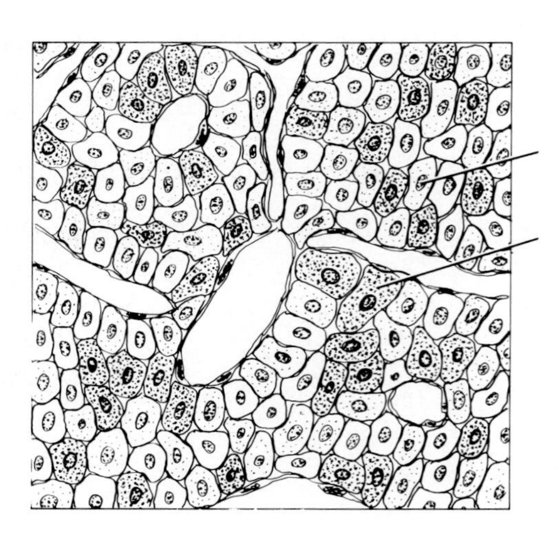

STUDENT NAME ______________________

LAB TIME/DATE ______________________

EXERCISE 25 Experiments on Hormonal Action

Effects of Hyperinsulinism

1. Briefly explain what was happening within the fish's system when the fish was immersed in the insulin solution.

____

____

2. What is the mechanism of the recovery process observed? ______________________

____

3. What would you do to help a friend who had inadvertently taken an overdose of insulin? ______________________

______________________ Why? ______________________

____

____

4. What is a glucose tolerance test? (Use an appropriate reference, as necessary, to answer this question.)

____

____

____

____

5. How does diabetic coma differ from insulin shock? ______________________

____

____

____

____

Effect of Epinephrine on the Heart

1. Based on your observations, what is the effect of epinephrine on the force and rate of the heartbeat?

____

2. What is the role of this effect in the "fight or flight" response? ______________________

____

Effect of Thyroid Hormone on Metabolic Rate

1. Relative to the measurement of oxygen consumption in rats, which group had the highest metabolic rate?

   ___

   Which group had the lowest metabolic rate? ___

   Correlate these observations with the pretreatment these animals received. ___

   ___

   ___

   Which group of rats was hyperthyroid? ___

   Which euthyroid? ___ Which hypothyroid? ___

2. Since oxygen used = carbon dioxide evolved, how were you able to measure the oxygen consumption in the experiments? ___

   ___

3. What did changes in the fluid levels in the manometer arms indicate? ___

   ___

4. The techniques used in this set of laboratory experiments probably allowed for several inaccuracies. One was the inability to control the activity of the rats. How would changes in their activity levels affect the results observed?

   ___

   ___

   Another possible source of error was the lack of control over the amount of food consumed by the rats in the 14-day period preceding the laboratory session. If each of the rats had been force-fed equivalent amounts of food in that 14-day period, which group (do you think) would have gained the most weight?

   ___ Which the least? ___ Explain your answers. ___

   ___

   ___

5. TSH, produced by the anterior pituitary, prods the thyroid gland to release thyroid hormone to the blood. Which group of rats can be assumed to have the *highest* blood levels of TSH? ___

   Which the lowest? ___ Explain your reasoning. ___

   ___

   ___

6. Use an appropriate reference to determine how each of the following factors modifies metabolic rate. Indicate increase by ↑ and decrease by ↓.

   increased exercise ___ aging ___ infection/fever ___

   small/slight stature ___ obesity ___ sex (♂ or ♀) ___

STUDENT NAME ____________________

LAB TIME/DATE ____________________

EXERCISE 26 Blood

Composition of Blood

1. Use the key to identify the cell type(s) or blood elements that fit the following descriptive statements.

Key:
a. red blood cell
b. megakaryocyte
c. eosinophil
d. basophil
e. monocyte
f. neutrophil
g. lymphocyte
h. formed elements
i. plasma

____________ most numerous leukocyte

____________, ____________, ____________ granulocytes

____________ also called an erythrocyte; anucleate formed element

____________, ____________ actively phagocytic leukocytes

____________, ____________ agranular leukocytes

____________ ancestral cell of platelets

____________ (a) through (g) are all examples of these

____________ number rises during parasite infections

____________ releases histamine; promotes inflammation

____________ many formed in lymphoid tissue

____________ transports oxygen

____________ primarily water, noncellular; the fluid matrix of blood

____________ increases in number during prolonged infections

____________, ____________, ____________,

____________, ____________ also called white blood cells

2. List three classes of nutrients normally found dissolved in plasma: ____________,

____________, and ____________.

Name two gases. ____________ and ____________

Name three ions. ____________, ____________, and ____________

3. Describe the consistency and color of the plasma you observed in the laboratory. ____________

____

4. What is the average life span of a red blood cell? How does its anucleate condition affect this life span?

___

___

___

5. From memory, describe the structural characteristics of each of the following blood cell types as accurately as possible, and note the percentage of each in the total white blood cell population.

eosinophils ___

___

neutrophils ___

___

lymphocytes ___

___

basophils ___

___

monocytes ___

___

Hematologic Tests

1. Broadly speaking, why are hematologic studies of blood so important in the diagnosis of disease?

___

___

2. Record information from the blood tests you conducted in the chart on the next page. Complete the chart by recording values for healthy male adults, and indicating the significance of high or low values for each test.

Test	Student Test Results	Normal Values (Healthy Male Adults)	Significance High Values	Low Values
total WBC count				
total RBC count				
hematocrit				
hemoglobin determination				
sedimentation rate				
coagulation time				

3. Why is a differential WBC count more valuable when trying to pin down the specific source of pathology than a total WBC count? ______________________________

____

____

4. Define *hematocrit.* ______________________________

5. If you had a high hematocrit, would you expect your hemoglobin determination to be high or low?

______________ Why? ______________________________

____

6. If your blood clumped with both anti-A and anti-B sera, your ABO blood type would be ______________

To what ABO blood groups could you give blood? ______________________________

From which ABO donor types could you receive blood? ______________________________

Which ABO blood type is most common? ______________ Least common? ______________

7. Explain why an Rh-negative person does not have a transfusion reaction on the first exposure to Rh-positive blood but *does* have a reaction on the second exposure. ____________________

____

What happens when an ABO blood type is mismatched for the first time? ____________________

____

8. Discuss the effect of each of the following factors on RBC count. Consult an appropriate reference, as necessary; and explain your reasoning.

   - athletic training (for example, running 4 to 5 miles per day for a period of 6 to 9 months) ____________________

     ____

     ____

   - a permanent move from sea level to a high-altitude area ____________________

     ____

     ____

   - lack of iron-containing foods in your diet ____________________

     ____

     ____

9. Provide the normal, or at least "desirable," range for plasma cholesterol concentration:

   ____________ mg/100 ml

10. Describe the relationship between high blood cholesterol levels and cardiovascular diseases such as hypertension, heart attacks, and strokes.

    ____

    ____

    ____

    ____

    ____

11. Correctly identify the blood pathologies described in column A by matching them with selections from column B:

	Column A	Column B
____________	abnormal increase in the number of WBCs	a. anemia
____________	abnormal increase in the number of RBCs	b. leukocytosis
____________	condition of too few RBCs or of RBCs with hemoglobin deficiencies	c. leukopenia
____________	abnormal decrease in the number of WBCs	d. polycythemia

STUDENT NAME ____________________

LAB TIME/DATE ____________________

EXERCISE 27

Anatomy of the Heart

Gross Anatomy of the Human Heart

1. Match the terms in the key to the descriptions provided below.

_______ location of the heart in the thorax

_______ superior heart chambers

_______ inferior heart chambers

_______ visceral pericardium

_______ "anterooms" of the heart

_______ equals cardiac muscle

_______ provide nutrient blood to the heart muscle

_______ lining of the heart chambers

_______ actual "pumps" of the heart

_______ drains blood into the right atrium

Key:

a. atria

b. coronary arteries

c. coronary sinus

d. endocardium

e. epicardium

f. mediastinum

g. myocardium

h. ventricles

2. What is the function of the fluid that fills the pericardial sac? ____________________

3. Why might a thrombus (blood clot) in the anterior descending branch of the left coronary artery cause sudden death? ____________________

4. An anterior view of the heart is shown here. Identify each numbered structure by writing its name on the correspondingly numbered line:

1. ____________________
2. ____________________
3. ____________________
4. ____________________
5. ____________________
6. ____________________
7. ____________________
8. ____________________
9. ____________________

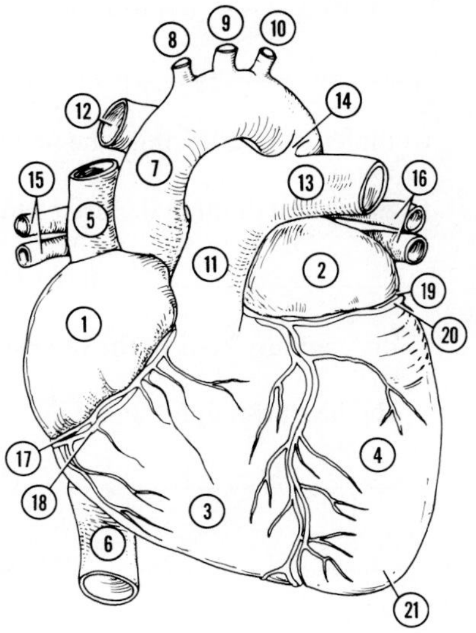

10. ____________________

11. ____________________

12. ____________________

13. ____________________

14. ____________________

15. ____________________

16. ____________________

17. ____________________

18. ____________________

19. ____________________

20. ____________________

21. ____________________

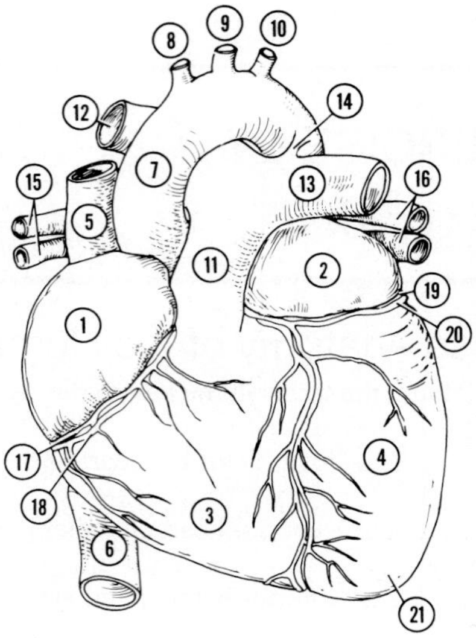

5. What is the function of the valves found in the heart? ____________________

____

6. Can the heart function with leaky valves? (Think! Can a water pump function with leaky valves?) __________

7. What is the role of the chordae tendineae? ____________________

____

8. Define:

angina pectoris ____________________

____

pericarditis ____________________

9. Differentiate clearly between the roles of the pulmonary and systemic circulations. ____________________

____

____

10. Complete the following scheme of circulation in the human body:

Right atrium through the tricuspid valve to the ____________________ through the ____________________ valve to the pulmonary trunk to the ____________________ to the capillary beds of the lungs to the ____________________ to the ____________________ of the heart through the ____________________ valve to the ____________________ through the ____________________ valve to the ____________________ to the systemic arteries to the

_________________ of the tissues to the systemic veins to the _________________ and

_________________ entering the right atrium of the heart.

11. If the mitral valve does not close properly, which circulation is affected? _________________

Dissection of the Sheep Heart

1. During the sheep heart dissection, you were asked initially to identify the right and left ventricles without cutting into the heart. During this procedure, what differences did you observe between the two chambers?

____

____

Knowing that structure and function are related, how would you say this structural difference reflects the relative functions of these two heart chambers? _________________

____

____

____

2. Semilunar valves prevent backflow into the _________________; AV valves prevent backflow into the _________________. Using your own observations, explain how the operation of the semilunar valves differs from that of the AV valves. _________________

____

____

____

____

3. Two remnants of fetal structures are observable in the heart—the ligamentum arteriosum and the fossa ovalis. What were they called in the fetus, where were they located, and what common purpose did they serve as functioning fetal structures?

____

____

____

____

Microscopic Anatomy of Cardiac Muscle

1. How would you distinguish cardiac muscle from skeletal muscle? _________________

____

____

____

____

2. What role does the unique structure of cardiac muscle play in its function? (Note: Before attempting a response, describe the unique anatomy.) ____

____

____

____

____

STUDENT NAME ____________________

LAB TIME/DATE ____________________

EXERCISE 28

Conduction System of the Heart and Electrocardiography

The Intrinsic Conduction System

1. List the elements of the intrinsic conduction system in order starting from the SA node.

   SA node ⟶ ____________________ ⟶ ____________________ ⟶

   ____________________ ⟶ ____________________

   Which of those structures is replaced when an artificial pacemaker is installed? ____________________

   At what structure in the transmission sequence is the impulse temporarily delayed? ____________________

   Why? ____________________

   ____

2. Even though cardiac muscle has an inherent ability to beat, the nodal system plays a critical role in heart physiology. What is that role? ____________________

   ____

   ____

3. How does the "all or none" law apply to normal heart operation? ____________________

   ____

   ____

Electrocardiography

1. Define *ECG*. ____________________

   ____

2. Draw an ECG wave form representing one heart beat. Label the P, QRS, and T waves; the P-R interval; the S-T segment, and the Q-T segment.

3. What changes from *baseline* were noted in the ECG recorded during running? ____________________

____

____

Explain. ____________________

____

____

In that recorded during breath holding? ____________________

____

Explain. ____________________

____

____

4. Describe what happens in the cardiac cycle in the following situations:

during the P wave ____________________

immediately before the P wave ____________________

immediately after the P wave ____________________

during the QRS wave ____________________

immediately after the QRS wave (S-T interval) ____________________

during the T wave ____________________

5. Define the following terms:

tachycardia ____________________

bradycardia ____________________

flutter ____________________

fibrillation ____________________

myocardial infarction ____________________

6. Which would be more serious, atrial or ventricular fibrillation? ____________________

Why? ____________________

____

7. Abnormalities of heart valves can be detected more accurately by auscultation than by electrocardiography. Why is this so? ____________________

____

____

STUDENT NAME ____________________

LAB TIME/DATE ____________________

EXERCISE 29 Anatomy of Blood Vessels

Microscopic Structure of the Blood Vessels

1. Use key choices to identify the blood vessel tunic described.

   Key: a. tunica intima b. tunica media c. tunica externa

   ____________ most internal tunic

   ____________ bulky middle tunic contains smooth muscle and elastin

   ____________ its smooth surface decreases resistance to blood flow

   ____________ tunic(s) of capillaries

   ____________, ____________, ____________ tunic(s) of arteries and veins

   ____________ is especially thick in elastic arteries

   ____________ most superficial tunic

2. Servicing the capillaries is the essential function of the organs of the circulatory system. Explain this statement.

   ____

   ____

   ____

3. Cross-sectional views of an artery and of a vein are shown here. Identify each; and on the lines beneath, note the structural details that enabled you to make these identifications:

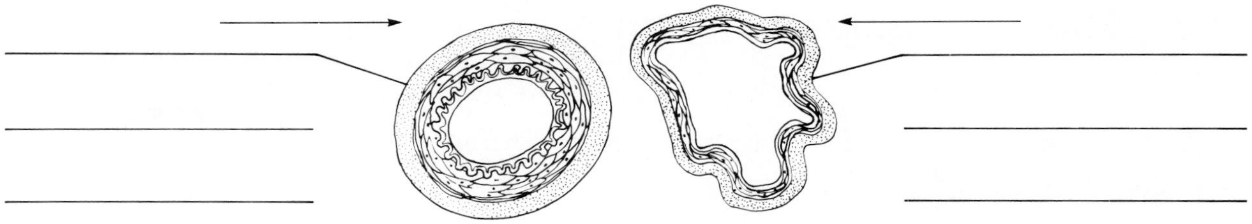

   ____________ ____________

   ____________ ____________

   ____________ ____________

   ____________ ____________

4. Why are valves present in veins but not in arteries? ____________________

   ____

   ____

5. Name two events *occurring within the body* that aids in venous return:

   ____

   and ____________________

6. Why are the walls of arteries proportionately thicker than those of the corresponding veins? ________________

____

____

Major Systemic Arteries and Veins of the Body

1. Use the key on the right to identify the arteries or veins described on the left.

_______, the arterial system has one of these; the venous system has two

_______ these arteries supply the myocardium

_______, _______ two paired arteries serving the brain

_______ longest vein in the lower limb

_______ artery on the dorsum of the foot checked after leg surgery

_______ serves the posterior thigh

_______ supplies the diaphragm

_______ formed by the union of the radial and ulnar veins

_______, _______ two superficial veins of the arm

_______ artery serving the kidney

_______ artery serving the liver

_______ artery that supplies the distal half of the large intestine

_______ drains the pelvic organs

_______ what the external iliac artery becomes on entry into the thigh

_______ major deep vein of the arm

_______ drains most of the small intestine

_______ join to form the inferior vena cava

_______ an arterial trunk that has three major branches, which run to the liver, spleen, and stomach

_______ major artery serving the tissues external to the skull

_______, _______, _______ three veins serving the leg

_______ artery generally used to take the pulse at the wrist

Key:

a. anterior tibial
b. basilic
c. brachial
d. brachiocephalic
e. celiac trunk
f. cephalic
g. common carotid
h. common iliac
i. coronary
j. deep femoral
k. dorsalis pedis
l. external carotid
m. femoral
n. greater saphenous
o. hepatic
p. inferior mesenteric
q. internal carotid
r. internal iliac
s. peroneal
t. phrenic
u. posterior tibial
v. radial
w. renal
x. subclavian
y. superior mesenteric
z. vertebral

Special Circulations

Pulmonary circulation:

1. Trace the pathway of a carbon dioxide gas molecule in the blood from the inferior vena cava until it leaves the bloodstream. Name all structures (vessels, heart chambers, and others) passed through en route.

____

____

____

2. Trace the pathway of an oxygen gas molecule from an alveolus of the lung to the right atrium of the heart. Name all structures through which it passes. ______________________________

____

____

3. Most arteries of the adult body carry oxygen-rich blood, and the veins carry oxygen-depleted, carbon dioxide-rich blood. What is different about the pulmonary arteries and veins? ____________________

____

____

4. How do the arteries of the pulmonary circulation differ structurally from the systemic arteries? What condition is indicated by this anatomic difference? ______________________________

____

____

Arterial supply of the brain and the circle of Willis:

1. What two paired arteries enter the skull to supply the brain?

______________________ and ______________________

2. Branches of the paired arteries just named cooperate to form a ring of blood vessels encircling the pituitary gland, at the base of the brain. What name is given to this communication network? ____________________

What is its function? ______________________________

____

3. What portion of the brain is served by the anterior and middle cerebral arteries? ____________________

____

Both the anterior and middle cerebral arteries arise from the ______________________ arteries.

4. Trace the pathway of a drop of blood from the aorta to the left occipital lobe of the brain, noting all structures through which it flows. ______________________________

____

Hepatic portal circulation:

1. What is the source of blood in the hepatic portal system? ______________________________

____

2. Why is this blood carried to the liver before it enters the systemic circulation? ______________________________

____

____

3. The hepatic portal vein is formed by the union of the ______________________________, which drains the

______________________, ______________________, ______________________,

______________________, and the ______________________, which drains the ____________

________ and ______________________. The ______________________ vein, which drains the lesser curvature of the stomach, empties directly into the hepatic portal vein.

4. Trace the flow of a drop of blood from the small intestine to the right atrium of the heart, noting all structures

encountered or passed through on the way. ______________________________

____

____

Fetal circulation:

1. The failure of two of the fetal bypass structures to become obliterated after birth can cause congenital heart disease, in which the youngster would have improperly oxygenated blood. Which two structures are these?

______________________ and ______________________

2. For each of the following structures, first note its function in the fetus; and then note what happens to it or what it is converted to after birth. Circle the blood vessel that carries the most oxygen-rich blood.

Structure	Function in Fetus	Fate
Umbilical artery		
Umbilical vein		
Ductus venosus		
Ductus arteriosus		
Foramen ovale		

3. What organ serves as a respiratory/digestive/excretory organ for the fetus? ______________________

STUDENT NAME ______________________________

LAB TIME/DATE ______________________________

EXERCISE 30

Human Cardiovascular Physiology—Blood Pressure and Pulse Determinations

Cardiac Cycle

1. Define the following terms:

   systole ______________________________

   diastole ______________________________

   cardiac cycle ______________________________

2. Using the key below, indicate the time interval occupied by the following events of the cardiac cycle.

   Key: a. 0.4 sec b. 0.3 sec c. 0.1 sec d. 0.8 sec

   ________ the length of the normal cardiac cycle

   ________ the quiescent period, or pause

   ________ time interval of atrial systole

   ________ ventricular contraction period

3. If an individual's heart rate is 80 beats/min, what is the length of the cardiac cycle? ________ What portion of the cardiac cycle is shortened by this more rapid heart rate? ______________________________

   ____

4. Answer the following questions, which concern events of the cardiac cycle:

   When are the AV valves closed? ______________________________

   Open ______________________________

   What event within the heart causes the AV valves to open? ______________________________

   What causes them to close? ______________________________

   When are the semilunar valves closed? ______________________________

   ____

   Open? ______________________________

   What event causes the semilunar valves to open? ______________________________

   ____

   To close? ______________________________

Are both sets of valves closed during any part of the cycle? ______

If so, when? ______

Are both sets of valves open during any part of the cycle? ______

At what point in the cardiac cycle is the pressure in the heart highest? ______

Lowest? ______

5. What two factors promote the movement of blood through the heart? ______

______ and ______

______

Auscultation of Heart Sounds

1. Complete the following statements:

The monosyllables describing the heart sounds are __1__. The first heart sound is a result of closure of the __2__ valves, whereas the second is a result of closure of the __3__ valves. The heart chambers that have just been filled when you hear the first heart sound are the __4__, and the chambers that have just emptied are the __5__. Immediately after the second heart sound, the __6__ are filling with blood, and the __7__ are empty.

1. ______
2. ______
3. ______
4. ______
5. ______
6. ______
7. ______

2. As you listened to the heart sounds during the laboratory session, what differences in pitch, length, and amplitude (loudness) of the two sounds did you observe? ______

______

______

3. Indicate where you would place your stethoscope to auscultate most accurately the following:

closure of the tricuspid valve ______

closure of the aortic semilunar valve ______

apical heartbeat ______

Which valve is heard most clearly when the apical heartbeat is auscultated? ______

4. No one expects you to be a full-fledged physician on such short notice; but on the basis of what you have learned about heart sounds, how might abnormal sounds be used to diagnose heart problems?

______

______

______

Palpation of the Pulse

1. Define *pulse.* ______________________________

____

2. Describe the procedure used to take the pulse. ______________________________

____

____

3. Identify the artery palpated at each of the following pressure points:

at the wrist ______________________ on the dorsum of the foot ______________________

in the front of the ear ______________________ at the side of the neck ______________________

4. When you were palpating the various pulse or pressure points, which appeared to have the greatest amplitude or tension? ______________________ Why do you think this was so? ______________________

____

5. Assume someone has been injured in an auto accident and is hemorrhaging badly. What pressure point would you compress to help stop bleeding from each of the following areas?

the thigh ______________________ the calf ______________________

the forearm ______________________ the thumb ______________________

6. How could you tell by simple observation whether bleeding is arterial or venous? ______________________

____

7. You may sometimes observe a slight difference between the value obtained from an apical pulse (beats/min) and that from an arterial pulse taken elsewhere on the body. What is this difference called?

____

Blood Pressure Determinations

1. Define *blood pressure.* ______________________________

____

2. Identify the phase of the cardiac cycle to which each of the following apply:

systolic pressure ______________________ diastolic pressure ______________________

3. What is the name of the instrument used to compress the artery and record pressures in the auscultatory method of determining blood pressure? ______________________

4. What are the sounds of Korotkoff? ______________________________

____

What causes the systolic sound? ______________________________

____

The disappearance of sound? ______________________________

5. Interpret 145/85/82. ______________________________

____

____

6. Assume the following BP measurement was recorded for an elderly patient with severe arteriosclerosis: 170/110/–. Explain the inability to obtain the third reading.

____

____

7. Define *pulse pressure*. ______________________________

Why is this measurement important? ______________________________

____

8. How do venous pressures compare to arterial pressures? ______________________________

Why? ______________________________

9. What maneuver to increase the thoracic pressure illustrates the effect of external factors on venous pressure?

______________ How was it performed? ______________________________

____

____

10. What might an abnormal increase in venous pressure indicate? (Think!) ______________________________

____

Effect of Various Factors on Blood Pressure and Heart Rate

1. What effect do the following have on blood pressure? (Indicate increase by I and decrease by D.)

   ______ increased diameter of the arterioles ______ hemorrhage

   ______ increased blood viscosity ______ arteriosclerosis

   ______ increased cardiac output ______ increased pulse rate

2. In which position (sitting, reclining, or standing) is the blood pressure normally the highest?

   ______ The lowest? ______

   What immediate changes did you observe when the subject stood up after being in the sitting or reclining position?

   ______

   ______

   What changes in the blood vessels might account for the change? ______

   ______

   ______

   After the subject stood for 3 minutes, what changes in blood pressure were observed? ______

   ______

   How do you account for this change? ______

3. What was the effect of exercise on blood pressure? ______

   On pulse? ______ Do you think these effects reflect changes in cardiac output *or* in peripheral resistance? ______

   ______

   Why are there normally no significant increases in diastolic pressure after exercise? ______

   ______

   ______

4. What effects of the following did you observe on blood pressure in the laboratory?

   nicotine ______ cold temperature ______

   What do you think the effect of heat would be? ______

   Why? ______

5. Differentiate between a hypo- and a hyperreactor relative to the cold pressor test. ______

   ______

   ______

Skin Color as an Indicator of Local Circulatory Dynamics

1. Describe normal skin color and the appearance of the veins in the subject's forearm before any testing was conducted. ______________________________

2. What changes occurred when the subject emptied his forearm of blood (by raising his arm and making a fist) and the flow was occluded with the cuff? ______________________________

   ____

   What changes occurred during venous congestion? ______________________________

   ____

3. What is the importance of collateral blood supplies? ______________________________

   ____

4. Explain the mechanism by which mechanical stimulation of the skin produced a flare. ______________________________

   ____

   ____

STUDENT NAME ______________________

LAB TIME/DATE ______________________

EXERCISE 31

Frog Cardiovascular Physiology

Refractory Period of Cardiac Muscle

1. Define *extrasystole.* ______________________

____

2. In responding to the following questions refer to the recordings you made during this exercise:

   What was the effect of stimulation of the heart during ventricular contraction? ______________________

   During ventricular relaxation (first portion)? ______________________

   During the pause interval? ______________________

   What does this indicate about the refractory period of cardiac muscle? ______________________

   ____

   Can cardiac muscle be tetanized? ________ Why or why not? ______________________

Physical and Chemical Factors Modifying Heart Rate

1. Describe the effect of thermal factors on the frog heart:

   cold ______________________ heat ______________________

2. What was the effect of vagal stimulation on heart rate? ______________________

   Which of the following factors cause the same (or very similar) heart rate–reducing effects? epinephrine, acetylcholine, atropine sulfate, pilocarpine, sympathetic nervous system activity, digitalis, potassium ions?

   ____

   Which of the factors listed above would reverse or antagonize vagal effects? ______________________

   ____

3. What is vagal escape? ______________________

   ____

   Why is vagal escape valuable in maintaining homeostasis? ______________________

   ____

4. Once again refer to your recordings. Did the administration of the following produce any changes in force of contraction (shown by peaks of increasing or decreasing height)? If so, explain the mechanism.

   epinephrine ____

   ____

   acetylcholine ____

   ____

   ____

   calcium ions ____

   ____

   ____

5. Excessive amounts of each of the following ions would most likely interfere with normal heart activity. Note the type of changes caused in each case.

   K^+ ____

   Ca^{2+} ____

   Na^+ ____

6. How does the Stannius ligature used in the laboratory produce heart block? ______________________

   ____

   ____

7. Define *partial heart block,* and describe how it was recognized in the laboratory. ______________________

   ____

   ____

8. Define *total heart block,* and describe how it was recognized in the laboratory. ______________________

   ____

   ____

   ____

9. What do your heart block experiment results indicate about the spread of impulses from the atria to the ventricles?

   ____

   ____

STUDENT NAME ______________________

LAB TIME/DATE ______________________

EXERCISE 32 The Lymphatic System and Immune Response

The Lymphatic System

1. Explain why the lymphatic system is a one-way system, whereas the blood vascular system is a two-way system.

____

____

____

2. How do lymphatic vessels resemble veins? ______________________

   How do lymphatic capillaries differ from blood capillaries? ______________________

   ____

3. What is the function of the lymphatic vessels? ______________________

   ____

4. What is lymph? ______________________

5. What factors are involved in the flow of lymphatic fluid? ______________________

   ____

6. What name is given to the terminal duct draining most of the body? ______________________

7. What is the cisterna chyli? ______________________

   ____

   How does the composition of lymph in the cisterna chyli differ from that in the general lymphatic stream?

   ____

8. Which portion of the body is drained by the right lymphatic duct? ______________________

   ____

9. Note three areas where lymph nodes are densely clustered: ______________________,

   ______________________, and ______________________

10. What are the two major functions of the lymph nodes? ______________________

    ____

11. The radical mastectomy is an operation in which a cancerous breast, surrounding tissues, and the underlying muscles of the anterior thoracic wall, plus the axillary lymph nodes, are removed. After such an operation, the arm usually swells, or becomes edematous, and is very uncomfortable—sometimes for months. Why?

    ____

The Immune Response

1. Define the following terms which relate to the operation of the immune system:

   Immunologic memory ____

   ____

   Specificity ____

   ____

   Recognition of self from nonself __________________________________

   ____

2. What is the function of B cells in the immune response? __________________________

   ____

   ____

3. What is the role of T cells? ____

   ____

   ____

   ____

   ____

Microscopic Anatomy of a Lymph Node

1. In the space below, make a rough drawing of the structure of a lymph node. Identify the cortex area, germinal centers, and medulla. For each identified area, note the cell type (T cell, B cell, or macrophage) most likely to be found there.

2. What structural characteristic ensures a *slow* flow of lymph through a lymph node? ______________

   ____

   Why is this desirable? ____

   ____

Antibodies and Tests for Their Presence

1. Describe the structure of the immunoglobulin monomer. ______________________________

____

____

2. Are the genes coding for one antibody entirely different from those coding for a different antibody?

   ________ Explain. ____

____

____

3. Rheumatoid arthritis is said to involve an autoimmune response. What *is* an autoimmune response?

____

____

4. When conducting the test to identify the presence of rheumatoid factor in plasma, a clumping, or agglutination, indicated a positive response (that is, the presence of the abnormal antibody). What event did the clumping reflect?

____

____

STUDENT NAME ______________________

LAB TIME/DATE ______________________

EXERCISE 33 Anatomy of the Respiratory System

Upper and Lower Respiratory Structures

1. Complete the labeling of the diagram of the upper respiratory structures (sagittal section).

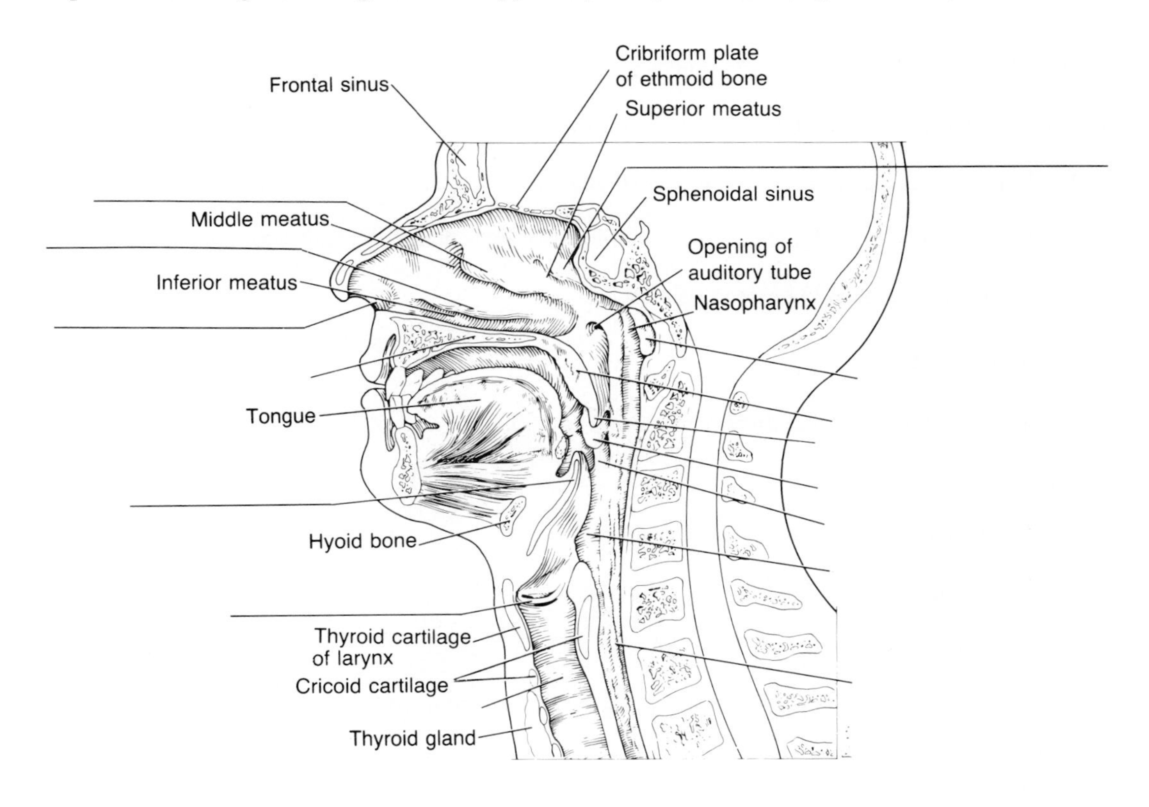

2. Two pairs of vocal folds are found in the larynx. Which pair are the true vocal cords? (Superior or inferior?)

____

3. What is the significance of the fact that the human trachea is reinforced with cartilage rings?

____

Of the fact that the rings are incomplete posteriorly? ______________________

____

4. Name the specific cartilages in the larynx that correspond to the following descriptions:

forms the Adam's apple ______________ shaped like a signet ring ______________

a "lid" for the larynx ______________ vocal cord attachment ______________

5. Trace a molecule of oxygen from the external nares to the pulmonary capillaries of the lungs: External nares →

____

____

____

6. What is the function of the pleural membranes? ______________________________

____

____

7. Name two functions of the nasal cavity mucosa: ______________________________

8. The following questions refer to the primary bronchi:

   Which is longer? __________ Larger in diameter? __________ More horizontal? __________

   The more common site for lodging of a foreign object that had entered the respiratory passageways? ________

9. Match the terms in column B to those in column A.

Column A		Column B
______	nerve that activates the diaphragm during inspiration	a. alveoli
______	"floor" of the nasal cavity	b. bronchioles
______	food passageway posterior to the trachea	c. conchae
______	flaps over the glottis during swallowing of food	d. epiglottis
______	contains the vocal cords	e. esophagus
______	part of the conducting pathway between the larynx and the primary bronchi	f. glottis
		g. larynx
______	pleural layer lining the walls of the thorax	h. palate
______	site from which oxygen enters the pulmonary blood	i. parietal pleura
______	autonomic nervous system nerve serving the thoracic region	j. phrenic nerve
______	opening between the vocal folds	k. primary bronchi
______	increases air turbulence in the nasal cavity	l. trachea
		m. vagus nerve
		n. visceral pleura

10. What portions of the respiratory system are referred to as anatomical dead space? ______________________

____

____

Why? ____

____

11. Define *external respiration:* ____

____

internal respiration: ____

____

12. On the diagram below identify alveolar epithelium, capillary endothelium, alveoli, and red blood cells, and bracket the respiratory membrane.

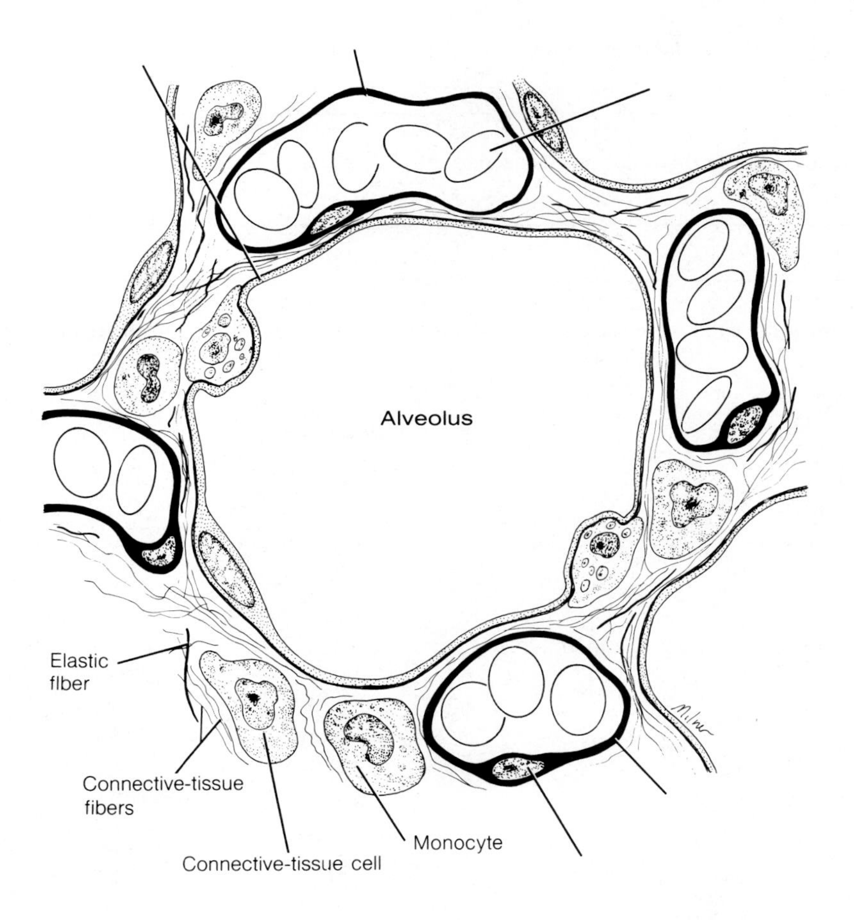

Sheep Pluck Demonstration

1. Does the lung inflate part by part or as a whole, like a balloon? ______________________________

   What happened when the pressure was released? ______________________________

   What type of tissue insures this phenomenon? ______________________________

Examination of Prepared Slides of Lung and Trachea Tissue

1. The tracheal epithelium is ciliated and has goblet cells. What is the function of each of these modifications?

   Cilia? ______________________________

   Goblet cells? ______________________________

2. The tracheal epithelium is said to be pseudostratified. Why? ______________________________

   ____

   ____

3. What structural characteristics of the alveoli make them an ideal site for the diffusion of gases?

   ____

   Why does oxygen move from the alveoli into the pulmonary capillary blood? ______________________________

   ____

   ____

4. If you observed pathologic lung sections, what were the conditions responsible and how did the tissue differ from normal lung tissue?

Slide Type	**Observations**

STUDENT NAME ____________________

LAB TIME/DATE ____________________

EXERCISE 34 Respiratory System Physiology

Mechanics of Respiration

1. For each of the following cases, check the column appropriate to your observations on the operation of the model lung.

	Diaphragm Pushed Up		Diaphragm Pulled Down	
Change	**Increased**	**Decreased**	**Increased**	**Decreased**
In internal volume of the bell jar (thoracic cage)				
In internal pressure				
In the size of the balloons (lungs)				
In direction of air flow	Into lungs	Out of lungs	Into lungs	Out of lungs

2. The activation of the diaphragm and the external intercostal muscles begins the inspiratory process. What results from the contractions of these muscles, and how is this accomplished? ____________________

____

____

____

3. What was the approximate increase in diameter of chest circumference during a quiet inspiration?

   ____________ inches During forced inspiration? ____________ inches

   What temporary physiologic advantage does the substantial increase in chest circumference during forced inspiration create? ____________________

   ____

   ____

4. The presence of a partial vacuum between the pleural membranes is integral to normal breathing movements. What would happen if an opening were made into the chest cavity, as with a puncture wound?

   ____

   How is this condition treated medically? ____________________

Respiratory Volumes and Capacities—Spirometry

1. Write the respiratory volume term and the normal value that is described by the following statements:

   volume of air present in the lungs after a forceful expiration ______________________

   volume of air that can be expired forcibly after a normal expiration ______________________

   volume of air that is breathed in and out during a normal respiration ______________________

   volume of air that can be inspired forcibly after a normal inspiration ______________________

   volume of air corresponding to TV + IRV + ERV ______________________

2. Record experimental respiratory volumes as determined in the laboratory.

   Average tidal volume ______________ ml   Average ERV ______________ ml

   Average IRV ______________ ml   Average VC ______________ ml

   Minute respiratory volume ______________ ml/min   % predicted VC ______________ %

3. Would your vital-capacity measurement differ if you performed the test while standing? ________ While lying down? ________ Explain. ______________________

   ____

Use of the Pneumograph to Determine Factors Influencing Rate and Depth of Respiration

1. Where are the neural control centers of respiratory rhythm? ______________ and ______________

2. Based on pneumograph reading of respiratory variation, what was the rate of quiet breathing?

   Initial testing ______________ breaths/min

   Record observations of how the initial pneumograph recording was modified during the various testing procedures described below. Indicate the respiratory rate, and include comments on the relative depth of the respiratory peaks observed.

Test Performed	**Observations**
Talking	
Yawning	
Laughing	
Standing	
Concentrating	
Swallowing water	
Coughing	
Lying down	
Running in place	

3. Student data:

breath-holding interval after a deep inhalation _______ sec   length of recovery period _______ sec

breath-holding interval after a forceful expiration _______ sec   length of recovery period _______ sec

After breathing quietly and taking a deep breath (which you held), was your urge to inspire or expire?

____

After exhaling and then breath holding, was the desire for inspiration or expiration? _______________

Explain these results. (*Hint:* what reflex is involved here?) _______________________________________

____

____

4. Observations after hyperventilation: ____

____

5. Length of breath holding after hyperventilation: _______ sec

Why does hyperventilation produce apnea or a reduced respiratory rate? ____________________________

____

____

6. Observations for rebreathing breathed air: ____

____

Why does rebreathing breathed air produce an increased respiratory rate? __________________________

____

7. What was the effect of running in place (exercise) on the duration of breath holding? _______________

Explain: ____

____

8. Relative to the test illustrating the effect of respiration on circulation: *(student data)*

Radial pulse before beginning test _______ /min   Radial pulse after testing _____________ /min

Relative pulse force before beginning test _____   Relative force of radial pulse after testing _____

Condition of neck and facial veins after testing ____

Explain: ____

____

9. Do the following factors generally increase (indicate with I) or decrease (indicate with D) the respiratory rate and depth?

increase in blood CO_2 _______   increase in blood pH _______

decrease in blood O_2 _______   decrease in blood pH _______

10. Blood CO_2 levels and blood pH are related. When blood CO_2 levels increase, does the pH increase or decrease?

______________ Explain why. ______________

____

____

Respiratory Sounds

1. Which of the respiratory sounds is heard during both inspiration and expiration? ______________

   Which is heard primarily during inspiration? ______________

2. Where did you best hear the vesicular respiratory sounds? ______________

Role of the Respiratory System in Acid-Base Balance of Blood

1. Define *buffer.* ______________

2. How successful was the laboratory buffer (pH 7) in resisting changes in pH when the acid was added?

   ____

   When the base was added? ______________

   How successful was the buffer in resisting changes in pH when the additional aliquots (3 more drops) of the acid and base were added to the original samples? ______________

3. What buffer system operates in blood plasma? ______________

   Which of its species resists a *drop* in pH? ______________

   Which resists a *rise* in pH? ______________

4. Explain how the carbonic acid–bicarbonate buffer system of the blood operates. ______________

   ____

   ____

   ____

   ____

   ____

STUDENT NAME ________________

LAB TIME/DATE ________________

EXERCISE 35

Anatomy of the Digestive System

General Histological Plan of the Alimentary Canal

The general anatomic features of the digestive tube have been presented. Fill in the table below to complete the information listed.

Wall layer	Subdivisions of the layer	Major functions
mucosa		
submucosa		
muscularis externa		
serosa or adventitia		

Organs of the Alimentary Canal

1. Match the items in column B with the descriptive statements in column A (continued on p. RS118):

Column A

_______ structure that suspends the small intestine from the posterior body wall

_______ fingerlike extensions of the intestinal mucosa that increase the surface area for absorption

_______ large collections of lymphoid tissue found in the submucosa of the small intestine

_______ deep folds of the mucosa and submucosa that extend completely or partially around the circumference of the small intestine

_______, _______ regions that break down foodstuffs mechanically

_______ mobile organ that manipulates food in the mouth and initiates swallowing

_______ conduit for both air and food

_______, _______, _______ three structures continuous with, and representing modifications of, the peritoneum

Column B

a. anus

b. appendix

c. esophagus

d. frenulum

e. greater omentum

f. hard palate

g. haustra

h. ileocecal valve

i. large intestine

j. lesser omentum

k. mesentery

________ the "gullet"; no digestive/absorptive function

________ folds of the gastric mucosa

________ sacculations of the large intestine

________ projections of the plasma membrane of a mucosal epithelial cell

________ valve at the junction of the small and large intestines

________ primary region of food and water absorption

________ membrane securing the tongue to the floor of the mouth

________ absorbs water and forms feces

________ area between the teeth and lips/cheeks

________ wormlike sac that outpockets from the cecum

________ initiates protein digestion

________ structure attached to the lesser curvature of the stomach

________ organ distal to the stomach

________ valve controlling food movement from the stomach into the duodenum

________ posterosuperior boundary of the oral cavity

________ site of the hepatopancreatic sphincter through which pancreatic secretions and bile pass

________ serous lining of the abdominal cavity wall

________ principal site for the synthesis of vitamin K by microorganisms

________ region containing two sphincters through which feces are expelled from the body

________ bone-supported anterosuperior boundary of the oral cavity

l. microvilli
m. oral cavity
n. parietal peritoneum
o. Peyer's patches
p. pharynx
q. plicae circulares
r. pyloric valve
s. rugae
t. small intestine
u. soft palate
v. stomach
w. tongue
x. vestibule
y. villi
z. visceral peritoneum

2. How is the muscularis externa of the stomach modified? ________________________________

____

How does this modification relate to the function of the stomach? ________________________

____

3. Correctly identify the three organs depicted in the diagrams on the opposite page.

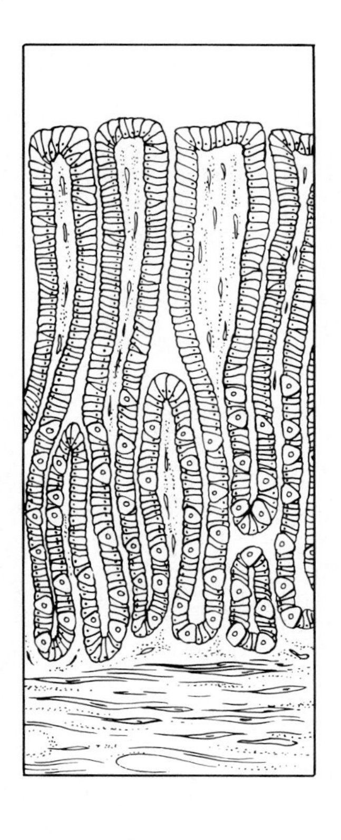

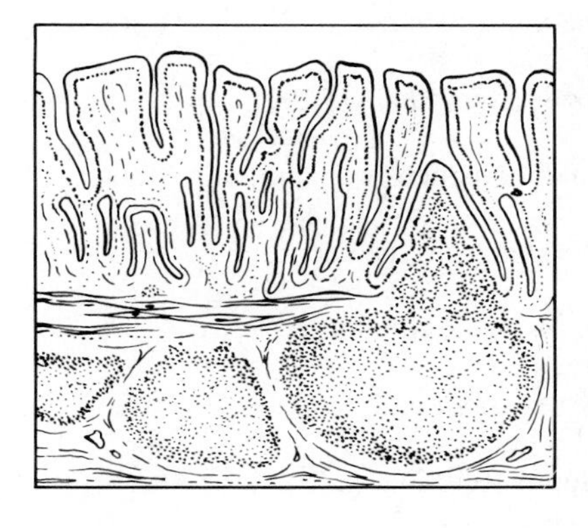

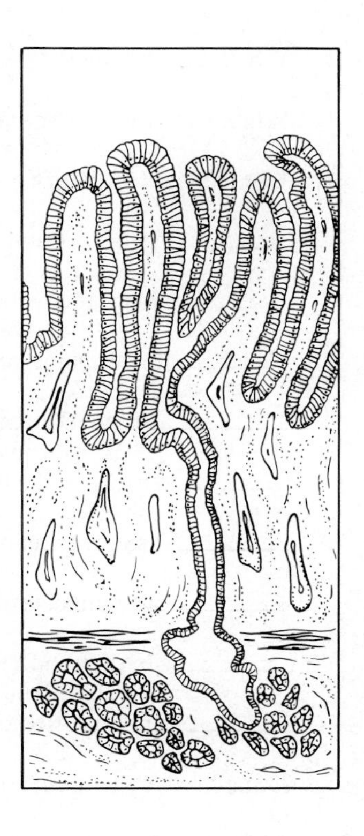

______________________ ______________________ ______________________

4. What transition in epithelium type exists at the cardiac-esophageal junction? ______________________

____

How do the epithelia of these two organs relate to their specific functions? ______________________

____

____

5. What cells of the stomach produce HCl? ______________ Pepsinogen? ______________

Accessory Digestive Organs

1. Various types of glands form a part of the alimentary tube wall or duct their secretions into it. Match the glands listed in column B with the function/locations described in column A.

Column A	Column B
________ produce mucus; found in the submucosa of the small intestine	a. Brunner's glands
________ produce a product containing amylase that begins starch breakdown in the mouth	b. gastric glands
________ produces a whole spectrum of enzymes and an alkaline fluid that is secreted into the duodenum	c. intestinal crypts
________ produces bile that it secretes into the duodenum via the common bile duct	d. liver
________ produce HCl and pepsinogen	e. pancreas
________ found in the mucosa of the small intestine; produce intestinal juice	f. salivary glands

2. Which of the salivary glands produces a secretion that is mainly serous? ____________________

3. What is the role of the gallbladder? ____________________

4. Use the key to identify each tooth area described below.

_______ visible portion of the tooth *in situ*

_______ material covering the tooth root

_______ hardest substance in the body

_______ attaches the tooth to bone and surrounding alveolar structures

_______ portion of the tooth embedded in bone

_______ forms the major portion of tooth structure; similar to bone

_______ form the dentin

_______ site of blood vessels, nerves, and lymphatics

_______ entire portion of the tooth covered with enamel

Key:
a. anatomic crown
b. cementum
c. clinical crown
d. dentin
e. enamel
f. gingiva
g. odontoblasts
h. periodontal ligament
i. pulp
j. root

5. In the human, the number of deciduous teeth is _______; the number of permanent teeth is _______.

6. The dental formula for permanent teeth is _______

Explain what this means: ____________________

____

7. What teeth are the "wisdom teeth"? ____________________

STUDENT NAME ______________________

LAB TIME/DATE ______________________

EXERCISE 36 Chemical Breakdown of Foodstuffs: Enzymatic Action

1. Match the following definitions with the proper key letters.

   Key: a. catalyst b. control c. enzyme d. substrate

   __________ increases the rate of a chemical reaction without becoming part of the product

   __________ provides a standard of comparison for test results

   __________ biologic catalyst; protein in nature

   __________ substance on which a catalyst works

2. List three characteristics of enzymes. ______________________

   ____

3. The enzymes of the digestive system are classed as hydrolases. What does this mean? ______________________

   ____

4. Fill in the following chart relative to the various digestive system enzymes encountered in this exercise.

Enzyme	Organ producing it	Site of action	Substrate(s)	Optimal pH
salivary amylase				
pepsin				
pancreatin				

5. Name the end products of digestion for the following types of foods:

   proteins __________ carbohydrates __________

   fats __________ and __________

6. You used several indicators or tests in the laboratory to determine the presence or absence of certain substances. Choose the correct test or indicator from the key to correspond to the conditions described below:

   Key: a. IKI (Lugol's iodine) b. Benedict's solution c. phenol red d. biuret test

   ______ used to test for the presence of protein, which was indicated by a violet color

   ______ used to test for the presence of starch, which was indicated by a blue-black color

   ______ used to test for the presence of fatty acids, which was evidenced by a color change from pink to yellow

   ______ used to test for the presence of reducing sugars (maltose, sucrose, glucose) as indicated by a blue to red color change

7. In the procedure concerning starch digestion by salivary amylase, how do you explain the fact that neither 0°C incubation nor conditions that involved a preliminary boiling of the enzyme preparation resulted in positive test results for the digestion of starch? ____

____

What conclusions can you draw when an experimental sample gives both a positive starch test and a positive Benedict's test after incubation? ____

____

Why was 37°C the optimal incubation temperature? ____

Why did very little, if any, starch digestion occur in test tubes 9 and 10? ____

____

Why was sample #1 completely negative for the presence of sugar? ____

8. In the procedure concerning pepsin digestion of protein, in which test tube did more protein hydrolysis occur? ________ Why? ____

____

Why did test tubes 3 and 5 yield negative results for digestion? ____

____

What functional relationship exists between HCl and pepsin? ____

____

Pepsin is a protein-digesting enzyme, and the structural material of cells is largely protein. Why doesn't the stomach digest itself? ____

____

9. In the procedure concerning pancreatic lipase digestion of fats and the action of bile salts, how did appearance of test tubes A and B differ? ____

____

How can you explain this difference? ____

____

Why did the phenol red indicator change from pink to yellow during the process of fat hydrolysis? ____

____

____

Why is bile not considered an enzyme? ____

____

What role does bile play in fat digestion? ____

____

10. The three-dimensional structure of a functional protein is altered by intense heat or excesses of pH even though peptide bonds may not break. Such inactivation is called denaturation, and denatured enzymes are nonfunctional.

Explain why. ____

____

____

What specific experimental conditions in the various procedures resulted in the denaturation of the enzymes?

____

11. Pancreatic and intestinal enzymes operate optimally at a pH that is slightly alkaline, yet the chyme entering the duodenum from the stomach is very acid. How is the proper pH for the functioning of the pancreatic-intestinal

enzymes assured? ____

____

12. Assume you have been chewing a piece of bread for 5 or 6 minutes. How would you expect its taste to change

during this interval? ____

Why? ____

13. Note the mechanism of absorption (passive or active transport) of the following food breakdown products, and indicate by a check mark (√) whether the absorption would result in their movement into the blood capillaries or the lymph capillaries (lacteals).

Substance	**Mechanism of absorption**	**Blood**	**Lymph**
monosaccharides			
fatty acids and glycerol			
amino acids			
water			
Na^+, Cl^-, Ca^{2+}			

14. People on a strict diet to lose weight begin to metabolize stored fats at an accelerated rate. How would this condition affect blood pH? ______________________________

15. Trace the pathway of a ham sandwich (ham = protein and fat; bread = starch) from the mouth to the site of absorption of its breakdown products, noting where digestion occurs and what specific enzymes are involved.

STUDENT NAME ______________________

LAB TIME/DATE ______________________

EXERCISE 37

Anatomy of the Urinary System

Gross Anatomy of the Human Urinary System

1. Complete the following statements:

The kidney is referred to as an excretory organ because it excretes __1__ wastes. It is also a major homeostatic organ because it maintains the electrolyte, __2__, and __3__ balance of the blood.

Urine is continuously formed by the __4__ and is routed down the __5__ by the mechanism of __6__ to a storage organ called the __7__. Eventually, the urine is conducted to the body __8__ by the urethra. In the male, the urethra is __9__ inches long and transports both urine and __10__. The female urethra is __11__ inches long and transports only urine.

Voiding or emptying the bladder is called __12__. Voiding has both voluntary and involuntary components. The voluntary sphincter is the __13__ sphincter. An inability to control this sphincter is referred to as __14.__

1. ______________________
2. ______________________
3. ______________________
4. ______________________
5. ______________________
6. ______________________
7. ______________________
8. ______________________
9. ______________________
10. ______________________
11. ______________________
12. ______________________
13. ______________________
14. ______________________

2. What is the function of the fat cushion that surrounds the kidneys in life? ______________________

____

3. Define *ptosis.* ______________________

____

4. Why is incontinence a normal phenomenon in the child under 1½ to 2 years old? ______________________

____

What events may lead to its occurrence in the adult? ______________________

____

____

5. Complete the labeling of the diagram to correctly identify the urinary system organs.

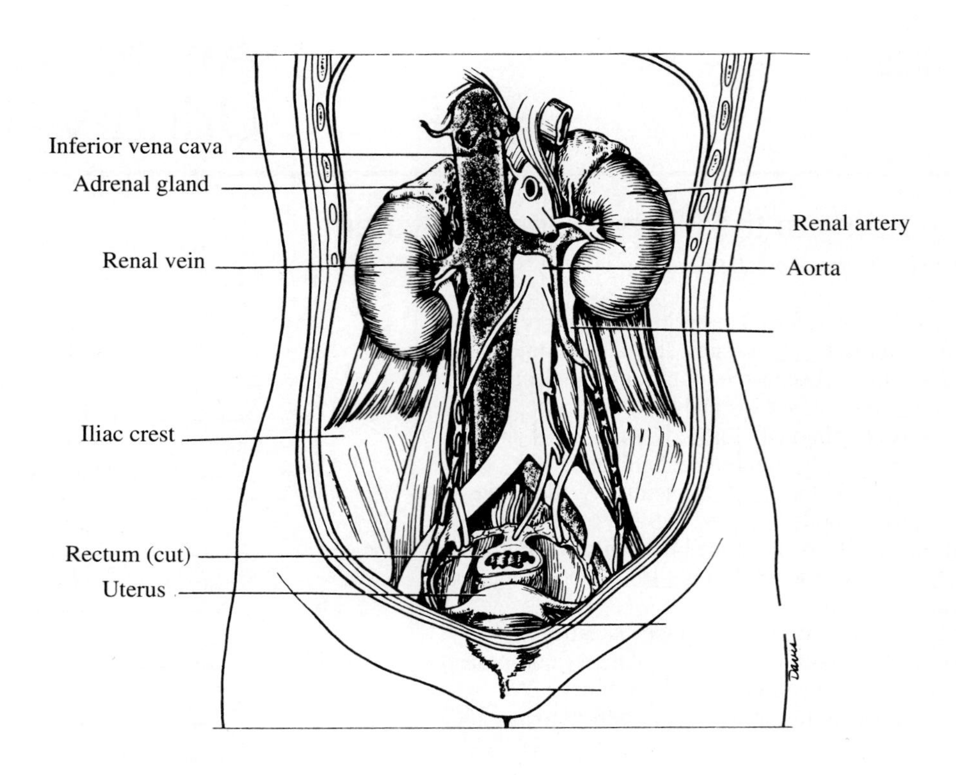

Gross Internal Anatomy of the Pig or Sheep Kidney

1. Match the appropriate structure in column B to its description in column A.

Column A		Column B
_______	smooth membrane, tightly adherent to the kidney surface	a. cortex
_______	portion of the kidney containing mostly collecting ducts	b. medulla
_______	portion of the kidney containing the bulk of the nephron structures	c. minor calyx
_______	superficial region of kidney tissue	d. renal capsule
_______	basinlike area of the kidney, continuous with the ureter	e. renal columns
_______	a cup-shaped extension of the pelvis that encircles the apex of a pyramid	f. renal pelvis
_______	area of cortical tissue running between the medullary pyramids	

Microscopic Anatomy of the Kidney and Bladder

1. Match each of the lettered structures on the diagram of the nephron (and associated renal blood supply) on the left with the terms on the right:

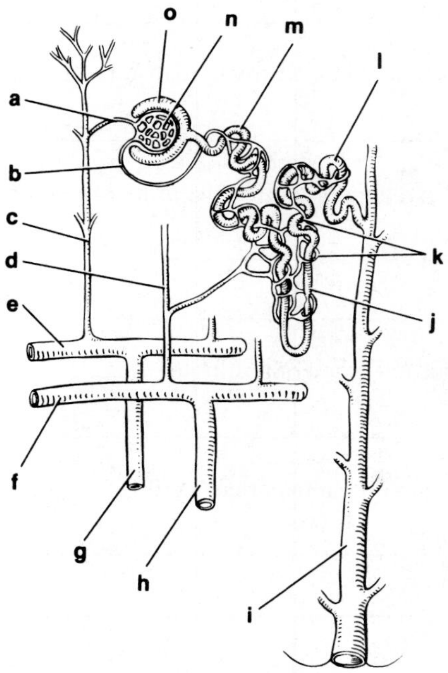

______ collecting tubule

______ glomerulus

______ peritubular capillaries

______ distal convoluted tubule

______ proximal convoluted tubule

______ interlobar artery

______ interlobular artery

______ arcuate artery

______ interlobular vein

______ efferent arteriole

______ arcuate vein

______ loop of Henle

______ afferent arteriole

______ interlobar vein

______ glomerular capsule

2. Using the terms provided in item 1, identify the following:

____________________ site of filtrate formation

____________________ primary site of tubular reabsorption

____________________ secondarily important site of tubular reabsorption

____________________ structure that conveys the processed filtrate (urine) to the renal pelvis

____________________ blood supply that directly receives substances from the tubular cells

____________________ Its inner (visceral) membrane forms part of the filtration membrane

3. Explain *why* the glomerulus is such a high-pressure capillary bed. ______________________________

____

How does its high pressure condition aid its function of filtrate formation? ______________________

____

4. What structural modification of certain tubule cells enhances their ability to reabsorb substances from the filtrate?

____

5. Explain the mechanism of tubular secretion and explain its importance in the urine formation process. ________

____

____

6. Compare and contrast the composition of blood plasma and glomerular filtrate. ____________________

____

7. Trace a drop of blood from the time it enters the kidney in the renal artery until it leaves the kidney through the renal vein. Renal artery → ____

____

____ → renal vein

8. Trace the anatomic pathway of a molecule of creatinine (metabolic waste) from the glomerular capsule to the urethra. Note each microscopic and/or gross structure it passes through in its travels. Name the subdivisions of the renal tubule. Glomerular capsule → ____

____

____ → urethra

9. What is important functionally about the specialized epithelium (transitional epithelium) in the bladder?

____

____

STUDENT NAME ____________________

LAB TIME/DATE ____________________

EXERCISE 38 Urinalysis

1. What is the normal volume of urine excreted in a 24-hour period? ______

   a. 0.1–0.5 liters b. 0.5–1.2 liters c. 1.0–1.8 liters

2. Assuming normal conditions, note whether each of the following substances would be (a) in greater relative concentration in the urine than in the glomerular filtrate, (b) in lesser concentration in the urine than in the glomerular filtrate, or (c) absent in both the urine and the glomerular filtrate.

   ______ water ______ amino acids ______ urea

   ______ phosphate ions ______ glucose ______ uric acid

   ______ sulfate ions ______ albumin ______ creatinine

   ______ potassium ions ______ red blood cells ______ pus (WBC)

   ______ sodium ions

3. Explain why urinalysis is a routine part of any good physical examination. ____________________

   ____

4. What substance is responsible for the normal yellow color of urine? ____________________

5. Which has a greater specific gravity: 1 ml of urine or 1 ml of distilled water? ____________________

   Explain. ____________________

   ____

6. Explain the relationship between the color, specific gravity, and volume of urine. ____________________

   ____

   ____

   ____

7. A microscopic examination of urine may reveal the presence of certain abnormal urinary constituents.

   Name three constituents that might be present if a urinary tract infection exists. ____________________,

   ____________________, and ____________________

8. How does a urinary tract infection influence urine pH? ____________________

   How does starvation influence urine pH? ____________________

9. Several specific terms have been used to indicate the presence of abnormal urine constituents. Identify each of the abnormalities described below by inserting a term from the list at the right that names the condition.

______________	presence of erythrocytes in the urine	a.	albuminuria
______________	presence of hemoglobin in the urine	b.	glycosuria
______________	presence of glucose in the urine	c.	hematuria
______________	presence of albumin in the urine	d.	hemoglobinuria
______________	presence of ketone bodies (acetone and others) in the urine	e.	ketonuria
		f.	pyuria
______________	presence of pus (white blood cells) in the urine		

10. What are renal calculi and what conditions favor their formation? ______________

____

11. All urine specimens become alkaline and cloudy on standing at room temperature. Explain. ______________

____

12. Glucose and albumin are both normally absent in the urine, but the reason for their exclusion differs. Explain the reason for the absence of glucose. ______________

____

The reason for the absence of albumin. ______________

____

13. Several conditions (both pathologic and nonpathologic) are named below. Using the key provided, characterize the probable abnormal constituents or conditions of the urinary product of each. More than one choice may be necessary to fully characterize the condition in most cases.

a. albumin
b. hemoglobin
c. blood cells
d. glucose
e. ketone bodies
f. bilirubin
g. pus
h. high specific gravity
i. low specific gravity
j. casts

______ glomerulonephritis

______ diabetes mellitus

______ pregnancy, exertion

______ hepatitis, cirrhosis of the liver

______ pyelonephritis

______ gonorrhea

______ starvation

______ diabetes insipidus

______ kidney stones

______ eating a 5-lb box of candy at one sitting

______ hemolytic anemias

______ cystitis (inflammation of the bladder)

14. Name the three major nitrogenous wastes found in the urine. ______________,

______________, and ______________

15. Explain the difference between organized and unorganized sediments. ______________

____

____

____

STUDENT NAME ___________________

LAB TIME/DATE ___________________

EXERCISE 39

Anatomy of the Reproductive System

Gross Anatomy of the Human Male Reproductive System

1. List the two principal functions of the testis: ___________________

2. Identify all indicated structures or portions of structures on the diagrammatic view of the male reproductive system below.

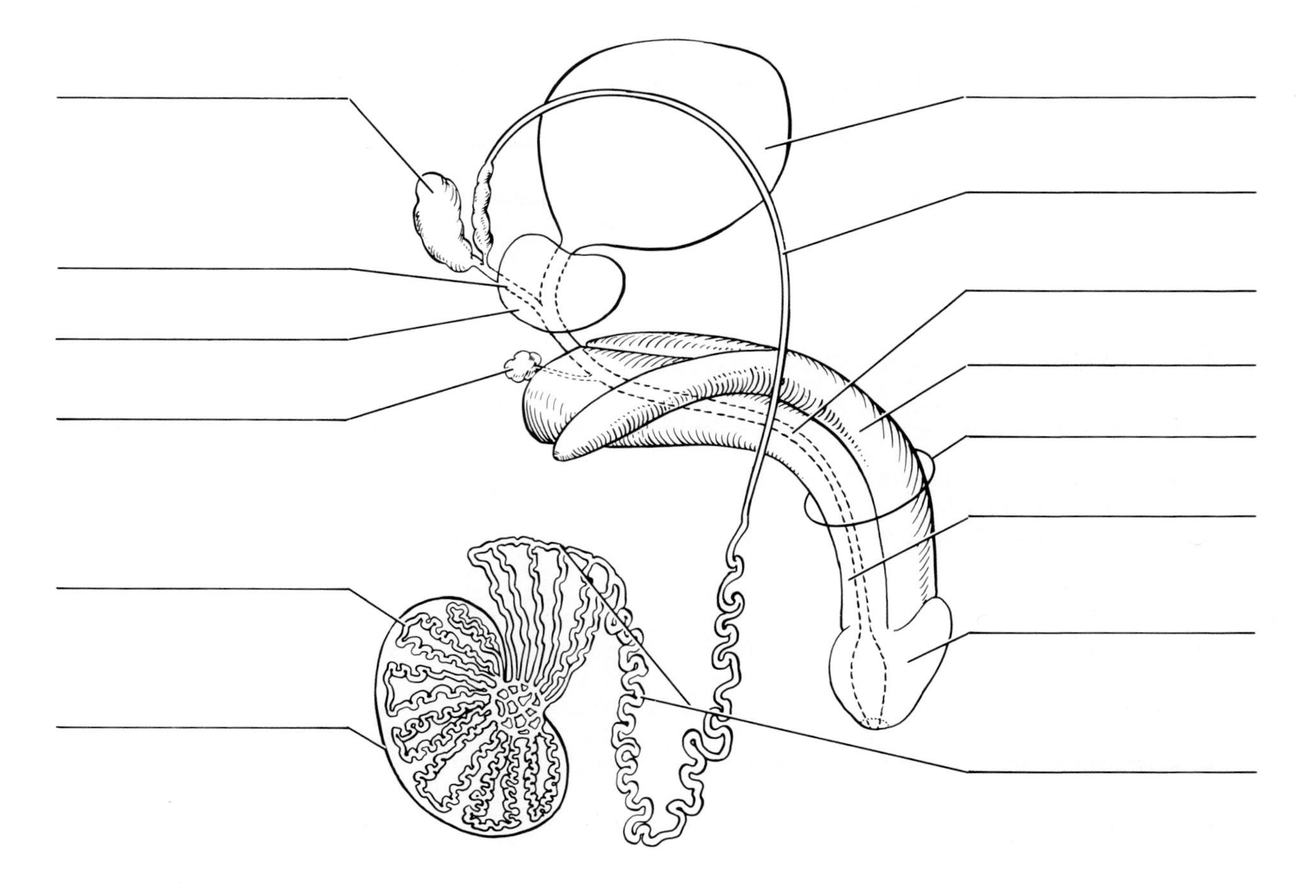

3. A common part of any physical examination of the male is palpation of the prostate gland. How is this accomplished? (Think!) ___________________

4. How might enlargement of the prostate gland interfere with urination or the reproductive ability of the male?

____

____

5. Match the terms in column B to the descriptive statements in column A.

	Column A		Column B
______	copulatory organ/penetrating device	a.	bulbourethral glands
______	site of sperm/androgen production	b.	epididymis
______	muscular passageway conveying sperm to the ejaculatory duct; in the spermatic cord	c.	glans penis
______	transports both sperm and urine	d.	membranous urethra
______	sperm maturation site	e.	penile urethra
______	location of the testis in adult males	f.	penis
______	loose fold of skin encircling the glans penis	g.	prepuce
______	portion of the urethra between the prostate gland and the penis	h.	prostate gland
______	empties a secretion into the prostatic urethra	i.	prostatic urethra
______	empties a secretion into the membranous urethra	j.	seminal vesicles
		k.	scrotum
		l.	testes
		m.	vas (ductus) deferens

6. Why are the testes located in the scrotum? ______

______

7. Describe the composition of semen and name all structures contributing to its formation. ______

______

______

8. Of what importance is the fact that seminal fluid is alkaline? ______

______

9. What structures comprise the spermatic cord? ______

______

Where is it located? ______

______

10. Using the following terms, trace the pathway of sperm from the testes to the urethra: rete testis, epididymis, seminiferous tubule, ductus deferens.

______ → ______ → ______ → ______

11. Using an appropriate reference, define cryptorchidism and discuss its significance.

______

______

Gross Anatomy of the Human Female Reproductive System

1. On the diagram of a frontal section of a portion of the female reproductive system seen below, identify all indicated structures.

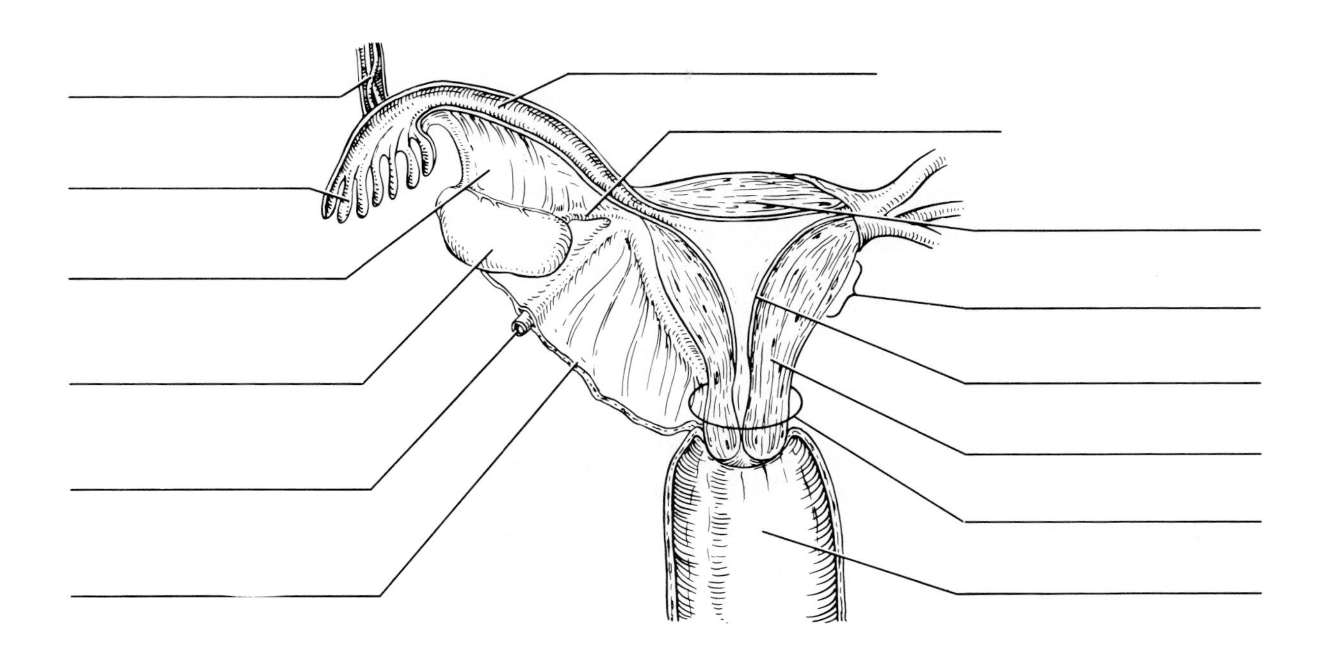

2. Identify the female reproductive system structures described below:

_________________________ site of fetal development

_________________________ copulatory canal

_________________________ "fertilized egg" typically formed here

_________________________ becomes erectile during sexual excitement

_________________________ duct extending superolaterally from the uterus

_________________________ partially closes the vaginal canal; a membrane

_________________________ produces eggs, estrogens, and progesterone

_________________________ fingerlike ends of the fallopian tube

3. Do any sperm enter the pelvic cavity of the female? Why or why not? _________________________

____

4. What is an ectopic pregnancy, and how can it happen? _________________________

____

____

____

5. Name the structures composing the external genitalia, or vulva, of the female. ______________________________

____

____

6. Put the following vestibular-perineal structures in their proper order from the anterior to the posterior aspect: vaginal orifice, anus, urethral opening, and clitoris.

   Anterior limit: ______________ → ______________ → ______________ → ______________

7. Name the male structure that is homologous to the female structures named below.

   labia majora ____________________ clitoris ____________________

8. Assume a couple has just consummated the sex act and the male's sperm have been deposited in the woman's vagina. Trace the pathway of the sperm through the female reproductive tract.

____

9. Define *ovulation:* ____

Microscopic Anatomy of Selected Reproductive Organs

1. The testis is divided into a number of lobes by connective tissue. Each of these lobes contains one to four

   ______________________________, which converge on a tubular region at the testis hilus called the

   ____________________.

2. What is the function of the cavernous bodies seen in the male penis? ______________________________

____

____

3. Name the three layers of the uterine wall from the inside out.

   ____________________, ____________________, ____________________

   Which of these is sloughed during menses? ______________________________

   Which contracts during childbirth? ______________________________

4. What is the function of the stereocilia exhibited by the epithelial cells of the mucosa of the epididymis? ________

____

STUDENT NAME ______________________

LAB TIME/DATE ______________________

EXERCISE 40 Physiology of Reproduction: Gametogenesis and the Female Cycles

Meiosis

1. The following statements refer to events occurring during mitosis and/or meiosis. For each statement, decide if the event occurs in (a) mitosis only, (b) meiosis only, or (c) both mitosis and meiosis.

   ________ dyads are visible

   ________ tetrads are visible

   ________ product is two diploid daughter cells

   ________ product is four haploid daughter cells

   ________ involves the phases prophase, metaphase, anaphase, and telophase

   ________ occurs throughout the body

   ________ occurs only in the ovaries and testes

   ________ provides cells for growth and repair

   ________ homologues synapse and chiasmata are seen

   ________ daughter cells are quantitatively and qualitatively different from the mother cell

   ________ daughter cells are genetically identical to the mother cell

   ________ chromosomes are replicated before the division process begins

   ________ provides cells for replication of the species

   ________ consists of two consecutive nuclear divisions, without chromosomal replication occurring before the second division

2. Describe the process of synapsis. ______________________

   ____

3. How does crossover introduce variability in the daughter cells? ______________________

   ____

   ____

4. Define *homologous chromosomes.* ______________________

   ____

Spermatogenesis

1. The cell types seen in the seminiferous tubules are listed in the key. Match the correct cell type(s) with the descriptions given below.

Key: a. primary spermatocyte
b. secondary spermatocyte
c. spermatogonium
d. Sertoli cell
e. spermatid
f. sperm

__________ primitive stem cell

__________ haploid

__________ provides nutrients to developing sperm

__________ product of meiosis II

__________ product of spermiogenesis

__________ product of meiosis I

2. Why are spermatids not considered functional gametes? __________

____

3. Define *spermatogenesis:* __________

Define *spermiogenesis:* __________

____

4. Draw a sperm below and identify the *acrosome, head, midpiece,* and *tail.* Then beside each label, note the composition and function of each of these sperm structures.

5. The life span of a sperm is very short. What anatomic characteristics might lead you to suspect this even if you didn't know its life span? __________

____

Oogenesis, the Ovarian Cycle, and the Menstrual Cycle

1. The sequence of events leading to germ cell formation in the female begins during fetal development. By the time the child is born, all the oogonia have been converted to __________.

In view of this fact, how does the total germ cell potential of the female compare to that of the male?

____

2. The female gametes develop in structures called *follicles.* What is a follicle? __________

____

How are primary and vesicular follicles anatomically different? __________

____

____

____

What is a corpus luteum? ____

____

3. What hormone is produced by the vesicular follicle? ______________________________

   By the corpus luteum? ____

4. Use the key to identify the cell type you would expect to find in the following structures.

   Key: a. oogonium   b. primary oocyte   c. secondary oocyte   d. ovum

   ________ forming part of the primary follicle in the ovary

   ________ in the uterine tube before fertilization

   ________ in the mature vesicular follicle of the ovary

   ________ in the uterine tube shortly after sperm penetration

5. The cellular product of spermatogenesis is four ________________; the final product of oogenesis is one ________________ and three ________________. What is the function of this unequal cytoplasmic division seen during oogenesis in the female? ______________________________

   ____

   ____

   What is the fate of the three tiny cells produced during oogenesis? ______________________

   Why? ____

6. The following statements deal with anterior pituitary and ovarian hormonal interrelationships. Name the hormone(s) described in each statement.

   ______________________ stimulates ovarian follicles to grow and to produce estrogen

   ______________________ ovulation occurs after its burstlike release

   ______________________ and ______________________ exert negative feedback on the anterior pituitary relative to FSH secretion

   ______________________ stimulates LH release by the anterior pituitary

   ______________________ stimulates the corpus luteum to produce progesterone and estrogen

   ______________________ maintains the hormonal production of the corpus luteum in a nonpregnant woman

7. Why does the corpus luteum deteriorate toward the end of the ovarian cycle? ______________________

   ____

8. For each statement below dealing with hormonal blood levels during the female ovarian and menstrual cycles, decide whether the condition in column A is usually (a) greater than, (b) less than, or (c) essentially equal to the condition in column B.

	Column A		Column B
______	amount of estrogen in the blood during menses	↔	amount of estrogen in the blood at ovulation
______	amount of progesterone in the blood on the fourteenth day	↔	amount of progesterone in the blood on the twenty-third day
______	amount of LH in the blood during menses	↔	amount of LH in the blood at ovulation
______	amount of FSH in the blood on day 6 of the cycle	↔	amount of FSH in the blood on day 20 of the cycle
______	amount of estrogen in the blood on the tenth day	↔	amount of progesterone in the blood on the tenth day

9. Ovulation and menstruation usually cease by the age of ____________.

10. What uterine tissue undergoes dramatic changes during the menstrual cycle? ____________

11. When during the female menstrual cycle would fertilization be unlikely? ____________

____

12. Assume that a woman could be an "on demand" ovulator like the rabbit, in which copulation stimulates the hypothalamic-anterior pituitary axis and causes LH release, and an oocyte was ovulated and fertilized on day 26 of her 28-day cycle. Why would a successful pregnancy be unlikely at this time?

____

____

13. The menstrual cycle depends on events within the female ovary. The stages of the menstrual cycle are listed below. For each, note its approximate time span and the related events in the uterus; and then to the right, record the ovarian events occurring simultaneously. Pay particular attention to hormonal events.

Menstrual Cycle Stage	Uterine Events	Ovarian Events

STUDENT NAME ______________________________

LAB TIME/DATE ______________________________

EXERCISE 41 Principles of Heredity

Introduction to the Language of Genetics

1. Match the key choices with the definitions given below.

Key: a. alleles
b. autosomes
c. dominant
d. genotype
e. heterozygous
f. homozygous
g. phenotype
h. recessive
i. sex chromosomes

______________ actual genetic makeup

______________ chromosomes determining maleness/femaleness

______________ situation in which an individual has identical alleles for a particular trait

______________ genes not expressed unless they are present in homozygous condition

______________ expression of a genetic trait

______________ situation in which an individual has different alleles making up his genotype for a particular trait

______________ genes for the same trait that may have different expressions

______________ chromosomes regulating most body characteristics

______________ the more-potent gene allele; masks the expression of the less-potent allele

Dominant-Recessive Inheritance

1. In humans, farsightedness is inherited by possession of a dominant gene. If a man who is homozygous for normal vision (aa) marries a woman who is heterozygous for farsightedness, what proportion of their children would be expected to be farsighted? ________%

2. A metabolic disorder called PKU is due to an abnormal recessive gene (p). Only homozygous recessive individuals exhibit this disorder. What percentage of the offspring will be anticipated to have PKU if the parents are Pp and pp? ________%

3. A man obtained 32 spotted and 10 solid-color rabbits from a mating of two spotted rabbits.

   Which trait is dominant? ______________ Recessive? ______________

   What is the probable genotype of the rabbit parents? ______________ × ________

4. Assume that the allele controlling brown eyes (B) is dominant over that controlling blue eyes (b) in human beings. (In actuality, eye color in humans is an example of multigene inheritance, which is much more complex than this.) A blue-eyed man marries a brown-eyed woman; and they have six children, all brown-eyed. What is the most likely genotype of the father? ________ Of the mother? ________ If the seventh child had *blue* eyes, what could you conclude about the parents' genotypes? ______________________________

Incomplete Dominance

1. Tail length on a bobcat is controlled by incomplete dominance. The alleles are T for normal tail length and t for tail-less. What name could/would you give to the tails of heterozygous (Tt) cats? ____________________

   How would their tail length compare with that of TT or tt bobcats? ____________________

2. If curly-haired individuals are genotypically CC, straight-haired individuals are cc, and wavy-haired individuals are heterozygotes (Cc), what percentage of the various phenotypes would be anticipated from a cross between a CC woman and a cc man?

   ________% curly ________% wavy ________% straight

Sex-Linked Inheritance

1. What does it mean when someone says a particular characteristic is sex-linked? ____________________

   ____

2. You are a male, and you have been told that hemophilia "runs in your genes." Whose ancestors, your mother's or your father's, should you investigate? ____________ Why? ____________________

   ____

   ____

3. An $X^C X^c$ female maries an $X^C Y$ man. Do a Punnet square for this match.

   What is the probability of producing a color-blind son? ____________

   A color-blind daughter? ____________

   A daughter that is a carrier for the color-blind gene? ____________

4. Why are consanguinous marriages (marriages between blood relatives) prohibited in most cultures?

   ____

   ____

Probability

1. What is the probability of having three daughters in a row? ____________________

2. A man and a woman, each of seemingly normal intellect, marry. Although neither is aware of the fact, each is a heterozygote for the allele for feeblemindedness. Is the allele for feeblemindedness dominant or recessive?

   ____

   What are the chances of their having one feebleminded child? ____________________

   What are the chances that all of their children (they plan a family of four) will be feebleminded?

   ____

Genetic Determination of Selected Human Characteristics

1. Look back at your data to complete this section. For each of the situations described here, determine if an offspring with the characteristics noted is possible with the parental genotypes listed. Check (√) the appropriate column.

		Possibility	
Parental Genotypes	**Phenotype of Child**	**Yes**	**No**
Jj × jj	Double-jointed thumbs		
FF × Ff	Straight little finger		
EE × ee	Detached ear lobes		
HH × Hh	Mid-digital hair		
$I^Ai \times I^Bi$	Type O blood		
$I^AI^B \times ii$	Type B blood		

2. You have dimples, and you would like to know if you are homozygous or heterozygous for this trait. You have six brothers and sisters. By observing your siblings, how could you tell, with some degree of certainty, that you are a heterozygote?

____

____

STUDENT NAME ____________________

LAB TIME/DATE ____________________

EXERCISE 42 Surface Anatomy Roundup

_______ 1. A blow to the cheek is most likely to break what superficial bone or bone part? (a) superciliary arches, (b) the philtrum, (c) zygomatic arch, (d) the tragus.

_______ 2. Rebound tenderness (a) occurs in appendicitis, (b) is whiplash of the neck, (c) is a sore foot from playing basketball, (d) occurs when the larynx falls back into place after swallowing.

_______ 3. The anatomical snuff box (a) is in the nose, (b) contains the styloid process of the radius, (c) is defined by tendons of the flexor carpi radialis and palmaris longus, (d) cannot really hold snuff.

_______ 4. Some landmarks on the body surface can be seen or felt, but others are abstractions that you must construct by drawing imaginary lines. Which of the following pairs of structures is abstract and invisible? (a) umbilicus and costal margin, (b) anterior superior iliac spine and natal cleft, (c) linea alba and linea semilunaris, (d) McBurney's point and midaxillary line, (e) philtrum and sternocleidomastoid.

_______ 5. Many pelvic organs can be palpated by placing a finger in the rectum or vagina, but only one pelvic organ is readily palpated through the skin. This is the (a) nonpregnant uterus, (b) prostate gland, (c) full bladder, (d) ovaries, (e) rectum.

_______ 6. A muscle that contributes to the posterior axillary fold is (a) pectoralis major, (b) latissimus dorsi, (c) trapezius, (d) infraspinatus, (e) pectoralis minor, (f) a and e.

_______ 7. Which of the following is not a pulse point? (a) anatomical snuff box, (b) inferior margin of mandible anterior to masseter muscle, (c) center of distal forearm at palmaris longus tendon, (d) medial bicipital furrow on arm, (e) dorsum of foot between the first two metatarsals.

_______ 8. Which pair of ribs inserts on the sternum at the sternal angle? (a) first, (b) second, (c) third, (d) fourth, (e) fifth.

_______ 9. The inferior angle of the scapula is at the same level as the spinous process of this vertebra: (a) C_5, (b) C_7, (c) T_3, (d) T_7, (e) L_4.

_______ 10. An important bony landmark that can be recognized by a distinct dimple in the skin is (a) posterior superior iliac spine, (b) styloid process of the ulna, (c) shaft of the radius, (d) acromion.

_______ 11. A nurse missed a patient's median cubital vein while trying to withdraw blood and then inserted the needle far too deeply into the cubital fossa. This error could cause any of the following problems, except this one: (a) paralysis of the ulnar nerve, (b) paralysis of the median nerve, (c) bruising the insertion tendon of the biceps brachii muscle, (d) blood spurting from the brachial artery.

_______ 12. Which of these organs is almost impossible to study with the techniques of surface anatomy? (a) heart, (b) lungs, (c) brain, (d) nose.

_______ 13. A preferred site for inserting an intravenous medication line into a blood vessel is (a) medial bicipital furrow on arm, (b) external carotid artery, (c) dorsal venous arch of hand, (d) popliteal fossa.

_______ 14. One listens for bowel sounds with a stethoscope that is placed (a) on the four quadrants of the abdominal wall; (b) in the triangle of auscultation; (c) in the right and left midaxillary line, just superior to the iliac crests; (d) inside the patient's bowels (intestines), on the tip of an endoscope.

15. Define palpation: ____________________

16. Explain how one locates the proper site for intramuscular injections into:

(a) Deltoid muscle: ____________________

(b) Ventral gluteal site: ______

17. Ashley, a pre–physical therapy student, was trying to locate the vertebral spinous processes on the flexed back of her friend Amber, but she kept losing count. Amber told her to check her count against several reliable "guideposts" along the way: the spinous processes of C_7, T_3, T_7, and L_4. Can you describe how to find each of these four particular vertebrae without having to count any vertebrae?

C_7: ______

T_3: ______

T_7: ______

L_4: ______

18. How does one find the midinguinal point? ______

19. Locate the standard points of surgical incision for reaching both the appendix and the gallbladder. ______

20. Gregory hit his funny bone. What nerve was hit against what bony process? ______

21. An athletic trainer was helping a college basketball player find the site of a pulled muscle. The trainer asked the athlete to extend her thigh at the hip forcefully, but she found this action too painful to perform. Then the trainer palpated her posterior thigh and felt some swelling of the muscles there. In simplest terms, which basic muscle group was injured? ______

22. Walking to her car after her sixty-fifth birthday party, Mrs. Schultz tripped on ice and fell forward on her outstretched palms. When she arrived at the emergency room, her right wrist and hand were bent like a fork handle. Dr. Jefferson felt that the styloid process of her radius was outside the anatomical snuff box and slightly proximal to the styloid process of the ulna. When he checked her elbow, he found that the olecranon process lay 2 cm proximal to the two epicondyles of the right humerus. Explain all these observations, and describe what had happened to Mrs. Schultz's limb. ______

APPENDIX A

The Metric System

Measurement	Unit and Abbreviation	Metric Equivalent	Metric to English Conversion Factor	English to Metric Conversion Factor
Length	1 kilometer (km)	= 1000 (10^3) meters	1 km = 0.62 mile	1 mile = 1.61 km
	1 meter (m)	= 100 (10^2) centimeters = 1000 millimeters	1 m = 1.09 yards 1 m = 3.28 feet 1 m = 39.37 inches	1 yard = 0.914 m 1 foot = 0.305 m
	1 centimeter (cm)	= 0.01 (10^{-2}) meter	1 cm = 0.394 inch	1 foot = 30.5 cm 1 inch = 2.54 cm
	1 millimeter (mm)	= 0.001 (10^{-3}) meter	1 mm = 0.039 inch	
	1 micrometer (μm) [formerly micron (μ)]	= 0.000001 (10^{-6}) meter		
	1 nanometer (nm) [formerly millimicron (mμ)]	= 0.000000001 (10^{-9}) meter		
	1 angstrom (Å)	= 0.0000000001 (10^{-10}) meter		
Area	1 square meter (m^2)	= 10,000 square centimeters	1 m^2 = 1.1960 square yards 1 m^2 = 10.764 square feet	1 square yard = 0.8361 m^2 1 square foot = 0.0929 m^2
	1 square centimeter (cm^2)	= 100 square millimeters	1 cm^2 = 0.155 square inch	1 square inch = 6.4516 cm^2
Mass	1 metric ton (t)	= 1000 kilograms	1 t = 1.103 ton	1 ton = 0.907t
	1 kilogram (kg)	= 1000 grams	1 kg = 2.205 pounds	1 pound = 0.4536 kg
	1 gram (g)	= 1000 milligrams	1 g = 0.0353 ounce 1 g = 15.432 grains	1 ounce = 28.35 g
	1 milligram (mg)	= 0.001 gram	1 mg = approx. 0.015 grain	
	1 microgram (μg)	= 0.000001 gram		
Volume (solids)	1 cubic meter (m^3)	= 1,000,000 cubic centimeters	1 m^3 = 1.3080 cubic yards 1 m^3 = 35.315 cubic feet	1 cubic yard = 0.7646 m^3 1 cubic foot = 0.0283 m^3
	1 cubic centimeter (cm^3 or cc)	= 0.000001 cubic meter = 1 milliliter	1 cm^3 = 0.0610 cubic inch	1 cubic inch = 16.387 cm^3
	1 cubic millimeter (mm^3)	= 0.000000001 cubic meter		
Volume (liquids and gases)	1 kiloliter (kl or kL)	= 1000 liters	1 kL = 264.17 gallons	1 gallon = 3.785 L 1 quart = 0.946 L
	1 liter (l or L)	= 1000 milliliters	1 L = 0.264 gallons 1 L = 1.057 quarts	
	1 milliliter (ml or mL)	= 0.001 liter = 1 cubic centimeter	1 ml = 0.034 fluid ounce 1 ml = approx. $\frac{1}{4}$ teaspoon 1 ml = approx. 15–16 drops (gtt.)	1 quart = 946 ml 1 pint = 473 ml 1 fluid ounce = 29.57 ml 1 teaspoon = approx. 5 ml
	1 microliter (μl or μL)	= 0.000001 liter		
Time	1 second (s)	= $\frac{1}{60}$ minute		
	1 millisecond (ms)	= 0.001 second		
Temperature	Degrees Celsius (°C)		°F = $\frac{9}{5}$°C + 32	°C = $\frac{5}{9}$(°F − 32)

B APPENDIX

Reagents/Solutions/ Special Preparations

This is a list of the solutions found in the laboratory manual. Most of the percent solutions are weight/volume (grams/100 milliliters). All solutions should be prepared with distilled water unless otherwise noted. Solutions containing glucose or sucrose should be refrigerated to inhibit bacterial growth.

Acetic Acid, 1%
For 500 milliliters, measure out 50 milliliters of 10% acetic acid. Add to a small amount of water. Add water to a final volume of 500 milliliters.

Agar Gel, 1.5%
Weigh out 15 grams of dried agar. Slowly add 1 liter of distilled water while heating. Bring slowly to a boil, stirring constantly until the agar dissolves. For immediate use, allow the agar to cool to about 45° C. Pour into petri dishes to solidify. Refrigerate in an inverted position. If the plates are to be kept for a longer time (more than one day), autoclave the agar solution in the flask, pour into sterile petri plates, allow the agar to solidify, invert the plates, and store in a refrigerator.

Albumin, 1%
Weigh out 1 gram of albumin powder. Add distilled water to a final volume of 100 milliliters; OR mix one part egg white with three parts distilled water.

Alpha-Amylase, 0.1%
Weigh out 0.1 gram alpha-amylase. Add distilled water to a final volume of 100 milliliters.

Ammonium Molybdate (($(NH_4)MoO_4$), Dilute, 0.01M
Weigh out 2.04 grams $(NH_4)MoO_4$ (Aldrich, #27,790-8). Add distilled water to a final volume of 1 liter. Caution! $(NH_4)MoO_4$ is a toxic irritant. Label TOXIC.

Atropine Sulfate in frog Ringer's solution) 5%
Weigh out 5 grams of atropine sulfate. Add frog Ringer's solution to a final volume of 100 milliliters. Caution! Atropine sulfate is toxic. Label TOXIC.

Barium Chloride, 10%
Weigh out 10 grams of barium chloride. Add distilled water to a final volume of 100 milliliters.

Benedict's Solution

- 173.0 grams sodium citrate
- 100.0 grams sodium carbonate, anhydrous
- 17.3 grams cupric sulfate (pure crystalline)

Add the citrate and carbonate salts to 700–800 milliliters distilled water. Heat to dissolve. Add the cupric sulfate to 100 milliliters distilled water. Heat to dissolve. Cool the solutions and then combine. Add distilled water to make 1 liter of solution.

Bile Solution, 2%
Weigh out 2 grams of bile salts. Add distilled water to a final volume of 100 milliliters.

Bleach (sodium hypochlorite) solution, 10%
Measure out 100 milliliters of household bleach. Add water to a final volume of 1 liter.

Calcium Chloride in frog Ringer's solution, 1%
Weigh out 1 gram of calcium chloride. Add frog Ringer's solution to a final volume of 100 milliliters.

Copper Sulfate ($CuSO_4$), 10%
Weigh out 10 grams of copper sulfate. Add distilled water to make 100 milliliters.

Digitalis in frog Ringer's solution, 2%
Weigh out 2 grams of digitoxin. Add frog Ringer's solution to a final volume of 100 milliliters.

Epinephrine, 1:10,000
Available in 1:1000 concentration. Dilute 1.0 milliliter of L-adrenaline chloride (1:1000) to 10 milliliters with distilled water or frog physiological saline. Caution! Epinephrine is toxic. Label TOXIC.

Epinephrine in frog Ringer's solution, 1%
Weigh out 1 grams of epinephrine. Add frog Ringer's solution to a final volume of 100 milliliters. Caution, epinephrine is toxic. Label TOXIC.

Epsom Salt Solution, 0.1%
For 500 milliliters, weigh out 0.5 grams of Epsom salts. Add water to a final volume of 500 milliliters.

Glucose, 10%
Weigh out 100 grams of glucose. Add distilled water to a final volume of 1 liter.

Glucose, 40%
Weigh out 40 grams of glucose and bring to 100 milliliters with distilled water. It may be necessary to heat the mixture to get the glucose into solution.

Hydrochloric Acid (HCl), 0.1N
Add 8 milliliters concentrated HCl to 900 milliliters distilled water. Add distilled water to a final volume of 1 liter; OR dilute 100 milliliters of 1N HCl to a final volume of 1 liter with distilled water.

Hydrochloric Acid (HCl), 1N
Add 80 milliliters concentrated HCl to 900 milliliters distilled water. Add water to make 1 liter of solution.

Hydrochloric Acid (HCl), Dilute, 3N
Add 258 milliliters of 36% HCl to 700 milliliters distilled water. Add water to a final volume of 1 liter.

Hydrochloric Acid (HCl), 0.01M
Add 0.8 milliliter concentrated HCl to 900 milliliters distilled water. Add distilled water to make 1 liter of solution; OR dilute 10 milliliters of 1N HCl to a final volume of 1 liter with distilled water.

Hydrochloric Acid (HCl), 0.01%
Add 0.27 milliliter of 1N HCl to 90 milliliters of distilled water. Add water to a final volume of 100 milliliters.

Hydrochloric Acid (HCl), 0.5%–1%
For a 1% HCl solution, add 27 milliliters 1N HCl to 60 milliliters distilled water. Add distilled water to make 100 milliliters. Dilute as needed for the experiment.

Insulin, 100 units/milliliter
Weigh out 10 milligrams of zinc stabilized insulin (25 IU/milligram dry weight, ICN #105707). Add water to a final volume of 25 milliliters.

Lugol's Iodine (IKI)
20 grams potassium iodide 4 grams iodine crystals Dissolve potassium iodide in 1 liter distilled water. Add the iodine crystals and stir to dissolve. Store in dark bottles.

Maltose, 1%
Weigh out 1 gram maltose. Add distilled water to a final volume of 100 milliliters.

Nitric Acid (HNO_3), 10%
For 100 milliliters, put 80 milliliters of distilled water into a graduated cylinder. Carefully add 14.3 milliliters of 69–71% nitric acid. Add water to a final volume of 100 milliliters. An alternative method is to put 50 milliliters of distilled water in a graduated cylinder, carefully add 10 milliliters of 69–71% nitric acid, and add water to a final volume of 70 milliliters.

Nitric Acid (HNO_3), Dilute, 3N
Add 183 milliliters of 69% HNO_3 to 700 milliliters distilled water. Add water to a final volume of 1 liter.

Pancreatin, 1.5%
Weigh out 1.5 grams pancreatin. Add distilled water to a final volume of 100 milliliters.
Pepsin, 2%
Weigh out 2 grams pepsin. Add distilled water to a final volume of 100 milliliters.
Physiologic Saline (Mammalian, 0.9%)
Weigh out 9 grams of NaCl. Add water to a final volume of 1 liter. Make fresh immediately prior to experiment.
Pilocarpine in frog Ringer's solution, 2.5%
Weigh out 2 grams of pilocarpine chloride. Add frog Ringer's solution to a final volume of 100 milliliters.
Potassium Chloride in frog Ringer's solution, 5%
Weigh out 5 grams of potassium chloride. Add frog Ringer's solution to a final volume of 100 milliliters.
Propylthiouracil (PTU), 0.02%
Weigh out 0.2 grams of 6-n-propylthiouracil (Aldrich, # H3,420-3). Add distilled water to make 1 liter of solution. Filter and store in light-resistant containers. (Note: If it is difficult to dissolve the PTU, add concentrated NaOH to adjust the pH to 8.0.) Caution! PTU is a possible carcinogen.
Quinine, 0.1%
For 500 milliliters, weigh out 0.5 gram of quinine sulfate. Add water to a final volume of 500 milliliters.
Rat Chow with Thyroid Extract
Grind up sufficient regular laboratory rat chow to feed the required number of animals for 2 weeks (approximately 40 grams chow/rat/day). Add 20 grams of dessicated thyroid powder (Carolina, #89-6150) for each 1000 grams of rat chow. Mix thoroughly.
Ringer's Solution, Frog
- 6.50 grams sodium chloride
- 0.14 grams potassium chloride
- 0.12 grams calcium chloride
- 0.20 grams sodium bicarbonate

Combine salts in flask and add distilled water to make 1 liter of solution.
Silver Nitrate ($AgNO_3$), 2.9 or 3%
Weigh out 2.9 grams (for 2.9%) or 3 grams (for 3%) of silver nitrate. Use caution; this is an oxidizing substance. Add distilled water to make 100 milliliters of solution. Store in light-resistant bottles. Make fresh for each use.
Sodium Chloride (NaCl) in frog Ringer's solution, 0.7%
Weigh out 0.7 grams of sodium chloride. Add frog Ringer's solution to a final volume of 100 milliliters.
Sodium Chloride (NaCl) , 0.9%
Weigh out 0.9 grams of sodium chloride. Add distilled water to a final volume of 100 milliliters.
Sodium Chloride (NaCl), 1.5%
Weigh out 1.5 grams NaCl. Add distilled water to a final volume of 100 milliliters.
Sodium Chloride (NaCl), 10%
Weigh out 10 grams NaCl and bring to 100 milliliters with distilled water. It may be necessary to heat the mixture to get the NaCl into solution.
Sodium Citrate, 5%
For 500 milliliters, weigh out 25 grams of sodium citrate. Add water to a final volume of 500 milliliters.
Sodium Hydroxide (NaOH), 1N
Weigh out 40 grams of sodium hydroxide. Dissolve in distilled water to make 1 liter; OR add distilled water to 100 milliliters of 10N NaOH to a final volume of 1 liter.
Sodium Hydroxide (NaOH), 10N
Weigh out 100 grams NaOH. Add distilled water to a final volume of 250 milliliters.
Sodium Hydroxide (NaOH), 0.2%
Dilute 1 milliliter of 20% NaOH to a final volume of 100 milliliters with distilled water.
Sodium Hydroxide (NaOH), 20%
Weigh out 20 grams NaOH. Add distilled water to a final volume of 100 milliliters
Starch Solution, Boiled, 0.1%
Weigh out 1 gram of starch for each 100 milliliters of solution Add distilled water to a final volume of 100 milliliters. Heat the mixture, stirring constantly until the starch goes into solution. Cool and filter. Refrigerate when not in use.
Starch Solution, Boiled, 1%
Add 1 gram of starch to 100 milliliters distilled water. Boil just until it changes from cloudy to translucent, cool and filter. Add a pinch of NaCl. Prepare fresh daily. Best results are obtained with potato starch from a biological supply house.
Sucrose, 5%
Weigh out 5 grams of sucrose. Add distilled water to a final volume of 100 milliliters.

Urine Solutions*

Urine, Artificial Normal Human
- 36.4 grams urea
- 15 grams sodium chloride
- 9.0 grams potassium chloride
- 9.6 grams sodium phosphate
- 4.0 grams creatinine
- 100 milligrams albumin

Add urea to 1.5 liters of distilled water. Mix until crystals dissolve. Add sodium chloride, potassium chloride, and sodium phosphate. Mix until solution is clear. The pH should be within the 5 to 7 pH range for normal human urine. (Adjust pH, if necessary, with 1N HCl or 1N NaOH.) Place a urine hydrometer in the solution and dilute with water to a specific gravity within the range of 1.015 to 1.025. This stock solution may be refrigerated for several weeks or frozen for months. Before use, warm to room temperature and add 4.0 grams creatinine and 100 milligrams of albumin for each 2 liters of solution.
Urine, Glycosuria
For a minimally detectable level of glucose, add a minimum of 600 milligrams of glucose to one liter of "normal" urine solution. For moderate to high glycosuria, add 2.5 to 5.0 grams of glucose to each liter of solution.
Urine, Hematuria
Add 1 milliliter of heparinized or defibrinated sheep blood to 1 liter of "normal" urine solution.
Urine, Hemoglobinuria
Add 2 milligrams of bovine hemoglobin to 1 liter of "normal" urine solution.
Urine, Hyposthenuria
Add distilled water to a sample of "normal" urine until the specific gravity approaches 1.005
Urine, Ketonuria
Add a minimum of 100 milligrams of acetoacetic acid or at least 1 milliliter of acetone to 1 liter of "normal" urine solution.
Urine, Leukocyte Presence
Add 100 to 200 units of pork or rabbit liver esterase to 100 milliliters of the "normal" urine solution. This test must be performed immediately after adding the enzyme.
Urine, pH Imbalance
Adjust "normal" urine to a pH of 4.0 to 4.5 with 1N HCl for acid urine. Adjust "normal" urine to a pH of 8 to 9 with 1N NaOH for alkaline urine.
Urine, Proteinuria
Add 300 milligrams or more of albumin per liter of "normal" urine solution. For severe renal damage, add 1 gram of albumin to each liter of solution.
Whole Spectrum Pathological Artificial Human Urine
Mix appropriate amounts of abnormal condition reagents to 1 liter of "normal" urine solution. For diabetes mellitus, use glycosuria and ketonuria solutions. For glomerular damage, use proteinuria, hemoglobinuria, and hematuria solutions.

* From "Artificial Urine for Laboratory Testing" by B. R. Shmaefsky, *American Biology Teacher* 52, no. 3 (1990), 170–172. Reprinted with permission.

APPENDIX C Predicted Vital Capacities for Males

Age	Height in centimeters																								
	146	148	150	152	154	156	158	160	162	164	166	168	170	172	174	176	178	180	182	184	186	188	190	192	194
16	3765	3820	3870	3920	3975	4025	4075	4130	4180	4230	4285	4335	4385	4440	4490	4540	4590	4645	4695	4745	4800	4850	4900	4955	5005
18	3740	3790	3840	3890	3940	3995	4045	4095	4145	4200	4250	4300	4350	4405	4455	4505	4555	4610	4660	4710	4760	4815	4865	4915	4965
20	3710	3760	3810	3860	3910	3960	4015	4065	4115	4165	4215	4265	4320	4370	4420	4470	4520	4570	4625	4675	4725	4775	4825	4875	4930
22	3680	3730	3780	3830	3880	3930	3980	4030	4080	4135	4185	4235	4285	4335	4385	4435	4485	4535	4585	4635	4685	4735	4790	4840	4890
24	3635	3685	3735	3785	3835	3885	3935	3985	4035	4085	4135	4185	4235	4285	4330	4380	4430	4480	4530	4580	4630	4680	4730	4780	4830
26	3605	3655	3705	3755	3805	3855	3905	3955	4000	4050	4100	4150	4200	4250	4300	4350	4395	4445	4495	4545	4595	4645	4695	4740	4790
28	3575	3625	3675	3725	3775	3820	3870	3920	3970	4020	4070	4115	4165	4215	4265	4310	4360	4410	4460	4510	4555	4605	4655	4705	4755
30	3550	3595	3645	3695	3740	3790	3840	3890	3935	3985	4035	4080	4130	4180	4230	4275	4325	4375	4425	4470	4520	4570	4615	4665	4715
32	3520	3565	3615	3665	3710	3760	3810	3855	3905	3950	4000	4050	4095	4145	4195	4240	4290	4340	4385	4435	4485	4530	4580	4625	4675
34	3475	3525	3570	3620	3665	3715	3760	3810	3855	3905	3950	4000	4045	4095	4140	4190	4225	4285	4330	4380	4425	4475	4520	4570	4615
36	3445	3495	3540	3585	3635	3680	3730	3775	3825	3870	3920	3965	4010	4060	4105	4155	4200	4250	4295	4340	4390	4435	4485	4530	4580
38	3415	3465	3510	3555	3605	3650	3695	3745	3790	3840	3885	3930	3980	4025	4070	4120	4165	4210	4260	4305	4350	4400	4445	4495	4540
40	3385	3435	3480	3525	3575	3620	3665	3710	3760	3805	3850	3900	3945	3990	4035	4085	4130	4175	4220	4270	4315	4360	4410	4455	4500
42	3360	3405	3450	3495	3540	3590	3635	3680	3725	3770	3820	3865	3910	3955	4000	4050	4095	4140	4185	4230	4280	4325	4370	4415	4460
44	3315	3360	3405	3450	3495	3540	3585	3630	3675	3725	3770	3815	3860	3905	3950	3995	4040	4085	4130	4175	4220	4270	4315	4360	4405
46	3285	3330	3375	3420	3465	3510	3555	3600	3645	3690	3735	3780	3825	3870	3915	3960	4005	4050	4095	4140	4185	4230	4275	4320	4365
48	3255	3300	3345	3390	3435	3480	3525	3570	3615	3655	3700	3745	3790	3835	3880	3925	3970	4015	4060	4105	4150	4190	4235	4280	4325
50	3210	3255	3300	3345	3390	3430	3475	3520	3565	3610	3650	3695	3740	3785	3830	3870	3915	3960	4005	4050	4090	4135	4180	4225	4270
52	3185	3225	3270	3315	3355	3400	3445	3490	3530	3575	3620	3660	3705	3750	3795	3835	3880	3925	3970	4010	4055	4100	4140	4185	4230
54	3155	3195	3240	3285	3325	3370	3415	3455	3500	3540	3585	3630	3670	3715	3760	3800	3845	3890	3930	3975	4020	4060	4105	4145	4190
56	3125	3165	3210	3255	3295	3340	3380	3425	3465	3510	3550	3595	3640	3680	3725	3765	3810	3850	3895	3940	3980	4025	4065	4110	4150
58	3080	3125	3165	3210	3250	3290	3335	3375	3420	3460	3500	3545	3585	3630	3670	3715	3755	3800	3840	3880	3925	3965	4010	4050	4095
60	3050	3095	3135	3175	3220	3260	3300	3345	3385	3430	3470	3500	3555	3595	3635	3680	3720	3760	3805	3845	3885	3930	3970	4015	4055
62	3020	3060	3110	3150	3190	3230	3270	3310	3350	3390	3440	3480	3520	3560	3600	3640	3680	3730	3770	3810	3850	3890	3930	3970	4020
64	2990	3030	3080	3120	3160	3200	3240	3280	3320	3360	3400	3440	3490	3530	3570	3610	3650	3690	3730	3770	3810	3850	3900	3940	3980
66	2950	2990	3030	3070	3110	3150	3190	3230	3270	3310	3350	3390	3430	3470	3510	3550	3600	3640	3680	3720	3760	3800	3840	3880	3920
68	2920	2960	3000	3040	3080	3120	3160	3200	3240	3280	3320	3360	3400	3440	3480	3520	3560	3600	3640	3680	3720	3760	3800	3840	3880
70	2890	2930	2970	3010	3050	3090	3130	3170	3210	3250	3290	3330	3370	3410	3450	3480	3520	3560	3600	3640	3680	3720	3760	3800	3840
72	2860	2900	2940	2980	3020	3060	3100	3140	3180	3210	3250	3290	3330	3370	3410	3450	3490	3530	3570	3610	3650	3680	3720	3760	3800
74	2820	2860	2900	2930	2970	3010	3050	3090	3130	3170	3200	3240	3280	3320	3360	3400	3440	3470	3510	3550	3590	3630	3670	3710	3740

APPENDIX C Predicted Vital Capacities for Females

Age	Height in centimeters																								
	146	148	150	152	154	156	158	160	162	164	166	168	170	172	174	176	178	180	182	184	186	188	190	192	194
16	2950	2990	3030	3070	3110	3150	3190	3230	3270	3310	3350	3390	3430	3470	3510	3550	3590	3630	3670	3715	3755	3800	3840	3880	3920
17	2935	2975	3015	3055	3095	3135	3175	3215	3255	3295	3335	3375	3415	3455	3495	3535	3575	3615	3655	3695	3740	3780	3820	3860	3900
18	2920	2960	3000	3040	3080	3120	3160	3200	3240	3280	3320	3360	3400	3440	3480	3520	3560	3600	3640	3680	3720	3760	3800	3840	3880
20	2890	2930	2970	3010	3050	3090	3130	3170	3210	3250	3290	3330	3370	3410	3450	3490	3525	3565	3605	3645	3695	3720	3760	3800	3840
22	2860	2900	2940	2980	3020	3060	3095	3135	3175	3215	3255	3290	3330	3370	3410	3450	3490	3530	3570	3610	3650	3685	3725	3765	3800
24	2830	2870	2910	2950	2985	3025	3065	3100	3140	3180	3220	3260	3300	3335	3375	3415	3455	3490	3530	3570	3610	3650	3685	3725	3765
26	2800	2840	2880	2920	2960	3000	3035	3070	3110	3150	3190	3230	3265	3300	3340	3380	3420	3455	3495	3530	3570	3610	3650	3685	3725
28	2775	2810	2850	2890	2930	2965	3000	3040	3070	3115	3155	3190	3230	3270	3305	3345	3380	3420	3460	3495	3535	3570	3610	3650	3685
30	2745	2780	2820	2860	2895	2935	2970	3010	3045	3085	3120	3160	3195	3235	3270	3310	3345	3385	3420	3460	3495	3535	3570	3610	3645
32	2715	2750	2790	2825	2865	2900	2940	2975	3015	3050	3090	3125	3160	3200	3235	3275	3310	3350	3385	3425	3460	3495	3535	3570	3610
34	2685	2725	2760	2795	2835	2870	2910	2945	2980	3020	3055	3090	3130	3165	3200	3240	3275	3310	3350	3385	3425	3460	3495	3535	3570
36	2655	2695	2730	2765	2805	2840	2875	2910	2950	2985	3020	3060	3095	3130	3165	3205	3240	3275	3310	3350	3385	3420	3460	3495	3530
38	2630	2665	2700	2735	2770	2810	2845	2880	2915	2950	2990	3025	3060	3095	3130	3170	3205	3240	3275	3310	3350	3385	3420	3455	3490
40	2600	2635	2670	2705	2740	2775	2810	2850	2885	2920	2955	2990	3025	3060	3095	3135	3170	3205	3240	3275	3310	3345	3380	3420	3455
42	2570	2605	2640	2675	2710	2745	2780	2815	2850	2885	2920	2955	2990	3025	3060	3100	3135	3170	3205	3240	3275	3310	3345	3380	3415
44	2540	2575	2610	2645	2680	2715	2750	2785	2820	2855	2890	2925	2960	2995	3030	3060	3095	3130	3165	3200	3235	3270	3305	3340	3375
46	2510	2545	2580	2615	2650	2685	2715	2750	2785	2820	2855	2890	2925	2960	2995	3030	3060	3095	3130	3165	3200	3235	3270	3305	3340
48	2480	2515	2550	2585	2620	2650	2685	2715	2750	2785	2820	2855	2890	2925	2960	2995	3030	3060	3095	3130	3160	3195	3230	3265	3300
50	2455	2485	2520	2555	2590	2625	2655	2690	2720	2755	2785	2820	2855	2890	2925	2955	2990	3025	3060	3090	3125	3155	3190	3225	3260
52	2425	2455	2490	2525	2555	2590	2625	2655	2690	2720	2755	2790	2820	2855	2890	2925	2955	2990	3020	3055	3090	3125	3155	3190	3220
54	2395	2425	2460	2495	2530	2560	2590	2625	2655	2690	2720	2755	2790	2820	2855	2885	2920	2950	2985	3020	3050	3085	3115	3150	3180
56	2365	2400	2430	2460	2495	2525	2560	2590	2625	2655	2690	2720	2755	2790	2820	2855	2885	2920	2950	2980	3015	3045	3080	3110	3145
58	2335	2370	2400	2430	2460	2495	2525	2560	2590	2625	2655	2690	2720	2750	2785	2815	2850	2880	2920	2945	2975	3010	3040	3075	3105
60	2305	2340	2370	2400	2430	2460	2495	2525	2560	2590	2625	2655	2685	2720	2750	2780	2810	2845	2875	2915	2940	2970	3000	3035	3065
62	2280	2310	2340	2370	2405	2435	2465	2495	2525	2560	2590	2620	2655	2685	2715	2745	2775	2810	2840	2870	2900	2935	2965	2995	3025
64	2250	2280	2310	2340	2370	2400	2430	2465	2495	2525	2555	2585	2620	2650	2680	2710	2740	2770	2805	2835	2865	2895	2925	2955	2990
66	2220	2250	2280	2310	2340	2370	2400	2430	2460	2495	2525	2555	2585	2615	2645	2675	2705	2735	2765	2800	2825	2860	2890	2920	2950
68	2190	2220	2250	2280	2310	2340	2370	2400	2430	2460	2490	2520	2550	2580	2610	2640	2670	2700	2730	2760	2795	2820	2850	2280	2910
70	2160	2190	2220	2250	2280	2310	2340	2370	2400	2425	2455	2485	2515	2545	2575	2605	2635	2665	2695	2725	2755	2780	2810	2840	2870
72	2130	2160	2190	2220	2250	2280	2310	2335	2365	2395	2425	2455	2480	2510	2540	2570	2600	2630	2660	2685	2715	2745	2775	2805	2830
74	2100	2130	2160	2190	2220	2245	2275	2305	2335	2360	2390	2420	2450	2475	2505	2535	2565	2590	2620	2650	2680	2710	2740	2765	2795

Reprinted with permission from Warren E. Collins, Inc., Braintree, Mass.

Index

NOTE: A *t* following a page number indicates tabular material and an *f* following a page number indicates a figure.